Travelling Waves and Periodic Oscillations in Fermi-Pasta-Ulam Lattices

Travelling Waves and Periodic Oscillations in Fermi-Pasta-Ulam Lattices

Alexander Pankov

The College of William and Mary, USA

Imperial College Press

Published by
Imperial College Press
57 Shelton Street
Covent Garden
London WC2H 9HE

Distributed by
World Scientific Publishing Co. Pte. Ltd.
5 Toh Tuck Link, Singapore 596224
USA office: 27 Warren Street, Suite 401-402, Hackensack, NJ 07601
UK office: 57 Shelton Street, Covent Garden, London WC2H 9HE

British Library Cataloguing-in-Publication Data
A catalogue record for this book is available from the British Library.

TRAVELLING WAVES AND PERIODIC OSCILLATIONS IN FERMI – PASTA – ULAM

Copyright © 2005 by Imperial College Press

All rights reserved. This book, or parts thereof, may not be reproduced in any form or by any means, electronic or mechanical, including photocopying, recording or any information storage and retrieval system now known or to be invented, without written permission from the Publisher.

For photocopying of material in this volume, please pay a copying fee through the Copyright Clearance Center, Inc., 222 Rosewood Drive, Danvers, MA 01923, USA. In this case permission to photocopy is not required from the publisher.

ISBN 1-86094-532-5

To Tanya

Preface

In the past two decades, there has been an explosion of interest to the study of wave propagation in spatially discrete nonlinear systems.

Probably, the most prominent example of such a system is the famous Fermi-Pasta-Ulam (FPU) lattice introduced in the pioneering work [Fermi et. al (1955)]. E. Fermi, J. Pasta and S. Ulam studied numerically the lattices of identical particles, i. e. monoatomic lattices, with cubic and quartic interaction potentials. These lattices are known today as α- and β-models, respectively. The aim of E. Fermi, J. Pasta and S. Ulam was to show the relaxation to equipartition of the distribution of energy among modes. Surprisingly enough, their numerical simulation yielded the opposite result. They observed that, at least at low energy, the energy of the system remained confined among the initial modes, instead of spreading towards all modes.

This work motivated a great number of further numerical and analytical investigations (for a relatively recent survey of the subject see [Poggi and Ruffo (1997)]). We mention here the so-called Toda lattice which is a completely integrable system. Due to the integrability, the dynamics of Toda lattice is well-understood (see [Toda (1989); Teschl (2000)]). Unfortunately, the Toda lattice is the only known completely integrable lattice of FPU type. Overwhelming majority of existing results concern either particular explicit solutions, both exact and approximate, or numerical simulation. Moreover, almost exclusively spatially homogeneous, i. e. monoatomic, lattices are under consideration, although inhomogeneous lattices (multiatomic lattices, lattices with impurities, etc.) are of great interest.

One of the first rigorous results about general FPU type lattices was obtained in [Friesecke and Wattis (1994)]. G. Friesecke and J. Wattis have proved the existence of solitary travelling waves in monoatomic FPU lattices under some general assumptions on the potential of interparticle interaction. The class of potentials includes the α- and β- models, the Toda potential, the Lennard-Jones potential, and others. The approach of G. Friesecke and J. Wattis is based on an appropriate constrained minimization procedure and the concentration compactness principle of P.-L. Lions [Lions (1984)]. In this approach the wave speed is unknown and is determined *a posteriori* through the corresponding Lagrange multiplier.

Later on D. Smets and M. Willem [Smets and Willem (1997)] considered the travelling wave problem as a problem with *prescribed* speed. Under another set of assumptions they have proved the existence of travelling waves for every prescribed speed beyond the speed of sound (naturally defined). The proof relies upon an appropriate version of the mountain pass theorem without Palais-Smale condition. In [Pankov and Pflüger (2000b)], K. Pflüger and the author revised the last approach considerably choosing periodic travelling waves as a starting point. The existence of periodic waves is obtained by means of the standard mountain pass theorem. Then one gets solitary waves in the limit as the wave lengths goes to infinity. This approach applies to many other problems (see, *e. g.* [Pankov and Pflüger (1999); Pankov and Pflüger (2000a)]).

We mention also the series of papers [Friesecke and Pego (1999); Friesecke and Pego (2002); Friesecke and Pego (2004a); Friesecke and Pego (2004b)], where near sonic solitary waves are studied. Under some generic assumptions on the potential of interaction near the origin the existence of such waves is obtained by means of perturbation from the standard Korteweg-de Vries (KdV) soliton. Many properties of near sonic waves are discussed including their dynamical stability.

Another line of development was originated by B. Ruf and P. Srikanth [Ruf and Srikanth (1994)] who considered time periodic motions of finite FPU type lattices not necessary consisting of identical particles. Similar problem for infinite lattices, still inhomogeneous, was studied in [Arioli and Chabrowski (1997); Arioli and Gazzola (1995); Arioli and Gazzola (1996); Arioli *et. al* (1996); Arioli and Szulkin (1997)] under more restrictive assumptions on the potential.

Another class of discrete media consists of chains of coupled nonlinear oscillators. One of the most known models of such kind is the so-called

Frenkel-Kontorova model introduced by Ya. Frenkel and T. Kontorova in 1938. As we have learned from [Braun and Kivshar (2004)], the same model have been appeared even before, in works by L. Prandtl and U. Dehlinger (1928-29). For physical applications of the Frenkel-Kontorova and related models we refer to [Braun and Kivshar (1998)] and [Braun and Kivshar (2004)]. Also chains of oscillators as systems that support breathers, *i. e.* spatially localized time periodic solutions, were studied in many works (see [Aubry (1997); James (2003); Livi *et. al* (1997); MacKay and Aubry (1994); Morgante *et. al* (2002)] and references therein). Some other mathematical results that concern time periodic solutions and travelling waves in such systems can be found in [Bak (2004); Bak and Pankov (2004); Bak and Pankov (to appear); Iooss and Kirschgässner (2000)].

Finally, we mention the third class of discrete systems of common interest – discrete nonlinear Schrödinger equations. Such equations are not considered here (see [Flach and Willis (1998); Hennig and Tsironis (1999); Kevrekidis and Weinstein (2003); Pankov and Zakharchenko (2001); Weinstein (1999)] and references therein).

Contents

The main aim of this book is to present rigorous results on time periodic oscillations and travelling waves in FPU lattices. Also we consider briefly similar results for chains of oscillators. Actually, we confine ourself in the circle of the results obtained by variational methods. Therefore, other approaches, like bifurcation theory and perturbation analysis, are not presented here. As we mentioned before, discrete nonlinear Schrödinger equations are outside the scope of the book.

In Chapter 1 we discuss general properties of equations that govern the dynamics of FPU lattices and chains of oscillators, with special attention paid to the well-posedness of the Cauchy problem. Also we remind here basic facts from the spectral theory of linear difference operators that are relevant to linear FPU lattices.

Chapter 2 deals with the existence of time periodic solutions in the lattices of FPU type. Since we employ global variational techniques, it is not natural to restrict the analysis to the case of spatially homogeneous, *i. e.* monoatomic, lattices. Instead, we allow periodic spatial inhomogeneities that means that we consider regular multiatomic lattices. We give complete proofs of all principal results. At the same time, for the results that require more technicalities we outline basic ideas and skip details. Also, skipping technical details, we present couple of results on the existence of time periodic solutions in some chains of nonlinear oscillators.

In Chapters 3 and 4 we study travelling waves in monoatomic FPU lattices. The first of them is devoted to waves with prescribed speed. This statement of problem seems to be most natural. Here we consider two types of travelling waves, periodic and solitary. In fact, we treat solitary waves as a limit case of periodic waves when the wavelength goes to infinity. In Chapter 4 we give some additional results. First of all, we present in details the approach of G. Friesecke and J. Wattis. This approach is technically more involved and, therefore, is postponed to the last chapter. Also we discuss here several other results, including exponential decay of solitary waves, as well as travelling waves in chains of oscillators.

Each chapter, except Chapter 3, ends with a special section devoted to various comments and open problems. Comments and open problems that concern travelling waves are put on the end of Chapter 4. Open problems we offer reflect author's point of view on what should be done next. Some of them are accessible by existing methods, while others are probably hard enough.

For reader's convenience we include four appendices. Their aim is to remind basic facts about functional spaces, concentration compactness, critical points and finite differences, and make the presentation more or less self-contained.

Audience
As audience we have researchers in mind. Although the book is formally self-contained, some acquaintance with variational methods and nonlinear analysis is recommended. Appropriate references are [Mawhin and Willem (1989); Rabinowitz (1986); Struwe (2000); Willem (1996)] (variational methods) and [Zeidler (1995a); Zeidler (1995b)] (nonlinear analysis). At the same time the present book is accessible to graduate students as well, especially in combinations with the books on variational methods listed above.

Acknowledgements
The present book was prepared during author's staying at Texas A&M University and the College of William and Mary as visiting professor. The

work was also supported in part by NATO, grant 970179.

The author thanks T. Bartsch, G. Papanicolaou, P. H. Rabinowitz, A. Szulkin, Z.-Q. Wang and M. Willem for many interesting discussions and valuable information.

Last but not least, I am deeply grateful to my wife Tanya for her generous support.

A. Pankov

Contents

Preface vii

1. Infinite Lattice Systems 1
   - 1.1 Equations of motion 1
   - 1.2 The Cauchy problem 7
   - 1.3 Harmonic lattices 10
   - 1.4 Chains of coupled nonlinear oscillators 16
   - 1.5 Comments and open problems 24

2. Time Periodic Oscillations 27
   - 2.1 Setting of problem 27
   - 2.2 Positive definite case 34
   - 2.3 Indefinite case 42
     - 2.3.1 Main result 42
     - 2.3.2 Periodic approximations 46
     - 2.3.3 Proof of main result 54
   - 2.4 Additional results 56
     - 2.4.1 Degenerate case 56
     - 2.4.2 Constrained minimization 58
     - 2.4.3 Multibumps 59
     - 2.4.4 Lattices without spatial periodicity 61
     - 2.4.5 Finite lattices 62
   - 2.5 Chains of oscillators 64
   - 2.6 Comments and open problems 70

3. Travelling Waves: Waves with Prescribed Speed 75

3.1	Statement of problem	75
3.2	Periodic waves	78
	3.2.1 Variational setting	78
	3.2.2 Monotone waves	81
	3.2.3 Nonmonotone and subsonic waves	85
3.3	Solitary waves	89
	3.3.1 Variational statement of the problem	89
	3.3.2 From periodic waves to solitary ones	93
	3.3.3 Global structure of periodic waves	101
	3.3.4 Examples	104
3.4	Ground waves: existence and convergence	105
	3.4.1 Ground waves: periodic case	105
	3.4.2 Solitary ground waves	109
	3.4.3 Monotonicity	112
3.5	Near sonic waves	114
	3.5.1 Amplitude estimate	114
	3.5.2 Nonglobally defined potentials	117

4. Travelling Waves: Further Results — 121

4.1	Solitary waves and constrained minimization	121
	4.1.1 Statement of problem	121
	4.1.2 The minimization problem: technical results	123
	4.1.3 The minimization problem: existence	133
	4.1.4 Proof of main result	140
	4.1.5 Lennard-Jones type potentials	143
4.2	Other types of travelling waves	146
	4.2.1 Waves with periodic profile functions	146
	4.2.2 Solitary waves whose profiles vanish at infinity	148
4.3	Yet another constrained minimization problem	150
4.4	Remark on FPU β-model	152
4.5	Exponential decay	154
4.6	Travelling waves in chains of oscillators	160
4.7	Comments and open problems	163

Appendix A Functional Spaces — 167

A.1 Spaces of sequences	167
A.2 Spaces of functions on real line	168

Appendix B Concentration Compactness — 173

Appendix C Critical Point Theory   177

    C.1 Differentiable functionals 177
    C.2 Mountain pass theorem 178
    C.3 Linking theorems 179

Appendix D Difference Calculus   183

Bibliography   185

Index   193

Chapter 1

Infinite Lattice Systems

1.1 Equations of motion

We consider a one dimensional chain of particles with nearest neighbor interaction. Equations of motion of the system read

$$m(n)\,\ddot{q}(n) = U'_{n+1}\bigl(q(n+1) - q(n)\bigr) - U'_n\bigl(q(n) - q(n-1)\bigr), \quad n \in \mathbb{Z}. \quad (1.1)$$

Here $q(n) = q(t,n)$ is the coordinate of n-th particle at time t, $m(n)$ is the mass of that particle, and U_n is the potential of interaction between n-th and $(n-1)$-th particles. We always assume that there are positive constants m_0 and M_0 such that

$$m_0 \leq m(n) \leq M_0$$

for every $n \in \mathbb{Z}$.

Equations (1.1) form an infinite system of ordinary differential equations which is a Hamiltonian system with the Hamiltonian

$$H = \sum_{n=-\infty}^{\infty} \left(\frac{p^2(n)}{2m(n)} + U_n\bigl(q(n+1) - q(n)\bigr) \right), \quad (1.2)$$

where $p(n) = m(n)\,\dot{q}(n)$ is the momentum of n-th particle.

Formally this statement is readily verified. However, to make it precise first one has to specify the phase space.

The simplest, but not so natural from the point of view of physics, choice of the configuration space is the space l^2 of two-sided sequences[1]

---

[1] For the definitions and notations of spaces of sequences see Appendix A.1.

$q = \{q(n)\}_{n \in \mathbb{Z}}$. This corresponds to the boundary condition

$$\lim_{n \to \pm\infty} q(n) = 0 \tag{1.3}$$

at infinity.

In this case the phase space is $l^2 \times l^2$ and Eq. (1.1) can be written as the first-order system

$$\dot{u} = J \nabla H(u),$$

where

$$u = \begin{pmatrix} q \\ p \end{pmatrix}, \quad J = \begin{pmatrix} 0 & I \\ -I & 0 \end{pmatrix} : l^2 \times l^2 \to l^2 \times l^2,$$

I is the identity operator and ∇H the functional gradient of H

$$\nabla H(u)(n) = \begin{pmatrix} U_n'\big(q(n) - q(n-1)\big) - U_{n+1}'\big(q(n+1) - q(n)\big) \\ p(n)/m(n) \end{pmatrix}.$$

Denote by G the nonlinear operator defined by

$$G(q)(n) = U_n'\big(q(n)\big), \quad n \in \mathbb{Z}, \tag{1.4}$$

where $q = \{q(n)\}$, and consider operators of right and left differences

$$(\partial^+ q)(n) := q(n+1) - q(n)$$

and

$$(\partial^- q)(n) := q(n) - q(n-1),$$

respectively. We suppose that G is a "good" nonlinear operator in l^2. Then

$$\nabla H(u) = \begin{pmatrix} -\partial^+ G(\partial^- q) \\ p/m \end{pmatrix}, \tag{1.5}$$

while Eq. (1.1) becomes a "divergence form" equation

$$m\ddot{q} = \partial^+ G(\partial^- q). \tag{1.6}$$

Note that ∂^+ and ∂^- are bounded linear operators in l^2 and

$$(\partial^+)^* = -\partial^-.$$

Another form of Eq. (1.6) is

$$m\ddot{q} = \partial^- G^+(\partial^+ q), \tag{1.7}$$

where
$$G^+(q)(n) = U'_{n+1}(q(n)). \tag{1.8}$$

However, more natural and most important choice of configuration space is the space $X = \tilde{l}^2$ that consists of two-sided sequences $q = \{q(n)\}_{n \in \mathbb{Z}}$ such that $\partial^+ q \in l^2$. Endowed with the norm
$$\|q\|_X = \left(\|\partial^+ q\|_{l^2}^2 + |q(0)|^2\right)^{1/2} = \left(\|\partial^- q\|_{l^2}^2 + |q(0)|^2\right)^{1/2},$$
X is a Hilbert space. Obviously,
$$\|\partial^- q\|_{l^2} = \|\partial^+ q\|_{l^2}.$$
Operators ∂^+ and ∂^- are linear bounded operators from the space X onto l^2 and have one dimensional kernel that consists of constant sequences.

Equation (1.1) (equivalently, (1.6)) is a Hamiltonian system on the phase space $\tilde{l}^2 \times l^2$. In this case the corresponding symplectic form [Marsden and Ratiu (1994)] is degenerate. Nevertheless, the Hamiltonian H defined by (1.2) is a conserved quantity provided $H(q,p)$ is C^1 on $\tilde{l}^2 \times l^2$. This can be verified by a direct calculation.

Now we introduce a reformulation of Eq. (1.6) in $\tilde{l}^2$ as an equation in l^2. Denote by
$$r(n) := q(n+1) - q(n),$$
i. e. $r = \partial^+ q$, the *relative displacements* of adjacent lattice sites and set
$$b(n) := a(n-1) = m(n)^{-1/2}.$$
Then Eq. (1.1) gives immediately
$$\ddot{r}(n) = a^2(n)\left[U'_{n+1}(r(n+1)) - U'_n(r(n))\right]$$
$$- a^2(n-1)\left[U'_n(r(n)) - U'_{n-1}(r(n-1))\right]. \tag{1.9}$$

Note that $r \in l^2$ whenever $q \in \tilde{l}^2$. In operator form, Eq.(1.9) reads
$$\ddot{r} = \partial^-\left[a^2 \partial^+ G(r)\right].$$
Also it can be written as (see Appendix D, Eq. (D.5))
$$\ddot{r} = \partial^+\left[b^2 \partial^- G^+(r)\right], \tag{1.10}$$

where
$$G^+(r)(n) = U'_{n+1}(r(n)).$$

Equation (1.10) is equivalent to the following first-order system

$$\dot{u} = F(u), \quad u = \begin{pmatrix} r \\ s \end{pmatrix}, \quad F(u) = \begin{pmatrix} \partial^+(bs) \\ b\partial^- G^+(r) \end{pmatrix}. \tag{1.11}$$

This is a Hamiltonian system

$$\dot{u} = J\nabla H(u), \tag{1.12}$$

where

$$J = \begin{pmatrix} 0 & \partial^+ b \\ -b\partial^- & 0 \end{pmatrix} \tag{1.13}$$

and

$$H(r,s) = \sum_{n=-\infty}^{\infty} \left[\frac{s(n)^2}{2} + U_{n+1}(r(n)) \right]. \tag{1.14}$$

In fact, here $s = bp = p/m^{1/2}$. The phase space of this system is $l^2 \times l^2$. It is readily verified that

$$(\partial^+ b)^* = -b\partial^-.$$

Certainly, $H(r,s)$ defined by (1.14) is a conserved quantity if H is C^1 on $l^2 \times l^2$.

Now let us discuss the relation between solutions of Eq. (1.6) and Eq. (1.10) (or (1.11)). Consider a solution $q = q(t,n)$ of Eq. (1.6) such that q is a C^1 function of t with values in $X = \tilde{l}^2$ and $\dot{q}$ is a C^1 function with values in l^2. Then $r = \partial^+ q$ and $s = ap$ are C^1 functions with values in l^2, and $u = (r,s)$ obviously solves (1.11). Moreover, the well-posedness, local or global in time, of the Cauchy problem for Eq. (1.6) in $\tilde{l}^2 \times l^2$ implies the same property for Eq. (1.11) (and (1.10)) in the space $l^2 \times l^2$.

Conversely, consider the Cauchy problem for (1.6), with

$$q|_{t=0} = q^{(0)} \in \tilde{l}^2, \quad \dot{q}|_{t=0} = q^{(1)} \in l^2.$$

Set

$$r^{(0)} = \partial^+ q^{(0)}, \quad s^{(0)} = m^{1/2} q^{(1)}.$$

Let $u = (r, s)$ be the solution of Cauchy problem for Eq. (1.11), with
$$u|_{t=0} = \left(r^{(0)}, s^{(0)}\right) \in l^2 \times l^2.$$
Define
$$q(t) = q^{(0)} + \int_0^t bs(\tau)\, d\tau.$$
Equation (1.11) implies that $\partial^+ q = r$. We have $\dot{q} = bs$ and, by Eq. (1.11),
$$\ddot{q} = b\dot{s} = b^2 \partial^- G^+(r) = b^2 \partial^- G^+(\partial^+ q).$$
Hence,
$$m\ddot{q} = \partial^- G^+(\partial^+ q)$$
which is another form (1.7) of Eq. (1.6). Obviously $\dot{q}|_{t=0} = q^{(1)}$. Thus, the two statements of problem, (1.6) and (1.11), are equivalent.

Now let us consider several examples. In these examples we always have $m(n) \equiv 1$, $U_n(r) \equiv U(r)$.

Example 1.1 Let
$$U(r) = \frac{c_0}{2} r^2, \quad c_0 > 0.$$
Then we obtain the *discrete wave equation*
$$\ddot{q} = c_0 \Delta q, \tag{1.15}$$
where $\Delta = \partial^+ \partial^-$ is the *discrete one-dimensional Laplacian*.

Example 1.2 If
$$U(r) = \frac{c_0}{2} r^2 + \frac{c_1}{3} r^3, \quad c_0 > 0, \quad c_1 \neq 0,$$
(cubic interaction), we obtain the so-called *Fermi-Pasta-Ulam α-model*.

Example 1.3 The *Fermi-Pasta-Ulam β-model* is defined by
$$U(r) = \frac{c_0}{2} r^2 + \frac{c_2}{4} r^4, \quad c_0 > 0, \quad c_2 > 0$$
(quartic interaction).

Example 1.4 The well-known *Toda lattice* [Toda (1989)] has the potential of interaction

$$U(r) = ab^{-1}(e^{-br} + br - 1).$$

This is an example of completely integrable Hamiltonian system [Toda (1989); Teschl (2000)].

Example 1.5 The *Lennard-Jones potential* [Friesecke and Wattis (1994)] is a singular potential defined by

$$U(r) = a\big[(d+r)^{-12} - 2d^{-6}(d+r)^{-6} + d^{-12}\big] = a\big[(d+r)^{-6} - d^{-6}\big]^2, \quad r > -d.$$

More generally, one considers the potentials

$$U(r) = a\big[(d+r)^{-k} - d^{-k}\big]^2,$$

with $a > 0$, $d > 0$ and positive integer k [Friesecke and Matthies (2002)].

In classical FPU lattices $m(n) \equiv m_0$ and U_n does not depend on $n \in \mathbb{Z}$. Such lattices are often called *monoatomic* lattices. More general class of FPU type lattices consists of systems with periodic dependence of $m(n)$ and U_n on $n \in \mathbb{Z}$, i. e. $m(n+N) = m(n)$, $U_{n+N} = U_n$. This is the class of *multiatomic*, or *N-atomic* lattices. If $N = 2$, such systems are called *diatomic* lattices.

Another interesting class of FPU type systems, *lattices with impurities*, appears when we assume that $m(n)$ and U_n are close to periodic for large values of $|n|$, i. e. $m = \overline{m} + m^{(0)}$ and $U_n(r) = \overline{U}_n(r) + U_n^{(0)}(r)$, where $\overline{m}(n)$ and $\overline{U}_n$ are N-periodic in n, while

$$\lim_{|n| \to \infty} m^{(0)}(n) = 0,$$

$$\lim_{|n| \to \infty} U_n^{(0)}(r) = 0.$$

In particular, if $m^{(0)}(n)$ and $U_n^{(0)}$ vanish for all, but finite, values of n, such a system can be viewed as a multiatomic lattice perturbed by replacing a finite number of original particles by particles of another sort, i. e. introducing an impurity.

1.2 The Cauchy problem

Consider the Cauchy problem for Eq. (1.12) or, equivalently, for Eq. (1.13) with the data

$$r|_{t=0} = r^{(0)} \in l^2, \quad \dot{r}|_{t=0} = r^{(1)} \in l^2, \tag{1.16}$$

or, respectively,

$$u|_{t=0} = u^{(0)} = \left(r^{(0)}, s^{(0)}\right) \in l^2 \times l^2. \tag{1.17}$$

We impose the following assumptions.

(i) *There exist $M_0 \geq m_0 \geq 0$ such that*

$$m_0 \leq m(n) \leq M_0, \quad n \in \mathbb{Z}.$$

(ii) *The potential $U_n(r)$ is a C^1 function on $\mathbb{R}$, $U_n(0) = U_n'(0) = 0$ and for every R there exists $C(R) > 0$ such that for all $n \in \mathbb{Z}$*

$$|U_n'(r_1) - U_n'(r_2)| \leq C(R)|r_1 - r_2|, \quad |r_1|, |r_2| \leq R. \tag{1.18}$$

Lemma 1.1 *Under assumption (ii) the operators G and G^+ (see Eqs. (1.4) and (1.8)) act in l^2 and are bounded locally Lipschitz continuous operators, i. e.*

$$\left\|G(q^{(1)}) - G(q^{(2)})\right\|_{l^2} \leq C(R) \left\|q^{(1)} - q^{(2)}\right\|_{l^2}, \quad \left\|q^{(1)}\right\|_{l^2}, \left\|q^{(2)}\right\|_{l^2} \leq R,$$

and similar inequalities for G^+.

Proof. Let $q \in l^2$ and $\|q\|_{l^2} \leq R$. Then $\|q\|_{l^\infty} \leq R$ and (1.18) implies that

$$\|G(q)\|_{l^2} \leq C(\|q\|_{l^\infty}) \|q\|_{l^2} \leq C(R) \|q\|_{l^2}.$$

Hence, G acts from l^2 into itself. The remaining parts of the lemma are similar. $\square$

Since $b = m^{-1/2}$ is a bounded sequence and the operators ∂^+ and ∂^- are bounded in l^2, the operator F in (1.11) acts from $l^2 \times l^2$ into itself and is a bounded locally Lipschitz continuous operator in that space. Therefore, the standard infinite dimensional version of the classical Picard theorem (see, e. g., [Dalec'kii and Krein (1974); Zeidler (1986); Zeidler (1995b)]) implies the following local well-posedness result

Theorem 1.1 *Under assumptions (i) and (ii), for every*

$$u^{(0)} = \left(r^{(0)}, s^{(0)}\right) \in l^2 \times l^2$$

there exists a unique solution

$$u = (r, s) \in C^1\left((-a, a), l^2 \times l^2\right)$$

of problem (1.13), (1.17). For any ball $B \subset l^2 \times l^2$ of initial data $u^{(0)}$ the number $a \in \mathbb{R}$ can be chosen independent on $u^{(0)}$ and for every $a' \in (0, a)$ the solution u as an element of $C^1\left([-a', a'], l^2 \times l^2\right)$ depends continuously on the initial data $u^{(0)} \in B$.

The problem of global existence is more delicate. A simple result of such kind is

Theorem 1.2 *Suppose that assumptions (i) and (ii), with (1.18) replaced by stronger inequality*

$$\left|U'_n(r_1) - U'_n(r_2)\right| \leq C\left|r_1 - r_2\right|, \quad r_1, r_2 \in \mathbb{R}, \tag{1.19}$$

are satisfied. Then for every $u^{(0)} \in l^2 \times l^2$ problem (1.13), (1.17) has a unique solution defined on $\mathbb{R}$.

This theorem is a particular case of a well-known result (see, e. g., [Dalec'kii and Krein (1974)], Theorem 1.2 of Chapter 8). However, it has a limited range of applications, since Eq. (1.19) means in particular that the potential U_n has at most quadratic growth at infinity.

Theorem 1.3 *Suppose that assumptions (i) and (ii) are satisfied and either*

(a) $U_n(r) \geq 0$ for all $n \in \mathbb{Z}$ and $r \in \mathbb{R}$,

or

(b) *there exists a nondecreasing continuous function $h(r)$, $r \geq 0$, such that* $\lim_{r \to \infty} h(r) = +\infty$ *and for every $n \in \mathbb{Z}$*

$$U_n(r) \geq h(|r|), \quad r \in \mathbb{R}.$$

Then for every $u^{(0)} = \left(r^{(0)}, s^{(0)}\right) \in l^2 \times l^2$ Cauchy problem (1.13), (1.17) has a unique global solution, i. e. solution defined for all $t \in \mathbb{R}$.

Proof. It is readily verified that the Hamiltonian $H(r,s)$ defined by Eq. (1.14) is a C^1 functional on $l^2 \times l^2$ and, hence, a conserved quantity. Thus, for any solution $u(t) = \bigl(r(t), s(t)\bigr)$ of problem (1.13), (1.17)

$$H\bigl(r(t), s(t)\bigr) = H\bigl(r^{(0)}, s^{(0)}\bigr).$$

In case (a) this implies that

$$\frac{1}{2}\|s(t)\|^2 \leq H\bigl(r^{(0)}, s^{(0)}\bigr).$$

Since $\dot{r} = \partial^+(bs)$, we obtain by integration that $\|r(t)\|$ remains bounded on any finite interval of existence of solution. This is enough to conclude that the solution is defined everywhere on $\mathbb{R}$ (see, e. g., [Reed and Simon (1975)], Theorem X.74).

Now we consider case (b). Fix $H_0 \geq 0$ such that $H\bigl(r^{(0)}, s^{(0)}\bigr) \leq H_0$. Conservation of the Hamiltonian implies that

$$\frac{1}{2} s(t,n)^2 + h\bigl(|r(t,n)|\bigr) \leq H_0.$$

Let $\bar{r} > 0$ be a solution of the equation

$$h(\bar{r}) = H_0.$$

Obviously, such a solution exists. Then $|r(t,n)| \leq \bar{r}$.

Now we introduce a modified potential defined as follows. Let $\psi(r)$ be an even function such that

$$\psi(r) = \begin{cases} 1 & \text{if } 0 \leq r \leq \bar{r}, \\ -r + \bar{r} + 1 & \text{if } \bar{r} \leq r \leq \bar{r}+1, \\ 0 & \text{if } r \geq \bar{r}. \end{cases}$$

Define $\widetilde{U}_n(r)$ by the formula

$$\widetilde{U}_n(r) = \int_0^r \Bigl[\psi(\varrho)\, U'_n(\varrho) + (1 - \psi(\varrho))\varrho\Bigr] d\varrho.$$

It is a simple exercise to verify that $\widetilde{U}_n$ satisfies assumption (1.19). Hence, the modified Hamiltonian satisfies the assumptions of Theorem 1.2. On the solution $u(t) = \bigl(r(t), s(t)\bigr)$, $\widetilde{H}$ coincides with H. Therefore, $u(t)$ extends to a global solution of the modified system. An elementary, but somewhat long, calculation shows that $\widetilde{U}_n(r) \geq \widetilde{h}(|r|)$, where

$$\widetilde{h}(r) = \begin{cases} h(r), & 0 \leq r \leq \overline{r}, \\ (\overline{r}+1-r)\, h(r) + \int_{\overline{r}}^{r} h(\varrho)\, d\varrho + \left(\dfrac{r^3}{3} - \overline{r}\dfrac{r^2}{2} + \dfrac{\overline{r}^3}{6}\right), & \overline{r} \leq r \leq \overline{r}+1, \\ r^2 + \int_{\overline{r}}^{\overline{r}+1} h(\varrho)\, d\varrho + \left[\dfrac{(\overline{r}+1)^2}{3} + \dfrac{\overline{r}^3}{6}\right], & r \geq \overline{r}+1. \end{cases}$$

Since

$$\widetilde{h}(r) \geq \widetilde{h}(\overline{r}) = h(\overline{r}) = H_0 \quad \text{for } r \geq \overline{r},$$

the argument from the beginning of the proof of case (b) shows that the extended solution satisfies $|r(t)| \leq \overline{r}$ and, therefore, is a solution of the original problem. $\square$

Remark 1.1 Theorems 1.1–1.3 imply (and are equivalent to) corresponding results on well-posedness of the Cauchy problem for Eq. (1.1) on $l^2 \times l^2$. Certainly, these parallel statements can be obtained directly by similar arguments.

Local Theorem 1.1 applies to all examples considered in Section 1.1 except the singular Lennard-Jones potential. Global existence for the β-model (Example 1.3), as well as for the Toda lattice (Example 1.4), follows immediately from Theorem 1.3. For the α-model (Example 1.2) we expect that global in time solutions do exist for some set of initial data, while for all other initial data the solutions blow up at finite time.

For the lattice with Lennard-Jones potential the existence of global solution holds true for all initial data $u^{(0)} = \left(r^{(0)}, s^{(0)}\right) \in l^2 \times l^2$ such that $r^{(0)}(n) > -d$ for all $n \in \mathbb{Z}$. In this case one can follow the proof of Theorem 1.3, case (b), modifying the potential near the singularity and behind it to reduce the problem to a nonsingular one.

1.3 Harmonic lattices

Here we consider *harmonic lattices*, i. e. lattices with quadratic interaction potential

$$U_n(r) = \frac{c(n)}{2}\, r^2.$$

Throughout this section we assume that

(h) there exist positive constants m_0, M_0 and C such that

$$0 < m_0 \leq m(n) \leq M_0$$

and

$$|c(n)| \leq C.$$

The dynamics of the harmonic lattice is governed by the equation

$$m\ddot{q} = \partial^+ c \partial^- q, \quad (1.20)$$

with the phase space $\tilde{l}^2 \times l^2$ (see Eq. (1.9)). In other words, we consider solutions such that q is a C^1 function with values in $\tilde{l}^2$, while $\dot{q}$ is a C^1 function with values in l^2.

Making use the change of unknown $x = m^{1/2}q$, we see that Eq. (1.20) is equivalent to the equation

$$\ddot{x} = m^{-1/2}\partial^+ c \partial^- m^{-1/2} x \quad (1.21)$$

considered in the phase space $\tilde{l}^2 \times l^2$. Let us introduce the operator A by the formula

$$-Ax = m^{-1/2}\partial^+ c \partial^- m^{-1/2} x. \quad (1.22)$$

Notice that the operators $-\partial^+$ and ∂^- are formally adjoint and the same holds true for the operators $-m^{-1/2}\partial^+$ and $\partial^- m^{-1/2}$.

Now Eq. (1.21) becomes

$$\ddot{x} = -Ax. \quad (1.23)$$

The operator A defined by (1.22) is a bounded linear operator in the space $\tilde{l}^2$ and, actually, the range of A lies in l^2. The last follows from the fact that the operator ∂^- maps $\tilde{l}^2$ into l^2, while the multiplication by any bounded sequence, as well as the operator ∂^+, leave l^2 invariant. Moreover, being restricted to the space l^2, the operator A is self-adjoint. In what follows the restriction of A to l^2 is still denoted by A.

The Cauchy problem for Eq.(1.20), with

$$q|_{t=0} = q^{(0)} \in \tilde{l}^2, \quad \dot{q}|_{t=0} = q^{(1)} \in l^2,$$

is equivalent to the Cauchy problem for Eq. (1.23), with

$$x|_{t=0} = x^{(0)} := m^{1/2} q^{(0)} \in \tilde{l}^2, \quad \dot{x}|_{t=0} = x^{(1)} := m^{1/2} q^{(1)} \in l^2. \quad (1.24)$$

It is well-known (see, e. g., [Dalec'kii and Krein (1974); Deimling (1977)]) that the solution of (1.23), (1.24) is given by

$$x = \cos\left(tA^{1/2}\right) x^{(0)} + A^{-1/2} \sin\left(tA^{1/2}\right) x^{(1)}, \qquad (1.25)$$

where the operator functions $\cos\left(tA^{1/2}\right)$ and $A^{-1/2} \sin\left(tA^{1/2}\right)$ are defined by means of power series expansions

$$\cos\left(tA^{1/2}\right) := \sum_{k=0}^{\infty} \frac{(-1)^k t^{2k}}{(2k)!} A^k,$$

$$A^{-1/2} \sin\left(tA^{1/2}\right) := \sum_{k=0}^{\infty} \frac{(-1)^k t^{2k+1}}{(2k+1)!} A^k.$$

Being considered in the space of all bounded linear operators in $\widetilde{l^2}$, these expansions are norm convergent. The same holds true with $\widetilde{l^2}$ replaced by l^2. Note that in the notations of these functions $A^{1/2}$ has no independent meaning.

Since $x^{(1)} \in l^2$, the function $A^{-1/2} \sin\left(tA^{1/2}\right)$ in the second term on the right hand part of Eq. (1.25) can be considered as a function of *self-adjoint* operator A in l^2. A direct calculation shows that

$$\cos\left(tA^{1/2}\right) - I = \left[\sum_{k=0}^{\infty} \frac{(-1)^{k+1} t^{2k+2}}{(2k+2)!}\right] A$$

$$= -\left[\int_0^t A^{-1/2} \sin\left(\tau A^{1/2}\right) d\tau\right] A.$$

Therefore, formula (1.25) becomes

$$x = x^{(0)} - \left[\int_0^t A^{-1/2} \sin\left(\tau A^{1/2}\right) d\tau\right] Ax^{(0)} + A^{-1/2} \sin\left(tA^{1/2}\right) x^{(1)}. \quad (1.26)$$

Another way to derive (1.26) is to look for the solution of (1.23), (1.24) in the form $x = x^{(0)} + u$, where u is a function with values in l^2, and reduce the problem to the nonhomogeneous Cauchy problem

$$\ddot{u} = -Au - Ax^{(0)}, \quad u|_{t=0} = 0, \quad \dot{u}|_{t=0} = x^{(1)},$$

in the space l^2, with the operator A being self-adjoint.

The advantage of Eq. (1.26) is that, since $x^{(1)}$ and $Ax^{(0)}$ are both in l^2, the solution is expressed in terms of the function $A^{-1/2} \sin\left(tA^{1/2}\right)$ of the *self-adjoint* operator A in l^2.

Actually, A is a second order difference operator of the form

$$(Ax)(n) = a(n)\,x(n+1) + a(n-1)\,x(n-1) + b(n)\,x(n) \qquad (1.27)$$

(Jacobi operator), where

$$a(n) = -\frac{c(n)}{\sqrt{m(n)\,m(n+1)}} \qquad (1.28)$$

and

$$b(n) = \frac{c(n) + c(n-1)}{m(n)}. \qquad (1.29)$$

This follows immediately from Eq. (1.22).

Looking for solutions to Eq. (1.23) in the form[2] $x = \exp(i\omega t)\,u(n)$, we arrive at the spectral problem

$$Au - \lambda u = 0, \qquad (1.30)$$

where $\lambda = \omega^2$. The spectral theory of Jacobi operators is well-understood (see, e. g., [Teschl (2000)] and references therein). Since, due to assumption (h), the operator A is a bounded self-adjoint operator in l^2, its spectrum $\sigma(H)$ is a bounded closed subset of $\mathbb{R}$. More precisely, one has (see, e. g., [Teschl (2000)], Lemma 1.8 and (1.151))

Proposition 1.1

(i) Let

$$a_+ = \sup_{n \in \mathbb{Z}} \Big[b(n) + \big|a(n)\big| + \big|a(n-1)\big| \Big],$$

$$a_- = \inf_{n \in \mathbb{Z}} \Big[b(n) - \big|a(n)\big| - \big|a(n-1)\big| \Big].$$

Then $\sigma(A) \subset [a_-, a_+]$.

(ii) If $c(n) > 0$ for all $n \in \mathbb{Z}$, then the operator A is nonnegative,

$$\sigma(A) \subset \big[0,\, 4\,\|c\|_{l^\infty}\,\|m^{-1}\|_{l^\infty}\big],$$

and 0 is in the essential spectrum $\sigma_{ess}(A)$.

---

[2]These are standing waves if $\omega \in \mathbb{R}$.

Now let us consider the case of harmonic N-atomic lattices, i. e. we assume that $m(n + N) = m(n)$ and $c(n + N) = c(n)$ for all $n \in \mathbb{Z}$. First, we introduce (a discrete version of) the *Floquet transform* as

$$(\mathcal{U}f)(n, \theta) = \sum_{k \in \mathbb{Z}} f(n - kN) e^{ix\theta}. \tag{1.31}$$

Sometimes this transform is called *Gelfand transform*. (For the theory of continuous Floquet transform and its applications we refer to [Kuchment (1993)]). The parameter θ is called *quasimomentum*. Observe that

$$(\mathcal{U}f)(n + N, \theta) = e^{i\theta}(\mathcal{U}f)(n, \theta). \tag{1.32}$$

This is the so-called *Floquet condition*. In physics literature this condition is called the *Bloch condition*. Also we have

$$(\mathcal{U}f)(n, \theta + 2\pi) = (\mathcal{U}f)(n, \theta), \tag{1.33}$$

i. e. the Floquet transform is 2π-periodic with respect to quasimomentum. Thus, the quasimomentum θ can be considered as the angle coordinate on the unit circle $\mathbb{S}^1$.

The periodicity assumption implies that the operator A commutes with the Floquet transform

$$(\mathcal{U}Af)(n, \theta) = A(\mathcal{U}f)(n, \theta). \tag{1.34}$$

The operator A on the right hand side of Eq. (1.34) acts on the function of the variable n satisfying Floquet condition (1.32), with θ being a parameter.

Let E denote the space of all complex valued functions on the set $I_N = \{0, 1, \ldots, N-1\}$ and E_θ the space of all functions u on $\mathbb{Z}$ satisfying the Floquet condition

$$u(n + N) = e^{i\theta}u(n), \quad n \in \mathbb{Z}. \tag{1.35}$$

Being endowed with the standard inner product, E becomes an N-dimensional Euclidian space. The restriction $u|_{I_N}$ defines an isomorphism between E_θ and E. The Floquet transform is a unitary operator

$$\mathcal{U} : l^2 \to L^2(\mathbb{S}^1; E).$$

The inverse operator $\mathcal{U}^{-1}$ is defined by the formula

$$(\mathcal{U}^{-1}v)(n) = \int_{\mathbb{S}^1} v(n, \theta) \frac{d\theta}{2\pi},$$

where $v(n, \theta)$ is extended from I_N to $\mathbb{Z}$ according to (1.35).

The operator A is well-defined on E_θ and is, in fact, a self-adjoint operator. Denote this operator by $A(\theta)$.

What we have described before, can be summarized in terms of direct integrals as

Theorem 1.4 *The Floquet transform provides a unitary equivalence between the operator A and the direct integral expansion*

$$\int_{\mathbb{S}^1}^{\oplus} A(\theta) \frac{d\theta}{2\pi}.$$

For the discussion of the notion of direct integral see, e. g., [Reed and Simon (1978)].

Due to the isomorphism between E_θ and E, the operators $A(\theta)$ can be considered as acting in E. Actually, $A(\theta)$ form a real analytic family of self-adjoint operators in the space E.

Let

$$\lambda_1(\theta) \leq \lambda_2(\theta) \leq \cdots \leq \lambda_N(\theta)$$

be the eigenvalues of $A(\theta)$. The functions $\lambda_j(\theta)$, $j = 1, 2, \ldots, N$, are continuous (actually, piecewise analytic) on $\mathbb{S}^1$. They are called *band functions* or *dispersion relations*. Let

$$\lambda_j^- = \min_{\theta \in \mathbb{S}^1} \lambda_j(\theta), \quad \lambda_j^+ = \max_{\theta \in \mathbb{S}^1} \lambda_j(\theta).$$

Theorem 1.5 *The spectrum $\sigma(A)$ is absolutely continuous and*

$$\sigma(A) = \bigcup_{j=1}^{N} [\lambda_j^-, \lambda_j^+]. \tag{1.36}$$

The intervals in (1.36) are called *spectral bands*, while the intervals $(\lambda_j^+, \lambda_{j+1}^-)$ are called *spectral gaps*. Some of the gaps may be empty (closed). In general, at most $(N-1)$ gaps open up.

The eigenfunctions of $A(\theta)$ are the generalized eigenfunctions of A called *Bloch eigenfunctions*. Corresponding solutions of Eq. (1.23)

$$x(t, n) = \exp(i\omega t)\, u(n), \quad \omega^2 = \lambda_j(\theta),$$

– so-called *Bloch waves* – have infinite energy. If $c \geq 0$, then $A \geq 0$ and $\sigma(A)$ is nonnegative. Therefore, all the eigenfrequencies are real and the Bloch waves are bounded. In the N-atomic case the Bloch waves are proper analoges of plane waves that occur in the monoatomic case.

For more details about spectral theory of periodic Jacobi operators, including generalized eigenfunction expansions, we refer to [Teschl (2000)].

Now we formulate a spectral result on harmonic lattices with impurities. Let $m = \overline{m} + m^{(0)}$ and $c = \overline{c} + c^{(0)}$, where $\overline{m}$ and $\overline{c}$ are N-periodic, while

$$\lim_{|n|\to\infty} m^{(0)}(n) = \lim_{|n|\to\infty} c^{(0)}(n) = 0.$$

Denote by $\overline{A}$ the operator (1.22) with the coefficients $\overline{m}$ and $\overline{c}$.

Theorem 1.6 $\sigma_{ess}(A) = \sigma(\overline{A})$. *If in addition* $|n|\, c^{(0)}(n) \in l^1$, *then the point spectrum of A is finite and is located in $\mathbb{R} \setminus \sigma(\overline{A})$, and the essential spectrum of A is absolutely continuous.*

See [Teschl (2000)], Theorem 7.11.

Finally, let us consider the case of monoatomic harmonic lattice. The problem reduces to the equation

$$\ddot{x}(n) = c_0^2\bigl(x(n+1) + x(n-1) - 2x(n)\bigr), \qquad (1.37)$$

where we assume that $c_0 > 0$. The so-called *plane wave* solutions are given by

$$x(t,n) = \exp\bigl[i(\varkappa n \pm \omega t)\bigr] = e^{\pm i\omega t} e^{i\varkappa n},$$

where the wavelength $\varkappa^{-1}$ and the frequency ω are related by the dispersion relation

$$\omega = \pm\sqrt{2}\, c_0 \sqrt{1 - \cos \varkappa} = \pm c_0 \sin \frac{\varkappa}{2}.$$

Note that plane waves are just Bloch waves in the monoatomic case.

For the spectral theory of differential and difference operators we refer to [Atkinson (1964); Berezanskii (1968); Berezin and Shubin (1991); Dunford and Schwartz (1988b); Edmunds and Evans (1987); Glazman (1966); Lanczos (1961); Levitan and Sargsyan (1975); Reed and Simon (1978); Schechter (1981)].

1.4 Chains of coupled nonlinear oscillators

Here we consider another lattice model – an infinite *chain of coupled nonlinear oscillators*.

Let $q(n)$ be a generalized coordinate of n-th oscillator and $U_n(r)$ the potential of the oscillator, i. e. the dynamics of n-th oscillator is given by

$$\ddot{q}(n) = -U'_n(q(n)),$$

provided the interaction between the oscillators is absent. Suppose that every oscillator interacts with two its neighbors via linear forces. Then the master equations of the system are

$$\ddot{q}(n) = -U'_n(q(n)) + a(n-1)(q(n-1)-q(n)) - a(n)(q(n)-q(n+1)), \quad (1.38)$$

where $n \in \mathbb{Z}$.

Let

$$U_n(r) = -\frac{c(n)}{2} r^2 + V_n(r),$$

where $V(0) = V'(0) = 0$. This means that $r = 0$ is a rest point of each oscillator. Set

$$b(n) = c(n) - a(n) - a(n-1).$$

Then Eq. (1.38) becomes

$$\ddot{q}(n) = a(n) q(n+1) + a(n-1) q(n-1) + b(n) q(n) - V'_n(q(n)), \quad n \in \mathbb{Z}. \quad (1.39)$$

We impose the following boundary condition at infinity

$$\lim_{n \to \pm\infty} q(n) = 0. \quad (1.40)$$

This condition means that at infinity the system is at rest.

In what follows we consider the case when the oscillators are anharmonic, i. e. $V_n \neq 0$. The case of harmonic chains is described by linear equation (1.23) with $m \equiv 1$ and, hence, reduces to the spectral theory of difference operators.

We impose the following two assumptions (*cf.* assumptions (*i*) and (*ii*) of Section 1.2):

(*i'*) the sequences $a(n)$ and $c(n)$ are bounded;
(*ii'*) the potential $V_n(r)$ is C^1 on $\mathbb{R}$, $V_n(0) = V'_n(0) = 0$ and for every $R > 0$ there exists $C = C(R) > 0$ such that

$$|V'_n(r_1) - V'_n(r_2)| \leq C|r_1 - r_2|, \quad |r_1 - r_2| \leq R, \quad (1.41)$$

for all $n \in \mathbb{Z}$.

The natural configuration space that incorporates boundary condition (1.40) is the space l^2 of two-sided sequences. The corresponding phase space is $l^2 \times l^2$.

Let us introduce the linear operator

$$(Aq)(n) = a(n)\,q(n+1) + a(n-1)\,q(n-1) + b(n)\,q(n)$$

and the nonlinear operator

$$B(q)(n) = -V_n'(q(n)).$$

Under assumptions (i') and (ii') A is a bounded self-adjoint operator in l^2, while the operator B is bounded and locally Lipschitz continuous. Equation (1.39) together with boundary condition (1.40) can be naturally written as an operator differential equation

$$\ddot{q} = Aq + B(q) \tag{1.42}$$

in the space l^2.

Equation (1.42) is a Hamiltonian system on $l^2 \times l^2$, with the Hamiltonian

$$H(q,p) = \frac{1}{2}\left[\|p\|^2 - (Aq, q)\right] + \sum_{n=-\infty}^{+\infty} V_n(q(n)), \tag{1.43}$$

where $p = \dot{q}$. Under assumptions (i') and (ii') $H(p, q)$ is a C^1 functional on $l^2 \times l^2$ and, hence, H is a conserved quantity.

Consider the Cauchy problem for (1.42) with the initial data

$$q|_{t=0} = q^{(0)} \in l^2, \quad \dot{q}|_{t=0} = q^{(1)} \in l^2. \tag{1.44}$$

Its local well-posedness under assumptions (i') and (ii') is a consequence of standard results (*cf.* Theorem 1.1). Exactly as in the case of FPU lattices (see Theorems 1.2 and 1.3) we have also the following results.

Theorem 1.7 *Assume (i') and (ii') with the constant C independent of R. Then for every $q^{(0)} \in l^2$ and $q^{(1)} \in l^2$ Cauchy problem (1.42), (1.43) has a unique solution defined for all $t \in \mathbb{R}$.*

Theorem 1.8 *In addition to assumptions (i') and (ii'), assume that the operator A is non-positive, i.e. $(Aq, q) \leq 0$ for all $q \in l^2$. Suppose also that one of the following two conditions holds:*

(a) $V_n(r) \geq 0$ for all $n \in \mathbb{Z}$ and $r \in \mathbb{R}$

or

(b) there exists a nondecreasing function $h(r)$, $r \geq 0$, such that $\lim_{r \to +\infty} h(r) = +\infty$ and $V_n(r) \geq h(|r|)$ for all $n \in \mathbb{Z}$ and $r \in \mathbb{R}$.

Then for every $q^{(0)} \in l^2$ and $q^{(1)} \in l^2$ problem (1.42), (1.43) has a unique solution defined for all $t \in \mathbb{R}$.

As it follows from the next proposition, under some additional condition one can skip the assumption of nonpositivity of A in Theorem 1.8.

Proposition 1.2 *Under assumptions of Theorem 1.8, case (b), without $(Aq, q) \leq 0$, suppose that*

$$\lim_{r \to +\infty} \frac{h(r)}{r^2} = +\infty.$$

Then problem (1.42), (1.43) has a unique global solution for every $q^{(0)} \in l^2$ and $q^{(1)} \in l^2$.

Proof. Write the potential U_n in the form

$$U_n(r) = -\frac{c(n) - 2\lambda}{2} r^2 + (V_n(r) - \lambda r^2),$$

with $\lambda > 0$ large enough. Then the new operator A that corresponds to the coefficients $a(n)$ and $c(n) - 2\lambda$ is non-positive. At the same time

$$V_n(r) - \lambda r^2 \geq h(|r|) - \lambda r^2 = h(|r|)\left(1 - \lambda \frac{r^2}{h(|r|)}\right).$$

This yields

$$V_n(r) - \lambda r^2 \geq k_1 h(|r|) - k_2,$$

with some $k_1 \in (0, 1)$ and $k_2 \geq 0$. Now it is enough to apply Theorem 1.8, with $h(r)$ replaced by $k_1 h(r) - k_2$. The proof is complete. $\square$

Now we consider few examples. In all these examples $a(n)$ and $U_n(r)$ do not depend on $n \in \mathbb{Z}$.

Example 1.6 Taking $a(n) \equiv a > 0$, $c(n) \equiv 0$ and

$$V_n(r) = 1 - \cos r,$$

one obtains the *Frenkel-Kontorova model*. The master equation of this model reads

$$\ddot{q}(n) = a\,(\Delta_d q)(n) - \sin q(n), \tag{1.45}$$

where

$$(\Delta_d q)(n) = q(n+1) + q(n-1) - 2q(n)$$

is the 1-dimensional discrete Laplacian. Equation (1.45) is a straightforward discretization of the famous sin-Gordon equation

$$q_{tt} - a q_{xx} + \sin q = 0.$$

The last equation is a completely integrable system (see, e. g. [Ablowitz and Segur (1981)]). At the same time its discrete counterpart, Eq. (1.45), is *not* completely integrable.

Example 1.7 When $a(n) \equiv a > 0$, $c(n) \equiv -m^2 \leq 0$ and

$$V_n(r) = -\frac{\lambda}{k+1} r^{k+1}, \quad k > 0 \text{ integer},$$

we obtain the *discrete nonlinear Klein-Gordon (DNKG) equation* $(m^2 > 0)$

$$\ddot{q}(n) = a\,(\Delta_d q)(n) - m^2 q(n) + \lambda q^k(n) \tag{1.46}$$

and *discrete nonlinear wave (DNW) equation* $(m^2 = 0)$

$$\ddot{q}(n) = a\,(\Delta_d)(n) + \lambda q^k(n). \tag{1.47}$$

The cubic ($k=3$) and quadratic ($k=2$) cases are of particular interest.

Example 1.8 If $a(n) \equiv a > 0$, $c(n) \equiv -m^2 \leq 0$ and

$$V_n(r) = -\frac{\lambda}{p+1} |r|^{p+1}, \quad p > 0,$$

we obtain the modified discrete Klein-Gordon $(m^2 > 0)$ and nonlinear wave $(m^2 = 0)$ equations

$$\ddot{q} = a\,\Delta_d q - m^2 q + \lambda |q|^{p-1} q$$

and

$$\ddot{q} = a\,\Delta_d q + \lambda |q|^{p-1} q,$$

respectively.

Example 1.9 If $a(n) \equiv a$, $c(n) = m^2 > 0$ and
$$V_n(r) = \frac{\lambda}{4} r^4, \quad \lambda > 0,$$
then we obtain the so-called *discrete φ^4-equation*
$$\ddot{q} = a\Delta_d q + m^2 q - \lambda q^3 = 0. \tag{1.48}$$

If $\lambda > 0$ and k is odd, then it follows from Theorem 1.8 that the Cauchy problem for equaions from Example 1.7 is globally well-posed. The same holds in Example 1.8 with $\lambda > 0$ and $p > 1$. This is so, because in those cases the operator
$$A = a\Delta_d - m^2$$
is a non-positive operator defined in l^2. The Cauchy problem for the quadratic DNKG equation with any $\lambda \neq 0$ has a unique global solution for *small initial data*, as it follows from Theorem 1.9 below.

Let
$$V_n(r) = -\frac{\lambda(n)}{3} r^3,$$
where $\lambda(n)$ is a bounded sequence. Also we suppose that the operator A is negative definite, *i. e.* there exists $\alpha_0 > 0$ such that
$$(Aq, q) \leq -\alpha_0 \|q\|_{l^2}^2 \quad \forall q \in l^2.$$
Set
$$\varphi(q) = -\frac{1}{2}(Aq, q) - \frac{1}{3}\sum_{n\in\mathbb{Z}} \lambda(n) q^3(n) =: \frac{1}{2}\varphi_0(q) + \frac{1}{3}\varphi_1(q).$$
Note that $\varphi_0(q)^{1/2}$ is an equivalent norm on l^2. We have that
$$H(p,q) = \frac{1}{2}\|p\|_{l^2}^2 + \varphi(q).$$
Since
$$|\varphi_1(q)| \leq c' \|q\|_{l^3}^3 \leq c'' \|q\|_{l^2}^3,$$
there exists a constant $\kappa > 0$ such that
$$|\varphi_1(q)|^{1/3} \leq \kappa\, \varphi_0(q)^{1/2}, \quad q \in l^2. \tag{1.49}$$

Let
$$\gamma := \inf \left\{ \sup_{\tau \geq 0} \varphi(\tau q) \ : \ q \in l^2, q \neq 0 \right\}. \tag{1.50}$$

We have that
$$\gamma \geq \frac{1}{6\kappa^6}. \tag{1.51}$$

Indeed,
$$\varphi(\tau q) = \frac{\tau^2}{2} \varphi_0(q) + \frac{\tau^3}{3} \varphi_1(q).$$

If $\varphi_1(q) \geq 0$, then
$$\sup_{\tau \geq 0} \varphi(\tau q) = +\infty.$$

If $\varphi_1(q) < 0$, then, maximizing the cubic polynomial $\varphi(\tau q)$, we obtain that
$$\sup_{\tau \geq 0} \varphi(\tau q) = \varphi\left(-\frac{\varphi_0(q)}{\varphi_1(q)} q\right) = \frac{1}{6} \frac{\varphi_0(q)^3}{\varphi_1(q)^2},$$

and the required follows from (1.49).

Now let us define
$$W = \{ q \in l^2 \ : \ 0 \leq \varphi(\tau q) < \gamma \quad \forall \tau \in [0,1] \}.$$

It is readily verified that the set W is star-shaped with respect to the origin, i. e. if $q \in W$, then $\theta q \in W$ for every $\theta \in [0,1]$.

Lemma 1.2 *For every $\varrho > 0$ such that*
$$\varrho \leq \frac{9}{4\kappa}, \quad \frac{\varrho}{2} + \frac{\kappa^2}{3} \varrho^{3/2} < \gamma,$$

the set W contains the open set
$$B = \{ q \in l^2 \ : \ \varphi_0(q) < \varrho \}.$$

Proof. By (1.49),
$$\frac{\tau^2}{2} \varphi_0(q) - \frac{\tau^3 \kappa^3}{3} \varphi_0(q)^{3/2} \leq \varphi(\tau q) \leq \frac{\tau^2}{2} \varphi_0(q) + \frac{\tau^3 \kappa^3}{3} \varphi_0(q)^{3/2}.$$

Thus, $\varphi(\tau q) \geq 0$ for every $\tau \in [0,1]$, provided
$$\frac{1}{2} - \frac{\tau \kappa^3}{3} \varphi_0(q)^{1/2} \geq 0 \quad \forall \tau \in [0,1].$$

Hence, if $\varphi_0(q) \leq 9/(4\kappa^2)$, the second condition for ϱ implies that $\varphi(\tau q) < \gamma$. □

Let
$$W_* = \{q \in l^2 : \varphi_0(q) + \varphi_1(q) > 0,\ \varphi(q) < \gamma\}.$$

By the continuity of the functionals φ_0, φ_1 and φ, the set W_* is open.

Lemma 1.3 *We have that $W = W_* \cup B$.*

Proof. Due to Lemma 1.2, it suffices to show that $W = W_* \cup \{0\}$.

Let $q \in W$, $q \neq 0$. If $\varphi_1(q) \geq 0$, then $\varphi_0(q) + \varphi_1(q) > 0$ and obviously $\varphi(q) < \gamma$.

If $\varphi_1(q) < 0$, then
$$\sup_{\tau \geq 0} \varphi(\tau q) = \varphi\left(-\frac{\varphi_0(q)}{\varphi_1(q)} q\right) \geq \gamma.$$

Hence, $-\varphi_0(q)/\varphi_1(q) > 1$ and $\varphi(q) < \gamma$. This implies that $q \in W_*$.
Conversely, let $q \in W_*$. If $\varphi_1(q) \geq 0$, then
$$\sup_{\tau \in [0,1]} \varphi(\tau q) = \varphi(q) < \gamma$$

and $q \in W$.

If $\varphi_1(q) < 0$, then the inequality $-\varphi_0(q)/\varphi_1(q) > 1$ implies that
$$\sup_{\tau \in [0,1]} \varphi(\tau q) = \varphi(q)$$

and we conclude. □

Since W_* and B are open set, the set W is open.

Lemma 1.4 *The set W is bounded.*

Proof. If $\varphi_1(q) \geq 0$, then $\varphi(q) \geq \varphi_0(q)/2$ and $\varphi_0(q) > 2\gamma$.

If $\varphi_1(q) < 0$, then Lemma 1.3 shows that $\varphi_1(q) > -\varphi_0(q)$. Hence, $\varphi(q) > \varphi_0(q)/6$ and $\varphi_0(q) < 6\gamma$. Thus, $W \subset \{q \in l^2 : \varphi_0(q) < 6\gamma\}$, *i. e.* W is contained in a bonded set and we conclude. □

Now we are ready to prove

Theorem 1.9 *Suppose that the operator A is negative definite and*
$$V_n(r) = -\frac{\lambda(n)}{3} r^4,$$

where the sequence $\lambda(n)$ is bounded. Let $q^{(0)} \in W$ and $q^{(1)} \in l^2$ satisfy

$$\frac{1}{2}\|q^{(1)}\|_{l^2}^2 + \varphi(q^{(0)}) < \gamma.$$

Then the Cauchy problem for (1.42) with the initial data $q^{(0)}$ and $q^{(1)}$ has a unique global solution.

Proof. The existence and uniqueness of the local solution $q(t)$ is obvious. To prove that $q(t)$ extends to a solution defined for all $t \in \mathbb{R}$ it is enough to show that $q(t)$ remains bounded.

We show that $q(t) \in W$. Assume not. Let $t_1 > 0$ be the smallest value of $t > 0$ for which $q(t) \notin W$. Then $q(t_1) \in \partial W$, the boundary of W. Since W is a star-shaped set, then $\theta q(t_1) \in W$ for all $\theta \in [0,1]$. Hence, $\varphi(\theta q(t_1)) < \gamma$. Passing to the limit as $\theta \to 1$, we obtain that $\varphi(q(t_1)) \leq \gamma$.

If $\varphi(q(t_1)) < \gamma$, then, by definition of W and the fact that $\varphi(\theta q(t_1)) < \gamma$ $\forall \theta \in [0,1)$, we conclude that $q(t_1) \in W$, a contradiction. Hence

$$\varphi(q(t_1)) = \gamma.$$

Since the Hamiltonian H is a conserved quantity, we have that

$$\gamma = \varphi(q(t_1)) \leq H(p(t_1), q_1(t_1)) = \frac{1}{2}\|\dot{q}(t_1)\|_{l^2}^2 + \varphi(q(t_1))$$
$$= H(q^{(0)}, q^{(1)}) = \frac{1}{2}\|q^{(1)}\|_{l^2}^2 + \varphi(q^{(0)})$$
$$< \gamma.$$

This contradiction shows that $q(t) \in W$ and, hence, $q(t)$ remains bounded for all $t > 0$.

Since Eq. (1.42) is invariant with respect to the inversion of time, the same holds for $t < 0$ and the proof is complete. □

Note that the assumptions on $q^{(0)}$ and $q^{(1)}$ are satisfied when the norms $\|q^{(0)}\|$ and $\|q^{(1)}\|$ are small enough.

1.5 Comments and open problems

In Sections 1.1–1.2 we have presented, in an appropriate form, more or less commonly known results. Theorem 1.3 is an extension of the global existence result given in [Friesecke and Pego (1999)].

For many nonlinear wave type equations global solution do exist for *small*, in an appropriate sense, initial data (see, *e. g.*, [Reed and Simon (1979); Reed (1976)]). The following problem remains open.

Problem 1.1 *Is it possible to find a norm in the phase space so that smallness of initial data would imply global existence for the α-model?*

The study of FPU lattices was initiated by E. Fermi, J. Pasta and S. Ulam in [Fermi *et. al* (1955)]. Since that time it is appeared an extensive physics literature on FPU type chains, as well as articles devoted to numerical simulation of lattice systems. See [Braun and Kivshar (1998); Braun and Kivshar (2004); Eilbeck (1991); Eilbeck and Flesch (1990); Flach and Willis (1998); Flytzanis *et. al* (1989); Peyrard *et. al* (1986); Rosenau (1989); Wattis (1993a); Wattis (1993b); Wattis (1996)], to mention a few. Applications of the Frenkel-Kontorova model in physics are discussed in [Braun and Kivshar (1998); Braun and Kivshar (2004)].

The contents of Section 1.4 is borrowed from [Bak and Pankov (to appear)]. In fact, Theorem 1.9 is a discrete counterpart of a result on nonlinear hyperbolic equations obtained in [Sattinger (1968)] (see also [Lions (1969)]).

We point out the following problems.

Problem 1.2 *Investigate global well-posedness of the Cauchy problem for DNKG and DNW equations in the case when $\lambda < 0$ and k is odd.*

Problem 1.3 *Study global well-posedness of Cauchy problem for the discrete φ^4-equation.*

Since DNKG and DNW equations are discrete counterparts of continuum KG and wave equations, it is natural to study the following

Problem 1.4 *When DNKG and DNW equations possess nonlinear scattering?*

Basic facts on nonlinear scattering theory (for classical continuum wave equations) can be found in [Reed (1976); Reed and Simon (1979)] and [Strauss (1989)].

Chapter 2

Time Periodic Oscillations

2.1 Setting of problem

Consider periodic in time solutions of system (1.1):

$$m(n)\ddot{q}(n) = U'_{n+1}\bigl(q(n+1) - q(n)\bigr) - U'_n\bigl(q(n) - q(n-1)\bigr), \qquad (2.1)$$

with the configuration space $\widetilde{l}^2$. Such solutions are often called *discrete breathers* (see, e. g. [Aubry (1997)]). Throughout this section we make use the following assumptions

(i) There exist $M_0 \geq m_0 > 0$ such that

$$m_0 \leq m(n) \leq M_0, \quad n \in \mathbb{Z}.$$

(ii) The potential U_n is a C^1 function on $\mathbb{R}$ and $U_n(0) = U'_n(0) = 0$.

Further restrictions will be imposed in subsequent sections. Note that for (local) well-posedness we need slightly more restrictive assumption (see Section 1.2, assumption (ii)).

Recall that (2.1) can be written as the following operator differential equation in $\widetilde{l}^2$ (see (1.6))

$$m\ddot{q} = \partial^+ G(\partial^- q), \qquad (2.2)$$

where

$$G(q)(u) = U'_n\bigl(q(n)\bigr),$$

and ∂^+ and ∂^- are the right and left differences, respectively.

Now we introduce a variational framework that serves T-periodic (in time) problem for Eq. (2.1). Denote by $\widetilde{H}$ the space of all T-periodic in t

functions

$$q(t) = \{q(t,n)\} \in H^1(0,T;\tilde{l}^2)$$

such that

$$\dot{q} = \{\dot{q}(t,n)\} \in L^2(0,T;l^2).$$

Endowed with the norm

$$\|q\|_{\tilde{H}} = \left[\|q\|^2_{L^2(0,T;\tilde{l}^2)} + \|\dot{q}\|^2_{L^2(0,T;l^2)}\right]^{1/2},$$

the space $\tilde{H}$ is a Hilbert space. A more explicit form for $\|q\|_{\tilde{H}}$ is

$$\|q\|_{\tilde{H}} = \left\{\int_0^T \|\dot{q}(t)\|^2_{l^2}\, dt + \int_0^T \|\partial^- q(t)\|^2_{l^2}\, dt + \int_0^T |q(t,0)|^2 dt\right\}^{1/2}.$$

Denote by H the subspace of $\tilde{H}$ that consists of all $q \in \tilde{H}$ such that

$$\int_0^T q(t,0)\, dt = 0. \tag{2.3}$$

This is a closed 1-codimensional subspace of $\tilde{H}$. The space H is considered as a Hilbert space with the norm

$$\|q\| = \left[\int_0^T \|\dot{q}(t)\|^2_{l^2}\, dt + \int_0^T \|\partial^- q(t)\|^2_{l^2}\, dt\right]^{1/2}$$

$$= \left\{\sum_{n\in\mathbb{Z}} \int_0^T \left[(\dot{q}(t,n))^2 + (q(t,n) - q(t,n-1))^2\right] dt\right\}^{1/2}. \tag{2.4}$$

This norm is equivalent to $\|\cdot\|_{\tilde{H}}$, because on H the term

$$\int_0^T |q(t,0)|^2 dt$$

can be estimated above by the term

$$\int_0^T |\dot{q}(t,0)|^2 dt.$$

On the space $\widetilde{H}$ we consider the functional

$$J(q) := \frac{1}{2} \int_0^T \|m^{1/2}\dot{q}(t)\|_{l^2}^2 \, dt - \sum_{n \in \mathbb{Z}} \int_0^T U_n\big(\partial^- q(t, n)\big) \, dt$$

$$= \sum_{n \in \mathbb{Z}} \left\{ \int_0^T \left[\frac{m(n)}{2} |\dot{q}(t,n)|^2 - U_n\big(q(t,n) - q(t, n-1)\big) \right] dt \right\}. \quad (2.5)$$

Proposition 2.1 *Assume (i) and (ii). Then J is a C^1 functional on $\widetilde{H}$.*

Proof. First we prove that the functional J is well-defined on $\widetilde{H}$. Let $q \in \widetilde{H}$. Since the embeddings

$$H^1(0, T) \subset C([0, T]) \subset L^\infty(0, T)$$

are continuous, we have[1] for every $n \in \mathbb{Z}$

$$\begin{aligned}
\|q(\cdot, n) - q(\cdot, n-1)\|_{L^\infty} &\leq c\Big(\|\dot{q}(\cdot, n) - \dot{q}(\cdot, n-1)\|_{L^2} \\
&\quad + \|q(\cdot, n) - q(\cdot, n-1)\|_{L^2} \Big) \\
&\leq c\Big(\|\dot{q}(\cdot, n)\|_{L^2} + \|\dot{q}(\cdot, n-1)\|_{L^2} \\
&\quad + \|q(\cdot, n) - q(\cdot, n-1)\|_{L^2} \Big).
\end{aligned}$$

Hence,

$$\sum_{n \in \mathbb{Z}} \|q(\cdot, n) - q(\cdot, n-1)\|_{L^\infty}^2 \leq c\|q\|^2 \leq c\|q\|_{\widetilde{H}}^2. \quad (2.6)$$

Assumption *(ii)* implies that there exists $r_0 > 0$ such that for all $n \in \mathbb{Z}$

$$|U_n(r)| \leq cr^2 \quad \text{if } |r| \leq r_0. \quad (2.7)$$

Due to (2.6) there exists $N_0 > 0$ such that

$$\|q(\cdot, n) - q(\cdot, n-1)\|_{L^\infty} \leq r_0,$$

provided $|n| > N_0$. Therefore, Eqs. (2.5), (2.6) and (2.7) yield

$$|J(q)| \leq c\|q\|_{\widetilde{H}}^2 + \sum_{|n| \leq N_0} \int_0^T \big|U_n\big(q(t,n) - q(t, n-1)\big)\big| \, dt < \infty.$$

---

[1] Everywhere in this chapter the L^p-norms are taken over the interval $(0, T)$ unless otherwise stated.

Now we prove that the Fréchet derivative of J exists and is continuous. For $h \in \widetilde{H}$, we have to prove that

$$\langle J'(q), h \rangle = \int_0^T \left(m\dot{q}(t), \dot{h}(t) \right) dt - \sum_{n \in \mathbb{Z}} \int_0^T U_n'\left(\partial^- g(t, n)\right) \partial^- h(t, n) \, dt$$

$$= \int_0^T \left(m\dot{q}(t), \dot{h}(t) \right) dt + \sum_{n \in \mathbb{Z}} \int_0^T \partial^+ U_n'\left(\partial^- q(t, n)\right) h(t, n) \, dt. \quad (2.8)$$

In more details,

$$\langle J'(q), h \rangle = \sum_{n \in \mathbb{Z}} \int_0^T \Big\{ m(n)\, \dot{q}(t, n)\, \dot{h}(t, n)$$
$$+ \Big[U_{n+1}'\big(q(t, n+1) - q(t, n)\big)$$
$$- U_n'\big(q(t, n) - q(t, n-1)\big) \Big] h(t, n) \Big\} dt.$$

The derivative of the quadratic part of J is easy to treat. Hence, it suffices to check that

$$\langle \Phi'(q), h \rangle = \sum_{n \in \mathbb{Z}} \int_0^T U_n'\left(\partial^- q(t, n)\right) \partial^- h(t, n) \, dt,$$

where

$$\Phi(q) := \sum_{n \in \mathbb{Z}} \int_0^T U_n\left(\partial^- q(t, n)\right) dt.$$

Thus, we have to prove that if $h \to 0$ in $\widetilde{H}$, then

$$\left| \sum_{n \in \mathbb{Z}} \int_0^T \Big[U_n\big(\partial^- q(t, n) + \partial^- h(t, n)\big) - U_n\big(\partial^- q(t, n)\big) \right.$$
$$\left. - U_n'\big(\partial^- q(t, n)\big) \partial^- h(t, n) \Big] dt \right| = o(\|h\|).$$

Using the Lagrange mean value theorem, we get, with some

$$\lambda_n = \lambda_n(t) \in (0, 1),$$

that the left-hand side here does not exceed

$$\sum_{n\in\mathbb{Z}} \left\| U'_n\bigl(\partial^- q(\cdot,n) + \lambda_n \partial^- h(\cdot,n)\bigr) - U'_n\bigl(\partial^- q(\cdot,n)\bigr) \right\|_{L^2} \left\| \partial^- h(\cdot,n) \right\|_{L^2}$$

$$\leq \|h\| \left[\sum_{n\in\mathbb{Z}} \left\| U'_n\bigl(\partial^- q(\cdot,n) + \lambda_n \partial^- h(\cdot,n)\bigr) - U'_n\bigl(\partial^- q(\cdot,n)\bigr) \right\|_{L^2}^2 \right]^{1/2}$$

$$=: \|h\| B(h).$$

Therefore, the result follows if $B(h) \to 0$ as $\|h\| \to 0$. By Eq. (2.6), given $\varepsilon > 0$ there exists an integer $N_\varepsilon > 0$ such that

$$\sum_{|n| \geq N_\varepsilon} \left\| \partial^- q(\cdot,n) \right\|_{L^\infty}^2 < \varepsilon.$$

If $\|h\|$ is small enough, we also have

$$\sum_{n \in \mathbb{Z}} \left\| \partial^- h(\cdot,n) \right\|_{L^\infty}^2 \leq \varepsilon.$$

Hence, by (ii),

$$\sum_{|n| \geq N_\varepsilon} \left\| U'_n\bigl(\partial^- q(\cdot,n) + \lambda_n \partial^- h(\cdot,n)\bigr) - U'_n\bigl(\partial^- q(\cdot,n)\bigr) \right\|_{L^2}^2$$

$$\leq C \sum_{|n| \geq N_\varepsilon} \left(\left\| \partial^- q(\cdot,n) + \lambda \partial^- h(\cdot,n) \right\|_{L^2}^2 + \left\| \partial^- q(\cdot,n) \right\|_{L^2}^2 \right)$$

$$\leq C\varepsilon,$$

with $C > 0$ independent of ε. Furthermore, U'_n is continuous, hence, uniformly continuous on compact sets and

$$\sum_{|n| < N_\varepsilon} \left\| U'_n\bigl(\partial^- q(\cdot,n) + \lambda_n \partial^- h(\cdot,n)\bigr) - U'_n\bigl(\partial^- q(\cdot,n)\bigr) \right\|_{L^2} \leq C\varepsilon,$$

provided $\|h\|$ is small enough, and we have proved the formula for $\Phi'(q)$.

To prove that J' is continuous, it suffices to check the continuity of Φ'. Let $q^{(k)} \to q$ in $\widetilde{H}$. Given $\varepsilon > 0$, by Eq. (2.6), for k large enough we have

$$\sum_{n \in \mathbb{Z}} \left\| \partial^- q^{(k)}(\cdot,n) - \partial^- q(\cdot,n) \right\|_{L^\infty}^2 \leq \varepsilon$$

and, for some N_ε,

$$\sum_{|n| \geq N_\varepsilon} \left\| \partial^- q(\cdot,n) \right\|_{L^\infty}^2 \leq \varepsilon.$$

By the uniform convergence of $q^{(k)}(t,n)$ on $[0,T]$, we have that

$$\sup_{\|h\|=1} \left| \sum_{|n|<N_\epsilon} \int_0^T \left[U_n'(\partial^- q^{(k)}(\cdot,n)) - U_n'(\partial^- q(\cdot,n)) \right] \partial^- h(\cdot,n)\, dt \right| \leq \varepsilon$$

provided k is large enough. Also we have

$$\sup_{\|h\|=1} \left| \sum_{|n|\geq N_\epsilon} \int_0^T \left[U_n'(\partial^- q^{(k)}(t,n)) - U_n'(\partial^- q(t,n)) \right] \partial^- h(t,n)\, dt \right|$$

$$\leq \sup_{\|h\|=1} \sum_{|n|\geq N_\epsilon} \int_0^T \left[\left| U_n'(\partial^- q^{(k)}(t,n)) \right| + \left| U_n'(\partial^- q(t,n)) \right| \right]$$
$$\times \left| \partial^- h(t,n) \right| dt$$

$$\leq C \sup_{\|h\|=1} \sum_{|n|\geq N_\epsilon} \int_0^T \left[|\partial^- q^{(k)}(t,n)| + |\partial^- q(t,n)| \right] |\partial^- h(t,n)|$$

$$\leq C \left[\sum_{|n|\geq N_\epsilon} \left(\|\partial^- q^{(k)}(\cdot,n)\|_{L^2}^2 + \|\partial^- q(\cdot,n)\|_{L^2}^2 \right) \right]^{1/2}$$

$$\leq C\varepsilon.$$

Hence,

$$\sup_{\|h\|=1} \left| \langle \Phi'(q^{(k)}) - \Phi'(q), h \rangle \right| \to 0,$$

and this completes the proof. □

Proposition 2.2 *Under assumptions (i) and (ii) any critical point $q \in \widetilde{H}$ of J is a solution of (2.1). Moreover, $q \in C^1(\mathbb{R}; \widetilde{l}^2)$ and $\dot{q} \in C^1(\mathbb{R}; l^2)$.*

Proof. Equation (2.8) shows that q is a weak solution of (2.2) (equivalently, (2.1)), i. e. a solution in the sense of (vector) distributions [Lions and Magenes (1972)]. Since $q \in H^1_{loc}(\mathbb{R}; \widetilde{l}^2)$, we have that $q \in C(\mathbb{R}; \widetilde{l}^2)$ (see Theorem A.1 and remarks right after it). Equation (2.1), understood in the sense of distributions, shows that $\ddot{q} \in C(\mathbb{R}; l^2)$ and the proof is complete. □

Now we reduce the problem to the subspace $H \subset \widetilde{H}$. Certainly, under assumptions (i) and (ii), the restriction $J|_H$ of the functional J to H is of class C^1 and we can consider critical points of J on H.

Proposition 2.3 *Assume (i) and (ii). Any critical point of $J|_H$ on H is a critical point of J on $\widetilde{H}$.*

Proof. Let H_0 denote the one-dimensional subspace of $\widetilde{H}$ generated by the vector $\mathbf{1} = (\ldots, 1, 1, 1, \ldots)$. It is easy to see that H_0 is a complement (not orthogonal) to H, i. e. $H_0 \cap H = \{0\}$ and $H + H_0 = \widetilde{H}$. It is trivial to verify that for every $q \in \widetilde{H}$ and $u \in H_0$

$$J(q + u) = J(q),$$

i. e. J is constant along H_0-direction. Therefore, the H_0-component of $J'(q)$ vanishes:

$$\langle J'(q), h \rangle = 0$$

for every $h \in H_0$. Hence, $J'(q) = 0$ provided the H-component of $J'(q)$ vanishes. This implies the required. □

Remark 2.1 Clearly, all critical points of J on $\widetilde{H}$ i. e. all solutions of Eq. (2.1), are of the form $q + u$, where $q \in H$ is a critical point of $J|_H$ (solution of (2.1) in H) and $u \in H_0$.

Now let us consider stationary (constant in time) solutions. In what follows we shall need the following

Proposition 2.4 *Under assumptions (i) and (ii), for any constant solution $q \in \widetilde{H}$ we have*

$$U'_n(\partial^- q(n)) = 0 \quad \text{for all } n \in \mathbb{Z}.$$

Proof. Suppose the contrary, that is, $\dot{q} \equiv 0$ and there exists $n_0 \in \mathbb{Z}$ such that

$$U'_{n_0}(\partial^- q(n_0)) = \gamma \neq 0.$$

Then Eq. (2.1) implies immediately that

$$U'_n(\partial^- q(n)) = \gamma \quad \text{for all } n \in \mathbb{Z}.$$

This is impossible, because $\partial^- q(n) \to 0$ as $n \to \pm\infty$ and, due to assumption (ii), $U'_n(r) \to 0$ as $r \to 0$ uniformly with respect to $n \in \mathbb{Z}$. □

Remark 2.2 Proposition 2.4 means that for any constant solution q no pair of neighbor particles undergoes any force and the displacement $\partial^- q(n)$ is a critical point of the interaction potential $U_n(r)$.

2.2 Positive definite case

Here we consider N-atomic lattices satisfying assumptions (i) and (ii) of Section 2.1 and

(iii) $U_n(r) = -\dfrac{c(n)}{2} r^2 + V_n(r), \quad c(n) > 0,$

where $V_n(0) = V_n'(0) = 0$ and $V_n'(r) = o(r)$ as $r \to 0$;

(iv) there exists $\theta > 2$ such that

$$V_n'(r)\, r \geq \theta V_n(r) \geq 0$$

and there exists $r_0 > 0$ such that $V_n(r) > 0$ if $|r| \geq r_0$;

(v) $m(n+N) = m(n)$, $c(n+N) = c(n)$ and $V_{n+N}(r) = V_n(r)$.

In the spatial periodicity assumption (v) we always assume that $N > 0$ is the minimal period.

Standard results on differential inequalities (see, e. g. [Hartman (2002)]) show that assumption (iv) implies

$$V_n(r) \geq d\,|r|^\theta - d_0, \quad d > 0,\ d_0 \geq 0. \tag{2.9}$$

In fact, one can prove that for every $d_0 > 0$ there exists $d > 0$ such that inequality (2.9) holds.

Let us also point out that $c(n) > 0$ and periodicity imply the existence of $c_0 > 0$ and $c_1 > 0$ such that $c_0 \leq c(n) \leq c_1$ for all $n \in \mathbb{Z}$.

Under the assumptions imposed here, the interaction potential $U_n(r)$ has a strict local maximum at 0 and admits at least two local minima. This means that the interparticle forces are repulsive-attractive, i. e. repulsive for small displacements and attractive for large displacements.

The main result of the section is

Theorem 2.1 *Assume that $(i)-(v)$ hold. Then there exists $T_0 > 0$ such that for every $T > T_0$ Eq. (2.1) admits a nonconstant T-periodic $\widetilde{l^2}$-valued solution.*

To prove Theorem 2.1, we are going to find a critical point of the functional J on the space H. Due to Propositions 2.2 and 2.3, any such critical point is a solution of (2.1).

Thus, we consider the restriction $J|_H$ of J to H. However, to simply the notation, we skip H here and still denote this restriction by J. In the case we consider, the functional J becomes

$$J(q) = \frac{1}{2}\int_0^T \|m^{1/2}\dot{q}(t)\|_{l^2}^2 dt + \frac{1}{2}\sum_{n\in\mathbb{Z}}\int_0^T c(n)\left[\partial^- q(t,n)\right]^2 dt$$
$$- \sum_{n\in\mathbb{Z}}\int_0^T V_n\bigl(\partial^- q(t,n)\bigr)\, dt. \tag{2.10}$$

Moreover, Eq. (2.8) gives

$$\langle J'(q), h\rangle = \int_0^T \bigl(m\,\dot{q}(t), \dot{h}(t)\bigr)\, dt + \sum_{n\in\mathbb{Z}}\int_0^T c(n)\,\partial^- q(t,n)\,\partial^- h(t,n)\, dt$$
$$- \sum_{n\in\mathbb{Z}}\int_0^T V_n'\bigl(\partial^- q(t,n)\bigr)\,\partial^- h(t,n)\, dt. \tag{2.11}$$

Let us introduce the operator $L : H \to H$ defined by

$$(Lq, h) = \int_0^T (m\,\dot{q}, \dot{h})_{\bar{l}^2}\, dt + \sum_{n\in\mathbb{Z}}\int_0^T c(n)\,\partial^- q(t,n)\,\partial^- h(t,n)\, dt \tag{2.12}$$

for all $h \in H$ and the functional

$$\Phi(q) := \sum_{n\in\mathbb{Z}}\int_0^T V_n\bigl(\partial^- q(t,n)\bigr)\, dt. \tag{2.13}$$

Then L is a bounded self-adjoint operator in H,

$$J(q) = \frac{1}{2}(Lq, q) - \Phi(q), \tag{2.14}$$

and

$$\langle J'(q), h\rangle = (Lq, h) - \langle \Phi'(q), h\rangle. \tag{2.15}$$

As the first step we apply the mountain pass theorem without (PS) (see Appendix C.2) to prove that the functional J on H admits a Palais-Smale sequence, *i. e.* a sequence $q^{(k)} \in H$ such that $J(q^{(k)})$ is convergent and $J'(q^{(k)}) \to 0$.

Proposition 2.5 *Under assumption (i)—(v), for every $T > 0$ the functional J on H admits a Palais-Smale sequence $q^{(k)}$. Moreover, $\|q^{(k)}\|$ is bounded from below and above by two positive constants and $J(q^{(k)})$ converges to the mountain pass level b.*

Proof. To prove the proposition we verify the assumptions of Theorem C.3.

From Eq. (2.10) we see that there exists $\alpha > 0$ such that

$$J(q) \geq \frac{\alpha}{2} \|q\|^2 - \sum_{n \in \mathbb{Z}} \int_0^T V_n(\partial^- q(t,n)) \, dt. \tag{2.16}$$

Assumption (iii) shows that given $\varepsilon > 0$ there exists $r_\varepsilon > 0$ such that

$$|V_n(r)| \leq \varepsilon r^2 \quad \text{if } |r| \leq r_\varepsilon.$$

Let $\varrho = r_\varepsilon / c^{1/2}$, where c is the constant from inequality (2.6). If $\|q\| = \varrho$, then (2.6) yields

$$\|\partial^- q(\cdot, n)\|_{L^\infty} \leq r_\varepsilon, \quad n \in \mathbb{Z}.$$

Hence,

$$\left|V_n(\partial^- q(t,n))\right| \leq \varepsilon \left(\partial^- q(t,n)\right)^2$$

and

$$\left|\sum_{n \in \mathbb{Z}} \int_0^T V_n(\partial^- q(t,n)) \, dt\right| \leq \varepsilon \int_0^T \|\partial^- q(t,n)\|_{l^2}^2 \, dt \leq \varepsilon \|q\|^2.$$

Taking ε small enough, we obtain from inequality (2.16) that

$$J(q) \geq \frac{\alpha}{4} \varrho^2 > 0 = J(0), \quad \|q\| = \varrho.$$

Now we fix q such that $q(t,0) = \varphi(t) \neq 0$ and $q(t,n) \equiv 0$ if $n \neq 0$. Then

$$J(\lambda q) = \frac{\lambda^2 m(0)}{2} \int_0^T \dot\varphi(t)^2 dt + \frac{\lambda}{2}\left(c(1) + c(0)\right) \int_0^T \varphi(t)^2 dt$$
$$- \int_0^T \left[V_1(\lambda \varphi(t)) + V_0(\lambda \varphi(t))\right] dt.$$

Inequality (2.9) implies that

$$J(\lambda q) \leq A\lambda^2 - B|\lambda|^\theta + B_0,$$

with $A > 0$, $B > 0$, $B_0 \geq 0$. Since $\theta > 2$,

$$J(\lambda q) < 0$$

for $\lambda > 0$ large enough.

Thus, Theorem C.3 applies and there exists a Palais-Smale sequence $q^{(k)}$ at the level b defined by Eq. (C.3), i. e.

$$J(q^{(k)}) \to b,$$

and $b \geq \alpha \varrho^2/4 > 0$.

We can assume that $J(q^{(k)}) \geq b/2 > 0$. Since the potentials V_n are nonnegative, Eq. (2.10) implies that

$$J(q) \leq K\|q\|^2, \quad q \in H,$$

and, therefore, $\|q^{(k)}\|$ is bounded below by a positive constant.

If remains to prove that $\|q^{(k)}\|$ is bounded above. Let $\varepsilon > 0$,

$$\varepsilon_k = \|J'(q^{(k)})\|$$

and k large enough so that

$$J(q^{(k)}) \leq b + \varepsilon/2.$$

Then

$$2b + \varepsilon + \varepsilon_k \|q^{(k)}\| \geq 2J(q^{(k)}) - \langle J'(q^{(k)}), q^{(k)} \rangle$$
$$= \sum_{n \in \mathbb{Z}} \int_0^T \left[V_n'(\partial^- q^{(k)}(t,n)) \partial^- q^{(k)}(t,n) \right.$$
$$\left. - 2V_n(\partial^- q^{(k)}(t,n)) \right] dt.$$

Due to assumption (iv), we get

$$2b + \varepsilon + \varepsilon_k \|q^{(k)}\| \geq \frac{\theta - 2}{\theta} \sum_{n \in \mathbb{Z}} \int_0^T V_n'(\partial^- q^{(k)}(t,n)) \partial^- q^{(k)}(t,n) \, dt.$$

Equation (2.11) shows that the right hand part here is equal to

$$\frac{\theta - 2}{\theta} \left[\int_0^T \|m^{1/2} \dot{q}^{(k)}(t)\|_{l^2}^2 \, dt \right.$$
$$\left. + \sum_{n \in \mathbb{Z}} \int_0^T c(n) [\partial^- q^{(k)}(t,n)]^2 dt - \langle J'(q^{(k)}), q^{(k)} \rangle \right]$$
$$\geq \frac{\theta - 2}{\theta} \left[\alpha \|q^{(k)}\|^2 - \varepsilon_k \|q^{(k)}\| \right].$$

Hence,

$$\alpha \frac{\theta-2}{\theta} \|q^{(k)}\|^2 \leq 2b + \varepsilon + \varepsilon_k \frac{\theta-1}{\theta} \|q^{(k)}\|,$$

which implies that $\|q^{(k)}\|$ is bounded above. □

Actually, in the proof of Proposition 2.5 we have shown that the functional J possesses the mountain pass geometry. However, it does not satisfy the Palais-Smale condition and, therefore, the classical mountain pass theorem, Theorem C.1, does not apply. Indeed, suppose that $q^{(k)} \in H$ is a precompact Palais-Smale sequence. Then the sequence $p^{(k)}$ defined by

$$p^{(k)}(t,n) = q^{(k)}(t,n+k) - \frac{1}{T} \int_0^T q^{(k)}(t,k)\, dt$$

is also a Palais-Smale sequence, but, obviously, no subsequence of $p^{(k)}$ converges.

Proof of Theorem 2.1. Given $T > 0$, we shall first show that there exists a critical point of J in H, hence, a T-periodic solution. Next we prove that this solution is not constant, provided T is large enough. The proof is divided into several steps.

Step 1. Let $q^{(k)}$ be a Palais-Smale sequence that exists according to Proposition 2.5 and

$$u^{(k)}(n) = \int_0^T \left[\dot{q}^{(k)}(t,n)^2 + \left(q^{(k)}(t,n+1) - q^{(k)}(t,n) \right)^2 \right] dt.$$

The real valued sequence $u^{(k)} = \{u^{(k)}(n)\}_{n \in \mathbb{Z}}$ belongs to l^1 and

$$\|u^{(k)}\|_{l^1} = \|q^{(k)}\|^2.$$

Due to Proposition 2.5,

$$0 < c_1 \leq \|q^{(k)}\|^2 \leq c_2.$$

Step 2. We prove that, up to a subsequence, there exists $\delta > 0$ and n_k such that

$$u^{(k)}(n_k) \geq \delta. \tag{2.17}$$

Indeed, if not, then $u^{(k)} \to 0$ in l^∞. In this case for k large enough the norm
$$\sup_{n \in \mathbb{Z}} \left\| q^{(k)}(\cdot, n+1) - q^{(k)}(\cdot, n) \right\|_{L^\infty}$$
is small enough. Since $V_n'(r) = o(r)$ at 0, Eq. (2.11) gives
$$\langle J'(q^{(k)}), q^{(k)} \rangle \geq m_0 \int_0^T \left\| \dot{q}^{(k)}(t) \right\|_{l^2}^2 dt + \frac{c_0}{2} \sum_{n \in \mathbb{Z}} \int_0^T \left[\partial^- q^{(k)}(t,n) \right]^2 dt$$
$$\geq \alpha \|q^{(k)}\|^2.$$

Since $J'(q^{(k)}) \to 0$, then $\|q^{(k)}\| \to 0$. This contradicts the inequality
$$0 < c_1 \leq \|q^{(k)}\|^2 \leq c_2,$$
hence, proves inequality (2.17).

Step 3. Due to the periodicity assumption, the sequence of the form
$$\widetilde{q}^{(k)}(t,n) = q^{(k)}(t, n + p_k N) - \frac{1}{T} \int_0^T q^{(k)}(t, p_k N) \, dt$$
with $p_k \in \mathbb{Z}$, is also a Palais-Smale sequence at the same mountain pass level. Making such a shift, we can assume that in (2.17) we have $0 \leq n_k < N$. Passing to a subsequence, we get the inequality
$$u^{(k)}(n_0) \geq \delta \qquad (2.18)$$
for some integer n_0, $0 \leq n_0 < N$.

Step 4. Since the sequence $q^{(k)}$ is bounded in the Hilbert space H, we can assume that $q^{(k)} \to q$ weakly in H. The compactness of Sobolev embedding implies that for every $n \in \mathbb{Z}$
$$\partial^- q^{(k)}(\cdot, n) \to \partial^- q(\cdot, n)$$
strongly in $L^\infty(0,T)$. This is local, or pointwise, convergence with respect to n. Since elements $h \in H$ such that
$$\partial^- h(t,n) \neq 0$$
for all, but finite, number of $n \in \mathbb{Z}$ are dense in H, this type of convergence of $q^{(k)}$ is enough to pass to the limit in J'. Indeed, in (2.11) the first two terms are continuous with respect to weak convergence of q, while

the last term is continuous with respect to local convergence stated above. Therefore,

$$J'(q) = 0$$

and q is a critical point of J. Passing to the limit in (2.18), we obtain that

$$\int_0^T \left[\dot{q}(t,n_0)^2 + \bigl(q(t,n_0+1) - q(t,n_0)\bigr)^2\right] dt \geq \delta > 0,$$

hence, q is a nonzero solution.

Now we are going to prove that the critical point q is below the mountain pass level. Let

$$g_n(r) = \frac{1}{2} V_n'(r) r - V_n(r).$$

Assumption (iv) shows that $g_n(r) \geq 0$, and we have that

$$b = \lim_{k \to \infty} J(q^{(k)}) = \lim_{k \to \infty} \left[J(q^{(k)}) - \frac{1}{2} \langle J'(q^{(k)}), q^{(k)} \rangle \right]$$

$$= \lim_{k \to \infty} \sum_{n \in \mathbb{Z}} \int_0^T g_n\bigl(\partial^- q^{(k)}(t,n)\bigr) dt.$$

Since $g_n(r) \geq 0$ and

$$\partial^- q^{(k)}(\cdot,n) \to \partial^- q(\cdot,n)$$

in $L^\infty(0,T)$ for every $n \in \mathbb{Z}$, we obtain that

$$b \geq \lim_{k \to \infty} \sum_{|n| \leq n_0} \int_0^T g_n\bigl(\partial^- q^{(k)}(t,n)\bigr) dt = \sum_{|n| \leq n_0} \int_0^T g_n\bigl(\partial^- q(t,n)\bigr) dt.$$

On the other hand,

$$J(q) = J(q) - \frac{1}{2} \langle J'(q), q \rangle = \sum_{n \in \mathbb{Z}} \int_0^T g_n\bigl(\partial^- q(t,n)\bigr) dt.$$

Since $g(r) \geq 0$ and n_0 is an arbitrary integer, we have that

$$J(q) \leq b, \tag{2.19}$$

where b is the mountain pass level.

Step 5. To complete the proof, we show that the solution q constructed above is not a constant, provided T is large enough.

Proposition 2.4 shows that for any nonzero constant solution q

$$J(q) = -T \sum U_n(\theta_n),$$

where the sum is extended to a finite number of indices and θ_n is a nontrivial critical point of U_n. Define

$$-d_0 = \min_{n \in \mathbb{Z}} \min_{r \in \mathbb{R}} U_n(r)$$

and

$$-d_1 = \max_{n \in \mathbb{Z}} \max_j U_n(\theta_{nj}),$$

where $\{\theta_{nj}\}$ is the set of all nontrivial critical points of U_n. Obviously, $d_0 \geq d_1$ and for any nonzero constant solution the critical value is not less than Td_1.

Due to (2.19), the required result will follow if we prove that the mountain pass value is less then Td_1. Given $\eta \in (0,1)$, let us consider the pass $\{q^{(\sigma)}, \sigma \in [0, \overline{\sigma}]\}$, where $q^{(\sigma)}$ is defined by

$$q^{(\sigma)}(t, n) = 0 \quad \text{if } n \neq 0$$

and

$$q^{(\sigma)}(t, 0) = \begin{cases} \sigma \sin\left(\dfrac{2\pi}{\eta T} t\right) & \text{if } 0 \leq t \leq \eta T, \\ 0 & \text{if } \eta T \leq t \leq T. \end{cases}$$

It is not difficult to verify that for all σ large enough, $J(q^{(\sigma)}) < 0$ (see the proof of Proposition 2.5).

By definition (see Appendix C.2), the mountain pass value does not exceed $\max J(q^{(\sigma)})$. Obviously,

$$\sum_{n \in \mathbb{Z}} \int_0^T U_n\big(\partial^- q^{(\sigma)}(t, n)\big)\, dt \geq -2\eta T d_0.$$

Then, for a suitable choice of η and for all T large we have

$$J(q^{(\sigma)}) \leq \frac{2\overline{\sigma}^2 \pi^2}{\eta^2 T} + 2\eta T d_0 < T d_1$$

and the proof is complete. $\square$

2.3 Indefinite case

2.3.1 *Main result*

In this section we consider the case when the quadratic part of the functional J is not positive definite. Here we keep assumptions (i), (ii), (iv) and (v). But now we allow $c(n)$ to be of arbitrary sign, i. e. we replace (iii) by

(iii') $U_n(r) = -\dfrac{c(n)}{2} r^2 + V_n(r)$, where $V_n(0) = V'_n(0) = 0$ and $V'_n(r) = o(r)$ as $r \to 0$.

This assumption means that the interaction potential $U_n(r)$ can be either of the same type as in Section 2.2, i. e. repulsive-attractive, or purely repulsive (the last case takes place if $c(n) \leq 0$).

We start with some properties of the operator L defined by (2.12).

If $c(n)$ vanishes for some $n = n_0$, then it is clear that the operator L is not invertible. Indeed, the vector $q \in H$ defined by

$$q(n) = \begin{cases} 0 & \text{if } n < n_0 \\ 1 & \text{if } n \geq n_0 \end{cases}$$

satisfies $Lq = 0$. This case will be considered separately later on, in Subsection 2.4.1. On the other hand, if $c(n) > 0$ for all $n \in \mathbb{Z}$, then L is positive definite and the spectrum $\sigma(L) \subset (0, +\infty)$. This is the case of Section 2.2.

Assume that $c(n) \neq 0$ for every $n \in \mathbb{Z}$ and $c(n)$ is negative for at least one value of n.

Let

$$A = \{n \in \mathbb{Z} \mid c(n) > 0\}.$$

Obviously $A \neq \mathbb{Z}$. Let Y be the subspace of constant functions satisfying

$$q(n) - q(n-1) = 0 \quad \text{for all } n \in A$$

and Z its orthogonal complement. The orthogonality condition means that $h \in Z$ if and only if

$$\int_0^T \partial^- h(t,n)\, dt = 0, \quad n \notin A.$$

Note that both spaces Y and Z are invariant with respect to L. Indeed, if $q \in Y$, then for every $h \in Z$ we have

$$(Lq, h) = \sum_{n \in \mathbb{Z}} \int_0^T \left[m(n)\, \dot{q}(t,n)\, \dot{h}(t,n) + c(n)\, \partial^- q(t,n)\, \partial^- h(t,n) \right] dt$$

$$= \sum_{n \in \mathbb{Z}} \int_0^T c(n)\, \partial^- q(n)\, \partial^- h(t,n)\, dt$$

$$= \left(\sum_{n \in A} + \sum_{n \notin A} \right) c(n)\, \partial^- q(n) \int_0^T \partial^- h(t,n)\, dt$$

$$= 0.$$

Hence, $Lq \in Y$. Since $Y \perp Z$, the subspace Z is also invariant.

Let

$$\alpha = \inf_{n \in A} c(n) > 0$$

(α is not defined when $A = \emptyset$) and

$$\beta = - \inf_{n \in \mathbb{Z}} c(n) > 0.$$

Lemma 2.1 *Assume that $c(n)$ is nowhere vanishing bounded sequence and $T < \pi/\sqrt{\beta}$. Then there exists $\lambda > 0$ such that*

$$(Lq, q) \leq -\lambda \|q\|^2, \quad q \in Y,$$

and

$$(Lq, q) \geq \lambda \|q\|^2, \quad q \in Z.$$

In particular, L is an invertible operator in H.

Proof. The case $q \in Y$ is trivial, since on Y

$$(Lq, q) = \sum_{n \notin A} c(n) \|\partial^- q(n)\|_{L^2}^2 = T^2 \sum_{n \notin A} c(n) |\partial^- q(n)|^2$$

and

$$\|q\|^2 = \sum_{n \notin A} \|\partial^- q(n)\|_{L^2}^2 = T^2 \sum_{n \notin A} |\partial^- q(n)|^2.$$

Recall that every $q \in Y$ is a constant function of t.

Let $q \in Z$. Since $q \perp Y$, then for all $n \notin A$ we have that

$$\int_0^T \partial^- q(t,n)\,dt = 0.$$

Therefore, for every $n \notin A$

$$\|\partial^- \dot{q}(\cdot,n)\|_{L^2} \geq \frac{2\pi}{T}\|\partial^- q(\cdot,n)\|_{L^2},$$

as it follows by means of an elementary Fourier series argument. Hence,

$$(Lq,q) = \sum_{n \in \mathbb{Z}} \int_0^T \left[\dot{q}(t,n)^2 + c(n)(\partial^- q(t,n))^2\right] dt$$

$$\geq \sum_{n \in A} \left(\|\dot{q}(\cdot,n)\|_{L^2}^2 + c(n)\|\partial^- q(\cdot,n)\|_{L^2}^2\right)$$

$$+ \sum_{n \notin A} \left(\|\dot{q}(\cdot,n)\|_{L^2}^2 - \left(\frac{\beta T^2}{4\pi^2} + \varepsilon\right)\|\partial^- \dot{q}(\cdot,n)\|_{L^2}^2 + \varepsilon\|\partial^- q(\cdot,n)\|_{L^2}^2\right).$$

Since

$$\|\partial^- \dot{q}(\cdot,n)\|_{L^2}^2 \leq 2\left(\|\dot{q}(\cdot,n)\|_{L^2}^2 + \|\dot{q}(\cdot,n-1)\|_{L^2}^2\right),$$

we have

$$(Lq,q) \geq \sum_{n \in A}\left[\left(1 - \left(\frac{\beta T^2}{2\pi^2} + 2\varepsilon\right)\right)\|\dot{q}(\cdot,n)\|_{L^2}^2 + \alpha\|\partial^- q(\cdot,n)\|_{L^2}^2\right]$$

$$+ \sum_{n \notin A, n+1 \notin A}\left[\left(1 - \left(\frac{\beta T^2}{\pi^2} + 4\varepsilon\right)\right)\|\dot{q}(\cdot,n)\|_{L^2}^2 + \varepsilon\|\partial^- q(\cdot,n)\|_{L^2}^2\right]$$

$$+ \sum_{n \notin A, n+1 \in A}\left[\left(1 - \left(\frac{\beta T^2}{2\pi^2} + \varepsilon\right)\right)\|\dot{q}(\cdot,n)\|_{L^2}^2 + \varepsilon\|\partial^- q(\cdot,n)\|_{L^2}^2\right].$$

Choosing $\varepsilon > 0$ small enough, we obtain

$$(Lq,q) \geq \lambda \|q\|^2,$$

with some positive λ. □

Remark 2.3 In [Arioli and Szulkin (1997)] it is shown that if $c(n)$ is independent of n and

$$\beta = -c(n) \geq \pi^2/T^2,$$

then $0 \in \sigma(L)$.

Now we state the main result of the section. Recall that we assume that not all $c(n)$ are positive.

Theorem 2.2 *Assume (i), (ii), (iii'), (iv) and (v), with $c(n) \neq 0$ for all $n \in \mathbb{Z}$. Then for all $T > 0$ system (2.1) admits a nonzero T-periodic solution $q \in H$. The solution is nonconstant if $c(n) < 0$ for all $n \in \mathbb{Z}$. If $c(n)$ takes both signs, then there exist $T_0 > 0$ and $T_1 > 0$, where T_0 depends on V_n and positive $c(n)$, while T_1 depends on $\min\{c(n)\}$, such that the solution is nonconstant, provided $T_0 < T_1$ and $T \in (T_0, T_1)$.*

The proof is given in Subsection 2.3.3.

Finally, we consider couple of examples.

Example 2.1 Consider a monoatomic chain with $c(n) = c < 0$ and

$$V(r) = \frac{d}{p}|r|^p, \quad d > 0.$$

In the case $p = 4$ this is the FPU β-model. The potential V satisfies all the assumptions of Theorem 2.2 and, hence, for every $T > 0$ there exists a nonconstant T-periodic solution, *i. e.* a breather.

Example 2.2 Consider a diatomoc chain ($N = 2$) with $c(n) < 0$ and

$$V_n(r) = \frac{d(n)}{p}|r|^p, \quad d > 0.$$

Theorem 2.2 gives the existence of breathers of arbitrary period $T > 0$. When $p = 4$ and

$$\varepsilon^2 = \frac{m(0)}{m(1)}$$

is sufficiently small, *i. e.* for a highly contrast FPU lattice, the existence of breathers was obtained in [Livi *et. al* (1997)]. These authors consider the system as a perturbation from the so-called anti-continuous limit[2], appropriately understood, with ε being a small parameter. In addition, their proof provides also the spatial exponential localization, *i. e.* that

$$|r(t,n)| := |q(t, n+1) - q(t,n)| \leq Ce^{-\alpha|n|},$$

with some $\alpha > 0$. We mention also recent paper [James and Noble (2004)] which extends the result of [Livi *et. al* (1997)] to the case of arbitrary mass ratio. This paper is based on a discrete spatial center manifold reduction.

---

[2]Probably, "decoupled limit" would be better.

2.3.2 Periodic approximations

In the proof of Theorem 2.2 we employ the so-called periodic approximations. Let k be a positive integer. We shall consider solutions of Eq. (2.1) that are T-periodic in time and kN-periodic in spatial variable $n \in \mathbb{Z}$. This means that we consider finite chains of particles with periodic boundary conditions.

Denote by $\widetilde{H}_k$ the Hilbert space of all functions $q = q(t,n)$, $t \in \mathbb{R}$, $n \in \mathbb{Z}$, T-periodic in t, kN-periodic in n and having finite norm

$$\|q\|_{k^\sim}^2 = \sum_{n=0}^{kN-1} \int_0^T \left[\dot{q}(t,n)^2 + \left(\partial^- q(t,n)\right)^2\right] dt + \int_0^T \left(q(t,0)\right)^2 dt.$$

The subspace $H_k \subset \widetilde{H}_k$ consists of all q such that

$$\int_0^T q(t,0)\, dt = 0.$$

This is a closed 1-codimensional subspace of $\widetilde{H}_k$, and on H_k the norm

$$\|q\|_k^2 = \sum_{n=0}^{kN-1} \int_0^T \left[\dot{q}(t,n)^2 + \left(\partial^- q(t,n)\right)^2\right] dt$$

is equivalent to $\|\cdot\|_{k^\sim}$. The corresponding inner product is denoted by $(\cdot,\cdot)_k$.

Let

$$J_k(q) := \sum_{n=0}^{kN-1} \int_0^T \left[\frac{1}{2} m(n)\, \dot{q}(t,n)^2 - U_n\!\left(\partial^- q(t,n)\right)\right] dt. \qquad (2.20)$$

As in Section 2.1, one can check that J_k is C^1 on $\widetilde{H}_k$, critical points of J_k are solutions of (2.1) satisfying the periodicity condition and critical points of the restriction of J_k to H_k are critical points of J_k. So, in what follows we consider the functional J_k on the space H_k. The derivative of J_k on H_k is given by the formula

$$\langle J_k'(q), h \rangle = \sum_{n=0}^{kN-1} \int_0^T \Big\{ m(n)\, \dot{q}(t,n)\, \dot{h}(t,n) + c(n)\, \partial^- q(t,n)\, \partial^- h(t,n)$$
$$- V_n'\!\left(\partial^- q(t,n)\right) \partial^- h(t,n) \Big\} dt. \qquad (2.21)$$

We have

$$J_k(q) = \frac{1}{2}(L_k q, q)_k - \Phi_k(q), \tag{2.22}$$

$$\langle J_k'(q), h \rangle = (L_k q, q)_k - \langle \Phi_k'(q), h \rangle, \tag{2.23}$$

where the self-adjoint operator L_k in H_k is defined in the similar way as L (see Eq. (2.11)) and

$$\Phi_k(q) := \sum_{n=0}^{kN-1} \int_0^T V_n(\partial^- q(t,n)) \, dt$$

(*cf.* Eq. (2.13)). Note that the norm of L_k is bounded by a constant independent of k.

Denote by Y_k the subspace of H_k that consists of constant (in t) functions $q \in H_k$ such that

$$q(n) - q(n-1) = 0 \quad \text{for all } n \in A,$$

and by Z_k the orthogonal complement of Y_k in H_k. The subspaces $Y_k \subset H_k$ and $Z_k \subset H_k$ are invariant subspaces of L_k. The statement of Lemma 2.1 is also valid for L_k, with λ independent of k, and Y and Z replaced by Y_k and Z_k, respectively.

Lemma 2.2 *The functional J_k on H_k satisfies the Palais-Smale condition.*

Proof. Let $q^{(j)}$ be a Palais-Smale sequence at the level b. Then

$$\begin{aligned}
b + 1 + \varepsilon_j \|q^{(j)}\|_k &\geq J_k(q^{(j)}) - \frac{1}{2}\langle J_k'(q^{(j)}), q^{(j)} \rangle \\
&\geq \left(\frac{1}{2} - \frac{1}{q}\right) \sum_{n=0}^{kN-1} \int_0^T V_n'(\partial^- q^{(j)}(t,n)) \\
&\quad \times \partial^- q^{(j)}(t,n) \, dt,
\end{aligned} \tag{2.24}$$

where $\varepsilon_j \to 0$. By the assumptions on the potentials, we have

$$|V_n'(r)|^2 \leq c_1 |r| |V_n'(r)| = c_1 r V_n'(r)$$

for all $|r| \leq 1$ and, trivially,

$$|V_n'(r)| \leq |r| |V_n'(r)| = r V_n'(r)$$

if $|r| \geq 1$.

Let
$$I_{n,j} = \{t \in [0,T] \mid |\partial^- q^{(j)}(t,n)| \leq 1\}$$

and $I_{n,j}^c = [0,T] \setminus I_{n,j}$. By Eq. (2.24), we have that

$$b + 1 + \varepsilon_j \|q^{(j)}\|_k \geq c_2 \left[\sum_{n=0}^{kN-1} \left(\int_{I_{n,j}} \left(V_n'(\partial^- q^{(j)}(t,n)) \right)^2 dt \right. \right.$$
$$\left. \left. + \int_{I_{n,j}^c} \left| V_n'(\partial^- q^{(j)}(t,n)) \right| dt \right) \right].$$

Therefore,

$$\sum_{n=0}^{kN-1} \int_{I_{n,j}} \left(V_n'(\partial^- q^{(j)}(t,n)) \right)^2 dt \leq c_2^{-1} \left(c + 1 + \varepsilon_j \|q^{(j)}\|_k \right),$$

$$\sum_{n=0}^{kN-1} \int_{I_{n,j}^c} \left| V_n'(\partial^- q^{(j)}(t,n)) \right| dt \leq c_2^{-1} \left(c + 1 + \varepsilon_j \|q^{(j)}\|_k \right).$$

Let $\overline{q}^{(j)}$ be the orthogonal projection of $q^{(j)}$ on Y_k and

$$\widetilde{q}^{(j)} = q^{(j)} - \overline{q}^{(j)}$$

the orthogonal projection of $q^{(j)}$ on Z_k. Since

$$\langle J_k'(q), h \rangle = (L_k q, h)_k - \langle \Phi_k'(q), h \rangle,$$

by the version of Lemma 2.1 for L_k, we have that

$$\lambda \|\overline{q}^{(j)}\|_k^2 \leq -(L_k \overline{q}^{(j)}, \overline{q}^{(j)})_k = -(L q^{(j)}, \overline{q}^{(j)})_k$$
$$= -\langle J_k'(q^{(j)}), \overline{q}^{(j)} \rangle - \langle \Phi_k'(q^{(j)}), \overline{q}^{(j)} \rangle.$$

Since, $J_k'(q^{(j)}) \to 0$, this implies that

$$\lambda \|\overline{q}^{(j)}\|_k^2 \leq \|\overline{q}^{(j)}\| - \sum_{n=0}^{kN-1} \int_0^T V_n'(\partial^- q^{(j)}(t,n)) \, \partial^- \overline{q}^{(j)}(t,n) \, dt.$$

The absolute value of the second term on the right does not exceed

$$\sum_{n=0}^{kN-1}\left[\left(\int_{I_{n,j}}V_n'(\partial^- q^{(j)}(t,n))^2 dt\right)^{1/2}\left(\int_{I_{n,j}}(\partial^-\overline{q}^{(j)}(t,n))^2 dt\right)^{1/2}\right.$$
$$\left.+\int_{I_{n,j}^c}\left|V_n'(\partial^- q^{(j)}(t,n))\right|dt\cdot\left\|\overline{q}^{(j)}(\cdot,n)\right\|_{L^\infty}\right]$$
$$\leq\left[\sum_{n=0}^{kN-1}\int_{I_{n,j}}V_n'(\partial^- q^{(j)}(t,n))^2 dt\right]^{1/2}\left[\sum_{n=0}^{kN-1}\left\|\partial^- q^{(j)}(\cdot,n)\right\|_{L^2}^2\right]^{1/2}$$
$$+\sup_n\left\|\overline{q}^{(j)}(\cdot,n)\right\|_{L^\infty}\sum_{n=0}^{kN-1}\int_{I_{n,j}^c}\left|V_n'(\partial^- q^{(j)}(t,n))\right|dt$$
$$\leq c_3\|\overline{q}^{(j)}\|_k\left[(b+1+\varepsilon_j\|q^{(j)}\|_k)^{1/2}+(b+1+\varepsilon_j\|q^{(j)}\|_k)\right],$$

because

$$\sup_n\left\|\overline{q}^{(j)}(\cdot,n)\right\|_{L^\infty}\leq c_0\|\overline{q}^{(j)}\|_k.$$

Thus, we have shown that

$$\lambda\|\overline{q}^{(j)}\|_k\leq 1+c_3\left[(b+1+\varepsilon_j\|q^{(j)}\|_k)^{1/2}+(b+1+\varepsilon_j\|q^{(j)}\|_k)\right].$$

Proceeding analogously for $\widetilde{q}^{(j)}$, we obtain

$$\lambda\|\widetilde{q}^{(j)}\|_k\leq 1+c_3\left[(b+1+\varepsilon_j\|q^{(j)}\|_k)^{1/2}+(b+1+\varepsilon_j\|q^{(j)}\|_k)\right].$$

Since

$$\|q^{(j)}\|_k^2=\|\overline{q}^{(j)}\|_k^2+\|\widetilde{q}^{(j)}\|_k^2,$$

we have that

$$\|q^{(j)}\|_k\leq C\left[1+(b+1+\varepsilon_j\|q^{(j)}\|_k)^{1/2}+(b+1+\varepsilon_j\|q^{(j)}\|_k)\right].$$

This inequality implies the boundedness of $q^{(j)}$.

Passing to a subsequence, we can assume that $q^{(j)}\to q$ weakly in H_k. The Sobolev compactness embedding implies that

$$\partial^- q^{(j)}(\cdot,n)\to\partial^- q(\cdot,n)$$

strongly in $L^\infty(0,T)$ for every $n \in \mathbb{Z}$. Moreover, $\overline{q}^{(j)} \to \overline{q}$ and $\widetilde{q}^{(j)} \to \widetilde{q}$ weakly in H_k, and

$$\partial^-\overline{q}^{(j)}(\cdot,n) \to \partial^-\overline{q}(\cdot,n),$$
$$\partial^-\widetilde{q}^{(j)}(\cdot,n) \to \partial^-\widetilde{q}(\cdot,n)$$

in $L^\infty(0,T)$ for all $n \in \mathbb{Z}$. This is enough to pass to the limit in

$$\langle J_k'(q^{(j)}), h \rangle = (L_k q^{(j)}, h)_k - \langle \Phi_k'(q^{(j)}), h \rangle, \quad h \in H_k,$$

and obtain that

$$\langle J_k'(q), h \rangle = (L_k q, h)_k - \langle \Phi_k'(q), h \rangle = 0, \quad h \in H_k, \qquad (2.25)$$

i. e. q is a critical point of J_k.

Now we have

$$(L_k \overline{q}^{(j)}, \overline{q}^{(j)})_k = (L_k q^{(j)}, \overline{q}^{(j)})_k = \langle J_k'(q^{(j)}), \overline{q}^{(j)} \rangle + \langle \Phi_k'(q^{(j)}), \overline{q}^{(j)} \rangle.$$

As before, passing to the limit, we get

$$\lim (L_k \overline{q}^{(j)}, \overline{q}^{(j)})_k = \langle \Phi_k'(q), \overline{q} \rangle = (L_k q, \overline{q})_k = (L_k \overline{q}, \overline{q})_k.$$

Similarly,

$$\lim (L_k \widetilde{q}^{(j)}, \widetilde{q}^{(j)})_k = (L_k \widetilde{q}, \widetilde{q})_k.$$

Note that

$$(L_k \overline{q}, \overline{q})_k - (L_k \widetilde{q}, \widetilde{q})_k = \left\| |L_k|^{1/2} q \right\|_k^2,$$

where $|L_k|^{1/2}$ is the square root of the absolute value of L_k defined by means of spectral decomposition, and $\left\| |L_k|^{1/2} g \right\|_k$ is a Hilbert norm on H_k equivalent to the original norm. Thus,

$$\left\| |L_k|^{1/2} q^{(j)} \right\|_k \to \left\| |L_k|^{1/2} q \right\|_k$$

and $q^{(j)} \to q$ weakly. This implies the convergence $q^{(j)} \to q$ in H_k and (PS) is proved. $\square$

Remark 2.4 An argument similar to the beginning of the proof of Lemma 2.2 shows that for critical points of J_k on H_k there exists an estimate of the form

$$\|q\|_k \leq \varphi(J_k(q)),$$

where $\varphi(r)$, $r \geq 0$, is a continuous function independent on k and such that $\varphi(0) = 0$.

Remark 2.5 Since for a critical point q we have

$$J_k(q) = J_k(q) - \frac{1}{2} \langle J'_k(q), q \rangle$$
$$= -\sum_{n=0}^{kN-1} \int_0^T \left[V_n(\partial^- q(t,n)) - \frac{1}{2} V'_n(\partial^- q(t,n)) \, \partial^- q(t,n) \right] dt,$$

assumption (iv) implies that any critical value is nonnegative. The same is true for J.

Lemma 2.3 *If $T < \pi/\sqrt{\beta}$, then there exists a constant $\varepsilon_0 > 0$ independent of k such that for any nontrivial critical point $q \in H_k$ of J_k we have*

$$\|q\|_k \geq \varepsilon_0.$$

Proof. According to assumptions (iii') and (v), there exists a continuous increasing function $\psi(r)$, $r \geq 0$ such that $\psi(0) = 0$ and

$$V'_n(r) \, r \leq \psi(|r|) \, r^2.$$

Let $q = \overline{q} + \widetilde{q}$, where $\overline{q} \in Y$ and $\widetilde{q} \in Z$. Then

$$\lambda \|\overline{q}\|_k^2 \leq \sum_{n=0}^{kN-1} \int_0^T \left| V'_n(\partial^- q(t,n)) \right| \left| \partial^- \overline{q}(t,n) \right| dt$$
$$\leq \sup_{n \in \mathbb{Z}} \psi\left(\left\| \partial^- q(\cdot, n) \right\|_{L^\infty} \right) \sum_{n=0}^{kN-1} \int_0^T \left| \partial^- \overline{q}(t,n) \right|^2 dt$$
$$\leq \psi\left(\sup_{n \in \mathbb{Z}} \left\| \partial^- q(\cdot, n) \right\|_{L^\infty} \right) \|\overline{q}\|_k^2$$
$$\leq \psi(\|q\|_k) \|q\|_k^2.$$

Similarly,

$$\lambda \|\widetilde{q}\|_k^2 \leq \psi(\|q\|_k) \|\widetilde{q}\|_k^2.$$

So, we obtain

$$\lambda \|q\|_k^2 \leq \psi(\|q\|_k) \|q\|_k^2,$$

because
$$\|q\|_k^2 = \|\bar{q}\|_k^2 + \|\tilde{q}\|_k^2.$$

Since $q \neq 0$, we get
$$\lambda \leq \psi(\|q\|_k),$$
hence, $\|q\|_k \geq \varepsilon_0 = \psi^{-1}(\lambda)$. □

Remark 2.6 Inequality $\|q\| \geq \varepsilon_0$ takes place also for any nontrivial critical point of J on H.

Now we are going to prove the existence of nontrivial critical points of J_k by means of linking theorem (see Theorem C.4).

Proposition 2.6 *If $T < \pi/\sqrt{\beta}$, then the functional J_k has a nontrivial critical point $q_k \in H_k$. Moreover, there exists a constant $C > 0$ independent of k such that $\|q_k\|_k \leq C$ and $J_k(q_k) \leq C$.*

Proof. Without loss of generality we can assume that $c(0) < 0$.
Let
$$N = \{q \in Z \mid \|q\|_k = \varrho_0\}.$$
From the assumptions it follows easily that
$$\Phi_k(q) = o(\|q\|_k^2) \quad \text{as} \quad \|q\|_k \to 0$$
uniformly in k. This together with Lemma 2.1 for L_k implies that there exists $\varrho_0 > 0$ independent of k such that
$$\inf_N J_k(q) > 0.$$

Let $z^0 \in Z$,
$$M = \{q = y + z^0 \mid y \in Y, \|u\| \leq \varrho_1, s \geq 0\}$$
and
$$M_0 = \{q = y + z^0 \mid y \in Y, \|y\|_k = \varrho_1 \text{ and } s \geq 0, \text{ or } \|u\| \leq \varrho_1, s = 0\},$$
i. e. $M_0 = \partial M$.

Now we make a particular choice of z^0, namely
$$z^0(t, n) = \begin{cases} \sin\left(\dfrac{2\pi}{T} t\right) & \text{if } n = 0, 1, \ldots, N-1, \\ 0 & \text{if } n = N, N+1, \ldots, kN-1, \end{cases}$$

and
$$z^0(t, n+kN) = z^0(t,n).$$

For $q = y + sz^0 \in M$, we have
$$J_k(q) = J_k(y+sz^0) = \frac{s^2}{2}(L_k z^0, z^0) + \frac{1}{2}(L_k y, y)$$
$$- \sum_{n=0}^{kN-1} \int_0^T V_n\big(\partial^-(y+sz^0)(t,n)\big)\,dt.$$

For every $d > 0$ there exists a constant $C_d > 0$ such that
$$V_n(r) \geq -d + C_d |r|^\theta.$$

Since y does not depend on t, this and Lemma 2.1 for L_k yield
$$J_k(q) \leq -\frac{\lambda}{2}\|y\|_k^2 + (L_k z^0, z^0)_k + Td - Cd\int_0^T \left|r_0 - s\cdot\sin\left(\frac{2\pi}{T}t\right)\right|^\theta d\theta$$
$$= -\frac{\lambda}{2}\|y\|_k^2 + \frac{2\pi^2 N}{T}s^2 \int_0^T \left(\cos\frac{2\pi}{T}t\right)^2 dt$$
$$+ c(0)s^2 \int_0^T \left(\sin\frac{2\pi}{T}t\right)^2 dt + Td$$
$$- Cd \int_0^T \left|r_0 - s\cdot\sin\left(\frac{2\pi}{T}t\right)\right|^\theta d\theta,$$

where $r_0 = \partial^- q(0)$. Since $c(0) < 0$ and for every r_0
$$\int_0^T \left|r_0 - s\cdot\sin\left(\frac{2\pi}{T}t\right)\right|^\theta d\theta \geq s^\theta \int_0^T \left|\sin\frac{2\pi}{T}t\right|^\theta d\theta,$$

we obtain that
$$J_k(q) \leq -\frac{\lambda}{2}\|y\|_k^2 + \frac{2\pi^2 N}{T}s^2 + Td - C_d' T s^\theta.$$

This inequality shows immediately that for ϱ_1 large enough, $J_k \leq 0$ on M_0. Moreover,
$$\sup_M J_k(q) \leq K := \max_{s \geq 0}\left[\frac{2\pi^2 N}{T}s^2 + Td - C_d' T s^\theta\right]. \qquad (2.26)$$

Applying the linking theorem (Theorem C.4), we obtain a nontrivial critical point $q_k \in H_k$, with $J_k(q_k) \leq K$. The estimate for $\|q_k\|_k$ follows from Remark 2.4. $\square$

2.3.3 Proof of main result

Now we are ready to prove Theorem 2.2. The idea is to obtain the solution as the limit of q_k as $k \to \infty$, where q_k is the solution of spatially kN-periodic problem found in Proposition 2.6.

Proof of Theorem 2.2. Step 1. Assume $T < \pi/\sqrt{\beta}$. We have that there exist $\varepsilon_0 > 0$ and a sequence n_k that

$$\|\partial^- q_k(\cdot, n_k)\|_{L^\infty} \geq \varepsilon_0 \qquad (2.27)$$

for all k. Indeed, if not, then

$$v_k := \sup_{n \in \mathbb{Z}} \|\partial^- q_k(\cdot, n)\|_{L^\infty} \to 0.$$

By assumptions (iii') and (v), there exists an increasing continuous function $\varphi(r)$, $r \geq 0$, such that $\varphi(0) = 0$ and

$$V'_n(r) r \leq \varphi(|r|) r^2.$$

We have

$$0 \leq J_k(q_k) = J_k(q_k) - \frac{1}{2} \langle J'_k(q_k), q_k \rangle$$

$$= \sum_{n=0}^{kN-1} \int_0^T \left[\frac{1}{2} V'_n\big(\partial^- q_k(t,n)\big) \partial^- q_k(t,n) - V_n\big(\partial^- q_k(t,n)\big) \right] dt$$

$$\leq \frac{1}{2} \sum_{n=0}^{kN-1} \int_0^T V'_n\big(\partial^- q_k(t,n)\big) \partial^- q_k(t,n) \, dt.$$

This implies that

$$0 \leq J_k(q_k) \leq \frac{1}{2} \varphi(v_k) \sum_{n=0}^{kN-1} \|\partial^- q_k(\cdot, n)\|_{L^2}^2 \leq C\varphi(v_k) \|q_k\|_k^2.$$

Since $\|q_k\|_k$ is bounded, we obtain that $J_k(q_k) \to 0$. Then, by Remark 2.4, $\|q_k\|_k \to 0$. This contradicts the conclusion of Lemma 2.3.

Note that, for any integer multiple p of N,

$$\tilde{q}_k(t, n) = q_k(t, n+p) - \frac{1}{T} \int_0^T q_k(t, p) \, dt$$

is also a critical point of J_k in H. Making such a shift, we can assume that $0 \leq n_k < N$ in (2.27). Passing to a subsequence, we can even assume that $n_k = n_0$ is independent of k.

Step 2. We still assume that $T < \pi\sqrt{\beta}$. Since $\|q_k\|_k$ is bounded, we can assume, passing to a subsequence, that $q_k(\cdot, n) \to q(\cdot, n)$ weakly in $H^1(0,T)$ and

$$\partial^- q_k(\cdot, n) \to \partial^- q(\cdot, n)$$

in $C([0,T])$ for all $n \in \mathbb{Z}$. Inequality

$$\|q_k\|_k \leq C$$

obviously implies that $q \in H$. It is also not difficult to show that q solves Eq. (2.1). Passing to the limit in (2.27) (remind that $n_k = n_0$) we obtain that

$$\|\partial^- q(\cdot, n_0)\|_{L^\infty} \geq \varepsilon_0,$$

hence, q is a nontrivial solution.

We have (see Remark 2.5)

$$J_k(q_k) = \sum_{n=-k[\frac{N}{2}]}^{k(N-[\frac{N}{2}])-1} \int_0^T \left(\frac{1}{2} V_n'(\partial^- q_k(t,n)) \partial^- q_k(t,n) - V_n(\partial^- q_k(t,n))\right) dt,$$

where $[x]$ denotes the integer part of x, and similarly

$$J(q) = \sum_{n=-\infty}^{\infty} \int_0^T \left(\frac{1}{2} V_n'(\partial^- q(t,n)) \partial^- q(t,n) - V_n(\partial^- q(t,n))\right) dt.$$

These two formulas together with the convergence $\partial^- q_k(\cdot, n) \to \partial^- q(\cdot, n)$ in $C([0,T])$ imply that

$$J(q) \leq \lim_{k \to \infty} J_k(q_k). \tag{2.28}$$

Step 3. Suppose that $c(n) < 0$ for all $n \in \mathbb{Z}$ and $T < \pi/\sqrt{\beta}$. Then Y is the subspace of all constant functions in H. It is easily verified that $J < 0$ on $Y \setminus \{0\}$. Since critical values of J are nonnegative, the solution we have just obtained is nonconstant.

If $T \geq \pi/\sqrt{\beta}$, then the operators L and L_k may not be invertible and the procedure employed above does not work in this case. However, for every integer $p > 0$ a T/p-periodic function is also T-periodic. Therefore, we have a nonconstant T-periodic solution for all $T > 0$.

Step 4. If $c(n) > 0$ for some value of n, then Eq. (2.1) admits non zero constant solutions. We set $T_1 = \pi/\sqrt{\beta}$. Now let us find $T_0 > 0$. As in the proof of Theorem 2.1, Step 5, critical values of nonzero constant solutions are not less that Td_1, where

$$-d_1 = \max_{n \in \mathbb{Z}} \max_j U_n(\theta_{nj})$$

and $\{\theta_{nj}\}$ is the set of all critical points of U_n. Calculating maximum in (2.26) and using (2.28), we obtain for the critical value $J(q)$ the estimate

$$J(q) \leq \alpha T^{-\frac{\theta+2}{\theta-2}} + Td,$$

where $\alpha > 0$ depends on d, but not on T and λ. Choose $d < \bar{d}$. Then, for sufficiently large T, say $T > T_0$,

$$J(q) \leq \alpha T^{-\frac{\theta+2}{\theta-2}} + Td < Td_1.$$

Hence, the solution q cannot be constant. This completes the proof. □

Remark 2.7 Since the conditions on T_0 and T_1 are independent of each other, there are potentials for which $T_0 < T_1$ and, hence, the problem has T-periodic solution for every $T \in (T_0, T_1)$. Since a T/p-periodic function ($p > 0$ integer) is T-periodic, we see that for every $T \in (pT_0, pT_1)$ there exists a nonconstant T-periodic solution. Thus, we have infinitely many bands for periods of nonconstant periodic solutions.

2.4 Additional results

2.4.1 *Degenerate case*

In this section we consider the case when $c(n)$ vanishes for some values of $n \in \mathbb{Z}$. In this case, in addition to assumptions (i), (ii), (iii'), (iv) and (v), we suppose that

(iii'') *for every $n \in \mathbb{Z}$ the potential $V_n(r)$ is strictly convex in a neighborhood of the origin.*

The situation is now more delicate because $0 \in \sigma(L)$. Nevertheless, we have the following result similar to Theorems 2.1 and 2.2.

Theorem 2.3 *Assume (i), (ii), (iii'), (iv) and $c(n) = 0$ for some value of n. In addition, we assume (iii''). Then for every $T > 0$ system (2.1) has a nonzero T-periodic solution $q \in H$. If $c(n) \leq 0$ for all $n \in \mathbb{Z}$, then the solution is nonconstant. If $c(n) \geq 0$ for all $n \in \mathbb{Z}$ and $c(n) > 0$ for at least one value of $n \in \mathbb{Z}$, then there exists $T_0 > 0$ such that the solution* is nonconstant provided $T > T_0$. If $c(n)$ change sign, then there exists $T_0 > 0$ and $T_1 > 0$, where T_0 depends on V_n and positive $c(n)$, while T_1 depends on $\min\{c(n)\}$, such that the solution is nonconstant if $T_0 < T_1$ and $T_0 < T < T_1$.

The proof goes along the same lines as in Section 2.3. In particular, it uses kN-periodic in n approximations and passage to the limit as $k \to \infty$.

The first step, the existence of spatially kN-periodic solutions, is based on the linking theorem. However, in this case the operator L_k (see Subsection 2.3.2) has a nontrivial kernel. Nevertheless, the functional J_k possesses the linking geometry.

Let K_k, Y_k, Z_k be the kernel, the negative and the positive subspaces of L_k, respectively. Assume that $T < \pi/\sqrt{\beta}$, with β defined in Section 2.3, and k is large enough. Then there exists $\lambda > 0$ independent of k such that

$$(L_k q, q)_k \leq -\lambda \|q\|_k^2, \quad q \in Y_k,$$

and

$$(L_k q, q)_k \geq \lambda \|q\|_k^2, \quad q \in Z_k.$$

Moreover, $K_k \oplus Y_k$ consists of constant in time functions. The space Y_k is trivial if $c(n) \geq 0$, but the kernel K_k is never trivial, since we assume that $c(n)$ vanishes for some $n \in \mathbb{Z}$. The linking geometry for J_k is generated by the space $K_k \oplus Y_k$ and its orthogonal complement Z_k. Even in the case when $c(n) \geq 0$, this geometry does not reduce to the mountain pass geometry.

The proof of (PS) condition for J_k is similar to that in Subsection 2.3.2. Thus, the linking theorem produces critical points $q_k \in H_k$ of J_k.

The passage to the limit is based on uniform estimates for $\|q_k\|_k$ and $J_k(q_k)$. This requires a long technical work and assumption (iii'') is used here.

For the detailed proof we refer to [Arioli and Szulkin (1997)].

Remark 2.8 At least in the nonnegative case $(c(n) \geq 0)$, assumption (iii'') can be weakened to the following: $V_n(r) = 0$ if and only if $r = 0$ [Arioli and Gazzola (1995)].

2.4.2 Constrained minimization

Here we sketch briefly another approach to the existence of periodic solutions, namely, constrained minimization. We consider system (2.1) with

$$U_n(r) = -\frac{c(n)}{2} r^2 + \frac{d(n)}{p} |r|^p, \quad p > 2. \tag{2.29}$$

We assume that the coefficients $m(n)$, $c(n)$ and $d(n)$ are N-periodic and strictly positive. Certainly, this case is covered by Theorem 2.1.

The functional J on H defined by (2.5) now reads

$$J(q) = \frac{1}{2} \int_0^T \|m\dot{q}(t)\|_{l^2}^2 \, dt + \sum_{n \in \mathbb{Z}} \int_0^T c(n) \big[\partial^- q(t,n)\big]^2 dt$$
$$- \frac{1}{p} \sum_{n \in \mathbb{Z}} \int_0^T d(n) |\partial^- q(t,n)|^p dt,$$

while for Φ defined by (2.13) we have

$$\Phi(q) = \frac{1}{2} \sum_{n \in \mathbb{Z}} \int_0^T d(n) |\partial^- q(t,n)|^p dt.$$

These are C^1 functionals on H.

Consider the following minimization problem:

$$\mu = \inf \left\{ \frac{1}{2}(Lq, q) \, : \, q \in H, \Phi(q) = 1 \right\}. \tag{2.30}$$

Theorem 2.4 *Under the assumption above, there exists a minimizer $q \in H$, $q \neq 0$, of problem (2.30). Moreover, there exists $T_0 > 0$ such that q is a nonconstant function of t if $T > T_0$.*

Under the assumptions imposed, the set $\{q \in H, \Phi(q) = 1\}$ is a C^1 manifold. The Lagrange multiplier rule yield the existence of $\lambda \in \mathbb{R}$ (Lagrange multiplier) such that, for the minimizer q,

$$(Lq, h) = \lambda \langle \Phi'(q), h \rangle, \quad \forall h \in H.$$

Taking $h = q$, we obtain that $\lambda > 0$. Using the homogeneity properties of (Lq, q) and $\Phi'(q)$ it is easy to scale out the Lagrange multiplier and obtain a solution of the original problem. Actually, for any minimizer q of (2.30), $\lambda^{\frac{1}{p-2}} q$ is a critical point of J, hence, a solution of (2.1). Thus, Theorem 2.4 produces a T-periodic solution of Eq. (2.1) in the case when the potentials are of the form (2.29). The proof can be found in [Arioli and Chabrowski (1997)].

2.4.3 *Multibumps*

Here we describe a multiplicity result for positive definite case obtained in [Arioli et. al (1996)]. Roughly speaking, the result states that, in a generic situation, the system possesses infinitely many nonconstant T-periodic solutions of multibump type, *i. e.* having most of their (finite) energy concentrated in a finite number of disjoint regions of the lattice.

Let us come back to the assumptions of Section 2.2. (In [Arioli et. al (1996)] it is supposed, in addition, that the derivatives V'_n are locally Lipschitz continuous, but this assumption is superfluous).

For all $\alpha \leq \beta$ we denote

$$J^\alpha = \{q \in H \mid J(q) \leq \alpha\}, \quad J_\beta = \{q \in H \mid J(q) \geq \beta\},$$
$$J^\alpha_\beta = \{q \in H \mid \beta \leq J(q) \leq \alpha\}.$$

Let

$$K = \{q \in H \setminus \{0\} \mid J'(q) = 0\}$$

be the set of all nonzero critical points of J in H,

$$K^\alpha = K \cap J^\alpha, \quad K_\beta = K \cap J_\beta, \quad K^\alpha_\beta = K^\alpha \cap K_\beta.$$

In Section 2.2 it is shown that the functional J possesses the mountain pass geometry, with mountain pass level $b > 0$, and there exists a nontrivial critical point $q \in H$ of J such that $J(q) \leq b$. Moreover, q is a nonconstant function of t if $T > T_0$.

The functional J is invariant under both a representation of $\mathbb{Z}$ denoted by $*$ and a representation of the unit circle $\mathbb{S}^1$ denoted by Ω. The first representation is defined by

$$* : \mathbb{Z} \times H \to H, \quad (k, q) \mapsto k * q,$$

where

$$(k * q)(t, n) = q(t, n + kN) - \frac{1}{T} \int_0^T q(t, kN)\, dt.$$

Essentially, this is the spatial shift adjusted to live H invariant. The representation Ω is just the shift of time variable

$$\Omega : \mathbb{S}^1 \times H \to H, \quad \Omega(\tau,q)(t,n) = q(t+\tau,n).$$

Let l be a positive integer,

$$\overline{k} = (k^{(1)}, k^{(2)}, \ldots, k^{(l)}) \in \mathbb{Z}^l$$

and

$$\overline{q} = (q^{(1)}, q^{(2)}, \ldots, q^{(l)}) \in H^l.$$

We set

$$\overline{k} * \overline{q} = \sum_{j=1}^{l} k^{(j)} * q^{(j)}.$$

For a sequence $\overline{k}_n = (k_n^{(1)}, k_n^{(2)}, \ldots, k_n^{(l)})$, by $\overline{k}_n \to \infty$ we mean that for $i \neq j$ we have

$$\left| k_n^{(i)} - k_n^{(j)} \right| \to \infty \quad \text{as} \quad n \to \infty.$$

A critical point q is said to be a *multibump solution* of type (l, ϱ), where l is a positive integer and $\varrho > 0$, if there exist

$$\overline{k} = (k^{(1)}, k^{(2)}, \ldots, k^{(l)}) \in \mathbb{Z}^l$$

and

$$\overline{q} = (q^{(1)}, q^{(2)}, \ldots, q^{(l)}), \quad q^{(j)} \in K \setminus \{0\},$$

such that q belongs to the ball (in H) of radius ϱ centered at $\overline{k} * \overline{q}$. This means that $\overline{k} * \overline{q}$ is an approximate solution of (2.1).

As it was pointed out in Section 2.2, due to $\mathbb{Z}$ invariance, the functional J does not satisfy the Palais-Smale condition. Nevertheless, the structure of Palais-Smale sequences is well-understood.

Theorem 2.5 *Under the assumptions of Section 2.2, let $q^{(j)} \in H$ be a Palais-Smale sequence for J at level $c > 0$. Then there exist a subsequence still denoted by $q^{(i)}$, l points $q^{(j)} \in K \setminus \{0\}$, and l sequences of integers $k_i^{(j)}$ ($j = 1, 2, \ldots, l$) such that*

$$\|q^{(i)} - \overline{k}_i * \overline{q}\| \to 0, \quad \sum_{j=1}^{l} J(q^{(i)}) = c$$

and $\overline{k}_i = (k_i^{(1)}, k_i^{(2)}, \ldots, k_i^{(l)}) \to \infty$.

Actually, the same result holds in the indefinite case of Section 2.3.

When we are interesting in multiplicity of solutions, it is natural to consider geometrically distinct solutions. We say that the solutions $q^{(1)} \in H$ and $q^{(2)} \in H$ are *geometrically distinct* if $q^{(2)}$ does not belong to the orbit of $q^{(1)}$ under the action of the group $\mathbb{Z} \times \mathbb{S}^1$, i. e. there exists no $(k, \tau) \in \mathbb{Z} \times \mathbb{S}^1$ such that

$$q^{(2)}(t) = (k * q^{(1)})(t + \tau).$$

Theorem 2.6 *Under assumptions of Section 2.2 let $T > T_0$, where T_0 is taken from Theorem 2.1. Then system (2.1) has infinitely many nonconstant geometrically distinct solutions. More precisely, assume that the following nondegeneracy condition holds:*

(ND) *there exist $\alpha > 0$ and a compact set $\mathcal{K} \subset H$ such that*

$$K^{b+\alpha} = \bigcup_{k \in \mathbb{Z}} k * \mathcal{K},$$

with $k_1 * \mathcal{K} \cap k_2 * \mathcal{K} = \emptyset$ if $k_1 \neq k_2$.

Then for every positive integer l, and for every $a > 0$ and $\varrho > 0$ system (2.1) admits infinitely many geometrically distinct multibump solutions of type (l, ϱ) in the set J_{lb-a}^{lb+a}.

For the proofs of Theorems 2.5 and 2.6 we refer to [Arioli *et. al* (1996)].

Note that if (ND) is not satisfied, then automatically J has infinitely many geometrically distinct critical points.

2.4.4 Lattices without spatial periodicity

First we consider the case when the lattice is harmonic at infinity.

Theorem 2.7 *Assume (i) of Section 2.1, (iv) of Section 2.2 and (iii') of Section 2.3. Suppose that*

$$-C_0 \leq c(n) \leq -c_0$$

for some positive c_0 and C_0, and that

$$\frac{1}{r} V_n'(r) \to 0 \quad as \quad |n| \to \infty$$

uniformly with respect to r in any bounded interval. Then for all $T > 0$ system (2.1) admits a T-periodic nonconstant solution.

The proof can be found in [Arioli and Szulkin (1997)]. Here we only point out that in this case the functional J satisfies the Palais-Smale condition.

The second result concerns lattices that are homogeneous at infinity.

Theorem 2.8 *Suppose that the potential $U_n(r)$ is given by Eq. (2.29). Assume that $m(n) \equiv 1$, there exists c_0, $\bar{c}$ and $\bar{d}$ such that either*

$$0 < c_0 \leq c(n) \leq \bar{c} \quad and \; 0 < \bar{d} < d(n)$$

or

$$0 < c_0 \leq c(n) < \bar{c} \quad and \;\; 0 < \bar{d} \leq d(n),$$

and

$$\lim c(n) = \bar{c}, \; \lim d(n) = \bar{d} \quad as \; |n| \to \infty.$$

Then for every $T > 0$ system (2.1) has a nontrivial T-periodic solution. The solution is nonconstant if T is large enough.

The proof given in [Arioli and Chabrowski (1997)] is based on the constrained minimization approach sketched in Subsection 2.4.2.

2.4.5 Finite lattices

We met already finite lattices with periodic boundary conditions in Subsection 2.3.2. They were used there to approximate infinite lattices. However, finite lattices are also interesting in their own rights and we mention here few results on such lattices.

We consider system (2.1) assuming that $m(n)$ is an N-periodic sequence and $U_{n+N} = U_n$ for all $n \in \mathbb{Z}$, where $N \in \mathbb{Z}$, $N > 0$. Under this assumption it is natural to consider spatially periodic solutions, *i. e.* solutions satisfying the periodic boundary condition

$$q(t, n + N) = q(t, n) \quad \text{for all } n \in \mathbb{Z}. \tag{2.31}$$

Theorem 2.9 *Assume (i), (ii), (iii'), (iv) and (v), with $c(n) \neq 0$ for all $n \in \mathbb{Z}$. Then for all $T > 0$ problem (2.1), (2.31) admits a nonzero T-periodic solution. The solution is nonconstant if $c(n) < 0$ for all $n \in \mathbb{Z}$. If $c(n) > 0$ for all $n \in \mathbb{Z}$, then there exists $T_0 > 0$ such that the solutions* is nonconstant if $T > T_0$. If $c(n)$ takes both sings, then there exist $T_0 > 0$ and $T_1 > 0$, where T_0 depends on V_n and positive $c(n)$, while T_1 depends on $\min\{c(n)\}$, such that the solution is nonconstant, provided $T_0 < T_1$ and $T \in (T_0, T_1)$.

The result follows basically from Proposition 2.6. Similar result holds for degenerate case when $c(n)$ may vanish. In this case we have to assume, in addition, (iii'') (cf. Subsection 2.4.1).

Another existence and multiplicity result for finite periodic lattices is the following

Theorem 2.10 *Assume that $U_n \in C^2(\mathbb{R})$ for all $n \in \mathbb{Z}$, $m(n+N) = m(n)$ and $U_{n+N} = U_n$. Suppose that there exists $\theta > 2$, $\delta_0 \geq -1$, $\delta_0 < \theta - 2$, $\alpha > 0$ and $r_0 > 0$ such that*

$$|U_n'(r)| \leq \alpha |r|^{1+\delta_0}, \quad r \leq -r_0, \tag{2.32}$$

$$r U_n'(r) \geq \theta U_n(r), \quad r \geq r_0. \tag{2.33}$$

Then for all $T > 0$ problem (2.1), (2.31) has infinitely many nonconstant T-periodic solution.

Assumptions (2.32) and (2.33) mean, roughly speaking, that the potential $U_n(r)$ is superquadratic at $+\infty$, while its growth rate at $-\infty$ is of lower order. Certainly, in this theorem one can switch the roles of $+\infty$ and $-\infty$.

Under some stronger assumptions the result of Theorem 2.10 was obtained in [Ruf and Srikanth (1994)] (without nonconstancy statement). The finial form of this theorem is a particular case of more general results found in [Tarallo and Terracini (1995)].

Example 2.3 The Toda potential (see Example 1.4)

$$U(r) = ab^{-1}(e^{-br} + br - 1)$$

satisfies the assumptions of Theorem 2.10.

Example 2.4 Let

$$U(r) = (r^+)^\alpha \pm (r^-)^\beta,$$

where $r^+ = \max[r, 0]$ and $r^- = r - r^+$. If $\alpha > \beta \geq -1$ and $\alpha > 2$, then Theorem 2.10 applies, with $\theta = \alpha$ and $\delta = \beta$.

Also one can consider finite lattices with fixed ends, i. e. solutions of (2.1) satisfying the Dirichlet boundary condition

$$q(t,0) = q(t, N+1) = 0.$$

Results similar to Theorems 2.9 and 2.10 can be also obtained in this case (see, e. g. [Ruf and Srikanth (1994)]).

2.5 Chains of oscillators

In this section we describe briefly the results on infinite chains of nonlinear oscillators obtained in [Bak and Pankov (2004); Bak (2004)]. We are looking for time periodic solutions of the equation

$$\ddot{q}(n) = a(n)\, q(n+1) + a(n-1)\, q(n-1) + b(n)\, q(n) - V'_n(q(n)), \quad (2.34)$$

where $n \in \mathbb{Z}$ (see Eq. (1.39)), with the following boundary condition

$$\lim_{n \to \pm\infty} q(t,n) = 0 \qquad (2.35)$$

at infinity. Actually, we consider (2.34) as a nonlinear operator differential equation

$$\ddot{q} = Aq + B(q)$$

in the Hilbert space l^2, where

$$(Aq)(n) = a(n)\, q(n+1) + a(n-1)\, q(n-1) + b(n)\, q(n), \quad n \in \mathbb{Z},$$

$$(Bq)(n) = -V'_n(q(n)), \quad n \in \mathbb{Z}.$$

The boundary condition is incorporated into the space l^2. In the notations of Section 1.4,

$$b(n) = c(n) - a(n) - a(n-1),$$

with $c(n)$ being the coefficient in the harmonic part of the potential U_n.

Note that problem (2.34)–(2.35) has a trivial solution $q \equiv 0$.

We consider nonhomogeneous, but spatially periodic, chains. Precisely, we impose the following assumptions.

(h_0) *The sequences $a(n)$ and $b(n)$ (equivalently, $a(n)$ and $c(n)$) and the sequence of potentials V_n are N-periodic in $n \in \mathbb{R}$, with $N \geq 1$.*

(h_1) The operator A is positive definite, i. e. there exists $\alpha_0 > 0$ such that

$$(Aq, q) \geq \alpha_0 \|q\|^2, \quad q \in l^2.$$

(h_2) For every $n \in \mathbb{Z}$ the potential V_n is C^1, $V_n(0) = V_n'(0) = 0$ and $V_n'(r) = 0(|r|)$ as $r \to 0$.

(h_3) There exists $\theta > 2$ such that

$$V_n'(r) r \geq \theta V_n(r) \geq 0,$$

and there exists $r_0 > 0$ such that $V_n(r) > 0$ if $|r| \geq r_0$.

Note that assumption (h_0) implies that the operator A is a bounded linear operator in l^2. Moreover, A is self-adjoint because the coefficients $a(n)$ and $b(n)$ are real. Assumption (h_1) means, roughly speaking, that the coefficient sequence $b(n)$ (equivalently, $c(n)$) is large comparably to $a(n)$.

We have the following result – an analog of Theorem 2.1.

Theorem 2.11 *Assume (h_0)–(h_3). Then there exists $T_0 > 0$ such that for every $T \geq T_0$ equation (2.34) has a nonconstant T-periodic in time l^2-valued solution.*

The proof of the theorem relies upon variational arguments. Skipping details, we present here the basic idea only.

Let X_T be the subspace in $H^1_{loc}(\mathbb{R}; l^2)$ that consists of T-periodic functions, i. e.

$$X_T := \{q \in H^1_{loc}(\mathbb{R}; l^2) \ : \ q(t+T) = q(t)\}.$$

This is a Hilbert space with the norm

$$\|q\|_T = \left(\|q\|^2_{L^2(0,T;l^2)} + \|\dot{q}\|^2_{L^2(0,T;l^2)}\right)^{1/2},$$

induced from $H^1(0, T; l^2)$.

Consider the functional

$$J(q) = \int_0^T \left[\frac{1}{2}\|\dot{q}\|^2_{l^2} + \frac{1}{2}(Aq(t), q(t)) - \sum_{n \in \mathbb{Z}} V_n(q(t,n))\right] dt.$$

Under assumptions imposed above, $J(q)$ is a C^1 functional on X_T and its critical points are exactly T-periodic l^2-valued weak solutions of Eq. (2.34).

It turns out to be that, due to positivity of A and superquadraticity of V_n at 0 and at infinity (assumptions (h_2) and (h_3)), the functional J possesses the mountain pass geometry. But, like the functional J in Section 2.1, it does not satisfy the Palais-Smale condition. Applying Theorem C.4, one obtains a Palais-Smale sequence $q^{(k)} \in X_T$. Moreover, $\|q^{(k)}\|_T$ is bounded below and above by positive constants, and $J(q^{(k)})$ converges to the mountain pass level. Passing to a subsequence, one can assume that $q^{(k)} \to q$ weakly in X_T. Moreover, due to the Sobolev embedding theorem (Theorem A.1), $q^{(k)}(\cdot, n) \to q(\cdot, n)$ in $L^\infty(\mathbb{R})$ for all $n \in \mathbb{Z}$. Passing to the limit one obtains that q is a critical point of J in X_T. Replacing $q^{(k)}$ by

$$\widetilde{q}^{(k)}(t, n) = q^{(k)}(t, n + m_k),$$

with appropriately chosen m_k, we can achieve that $q \neq 0$. The final step is to show that q is nonconstant in t, provided T is large enough. This approach is similar to that of Section 2.2, but the details require a number of technical changes.

Another approach to the same result, realized in [Bak and Pankov (2004)], is similar to that presented in Section 2.3, and is based on periodic approximations. The last means that we consider the periodic boundary condition

$$q(t, n + kN) = q(t, n) \qquad (2.36)$$

instead of (2.35).

More precisely, let E_k be the space of all kN-periodic sequences $q(n)$. Endowed with the norm

$$\left(\sum_{n=0}^{kN-1} q(n)^2 \right)^{1/2},$$

this is a kN-dimensional Hilbert space. Let $X_{T,k}$ be the space of all functions in $H^1_{loc}(\mathbb{R}; E_k)$ that are T-periodic in t, i. e.

$$X_{T,k} := \{ q \in H^1_{loc}(\mathbb{R}; E_k) \ : \ q(t + T) = q(t) \}.$$

This is a Hilbert space with respect to the $H^1(0, T; E_k)$-norm, denoted here by $\| \cdot \|_{T,k}$.

Due to the periodicity of coefficients, the operator A acts also in all E_k. Moreover, A is positive definite in E_k with the same constant α_0 as in (h_1). Actually, the spectral theory of periodic difference operators (see

Section 1.3 and, for detailed presentation, [Teschl (2000)]) tells us that the spectrum of A in E_k is contained in the spectrum of A in l^2.

Let

$$J_k(q) = \int_0^T \left[\frac{1}{2} \|\dot{q}(t)\|_{E_k}^2 + \frac{1}{2}(Aq, q)_{E_k} - \sum_{n=0}^{kN-1} V_n(q(t,n)) \right] dt.$$

The functional J_k is a C^1 functional on $X_{T,k}$ and its critical points are T-periodic in time solutions of problem (2.34), (2.36).

Like J, the functional J_k possesses the mountain pass geometry. But, in contrast to J, J_k satisfies also the Palais-Smale condition. This follows from the fact that E_k is finite dimensional and the Sobolev embedding theorem. Hence, the mountain pass theorem (Theorem C.1) provides a nontrivial critical point $q^{(k)}$ which is a solution of (2.34), (2.36). Actually, $q^{(k)}$ is a nonconstant in time solution, provided T is large enough.

As in the case of Palais-Smale sequence of mountain pass type for J, $\|q^{(k)}\|_{T,k}$ is bounded below and above by positive constants. Therefore, passing to a subsequence, we can assume that $q^{(k)}(\cdot, n) \to q(\cdot, n)$ weakly in $H^1(0,T)$ and strongly in $L^\infty(\mathbb{R})$ for all $n \in \mathbb{Z}$. The uniform bound for $\|q^{(k)}\|_{T,k}$ implies that $q \in X_T$. Moreover, q is a critical point of J. Replacing $q^{(k)}$ by $\tilde{q}^{(k)}(\cdot, n + m_k)$, with appropriate m_k, we can obtain a nonzero limit q. In fact, q is nonconstant in time if T is large enough.

Remark 2.9 Actually, in Theorem 2.11 a nontrivial solution exists for all $T > 0$. But, if $T < T_0$, this solution may be independent of t, i. e. a stationary solution. It follows from [Pankov and Zakharchenko (2001)] that under the assumptions of Theorem 2.11 nonzero stationary solutions do exist.

In the particular case when

$$V_n(r) = \frac{d(n)}{p} |r|^p,$$

where $p > 2$ and $d(n) > 0$ is an N-periodic sequence, to solve (2.34) in the space of T-periodic functions one can use a constrained minimization similar to that discussed in Subsection 2.4.2. In this case Eq. (2.34) reads

$$\ddot{q}(n) = a(n)\, q(n+1) + a(n-1)\, q(n-1) + b(n)\, q(n)$$
$$- d(n)|q(n)|^{p-2} q(n). \qquad (2.37)$$

Let

$$Q(v) := \frac{1}{2} \int_0^T \left[\|\dot v\|_{l^2}^2 + (Av, v) \right] dt$$

and

$$\Psi(v) := \frac{1}{p} \int_0^T \left(\sum_{n \in \mathbb{Z}} d(n) |v(t,n)|^p \right) dt.$$

Then

$$J(v) = Q(v) - \Psi(v).$$

The functional Q is a continuous quadratic functional on X_T, while Ψ is C^1 on X_T.

Given $\alpha > 0$, consider the following minimization problem

$$I_\alpha = \inf \{ Q(v) \, : \, v \in X_T, \Psi(v) = \alpha \}. \qquad (2.38)$$

Let u be a point of minimum in (2.37) (if it exists). Obviously, $u \neq 0$. Since both the functionals Q and Ψ are C^1, then there exists a Lagrange multiplier $\lambda \in \mathbb{R}$ such that

$$Q'(u) = \lambda \Psi'(u),$$

or, explicitly,

$$\int_0^T \left[(\dot u(t), \dot h(t)) + (Au(t), h(t)) \right] dt$$
$$= \lambda \int_0^T \left[\sum_{n \in \mathbb{Z}} d(n) |u_n(t,n)|^{p-2} u_n(t,n) \, h(t,n) \right] dt \qquad (2.39)$$

for all $h \in X_T$. Testing with $h = u$, we obtain that

$$\lambda = \frac{2 I_\alpha}{\alpha p} > 0$$

and a straightforward calculation shows that

$$q = \lambda^{\frac{1}{p-2}} u$$

is a solution to (2.38).

The following result is proved in [Bak (2004)].

Theorem 2.12 *For every $\alpha > 0$ problem (2.38) has a solution $u \in X_T$. Furthermore, if T is sufficiently large, then $u \neq const$.*

Avoiding technical details, we explain only the basic idea of the proof. The functionals Q and Ψ satisfy

$$Q(\tau v) = \tau^2 Q(v), \quad \Psi(\tau v) = \tau^p \Psi(v), \quad \tau > 0.$$

This implies immediately that problems (2.38), with different values of α, are equivalent and

$$I_\alpha = \alpha^{2/p} I_1. \tag{2.40}$$

Let $u^{(k)}$ be a minimizing sequence for (2.38), i. e. $u^{(k)} \in X_T$, $\Psi(u^{(k)}) = \alpha$ and $Q(u^{(k)}) \to \alpha$. It is readily verified that the sequence $u^{(k)}$ is bounded in X_T. We define an element $w^{(k)} \in l^1$ by

$$w^{(k)}(n) = \frac{d(n)}{p} \int_0^T |u^{(k)}(t,n)|^p dt.$$

Then $\|w^{(k)}\|_{l^1} = \alpha$. Applying concentration compactness Lemma B.3, we obtain that w^k satisfies one of the statement (i)–(iii) of that lemma.

Vanishing (statement (ii)) is impossible because otherwise $\alpha = \|w^{(k)}\|_{l^1}$ would go to zero.

To rule out dichotomy (statement (iii)) one uses the inequality (subadditivity inequality)

$$I_\lambda < I_\alpha + I_{\lambda-\alpha}, \quad \lambda \in (0, \alpha),$$

that follows from (2.40).

Thus, we see that $w^{(k)}$ concentrates, i. e. satisfy property (iii) of Lemma B.3. This permits us to conclude that, after appropriate spatial shifts and passing to a subsequence, $u^{(k)} \to u \in X_T$ weakly in X_T and strongly in $L^p(0,T;l^p)$. The functional Ψ is, obviously, continuous on $L^p(0,T;l^p)$, hence, $\Psi(u) = \alpha$. The quadratic functional Q is positive definite and weakly lower semicontinuous. Therefore,

$$Q(u) \leq \lim Q(u^{(k)}) = I_\alpha.$$

Hence, $Q(u) = I_\alpha$ and we conclude.

Example 2.5 Consider a chain of identical oscillators with the coupling constant $a \equiv a(n)$, $c(n) \equiv c$ and the anharmonic potential $V_n(r) \equiv V(r)$

that satisfies (h_2) and (h_3), for instance, the potential

$$V(r) = \frac{d}{p}|r|^p$$

with $p > 2$ and $d > 0$. Then the equation of motion reads

$$\ddot{q} = a\Delta_d q + cq - d|q|^{p-2}q, \qquad (2.41)$$

where

$$(\Delta_d q)(n) = q(n+1) + q(n-1) - 2q(n)$$

is the one-dimensional discrete Laplacian. If $c > 0$, $a > 0$ and $p = 4$, this is the discrete φ^4-equation. It is well-known (and easy to verify) that

$$0 \geq (\Delta_d q, q) \geq -4\|q\|_{l^2}^2,$$

and 4 is the sharp constant. Therefore, for

$$A = a\Delta_d + c$$

we have that

$$(Aq, q) \geq (c - 4a)\|q\|_{l^2}^2.$$

Hence, A is positive defined if and only if $a < c/4$. Thus, if $a < c/4$, then equation (2.41) possesses nonconstant T-periodic solutions (breathers) whenever T large enough. In particular, this holds for the discrete φ^4-equation with $a < c/4$.

Example 2.6 When $a > 0$ and $c = -m^2 < 0$ Eq. (2.41) becomes the modified nonlinear discrete Klein-Gordon equation. In this case the operator $A = a\Delta_d - m^2$ is negative defined. The potential V satisfies (h_2) and (h_3) if $d > 0$. If $d < 0$, then $-V$ satisfies (h_2) and (h_3). In any case Theorem 2.11 does not apply.

2.6 Comments and open problems

Variational approach to time periodic solutions of FPU type system was introduced first in [Ruf and Srikanth (1994)]. These authors considered finite lattices with Toda-like potentials satisfying some one side conditions. They have obtained a number of results on existence of infinitely many T-periodic solutions. The approach of [Ruf and Srikanth (1994)] is based on minimax methods and a relative $\mathbb{S}^1$-index. Another approach to similar problems based on a linking type arguments was suggested in [Tarallo and Terracini (1995)], and the results of [Ruf and Srikanth (1994)] were extended considerably. A result of such type, Theorem 2.10, is presented in Subsection 2.4.5.

Those results motivated subsequent works on infinite lattices [Arioli and Chabrowski (1997); Arioli and Gazzola (1995); Arioli and Gazzola (1996); Arioli et. al (1996); Arioli et. al (2003); Arioli and Szulkin (1997)]. However, the following problem is still open.

Problem 2.1 *Find any result similar to Theorem 2.10, i. e. under asymmetric assumptions on the potentials, in the case of infinite lattices.*

Since Theorem 2.10 applies to any period of the form kN, $k \geq 1$ an integer, a natural idea is to use periodic approximations in the spirit of Section 2.3. The key point here is to obtain appropriate uniform, with respect to k, *a priori* estimates for time periodic solutions that are spatially kN-periodic. A closely related problem is

Problem 2.2 *Extend the results of Sections 2.2–2.3 to the case of not everywhere defined potentials, like the singular Lennard-Jones potential.*

The results of Sections 2.1–2.3 are borrowed from [Arioli and Gazzola (1995); Arioli and Gazzola (1996); Arioli and Szulkin (1997)]. However, the proof of Theorem 2.2 is new.

The approach developed in this chapter seems to be quite general and should work for more general lattices. For instance, let us suggest the following two problems.

Problem 2.3 *Extend the results of this chapter to the case of N-dimensional lattices.*

Problem 2.4 *Study time periodic solutions for lattices with second neighbor interactions.*

We mention here the paper [Srikanth (1998)] that deals with finite two dimensional lattices in the spirit of [Ruf and Srikanth (1994)].

The following problem seems to be interesting for both applications and mathematical development.

Problem 2.5 *Study time periodic oscillations of infinite spatially periodic lattices with impurities both in positive definite and indefinite cases.*

More precisely, this means that

$$m(n) = \overline{m}(n) + m^0(n),$$
$$c(n) = \overline{c}(n) + c^0(n)$$

and

$$V_n(r) = \overline{V}_n(r) + V_n^0(r),$$

where $\overline{m}(n)$, $\overline{c}(n)$ and $\overline{V}_n(r)$ are N-periodic in n, while $m^0(n)$, $c^0(n)$ and V_n^0 tend to 0 as $|n| \to \infty$. Theorem 2.8 could be considered as a prototype of results we expect to exist.

Another interesting class of solutions consists of so-called *homoclinics*, i. e. solutions that satisfy the boundary condition

$$q(-\infty, n) = q(+\infty, n) = 0.$$

Formally, this corresponds to $T = +\infty$. The following problem is of interest.

Problem 2.6 *Find homoclinic solutions to Eq. (2.1) with superquadratic potentials (the indefinite case is especially interesting).*

For results on homoclinics for finite dimensional second order Hamiltonian systems see [Omana and Willem (1992); Rabinowitz (1990); Rabinowitz (1996); Rabinowitz and Tanaka (1991)] and references therein.

The class of asymptotically linear systems of the form (3.1), i. e. the systems with

$$U_n'(r) = \begin{cases} -c^0(n)\,r + o(r) & \text{as } r \to 0 \\ -c^\infty(u)\,r + o(r) & \text{as } r \to \infty, \end{cases}$$

is not covered by existing theory. So, we offer

Problem 2.7 *Study time periodic and homoclinic solutions to Eq. (2.1) in the case of asymptotically linear nonlinearities.*

Time periodic solutions of finite dimensional asymptotically linear Hamiltonian systems are considered in [Mawhin and Willem (1989)]. For homoclinics we refer to [Szulkin and Zou (2001)].

Multibump type solutions are shown to exist for a variety of problems (see, e. g. [Coti Zelati and Rabinowitz (1991); Coti Zelati and Rabinowitz (1991); Coti Zelati and Rabinowitz (1992); Rabinowitz (1993); Rabinowitz

(1996)]). However, in the case of lattice systems Theorem 2.6 seems to be the only known result of such kind.

Problem 2.8 *Find multibump solutions in the indefinite case.*

We expect that the idea of gluing on Nehari's manifold [Li and Wang (2001)] together with the techniques of generalized Nehari's manifold suggested in [Pankov (2004)] should work in this case. Certainly, this is not the only possible approach to the last problem.

Discrete systems that consist of nonlinear oscillators, like Frenkel-Kontorova model, discrete Klein-Gordon and φ^4 equations, *etc.*, and their time periodic solutions (breathers) attracted a great attention in low dimensional nonlinear physics (see, *e. g.* [Aubry (1997); Braun and Kivshar (1998); Braun and Kivshar (2004); Flach and Willis (1998)] and references therein).

The results of Section 2.5 are borrowed from [Bak and Pankov (2004); Bak (2004)]. To our best knowledge, these are the first results of such kind in the case on spatially nonhomogeneous systems of the form (2.34). Certainly, the results are far to be complete. First we mention

Problem 2.9 *Find and classify time periodic solutions that are at rest at spatial infinity to systems of the form (2.34) with potentials that possess many local maxima, like the Frenkel-Kontorova potential.*

In view of Example 2.6 it is interesting to study the case when the operator A is not positive definite and either V_n or $-V_n$ satisfies (h_2) and (h_3). The operator

$$\mathcal{L}q = -\ddot{q} + Aq,$$

with T-periodic boundary conditions, is a self-adjoint operator in $L^2(0,T;l^2)$. Due to spatial periodicity, its spectrum $\sigma(\mathcal{L})$ possesses the band structure. Therefore, generically $\sigma(\mathcal{L})$ may have gaps. Suppose that 0 lies in a gap of $\sigma(\mathcal{L})$. Then it is very likely that the approach of [Pankov (2004)] allows to find nontrivial critical points of the functional J. However, it is not clear *a priori* that the solution obtained is not constant.

Notice that if A is positive definite, then $\mathcal{L}$ is positive definite. If, in addition, $rV(r) \leq 0$, then it is easily seen that the only time periodic solution is 0 (*cf.* Section 4.4). So, this case is trivial.

Problem 2.10 *Do nonconstant T-periodic solutions exist in the case when 0 lies in a gap of $\sigma(\mathcal{L})$ and either V_n or $-V_n$ satisfies (h_2) and (h_3)?*

Certainly, similar problem makes sense in the case of FPU lattices. Again, the case when the quadratic part is positive definite, while $rV(r) \leq 0$, is trivial.

Due to the analogy with *gap solutions* [Aceves (2000); de Sterke and Sipe (1994); Pankov (2004)], solutions of such kind can be called *gap breathers*.

Commonly, breathers are considered as spatially exponentially localized objects. This important property is not yet studied for the solutions obtained in this chapter.

Problem 2.11 *Are the solutions obtained in Theorems 2.1, 2.2, 2.5 and 2.11 exponentially localized in the spatial variable?*

Exponential localization means that

$$|r(t,n)| := |q(t,n+1) - q(t,n)| \leq Ce^{-\alpha|n|}$$

in the case of FPU type systems and

$$|q(t,n)| \leq Ce^{-\alpha|n|}$$

in the case of chains of oscillators, with $\alpha > 0$ and $C > 0$. In the degenerate case of Subsection 2.4.1 we cannot expect exponential localization. Instead, we conjecture power localization, *i. e.*

$$|r(t,n)| \leq C|n|^{-\alpha}, \quad \alpha > 0, C > 0.$$

Finally, we would like to point out probably the most important problem that concerns time periodic oscillations of lattices systems.

Problem 2.12 *Investigate stability properties of time periodic solutions of spatially periodic infinite lattices and lattices with impurities for both FPU systems and chains of oscillators.*

This problem seems to be completely open. Here it could be useful the concept of *Howland semigroup* (see, e. g., [Chicone and Latushkin (1999); Howland (1974); Howland (1979)]).

Chapter 3

Travelling Waves: Waves with Prescribed Speed

3.1 Statement of problem

In this chapter we consider monoatomic FPU lattices. Without loss of generality one can assume that all (identical) particles are of unit mass and, therefore, the corresponding equations of motion read

$$\ddot{q}(n) = U'\bigl(q(n+1) - q(n)\bigr) - U'\bigl(q(n) - q(n-1)\bigr), \quad n \in \mathbb{Z}, \qquad (3.1)$$

where U is the potential of interaction between two adjacent particles (see Eq. (1.1)).

Travelling wave is a solution of the form

$$q(t, n) = u(n - ct), \qquad (3.2)$$

where the function $u(t)$, $t \in \mathbb{R}$, is called the *profile function*,[1] or simply *profile*, of the wave and the constant c, the *speed* of the wave, is always assumed to be positive. *In the following we often do not distinguish the profile function and the wave itself.* These are waves travelling to the right with the speed c. Making use Ansatz (3.2), we obtain the following equation for the profile function

$$c^2 u''(t) = U'\bigl(u(t+1) - u(t)\bigr) - U'\bigl(u(t) - u(t-1)\bigr). \qquad (3.3)$$

This is a forward-backward differential-difference equation.

One can rewrite Eq. (3.3) in terms of relative displacements. Let

$$r(t) := u(t+1) - u(t).$$

---

[1] We denote by t the independent variable of the profile function, but this *does not* mean time.

This function is exactly the *profile function for relative displacements* because

$$q(t, n+1) - q(t, n) = r(n + ct).$$

For the function $r(t)$ we have the equation

$$c^2 r''(t) = U'(r(t+1)) + U'(r(t-1)) - 2U'(r(t)). \tag{3.4}$$

Note that the relative displacement profile $r(t)$ is related to the profile function $u(t)$ also by the following formula

$$r(t) = \int_t^{t+1} u(s)\,ds. \tag{3.5}$$

In the following we shall consider two types of travelling waves: solitary waves and periodic waves.

A *solitary travelling wave* is a travelling wave such that its relative displacement profile function $r(t)$, or the derivative $u'(t)$ of the profile function (see Eq. (3.5)), vanishes at infinity.

A *periodic travelling wave* is a travelling wave such that its relative displacement profile function $r(t)$ (equivalently, $u'(t)$) is a periodic function of $t \in \mathbb{R}$, say, $2k$-periodic. In general, the profile function $u(t)$ of such wave is *not periodic*. In fact, it is of the form

$$u(t) = \langle u' \rangle t + \text{ periodic function},$$

where

$$\langle u' \rangle = \int_{-k}^{k} u'(s)\,ds$$

is the mean value of the $2k$-periodic function $u'(t)$.

Remark that for every solution $u(t)$ of Eq. (3.3) $u(t + \alpha) + \beta$ is also a solution of Eq. (3.3) for all $\alpha \in \mathbb{R}$ and $\beta \in \mathbb{R}$. Therefore, in general travelling waves form two-parametric families.

Notice also that looking for a wave of the form

$$q(n, t) = u(n + \dot{c}t), \quad c > 0$$

travelling to the left with speed $c > 0$ we obtain the same equation (3.3) for the profile function $u(t)$. Therefore, if $u(t)$ is the profile function for a wave travelling to the right, then there is also a wave with the same profile travelling to the left with the same speed, and *vice versa*.

Let $u(t)$ be the profile function of $2k$-periodic travelling wave ($k \in \mathbb{R}$, $k > 0$). For the corresponding lattice solution $q(n,t)$ the function

$$\dot{q}(t,n) = -cu'(n-ct)$$

is $2k/c$-periodic in time, but only almost periodic with respect to the spatial variable n in general. The function $\dot{q}(t,n)$ is periodic in n only in the case of rational k/c.

It is convenient to represent the interaction potential $U(r)$ in the form

$$U(r) = \frac{a}{2}r^2 + V(r), \qquad (3.6)$$

where $V(r)$ is the anharmonic part of the potential.

In the case of harmonic potential ($V(r) = 0$), Eq. (3.4) becomes

$$c^2 r''(t) = a\big(r(t+1) + r(t-1) - 2r(t)\big).$$

Making use the Fourier transform

$$\widehat{r}(\xi) = \frac{1}{\sqrt{2\pi}} \int_{-\infty}^{+\infty} e^{-i\xi t} r(t)\, dt,$$

we obtain

$$-\xi^2 c^2 \widehat{r}(\xi) = 2a\left(\cos\xi - 1\right)\widehat{r}(\xi).$$

Hence $\widehat{r}(\xi) = 0$ almost everywhere, and thus $r \equiv 0$. Therefore, in the harmonic case nontrivial solitary waves do not exist.

Now let $a = c_0^2$ ($c_0 > 0$) and $V(r) = 0$. Then, as it is mentioned on the end of Section 1.3, Eq. (1.37) for relative displacements, allows plane wave solutions

$$r(t,n) = \exp\big[i(\varkappa n \pm \omega t)\big],$$

with the dispersion relation[2]

$$\omega = \pm 2c_0 \sin\frac{\varkappa}{2}.$$

Such solutions are non localized travelling waves. For either branch of this dispersion relation, the group velocity for linear wave propagation satisfies

$$\left|\frac{d\omega}{d\varkappa}\right| = \left|c_0 \cos\frac{k}{2}\right| \le c_0.$$

---

[2] Here $r(t,n)$ is the relative displacement of adjacent lattice sites, *not* the relative displacement profile function $r(t)$.

This is the reason to name $c_0 > 0$ the *speed of sound*.

In what follows we are interesting, primarily, in monotone waves, increasing and decreasing. This means that the corresponding wave profile is a monotone function or, equivalently, the relative displacement profile is either positive or negative. From the point of view of physics, increasing waves are *expansion waves*, while decreasing waves are *compression waves*. Also we present few results on nonmonotone waves.

3.2 Periodic waves

3.2.1 *Variational setting*

Consider $2k$-periodic travelling waves. This means that we are looking for solutions of Eq. (3.3) with the boundary condition

$$u'(t+2k) = u'(t). \tag{3.7}$$

Let X_k be the Hilbert space defined by

$$X_k := \{u \in H^1_{loc}(\mathbb{R}) \ : \ u'(t+2k) = u'(t), u(0) = 0\},$$

with the inner product

$$(u,v)_k := \int_{-k}^{k} u'(t)v'(t)\,dt$$

and corresponding norm $\|u\|_k = (u,u)_k^{1/2}$. Since every function that belongs to $H^1_{loc}(\mathbb{R})$ is continuous (see Theorem A.1), the condition $u(0) = 0$ is meaningful. By $\|\cdot\|_{k,*}$ we denote the dual norm on X_k^*, the dual space to X_k.

Actually, X_k is a 1-codimensional subspace of the Hilbert space

$$\widetilde{X}_k := \{u \in H^1_{loc}(\mathbb{R}) \ : \ u'(t+2k) = u'(t)\},$$

with

$$\int_{-k}^{k} u'(t)\,v'(t)\,dt + u(0)\,v(0)$$

as the inner product.

On $\widetilde{X}_k$ we define the operator A by

$$(Au)(t) := u(t+1) - u(t) = \int_t^{t+1} u'(s)\, ds. \tag{3.8}$$

Obviously, A acts from $\widetilde{X}_k$ into itself.

Lemma 3.1 *The operator A is a linear bounded operator from X_k into $L^2(-k,k) \cap L^\infty(-k,k)$ satisfying*

$$\|Au\|_{L^\infty(-k,k)} \le \|u\|_k$$

and

$$\|Au\|_{L^2(-k,k)} \le \|u\|_k.$$

Proof. By the Cauchy-Schwatz inequality we have

$$|Au(t)| = \left| \int_t^{t+1} u'(s)\, ds \right| \le \left(\int_t^{t+1} |u'(s)|^2 ds \right)^{1/2} \le \|u\|_k$$

and

$$\int_{-k}^k |Au(t)|^2 dt \le \int_{-k}^k \int_t^{t+1} |u'(s)|^2 ds\, dt \le \|u\|_k^2.$$

$\square$

Consider the functional

$$J_k(u) := \int_{-k}^k \left[\frac{c^2}{2} u'(t)^2 - U(Au(t)) \right] dt.$$

We always assume that

(i) the potential U is C^1 on $\mathbb{R}$ and $U(0) = U'(0) = 0$.

Proposition 3.1 *Under assumption (i) the functional J_k is well-defined on X_k. Moreover, J_k is C^1 and*

$$\langle J'_k(u), h \rangle = \int_{-k}^k \left[c^2 u'(t) h'(t) - U'(Au(t)) Ah(t) \right] dt.$$

Proof. We have

$$J_k(u) = \frac{c^2}{2} (u,u)_k - \Phi_k(u),$$

where
$$\Phi_k(u) := \int_{-k}^{k} U(Au(t))\, dt.$$

Hence, we have to consider only the functional Φ_k.

Since for any $u \in X_k$ the function Au is continuous, $\Phi_k(u) < \infty$. A straightforward calculation shows that the Gateaux derivative of Φ_k exists and is given by
$$\langle \Phi_k'(u), h \rangle = \int_{-k}^{k} U'(Au(t)) Ah(t)\, dt.$$

Finally, we check the continuity of Φ_k'.

Let $\|h\|_k \leq 1$ and $u_n \to u$ in X_k. Then $Au_n \to Au$ uniformly on $[-k, k]$ and
$$\begin{aligned}
\left| \langle \Phi_k'(u_n) - \Phi_k'(u), h \rangle \right| &\leq \|Ah\|_{L^1} \cdot \|U'(Au_n) - U'(Au)\|_{L^\infty} \\
&\leq (2k)^{1/2} \|Ah\|_{L^2} \cdot \|U'(Au_n) - U'(Au)\|_{L^\infty} \\
&\leq (2k)^{1/2} \|U'(Au_n) - U'(Au)\|_{L^\infty},
\end{aligned}$$
where all L^p-norms are taken over $[-k, k]$. Since $U'(Au_n) \to U'(Au)$ uniformly on $[-k, k]$, we conclude. $\square$

Proposition 3.2 *Under assumption (i) any critical point of J_k is a classical, i. e. C^2, solution of Eq. (3.3) satisfying (3.7).*

Proof. Let $\varphi(t)$ be any $2k$-periodic C^∞ function. Then
$$h(t) = \varphi(t) - \varphi(0) \in X_k.$$

If u is a critical point of J_k, we have
$$\begin{aligned}
0 &= \int_{-k}^{k} \left[c^2 u'(t) h'(t) - U'(u(t+1) - u(t))(h(t+1) - h(t)) \right] dt \\
&= \int_{-k}^{k} \left[c^2 u'(t) \varphi'(t) - U'(u(t+1) - u(t))(\varphi(t+1) - \varphi(t)) \right] dt \\
&= \int_{-k}^{k} \left[c^2 u'(t) \varphi'(t) - \left[U'(u(t) - u(t-1)) - U'(u(t+1) - u(t)) \right] \right] \varphi(t)\, dt.
\end{aligned}$$

Hence u is a weak solution of (3.3). Since $u(t)$ and $U'(r)$ are continuous, Eq. (3.3) implies that $u \in C^2$. $\square$

Remark 3.1 Taking the test function $h(t) \equiv t$, we obtain that any critical point of J_k satisfies the following additional identity

$$c^2[u(k) - u(-k)] = \int_{-k}^{k} U'\bigl(u(t+1) - u(t)\bigr) dt.$$

3.2.2 Monotone waves

Here we present a result on existence of *monotone* supersonic waves. We impose the following conditions:

(i') $U(r) = \dfrac{c_0^2}{2} r^2 + V(r)$, where $c_0 \geq 0$, V is C^1 on $\mathbb{R}$, $V(0) = V'(0) = 0$ and $V'(r) = o(|r|)$ as $r \to 0$,

and either

(ii^+) there exist $r_0 > 0$ and $\theta > 2$ such that $V(r_0) > 0$ and, for $r \geq 0$, we have

$$0 \leq \theta V(r) \leq r V'(r),$$

or

(ii^-) there exist $r_0 < 0$ and $\theta > 2$ such that $V(r_0) > 0$ and, for $r \leq 0$, we have

$$0 \leq \theta V(r) \leq r V'(r).$$

Assumption (ii^+) can be written as the differential inequality

$$r^{\theta+1} \frac{d}{dr}\bigl(r^{-\theta} V(r)\bigr) \geq 0, \quad r > 0.$$

Integration shows that

$$V(r) \geq a_0 r^\theta \quad \text{for } r > r_0,$$

with $a_0 = r_0^{-\theta} V(r_0)$. Together with assumption (i') this implies that

$$V(r) \geq a_1(r^\theta - r^2), \quad r \geq 0. \tag{3.9}$$

Similarly in the case of (ii^-) the last inequality holds for $r \leq 0$.

We are interesting in $2k$-periodic travelling waves having either nondecreasing or nonincreasing profile.

Theorem 3.1 *Assume (i') and $k \geq 1$.*

(a) Under assumption (ii^+) for every $c > c_0$ there exists a nontrivial nondecreasing $2k$-periodic travelling wave $u_k \in X_k$.

(b) Under assumption (ii^-) for every $c > c_0$ there exists a nontrivial nonincreasing $2k$-periodic travelling wave $u_k \in X_k$.

Moreover, in both cases there exist constants $\delta > 0$ and $M > 0$, independent of k, such that the critical value $J_k(u_k)$ satisfies

$$0 < \delta \leq J_k(u_k) \leq M.$$

The proof is based on a version of mountain pass theorem (see Theorem C.2).

Let

$$(Pu)(t) = \int_0^t |u'(s)|\, ds.$$

It is readily verified that P maps continuously the space X_k into itself and PX_k consists of nondecreasing functions.

Since we are looking for monotone waves, we can assume that $V(r) \equiv 0$ for $r < 0$ in case (a) and $V(r) = 0$ for $r > 0$ in case (b). (Equally well one can assume in both cases that $V(r)$ is an even function). In particular, this means that the modified potential satisfies

$$0 \leq \theta V(r) \leq r V'(r) \quad \text{for all } r \in \mathbb{R}.$$

In the following we consider case (a) only. Case (b) is similar.

Lemma 3.2 *Under assumptions of Theorem 3.1 there exists $\delta > 0$ and $\varrho > 0$ such that $J_k(u) \geq \delta$ if $\|u\|_k = \varrho$. Furthermore, there exists $e_k \in PX_k$ such that $\|e_k\|_k > \varrho$ and $J_k(e_k) = J_1(e_1) \leq 0$.*

Proof. Assumption (i') implies that, given $\varepsilon > 0$, there exists $\varrho > 0$ such that

$$|V(r)| \leq \varepsilon r^2$$

if $|r| \leq \varrho$. If $\|u\|_k \leq \varrho$, then, by Lemma 3.1, $\|Au\|_{L^\infty} \leq \varrho$, and

$$J_k(u) \geq \int_{-k}^{k} \left[\frac{c^2}{2}|u'|^2 - \frac{c_0^2}{2}|Au|^2 - \varepsilon |Au|^2 \right] dt \geq \frac{c^2 - c_0^2 - 2\varepsilon}{2} \|u\|_k^2.$$

Choosing ε small enough, we obtain the first statement of the lemma.

To construct the function e_k, we first choose a function $v \in PX_1$ such that $v(t) = 0$ for $0 \leq t \leq 1$, $v'(-1) = 0$ and $Av(t_0) > 0$ for some t_0. Note that v' can be nonzero only on intervals of the form $(2l-1, 2l)$, $l \in \mathbb{Z}$. Since $Av \geq 0$, by (3.9) we have that

$$J_1(\tau v) \leq \tau^2 \int_{-1}^{1} \left[\frac{c^2}{2} |v'|^2 + \left(a_1 - \frac{c_0^2}{2}\right) |Av|^2 \right] dt - \tau^\theta a_1 \int_{-1}^{1} |Av|^\theta dt$$

$$\leq \tau^2 \left(\frac{c^2}{2} \|v\|_1^2 + a_1 \|Av\|_{L^2}^2 \right) - \tau^\theta a_1 \|Av\|_{L^\theta}^\theta.$$

Since $\theta > 2$, we obtain that $J_1(\tau v) \to -\infty$ as $\tau \to +\infty$. Hence, we can fix $e_1 = \tau_0 v$ satisfying $J_1(e_1) \leq 0$ and $\|e_1\|_1 > \varrho$.

Now we define $e_k \in X_k$ by $e_k(t) = e_1(t)$ if $|t| \leq 1$ and $e'_k(t) = 0$ if $1 \leq t \leq k$. (Extending e'_k to $\mathbb{R}$ by $2k$-periodicity, we define $e_k \in X_k$ uniquely). Obviously, e'_k can be nonzero only on intervals $(2kl - 1, 2kl)$, $l \in \mathbb{Z}$. We see immediately that $\|e_k\|_k = \|e_1\|_1$. Moreover,

$$(Ae_k)(t) = e_k(t+1) - e_k(t) = \begin{cases} e_1(t+1) - e_1(t), & t \in [-2, 0], \\ 0, & t \in [-k, k] \setminus [-2, 0]. \end{cases}$$

Consequently,

$$\int_{-k}^{k} V(Ae_k(t)) \, dt = \int_{-2}^{0} V(Ae_k(t)) \, dt = \int_{-2}^{-1} V(e_1(t+1) - e_1(t)) \, dt$$

$$+ \int_{-1}^{0} V(e_1(t+1) - e_1(t)) \, dt$$

$$= \int_{0}^{1} V(e_1(t-1) - e_1(t-2)) \, dt$$

$$+ \int_{-1}^{0} V(e_1(t+1) - e_1(t)) \, dt$$

$$= \int_{-1}^{1} V(Ae_1(t)) \, dt$$

because the difference $e_1(t+1) - e_1(t)$ is 2-periodic. In particular, we have that $J_k(e_k) = J_1(e_1) \leq 0$. □

Remark 3.2 In fact, $J_k(se_k) = J_1(se_1)$ for all $s \in \mathbb{R}$.

Lemma 3.3 *Assume* (i') *and* $c > c_0$. *If the potential satisfies*

$$\theta V(r) \leq rV'(r), \quad r \in \mathbb{R},$$

with $\theta > 2$, *then the functional* J_k *satisfies the Palais-Smale condition.*

Proof. Let $u_n \in X_k$ be a Palais-Smale sequence at the level a. Then, for n large enough, $\|J_k'(u_n)\|_{k,*} \leq 1$ and $|J_k(u_n)| \leq a+1$. Hence,

$$a+1+\frac{1}{\theta}\|u_n\| \geq J_k(u_n) - \frac{1}{\theta}\langle J_k'(u_n), u_n\rangle$$
$$= \left(\frac{1}{2} - \frac{1}{\theta}\right)\int_{-k}^{k} \left(c^2|u_n'|^2 - c_0^2|Au_n|^2\right)dt$$
$$+ \int_{-k}^{k}\left[\frac{1}{\theta}V'(Au_n)Au_n - V(Au_n)\right]dt.$$

Due to the assumption on $V(r)$ the second integral is nonnegative and, by Lemma 3.1, we obtain that

$$a+1+\frac{1}{\theta}\|u_n\|_k \geq \left(\frac{1}{2} - \frac{1}{\theta}\right)(c^2 - c_0^2)\|u_n\|_k^2.$$

Hence, the sequence u_n is bounded in X_k.

The boundedness of u_n implies that, up to a subsequence, $u_n \to u$ weakly in X_k, hence, $Au_n \to Au$ weakly in $\widetilde{X}_k$ and, by the compactness of Sobolev embedding (Theorem A.1), in $L^2(-k,k)$. A straightforward calculation shows that

$$c^2\|u_n - u\|_k^2 = \int_{-k}^{k} c^2(u_n' - u')^2 dt$$
$$= \langle J_k'(u_n) - J_k'(u), u_n - u\rangle + c_0^2\|Au_n - Au\|_{L^2}^2$$
$$+ \int_{-k}^{k}[V'(Au_n) - V'(Au)](Au_n - Au)\,dt.$$

The first and the second terms on the right obviously converge to 0. By Lemma 3.1, Au_n is bounded in $L^\infty(-k,k)$. Hence, $[V'(Au_n) - V'(Au)]$ is bounded in $L^\infty(-k,k)$. Since, $Au_n \to Au$ strongly in $L^2(-k,k)$, the last integral term also tends to zero. Therefore, we conclude that $\|u_n - u\|_k \to 0$, which proves the lemma. □

Proof of Theorem 3.1. Case (a). To prove the existence we apply Theorem C.2. Due to Lemmas 3.2 and 3.3, the only we have to verify is that $J_k(Pu) \leq J(u)$ for every $u \in X_k$. We have

$$(APu)(t) = \int_t^{t+1}(Pu)'(s)\,ds = \int_t^{t+1}|u'(s)|\,ds \geq \left|\int_t^{t+1}u'(s)\,ds\right|.$$

Hence,

$$(APu)(t) \geq |(Au)(t)| \geq (Au)(t).$$

Since the modified potential $V(r)$ is nondecreasing on $\mathbb{R}$, then

$$\begin{aligned} J_k(Pu) &= \int_{-k}^{k} \left[c^2|(Pu)'|^2 - c^2(APu)^2 - V(APu) \right] dt \\ &= \int_{-k}^{k} \left[c^2|u'|^2 - c_0^2(APu)^2 - V(APu) \right] dt \\ &\leq \int_{-k}^{k} \left[c^2|u'|^2 - c_0^2(Au)^2 - V(Au) \right] dt \\ &= J_k(u). \end{aligned}$$

Theorem C.2 implies the existence of a nontrivial critical point $u_k \in PX_k$ of J_k such that $J_k(u_k) \geq \delta$, with $\delta > 0$ from Lemma 3.2, and

$$J_k(u_k) \leq \max_{s \in [0,1]} J_k(se_k).$$

By Remark 3.2, $J_k(se_k) = J_1(se_1)$ and

$$J_k(u_k) \leq M := \max_{s \in [0,1]} J_1(se_1).$$

Case (b) is similar, with P replaced by $-P$. $\square$

Remark that the assumption $c > c_0$ means that the wave speed is greater then the speed of sound and, hence, the wave is supersonic.

Remark 3.3 It is not clear whether monotone waves obtained in Theorem 3.1 are strictly monotone.

3.2.3 *Nonmonotone and subsonic waves*

First we present a version of Theorem 3.1 that concerns the existence of not necessary monotone waves.

Theorem 3.2 *Assume that*

(i'') $U(r) = \dfrac{a}{2} u^2 + V(r)$, *where V is C^1 on $\mathbb{R}$, $V(0) = V'(0) = 0$ and $V'(r) = o(|r|)$ as $r \to 0$,*

(ii') *for some $r_0 \in \mathbb{R}$ we have that $V(r_0) > 0$ and there exists $\theta > 2$ such that*

$$\theta V(r) \leq rV'(r), \quad r \in \mathbb{R}.$$

Let $c^2 > \max(a, 0)$. Then for every $k \geq 1$ there exists a nontrivial $2k$-periodic travelling wave $u_k \in X_k$. Moreover, the corresponding critical value $J_k(u_k)$ satisfies

$$0 < \delta \leq J_k(u_k) \leq M,$$

with $\delta > 0$ and $M > 0$ independent of k.

Proof. The proof is similar to that of Theorem 3.1. Just use the standard mountain pass theorem (Theorem C.1) instead of Theorem C.2. □

Now let us come back to assumption (i') and consider the case when $0 < c \leq c_0$, i. e. the case of *subsonic waves*.

Theorem 3.3 *Assume (i'), (ii^+) and (ii^-). Then for every $k \geq 1$ and $c \in (0, c_0]$ there exists a $2k$-periodic travelling wave $u_k \in X_k$.*

Proof. We sketch briefly how to apply the linking theorem (see Theorem C.4). We have to check that the functional J_k possesses the linking geometry and satisfies the Palais-Smale condition.

Note first that the space X_k splits into the orthogonal sum of the one-dimensional subspace generated by the function $h_0(t) = t$ and the space $H^1_{k,0}$ of all $2k$-periodic functions from X_k with zero mean value. Consider the operator L defined by

$$Lu = c^2 u'' - c_0^2 Au,$$

with $2k$-periodic boundary conditions. Elementary Fourier analysis shows that L is a self-adjoint operator in $L^2(-k, k)$ bounded below and that L has discrete spectrum which accumulates at $+\infty$. The eigenvalues and eigenfunctions can be calculated explicitly, but we do not use this fact. We mention only that all eigenvalues, λ_j, with nonconstant eigenfunctions are double. Denote by $h_j^\pm \in H^1_{k,0}$ linearly independent pairs of eigenfunctions with the eigenvalues λ_j.

Let Z be the subspace of $H^1_{k,0}$ generated by the functions $h_j^\pm$ with $\lambda_j > 0$ and Y be the subspace of X_k generated by the functions $h_j^\pm$ with $\lambda_j \leq 0$ and the function h_0. Note that $\dim Y < \infty$. It is readily verified that $Y \perp Z$ and $X_k = Y \oplus Z$. A straightforward verification shows that

$$Q_k(y + z) = Q_k(y) + Q_k(z), \quad y \in Y, \; z \in Z,$$

where Q_k is the quadratic part of the functional J_k,

$$Q_k(u) = \frac{1}{2} \int_{-k}^{k} \left(c^2 |u'|^2 - c_0^2 |Au|^2 \right) dt.$$

Since on Z the quadratic form Q_k is positive definite, i. e.

$$Q_k(u) \geq \alpha \|u\|_k^2,$$

with $\alpha > 0$, the argument from the beginning of the proof of Lemma 3.2 shows that

$$J_k(u) \geq \delta > 0$$

on

$$N = \{u \in Z : \|u\|_k = r\}$$

provided $r > 0$ is small enough.

Now we fix any $z \in Z$, $\|z\|_k = 1$, and set

$$M = \{u = y + \lambda z : y \in Y, \|u\|_k \leq \varrho, \lambda \leq 0\}.$$

We have to prove that $J_k(u) \leq 0$ on $M_0 = \partial M$ provided ϱ is large enough. Recall that

$$M_0 = \{u = y + \lambda z : y \in Y, \|u\|_R = \varrho \text{ and } \lambda \geq 0, \text{ or } \|u\|_k \leq \varrho \text{ and } \lambda = 0\}.$$

We have

$$J_k(y + \lambda z) = Q_k(y) + \lambda^2 Q_k(z) - \int_{-k}^{k} V(A(y + \lambda z)) \, dt.$$

Due to assumptions (i'), (ii^+) and (ii^-), there exists a constant $C > 0$ such that

$$V(r) \geq C|r|^\theta - 1.$$

Therefore,

$$J_k(y + \lambda z) \leq \lambda^2 \alpha_0 + 2k - C\|y + \lambda z\|_{L^\theta}^\theta,$$

where $\alpha_0 = Q_k(z)$. Since

$$\varrho^2 = \|y + \lambda z\|_k^2 = \|y\|^2 + \lambda^2,$$

we have $\lambda^2 \leq \varrho^2$. Furthermore, on finite dimensional spaces all norms are equivalent. Hence,

$$\|y + \lambda z\|_{L^\theta} \geq c\|y + \lambda z\|_k = c\varrho$$

and

$$J_k(y + \lambda z) \leq \alpha_0 \varrho^2 + 2k - C\varrho^\theta.$$

Since $\theta > 2$, the right hand part here is negative if ϱ is large enough. Hence, $J_k(y + \lambda z) \leq 0$. If $u \in M_0$, $\|u\|_k \leq \varrho$ and $\lambda = 0$, then $u = y \in Y$ and, obviously, $J_k(u) \leq 0$. Thus, we see that J_k possesses the linking geometry.

Finally, we verify the Palais-Smale condition (PS). Let $u_n \in X_k$ be a Palais-Smale sequence at some level a. Choose $\beta \in (\theta^{-1}, 2^{-1})$. For n large we have

$$a + 1 + \beta\|u_n\| \geq J_k(u_n) - \beta\langle J'_k(u_n), u_n\rangle$$

$$= \int_{-k}^{k} \left[\left(\frac{1}{2} - \beta\right)c^2(u'_n)^2 - \left(\frac{1}{2} - \beta\right)c_0^2(Au_n)^2\right.$$

$$\left. + \beta V'(Au_n) Au_n - V(Au_n)\right] dt$$

$$\geq \left(\frac{1}{2} - \beta\right)c^2\|u_n\|^2 - \left(\frac{1}{2} - \beta\right)c_0^2\|Au_n\|_{L^2}^2$$

$$+ (\beta\theta - 1)\int_{-k}^{k} V(Au_n)\, dte$$

$$\geq \left(\frac{1}{2} - \beta\right)c^2\|u_n\|_k^2 - \left(\frac{1}{2} - \beta\right)c_0^2\|Au_n\|_{L^2}^2$$

$$+ C(\beta\theta - 1)\|Au_n\|_{L^\theta}^\theta - C_0. \tag{3.10}$$

Since $\theta > 2$, we have

$$\|Au_n\|_{L^2}^2 \leq C\|Au_n\|_{L^\theta}^2 \leq K(\varepsilon) + \varepsilon\|Au_n\|_{L^\theta}^\theta,$$

where $K(\varepsilon) \to \infty$ as $\varepsilon \to 0$. Choosing ε small enough, we see that the L^θ term in (3.10) absorbs the L^2 term and

$$a + 1 + \beta\|u_n\|_k \geq \left(\frac{1}{2} - \beta\right)c^2\|u_n\|_k^2 + C(\beta\theta - 1)\|Au_n\|_{L^\theta}^\theta - C_0.$$

Since $\beta\theta > 1$,
$$a + 1 + \beta\|u_n\|_k \geq \left(\frac{1}{2} - \beta\right) c^2 \|u_n\|_k^2 - C_0.$$

This implies that u_n is bounded. Now the relative compactness of u_n in X_k follows exactly as in the proof of Lemma 3.3. □

Remark 3.4 The linking geometry in Theorem 3.3 is not uniform with respect to k. Therefore, we cannot derive any uniform bound for solutions obtained in that theorem.

3.3 Solitary waves

3.3.1 *Variational statement of the problem*

In a sense, the case of solitary waves is a limit case of the setting considered in Section 3.2 when $k = \infty$. Let X be the Hilbert space
$$X := \{u \in H^1_{loc}(\mathbb{R}) \,:\, u' \in L^2(\mathbb{R}), u(0) = 0\}$$
endowed with the inner product
$$(u, v) = \int_{-\infty}^{+\infty} u'(t)\, v'(t)\, dt$$
and corresponding norm $\|u\| = (u,u)^{1/2}$. Note that the condition $u(0) = 0$ in the definition of X is meaningful because every element of $H^1_{loc}(\mathbb{R})$ is a continuous function. As usual $\|\cdot\|_*$ stands for the dual norm on the dual space X^*. The space X is a 1-dimensional subspace of the Hilbert space
$$\widetilde{X} := \{u \in H^1_{loc}(\mathbb{R}) \,:\, u' \in L^2(\mathbb{R})\},$$
with the inner product
$$u(0)\,v(0) + \int_{-\infty}^{+\infty} u'(t)\, v'(t)\, dt.$$

It is readily verified that the linear operator A defined by
$$(Au)(t) := u(t+1) - u(t) = \int_t^{t+1} u'(s)\, ds$$
acts from $\widetilde{X}$ into itself.

The following statement is similar to Lemma 3.1.

Lemma 3.4 *The operator A acts from $\widetilde{X}$ into $L^2(\mathbb{R}) \cap L^\infty(\mathbb{R})$. Moreover, for $u \in \widetilde{X}$ the function Au is continuous, $(Au)(\pm\infty) = 0$, i. e.*

$$\lim_{t \to \pm\infty} (Au)(t) = 0,$$

$$\|Au\|_{L^\infty(\mathbb{R})} \leq \|u\|$$

and

$$\|Au\|_{L^2(\mathbb{R})} \leq \|u\|.$$

Proof. The proof is similar to the proof of Lemma 3.1. The only novelty concerns $(Au)(\pm\infty) = 0$. We have that

$$|(Au)(t)| \leq \int_t^{t+1} |u'(s)|\, ds \leq \left(\int_t^{t+1} |u'(s)|^2 ds \right)^{1/2},$$

and this implies the required since $u' \in L^2(\mathbb{R})$. $\square$

We remark also that the map

$$(Pu)(t) = \int_0^t |u'(s)|\, ds$$

acts continuously from X into itself.

On the space X we consider the functional

$$J(u) := \int_{-\infty}^{\infty} \left[\frac{c^2}{2} u'(t) - U(Au(t)) \right] dt.$$

We impose the following assumption:

(i''') *the potential U is C^1 on $\mathbb{R}$, $U(0) = U'(0) = 0$ and for some $R > 0$*

$$\sup_{|u| \leq R} \left| \frac{U'(r)}{r} \right| < \infty. \tag{3.11}$$

Actually, (3.11) is the restriction on the behavior of V' near $r = 0$. Hence, if it holds for some $R > 0$, then so is for every $R > 0$. Note that (i''') is slightly stronger than (i).

Proposition 3.3 *Under assumption (i''') the functional J is well-defined on X. Moreover, J is C^1 and*

$$\langle J(u), h \rangle = \int_{-\infty}^{+\infty} \left[c^2 u'(t) h'(t) - U'(Au(t)) Ah(t) \right] dt.$$

Proof. Since
$$J(u) = \frac{c^2}{2}(u,u) - \Phi(u),$$
where
$$\Phi(u) = \int_{-\infty}^{+\infty} U(Au(t))\,dt,$$
we have to consider only the functional Φ.

Due to assumption (i''') (in particular, we use here (3.11)), for every $R > 0$
$$\sup_{|r|\leq R}\left|\frac{U(r)}{r^2}\right| < \infty.$$

By Lemma 3.4, there exists a constant $C > 0$ depending on $\|Au\|_{L^\infty}$, hence, on $\|u\|$, such that
$$\left|U(Au(t))\right| \leq C|Au(t)|^2.$$

This implies immediately that $\Phi(u)$ is finite.

A direct calculation shows that the Gateaux derivative of Φ exists and is given by
$$\langle \Phi'(u), h\rangle = \int_{-\infty}^{+\infty} U'(Au(t))Ah(t)\,dt.$$

To complete the proof we have to verify that Φ' is continuous. Let $\|h\| \leq 1$ and $u_n \to u$ in X. Then $Au_n \to Au$ in X, in $L^2(\mathbb{R})$ and also in $L^\infty(\mathbb{R})$. We have, by Lemma 3.4,
$$\left|\langle \Phi'(u_n) - \Phi'(u), h\rangle\right| \leq \|Ah\|_{L^2}\|U'(Au_n) - U'(Au)\|_{L^2}$$
$$\leq \|U'(Au_n) - U'(Au)\|_{L^2}.$$

Here L^p-norms are taken over $\mathbb{R}$. Assumption (i''') implies that there exists a constant $C > 0$ such that
$$|U'(r)| \leq C|r|, \quad |r| \leq R = \max\left(\|Au\|_{L^\infty}, \|Au_n\|_{L^\infty}\right).$$

Now

$$\|U'(Au_n) - U'(Au)\|_{L^2}^2 \leq \int_{-a}^{a} \left|U'(Au_n(t)) - U'(Au(t))\right|^2 dt$$

$$+ \int_{|t| \geq a} \left[\left|U'(Au_n(t))\right|^2 + \left|U'(Au(t))\right|^2\right] dt$$

$$\leq \int_{-a}^{a} \left|U'(Au_n(t)) - U'(Au(t))\right|^2 dt$$

$$+ C \int_{|t| \geq a} \left[\left|Au_n(t)\right|^2 + \left|Au(t)\right|^2\right] dt.$$

Let $\varepsilon > 0$. Since $Au_n \to Au$ in $L^2(\mathbb{R})$, we can choose $a > 0$ independent of n and such that the second integral above is less than ε. Since $Au_n \to Au$ uniformly on $[-a, a]$, then the first integral above is less than ε, provided n is large enough. Hence, for n large

$$\|U'(Au_n) - U'(Au)\|_{L^2}^2 \leq \varepsilon + C\varepsilon,$$

and we conclude. $\square$

Proposition 3.4 *Under assumption (i''') a function $u \in X$ is a critical point of J if and only if u is a classical, i. e. $u \in C^2(\mathbb{R})$, solution of Eq. (3.3).*

Proof. Let $\varphi \in C_0^\infty(\mathbb{R})$. Then $h(t) = \varphi(t) - \varphi(0) \in X$. If u is a critical point of J, we see as in the proof of Proposition 3.2 that u is a weak solution of (3.3) and, moreover, $u \in C^2(\mathbb{R})$. Hence, u is a classical solution.

Conversely, if $u \in X$ is a classical solution of Eq. (3.3), then

$$\langle J(u), h \rangle = \int_{-\infty}^{\infty} \left[c^2 u'(t) h'(t) - U'(Au(t)) Ah(t)\right] dt = 0$$

for every $h \in X$ such that $h' \in C_0^\infty(\mathbb{R})$. Since $C_0^\infty(\mathbb{R})$ is dense in $L^2(\mathbb{R})$, we conclude. $\square$

The next proposition deals with simple general properties of solitary waves. See Section 4.5 for further information.

Proposition 3.5 *Assume (i'''). Then for any travelling wave $u \in X$*

(a) $u'(\infty) = \lim_{t \to \infty} u'(t) = 0$;

(b) *for the displacement profile $r(t) = u(t+1) - u(t)$ we have*

$$r(\infty) = \lim_{t \to \infty} r(t) = 0.$$

Proof. Since $Au(t) \in L^\infty(\mathbb{R})$, assumption (i''') implies that the right hand part of Eq. (3.3) belongs to $L^2(\mathbb{R})$. Hence, $u'' \in L^2(\mathbb{R})$ and $u' \in H^1(\mathbb{R})$. By the Sobolev embedding theorem (Theorem A.1), $u' \in C_0(\mathbb{R})$ and statement (a) is proved.

Statement (b) follows from Lemma 3.4. □

3.3.2 *From periodic waves to solitary ones*

To pass to the limit as $k \to \infty$, we need some preliminary results. Actually, we shall consider such limits along sequences $k_n \to \infty$. Therefore, to simplify the notation every time when we use an expression like "a sequence u_k" this means that there is a sequence $k_n \to \infty$ and $u_k = u_{k_n}$. So, this *does not* mean that k is integer.

Lemma 3.5

(a) *Assume (i''), (ii') and $c^2 > \max(a, 0)$. Then there exists a constant $\varepsilon_1 > 0$ independent of k such that for any nontrivial critical points $u_k \in X_k$ of J_k and $u \in X$ of J*

$$\varepsilon_1 \leq (c^2 - a)\|u_k\|_k^2 \leq \frac{2\theta}{\theta - 2} J_k(u_k)$$

and

$$\varepsilon_1 \leq (c^2 - a)\|u\|^2 \leq \frac{2\theta}{\theta - 2} J(u).$$

(b) *Under assumptions (i'), (ii^+) (resp., (ii^-)) and $c > c_0$ the statement of part (a) holds for nontrivial critical points $u_k \in PX_k$ (resp., $u_k \in -PX_k$) and $u \in PX$ (resp., $u \in -PX$) of J_k and J, with $a = c_0^2$.*

Proof. Part (b) follows from part (a), with $a = c_0^2$, if we modify $V(r)$ so that the new potential coincides with $V(r)$ for $r > 0$ (resp., $r < 0$) and vanishes for $r < 0$ (resp., $r > 0$).

Let us prove the upper bound for J_k. By assumption (ii'),

$$\begin{aligned} J_k(u_k) &= J_k(u_k) - \frac{1}{2} \langle J_k'(u_k), u_k \rangle \\ &= \int_{-k}^{k} \left[\frac{1}{2} V'(Au_k) Au_k - V(Au_k) \right] dt \\ &\geq \frac{\theta - 2}{2} \int_{-k}^{k} V(Au_k) dt. \end{aligned}$$

Hence,

$$(c^2 - a)\|u_k\|_k^2 = 2J_k(u_k) + 2\int_{-k}^{k} V(Au_k)\,dt$$

$$\leq \frac{2\theta}{\theta - 2} J_k(u_k).$$

To obtain the lower bound, we assume on the contrary that there exists a sequence of nontrivial critical points $u_{k_n} \in X_{k_n}$ such that $\|u_{k_n}\|_{k_n} \to 0$. (It is not necessary that $k_n \to \infty$). By Lemma 3.1, $\|Au_{k_n}\|_{L^\infty(-k_n,k_n)} \to 0$ and assumption (i') implies that

$$\left|V'(Au_{k_n}(t))\,Au_{k_n}(t)\right| \leq \varepsilon_n \left|Au_{k_n}(t)\right|^2,$$

where $\varepsilon_n \to 0$ as $n \to \infty$. Since

$$\langle J_{k_n}(u_{k_n}), u_{k_n}\rangle = 0,$$

we have

$$c^2\|u_{k_n}\|_{k_n}^2 = \int_{-k_n}^{k_n} \left[a|Au_{k_n}|^2 + V'(Au_{k_n})\,Au_{k_n}\right] dt$$

$$\leq (a + \varepsilon_n)\|u_{k_n}\|_{k_n}^2.$$

Since $c^2 > a$, this is a contradiction and the lemma is proved.

The case of J is similar. □

Remark 3.5 In the proof of Lemma 3.5 we have only used the fact that $\langle J_k(u_k), u_k\rangle = 0$ (resp., $\langle J(u), u\rangle = 0$). Hence, the statements of that lemma remain true for nonzero elements $u_k \in X_k$ (resp., $u \in X$) satisfying the last identities.

Lemma 3.6 *Assume (i'') and $c^2 > \max(a,0)$. Let $u_k \in X_k$ be a sequence such that $\|J'_k(u_k)\|_{k,*} \to 0$ and $\|u_k\|_k$ is bounded. Then either $\|u_k\|_k \to 0$, or for any $r > 0$ there exist $\eta > 0$, a subsequence of u_k, still denoted by u_k, and $\zeta_k \in \mathbb{R}$ such that*

$$\int_{\zeta_k - r}^{\zeta_k + r} |Au_k(t)|^2 dt \geq \eta.$$

Proof. Assume that

$$\limsup_{k \to \infty} \sup_{\zeta \in \mathbb{R}} \int_{\zeta - r}^{\zeta + r} |Au_k(t)|^2 dt = 0.$$

Let $\varphi_k \in C_0^\infty(\mathbb{R})$ be a function such that
$$0 \leq \varphi_k(t) \leq 1, \quad \varphi_k(t) = 1 \quad \text{if } |t| \leq k,$$
$$\varphi_k(t) = 0 \quad \text{if } |t| \geq k+1$$
and $|\varphi_k'(t)| \leq C$, where $C > 0$ is independent of k. We set
$$f_k(t) = \varphi_k(t) A u_k(t).$$
It is readily verified that $f_k \in H^1(\mathbb{R})$, $\|f_k\|_{H^1}$ is bounded and
$$\lim_{k \to \infty} \sup_{\zeta \in \mathbb{R}} \int_{\zeta-r}^{\zeta+r} |f_k(t)|^2 dt = 0.$$
Then Lemma B.2 implies that $\|f_k\|_{L^p(\mathbb{R})} \to 0$ for all $p > 2$. Since
$$\|A u_k\|_{L^p(-k,k)} \leq \|f_k\|_{L^p(\mathbb{R})},$$
we see that
$$\|A u_k\|_{L^p(-k,k)} \to 0$$
for all $p > 2$.

Let $\varepsilon_k = \|J_k'(u_k)\|_{k,*} \to 0$. Then we have
$$\langle J_k'(u_k), u_k \rangle_k = \int_{-k}^{k} [c^2 (u_k')^2 - a|A u_k|^2 - V'(A u_k) A u_k] \, dt \leq \varepsilon_k \|u_k\|_k.$$
By Lemma 3.1,
$$\|A u_k\|_{L^\infty(-k,k)} \leq C.$$
Fix any $p > 2$. Then assumption (i'') implies that for every $\varepsilon > 0$ there exists a constant $C_\varepsilon > 0$ such that
$$|V'(r)r| \leq \varepsilon r^2 + C_\varepsilon |r|^p, \quad |r| \leq C.$$
Now we have that
$$c^2 \|u_k\|_k^2 \leq \int_{-k}^{k} [a|A u_k|^2 + V'(A u_k) A u_k] \, dt + \varepsilon_k \|u_k\|_k$$
$$\leq \int_{-k}^{k} [(a+\varepsilon)|A u_k|^2 + C_\varepsilon |A u_k|^p] \, dt + \varepsilon_k \|u_k\|_k$$
$$= (a+\varepsilon) \|A u_k\|_{L^2(-k,k)}^2 + C_\varepsilon \|A u_k\|_{L^p(-k,k)}^p + \varepsilon_k \|u_k\|_k.$$

If $a < 0$ we can take $\varepsilon > 0$ small enough so that $a + \varepsilon < 0$ and obtain

$$c^2 \|u_k\|_k^2 \leq C_\varepsilon \|Au_k\|_{L^p(-k,k)}^p + \varepsilon_k \|u\|_k$$

which implies that $\|u_k\|_k \to 0$.

If $a > 0$, using Lemma 3.1 we obtain that

$$(c^2 - a - \varepsilon) \|u_k\|_k^2 \leq C_\varepsilon \|Au_k\|_{L^p(-k,k)}^p + \varepsilon_k \|u\|_k.$$

Since $c^2 > a$, we can choose $\varepsilon > 0$ so small that $c^2 - a - \varepsilon > 0$ and complete the proof as above. □

Proposition 3.6

(a) *Assume* (i''), (ii') *and* $c^2 > \max(a, 0)$. *Let* $u_k \in X_k$ *be a sequence of nontrivial critical points of* J_k *such that the critical values* $J_k(u_k)$ *are uniformly bounded. Then there exist a nontrivial critical point* $u \in X$ *of* J *and a sequence* $\zeta_k \in \mathbb{R}$ *such that a subsequence of* $u_k(\cdot + \zeta_k) - u_k(\zeta)$ *converges to* u *uniformly on compact intervals together with first and second derivatives, i. e. in* $C^2(\mathbb{R})$.

(b) *Under assumptions* (i'), (ii^+) *(resp.,* (ii^-)*) and* $c^2 > c_0^2$ *the same statement holds for nontrivial critical points* $u_k \in PX_k$ *(resp.,* $u_k \in -PX_k$*), with* $u \in PX$ *(resp.,* $u \in -PX$*).*

Proof. By Lemma 3.5, $\|u_k\|_k$ is bounded above and below by two positive constants. This means that $\|u_k\|_k$ does not converge to 0 and Lemma 3.6 shows that, along a subsequence,

$$\int_{\zeta_k - r}^{\zeta_k + r} |Au_k|^2 dt \geq \eta \tag{3.12}$$

for some $r > 0$, $\eta > 0$ and $\zeta_k \in \mathbb{R}$.

Let $\widetilde{u}_k(t) = u_k(t + \zeta_k) - u_k(\zeta_k)$. Then $\|\widetilde{u}_k\|_k = \|u\|_k$ and since J_k is invariant under translations and adding constants, $J_k(\widetilde{u}_k) = J_k(u_k)$ and $J_k'(\widetilde{u}_k) = 0$. Moreover, since $\|\widetilde{u}_k\|_k$ is bounded, there exist a subsequence, still denoted by $\widetilde{u}_k$, that converges to a function $u \in H^1_{loc}(\mathbb{R})$ weakly in $H^1_{loc}(\mathbb{R})$, i. e. weakly in $H^1(a, b)$ for every finite interval (a, b).

First, we check that $u \in \widetilde{X}$. Indeed, we have $\widetilde{u}_k' \to u'$ weakly in $L^2_{loc}(\mathbb{R})$. Hence, for every $a < b$

$$\int_a^b |u'(t)|^2 dt \leq \liminf \int_a^b |\widetilde{u}_k'(t)|^2 dt \leq \liminf \|\widetilde{u}_k\|_k^2 \leq C.$$

Passing to the limit as $a \to -\infty$ and $b \to +\infty$, we obtain the result.

By the compactness of Sobolev embedding, $A\widetilde{u}_k \to Au$ strongly in $L^\infty_{loc}(\mathbb{R})$, i. e. uniformly on finite intervals, and in $L^2_{loc}(\mathbb{R})$. This and Eq. (3.12) shows that

$$\int_{-r}^{r} |Au|^2 dt \geq \eta > 0,$$

hence, $u \neq 0$.

Since $A\widetilde{u}_k \to Au$ in $L^\infty_{loc}(\mathbb{R})$, then $U'(Au_k) \to U'(Au)$ in $L^\infty_{loc}(\mathbb{R})$. Let

$$\varphi \in C^\infty(\mathbb{R}), \quad \varphi(0) = 0, \quad \varphi' \in C_0^\infty(\mathbb{R}).$$

For k large, supp $A\varphi \subset [-k, k]$. For such k, let $\varphi_k \in X_k$ be the primitive function of the $2k$-periodic extension of $\varphi'|_{[-k,k]}$. Then

$$\langle J'(u), \varphi \rangle = \int_{-\infty}^{\infty} \left[c^2 u' \varphi' - U'(Au) A\varphi \right] dt$$

$$= \int_{\text{supp } A\varphi} \left[c^2 u' \varphi' - U'(Au) A\varphi \right] dt$$

$$= \lim_{k \to \infty} \int_{\text{supp } A\varphi} \left[c^2 \widetilde{u}'_k \varphi' - U'(A\widetilde{u}_k) A\varphi \right] dt$$

$$= \lim_{k \to \infty} \int_{-k}^{k} \left[c^2 \widetilde{u}'_k - U'(A\widetilde{u}_k) A\varphi \right] dt$$

$$= 0.$$

Hence, u is a nontrivial solution of Eq. (3.3).

The right hand side of Eq. (3.3) for $\widetilde{u}_k$ converges in $L^\infty_{loc}(\mathbb{R})$ to the right hand side of that equation for u. Therefore, $\widetilde{u}''_k \to u''$, hence, $\widetilde{u}'_k \to u'$ and $\widetilde{u}_k \to u$, in $L^\infty_{loc}(\mathbb{R})$. In particular, $u(0) = 0$ and $x \in X$. This proves part (a).

Part (b) follows from part (a). Just replace $V(r)$ by a function that coincides with $V(r)$ for $r > 0$ (resp., $r < 0$) and vanishes for $r < 0$ (resp., $r > 0$), and note that the limit of a sequence of monotone functions is a monotone function. □

Remark 3.6 Proposition 3.6 is still valid if, instead of a sequence of critical points, we consider a sequence $u_k \in X_k$ such that $\|J'_k(u_k)\|_{k,*} \to 0$ and $J_k(u_k)$ is a bounded sequence.

Combining Proposition 3.6 with Theorems 3.1 and 3.2, we obtain the following existence results.

Theorem 3.4

(a) *Assume* (i'), (ii^+) *and* $c > c_0$. *Then there exists a nontrivial nondecreasing solitary wave* $u \in X$.

(b) *Assume* (i'), (ii^-) *and* $c > c_0$. *Then there exists a nontrivial nonincreasing solitary wave* $u \in X$.

Theorem 3.5 *Assume* (i''), (ii') *and* $c^2 > \max(a, 0)$. *Then there exists a nontrivial solitary wave* $u \in X$.

The next result is an important supplement to Theorem 3.4 (*cf.*, however, Remark 3.3).

Proposition 3.7 *Under the assumptions of Theorem 3.4 suppose, in addition, that* $r \cdot U'(r) > 0$ *for all* $r > 0$ *(resp.* $r < 0$*). Then any nontrivial monotone wave* $u \in X$ *is strictly monotone. In particular, this is so if either* $c_0 > 0$, *or* $r \cdot V'(r) > 0$ *for all* $r > 0$ *(resp.* $r < 0$*).*

Proof. We consider only the case of nondecreasing waves. The case of nonincreasing waves is similar.

Let $u \in X$ be a nontrivial nondecreasing solitary wave. Then there exists $t_0 \in \mathbb{R}$ such that $u'(t_0) > 0$. Hence, if $t_0 - 1 \leq t \leq t_0$, then

$$Au(t) = \int_t^{t+1} u'(s)\, ds > 0 \qquad (3.13)$$

because the interval of integration contains the point t_0.

In Section 4.5, Theorem 4.10 and Remark 4.6, it is shown that $r(t) = Au(t)$, $u'(t)$ and $u''(t)$ decay exponentially at infinity.[3] Integrating Eq. (3.3) over the interval $(-\infty, t)$, we obtain that

$$u'(t) = \frac{1}{c^2} \int_{t-1}^{t} U'(Au(s))\, ds.$$

Together with (3.13), this shows that $u'(t) > 0$ whenever $t_0 - 1 \leq t \leq t_0 + 1$.

Iterating this argument n times, we obtain that the derivative $u'(t)$ is positive on the interval $[t_0 - n, t_0 + n]$. This completes the proof. □

In Theorems 3.4 and 3.5 solitary waves, *i. e.* critical points of J, are obtained as local limits of mountain pass critical points of J_k. Note that the functional J itself also possesses the mountain pass geometry.

---

[3] Note that Section 4.5 is independent on the previous material.

Confining ourself to the setting of Theorem 3.5, let

$$\alpha_k = \inf_{\gamma \in \Gamma_k} \max_{s \in [0,1]} J_k(\gamma(s))$$

and

$$\alpha = \inf_{\gamma \in \Gamma} \max_{s \in [0,1]} J(\gamma(s))$$

be mountain pass values of J_k and J, respectively. Here

$$\Gamma_k = \Big\{ \gamma \in C([0,1]; X_k) \, : \, \gamma(0) = 0, \gamma(1) = e_k \Big\},$$

$$\Gamma = \Big\{ \gamma \in C([0,1]; X) \, : \, \gamma(0) = 0, \gamma(1) = e \Big\},$$

$e \in X$ and $e \in X_k$ are arbitrary elements such that $J(e) < 0$ and $J_k(e_k) < 0$. Note that α is not necessary a critical value of J.

Proposition 3.8 *Under assumptions of Theorem 3.5 we have*

$$J(u) \leq \limsup \alpha_k \leq \alpha \leq \inf_{v \in X \setminus \{0\}} \sup_{\tau > 0} J(\tau v).$$

Proof. In Theorem 3.5 the critical point $u \in X$ is obtained as the limit of an appropriate sequence $u_k \in X_k$ of mountain pass critical points of the functionals J_k in the topology of $C^1(\mathbb{R})$, i. e. uniformly on finite interval together with first derivatives.

Let $k_0 \geq 1$ and $k \geq k_0$. Then

$$J_k(u_k) = J_k(u_k) - \frac{1}{2} \langle J_k(u_k), u_k \rangle = \int_{-k}^{k} g(Au_k)\, dt,$$

where

$$g(r) = \frac{1}{2} V'(r)\, r - V(r).$$

Due to assumption (ii'), $g(r) \geq 0$. Hence,

$$J_k(u_k) \geq \int_{-k_0}^{k_0} g(Au_k)\, dt.$$

Since $Au_k \to Au$ in $L^\infty_{loc}(\mathbb{R})$, we obtain that

$$\limsup J_k(u_k) = \limsup \alpha_k \geq \int_{-k_0}^{k_0} g(Au)\, dt.$$

Passing to the limit as $k_0 \to \infty$ yield

$$\int_{-\infty}^{+\infty} g(Au)\, dt \leq \limsup \alpha_k.$$

Since

$$J(u) = J(u) - \frac{1}{2}\langle J'(u), u\rangle = \int_{-\infty}^{+\infty} g(Au)\, dt,$$

we get

$$J(u) \leq \limsup \alpha_k.$$

Now let $\gamma \in \Gamma$. Consider the pass $\gamma_k \in \Gamma_k$ uniquely defined by:

$$\frac{d}{dt}(\gamma_k(s)(t)) = \begin{cases} \dfrac{d}{dt}(\gamma(s)(t)), & |t| \leq k-1, \\ 0, & k-1 < |t| \leq k. \end{cases}$$

(Here $s \in [0,1]$ is the parameter of the pass). Then it is not difficult to see that

$$\lim_{k\to\infty} J_k(\gamma_k(s)) = J(\gamma(s))$$

uniformly with respect to $s \in [0,1]$. This immediately implies the inequality

$$\limsup \alpha_k \leq \alpha.$$

Finally, for every $v \in X \setminus \{0\}$ there exists $\tau_0 > 0$ such that

$$J(\tau v) < 0$$

for all $\tau \geq \tau_0$. Considering the path that consists of the segment $\{\tau v : 0 \leq \tau \leq \tau_0\}$ and a path joining the points $\tau_0 v$ and e inside the domain where J is negative, we obtain the last inequality of the proposition. □

Similar result takes place in the setting of Theorem 3.4. The only we need is to choose e and e_k to be increasing (respectively, decreasing) and take the infimum in the last inequality of Proposition 3.8 among all nonzero $v \in PX$ (respectively, $v \in -PX$).

3.3.3 Global structure of periodic waves

The convergence result of Proposition 3.6 shows, in particular, that locally a periodic wave profile looks like an appropriate single solitary wave profile. Here we present a refined version of that result, which shows that a periodic wave profile, with large period, on the wavelength interval looks like a sum of finite number of solitary profiles. Actually, it is convenient to consider not exact periodic waves, *i. e.* critical points of J_k, but approximate waves, *i. e.* sequences $u_k \in X_k$ such that $\|J_k(u_k)\|_{k,*} \to 0$ as $k \to \infty$.

Proposition 3.9 *In addition to assumptions* (i'') *and* (ii'), *with* $c^2 > \max(a, 0)$, *suppose that* $V \in C^2(\mathbb{R})$. *Let* $u_k \in X_k$ *be a sequence satisfying* $\|J'_k(u_k)\|_{k,*} \to 0$ *and* $J_k(u_k) \to \alpha > 0$ *as* $k \to \infty$. *Then there exist a finite number of critical points* $u^i \in X$ *of* J, $i = 1, 2, \ldots, l$, *and* $\zeta_k^i \in \mathbb{R}$, $i = 0, 1, \ldots, l-1, l$, *with* $\zeta_k^l = 0$, *such that*

$$\sum_{i=1}^{l} J(u^i) = \alpha$$

and, as $k \to \infty$ *along a sequence,*

$$\left\| \left[u_k(\cdot + \zeta_k^0) - \sum_{i=1}^{l} u^i(\cdot + \zeta_k^i) \right]' \right\|_{L^2(-k,k)} \to 0.$$

Proof. By Remark 3.6, there exist a critical point $u = u^1 \in X$ of J and $\zeta_k \in \mathbb{R}$ such that, along a subsequence,

$$\widetilde{u}_k(t) = u_k(t + \zeta_k) - u_k(\zeta_k)$$

converges to u in the sense of Proposition 3.6.

Let $\varphi_k \in C^\infty(\mathbb{R})$ be a sequence with the following properties:

$$\varphi_k(0) = 0, \quad \varphi'_k \in C_0^\infty(\mathbb{R}), \quad \operatorname{supp} \varphi_k \subset [-k+1, k-1]$$

and $\varphi_k \to u$ strongly in X. Obviously, such a sequence exists. Since J is C^1, then $J'(\varphi_k) \to 0$ and $J(\varphi_k) \to J(u)$. Denote by $\widetilde{\varphi}_k \in X_k$ the function such that $(\widetilde{\varphi}_k)'$ is the $2k$-periodic extension of $\varphi'_k|_{[-k,k]}$ to $\mathbb{R}$. For any $\eta \in X_k$ we set

$$\eta_k(t) := (T_k \eta)(t) := \begin{cases} \eta(t), & -k-1 \leq t \leq k+1, \\ \eta(k+1), & t \geq k+1, \\ \eta(-k-1), & t < -k-1. \end{cases}$$

This is a kind of cut-off. The function η_k is well-defined because η is continuous. It is easy that $T_k \eta \in X$ and

$$\|T_k\eta\| \le 2\|\eta\|_k.$$

We have

$$\left|\langle J'_k(\widetilde{\varphi}_k), \eta\rangle\right| = \left|\langle J'(\varphi_k), \eta_k\rangle\right| \le \|J'(\varphi_k)\|_* \|\eta_k\|$$
$$\le 2 \|J'(\varphi_k)\|_* \|\eta\|_k.$$

Therefore, $\|J'_k(\widetilde{\varphi}_k)\|_{k,*} \to 0$.

Setting $v_k = \widetilde{u}_k - \widetilde{\varphi}_k$, we want to show that $\|J'_k(v_k)\|_{k,*} \to 0$. First, we note that, for $\eta \in X_k$,

$$\langle J'_k(v_k), \eta\rangle = \langle J'_k(\widetilde{u}_k) - J'_k(\widetilde{\varphi}_k), \eta\rangle$$
$$+ \int_{-k}^{k} \left[V'(A\widetilde{u}_k) - V'(A\varphi_k) - V'(A\widetilde{u}_k - A\varphi)\right] A\eta \, dt. \quad (3.14)$$

Since $\varphi_k \to u$ in X, for any $\varepsilon > 0$ we can find $k_0 > 0$ such that

$$|A\varphi_k(t)| \le \varepsilon$$

for $t \notin B := [-k_0, k_0]$ and

$$\int_{\mathbb{R}\setminus B} |A\varphi_k|^2 dt \le \varepsilon^2.$$

Let $B_k = [-k, k] \setminus B$. Since $A\varphi_k$ is bounded in $L^\infty(\mathbb{R})$, assumption (i'') yields

$$\left|V'(A\varphi_k(t))\right| \le C|A\varphi_k(t)|,$$

with $C > 0$ independent of k. Then

$$\left|\int_{B_k} V'(A\varphi_k) A\eta\right| \le C \left(\int_{\mathbb{R}\setminus B} |A\varphi_k|^2 dt\right)^{1/2} \|A\eta\|_{L^2(-k,k)} \le C\varepsilon \|\eta\|_k.$$

Using the fact that $V \in C^2$ and $A\widetilde{u}_k$ and $A\varphi_k$ are bounded in $L^\infty(\mathbb{R})$, we obtain

$$\left|V'(A\widetilde{u}_k - A\varphi_k) - V'(A\widetilde{u}_k)\right| = \left|V''(A\widetilde{u}_k - \theta A\varphi_k) A\varphi_k\right| \le C|A\varphi_k|,$$

with $\theta \in [0,1]$. Hence,

$$\left|\int_{B_k} [V'(A\widetilde{u}_k - A\varphi_k) - V'(A\widetilde{u}_k)] A\eta\, dt\right| \le C \int_{B_k} |A\varphi_k| |A\eta|\, dt$$
$$\le C \left(\int_{B_k} |A\varphi_k|^2 dt\right)^{1/2} \|\eta\|_k$$
$$\le C\varepsilon \|\eta\|_k.$$

Since $A\widetilde{u}_k \to Au$ in $L^{\infty}_{loc}(\mathbb{R})$, $A\varphi_k \to Au$ in $L^{\infty}(\mathbb{R})$ and $V \in C^1(\mathbb{R})$, we see that

$$\left|\int_B [V'(A\widetilde{u}_k) - V'(A\varphi_k) - V'(A\widetilde{u}_k - A\varphi_k)] A\eta\, dt\right| \le C\varepsilon \|\eta\|_k,$$

provided k is large enough. Thus, given $\varepsilon > 0$ the integral term in Eq. (3.14) is estimated by $C\varepsilon\|\eta\|_k$ provided k is large enough. This means that $\|J'_k(v_k)\|_{k,*} \to 0$.

Now a straightforward calculation shows that

$$J_k(\widetilde{u}_k) = J_k(v_k) + J_k(\widetilde{\varphi}_k) + (v_k, \widetilde{\varphi}_k)_k$$
$$- \int_{-k}^{k} [V(Av_k - A\varphi_k) - V(Av_k) - V(A\varphi_k)]\, dt. \quad (3.15)$$

The integral term here can be estimated as above and, hence, tends to zero. A trivial calculation shows that

$$(v_k, \widetilde{\varphi}_k)_k = (T_k v_k, \varphi_k).$$

Since $v'_k \to 0$ in $L^{\infty}_{loc}(\mathbb{R})$, so is $(T_k v_k)'$. Moreover, $T_k v_k \to 0$ weakly in X because $T_k v_k$ is bounded in X. Since φ_k is strongly convergent in X, we have that $(T_k v_k, \varphi_k) \to 0$, hence, $(v_k, \widetilde{\varphi}_k)_k \to 0$. Now Eq. (3.15) implies that

$$\alpha = \lim J_k(\widetilde{u})_k = J(u) + \lim J_k(v_k),$$

i. e. $J_k(v_k) \to \alpha - J(u)$. Moreover, u is a nontrivial critical point of J, hence, $J(u) > 0$, by Lemma 3.5.

Since $\|J'_k(v_k)\|_{k,*} \to 0$, the beginning of the proof of Proposition 3.6 shows that either $J_k(v_k) \to 0$ and $\|v_k\|_k \to 0$, or $J_k(v_k)$ and $\|v_k\|_k$ is bounded below by a positive constant independent of k. In the first case we are done, with $l = 1$. In the second case, by Lemma 3.5,

$$0 < \varepsilon_2 := \frac{\theta - 2}{2(\theta - 2)} \varepsilon_1 \le J(u) < \alpha,$$

with ε_1 from Lemma 3.5. Note that ε_2 is independent of u_k and u. Now we can repeat the arguments above, with u_k replaced by v_k and α replaced by $\alpha - J(u) < \alpha$, to get u^2. After a finite number, not greater that α/ε_2, of such steps we conclude. □

3.3.4 Examples

Here we consider several examples.

Example 3.1 In the case of Toda lattice the potential is given by

$$U(r) = ab^{-1}\left(e^{-br} + br - 1\right), \quad ab > 0.$$

If $b > 0$ and $c > ab$, then for every $k \geq 1$, Eq. (3.1) has a nontrivial nonincreasing $2k$-periodic travelling wave and also a nontrivial decreasing solitary wave. If $b < 0$ and $c > ab$, then there exist a nontrivial nondecreasing $2k$-periodic travelling wave and a nontrivial increasing solitary wave. Certainly, these results are known [Toda (1989)].

Example 3.2 Consider the potential

$$U(r) = \frac{c_0^2}{2}r^2 + \frac{c_1}{2n+1}r^{2n+1}, \quad c_0 > 0, c_1 \neq 0.$$

If $n = 1$, this is the FPU α-model. Let $c > c_0$. If $c_1 > 0$, then there exist nontrivial nondecreasing travelling waves, both $2k$-periodic, with $k \geq 1$, and solitary. If $c_1 < 0$, then there exist nontrivial $2k$-periodic and solitary travelling waves, nonincreasing in this case. Actually, due to Proposition 3.7, the solitary waves are strictly monotone.

Example 3.3 Let

$$U(r) = \frac{c_0^2}{2}r^2 + \frac{c_1}{2n}r^{2n}, \quad c_0 > 0, c_1 > 0.$$

In the case when $n = 2$ we obtain the FPU β-model. Note that the potential $U(r)$ is even. If $c > c_0$, then for every $k \geq 1$, there exists a pair $\pm u_k \in X_k$ of nontrivial travelling waves, one nondecreasing and the other nonincreasing. Also there exists a pair $\pm u \in X$ of nontrivial strictly monotone solitary waves. If $0 < c \leq c_0$, then for every $k \geq 1$ there exists a pair $\pm u_k \in X_k$ of $2k$-periodic subsonic travelling waves with the speed c.

Example 3.4 Consider the potential

$$U(r) = \frac{a}{2}r^2 + \sum_{i=3}^{m} a_i |r|^i + \sum_{i=m+1}^{n} b_i |u|^i,$$

with $a_i \leq 0$, $b_i \geq 0$ and $b_n > 0$. If $c^2 > \max(0,a)$, then Theorems 3.2 and 3.5 apply and, hence, we obtain nontrivial periodic and solitary travelling waves, not necessary monotone.

3.4 Ground waves: existence and convergence

3.4.1 Ground waves: periodic case

Consider a travelling wave that has the profile $u \in X_k$, i. e. a $2k$-periodic wave in our terminology. We say that u is a *periodic ground wave* if the corresponding critical value $J_k(u)$ is minimal possible among all nontrivial critical values of the functional J_k.

First we study general, not necessary monotone waves, in the setting of Subsection 3.2.3 and postpone the discussion of monotonicity to Subsection 3.4.3. Since the functional J_k satisfies the Palais-Smale condition, the existence of ground waves in this case is almost trivial and we have

Proposition 3.10 *Assume (i'') and (ii') from Theorem 3.2. Let $c^2 > \max(a,0)$. Then for every $k \geq 1$ there exists a $2k$-periodic ground wave $u_k \in X_k$. Moreover, any $2k$-periodic ground wave u_k satisfies*

$$\delta \leq J_k(u_k) \leq M,$$

with $\delta > 0$ and $M > 0$ independent of k.

Proof. Fixed $k \geq 1$, let α be the infimum of nontrivial critical values of J_k. By Theorem 3.2, $0 < \alpha < +\infty$. Then there exists a sequence $u^{(n)}$ of critical points of J_k such that the sequence $J_k(u^{(n)})$ of corresponding critical values converges to α. Obviously, $u^{(n)}$ form a Palais-Smale sequence. Since J_k satisfies the Palais-Smale condition, the sequence $u^{(n)}$ is precompact in X_k and any its limit point is a critical point of J_k at the level α, hence, a ground wave. □

Remark 3.7 In addition to the assumptions of Proposition 3.10, suppose that the interaction potential is even. This implies that the functional J_k is even. Using standard multiplicity results for even functionals [Rabinowitz (1986); Willem (1996)], one can show that, for every $k \geq 1$, there exists an infinite sequence $u^{(n)} \in X_k$ of critical points, such that the critical values $J_k(u^{(n)})$ go to infinity. We do not present the details because no counterpart of this result is known in the case of solitary waves.

Now we are going to present a more explicit description of ground waves. With this aim we impose the following assumption

(N) $V \in C^2$ and there exists $\mu \in (0,1)$ such that

$$0 < r^{-1} V'(r) \leq \mu V''(r), \quad r \neq 0.$$

An elementary integration by parts shows that (N) implies (ii'), with

$$\theta = \frac{1+\mu}{\mu} > 2.$$

Moreover, the function $V'(r)/r$ is monotone decreasing from $+\infty$ to 0 on $(-\infty, 0)$ and monotone increasing from 0 to $+\infty$ on $(0, +\infty)$.

Let us introduce the functional

$$I_k(u) := \langle J_k'(u), u \rangle = \int_{-k}^{k} \left[c^2 (u')^2 - a(Au)^2 - V'(Au)\, Au \right] dt. \quad (3.16)$$

This is the so-called *Nehari functional* of J_k. The same argument as in the proof of Proposition 3.1 shows that, under assumptions (i'') and (N), the functional I_k is a C^1 functional on X_k and

$$\langle I_k'(u), h \rangle = \int_{-k}^{k} \left[2c^2 u' h' - 2a\, Au\, Ah - V''(Au)\, Au\, Ah - V'(Au)\, Ah \right] dt \quad (3.17)$$

for all $h \in X_k$.

Now we define the *Nehari manifold* by

$$S_k = \{ u \in X_k \setminus \{0\} : I_k(u) = 0 \}. \quad (3.18)$$

Assumption (N) implies that S_k is a C^1 submanifold of codimension 1 in X_k. Indeed, this follows from the implicit function theorem. For, we have to check only that $I_k'(u) \neq 0$ for every $u \in S_k$. But, for any such u, a straightforward calculation shows that

$$\langle I'_k(u), u \rangle = \int_{-k}^{k} \left[c^2 (u')^2 - a(Au)^2 - V''(Au)(Au)^2 \right] dt$$

$$\leq \int_{-k}^{k} \left[c^2 (u')^2 - a(Au)^2 - \mu^{-1} V'(Au) Au \right] dt$$

$$= \int_{-k}^{k} \left[c^2 (u')^2 - a(Au)^2 - V'(Au) Au - (\mu^{-1} - 1) V'(Au) Au \right] dt$$

$$= -(\mu^{-1} - 1) \int_{-k}^{k} V'(Au) Au \, dt,$$

where we have used the definition of S_k and assumption (N). Since $u \neq 0$, the last integral above is positive. Hence,

$$\langle I'_k(u), u \rangle < 0$$

and $I'_k(u) \neq 0$. This also shows that if $u \in S_k$, then the line $\mathbb{R}u$ is transverse to the tangent space $\ker I'_k(u)$ to the manifold S_k at the point u.

Another important property of the Nehari manifold S_k is that, under our assumption, for any nonzero $u \in X_k$ there exists a unique $\tau > 0$ such that $\tau u \in S_k$. In particular, $S_k \neq \emptyset$. Indeed, for $\tau > 0$ we have

$$I_k(\tau u) = \tau^2 \left[c^2 \|u\|_k^2 - a\|Au\|_{L^2(-k,k)}^2 - \tau^{-1} \int_{-k}^{k} V'(\tau Au) Au \, dt \right].$$

Since

$$c^2 \|u\|_k^2 - a\|Au\|_{L^2(-k,k)}^2 > 0,$$

the monotonicity property of $V'(r)/r$ implies the required.

We consider the restriction $J_k|_{S_k}$ of the functional J_k to the Nehari manifold S_k. Note that

$$J_k(u) = \int_{-k}^{k} \left[\frac{1}{2} V'(Au) Au - V(Au) \right] dt \quad \text{on } S_k. \tag{3.19}$$

Proposition 3.11 *Under assumptions (i'') and (N), let $c^2 > \max(a, 0)$. Then any critical point of $J_k|_{S_k}$ is a critical point of J_k.*

Proof. A straightforward calculation shows that, for any $u \in X_k$,

$$\frac{d}{d\tau} J_k(\tau u)\big|_{\tau=1} = I_k(u).$$

Therefore, if $u \in S_k$, then $J'_k(u)$ vanishes on the line $\mathbb{R}u$. But $J'_k(u)$ vanishes also on the tangent space $\ker I'_k(u)$ to S_k at the point u because u is a critical point of $J_k|_{S_k}$. Since the subspaces $\ker I'_k(u)$ and $\mathbb{R}u$ a transverse, then $J'_k(u) = 0$ and u is a critical point of entire functional J_k. □

Remark 3.8 Obviously, all nontrivial critical points of J_k belong to S_k.

Due to Proposition 3.11, it is natural to consider the following minimization problem

$$m_k = \inf\{J_k(u) \; : \; u \in S_k\}. \tag{3.20}$$

Equation (3.19) together with (ii'), a consequence of (N), shows that $m_k \geq 0$. In fact, we have

Lemma 3.7 *Under assumptions (i'') and (N), let $c^2 > \max(a, 0)$. Then there exist constants $\varepsilon_1 > 0$ and $\varepsilon_2 > 0$, independent of k, such that $\|u\|_k \geq \varepsilon_1$ and $J_k(u) \geq \varepsilon_2$ for all $u \in S_k$. In particular, $m_k \geq \varepsilon_2 > 0$.*

Proof. Similar to the proof of Lemma 3.6 (see Remark 3.5). □

Now we summarize geometric properties of the Nehari manifold. The manifold S_k separates two open subsets $D_k^+ \cup \{0\}$ and D_k^-, i. e.

$$X_k = \left(D_k^+ \cup \{0\}\right) \cup D_k^- \cup S_k$$

and

$$S_k = \partial(D_k^+ \cup \{0\}) = \partial D_k^-,$$

where

$$D_k^+ = \{u \in X_k \; : \; I_k(u) > 0\}$$

and

$$D_k^- = \{u \in X_k \; : \; I_k(u) < 0\}.$$

Moreover, $D_k^+ \cup \{0\}$ is a star-shaped set that contains a ball centered at the origin. The functional J_k restricted to any transverse ray $\{\tau v, \tau > 0\}$, $v \in S_k$, attains its maximum exactly at $v \in S_k$.

Since all nontrivial critical points of J_k lie in S_k, the importance of problem (3.20) is that any minimum point is a $2k$-periodic ground wave.

Now let e_k be the function from Lemma 3.2, $J_k(e_k) \leq 0$. Denote by α_k the mountain pass critical value

$$\alpha_k := \inf_{\gamma \in \Gamma_k} \max_{t \in [0,1]} J_k(\gamma(y))$$

of J_k, where

$$\Gamma_k := \left\{ \gamma \in C([0,1]; X_k) \; : \; \gamma(0) = 0, \gamma(1) = e_k \right\}.$$

This is exactly the critical value from Theorem 3.2.

Lemma 3.8 *Assume (i'') and (N), with $c^2 > \max(a, 0)$. Then $\alpha_k = m_k$ and any mountain pass critical point is a solution of (3.20).*

Proof. Since α_k is a critical value with the critical point $u_k \in S_k$, then $m_k \leq \alpha_k$.

Let $v_0 \in X_k$, $\|v_0\|_k = 1$. Then, as in the proof of Lemma 3.2, we see that there exists $\tau_0 > 0$ such that $J_k(\tau v) < 0$ for all $\tau \geq \tau_0$ and all $v \in \mathrm{span}\{v_0, e_k\}$, with $\|v\|_k = 1$. This means that one can join $\tau_0 v_0$ and e_k by a path γ' such that J_k is negative on γ'. Combining γ' with the segment $\{\tau v_0, 0 \leq \tau \leq \tau_0\}$, we obtain the path that belongs to Γ_k. This implies that

$$\alpha_k \leq \alpha_k' = \inf_{v \in X_k \setminus \{0\}} \sup_{\tau > 0} J_k(\tau v).$$

Now observe that, for any $v \in X_k$, $v \neq 0$, the function $J_k(\tau v_0)$ of $\tau > 0$ attains its maximum value at some point $\tau_0 > 0$ and we have

$$0 = \frac{d}{dt} J_k(\tau v)|_{\tau = \tau_0} = \frac{1}{\tau_0} I_k(\tau_0 v).$$

Hence, $\tau_0 v \in S_k$. This shows that

$$\alpha_k' = m_k = \inf\{J_k(u) \; : \; u \in S_k\}$$

and we conclude. $\square$

3.4.2 Solitary ground waves

Here we consider *solitary ground waves*, i. e. nontrivial solitary travelling waves with profile $u \in X$ such that the critical value $J(u)$ is minimal possible among all nontrivial critical values of J. Since the functional J does not satisfy the Palais-Smale condition, the existence of such waves is not trivial and to prove it we employ the Nehari manifold approach. Therefore, we still assume (N) and (i'), with $c^2 > \max(a, 0)$.

The Nehari functional for J reads

$$I(u) := \langle J'(u), u \rangle = \int_{-\infty}^{+\infty} \left[c^2 (u')^2 - a(Au)^2 - V'(Au)\, Au \right] dt \qquad (3.21)$$

and the Nehari manifold is defined by

$$S := \{ u \in X \setminus \{0\} \ : \ I(u) = 0 \}. \qquad (3.22)$$

Note that

$$J(u) = \int_{-\infty}^{\infty} \left[\frac{1}{2} V'(Au)\, Au - V(Au) \right] dt =: \Psi(u) \quad \text{on } S. \qquad (3.23)$$

Under assumptions (i'') and (N) geometric properties of I and S are the same as in the case of I_k and S_k. In particular, S is a C^1-submanifold of X of codimension 1. All nontrivial critical points of J belong to S. Moreover, we have

Proposition 3.12 *Under assumptions (i'') and N, let $c^2 > \max(a, 0)$. Then any critical point of $J|_S$ is a critical point of J.*

Proof uses the same argument as the proof of Proposition 3.11. □

As in the case of S_k, for every $u \in X$, $u \neq 0$, there exists a unique $\tau = \tau(u) > 0$ such that $\tau u \in S$. Moreover, $\tau(u)$ depends continuously on $u \in X$.

Consider the minimization problem

$$m = \inf\{ J(u) \ : \ u \in S \} \qquad (3.24)$$

analogous to problem (3.20). Due to Proposition 3.12, any minimizer of (3.24) if it exists is a ground wave.

Let e_1 be the function from Lemma 3.2. Denote by $e \in X$ the function defined by $e'|_{[-1,1]} = e'_1|_{[-1,1]}$ and $e'(t) = 0$ if $|t| > 1$. Then $J(e) \leq 0$. Let α be the mountain pass value

$$\alpha := \inf_{\gamma \in \Gamma} \max_{s \in [0,1]} J(\gamma(s)),$$

where

$$\Gamma := \left\{ \gamma \in C([0,1]; X) \ : \ \gamma(0) = 0, \gamma(1) = e \right\}.$$

Note that *a priori* it is not known that α is a critical value of J because J does not satisfy (PS).

Lemma 3.9 *Under assumptions of Proposition 3.12 we have that $\alpha = m$.*

Proof is identical to the proof of Lemma 3.8. □

Now we are ready to prove the main result of this section that gives, in particular, the existence of solitary ground waves.

Theorem 3.6 *Assume (i'') and N, with $c^2 > \max(a, 0)$. Let $u_k \in X_k$ be a sequence of periodic ground waves. Then there exist a sequence $\zeta_k \in \mathbb{R}$ and a solitary ground wave $u \in X$ such that, along a subsequence $k \to \infty$, the functions $\widetilde{u}_k = u_k(\cdot + \zeta_k) - u_k(\zeta)$ satisfy*

$$\lim_{k \to \infty} \|\widetilde{u}_k' - u'\|_{L^2(-k,k)} = 0. \tag{3.25}$$

Moreover, $m = J(u)$ and $m_k \to m$.

Proof. The existence of periodic ground waves follows from Proposition 3.10. The same proposition and Lemma 3.6 imply that $\|u_k\|_k$ is bounded. Now, applying Proposition 3.6, we see that there exist ζ_k and a nontrivial critical point $u \in X$ of J such that $\widetilde{u}_k \to u$ in $H^1_{loc}(\mathbb{R})$.

To prove that the limit u is actually a ground wave, first note that for any $v \in S$ and $\varepsilon > 0$ there exists $v_k \in S_k$ such that

$$J_k(v_k) \le J(v) + \varepsilon$$

if k is large enough. To show this, take a sequence $\varphi_k \in C^\infty(\mathbb{R})$ such that $\operatorname{supp}\varphi_k' \subset [-k+1, k-1]$ and $\varphi_k \to v$ in X. Denote by $\overline{\varphi}_k \in X_k$ the function such that $(\overline{\varphi}_k)' = \varphi_k'$ on $[-k+1, k-1]$. Let $\tau_k > 0$ be a unique number such that $v_k = \tau_k \overline{\varphi}_k \in S_k$. In fact, we have also that $\tau_k \varphi_k \in S$. Hence, $\tau_k \to 1$ and $\tau_k \varphi_k \to v$. Therefore,

$$J_k(v_k) = J(\tau_k \varphi_k) \to J(v)$$

and we obtain the required.

As consequence, since $m_k \le J_k(v_k)$, we have that

$$\limsup_{k \to \infty} m_k \le J(v) + \varepsilon$$

for all $v \in S$. Hence,

$$\limsup_{k \to \infty} m_k \le m.$$

Let

$$g(r) = \frac{1}{2} V'(r)\, r - V(r).$$

Since $g(r) \geq 0$, then, by (3.19),

$$m_k = \int_{-k}^{k} g(A\tilde{u}_k) \, dt \geq \int_{B} g(A\tilde{u}_k) \, dt$$

for any bounded interval B and k large enough. We know that $Au_k \to Au$ in $L^\infty_{loc}(\mathbb{R})$. Therefore, $g(A\tilde{u}_k) \to g(Au)$ in $L^\infty_{loc}(\mathbb{R})$, and the integral in the right hand part of the inequality above converges to

$$\int_{B} g(Au) \, dt.$$

Since B is an arbitrary interval, this implies that

$$\limsup_{k \to \infty} m_k \geq \int_{-\infty}^{+\infty} g(Au) \, dt.$$

As u is a critical point of J, we have that $u \in S$ and, by (3.23),

$$\int_{-\infty}^{+\infty} g(Au) = J(u) \geq m.$$

This implies that

$$\limsup_{k \to \infty} m_k \geq J(u) \geq m$$

and therefore

$$\lim_{k \to \infty} m_k = m = J(u).$$

Thus u is a solitary ground wave.

The convergence statement (3.25) follows from Proposition 3.9. Indeed, since $J(u)$ is the lowest nontrivial critical point of J, we have that $l = 1$ in that proposition and we are done. □

3.4.3 Monotonicity

Now we consider the case of even interaction potential.

Proposition 3.13 *Assume (i''), with $a = c_0^2$, and (N). Suppose that $V(r)$ is an even function and $c > c_0$. Then every ground wave, periodic or solitary, is either nondecreasing or nonincreasing.*

Proof. We consider the case of solitary waves. The remaining case of periodic waves is quite similar.

Let $u \in X$ be a ground solitary wave. Since the potential is even, the function $-u$ is also a ground wave. Remind that

$$Pu(t) = \int_0^t |u'(s)|\, ds.$$

Suppose that u is not monotone. Then $Pu \neq \pm u$. By the definitions of operators A and P,

$$(APu)(t) = \int_t^{t+1} |u'(s)|\, ds$$

and

$$|Au(t)| \leq (APu)(t).$$

Moreover, since $u(t)$ is not a monotone function, the last inequality is strict for all t in an open subset of $\mathbb{R}$.

By Theorem 3.6, $u \in S$ is a minimizer of problem (3.24). Since $(Pu)' = |u'|$, we have that

$$0 = I(u) = \int_{-\infty}^{+\infty} \left[c^2(u')^2 - c_0^2|Au|^2 - V'(Au)\, Au\right] dt$$

$$= \int_{-\infty}^{+\infty} \left\{c^2[(Pu)']^2 - c_0^2|Au|^2 - V'(|Au|)\, |Au|\right\} dt$$

$$> \int_{-\infty}^{+\infty} \left\{c^2|(Pu)'|^2 - c_0^2|APu|^2 - V'(APu)\,(APu)\right\} dt$$

$$= I(Pu).$$

We have used here the strict monotonicity of $V'(r)r$. Thus, $Pu \notin S$, but there exists $\tau > 0$ such that $v = \tau Pu \in S$. Moreover, since $I(Pu) < 0$, we have that $\tau < 1$.

A straightforward calculation shows that

$$\left[\frac{1}{2}V(r)\,r - V(r)\right]' = \frac{r}{2}[V''(r) - r^{-1}V'(r)].$$

Hence, due to assumption (N), the function

$$g(r) := \frac{1}{2}V(r)\,r - V(r)$$

is decreasing if $r \leq 0$ and increasing if $r \geq 0$. Therefore,

$$J(u) = \Psi(u) = \int_{-\infty}^{+\infty} g(Au)\,dt = \int_{-\infty}^{+\infty} g(|Au|)\,dt > \int_{-\infty}^{+\infty} g(APu)\,dt$$
$$= \Psi(Pu).$$

Further,

$$\Psi(v) = \int_{-\infty}^{+\infty} g(\tau APu)\,dt < \int_{-\infty}^{+\infty} g(APu)\,dt = \Psi(Pu).$$

Hence,

$$\Psi(v) < J(u) = m.$$

Since $v \in S$, this mean that u is not a minimizer of (3.24). Contradiction. Thus, we have proved that u is nondecreasing. $\square$

Remark 3.9 Actually, in Proposition 3.14 solitary ground waves are strictly monotone as it follows from Proposition 3.8.

3.5 Near sonic waves

3.5.1 Amplitude estimate

Here we consider travelling waves with the speed $c > c_0$ close to the speed of sound c_0. We start with the following statement which is a key point of the present section.

Proposition 3.14 *Assume (i'), (ii^+) and (ii^-). In addition, suppose that there exist $a_0 > 0$ and $q > 2$ such that*

$$V(r) \geq a_0 |r|^q, \quad r \in \mathbb{R}. \tag{3.26}$$

Let $c > c_0$ and $u_k = u_{k,c} \in X_k$ (resp. $u = u_c \in X$) denote any nontrivial $2k$-periodic (resp. solitary) travelling wave with critical value not greater than the mountain pass level of J_k (resp. J). Then there exist constants $\bar{c}$ and C such that for every $c \in (c_0, \bar{c}]$

$$\|u_k\|_k^2 \leq C(c^2 - c_0^2)^{\frac{q+4}{q-2}},$$

provided k is large enough $(k \geq k_0(c))$, and

$$\|u\|^2 \leq C(c^2 - c_0^2)^{\frac{q+4}{q-2}}.$$

Proof. We present the proof in the case when $c_0 > 0$. The case $c_0 = 0$ is similar and simpler.

By Proposition 3.8, for every $v \in X$, $v \neq 0$,
$$J(u) \leq \sup_{\tau > 0} J(\tau v) := \varphi(v).$$

We have that
$$J(\tau v) \leq \frac{\tau^2}{2} \int_{-\infty}^{+\infty} \left[c^2(v')^2 - c_0^2 |Av|^2 \right] dt - a_0 \tau^q \int_{-\infty}^{+\infty} |Av|^q dt$$
$$= \tau^2 Q(v) - a_0 \tau^q \Phi(v). \tag{3.27}$$

Now we make a particular choice of v. Namely, let $\varepsilon \in (0, 1/2]$ and $v = v_\varepsilon$ be a function such that its Fourier transform $\widehat{v}$ is given by
$$\widehat{v}(\xi) = \begin{cases} 1 & \text{if } |\xi| \leq 2\varepsilon \\ 0 & \text{if } |\xi| > 2\varepsilon. \end{cases}$$

An elementary calculation shows that
$$v(t) = \frac{1}{\sqrt{2\pi}} \int_{-\infty}^{+\infty} e^{it\xi} \widehat{v}(\xi) \, d\xi = \frac{2}{\sqrt{2\pi}} \frac{\sin 2\varepsilon t}{t}.$$

(In order to work in the space X, we have to consider $v(t) - v(0)$ instead of $v(t)$, but this does not change Q and Φ).

Change of variable $x = 2\varepsilon t$ shows that
$$\Phi(v) = \frac{2^q}{\sqrt{(2\pi)^q}} (2\varepsilon)^{q-1} \int_{-\infty}^{+\infty} \left| \frac{\sin(x+1)}{x+1} - \frac{\sin x}{x} \right|^q dx = a_1 \varepsilon^{q-1}, \tag{3.28}$$

where $a_1 > 0$.

Next we estimate $Q(v)$. Integrating by parts, we obtain
$$Q(v) = \frac{1}{2} \left(\int_{-\infty}^{+\infty} c^2 v' \cdot v' dt - \int_{-\infty}^{+\infty} c_0^2 Av \cdot Av \, dt \right)$$
$$= \frac{1}{2} \int_{-\infty}^{+\infty} \left[-c^2 v'' + c_0^2 (A^* Av) \right] \cdot v \, dt$$
$$= \frac{1}{2} \int_{-\infty}^{+\infty} Lv \cdot v \, dt,$$

where
$$(A^* v)(t) = v(t) - v(t-1)$$

is the operator formally adjoint to A and
$$(Lv)(t) := -c^2 v''(t) + c_0^2 (A^* A v)(t)$$
$$= -c^2 v''(t) + c_0^2 (v(t+1) + v(t-1) - 2v(t)).$$

The operator L is a pseudodifferential operator with the symbol
$$\sigma_L(\xi) = c^2 \xi^2 - 4c_0^2 \sin^2 \frac{\xi}{2}.$$

This follows from easy formulas
$$\sigma_A(\xi) = e^{i\xi} - 1,$$
$$\sigma_{A^*}(\xi) = 1 - e^{-i\xi}$$

and
$$\sigma_{A^* A}(\xi) = (1 - e^{-i\xi})(e^{i\xi} - 1) = -4 \sin^2 \frac{\xi}{2}.$$

We have that
$$Q(v) = \frac{1}{2} \int_{-\infty}^{+\infty} Lv \cdot v \, dt = \frac{1}{2} \int_{-\infty}^{+\infty} \left(c^2 \xi^2 - 4c_0 \sin^2 \frac{\xi}{2} \right) \hat{v}^2(\xi) \, d\xi$$
$$= \frac{1}{2} \int_{-2\varepsilon}^{2\varepsilon} \left(c^2 \xi^2 - 4c_0^2 \sin^2 \frac{\xi}{2} \right) d\xi.$$

(The last formula can be also obtained by means of a straightforward calculation). For $|x| \leq \varepsilon$
$$\sin^2 x \geq \left(|x| - \frac{x^2}{2} \right)^2 = x^2 - |x|^3 + \frac{x^4}{4} \geq x^2(1 - |x|) \geq (1 - \varepsilon) x^2.$$

Therefore, if $|\xi| \leq 2\varepsilon$, then
$$c^2 \xi^2 - 4c_0^2 \sin^2 \frac{\xi}{2} \leq c^2 \xi^2 - c_0^2 (1 - \varepsilon) \xi^2 = [c^2 - c_0^2 + \varepsilon c_0^2] \xi^2.$$

Choosing
$$\varepsilon = \frac{c^2 - c_0^2}{c_0^2}, \qquad (3.29)$$

we obtain that
$$Q(v) \leq (c^2 - c_0^2) \int_{-2\varepsilon}^{2\varepsilon} \xi^2 d\xi = \frac{8}{3}(c^2 - c_0^2)\varepsilon^3 = a_2(c^2 - c_0^2)^4, \qquad (3.30)$$

where $a_2 > 0$ is independent of c.

Equations (3.27)–(3.30) yield
$$\varphi(v) \le a_2(c^2 - c_0^2)^4 \tau^2 - a_3(c^2 - c_0^2)^{q-1} \tau^q.$$

Elementary calculus shows that the right hand part here attains its maximum value at the point
$$\tau_0 = \left[\frac{2a_2}{qa_3}(c^2 - c_0^2)^{5-q}\right]^{\frac{1}{q-2}} = a_4(c^2 - c_0^2)^{\frac{5-q}{q-2}}.$$

Hence,
$$\varphi(v) \le a_2 a_4^2 (c^2 - c_0^2)^{\frac{2(5-q)}{q-2}+4} = a_5(c^2 - c_0^2)^{\frac{2(q+1)}{q-2}}. \tag{3.31}$$

Now Lemma 3.6 shows that
$$\|u_c\|^2 \le a_6(c^2 - c_0^2)^{\frac{2(q+1)}{q-2}-1} = a_6(c^2 - c_0^2)^{\frac{q+4}{q-2}}$$

and the second inequality of the proposition is proved.

Proposition 3.8 shows us that
$$J_k(u_k) \le \alpha_k \le 2\varphi(v),$$

provided k is large enough, and the first inequality of the proposition follows from inequality (3.31) and Lemma 3.6 as above. □

Remark 3.10 Arguing as in proof of Proposition 3.14, with J replaced by J_k, and taking as the test function $v(t) = t$, considered as an element of X_k, one can obtain a similar estimate for $\|u_k\|_k$ for all $k \ge 1$, but with C dependent on k.

3.5.2 Nonglobally defined potentials

Here we consider the case when the potential is of the form
$$U(r) = \frac{c_0^2}{2} r^2 + V(r),$$

where $c_0 \ge 0$ and $V(r)$, the unharmonic part, is defined in a neighborhood of $r = 0$. We shall assume a local version, (i'_{loc}), of (i'), i. e. that $V \in C^1$ in neighborhood of 0, $V(0) = V'(0) = 0$, and $V'(r) = o(|r|)$ as $r \to 0$. We use also local versions of (ii^+) and (ii^-) refereed to as (ii^+_{loc}) and (ii^-_{loc}), respectively, i. e. in those assumptions we restrict r_0 and r to a neighborhood of 0.

Theorem 3.7 *Assume (i'_{loc}) and (ii^+_{loc}) (resp., (ii^-_{loc})). Suppose, in addition, that inequality (3.26), with $a_0 > 0$ and $q > 2$, takes place in a right (resp., left) neighborhood of 0. Then there exists $\bar{c} > c_0$ such that, given $c \in (c_0, \bar{c})$, for every k large enough ($k \geq k_0(c)$) there exists a nondecreasing (resp. nonincreasing) travelling wave $u_k \in X_k$ and an increasing (resp., decreasing) solitary travelling wave $u \in X$, and, up to a subsequence and shifts, $u_k \to u$ in $C^2(\mathbb{R})$.*

Proof. Passing to the symmetrized potential, we may assume without loss of generality that the potential is even and satisfies both (ii^+_{loc}) and (ii^-_{loc}), and inequality (3.26) holds in a whole neighborhood of 0. Thus, we can treat increasing and decreasing waves simultaneously.

Now let us introduce a new even potential $\widetilde{V}$ defined on whole $\mathbb{R}$. Fix $\delta_1 > \delta_0 > 0$ such that $V(r)$ is defined on $[-\delta_1, \delta_1]$ and let $q_1 = \max[q, \theta]$. Let

$$F(r) = dr^{q_1}.$$

We fix $d > 0$ so that $F(r) \geq V(r)$ on $[\delta_0, \delta_1]$ (obviously, this is possible). Since $\theta \geq q_1$, we have that

$$\theta F(r) \leq q_1 F(r) = F'(r) r.$$

Choose a smooth function $\eta(r)$ such that

$$\eta(r) = \begin{cases} 1 & \text{for } 0 \leq r \leq \delta_0 \\ 0 & \text{for } r \geq \delta_1 \end{cases}$$

and $\eta'(r) \leq 0$. Define the potential $\widetilde{V}$ by

$$\widetilde{V}(r) = \eta(r) V(r) + (1 - \eta(r)) F(r)$$

(we mean that $\eta(r) V(r) = 0$ if $r \geq \delta_1$) and $\widetilde{V}(-r) = \widetilde{V}(r)$. We have that

$$\widetilde{V}'(r) r = \eta(r) V'(r) r + (1 - \eta(r)) F'(r) r + \eta'(r) (V(r) - F(r)) r,$$

for $r \geq 0$. Since $F(r) \geq V(r)$ on $[\delta_0, \delta_1]$, we obtain

$$\widetilde{V}'(r) r \geq \eta(r) V'(r) r + (1 - \eta(r)) F'(r) r$$
$$\geq \eta(r) \theta V(r) + (1 - \eta(r)) \theta F(r)$$
$$= \theta \widetilde{V}(r).$$

It is also easy to see that $\widetilde{V}$ satisfies (3.26), with another constant $a_0 > 0$.

Applying Theorems 3.1 and 3.6, we obtain travelling waves u_k and u with desired convergence property for the system with the potential $\widetilde{V}$. By Proposition 3.14, if c is close enough to c_0 (and k is large enough), then

$$\|u_k\|_k \leq \delta_1, \quad \|u\| \leq \delta_1.$$

Hence

$$\|Au_k\|_{L^\infty(-k,k)} \leq \delta_1, \quad \|Au\|_{L^\infty(\mathbb{R})} \leq \delta_1.$$

Therefore, $\widetilde{V}(Au_k) = V(Au_k)$ and $\widetilde{V}(Au) = V(Au)$, i.e. u_k and u are solutions of the original system. □

Now we show that assumptions (ii_{loc}^+) and (ii_{loc}^-) are generic.

Proposition 3.15 *Suppose that V is C^n, $n > 2$, in a neighborhood of 0 and $V(0) = V'(0) = \ldots = V^{(n-1)}(0) = 0$, $V^{(n)}(0) = a_0 \neq 0$.*

(a) *If n is odd and $a_0 > 0$ (resp., $a_0 < 0$), then V satisfies (ii_{loc}^+) (resp., (ii_{loc}^-)), with $\theta < n$ sufficiently close to n.*

(b) *If n is even and $a_0 > 0$, then V satisfies both (ii_{loc}^+) and (ii_{loc}^-), with $\theta < n$ sufficiently close to n.*

Proof. We consider case (b) only (case (a) is similar).

By Taylor's formula, we have near $r = 0$

$$V(r) = \frac{a_0}{n!} r^n + r^n \varphi(r),$$
$$V'(r) = \frac{a_0}{(n-1)!} r^{n-1} + r^{n-1} \psi(r),$$

where $\varphi(r) = o(1)$ and $\psi(r) = o(1)$. Then, in a neighborhood of 0, we have

$$\theta V(r) = r^n \left(\frac{\theta a_0}{n!} + \theta \varphi(r) \right) \leq r^n (\theta + \varepsilon) \frac{a_0}{n!}.$$

Similarly,

$$V'(r) r = r^n \left(\frac{a_0}{(n-1)!} + \psi(r) \right) \geq r^n \frac{a_0}{(n-1)!} (1 - \varepsilon)$$

near the origin. Taking $\varepsilon > 0$ such that

$$\theta + \varepsilon = n(1 - \varepsilon),$$

i. e.

$$\varepsilon = \frac{n - \theta}{n + 1},$$

we obtain the result. □

Example 3.5 In the case of generalized Lennard-Jones potential (see Example 1.5)

$$U(r) = a\big[(d+r)^{-k} - d^{-k}\big]^2,$$

where $a > 0$, $d > 0$ and k is a positive integer, a straightforward calculation shows that

$$c_0^2 = U''(0) = 2ad^{-2(k+1)}k^2$$

and $V(0) = V'(0) = V''(0) = 0$, $V'''(0) < 0$. Proposition 3.15 and Theorem 3.7 imply now the existence of decreasing solitary waves for every speed $c > c_0$ sufficiently close to c_0. Also there exist corresponding periodic waves of large periods.

Chapter 4

Travelling Waves: Further Results

4.1 Solitary waves and constrained minimization

4.1.1 Statement of problem

Here we present another approach to solitary waves borrowed from [Friesecke and Wattis (1994)]. Consider the following problem.

(FW) *Minimize the* average kinetic energy

$$\mathbf{T}(u) := \frac{1}{2} \int_{-\infty}^{+\infty} u'(t)^2 dt$$

subject to the constraint that the average potential energy *is fixed*

$$\mathbf{U}(u) := \int_{-\infty}^{+\infty} U\big(u(t+1) - u(t)\big) dt = K,$$

where $K > 0$ is a given number.

Problem (FW) is set up on the space X.

The travelling wave equation (3.3) is easily recognized as the Euler-Lagrange equation of problem (FW), with the inverse square, c^{-2}, of the wave speed corresponding to the Lagrange multiplier.

Now let us summarize the main result of this section.

Theorem 4.1 *Let $U \in C^2(\mathbb{R})$, $U \geq 0$ in a neighborhood of zero, $U(0) = 0$ and U superquadratic on at least one side, i. e. $r^{-2}U(r)$ increases strictly with $|r|$ for all $r \in \Lambda$, where either $\Lambda = (-\infty, 0)$ or $\Lambda = (0, \infty)$. Then there exists $K_0 \geq 0$ such that for every $K > K_0$ the system possesses a nontrivial solitary travelling wave $u = u_K \in X$, with $\mathbf{U}(u_K) = K$ and some speed $c = c_K > 0$. The waves u_K have the following properties:*

(p1) *They are monotone functions, increasing if $\Lambda = (0, \infty)$ and decreasing if $\Lambda = (-\infty, 0)$.*

(p2) *$u_K \in C^2(\mathbb{R})$, and $u_K(t+1) - u_K(t) \to 0$ and $u'_K(t) \to 0$ as $t \to \pm\infty$ (localization).*

(p3) *They are supersonic, i. e. $c_K^2 > c_0^2 = U''(0)$.*

(p4) *They are solutions of problem (FW), with $\mathbf{U}$ replaced by the symmetrized energy*

$$\widetilde{\mathbf{U}}(u) := \int_{-\infty}^{+\infty} \left[\widetilde{U}(u(t+1)) - \widetilde{U}(u(t))\right] dt,$$

where

$$\widetilde{U}(r) = \begin{cases} U(r), & r \in \Lambda \\ U(-r), & r \notin \Lambda. \end{cases}$$

If in addition U satisfies either

$$U''(0) = 0 \tag{4.1}$$

or the following nondegeneracy condition at zero:

$$U(r) = \frac{1}{2} U''(0) r^2 + \varepsilon |r|^p + o(|r|^p), \quad r \in \Lambda, r \to 0, \tag{4.2}$$

for some $\varepsilon > 0$ and $2 < p < 6$, then K_0 can be taken to be zero.

Property (p2) follows immediately from Proposition 3.5, since assumption (i''') is obviously satisfied. From the results of Section 4.5 it follows that travelling waves obtained in Theorem 4.1 are exponentially localized. The proof of the main part of the theorem is given in Subsections 4.1.2–4.1.4.

Remark 4.1 In addition to the assumptions of Theorem 4.1, suppose that $U \in C^3$ near 0 and

$$U'''(0) \begin{cases} > 0 & \text{if } \Lambda = (0, \infty) \\ < 0 & \text{if } \Lambda = (-\infty, 0). \end{cases}$$

Then it is obvious that (4.2) holds.

4.1.2 *The minimization problem: technical results*

First, note that **T** and **U** are well-defined C^2 functionals on X (*cf.* Proposition 3.3). Let

$$\mathcal{A}_K := \{u \in X \,:\, \mathbf{U}(u) = K\}.$$

We consider the minimization problem (*FW*) and denote by T_K its minimum value

$$T_K := \inf\{\mathbf{T}(u) \,:\, u \in \mathcal{A}_K\}.$$

Lemma 4.1 *Suppose that $U \in C^2(\mathbb{R})$, $U(0) = 0$, $U \geq 0$ on $\mathbb{R}$ and $U(r) > 0$ for $r \in (-\delta, \delta) \setminus \{0\}$ for some $\delta > 0$. For every $K \geq 0$ the set $\mathcal{A}_K$ is nonempty and, hence, the minimum value T_K is well-defined. Moreover, T_K is a monotone nondecreasing and continuous function of $K \in [0, \infty)$, and $T_K > 0$ for all $K > 0$.*

Proof. Pick $\lambda_0 \in (0, \delta)$. Then $U(\lambda_0) > 0$. Let us introduce the the function

$$v_{\lambda,l}(t) := \begin{cases} 0, & t \leq 0 \\ \lambda t, & 0 \leq t \leq l \\ \lambda l, & t \geq l, \end{cases}$$

with $\lambda \geq 0$ and $l \geq 1$. Obviously $v_{\lambda,l} \in X$ and

$$\mathbf{U}(v_{\lambda,l}) = (l-1)U(\lambda) + 2\int_0^1 U(\lambda t)\,dt.$$

In particular, we have that $\mathbf{U}(v_{\lambda_0,l}) \to \infty$ as $l \to \infty$. The functional **U** is continuous, while $v_{\lambda,l} \in X$ depends continuously on λ and l. Therefore, the continuous function $\mathbf{U}(v_{\lambda,1})$ of $\lambda \in [0, \lambda_0]$ attains each intermediate value in the segment $[0, \mathbf{U}(v_{\lambda_0,1})]$, while the continuous function $\mathbf{U}(v_{\lambda_0,l})$ of $l \geq 1$ attains each intermediate value in $[\mathbf{U}(v_{\lambda_0,1}), \infty)$. This proves the first statement of the lemma.

Now let us show that T_K is nondecreasing with K. Let $\alpha \leq K$ and $v \in \mathcal{A}_K$. Since

$$\mathbf{U}(\lambda v) = \begin{cases} 0 & \text{if } \lambda = 0 \\ K & \text{if } \lambda = 1, \end{cases}$$

there exists $\lambda_0 \in [0, 1]$ such that $\mathbf{U}(\lambda_0 v) = \alpha$. Hence

$$T_\alpha \leq \mathbf{T}(\lambda_0 v) = \lambda_0^2 \mathbf{T}(v) \leq \mathbf{T}(v).$$

Since v is an arbitrary element of $\mathcal{A}_K$, we obtain that $T_\alpha \leq T_K$, i. e. T_K increases with K.

Next, we prove that T_K depends continuously on K. In view of monotonicity of T_K, it suffice to show that there exist $\eta(\delta) \to 0$ as $\delta \to 0$ such that

$$T_{K+\delta} - T_K \leq \eta(\delta)$$

for all $K \geq 0$ and $\delta > 0$. Fix K and δ. Given $\varepsilon > 0$, pick $v_K \in \mathcal{A}_K$ such that

$$\mathbf{T}_K(v_K) \leq T_K + \varepsilon,$$

and consider the function

$$v(t) := \begin{cases} v_K(t), & t \leq a \\ v_K(t) + v_{\lambda,1}(t - a - 1), & t \geq a, \end{cases}$$

with $v_{\lambda,1}$ defined before, and λ and $a > 0$ to be specified later. Then $v \in X$ and

$$\mathbf{U}(v) = \int_{-\infty}^{a-1} U(Av_K)\, dt + \int_{a-1}^{a} U(Av)\, dt + \mathbf{U}(v_{\lambda,1}).$$

Here we use the notation

$$(Av)(t) = v(t+1) - v(t)$$

introduced in Section 3.2. Clearly the first term tends to K and the second term to zero as $a \to \infty$. Thus

$$\mathbf{U}(v) \to K + \mathbf{U}(v_{\lambda,1}) = K + 2\int_0^1 U(\lambda t)\, dt$$

as $a \to \infty$. Assuming that

$$0 < \delta < \frac{1}{2}\mathbf{U}(v_{\lambda_0,1}),$$

we define

$$\lambda_1(\delta) := \inf\{\lambda > 0 : \mathbf{U}(v_{\lambda,1}) = 2\delta\}.$$

By continuity of the functional $\mathbf{U}$, there exists $a \in \mathbb{R}$ and $\lambda \leq \lambda_1(\delta)$ such that

$$\mathbf{U}(v) = K + \delta.$$

A straightforward calculation shows that

$$\mathbf{T}(v) \leq \mathbf{T}(v_K) + \mathbf{T}(v_{\lambda,1}).$$

Therefore,

$$T_{K+\delta} - T_K \leq \mathbf{T}(v) - T_K \leq \mathbf{T}(v) - \mathbf{T}(v_K) + \varepsilon \leq \mathbf{T}(v_{\lambda,1}) + \varepsilon.$$

But

$$\mathbf{T}(v_{\lambda,1}) = \frac{1}{2}\lambda^2 \leq \frac{1}{2}\lambda_1(\delta)^2$$

and, hence,

$$T_{K+\delta} - T_K \leq \frac{1}{2}\lambda_1(\delta)^2 + \varepsilon.$$

We see also that $\lambda_1(\delta)$ is independent of K and ε, and tends to zero as $\delta \to 0$, because $U > 0$ near zero. This proves the continuity of T_K.

It remains to show that $T_K > 0$ if $K > 0$. Let $v \in X$ and $\|v\| \leq C$. By Lemma 3.5, $|Av(t)| \leq C$. Choose $C_1 > 0$ such that $|U''(r)| \leq C_1$ on $[-C,C]$. Since $U'(0) = 0$, we have that

$$\int_{-\infty}^{+\infty} |U(Av)|\, dt \leq C_1 \int_{-\infty}^{+\infty} |Av|^2 dt \leq C_1 \|v\|^2. \tag{4.3}$$

Choose $v \in \mathcal{A}_K$ such that $\mathbf{T}(v) \leq T_K + \varepsilon$, with $0 < \varepsilon < 1$. Since

$$\|v\|^2 = 2\mathbf{T}(v),$$

taking $C = 2(T_K + 1)$, we obtain from (4.3) that

$$K \leq 2C_1 \mathbf{T}(v).$$

This implies that $T_K > 0$ and we conclude. $\square$

To address the question of existence of minimizers we use the concentration compactness principle of P.-L. Lions [Lions (1984)] in the following form (*cf.* Lemma B.1).

Lemma 4.2 *Under the assumptions of Lemma 4.1, let u^n be a sequence in $H^1_{loc}(\mathbb{R})$ such that $(u^n)'$ is bounded in $L^2(\mathbb{R})$ and*

$$\int_{-\infty}^{+\infty} U^n dt = K,$$

where $K > 0$ is fixed and $U^n(t) = U(Au^n(t))$. Then there exists a subsequence, still denoted U^n, satisfying one of the following three possibilities:

(a) *(compactness)* There exists $y_n \in \mathbb{R}$ such that $U^n(\cdot + y_n)$ is tight, i. e. for every $\varepsilon > 0$ there exists $R > 0$ such that

$$\left(\int_{-\infty}^{y_n - R} + \int_{y_n + R}^{+\infty} \right) U^n dt \leq \varepsilon.$$

(b) *(vanishing)* For every $R > 0$

$$\lim_{n \to \infty} \sup_{y \in \mathbb{R}} \int_{y-R}^{y+R} U^n dt = 0.$$

(c) *(dichotomy)* There exists $\alpha \in (0, K)$ and $u_1^n, u_2^n \in H^1_{loc}(\mathbb{R})$, with the derivatives in $L^2(\mathbb{R})$, such that

$$\mathrm{dist}\left(\mathrm{supp}\,(u_1^n)', \mathrm{supp}\,(u_2^n)' \right) \to \infty,$$

$$\int_{-\infty}^{+\infty} ((u^n)')^2 dt - \int_{-\infty}^{+\infty} ((u_1^n)')^2 dt - \int_{-\infty}^{+\infty} ((u_2^n)')^2 dt \geq 0,$$

and for every $\varepsilon > 0$ there exists $n_\varepsilon \geq 1$ with the properties that for $n \geq n_\varepsilon$

$$\| U^n - (U_1^n + U_2^n) \|_{L^1(\mathbb{R})} \leq \varepsilon,$$

$$\left| \int_{-\infty}^{+\infty} U_1^n dt - \alpha \right| \leq \varepsilon$$

and

$$\left| \int_{-\infty}^{+\infty} U_2^n dt - (K - \alpha) \right| \leq \varepsilon,$$

where $U_j^n(t) = U(Au_j^n(t))$, $j = 1, 2$.

Proof. The concentration function

$$Q_n(R) := \sup_{y \in \mathbb{R}} \int_{y-R}^{y+R} U^n dt$$

is a monotone nondecreasing function of $R \geq 0$, and $Q_n(R) \leq K$ for all $R \geq 0$. Passing to a subsequence, we may assume that $Q_n(R) \to Q(R)$ pointwise, and the function $Q(R)$ is monotone nondecreasing and $Q(R) \leq K$. Let

$$\alpha := \lim_{R \to \infty} Q(R) \in [0, K].$$

We note that $\alpha = 0$ implies (a) and $\alpha = K$ implies (b). This follows also from Lemma B.1. Case (c) that corresponds to $\alpha \in (0, K)$ does not follow immediately from Lemma B.1 and require more work.

Suppose that $\alpha \in (0, K)$. Fix $\varepsilon > 0$ and choose $R > 0$ such that $Q(R-1) > \alpha - \varepsilon$. Then, for n large, $Q_n(R-1) > \alpha - \varepsilon$ and, hence, there exists $y_n \in R$ such that

$$\int_{y_n - (R-1)}^{y_n + (R-1)} U^n dt > \alpha - \varepsilon. \tag{4.4}$$

Furthermore, since $\lim_{R \to \infty} Q(R) = \alpha$, we can find $R^n \to \infty$ such that

$$Q_n(R^n + 1) < \alpha + \varepsilon. \tag{4.5}$$

Now for

$$R_1^n \in [R, \tfrac{1}{3}(R^n - R)], \quad R_2^n \in [R + \tfrac{2}{3}(R^n - R), R^n]$$

(R_1^n and R_2^n to be specified later), define

$$u_1^n(t) := \begin{cases} u^n(y_n - R_1^n), & t \leq y_n - R_1^n, \\ u^n(t), & y_n - R_1^n \leq t \leq y_n + R_1^n, \\ u^n(y_n + R_1^n), & t \geq y_n + R_1^n, \end{cases}$$

$$u_2^n(t) := \begin{cases} u^n(t) - u^n(y_n - R_2^n), & t \leq y_n - R_2^n, \\ 0, & y_n - R_2^n \leq t \leq y_n + R_2^n, \\ u(t) - u^n(y_n + R_2^n), & t \geq y_n + R_2^n. \end{cases}$$

Then it is readily verified that

$$\int_{-\infty}^{+\infty} ((u^n)')^2 dt - \int_{-\infty}^{+\infty} ((u_1^n)')^2 dt - \int_{-\infty}^{+\infty} ((u_2^n)')^2 dt$$
$$= \left(\int_{y_n - R_2^n}^{y_n - R_1^n} + \int_{y_n + R_2^n}^{y_n + R_1^n} \right) ((u^n))^2 dt \geq 0$$

and $\text{dist}\left(\text{supp}\,(u_1^n)', \text{supp}\,(u_2^n)'\right) \to \infty$.

It remains to verify the other statements of (c). This can be achieved by an appropriate choice of R_1^n and R_2^n. First, note that for every $u \in H^1_{loc}(\mathbb{R})$, with $\|u'\|_{L^2(\mathbb{R})} \leq C$, one has

$$U(Au(t)) \leq C_1 (Au(t))^2 \leq C_1 \int_t^{t+1} (u')^2 dt, \tag{4.6}$$

with an appropriate constant $C_1 > 0$ depending on C.

For notational convenience let us drop the superscript n. Using (4.6), we have that

$$\int_{y_n-R_1-1}^{y_n-R_1} U_1 \, dt \leq C_1 \int_{y_n-R_1-1}^{y_n-R_1} \left(\int_t^{t+1} (u'_1(s))^2 ds \right) dt$$

$$= C_1 \int_{y_n-R_1-1}^{y_n-R_1} \left(\int_{\max[t,y_n-R_1]}^{t+1} (u'_1(s))^2 ds \right) dt$$

$$\leq C_1 \int_{y_n-R_1}^{y_n-R_1+1} (u'(t))^2 dt.$$

Similarly,

$$\int_{y_n+R_1-1}^{y_n+R_1} U_1 \, dt \leq C_1 \int_{y_n+R_1-1}^{y_n+R_1} (u'(t))^2 dt,$$

$$\int_{y_n-R_2-1}^{y_n-R_2} U_2 \, dt \leq C_1 \int_{y_n-R_2-1}^{y_n-R_2} (u'(t))^2 dt$$

and

$$\int_{y_n+R_2-1}^{y_n+R_2} U_2 \, dt \leq C_1 \int_{y_n+R_2}^{y_n+R_2+1} (u'(t))^2 dt.$$

Since

$$\int_{-\infty}^{\infty} (u'(t))^2 dt \leq C^2,$$

we obtain that

$$\min_{R_1 \in [R, R+\frac{1}{3}(R^n-R)]} \left(\int_{y_n-R_1-1}^{y_n-R_1} + \int_{y_n+R_1-1}^{y_n+R_1} \right) U_1 \, dt \leq \delta(n)$$

and
$$\min_{R_2 \in \left[R + \frac{2}{3}(R^n - R), R\right]} \left(\int_{y_n - R_2 - 1}^{y_n - R_2} + \int_{y_n + R_2 - 1}^{y_n + R_2} \right) U_2 \, dt \leq \delta(n),$$

where
$$\delta(n) = C_1 C^2 \frac{1}{\frac{1}{3}(R^n - R) - 1} \to 0$$

as $n \to \infty$.

Choosing R_1 and R_2 so that the above minima are attained at R_1 and R_2 respectively, we have that

$$\int_{-\infty}^{+\infty} |U - U_1 - U_2| \, dt = \left(\int_{y_n - R_1 - 1}^{y_n - R_1} + \int_{y_n + R_1 - 1}^{y_n + R_1} \right) |U - U_1| \, dt$$
$$+ \left(\int_{y_n - R_2 - 1}^{y_n - R_2} + \int_{y_n + R_2 - 1}^{y_n + R_2} \right) |U - U_2| \, dt$$
$$+ \left(\int_{y_n - R_2}^{y_n - R_1 - 1} + \int_{y_n + R_1}^{y_n + R_1 - 1} \right) U \, dt$$
$$\leq 2\delta(n) + \int_{y_n - R^n - 1}^{y_n + R^n + 1} U \, dt - \int_{y_n - R + 1}^{y_n + R - 1} U \, dt$$
$$\leq 2\delta(n) + (\alpha + \varepsilon) - (\alpha - \varepsilon)$$
$$= 2\delta(n) + 2\varepsilon.$$

Moreover,

$$\int_{-\infty}^{+\infty} U_1 \, dt = \left(\int_{y_n - R_1 - 1}^{y_n - R_1} + \int_{y_n + R_1 - 1}^{y_n + R_1} \right) U_1 \, dt + \int_{y_n - R_1}^{y_n + R_1 - 1} U \, dt \quad (4.7)$$

and

$$\int_{-\infty}^{+\infty} U_2 \, dt = \left(\int_{y_n - R_2 - 1}^{y_n - R_2} + \int_{y_n + R_2 - 1}^{y_n + R_2} \right) U_2 \, dt$$
$$+ \left(\int_{-\infty}^{y_n - R_2 - 1} + \int_{y_n + R_2}^{\infty} \right) U \, dt. \quad (4.8)$$

The first terms in both (4.7) and (4.8) do not exceed $\delta(n)$. Due to (4.4) and (4.5), the second terms lie, respectively, in the interval

$$\left[\int_{y_n-R+1}^{y_n+R-1} U\,dt, \int_{y_n-R^n-1}^{y_n+R^n+1} U\,dt\right] \subset (\alpha - \varepsilon, \alpha + \varepsilon)$$

and in

$$\left[K - \int_{y_n-R^n-1}^{y_n+R^n+1} U\,dt, K - \int_{y_n-R+1}^{y_n+R-1} U\,dt\right] \subset ((K-\alpha)-\varepsilon, (K-\alpha)+\varepsilon).$$

Since $\delta(n) \to 0$ as $n \to 0$, (c) follows, and the proof is complete. $\square$

Lemma 4.3 *Under the assumptions of Lemma 4.1 let $K > 0$ be fixed. Then the following two statements are equivalent:*

1) *Dichotomy does not occur, i. e. no minimizing sequence $u_n \in \mathcal{A}_K$ of $\mathbf{T}$ satisfies (c) of Lemma 4.2.*
2) *Subadditivity inequality*

$$T_K < T_\alpha + T_{K-\alpha} \quad \text{for all } a \in (0, K) \tag{4.9}$$

holds.

Proof. First suppose that (4.9) holds. Let $u^n \in \mathcal{A}_K$ be a sequence satisfying (c), with some $\alpha \in (0, K)$. Then defining $\alpha_n := \mathbf{U}(u_1^n)$ and $\beta_n := \mathbf{U}(u_2^n)$, letting $n \to \infty$ and using the continuity of T_K (Lemma 4.1) we obtain

$$T_K \geq \liminf_{n \to \infty}\left[\mathbf{T}(u_1^n) + \mathbf{T}(u_2^n)\right] \geq \liminf_{n \to \infty}(T_{\alpha_n} + T_{\beta_n})$$
$$= T_\alpha + T_{K-\alpha}$$

that contradicts (4.9).

Now assume that (4.9) does not hold, i. e. there exists $\alpha \in (0, K)$ such that $T_K \geq T_\alpha + T_{K-\alpha}$. It suffices to construct a minimizing sequence $u^n \in \mathcal{A}_K$ satisfying (c). Pick minimizing sequences $u_\alpha^n \in \mathcal{A}_\alpha$ and $u_\beta^n \in \mathcal{A}_{K-\alpha}$. Arguing similarly to the proof of the continuity of T_K in Lemma 4.1, we may assume without loss that $\operatorname{supp}(u_\alpha^n)'$ and $\operatorname{supp}(u_{K-\alpha}^n)'$ are contained in some interval $(-R_n, R_n)$. Then

$$u^n(t) := u_\alpha^n(t + R_n + n) - u_{K-\alpha}^n(t - R_n - n) + C_n,$$

where C_n is chosen so that $u^n(0) = 0$, has the required property. $\square$

Lemma 4.4 *Suppose that U is as in Lemma 4.1. Let $K > 0$ be fixed. Then the following three statements are equivalent:*

1) *Vanishing does not occur, i. e. no minimizing sequence $u^n \in \mathcal{A}_K$ satisfies (b) of Lemma 4.2.*
2) *There exists $\varepsilon = \varepsilon(K) > 0$ such that for every minimizing sequence $u^n \in \mathcal{A}_K$*

$$\liminf_{n \to \infty} \|Au^n\|_{L^\infty(\mathbb{R})} > \varepsilon.$$

3) *The following energy inequality*

$$U''(0) T_K < K \qquad (4.10)$$

holds.

Proof. For $\varepsilon > 0$, let

$$T_{K,\varepsilon} := \inf \{ \mathbf{T}(u) \ : \ u \in \mathcal{A}_k, \|Au\|_{L^\infty(\mathbb{R})} \leq \varepsilon \}.$$

Since the function $v_{\varepsilon,L}$ defined in the proof of Lemma 4.1 belongs to the above set, provided $\varepsilon > 0$ is small enough and $L = L(\varepsilon)$ is appropriately chosen, the $T_{K,\varepsilon}$ is well-defined at least for all small $\varepsilon > 0$. We shall show that

$$\lim_{\varepsilon \to 0} T_{K,\varepsilon} = \frac{K}{U''(0)}, \qquad (4.11)$$

where the limit is to be understood as $+\infty$ in case $U''(0) = 0$.

First we estimate this limit from bellow. Let $u^n \in \mathcal{A}_K$ be an arbitrary sequence such that $\|Au^n\|_{L^\infty(\mathbb{R})} \to 0$. Then

$$K = \int_{\mathbb{R}} U(Au^n) \, dt$$
$$\leq \frac{1}{2} \sup_{r \in [-a_n, a_n]} |U''(r)| \int_{-\infty}^{+\infty} \int_t^{t+1} ((u^n)')^2 \, ds \, dt$$
$$= \sup_{r \in [-a_n, a_n]} |U''(r)| \, \mathbf{T}(u^n),$$

where $a_n = \|Au^n\|_{L^\infty(\mathbb{R})} \to 0$. Therefore,

$$K \leq U''(0) \liminf_{n \to \infty} \mathbf{T}(u^n)$$

and, in particular,
$$K \leq U''(0) \liminf_{\varepsilon \to 0} \mathbf{T}_{K,\varepsilon}.$$

To obtain an upper bound we use the function $v_{\lambda,l}$ defined in the proof of Lemma 4.1. Recall that
$$\mathbf{T}(v_{\lambda,l}) = \frac{1}{2}\lambda^2 l$$

and
$$\mathbf{U}(v_{\lambda,l}) = (l-1)U(\lambda) + 2\int_0^l U(\lambda t)\,dt.$$

Pick λ_0 so small that $\mathbf{U}(v_{\lambda,1}) \leq K$ for all $\lambda \in (0, \lambda_0)$. Then for every $\lambda \in (0, \lambda_0)$ there exists $l = l(\lambda) \geq 1$ such that $\mathbf{U}(v_{\lambda,l}) = K$. Furthermore, it is easy to verify that $l(\lambda) \to \infty$ as $\lambda \to 0$. Now

$$\frac{(l-1)U(\lambda)}{\frac{1}{2}\lambda^2 l} \leq \frac{K}{\mathbf{T}(v_{\lambda,l})} \leq \frac{(l-1)U(\lambda) + 2\int_0^1 U(\lambda t)\,dt}{\frac{1}{2}\lambda^2 l}. \qquad (4.12)$$

Since
$$U(\lambda) = \frac{1}{2}U''(0)\lambda^2 + o(\lambda^2)$$

and $l = l(\lambda) \to \infty$ as $\lambda \to 0$, both the left-hand part and the right-hand part of (4.12) tend to $U''(0)$ as $\lambda \to 0$. Hence,
$$K = U''(0)\lim_{\lambda \to 0}\mathbf{T}(v_{\lambda,l}) \geq U''(0)\limsup_{\varepsilon \to 0} T_{K,\varepsilon}$$

and we obtain (4.11). This proves the equivalence 2) $\Leftrightarrow$ 3).

It remains to prove that 1) $\Leftrightarrow$ 2). The implication "$\Rightarrow$" is trivial. To verify "$\Leftarrow$" it suffices to show that, given $\varepsilon > 0$ and $C > 0$, there exist $\varepsilon_1 = \varepsilon_1(\varepsilon, C) > 0$ and $R = R(\varepsilon, C) > 0$ such that all $u \in X$, with $\|u\| \leq C$ and $\|Au\|_{L^\infty(\mathbb{R})} \geq \varepsilon$, satisfy
$$\sup_{y \in \mathbb{R}} \int_{y-R}^{y+R} U(Au)\,dt > \varepsilon_1.$$

Pick $y \in \mathbb{R}$ such that $U(Au(y)) \geq \varepsilon$. Obviously, such a y exists. By Lemma 3.5,
$$\|Au\|_{L^\infty(\mathbb{R})} \leq C.$$

Letting
$$C_0 := \sup_{r \in [-C,C]} |U'(r)|,$$

we obtain for $t \geq y$

$$\left|U(Au(t)) - U(Au(y))\right| \leq C_0 |(Au)(t) - (Au)(y)|$$
$$\leq C_0 \int_y^t \left(|u'(s+1)| + |u'(s)|\right) ds$$
$$\leq 2CC_0\sqrt{t-y}.$$

Hence,

$$U(Au(t)) \geq \varepsilon - 2CC_0\sqrt{t-y}. \tag{4.13}$$

This implies the required, with

$$R = \left(\frac{\varepsilon}{2CC_0}\right)^2$$

and

$$\varepsilon_1 = \int_0^R (\varepsilon - 2CC_0\sqrt{s})\, ds.$$

The proof is complete. □

4.1.3 The minimization problem: existence

The technicalities of the previous subsection give the possibility to obtain the existence of solution to problem (FW). We start with the following

Proposition 4.1 *Let U be as in Lemma 4.1 and $K > 0$. Assume that the subadditivity inequality (4.9) and the energy inequality (4.10) hold. Then $\mathbf{T}$ attains its minimum on $\mathcal{A}_K$.*

Proof. This is a consequence of standard arguments in the calculus of variations and of Lemmas 4.2–4.4. Take any minimizing sequence $u^n \in \mathcal{A}_K$ for the functional $\mathbf{T}$. Lemmas 4.2–4.4 show that, passing to a subsequence, one can assume that u^n satisfies property (a) of Lemma 4.2. Since $\mathbf{T}$ and $\mathbf{U}$ are invariant under translations and under adding constants, we can replace $u^n(t)$ by $u^n(t + y_n) - u^n(y_n)$. Hence, we can assume that $y_n = 0$.

Obviously, u^n is bounded in X and, passing to a subsequence, we can assume that u^n converges weakly to $u \in X$. Since the norm in a Hilbert space is weakly lower semicontinuous, we have that

$$\mathbf{T}(u) \leq \inf \{\mathbf{T}(v) \,:\, v \in \mathcal{A}_k\} = T_K.$$

It remains to prove that $\mathbf{U}(u) = K$. By property (a), with $y_n = 0$, it is enough to show that for every $R > 0$

$$\lim_{n \to \infty} \int_{-R}^{R} U(Au^n)\, dt = \int_{-R}^{R} U(Au)\, dt,$$

but this follows immediately from the compactness of the Sobolev embedding $H^1(-R, R) \subset L^\infty(-R, R)$ (see Theorem A.1) which implies that $u^n \to u$ strongly in $L^\infty(-R, R)$ for every $R > 0$. □

The next step is to study for which potentials the subadditivity inequality and the energy inequality hold. For, we shall need the following elementary lemma that goes back to [Lions (1984)].

Lemma 4.5 *Let $h : [0, K] \to \mathbb{R}$. Suppose that*

$$h(\theta \alpha) \leq \theta h(\alpha), \quad 0 < \alpha < \frac{K}{2}, 1 < \theta \leq \frac{K}{\alpha},$$

and

$$h(\theta \alpha) < \theta h(\alpha), \quad \frac{K}{2} \leq \alpha \leq K, 1 < \theta \leq \frac{K}{\alpha}.$$

Then

$$h(K) < h(\alpha) + h(K - \alpha)$$

for all $\alpha \in (0, K)$.

Proof. Assume, without loss of generality, that $\alpha \geq K/2$. Then

$$h(K) < \frac{K}{\alpha} h(\alpha)$$

and

$$h(K) \leq \frac{K}{K - \alpha} h(K - \alpha).$$

Hence,

$$h(\alpha) + h(K - \alpha) > \left(\frac{K}{\alpha} + \frac{K - \alpha}{K}\right) h(K) = h(K)$$

and we conclude. □

Proposition 4.2 *Let U be as in Lemma 4.1. Suppose that U is superquadratic in the sense that $r^{-2}U(r)$ increases strictly with $|r|$. Then there exists $K_0 \geq 0$ such that for $K > K_0$ subadditivity inequality (4.9) and energy inequality (4.10) hold. In particular, by Proposition 4.1, $\mathbf{T}$ attains its minimum value on $\mathcal{A}_K$ for all $K > K_0$.*

Proof. Since $r^{-2}U(r)$ increases strictly with $|r|$ and tends to $U''(0)/2$ as $r \to 0$, we have that

$$U(r) > \frac{1}{2} U''(0) r^2 \qquad (4.14)$$

for all $r \neq 0$. Fix $\lambda > 0$ and consider the function $v_{\lambda,l}$ defined in the proof of Lemma 4.1. We have that

$$\mathbf{U}(v_{\lambda,l}) = (l-1) U(\lambda) + 2 \int_0^1 U(\lambda t)\, dt > (l-1) U(\lambda)$$

and

$$U''(0)\, \mathbf{T}(v_{\lambda,l}) = \frac{1}{2} U''(0)\, \lambda^2 l.$$

Hence, there exists $l_0 \geq 1$ such that for $l \geq l_0$ we have, using (4.14), that

$$(l-1) U(\lambda) > \frac{1}{2} U''(0) \lambda^2 l.$$

Therefore, for all $l \geq l_0$

$$\mathbf{U}(v_{\lambda,l}) > U''(0)\, \mathbf{T}(v_{\lambda,l}).$$

Let $K_0 = \mathbf{U}(v_{\lambda,l_0})$. Then the last inequality implies immediately that

$$K > U''(0)\, T_K$$

for all $K \geq K_0$. The energy inequality is proved.

Now we claim that, with this definition of K_0, the subadditivity holds for $K \geq 2K_0$. We check that for all $K \geq 2K_0$ the function $h(\alpha) = T_\alpha$ satisfies the assumption of Lemma 4.5. Fix $\alpha \in (0, K)$ and $\theta \in (1, K/\alpha]$. We consider the case $\alpha \geq K/2$ only, the other case being easier.

By Lemma 4.4 and the boundedness of minimizing sequences, there exist $\varepsilon > 0$ and $C > 0$ such that

$$T_\alpha = \inf \{\mathbf{T}(u) \,:\, u \in \mathcal{A}_{\alpha,\varepsilon,C}\},$$

where
$$\mathcal{A}_{\alpha,\varepsilon,C} = \{u \in X \,:\, \mathbf{U}(u) = \alpha, \|Au\|_{L^\infty(\mathbb{R})} \geq \varepsilon, \|u\| \leq C\}.$$

By (4.14), there exists $\alpha_0 > 0$ such that
$$\int_{\{t\,:\,|Au(t)|\geq \varepsilon/2\}} U\bigl(Au(t)\bigr)\,dt \leq \alpha_0$$
for all $u \in \mathcal{A}_{\alpha,\varepsilon,C}$. Let
$$\theta_0 := \min\left\{\frac{U(\lambda r)}{\lambda^2 U(r)} \,:\, |r| \in \left[\frac{\varepsilon}{2}, C\right], \lambda \in \left[1, \sqrt{\theta}\right]\right\}.$$

By the superquadraticity of U, we have that $\theta_0 > 1$.

Now take $v \in \mathcal{A}_{\alpha,\varepsilon,C}$. Since $\mathbf{U}(\lambda v) = \alpha$ at $\lambda = 1$ and
$$\mathbf{U}(\lambda v) = \mathbf{U}(\sqrt{\theta} v) \geq \theta\,\mathbf{U}(v) = \theta \alpha$$
at $\lambda = \sqrt{\theta}$, there exists $\lambda = \lambda(\theta, v) \in \left[1, \sqrt{\theta}\right]$ such that
$$\mathbf{U}(\lambda v) = \theta \alpha.$$

In fact, $\lambda < \sqrt{\theta}$. Indeed, we have that
$$\theta \alpha = \mathbf{U}(\lambda v) = \int_{\{|Av(t)|<\varepsilon/2\}} U\bigl(\lambda Av(t)\bigr)\,dt + \int_{\{|Av(t)|\geq \varepsilon/2\}} U\bigl(\lambda Av(t)\bigr)\,dt$$
$$\geq \lambda^2 \int_{\{|Av(t)|<\varepsilon/2\}} U\bigl(Av(t)\bigr)\,dt + \theta_0 \lambda^2 \int_{\{|Av(t)|\geq \varepsilon/2\}} U\bigl(Av(t)\bigr)\,dt.$$

Denoting by I_ε the second integral in the right-hand side above, we see that the first integral is equal to $\alpha - I_\varepsilon$. Since $I_\varepsilon \geq \alpha_0$, we obtain that
$$\theta \alpha \geq \lambda^2(\alpha - I_\varepsilon) + \lambda^2 \theta_0 I_\varepsilon = \lambda^2\bigl[\alpha + (\theta_0 - 1) I_\varepsilon\bigr]$$
$$\geq \lambda^2\bigl[\alpha + (\theta_0 - 1)\alpha_0\bigr].$$

Hence,
$$\lambda^2 \leq \theta \frac{\alpha}{\alpha + (\theta_0 - 1)\alpha_0} =: \lambda_0^2 < \theta.$$

Consequently,
$$T_{\theta\alpha} \leq \inf\left\{\mathbf{T}(\lambda(\theta,v)\,v) \,:\, v \in \mathcal{A}_{\alpha,\varepsilon,C}\right\} \leq \lambda_0^2 \inf\left\{\mathbf{T}(v) \,:\, v \in \mathcal{A}_{\alpha,\varepsilon,C}\right\}$$
$$= \lambda_0^2 T_\alpha < \theta T_\alpha.$$

This completes the proof of the subadditivity inequality. □

Remark 4.2 If $U''(0) = 0$, then in the proof of Proposition 4.2 one can take arbitrary $K_0 > 0$. Therefore, in this case the energy inequality holds for all $K > 0$. Now the proof of Proposition 4.2 shows that the subadditivity inequalities also holds for all $K > 0$.

Remark 4.3 In [Friesecke and Wattis (1994)] it is shown that if U is subquadratic, i. e. $r^{-2}U(r)$ is nonincreasing with $|r|$, then, for every $K > 0$, neither energy inequality nor the subadditivity inequalities hold, and the minimum of $\mathbf{T}$ on $\mathcal{A}_K$ is not attained.

Now we discuss the nondegeneracy assumption (4.2).

Proposition 4.3 *Let U be as in Lemma 4.1 and superquadratic. In addition, let U satisfy nondegeneracy condition (4.2), with $\varepsilon > 0$ and $p \in (2,6)$. Then subadditivity and energy inequalities (4.9) and (4.10) hold for all $K > 0$. Therefore, by Proposition 4.1, the functional $\mathbf{T}$ attains its minimum on $\mathcal{A}_K$ for all $K > 0$.*

Proof. As in the proof of Proposition 4.2, it suffices to prove that the energy inequality holds for all $K > 0$. Let

$$w_{\lambda,\beta} = \lambda \tanh\left[\beta\left(z - \frac{1}{2}\right)\right] + C,$$

where C is a constant such that $w_{\lambda,\beta}(0) = 0$ so that $w_{\lambda,\beta} \in X$.

First we calculate $\mathbf{T}(w_{\lambda,\beta})$. A straightforward calculation shows that

$$\varphi(t) := Aw_{\lambda,\beta}(t) = \frac{2\lambda \tanh\left(\frac{\beta}{2}\right) \operatorname{sech}^2(\beta t)}{1 - \tanh^2(\beta t) \tanh^2\left(\frac{\beta}{2}\right)}.$$

We expand φ using

$$2 \tanh\left(\frac{\beta}{2}\right) = \beta - \frac{1}{12}\beta^3 + O(\beta^5)$$

and

$$\frac{1}{1-x} = 1 + x + O(x^2).$$

We have

$$\varphi = \lambda \left[\beta - \frac{1}{12}\beta^3 + O(\beta^5)\right]\left[1 + \frac{1}{4}\beta^2 \tanh^2(\beta t) + O(\beta^4)\right] \operatorname{sech}^2(\beta t)$$

$$= \lambda \left[\beta + \frac{1}{4}\beta^3 \tanh^2(\beta t) - \frac{1}{12}\beta^3 + O(\beta^5)\right] \operatorname{sech}^2(\beta t)$$

as $\beta \to 0$. Since
$$w'_{\lambda,\beta} = \lambda\beta \operatorname{sech}^2(\beta t),$$
we obtain that
$$\mathbf{T}(w_{\lambda,\beta}) = \frac{1}{2}\lambda^2\beta \int_{-\infty}^{+\infty} \operatorname{sech}^4(x)\,dx.$$

Now we show that, given $K > 0$, there exists a $\lambda = \lambda(\beta)$ such that
$$\mathbf{U}(w_{\lambda,\beta}) = K.$$

To calculate $\mathbf{U}(w_{\lambda,\beta})$, we need first to estimate the term $\beta\lambda(\beta)$. By (4.14), $\lambda(\beta)$ does not exceed the solution $\lambda_0(\beta)$ of the equation
$$\int_{-\infty}^{+\infty} \frac{1}{2} U''(0) \bigl(Aw_{\lambda_0,\beta}(t)\bigr)^2 dt = K.$$

The left-hand part simplifies to
$$\frac{1}{2} U''(0) \lambda_0^2 \bigl[\beta^2 + O(\beta^4)\bigr] \int_{-\infty}^{+\infty} \operatorname{sech}^4(\beta t)\,dt$$
$$= \frac{1}{2} U''(0) \lambda_0^2 \beta \bigl[1 + O(\beta^2)\bigr] \int_{-\infty}^{+\infty} \operatorname{sech}^4(x)\,dx.$$

Let $r = r(\beta) := \lambda_0(\beta)\beta$ and
$$C_1 = \frac{1}{2} U''(0) \int_{-\infty}^{+\infty} \operatorname{sech}^4 x\,ds.$$

We obtain the equation for r:
$$C_1(1 + O(\beta^2))r = K.$$

Hence,
$$r(\beta) = \frac{K}{C_1(1 + O(\beta^2))} = \frac{K}{C_1}(1 + O(\beta^2)),$$

$$\lambda_0(\beta) = \sqrt{\frac{r(\beta)}{\beta}} = \sqrt{\frac{K}{C_1\beta}} + O(\beta^{1/2}),$$

and
$$\beta\lambda(\beta) \leq \beta\lambda_0(\beta) = O(\beta^{1/2}).$$

Knowing that the order of magnitude of λ is at most $\beta^{-1/2}$, we can expand $\mathbf{U}(w_{\lambda,\beta})$. For $\varphi = Aw_{\lambda,\beta}$, we have

$$\begin{aligned}U(\varphi) &= \frac{1}{2}U''(0)\,\varphi^2 + \varepsilon|\varphi|^p + o(|\varphi|^p) \\ &= \frac{1}{2}\lambda^2 U''(0)\left[\beta^2 + \frac{1}{2}\beta^4\tanh^2(\beta t) - \frac{1}{6}\beta^4 + O(\beta^6)\right]\operatorname{sech}^2(\beta t) \\ &\quad + \varepsilon\lambda^p\left[\beta^p + o(\beta^p)\right]\operatorname{sech}^2(\beta t) + o\!\left(\lambda^p\beta^p\operatorname{sech}^{2p}(\beta t)\right).\end{aligned}$$

This gives us the following expression for $\mathbf{U}(w_{\lambda,\beta})$:

$$\begin{aligned}\mathbf{U}(w_{\lambda,\beta}) &= \frac{1}{2}\lambda^2\beta\, U''(0)\left[\int_{-\infty}^{+\infty}\operatorname{sech}^4 x\,dx - \beta^2 C_2 + O(\beta^2)\right] \\ &\quad + \varepsilon\lambda^p\beta^{p-1}\int_{-\infty}^{+\infty}\operatorname{sech}^{2p} x\,dx + o\!\left(\beta^{\frac{p}{2}-1}\right),\end{aligned}$$

where

$$C_2 = \int_{-\infty}^{+\infty}\left(\frac{1}{2}\tanh^2 x - \frac{1}{2}\right)\operatorname{sech}^4 x\,dx.$$

Since $\mathbf{U}(w_{\lambda,\beta}) = K$, we obtain that

$$\lambda(\beta) = \sqrt{\frac{K}{C_1\beta}} + o(\beta^{-1/2}).$$

Finally, we have to check the inequality

$$U''(0)\,\mathbf{T}(w_{\lambda,\beta}) < \mathbf{U}(w_{\lambda,\beta}). \tag{4.15}$$

We obtain that

$$\begin{aligned}\mathbf{U}(w_{\lambda,\beta}) - U''(0)\mathbf{T}(w_{\lambda,\beta}) &= \frac{1}{2}\lambda^2\beta^3 U''(0)\left[C_2 + O(\beta^2)\right] \\ &\quad + \varepsilon\lambda^p\beta^{p-1}\int_{-\infty}^{+\infty}\operatorname{sech}^{2p} x\,dx + o\!\left(\beta^{\frac{p}{2}-1}\right) \\ &= a_1\beta^2 + o(\beta^2) + a_2\beta^{\frac{p}{2}-1} + o\!\left(\beta^{\frac{p}{2}-1}\right),\end{aligned}$$

where a_1, a_2 are constants, depending on K, C_1 and C_2, but independent of β and λ, and $a_2 > 0$. On the other hand, one calculates $C_2 = -4/45$ and, hence, $a_1 < 0$. This implies immediately that (4.15) is valid for β small enough, provided $p/2 - 1 < 2$, i. e. $p < 6$. The proof is complete. $\square$

4.1.4 Proof of main result

First we show that the minimizers of problem (FW) solve the travelling wave equation.

Lemma 4.6 *Let U be as in Lemma 4.1. Suppose in addition that $U(r)$ is increasing for $r > 0$ and decreasing for $r < 0$. Let $K > 0$ and let $u \in \mathcal{A}_K$ be a minimizer of $\mathbf{T}$ on $\mathcal{A}_K$. Then $u \in C^2(\mathbb{R})$ and satisfies the travelling wave equation (3.3) with some $c > 0$.*

Proof. The quadratic functional $\mathbf{T}$ is continuous on X, hence, of class C^2. It is not difficult to verify that $\mathbf{U}$ is also a C^2 functional (*cf.* the proof of Proposition 3.3), and, for every $h \in X$,

$$\langle \mathbf{U}'(v), h \rangle = \int_{-\infty}^{+\infty} U'(Av(t)) Ah(t)\, dt.$$

Since $U'(r)\, r > 0$ for all $r \neq 0$, we have that

$$\langle \mathbf{U}'(v), v \rangle > 0$$

for all $v \in X$, $v \neq 0$. Hence, $\mathbf{U}'(v) \neq 0$. By the implicit function theorem, this shows that $\mathcal{A}_K$ is a C^1 submanifold of X.

Applying the Lagrange multiplier rule, we obtain that there exists $\lambda \in \mathbb{R}$ such that

$$\mathbf{T}'(u) = \lambda \mathbf{U}'(u),$$

or, in more details,

$$\int_{-\infty}^{+\infty} u'(t) h'(t)\, dt = \lambda \int_{-\infty}^{+\infty} U'(Au(t))\, A(h)\, dt \qquad (4.16)$$

for all $h \in X$. Taking $h = u$, we see that $\lambda > 0$. Equation (4.16) means that u is a weak solution of (3.3), with $c = \lambda^{-1/2}$ (see Proposition 3.4). Proposition 3.4 implies also that $u \in C^2$. The proof is complete. $\square$

Now we show that, in the case of symmetric potential, the minimizers constructed in Subsection 4.1.2 are monotone functions.

Lemma 4.7 *Suppose that the interaction potential U satisfies the assumptions of Lemma 4.1 and, in addition, U is symmetric, i. e. $U(r) = U(-r)$, and superquadratic. Let $K > 0$ and $u \in \mathcal{A}_K$ be a minimizer of $\mathbf{T}$ on $\mathcal{A}_K$. Then u is a monotone function, i. e. either $Au \geq 0$ or $Au \leq 0$.*

Proof. Consider the function

$$\widetilde{u}(t) := Pu(t) = \int_0^t |u'(s)|\, ds.$$

Let $\varphi = Au$ and $\widetilde{\varphi} = A\widetilde{u}$.

We claim that

$$\widetilde{\varphi}(t) = |\varphi(t)|. \tag{4.17}$$

It is clear that $\widetilde{\varphi} \geq |\varphi|$. Suppose for contradiction that $\widetilde{\varphi}(t_0) > |\varphi(t_0)|$ at some point t_0. By continuity of φ and $\widetilde{\varphi}$, we have that $\widetilde{\varphi} > |\varphi|$ on some neighborhood of t_0. By symmetry and strict monotonicity of U, on this neighborhood $U(\widetilde{u}) > U(u)$ and, hence, $\mathbf{U}(\widetilde{u}) > \mathbf{U}(u)$. As consequence, there exists $\lambda \in (0,1)$ such that $\mathbf{U}(\lambda\widetilde{u}) = \mathbf{U}(u) = K$, so that $\lambda\widetilde{u} \in \mathcal{A}_K$. But, by definition of $\widetilde{u}$, $\mathbf{T}(\widetilde{u}) = \mathbf{T}(u)$, and we have that

$$\mathbf{T}(\lambda\widetilde{u}) = \lambda^2 \mathbf{T}(\widetilde{u}) < \mathbf{T}(\widetilde{u}) = \mathbf{T}u$$

which contradicts the fact that u is a minimizer and proves (4.17).

Now suppose that u is not monotone. By Lemma 4.6, $u \in C^2(\mathbb{R})$. Therefore, there exist $a, b \in \mathbb{R}$ such that $u'(a) < 0$ and $u'(b) > 0$. For definiteness, we assume that $a < b$. By (4.17),

$$\int_t^{t+1} |u'(s)|\, ds = \left| \int_t^{t+1} u'(s)\, ds \right|$$

for all $t \in \mathbb{R}$. Hence, on each interval $[t, t+1]$ of length one, either $u' \geq 0$ or $u' \leq 0$. This implies that there exist $t_1, t_2 \in (a, b)$ such that $t_2 - t_1 \geq 1$ and $u' \equiv 0$ on $[t_1, t_2]$. Choose t_1 and t_2 so that, in addition,

$$\operatorname{dist}\Big(t_i, \{t \in \mathbb{R} : u'(t) \neq 0\}\Big) = 0, \quad i = 1, 2. \tag{4.18}$$

Let

$$w(t) := \begin{cases} \widetilde{u}(t), & t \leq t_1 \\ \widetilde{u}(t + (t_2 - t_1)), & t > t_1. \end{cases}$$

Since $t_2 - t_1 \geq 1$, we have that

$$\mathbf{U}(w) - \mathbf{U}(\widetilde{u}) = \int_{t_1-1}^{t_1} U\Big(\widetilde{u}\big(t+1+(t_2-t_1)\big) - \widetilde{u}(t)\Big)\, dt$$
$$- \int_{t_1-1}^{t_1} U\big(\widetilde{u}(t_1) - \widetilde{u}(t)\big)\, dt$$
$$- \int_{t_2-1}^{t_2} U\big(\widetilde{u}(t+1) - \widetilde{u}(t_1)\big)\, dt$$
$$= \int_{t_1-1}^{t_1} \Big[U\big(r_1(t) + r_2(t)\big) - U\big(r_1(t)\big) - U\big(r_2(t)\big)\Big]\, dt,$$

where

$$r_1(t) = \widetilde{u}(t_1) - \widetilde{u}(t) \geq 0,$$
$$r_2(t) = \widetilde{u}\big(t+1+(t_2-t_1)\big) - \widetilde{u}(t_1) \geq 0.$$

By superquadraticity of U, we have that

$$U(r_1 + r_2) \geq U(r_1) + U(r_2)$$

for all $r_1, r_2 \geq 0$, with equality if and only if one of the r_i is zero. Now (4.18) implies the existence of $t_0 \in [t_1-1, t_1]$ such that both $r_1(t_0) > 0$ and $r_2(t_0) > 0$. Hence, $\mathbf{U}(w) > \mathbf{U}(\widetilde{u})$. Now there exists $\lambda \in (0,1)$ such that $\mathbf{U}(\lambda w) = \mathbf{U}(\widetilde{u}) = K$ and

$$\mathbf{T}(\lambda w) = \lambda^2 \mathbf{T}(w) < \mathbf{T}(w) = \mathbf{T}(\widetilde{u}) = \mathbf{T}(u),$$

that contradicts the fact that u is a minimizer. The proof is complete. $\square$

Proof of Theorem 4.1. Applying Propositions 4.2, 4.3, Remark 4.2 and Lemma 4.6 to the symmetrized potential, we obtain in this case all the statements of Theorem 4.1, except (p3) and (p1). The last holds in a weaker form: u is a nondecreasing (resp. nonincreasing) function. By Lemma 4.7, the minimizer u we consider is a monotone function and, therefore, either u or $-u$ is a solution of the original problem.

Property (p1), i. e. strict monotonicity, follows from Proposition 3.8.

If remains to prove (p3), i. e. the fact that the wave is supersonic. Since $r^{-2}U(r)$ increases with $|r|$, differentiating we obtain

$$\frac{1}{2} U'(r)\, r \geq U(r), \quad r \in \Lambda.$$

Now Eq. (4.16), with $h = u$, $c^2 = \lambda^{-1}$, and the energy inequality, which holds by Proposition 4.2, imply that

$$c^2 = \frac{1}{2\mathbf{T}(u)} \int_{-\infty}^{+\infty} U'(Au) \, Au \, dt \geq \frac{\mathbf{U}(u)}{\mathbf{T}(u)} > U''(0).$$

This completes the proof of the theorem. □

4.1.5 Lennard-Jones type potentials

Here we discuss singular potentials, like the Lennard-Jones potential (see Examples 1.5 and 3.5)

$$U(r) = a\left[(d+r)^{-6} - d^{-6}\right]^2, \quad r > -d,$$

where $a > 0$ and $d > 0$. We start with the following result that concerns the case of solitary waves with small averaged potential energy.

Theorem 4.2 *Let*

$$U(r) = U_0(r) \quad \text{if } r > -d,$$

with some $d > 0$, and

$$U(r) = +\infty \quad \text{if } r \leq -d.$$

Assume that $U_0 \in C^2(-d, \infty)$, $U_0 \geq 0$, $U_0(0) = 0$, U_0 is superquadratic on $(-d, 0)$, $U_0''(0) \neq 0$ and that (4.2) holds. Then there exists

$$K_0 \geq \frac{1}{2} U''(0) \, d^2$$

such that for every $K \in (0, K_0)$ the system admits a solitary wave $u = u_K \in X$ with the averaged potential energy $\mathbf{U}(u) = K$, with the property

$$-d \leq Au_K(t) \leq 0, \quad t \in \mathbb{R},$$

and properties (p2)–(p4) of Theorem 4.1.

Proof. The proof relies on a cutoff argument which makes Theorem 4.1 applicable. Fix $\varepsilon \in (0, d/2)$ and choose a cutoff function $\eta \in C^\infty(\mathbb{R})$ such that

$$\eta = \begin{cases} 0 & \text{on } (-\infty, -d+\varepsilon) \\ 1 & \text{on } (-d+2\varepsilon, \infty) \end{cases}$$

and $\eta' \geq 0$.

Let p be the exponent from assumption (4.2). Choosing $C = C_\varepsilon$ so large that

$$W_\varepsilon(r) := \frac{1}{2} U''(0) r^2 + C|r|^p > U(r)$$

on $(-d+\varepsilon, 0)$, we define a new potential on the whole of $\mathbb{R}$ by

$$U_\varepsilon(r) := \begin{cases} W_\varepsilon(r), & r \leq -d \\ \eta(r) U(r) + (1-\eta(r)) W_\varepsilon(r), & r > -d. \end{cases}$$

The superquadraticity of U and W, and the fact that

$$\eta'(U - W) \leq 0 \quad \text{on } (-\infty, 0)$$

imply immediately that

$$\left(\frac{U_\varepsilon(r)}{r^2}\right)' < 0, \quad r < 0,$$

which means that U_ε is superquadratic on $(-\infty, 0)$. Also it is clear that U_ε satisfies (4.2). Hence, Theorem 4.1 applies and for all $K > 0$ the system with the modified potential U_ε possesses a travelling wave $u = u_K = u_{\varepsilon, K}$ such that $\mathbf{U}_\varepsilon(u) = K$.

Now we shall show that, for

$$K < \frac{1}{2} U''(0) d^2$$

and appropriately chosen ε,

$$\varphi(t) = Au(t) > -d + 2\varepsilon, \quad t \in \mathbb{R}.$$

Since $U_\varepsilon = U$ on $(-d + 2\varepsilon, \infty)$, this completes the proof.

By Lemma 3.5 and energy inequality (4.10),

$$\|\varphi\|_{L^\infty(\mathbb{R})} \leq \|u\| = (2\mathbf{T}_{\varepsilon,K})^{1/2} < \left(\frac{2K}{U''(0)}\right)^{1/2},$$

where $\mathbf{T}_{\varepsilon,K}$ is the minimum of $\mathbf{T}$ subject to the modified constraint. Hence, given

$$K < \frac{1}{2} U''(0) d^2,$$

one can take

$$\varepsilon \leq \frac{1}{2}\left[d - \left(\frac{2K}{U''(0)}\right)^{1/2}\right]$$

to obtain $\varphi(t) \geq -d + 2\varepsilon$. □

The following results are obtained in [Friesecke and Matthies (2002)].

Theorem 4.3 *In addition to the assumptions of Theorem 4.2, suppose that $U_0 \in C^3((-d, \infty))$,*

$$U_0'''(r) < 0 \text{ on } (-d, 0], \quad U_0(r) < U_0(-r) \text{ on } (0, d)$$

and

$$U_0(r) \geq a(r+d)^{-1} \tag{4.19}$$

for some $a > 0$ and all r close to $-d$. Then the conclusion of Theorem 4.2 holds for all $K > 0$ and $u = u_K$ is a solution of problem (FW), with the original (not symmetrized) potential.

We skip the proof and mention only that (4.2) follows from $U_0'''(0) < 0$. The assumption $U_0(r) < U_0(-r)$ on $(0, d)$ is needed to show that $u = u_K$ obtained as a solution of symmetrized problem (FW) is, actually, a solution of unsymmetrized problem (FW). Assumption (4.19) is essential to prove that all $K > 0$ are allowed.

Theorem 4.4 *Under the assumptions of Theorem 4.3, let $u_K \in X$, $K > 0$, be any solution obtained by means of problem (FW), and*

$$u_\infty(t) = \begin{cases} 0, & t \leq 0 \\ -dx, & 0 \leq t \leq 1 \\ -d, & t \geq 1. \end{cases}$$

Then there exist constants $a_K, b_K \in \mathbb{R}$ such that for

$$\tilde{u}_K(t) = b_K + u_K(t - a_K)$$

one has

$$\tilde{u}_K \to u_\infty \text{ in } L^\infty(\mathbb{R})$$

and[1]

$$(\tilde{u}_K)' \to (u_\infty)' \text{ in } L^p(\mathbb{R}) \text{ for all } p \in [1, \infty),$$

as $K \to \infty$. Furthermore, the corresponding wave speed c_K satisfies

$$\lim_{K \to \infty} c_K = \infty.$$

The classical Lennard-Jones potential, as well as its generalized version mentioned in Examples 1.5 and 3.5, satisfies the assumptions of Theorems 4.2 and 4.3.

4.2 Other types of travelling waves

4.2.1 Waves with periodic profile functions

In Section 3.2 we have considered travelling waves whose relative displacement profile is periodic, i. e.

$$r(t) = r(t + 2k),$$

or, equivalently,

$$u'(t) = u'(t + 2k),$$

where $k \geq 1$. In the present subsection we impose the boundary condition

$$u(t) = u(t + 2k) \tag{4.20}$$

which means that the wave profile itself is $2k$-periodic.

Let

$$Y_k := \{u \in H^1_{loc}(\mathbb{R}) \; : \; u(t + 2k) = u(t), u(0) = 0\}.$$

Endowed with the norm

$$\|u\|_k = \|u'\|_{L^2(-k,k)},$$

the space Y_k is a Hilbert space and, obviously, this is a closed subspace of the space X_k introduced in Section 3.2. Moreover, $u \in X_k$ belongs to Y_k if

---

[1] It follows from Theorem 4.10 and Remark 4.6 that u'_k decay exponentially and, hence, belongs to $L^p(\mathbb{R})$ for all $p \geq 1$.

and only if its derivative u' has zero mean value:

$$\langle u' \rangle := \int_{-k}^{k} u'(t)\, dt = 0.$$

Therefore, Y_k is 1-codimensional subspace of X_k. The orthogonal complement is exactly the subspace of X_k generated by the function $h_0(t) = t$.

We impose the following assumptions already used in Subsection 3.2.3:

(i'') $U(r) = \dfrac{a}{2} r^2 + V(t)$, where $V \in C^1(\mathbb{R})$, $V(0) = V'(0)$ and $V'(r) = o(r)$ as $r \to 0$,

(ii') For some $r_0 \in \mathbb{R}$ we have that $V(r_0) > 0$ and there exists $\theta > 2$ such that

$$\theta V(r) \leq r V'(r), \quad r \in \mathbb{R}.$$

Consider the functional (see Subsection 3.2.1)

$$J_k(u) = \int_{-k}^{k} \left[\frac{c^2}{2} u'(t)^2 - U(Au(t)) \right] dt$$

restricted to the subspace Y_k. We keep the notation J_k for the restricted functional. Since in the proof of Proposition 3.2 we have used only $2k$-periodic C^∞ test functions, we see that any critical point the restriction of J_k to Y_k is, in fact, a solution of Eq. (3.3) that satisfies (4.20).

Theorem 4.5 *Assume (i'') and (ii'). Let $c > \max(a, 0)$. Then for every $k \geq 1$ there exists a nontrivial travelling wave $w_k \in Y_k$. Moreover, there exist constants $\delta > 0$ and $M > 0$ such that*

$$\delta \leq J_k(w_k) \leq M$$

and

$$\delta \leq \|w_k\|_k \leq M.$$

Proof relies upon the standard mountain pass theorem (Theorem C.1) and goes along the same lines as the proof of Theorem 3.1, with only minor modifications. □

Note that since $\langle w_k' \rangle = 0$, the waves obtained are definitely not monotone, even when $a = c_0^2 > 0$. In the last case by Theorem 3.1, there exist two monotone waves, nondecreasing and nonincreasing, with periodic relative displacement profile and, by Theorem 4.5, an additional nonmonotone wave that has a periodic profile function.

4.2.2 Solitary waves whose profiles vanish at infinity

In contrast to the case of Section 3.3, here we impose the boundary condition

$$u(\infty) = 0 \tag{4.21}$$

that means that the wave profile vanishes at infinity.

Let Y be the closure of $C_0^\infty(\mathbb{R})$ with respect to the norm

$$\|u\| = \left(\int_{-\infty}^{+\infty} u'(t)^2 dt \right)^{1/2}.$$

Obviously, Y is a closed subspace of the space $\widetilde{X}$ defined in Subsection 3.3.1. One can think of functions from Y as satisfying boundary condition (4.21) in some generalized sense.

Consider the functional (see Subsection 3.3.1)

$$J(u) = \int_{-\infty}^{+\infty} \left[\frac{c^2}{2} u'(t)^2 - U(Au(t)) \right] dt$$

restricted to the space Y. The restriction of J to Y is still denoted by J. Under assumptions (i''') and (ii') (see Subsection 4.2.1), J is a C^1 functional on Y. Moreover, every critical point of J in Y is a C^2 solution of Eq. (3.3). Indeed, since $C_0^\infty(\mathbb{R}) \subset Y$, the proof of Proposition 3.5 shows that u is a weak solution of (3.3) and, in fact, a classical solution.

Now we have the following existence result.

Theorem 4.6 *Under assumptions (i''') and (ii'), with $c > \max(a, 0)$, there exists a nontrivial travelling wave $w \in Y$.*

Proof. As we have pointed out in Subsection 3.3.2, the functional J possesses the mountain pass geometry in the space X. But since there exists, obviously, a vector $e \in Y$ such that $J(e) < 0$, the same holds in Y. By the mountain pass theorem without Palais-Smale condition (see Theorem C.4), there exists a sequence $w_n \in Y$ such that

$$J(w_n) \to b$$

and

$$J'(w_n) \to 0 \quad \text{in } Y^* \quad \text{(the dual space)},$$

where b is the mountain pass level, *i. e.* w_n is a Palais-Smale sequence.

As usual, assumption (ii') permits us to conclude that the sequence w_n is bounded in Y. Moreover, as in Lemma 3.5, we deduce that $\|w_n\|$ is bounded below by a positive constant. Hence, $\|w_n\|$ does not converge to 0. Therefore, we can assume that w_n converges to w weakly in Y. Arguing as in Lemma 3.6, we obtain that, for any $r > 0$ there exists $\eta > 0$ such that along a subsequence,

$$\int_{\zeta_n - r}^{\zeta_n + r} |Aw_n(t)|^2 dt \geq \eta,$$

with some $\zeta_n \in \mathbb{R}$. Replacing $w_n(t)$ by $w_n(t - \zeta_n)$, we obtain that

$$\int_{-r}^{r} |Aw_n(t)|^2 dt \geq \eta$$

and the new w_n still form a Palais-Smale sequence. By the compactness of Sobolev embedding, $Aw_n \to Aw$ in $L^\infty_{loc}(\mathbb{R})$, *i. e.* uniformly on compact intervals and we deduce that

$$\int_{-r}^{r} |Aw(t)|^2 dt \geq \eta,$$

hence, $w \neq 0$.

Testing $J'(w_n)$ with an arbitrary function $\varphi \in C_0^\infty(\mathbb{R})$, we obtain, as in the proof of Proposition 3.6, that $J'(w) = 0$, *i. e.* $w \neq 0$ is a critical point of J in Y. $\square$

Remark 4.4 As it follows from Theorem 4.10 and Remark 4.6, at least in the case when $a = c_0^2 \geq 0$ the solution $w \in Y$ obtained in Theorem 4.6 decays exponentially at infinity and, hence, satisfies (4.21).

Note that for the solution w obtained in Theorem 4.6 we have that $0 < J(w) \leq b$. It is easy to verify that the arguments of the proof of Proposition 3.14 work in the case we consider as well. Therefore, under the additional assumption that $V(r) \geq a_0 |r|^q$, with $q > 2$, we obtain the estimate

$$\|w\|^2 \leq C(c^2 - c_0^2)^{\frac{q+4}{q-2}}$$

provided $c > c_0$ is close to c_0 (here $a = c_0^2 > 0$). Repeating the proof of Theorem 3.7, we obtain

Theorem 4.7 *Suppose that $V(r)$ is defined in a neighborhood of 0 and satisfies in that neighborhood assumptions (i'') and (ii'), with $a = c_0^2 \geq 0$. Assume, in addition, that for $|r|$ small*

$$V(r) \geq a_0 |r|^q,$$

with $a_0 > 0$ and $q > 2$. Then there exists $\bar{c} > c_0$ such that for every $c \in (c_0, \bar{c})$ there exists a nontrivial travelling wave $w \in X$ with the speed c.

Remark 4.5 Similar result holds for waves considered in Subsection 4.2.1.

Note that Theorem 4.6 applies immediately to FPU α- and β-models (see Examples 1.2, 1.3 and 3.2, 3.3), while Theorem 4.7 works is the case of Lennard-Jones type potentials (Examples 1.5 and 3.5). Thus we obtain in those cases additional nonmonotone waves.

4.3 Yet another constrained minimization problem

Consider a particular case when the potential is given by

$$V(r) = \frac{c_0^2}{2} + \frac{d}{p}|r|^p, \quad p > 2, d > 0.$$

In this case one can obtain travelling waves solving another constrained minimization problem.

Let

$$Q_k(u) = \frac{1}{2} \int_{-k}^{k} \left[c^2 (u'(t))^2 - c_0^2 (Au(t))^2 \right] dt$$

and

$$\Psi_k(u) = \frac{d}{p} \int_{-k}^{k} |Au(t)|^p dt.$$

These are well-defined C^1 functionals on the space X_k. Similarly, the functionals

$$Q(u) = \frac{1}{2} \int_{-\infty}^{+\infty} \left[c^2 (u'(t))^2 - c_0^2 (Au(t))^2 \right] dt$$

and
$$\Psi(u) = \frac{d}{p}\int_{-\infty}^{+\infty} |Au(t)|^p dt$$
are of class C^1 on X. Obviously,
$$J_k(u) = Q_k(u) - \Psi_k(u)$$
and
$$J(u) = Q(u) - \Psi(u),$$
where J_k and J are introduced in Sections 3.2 and 3.3, respectively.

Given $\alpha > 0$, consider the following two minimization problems:
$$I_k(\alpha) = \inf\{Q_k(v) \, : \, v \in X_k, \Psi_k(v) = \alpha\} \qquad (4.22)$$
and
$$I(\alpha) = \inf\{Q(v) \, : \, v \in X, \Psi(v) = \alpha\}. \qquad (4.23)$$

Since Ψ_k and Ψ are positive homogeneous of degree $p > 2$, and Q_k and Q are positive homogeneous of degree 2, it is readily verified that problems (4.22) (resp., (4.23)) with different values of $\alpha > 0$, are equivalent. Moreover,
$$I_k(\alpha) = \alpha^{2/p} I_k(1)$$
and
$$I(\alpha) = \alpha^{2/p} I(1).$$

We have the following result

Theorem 4.8 *Suppose that $c > c_0 \geq 0$. Then there exists a solution $v_k \in X_k$ $k \geq 1$, (resp., $v \in X$) of problem (4.22) (resp., (4.23)).*

To prove the result for problem (4.22) one considers an arbitrary minimizing sequence which turns out to be bounded in X_k. The passage to the limit is straightforward and uses the compactness of the Sobolev embedding.

In the case of problem (4.23) one has to employ, in addition, a concentration compactness argument based on Lemma B.1. Alternatively, one can obtain a solutions as the limit of appropriately shifted solutions v_k of the problem (4.22) as $k \to \infty$.

Since the functionals Q_k and Ψ_k, as well as Q and Ψ, are C^1, there exist Lagrange multipliers λ_k and λ such that

$$Q'_k(v_k) = \lambda_k \Psi'(v_k)$$

and

$$Q'(v) = \lambda \Psi'(v).$$

Actually,

$$\lambda_k = \frac{I_k(\alpha)}{\alpha p} > 0$$

and

$$\lambda = \frac{I(\alpha)}{\alpha p} > 0.$$

These Lagrange multiplies can be scaled out, using the homogeneity properties of the problems. Precisely, letting

$$u_k = \lambda_k^{\frac{1}{p-2}} v_k$$

and

$$u = \lambda^{\frac{1}{p-2}} v,$$

we obtain travelling waves $u_k \in X_k$ and $u \in X$.

4.4 Remark on FPU β-model

Let us consider the FPU β-model, with the interaction potential

$$U(r) = \frac{c_0^2}{2} r^2 + \frac{d}{4} r^4 = \frac{c_0^2}{2} r^2 + V(r),$$

in the case when $d < 0$. In [Friesecke and Wattis (1994)] it is conjectured that nontrivial solitary waves do not exist in this case. The conjecture trivially holds true, as a consequence of following

Proposition 4.4 *Suppose that*

$$U(r) = \frac{c_0^2}{2} r^2 + V(r)$$

where $V \in C^1$, $V(0) = V'(0) = 0$ and $V(r)r \leq 0$. If $c > c_0 \geq 0$, then nontrivial travelling waves, both periodic and solitary, do not exist.

Proof. For the functional J_k introduced in Section 3.2 we have that

$$\langle J_k'(u), u \rangle = \int_{-k}^{k} \left[c^2 (u'(t))^2 - c_0^2 (Au(t))^2 - V'(Au(t)) Au(t) \right] dt$$

$$\geq \int_{-k}^{k} \left[c^2 (u'(t))^2 - c_0^2 (Au(t))^2 \right] dt$$

$$= c^2 \|u\|_k^2 - c_0^2 \|Au\|_{L^2(-k,k)}^2.$$

By Lemma 3.1

$$\langle J_k'(u), u \rangle \geq (c - c_0)^2 \|u\|_k^2.$$

Therefore, the only critical point of J_k is the point of minimum $u = 0$, and there is no nontrivial travelling wave $u \in X_k$.

The same reasoning shows that the functional J has the only critical point $u = 0$ and nontrivially solitary waves do not exist. □

However, if $0 < c \leq c_0$, the FPU β-model with $d < 0$ possesses periodic travelling waves. More generally, we have

Theorem 4.9 *Suppose that U,*

$$U(r) = \frac{c_0^2}{2} r^2 + V(r),$$

where $V \in C^1(\mathbb{R})$, $V(0) = V'(0) = 0$ and $V'(r) = o(r)$ near $r = 0$, and suppose that there exists $\theta > 2$ such that

$$0 \geq \theta V(r) \geq r V'(r)$$

and $V(r) < 0$ if $|r| \geq r_0$ for some $r_0 > 0$, i. e. $-V$ satisfies assumptions (ii^+) and (ii^-) from Section 3.2. Then for every $c \in (0, c_0]$ and every $k \geq 1$ there exists a $2k$-periodic travelling wave $u_k \in X_k$.

The proof goes along the same lines as in the case of Theorem 3.3. We consider the functional

$$-J_k(u) = -\int_{-k}^{k} \left[\frac{c^2}{2} (u'(t))^2 - \frac{c_0^2}{2} (Au(t))^2 - V(A(t)) \right] dt$$

and apply the generalized linking theorem of [Benci and Rabinowitz (1979)] (Theorem C.5).

Employing the notation used in the proof of Theorem 3.3, we denote by Z the subspace of X_k generated by $h_j^\pm$ with $\lambda_j < 0$, while Y stands now for the subspace generated by $h_j^\pm$ with $\lambda_j \geq 0$ and the function h_0. So, the quadratic part of $-J_k$ is nonpositive on Y and positive on Z. Clearly, $Y \perp Z$ and $X_k = Y \oplus Z$, but now $\dim Y = \infty$ and $\dim Z < \infty$. This is the reason to use Theorem C.5 instead of Theorem C.4.

The verifications of Palais-Smale condition and of linking geometry are quite similar to the corresponding points of the proof of Theorem 3.3.

4.5 Exponential decay

Here we prove that solitary waves obtained before have exponentially decaying relative displacement profile

$$r(t) = Au(t) = u(t+1) - u(t).$$

Being rewritten in terms of r, Eq. (3.3) becomes

$$c^2 r''(t) = -A^* A U'(r(t)), \qquad (4.24)$$

where

$$A^* v(t) = v(t) - v(t-1)$$

is the operator formally adjoint to A, and

$$A^* A v(t) = (v(t+1) + v(t-1) - 2v(t)).$$

Separating harmonic and unharmonic parts of the potential U, we impose the following assumption:

(h) $U(r) = \dfrac{c_0^2}{2} r^2 + V(r)$, where $V \in C^1$ in a neighborhood of 0, $V(0) = V'(0) = 0$ and $V'(r) = o(r)$ as $r \to 0$.

So, we allow locally defined potentials.

Now we write Eq. (4.24) in the following form

$$Lr = -A^* A V'(r), \qquad (4.25)$$

where

$$\begin{aligned} Lv(t) &= -c^2 v''(t) + c_0^2 (A^* A) v(t) \\ &= -c^2 v''(t) + c_0^2 (v(t+1) + v(t-1) - 2v(t)). \end{aligned}$$

The operator L is a pseudodifferential operator with the symbol

$$\sigma_L(\xi) = c^2\xi^2 - 4c_0^2 \sin^2 \frac{\xi}{2},$$

while

$$\sigma_A(\xi) = e^{i\xi} - 1,$$

$$\sigma_{A^*}(\xi) = 1 - e^{-i\xi}$$

and

$$\sigma_{A^*A}(\xi) = (1 - e^{-i\xi})(e^{i\xi} - 1) = -4\sin^2\frac{\xi}{2}$$

(*cf.* Subsection 3.5.1).

Making use the Fourier transform, we obtain from Eq. (4.25)

$$\sigma_L(\xi)\hat{r}(\xi) = -\sigma_{A^*A}(\xi)\widehat{[V'(r)]}(\xi),$$

i. e.

$$r = TV'(r), \qquad (4.26)$$

where

$$\sigma_T(\xi) = -\frac{\sigma_{A^*A}(\xi)}{\sigma_L(\xi)} = \frac{4\sin^2(\xi/2)}{c^2\xi^2 - 4c_0^2\sin^2(\xi/2)}.$$

(Formally, $T = -L^{-1}A^*A$). Actually, this means that T is a convolution operator with the integral kernel $K(x)$ such that $\hat{K}(\xi) = \sigma_T(\xi)$.

To study Eq. (4.26) we need the following

Lemma 4.8 *Let $f(t)$ and $g(t)$ be bounded non-negative functions on $\mathbb{R}$, with $\lim_{t\to\pm\infty} g(t) = 0$. Suppose that*

$$f(t) \leq \int_{-\infty}^{+\infty} e^{-\beta|t-\tau|} g(\tau) f(\tau) \, d\tau,$$

with $\beta > 0$. Then for every $\alpha \in (0,\beta)$ there exists a constant $C = C_\alpha$ such that

$$f(t) \leq Ce^{-\alpha|t|}.$$

Proof. It suffices to prove the desired estimate for $t > 0$ because the case $t < 0$ reduces to the previous one by replacing t by $-t$.

For any integer $n > 0$ let

$$L_n = \int_{-\infty}^{n} e^{-\beta(n-\tau)} g(\tau) f(\tau) \, d\tau,$$

$$F_n = \sup_{t \geq n} f(t), \quad G_n = \sup_{t \geq n} g(t).$$

For $t \geq n$ we have that

$$f(t) \leq \int_{-\infty}^{n} e^{-\beta(t-\tau)} g(\tau) f(\tau) \, d\tau + \int_{n}^{+\infty} e^{-\beta|t-\tau|} g(\tau) f(\tau) \, d\tau$$

$$\leq L_n + F_n G_n \int_{n}^{+\infty} e^{-\beta|t-\tau|} d\tau \leq L_n + F_n G_n \int_{-\infty}^{+\infty} e^{-\beta|\tau|} d\tau$$

$$= L_n + \frac{2}{\beta} F_n G_n.$$

From there we deduce immediately that

$$F_n \leq L_n + \frac{2}{\beta} F_n G_n. \qquad (4.27)$$

Next we estimate L_{n+1}:

$$L_{n+1} = e^{-\beta} \int_{-\infty}^{n} e^{-\beta(n-\tau)} g(\tau) f(\tau) \, d\tau$$

$$+ \int_{n}^{n+1} e^{-\beta(n+1-\tau)} g(\tau) f(\tau) \, d\tau$$

$$= e^{-\beta} L_n + F_n G_n. \qquad (4.28)$$

Since, by assumption, $G_n \to 0$ as $n \to +\infty$, then there exists an integer $M > 0$ such that for $n \geq M$ we have

$$2 G_n \leq \min\left[\frac{\beta}{2}, e^{-\alpha} - e^{-\beta}\right]. \qquad (4.29)$$

Now (4.27) implies immediately that

$$F_n \leq L_n + \frac{1}{2} F_n.$$

Hence,

$$F_n \leq 2 L_n.$$

This together with inequalities (4.28) and (4.29) gives us

$$L_{n+1} \le L_n(e^{-\beta} + 2G_n) \le L_n e^{-\alpha}$$

The last inequality implies easily that, for $n \ge M$,

$$L_n \le e^{-\alpha n} e^{\alpha M} L_M = K e^{-\alpha n}.$$

Therefore,

$$F_n \le 2K e^{-\alpha n}$$

and this implies the required. Indeed, for $t \in [n, n+1]$, with $n \ge M$, we have that

$$f(t) \le F_n \le 2K e^{-\alpha n} = 2K e^{\alpha(t-n)} e^{-\alpha t} \le (2K e^{\alpha}) e^{-\alpha t}$$

and the proof is complete. □

Now we obtain an exponential bound for the kernel $K(x)$.

Lemma 4.9 *Suppose that $c^2 > c_0^2 > 0$. Then there exists $\beta_0 > 0$ such that for every $\beta \in (0, \beta_0)$*

$$|K(x)| \le C e^{-\beta |x|},$$

with some $C > 0$.

Proof. We start with

$$\widehat{K}(\xi) = \frac{4\sin^2(\xi/2)}{c^2 \xi^2 - 4c_0^2 \sin^2(\xi/2)} = \frac{1}{c_0^2} \frac{\operatorname{sinc}^2(\xi/2)}{d^2 - \operatorname{sinc}^2(\xi/2)},$$

where $d^2 = c^2/c_0^2 > 1$ and $\operatorname{sinc} z = z^{-1} \sin z$, and study $\widehat{K}(\xi)$ for complex values of ξ. Obviously, $\widehat{K}(\xi)$ is a meromorphic function of ξ. Since $d^2 > 0$, the point $\xi = 0$ is a regular point of $\widehat{K}(\xi)$.

The poles of $\widehat{K}(\xi)$ are exactly the roots of the equation

$$d^2 - \operatorname{sinc}^2 \frac{\xi}{2} = 0. \qquad (4.30)$$

The roots possesses the following symmetries: if ξ is a root, then $-\xi$, $\overline{\xi}$ and $-\overline{\xi}$ are also. Therefore, it suffice to describe roots of Eq. (4.30) in the first quadrant.

Let $\xi = a + ib$, $a \geq 0$, $b \geq 0$. The identity

$$\left|\frac{\xi}{2}\right|^2 \operatorname{sinc}\frac{\xi}{2} = \frac{1}{2}a\sin\frac{a}{2}\cosh\frac{b}{2} + \frac{1}{2}b\cos\frac{a}{2}\sinh\frac{b}{2}$$
$$+ i\left(\frac{a}{2}\cos\frac{a}{2}\sinh\frac{b}{2} - \frac{b}{2}\sin\frac{a}{2}\cosh\frac{b}{2}\right)$$

shows that in the case $a \geq 0$ and $b \geq 0$ Eq. (4.30) is equivalent to: *either*

$$a = 0, \quad \frac{\sinh(b/2)}{b/2} = d, \tag{4.31}$$

or

$$\frac{\tan(a/2)}{a/2} = \frac{\tanh(b/2)}{b/2}, \quad \frac{\sinh(b/2)}{b/2} = \frac{d}{|\cos(a/2)|}. \tag{4.32}$$

Since $d > 1$, the second equation in (4.31) has a unique solution β_0. The second equation in (4.32) shows that for all other roots $b \geq \beta_0$.

Now we compute the residues of $\widehat{K}(\xi)$ at the poles $\pm i\beta_0$. They are $\mp i\alpha_0$ respectively and since $d = \operatorname{sinc}(i\beta_0/2)$, we obtain

$$\alpha_0 = i \lim_{\xi \to i\beta_0} (\xi - i\beta_0)\widehat{K}(\xi) = i \lim_{\xi \to i\beta_0} \frac{d^2(\xi - i\beta_0)}{2d(d - \operatorname{sinc}(\xi/2))}$$
$$= \frac{1}{2}\frac{d^2\beta_0^2}{\sinh\beta_0 - d^2\beta_0} > 0.$$

We write

$$K(x) = K_0(x) + K_1(x),$$

where

$$\widehat{K}_0(\xi) = -\frac{i\alpha_0}{\xi - i\beta_0} + \frac{i\alpha_0}{\xi + i\beta_0} = \frac{2\alpha_0\beta_0}{\xi^2 + \beta_0^2}$$

and

$$\widehat{K}_1(\xi) = \widehat{K}(\xi) - \widehat{K}_0(\xi).$$

The inverse Fourier transform gives

$$K_0(x) = \alpha_0 e^{-\beta_0|x|}.$$

The remainder $\widehat{K}_1$ has no singularities in the strip

$$S = \{\xi \in \mathbb{C} : |\operatorname{Im}\xi| < \beta_0\}.$$

From the form of $\widehat{K}$ and $\widehat{K}_0$ it is clear that

$$|\widehat{K}_1(\xi)| \leq \frac{C}{1+|\xi|^2}, \quad \xi \in S. \tag{4.33}$$

Now we estimate $K_1(x)$. Fix $\beta \in (-\beta_0, \beta_0)$. Applying the Fourier inversion formula and shifting the contour of integration by $\xi = \zeta + i\beta$, we have, using (4.33), that

$$\left|e^{\beta x} K_1(x)\right| = \left|\frac{1}{\sqrt{2\pi}} \int_{-\infty}^{+\infty} \widehat{K}_1(\zeta + i\beta) e^{i\zeta x} \, dx\right|$$

$$\leq C \int_{-\infty}^{+\infty} \frac{d\zeta}{1+\zeta^2} = \pi C$$

for any real x. This implies the required. $\square$

Now we obtain

Theorem 4.10 *Assume (h) with $c^2 > c_0^2 > 0$. Let $u \in X$ be a travelling wave and $r = Au$ its relative displacement profile. Then for every $\alpha \in (0, \beta_0)$, with β_0 from Lemma 4.9, there exists a constant $C = C_\alpha$ such that*

$$|r(t)| \leq C e^{-\alpha|t|}.$$

Moreover, $u''(t)$, $u'(t)$ and $u(t) - u(\pm\infty)$ also decay exponentially as $t \to \pm\infty$.

Proof. From Eq. (4.26) we obtain that

$$|r(t)| \leq \int_{-\infty}^{+\infty} |K(t-\tau)| g(\tau) |r(\tau)| \, d\tau,$$

where

$$g(\tau) = \left|\frac{V'(r(t))}{r(t)}\right|.$$

Assumption (h) implies that $g(t) \to 0$ as $t \to \pm\infty$. By Lemma 4.9,

$$|r(t)| \leq C \int_{-\infty}^{+\infty} e^{-\beta|t-\tau|} g(\tau) |r(\tau)| \, d\tau$$

for every $\beta \in (0, \beta_0)$. Applying Lemma 4.8, we obtain the first assertion of the theorem.

Exponential decay of u'' follows immediately from Eq. (3.3), assumption (h) and the first part of the theorem.

Since $u' \in L^2(\mathbb{R})$, we have that

$$u'(t) = \int_{-\infty}^{t} u''(s)\,ds = -\int_{-t}^{+\infty} u''(s)\,ds.$$

Together with exponential decay of u'', this gives that $u'(t)$ is an exponentially decaying function. Repeating this argument, we obtain the exponential decay of $u(t) - u(\pm\infty)$. The existence of $u(\in \pm\infty)$ follows immediately from the exponential decay of u'. The proof is complete. $\square$

Remark 4.6 If $c_0 = 0$, the situation simplifies considerably and we obtain the same result as in Theorem 4.10, with $\beta_0 = +\infty$, i. e.

$$|r(t)| \le C_\alpha e^{-\alpha|t|}$$

for every $\alpha > 0$. We expect that the solutions still decay exponentially if we replace c_0^2 in assumption (h) by $a < 0$ and suppose that $c^2 > 0$.

Remark 4.7 Thus, the solitary travelling waves obtained in Theorems 3.4–3.7, 4.1–4.3 decay exponentially.

Remark 4.8 Under assumptions of Theorem 4.10, or Remark 4.6, we have the following identity

$$c^2\left[u(+\infty) - u(-\infty)\right] = \int_{-\infty}^{+\infty} U'\bigl(u(t+1) - u(t)\bigr)\,dt$$

which is a counterpart of identity from Remark 3.1.

4.6 Travelling waves in chains of oscillators

The approach developed in Chapter 3 applies, with minor modifications, in the case of travelling waves in a homogeneous chain of nonlinear oscillators governed by Eq. (1.38). An equivalent equation of motion is (1.39). In this section we present results by S. Bak and the author (unpublished).

We consider a homogeneous chain of nonlinear oscillators with the potential

$$U(r) = -\frac{c_0}{2}r^2 + V(r)$$

and linear coupling between nearest neighbors (a stands for the coupling constant). The equation of motion becomes

$$\ddot{q}(n) = a\Delta_d q(n) + c_0\, q(n) - V'\big(q(n)\big), \qquad (4.34)$$

where $q(n) = q(t,n)$ and

$$(\Delta_d q)(n) = q(n+1) + q(n-1) - 2q(n)$$

is the one-dimensional discrete Laplacian.

Making use the travelling wave Ansatz

$$q(t,n) = u(n - ct),$$

we obtain the equation

$$c^2 u''(t) = a\big(u(t+1) + u(t-1) - 2u(t)\big) + c_0 u(t) - V'\big(u(t)\big) \qquad (4.35)$$

governing travelling waves. This equation has, actually, a variational structure.

Let

$$H_k^1 = \{u \in H_{loc}^1(\mathbb{R}) \ :\ u(t+2k) = u(t)\}$$

endowed with the norm

$$\|u\|_k = \left(\|u\|_{L^2(-k,k)}^2 + \|u'\|_{L^2(-k,k)}^2\right)^2,$$

i. e. the Sobolev space of $2k$-periodic functions, and let $H^1(\mathbb{R})$ be the standard Sobolev space on $\mathbb{R}$. On these spaces we consider the functionals

$$J_k(u) = \int_{-k}^{k} \left[\frac{c^2}{2} u'(t)^2 - \frac{a}{2}\big(Au(t)\big)^2 + \frac{c_0^2}{2} u(t)^2 - V\big(u(t)\big)\right] dt \qquad (4.36)$$

and

$$J(u) = \int_{-\infty}^{+\infty} \left[\frac{c^2}{2} u'(t)^2 - \frac{a}{2}\big(Au(t)\big)^2 + \frac{c_0^2}{2} u(t)^2 - V\big(u(t)\big)\right] dt, \qquad (4.37)$$

respectively, where

$$Au(t) := u(t+1) - u(t) = \int_t^{t+1} u(s)\, ds.$$

We assume that

(h_1) the function V is C^1, $V(0) = V'(0) = 0$ and $V'(r) = o(r)$ as $r \to 0$,
(h_2) there exist $r_0 \in \mathbb{R}$ and $\theta > 2$ such that $V(r_0) > 0$ and

$$\theta V(r) \leq r V'(r), \quad r \in \mathbb{R}.$$

The functionals J_k and J are well-defined C^1 functionals on H_k^1 and $H^1(\mathbb{R})$, respectively. One can verify that their critical points are weak solutions of Eq. (4.35). In fact, they are C^2 solutions (cf. Propositions 3.4 and 3.5). Moreover, by Theorem A.1 in the case of functional J any its critical point $u \in H^1(\mathbb{R})$ satisfies

$$u(\infty) := \lim_{t \to \infty} u(t) = 0.$$

Under the additional assumption that $c^2 > \max(a, 0)$ and $c_0 > 0$, the quadratic parts of J_k and J are positive defined, and the functionals themselves possess the mountain pass geometry. Moreover, the functional J_k satisfies also the Palais-Smale condition (here the compactness of Sobolev embedding plays a crucial role). Thus, applying the mountain pass theorem (Theorem C.1), we obtain the existence of periodic travelling waves in the system considered:

Theorem 4.11 *Assume* (h_1) *and* (h_2). *Suppose that* $c_0 > 0$. *Then for every* $k \geq 1$ *and* $c^2 > \max(a, 0)$ *there exists a nontrivial $2k$-periodic solution* $u_k \in H_k^1$ *of Eq. (4.35), i. e. a $2k$-periodic travelling wave with the speed c.*

Notice that, in contrast to Section 3.2, the waves considered in this theorem have periodic profile functions.

Arguing as in Section 3.3, one can pass to the limit as $k \to \infty$ and obtain a nontrivial solitary wave. We have

Theorem 4.12 *Assume* (h_1) *and* (h_2) *and suppose that* $c_0 > 0$. *Then for every* $c^2 > \max(a, 0)$ *there exists a nontrivial solution* $u \in X$ *of Eq. (4.35), i. e. a solitary travelling wave with the speed c. Moreover, the solution u decays exponentially:*

$$|u(t)| \leq C e^{-\alpha |t|},$$

with some $C > 0$ *and* $\alpha > 0$.

Exponential decay can be obtained using exponential bound for the Green function of the operator

$$(Lu)(t) = -c^2 u'' + a\big(u(t+1) + u(t-1) - 2u(t)\big) + c_0 u(t),$$

which is easy to derive by means of Fourier transform.

Remark 4.9 Using linking instead of mountain pass, one can extend the result of Theorem 4.11 to the case when $c^2 > 0$, and a and c are arbitrary real number. However, it is very likely that nontrivial solitary waves do not exist in this case.

4.7 Comments and open problems

Travelling solitary waves in FPU lattices were observed numerically long time ago (see, e. g. [Eilbeck (1991); Eilbeck and Flesch (1990); Flytzanis et. al (1989); Hochstrasser et. al (1989); Peyrard et. al (1986)]). Another area of work in this field deals with the so-called continuum approximations. Here the equation of motion reduces to a patrial differential equation and the equation governing travelling waves becomes an ordinary differential equation. The results are of perturbative nature and concern the waves with near sonic speed (see [Flytzanis et. al (1989); Friesecke and Pego (1999); Peyrard et. al (1986); Rosenau (1989); Wattis (1993a)]).

The first rigorous result on the existence of solitary waves on general FPU lattices was obtained in [Friesecke and Wattis (1994)] (we present this approach in Section 4.1). Note that this approach is based on appropriate constrained minimization problem and relies essentially on the concentration compactness method introduced in [Lions (1984)]. Similar kind of constrained minimization was used before in [Valkering (1978)] to establish the existence of periodic travelling waves for some class of Lennard-Jones type potentials. Certainly, in the last case the problem simplifies considerably and no concentration compactness is needed. The one dimensional Sobolev embedding theorem is enough for that purpose.

Later, in [Smets and Willem (1997)], the existence of solitary waves with prescribed speed was considered and the techniques based on the mountain pass theorem without (PS) condition was introduces. Another approach to the same problem was proposed in [Pankov and Pflüger (2000b)]. This approach relies on a combination of periodic approximations and the standard mountain pass theorem and is technically simpler than that of [Smets and Willem (1997)]. In addition, it gives a local convergence of periodic waves to solitary ones. Moreover, employing the Nehari manifold method originated in [Nehari (1960); Nehari (1961)] one obtains an improved convergence of periodic ground waves to ground solitary waves. The notion of ground wave was also introduced in [Pankov and Pflüger (2000b)]. Note that the existence of ground solitary waves is not obvious at all. This approach is discussed in Sections 3.1–3.4 in full details. Remark that Theorem 3.3 is new. We conjecture that under assumptions of Theorem 3.3, nontrivial subsonic solitary waves do not exist.

Problem 4.1 *Prove, or disprove, this conjecture.*

In Proposition 3.8 we have shown that under natural assumptions monotone solitary waves are strictly monotone. However, the following problem remains open.

Problem 4.2 *In the setting of Section 3.2, are monotone periodic waves strictly monotone?*

The results of Section 3.5 are obtained by the author and nowhere published before. These results deal with the near sonic limit $c \to c_0$. On the other hand, the following problem is completely open.

Problem 4.3 *Study the behavior of ground travelling waves, periodic and solitary, in the high speed limit $c \to \infty$.*

Another result on near sonic waves is obtained in [Friesecke and Pego (1999)]. Those authors employ a technique that relies on perturbation from an appropriate continuous limit, in fact, KdV-equation. Actually, under certain assumptions they show that for $c > c_0$ close to c_0 there exists a unique, up to translations and adding constants, solitary wave $u(t)$ such that $u'(t) > 0$ ($u'(t) < 0$) and u' is an even function, *i. e.* the wave itself is monotone and symmetric. In this connection we offer the following

Problem 4.4 *Under assumptions of Section 3.2, is an increasing (resp., decreasing) 2k-periodic travelling wave unique up to translations and adding constants? Same question about monotone solitary waves. What's about uniqueness of ground waves in the setting of Section 3.4? Are all such waves symmetric?*

We mention also the paper [Iooss (2000)], where bifurcation analysis of travelling waves on FPU type lattices is carried out.

As we already mentioned, in Section 4.1 we present the results of [Friesecke and Wattis (1994)], while Theorems 4.3 and 4.4 are borrowed from [Friesecke and Matthies (2002)]. Note also that a result similar to

Theorem 4.4 is obtained in [Treschev (2004)], where the method of integral equations is employed. Theorem 4.4 deals with high energy limit of solitary waves when the averaged potential energy $K \to \infty$ in the case of Lennard-Jones type potentials.

Problem 4.5 *What can one say about high energy limit for systems like FPU α- and β-models?*

Another question in this direction is the following. Let u_K be a solitary wave with the average potential energy K and c_K its wave speed. Suppose that $K_0 = 0$ in Theorem 4.1.

Problem 4.6 *Is it true that $c_K \to c_0$ as $K \to 0$, where $c_0 = \sqrt{V''(0)}$ is the speed of sound?*

Numerical simulation [Friesecke and Matthies (2002)] supports that this should be true.

The results of Section 4.2 are, actually, modifications of those presented in Sections 3.2 and 3.3. They shows that different types of travelling waves may occur in FPU lattices. If the potential of interaction has more complicated shape, the situation is completely unclear. For instance, suppose that $V(r)$ has at least two local (or global) maxima at $r = r_1$ and $r = r_2$, and a local minimum point r_0 in between. Applying Theorem 3.7, we see that, under some mild assumptions on the potential near $r = r_0$, for every $c > c_0 = V'(0)$ and c close to c_0 there exists a solitary wave that looks like $r_0 t + c_\pm$ near $\pm\infty$, with some constants $c_\pm$. However, the following problem is completely open.

Problem 4.7 *Does there exist a travelling wave that connects the equilibria $r = r_1$ and r_2, i. e. looks like $r_1 t + c_-$ (resp., $r_2 t + c_-$) at negative infinity and $r_2 t + c_+$ (resp., $r_1 t + c_+$) at positive infinity?*

Such waves can be considered as *travelling transition layers*. Problem 4.7 is inspired by the results of [Rabinowitz and Stredulinsky (2003)].

The result on exponential decay of solitary waves (Section 4.5) is a simplified and less precise version of that obtained in [Friesecke and Pego (1999)] in which one can find more information.

Systems of nonlinear oscillators considered in Section 4.6 seem to be similar to and even simpler than FPU lattices. Nevertheless, not so many rigorous results about travelling waves in such systems are known. We mention here the paper [Iooss and Kirschgässner (2000)] in which bifurcation of such waves is considered. The following problem seems to be interesting.

Problem 4.8 *Do there exist travelling waves with different behavior at $\pm\infty$ in the case when the anharmonic potential has many (local) extreme points? A prototypical example is the Frenkel-Kontorova potential.*

In Chapters 3 and 4 we consider only monoatomic lattices. However, N-atomic FPU lattices also may support travelling waves. Some results on periodic waves in diatomic lattices can be found in [Georgieva et. al (1999); Georgieva et. al (2000)], but the problem is still not studied in details, especially in the case of solitary waves. Also we believe that the variational approach presented here can be extended to the case of two- and N-dimensional lattices, as well as to the case of lattices with second nearest neighbor interaction.

Finally, we would like to point out one of the most challenging problems that concern lattice travelling waves – stability. In the series of papers [Friesecke and Pego (1999); Friesecke and Pego (2002); Friesecke and Pego (2004a); Friesecke and Pego (2004b)] G. Friesecke and R. Pego develop an interesting approach to this problem and obtain some results about stability of FPU solitary waves, in particular, near sonic waves. Nevertheless, the problem is still far to be understood well.

Appendix A

Functional Spaces

We recall here definitions of basic functional spaces just to fix the notation. We consider spaces of real valued functions. However, each such a space has its natural "complex valued" counterpart for which we use the same notation.

A.1 Spaces of sequences

Two-sided sequences of real numbers are considered as real valued functions on the set of integers. We write

$$u = \{u(n)\}_{n \in \mathbb{Z}} = \{u(n)\}$$

for such a sequence. Let $p \geq 1$. We denote by l^p the vector space of all sequences $u = \{u(n)\}$ such that the norm

$$\|u\|_{l^p} = \left[\sum_{n=-\infty}^{\infty} |u(n)|^p\right]^{1/p}$$

is finite. By l^∞ we denote the vector space of all bounded sequences endowed with the norm

$$\|u\|_{l^\infty} = \sup_{n \in \mathbb{Z}} |u(n)|.$$

It is known that l^p, $1 \leq p \leq \infty$, is a Banach space. The space l^p is reflexive if $1 < p < \infty$, and its dual space, $(l^p)'$, is identified with $l^{p'}$,

$$\frac{1}{p} + \frac{1}{p'} = 1,$$

via the bilinear form

$$(u, v) = \sum_{n=-\infty}^{\infty} u(n)\, v(n). \qquad (A.1)$$

This means that every $v \in l^{p'}$ generates a bounded linear functional

$$f_v(u) = (u, v)$$

and the map $v \mapsto f_v$ is a linear isometric isomorphism of $l^{p'}$ onto $(l^p)'$ provided $1 < p < \infty$. In fact, one has the following Hölder inequality

$$|(u, v)| \leq \|u\|_{l^p} \cdot \|v\|_{l^{p'}} \qquad (A.2)$$

for all $p \in [1, \infty]$ (we set $1' = \infty$ and $\infty' = 1$). The space l^2 is a Hilbert space with respect to the inner product defined by Eq. (A.1).

By l_0 we denote the vector space of all finite sequences, i. e. sequences $u = \{u(n)\}$ such that

$$\operatorname{supp} u = \{n \in \mathbb{Z} : u(n) \neq 0\}$$

is a finite set. Obviously, l_0 is a dense subspace of l^p, $1 \leq p < \infty$. Recall that

$$l^p \subset l^q, \quad 1 \leq p \leq q \leq \infty,$$

and the embedding is continuous. Moreover, the embedding is dense if $q < \infty$.

A.2   Spaces of functions on real line

We denote by $C([a, b])$ the Banach space of all continuous functions endowed with the standard *supremum* norm

$$\|f\|_C = \sup_{x \in [a, b]} |f(x)|,$$

while $C((a, b))$ stands for the vector space of all continuous functions on the open interval (a, b). The last space is a Fréchet space with the topology of uniform convergence on compact subsets of (a, b).

For any natural number n we denote by $C^n([a, b])$ the space of all n times continuously differentiable functions on $[a, b]$. This is a Banach space with the norm

$$\|f\|_{C^n} = \sum_{k=0}^{n} \|f^{(n)}\|_C.$$

Denote by $C^n((a,b))$ the Fréchet space of all n times continuously differentiable functions on (a,b). We set

$$C^\infty([a,b]) = \bigcap_{n \geq 1} C^n([a,b])$$

and

$$C^\infty((a,b)) = \bigcap_{n \geq 1} C^n((a,b)).$$

These are the spaces of all infinitely differentiable functions on $[a,b]$ and (a,b), respectively. We also denote by $C_0^\infty((a,b))$ the vector space of all compactly supported infinitely differentiable functions on (a,b).

We denote by $C_b(\mathbb{R})$ the Banach space of all bounded continuous functions on $\mathbb{R}$ equipped with the norm

$$\|f\|_{C_b} = \sup_{x \in \mathbb{R}} |f(x)|.$$

Let us also denote by $C_0(\mathbb{R})$ a closed subspace of $C_b(\mathbb{R})$ that consist of all functions vanishing at infinity, i. e.

$$\lim_{x \to \infty} f(x) = 0.$$

Let $L^p(a,b)$, $1 \leq p < \infty$, be the Banach space of all Lebesgue measurable functions on (a,b) with finite norm

$$\|f\|_{L^p} = \left(\int_a^b |f(x)|^p dx \right)^{1/p}.$$

By $L^\infty(a,b)$ we denote the Banach space of all essentially bounded measurable functions, endowed with the norm

$$\|f\|_{L^\infty} = \operatorname*{ess\,sup}_{x \in (a,b)} |f(x)|.$$

If $1 < p < \infty$, the space $L^p(a,b)$ is reflexive and its dual space is identified with the space $L^{p'}(a,b)$, $p^{-1} + (p')^{-1} = 1$. The duality between these two spaces is given by the following bilinear form

$$(f,g) = \int_a^b f(x)\,g(x)\,dx. \tag{A.3}$$

Recall that

$$|(f,g)| \leq \|f\|_{L^p} \cdot \|g\|_{L^{p'}} \tag{A.4}$$

(the Hölder inequality) for all $p \in [1, \infty]$. The space $L^2(a,b)$ is a Hilbert space with the inner product (A.3).

If (a,b) is a finite interval, then

$$L^p(a,b) \subset L^q(a,b), \quad 1 \leq q \leq p \leq \infty,$$

and the embedding is continuous and dense. Moreover, the embedding $C([a,b]) \subset L^p(a,b)$ is continuous. It is also dense provided $p < \infty$.

Local Lebesgue spaces are defined as follows. The space $L^p_{loc}(a,b)$, $1 \leq p \leq \infty$, consists of all measurable functions f such that the restriction of f to every relatively compact subinterval $(\alpha, \beta) \subset (a,b)$ belongs to $L^p(\alpha, \beta)$. Endowed with the topology of L^p-convergence on relatively compact subintervals, $L^p_{loc}(a,b)$ is a Fréchet space.

Now let us recall basic facts about *Sobolev spaces*. Denote by $H^1(a,b)$ the space of all functions $u \in L^2(a,b)$ such that the weak derivative, u', of u belongs to $L^2(a,b)$. Recall that a function v is called the *weak derivative* of u if

$$\int_a^b u\,\varphi'\,dx = -\int_a^b v\,\varphi\,dx$$

for all $\varphi \in C_0^\infty((a,b))$. The space $H^1(a,b)$ is a Hilbert space with the norm

$$\|u\|_{H^1} = \left(\|u\|_{L^2}^2 + \|v'\|_{L^2}^2\right)^{1/2}. \tag{A.5}$$

The corresponding inner product in $H^1(a,b)$ is

$$(u_1, u_2)_{H^1} = (u_1, u_2) + (u'_1, u'_2).$$

If (a,b) is a finite interval, $H^1(a,b)$ coincides with the closure of $C^1([a,b])$ with respect to the norm defined by (A.5). The closure of $C_0^\infty((a,b))$ with respect to (A.5) is denoted by $H_0^1(a,b)$. This is a closed linear subspace of $H^1(a,b)$. Recall that $H^1(\mathbb{R}) = H_0^1(\mathbb{R})$.

The local space $H^1_{loc}(\mathbb{R})$ consists of all functions $u \in L^2_{loc}(\mathbb{R})$ such that for any $\varphi \in C_0^\infty(\mathbb{R})$ the function $\varphi \cdot f$ belongs to $H^1(\mathbb{R})$. In other words, $u \in L^2_{loc}(\mathbb{R})$ belongs to $H^1_{loc}(\mathbb{R})$ if and only if the restriction of u to each finite interval (a,b) is an element of $H^1(a,b)$. The space $H^1_{loc}(\mathbb{R})$ is a Fréchet space with the topology of H^1 convergence on any finite interval.

The following result is a particular case of the *Sobolev embedding theorem*.

Theorem A.1

(a) Let (a,b) be a finite interval. Any function $u \in H^1(a,b)$ coincides almost everywhere with an absolutely continuous function on $[a,b]$. The natural embedding $H^1(a,b) \subset C([a,b])$ is compact.

(b) $H^1(\mathbb{R})$ is embedded continuously into $L^p(\mathbb{R})$, $2 \leq p \leq \infty$ and $C_0(\mathbb{R})$. In particular,

$$\lim_{t \to \pm\infty} u(t) = 0$$

for every $u \in H^1(\mathbb{R})$.

The assertion concerning $C_0(\mathbb{R})$ is probably less known. Therefore, we sketch its proof. Let $u \in H^1(\mathbb{R})$. Since $u \in L^2(\mathbb{R})$, there exists a sequence $t_n \to +\infty$ such that $u(t_n) \to 0$. For $x \geq t_n$ the identity

$$\frac{1}{2}|u(x)|^2 - \frac{1}{2}|u(t_n)|^2 = \frac{1}{2}\int_{t_n}^x (u^2(t))' dt = \int_{t_n}^x u(t)\, u'(t)\, dt$$

yields

$$|u(x)|^2 \leq |u(t_n)| + 2\left(\int_{t_n}^x u^2(t)\, dt\right)^{1/2} \left(\int_{t_n}^x (u'(t))^2 dt\right)^{1/2}$$

and we obtain that

$$\lim_{x \to +\infty} u(x) = 0.$$

Similarly,

$$\lim_{x \to -\infty} u(x) = 0.$$

Remark A.1 Certainly, for any finite interval the embedding

$$H^1(a,b) \subset L^p(a,b), \quad 1 \leq p < \infty,$$

is compact.

The definitions of the functional spaces we consider extend naturally to the case of vector valued functions. In particular, for any Banach space E, there are well-defined spaces $l^p(E)$, $C([a,b],E)$, $C^n([a,b],E)$, $C_b(\mathbb{R},E)$, $L^p(a,b;E)$ of E-valued functions. One considers also the space $H^1(a,b;H)$ of H-valued functions, where H is a Hilbert space. The statements of Theorem A.1 are still valid in this case, with only one exception: the embedding

$$H^1(a,b;H) \subset C([a,b],H)$$

is compact if and only if the space H is finite dimensional.

References: [Adams and Fournier (2003); Dunford and Schwartz (1988a); Evans and Gariepy (1992); Lions and Magenes (1972)].

Appendix B

Concentration Compactness

Concentration compactness is a powerful techniques for studying variational problems without compactness introduced in [Lions (1984)]. We sketch here few technical lemmas.

Lemma B.1 *Suppose that $\{\rho_k\}$ is a sequence of nonnegative functions in $L^1(\mathbb{R})$ such that $\|\rho_k\|_{L^1} = \lambda > 0$. Then, after passing to a subsequence, one of the three following statements holds true:*

(i) *(concentration) there is a sequence $\{y_k\} \subset \mathbb{R}$ such that for every $\varepsilon > 0$ there exists $R > 0$ with the property that*

$$\int_{y_k-R}^{y_k+R} \rho_k(x)\,dx \geq \lambda - \varepsilon;$$

(ii) *(vanishing)*

$$\limsup_{k \to \infty} \sup_{y \in \mathbb{R}} \int_{y_k-R}^{y_k+R} \rho_k(x)\,dx = 0$$

for all $R > 0$;

(iii) *(dichotomy) there exist $\alpha \in (0, \lambda)$ and sequences of compactly supported nonnegative functions $\{\rho_k^{(1)}\}, \{\rho_k^{(2)}\} \subset L^1(\mathbb{R})$ such that*

$$\operatorname{dist}\left[\operatorname{supp}\left(\rho_k^{(1)}\right), \operatorname{supp}\left(\rho_k^{(2)}\right)\right] \to \infty \text{ as } k \to \infty,$$

$$\lim_{k \to \infty} \left\|\rho_k - \left(\rho_k^{(1)} + \rho_k^{(2)}\right)\right\|_{L^1} = 0,$$

$$\lim_{k \to \infty} \left\|\rho_k^{(1)}\right\|_{L^1} = \alpha$$

and
$$\lim_{k\to\infty} \|\rho_k^{(2)}\|_{L^1} = \lambda - \alpha.$$

Proof. See [Lions (1984); Chabrowski (1997)].

Lemma B.2 *Let $r > 0$. If $\{u_k\}$ is a bounded sequence in $H^1(\mathbb{R})$ and if*
$$\sup_{y\in\mathbb{R}} \int_{y-r}^{y+r} |u_k(x)|^2 dx \to 0, \quad k \to \infty,$$
then $u_k \to 0$ in $L^p(\mathbb{R})$ for all $p > 2$.

Proof. See [Lions (1984)].

Now we present discrete versions of Lemmas B.1 and B.2 (see [Pankov and Zakharchenko (2001)]).

Lemma B.3 *Let $\{v_k\}$ be a sequence of nonnegative elements of l^1 such that $\|v_k\|_{l^1} \to \lambda > 0$. Then there exists a subsequence, still denoted by v_k, such that one of the three following possibilities holds:*

(i) *(concentration) there is a sequence $\{m_k\}$ of integers such that for every $\varepsilon > 0$ there exists an integer $r > 0$ with the property that*
$$\sum_{n=m_k-r}^{m_k+r} v_k(n) \geq \lambda - \varepsilon;$$

(ii) *(vanishing)*
$$\lim_{k\to\infty} \|v_k\|_{l^\infty} = 0;$$

(iii) *(dichotomy) there exist $\alpha \in (0, \lambda)$ and two sequences $\{v_k^{(1)}\}$, $\{v_k^{(2)}\} \subset l_0$ such that*
$$\text{dist}\left[\text{supp}\,(v_k^{(1)}), \text{supp}\,(v_k^{(2)})\right] \to \infty \text{ as } k \to \infty,$$
$$\lim_{k\to\infty} \left\|v_k - (v_k^{(1)} + v_k^{(2)})\right\|_{l^1} = 0,$$
$$\lim_{k\to\infty} \|v_k^{(1)}\|_{l^1} = \alpha$$

and
$$\lim_{k\to\infty} \|v_k^{(2)}\|_{l^1} = \lambda - \alpha.$$

Lemma B.4 *Let $\{v_k\}$ be a bounded sequence in l^2 such that*
$$\lim_{k\to\infty} \|v_k\|_{l_\infty} = 0.$$
Then $v_k \to 0$ in l^p, $2 < p \leq \infty$.

Appendix C

Critical Point Theory

C.1 Differentiable functionals

Let φ be a real valued functional on a Banach space X. We say that φ has a *Gateaux derivative* $f \in X'$ at $u \in X$ if for every $h \in X$

$$\lim_{t \to \infty} \frac{1}{t} \left[\varphi(u + th) - \varphi(u) - \langle f, th \rangle \right] = 0.$$

Here, as usual $\langle f, h \rangle$ stands for the value of the linear functional $f \in X'$ at $h \in X$, i. e. $\langle \cdot, \cdot \rangle$ is the canonical bilinear form on $X' \times X$. The Gateaux derivative of f at u is denoted by $\varphi'(u)$.

The functional φ has a *Fréchet derivative* $f' \in X$ at $u \in X$ if

$$\lim_{h \to 0} \frac{1}{\|h\|} \left[\varphi(u + h) - \varphi(u) - \langle f, h \rangle \right] = 0. \tag{C.1}$$

Any Fréchet derivative is a Gateaux derivative. Therefore, for Fréchet derivatives we use the same notation as for Gateaux derivatives.

If X is a Hilbert space with the inner product $(\cdot, \cdot)$ and φ has a Gateaux derivative at $u \in X$, the *gradient*, $\nabla \varphi(u)$, is defined by

$$\left(\nabla \varphi(u), h \right) = \langle \varphi'(u), h \rangle \quad \forall h \in X.$$

The functional φ is said to be C^1 if the Fréchet derivative φ' exists and is continuous, i. e. the map $u \mapsto \varphi'(u)$ from X into X' is continuous. Actually, if φ has a continuous Gateaux derivative on X, then φ is C^1.

We say that the functional φ of class C^1 has a *second Gateaux derivative*

$L \in L(X, X')$ at $u \in X$ if, for every $h, v \in X$,

$$\lim_{t \to 0} \frac{1}{t} \langle \varphi'(u+th) - \varphi'(u) - tLh, v \rangle = 0.$$

The second Gateaux derivative at u is denoted by $\varphi''(u)$. Here $L(E, F)$ denotes the Banach space of all linear bounded operators from E into F.

The functional φ has a *second Fréchet derivative* $L \in L(X, X')$ at $u \in X$ if

$$\lim_{h \to 0} \frac{1}{\|h\|} \left(\varphi'(u+h) - \varphi'(u) - Lh \right) = 0.$$

Certainly, any second Fréchet derivative is a second Gateaux derivative and we apply the same notation $\varphi''(u)$.

The functional φ is said to be C^2 if the second Fréchet derivative $\varphi''(u)$ exists and is continuous in $u \in X$. In fact, it is enough to assume that the second Gateaux derivative exists and is continuous.

A point $u \in X$ is called a *critical point* of φ if φ has the Gateaux derivative $\varphi'(u)$ and $\varphi'(u) = 0$. The value of the functional φ at a critical point is called a *critical value* of φ.

C.2 Mountain pass theorem

Let φ be a C^1 functional on a Banach space X. We say that φ satisfies the *Palais-Smale condition* (for shortness, *condition (PS)*) if any sequence $\{u_n\}$ in X such that the sequence $\{\varphi(u_n)\}$ is bounded and $\varphi'(u_n) \to 0$ contains a convergent subsequence.

The simplest mountain pass theorem reads (see [Rabinowitz (1986); Willem (1996)])

Theorem C.1   *Let φ be a C^1 functional on a Banach space X. Assume that there exist $e \in X$ and $r > 0$ such that $\|e\| > r$,*

$$\beta := \inf_{\|u\|=r} \varphi(u) > \varphi(0) \geq \varphi(e), \qquad (C.2)$$

and φ satisfies the Palais-Smale condition. Let

$$b := \inf_{\gamma \in \Gamma} \max_{t \in [0,1]} \varphi(\gamma(t)) \qquad (C.3)$$

where
$$\Gamma := \left\{\gamma \in C([0,1], X) : \gamma(0) = 0, \gamma(1) = e\right\}. \tag{C.4}$$
Then b is a critical value of φ and $b \geq \beta$.

The following theorem of mountain pass type can be found in [Berestycki et. al (1995)].

Theorem C.2 *Under the assumptions of Theorem C.1 let $P : X \to X$ be a continuous mapping such that*
$$\varphi(Pu) \leq \varphi(u) \quad \text{for all } u \in X,$$
$P(0) = 0$ and $P(e) = e$. Then there exists a critical point $u \in \overline{PX}$ (the closure of PX) of φ with the critical value b.

Remark C.1 Typically, in applications the functional φ has a local *minimum* at 0, a trivial critical point. In this case Theorem C.1 gives the existence of a nontrivial critical point.

There is also a version of the mountain pass theorem without condition (PS).

Theorem C.3 *Suppose that all the assumptions of Theorem C.1 except condition (PS) are satisfied. Let b be defined by (C.3), (C.4). Then there exists a sequence $\{u_n\} \subset X$ such that*
$$\lim_{n \to \infty} \varphi(u_n) = b$$
and
$$\lim_{n \to \infty} \varphi'(u_n) = 0,$$
i. e. $\{u_n\}$ is a Palais-Smale sequence at the level b.

We say that the functional φ possesses the *mountain pass geometry* if φ satisfies the assumptions of Theorem C.1 except (PS). The number defined by Eq. (C.4) is called the *mountain pass value* (level).

C.3 Linking theorems

Let H be a Hilbert space decomposed into the direct orthogonal sum $H = Y \oplus Z$. Let $\varrho > r > 0$ and let $z \in Z$ be such that $\|z\| = r$. Define
$$M := \left\{u = y + \lambda z : y \in Y, \|u\| \leq \varrho, \lambda \geq 0\right\}$$

and

$$M_0 := \{u = y + \lambda z : y \in Y, \|u\| = \varrho \text{ and } \lambda \geq 0, \text{ or } \|u\| \leq \varrho \text{ and } \lambda = 0\},$$

i. e. M_0 is the boundary ∂M of M. Let

$$N := \{u \in Z : \|u\| = r\}.$$

Consider a functional φ on H and suppose that

$$\beta := \inf_{u \in N} \varphi(u) > \alpha := \sup_{u \in M_0} \varphi(u). \tag{C.5}$$

In this situation we say that the functional φ possesses the *linking geometry*.

Theorem C.4 *Suppose that the functional φ of class C^1 satisfies the Palais-Smale condition and possesses the linking geometry. Let*

$$b := \Big\{ \inf_{\gamma \in \Gamma} \sup_{u \in M} \varphi(\gamma(u)) \Big\}, \tag{C.6}$$

where

$$\Gamma := \{\gamma \in C(M; H) : \gamma = id \text{ on } M_0\}.[2] \tag{C.7}$$

If $\dim Y < \infty$, then b is a critical value of φ and

$$\beta \leq b \leq \sup_{u \in M} \varphi(u).$$

Remark C.2 The additional assumption $\dim Y < \infty$ in Theorem C.4 implies that the *suprema* in (C.5) and (C.6) are attained and, hence, can be replaced by *maxima*.

Under further restrictions on the functional φ one can drop the assumption $\dim Y < \infty$ in the previous theorem.

Theorem C.5 *Suppose that the functional φ possesses the linking geometry and satisfies the following assumptions:*

(i) $\varphi(u) = \dfrac{1}{2}(Au, u) + b(u)$, *where* $u = u_1 + u_2 \in Y \oplus Z$, $Au = A_1 u_1 + A_2 u_2$, $A_1 : Y \to Y$ *and* $A_2 : Z \to Z$ *are linear bounded self-adjoint operators;*

(ii) *the functional φ is weakly continuous and uniformly differentiable on bounded sets, i. e. the limit in (C.1) is uniform in $u \in B$ for every bounded set $B \subset H$;*

---

[2] As usual, *id* denotes the identity map.

(iii) if $\{u_n\} \in H$ is a sequence such that $\varphi(u_n)$ is bounded above and $\varphi'(u_n) \to 0$, then it is bounded.

Then b defined by (C.6) and (C.7) is a critical value and

$$\beta \leq b \leq \sup_{u \in M} \varphi(u).$$

For the proof see [Benci and Rabinowitz (1979)].

Remark C.3 There is an extension of Theorem C.5 in the spirit of Theorem C.3 (see [Kryszewski and Szulkin (1998)] and [Willem (1996)]; a simplified proof is contained in [Bartsch and Ding (1999)]).

Appendix D

Difference Calculus

Let l be the vector space of all real, or complex, two-sided sequences, *i. e.* functions on $\mathbb{Z}$. *Difference operations* are linear maps from l into itself. The simplest are *left* and *right shifts* defined by

$$(S^-u)(n) := u(n-1) \tag{D.1}$$

and

$$(S^+u)(n) := u(n+1), \tag{D.2}$$

respectively.

Operations of *left* and *right differences* are defined by

$$(\partial^- u)(n) := u(n) - u(n-1) \tag{D.3}$$

and

$$(\partial^+ u)(n) := u(n+1) - u(n), \tag{D.4}$$

respectively. In other words,

$$\partial^+ = I - S^-$$

and

$$\partial^- = S^+ - I.$$

Here I stands for the identity operation.

The following identity is useful

$$\partial^-(a\partial^+ u) = \partial^+\left[(S^-a)\partial^- u\right], \tag{D.5}$$

where $a \in l$. We mention also the product rules

$$\partial^+(fg)(n) = f(n)\left(\partial^+g\right)(n) + g(n+1)\left(\partial^+f\right)(n),$$
$$\partial^-(fg)(n) = f(n)\left(\partial^-g\right)(n) + g(n-1)\left(\partial^-f\right)(n).$$

We do not consider general difference operations here. Instead, we restrict ourself to the case of *divergence form difference operations of second order*. These are operations of the form

$$Ru = \partial^-\left(a\partial^+u\right) + bu \tag{D.6}$$

where $a \in l$ and $b \in l$ are given sequences (coefficients). Alternatively,

$$Ru = \partial^+\left[(S^-a)\partial^-u\right] + bu, \tag{D.7}$$

as it follows from Eq. (D.5).

Abel's summation by parts formula reads

$$\sum_{j=m}^{n} g(j)\left(\partial^+f\right)(j) = g(n)\,f(n+1) - g(m-1)\,f(m)$$
$$-\sum_{j=m}^{n} \left(\partial^-g\right)(j)\,f(j). \tag{D.8}$$

It implies that the operations ∂^+ and $-\partial^-$ are formally adjoint, *i. e.*

$$\sum_{j} g(j)\left(\partial^+f\right)(j) = \sum_{j}\left(-\partial^-g\right)(j)\,f(j)$$

for all $f \in l$ and $g \in l_0$ (see Appendix A.1 for the definition of l_0). As consequence, if the coefficient sequences in Eq. (D.6) are real, then the operation R is formally self-adjoint, *i. e.*

$$\sum_{j} g(j)\,(Rf)(j) = \sum_{j}(Rg)(j)\,f(j)$$

for all $f \in l$ and $g \in l_0$.

The operation

$$\Delta := \partial^+\partial^- = \partial^-\partial^+$$

is called (one dimensional) *discrete Laplacian*.

References: [Atkinson (1964); Gel'fond (1971); Samarskii (2001); Teschl (2000)].

Bibliography

Ablowitz, M. J. and Segur, H. (1981). *Solitons and the Inverse Scattering Transform*, SIAM, Philadelphia.

Aceves, A. B. (2000). Optical gap solitons: Past, present, and future; theory and experiments, *Chaos*, **10**, pp. 584–589.

Adams, R. A. and Fournier, J. J. F. (2003). *Sobolev Spaces, 2nd Ed.*, Acad. Press, Amsterdam e.a.

Ambrosetti, A. and Prodi, G. (1995). *A Primer of Nonlinear Analisis (Corrected reprint of the 1993 original)*, Cambridge Univ. Press, Cambridge.

Arioli, G. and Chabrowski, J. (1997). Periodic motions of a dynamical system consisting of an infinite lattice of particles, *Dyn. Syst. Appl.*, **6**, pp. 387–395.

Arioli, G. and Gazzola, F. (1995). Existence and numerical approximation of periodic motions of an infinite lattice of particles, *Z. Angew. Math. Phys.*, **46**, pp. 898–912.

Arioli, G. and Gazzola, F. (1996). Periodic motions of an infinite lattice of particles with nearest neigbour interaction. *Nonlin. Anal., TMA*, **26**, pp. 1103–1114.

Arioli, G., Gazzola, F. and Terracini, S. (1996). Multibump periodic motions of an infinite lattice of particles, *Math. Z.*, **223**. pp. 627–642.

Arioli, G., Koch, H. and Terracini, S. (2003). *The Fermi-Pasta-Ulam model: Periodic solutions*, preprint.

Arioli, G. and Szulkin, A.(1997). Periodic motions of an infinite lattice of particles: the strongly indefinite case, *Ann. Sci. Math. Québec*, **22**, pp. 97–119.

Arnol'd, V. I. (1997). *Mathematical Methods of Classical Mechanics, 4th printing*, Springer, New York.

Atkinson, F.V. (1964). *Discrete and Continious Boundary Problems*, Acad. Press, New York – London.

Aubry, S. (1997). Breathers in nonlinear lattices: Existence, linear stability and quantization, *Physica D*, **103**, pp. 201–250.

Bak, S. (2004). Constrained minimization method in the problem on oscillations of a chain of nonlinear oscillators, *Math. Phys., Anal., Geom.*, **11**, no 3, pp. 263–273 (in Russian).

Bak, S. and Pankov, A. (2004). On periodic oscillations of an infinite chain of linearly coupled nonlinear oscillators, *Repts Net. Acad. Sci. Ukraine*, no 3 (in Russian).

Bak, S. and Pankov, A. (to appear). On dynamical equations of a system of linearly coupled nonlinear oscillators, *Ukr. Math. Journ.* (in Russian).

Bartsch, T. and Ding, Y. (1999). On a nonlinear Schrödinger equation with periodic potential, *Math. Ann.*, **313**, pp. 15–37.

Benci, V. and Rabinowitz, P. H. (1979). Critical point theorems for indefinite functionals, *Invent. Math.*, **52**, pp. 241–273.

Berestycki, H., Capuzzo-Dolcetta, I. and Nirenberg, L. (1995). Variational methods for indefinite superlinear homogeneous elliptic problems, *Nonlin. Diff. Equat. Appl.*, **2**, pp. 553–572.

Berezanskii, Y. M. (1968). *Expansions in Eigenfunctions of Selfadjoint Operators*, Amer. Math. Soc., Providence, R. I.

Berezin, F. A. and Shubin, M. A. (1991). *The Schödinger Equation*, Kluwer, Dordrecht.

Berger, M. S. (1977). *Nonlinearity and Functional Analysis: Lectures on Nonlinear problems in Mathematical Analysis*, Acad. Press, New York.

Braun, O. M. and Kivshar, Y. S. (1998). Nonlinear dynamics of the Frenkel-Kontorova model, *Physics Repts.*, **306**, pp. 1-108.

Braun, O. M. and Kivshar, Y. S. (2004). *The Frenkel-Kontorova model*, Springer-Verlag, Berlin.

Bronski, J. S., Segev, M. and Weinstein, M. I. (2001). Mathematical frontiers in optical solitons, *Proc. Nat. Acad. Sci. USA*, **98**, pp. 12872–12873.

Chabrowski, J. (1997). *Variational Methods for Potential Operator Equations*, de Gruyter, Berlin.

Chernoff, P. R. and Marsden, J. E. (1974). *Properties of Infinite Dimentional Hamiltonian Systems*, Lecture Notes Math., **425**, Springer, Berlin.

Chicone, C. and Latushkin Y. (1999). *Evolution Semigroups in Dynamical Systems and Differential Equations*, Amer. Math. Soc., Providence, R. I.

Coti Zelati, V. and Rabinowitz, P. H. (1991). Homoclinic orbits for second order Hamiltonian systems possessing superquadratic potentials, *J. Amer. Math. Soc.*, **4**, pp. 693–727.

Coti Zelati, V. and Rabinowitz, P. H. (1992). Homoclinic type solutions for a semilinear elliptic PDE on $\mathbb{R}^n$, *Comm. Pure Appl. Math.*, **45**, pp. 1217–1269.

Dalec'kii, Ju. L. and Krein, M. G. (1974). *Stability of Solutions of Differential Equations in Banach Spaces*, Amer. Math. Soc., Providence, R. I.

Deimling, K. (1977). *Ordinary Differential Equations in Banach Spaces*, Lect. Notes Math., **596**, Springer, Berlin.

Deift, P., Kriecherbauer, T. and Venakides S. (1995). Forced lattice vibrations, I, II, *Commun. Pure Appl. Math.*, **48**, pp. 1187–1249, pp. 1251–1298.

de Sterke, C. M. and Sipe, J. E. (1994). Gap solitons, *Progress in Optics, vol. 33* (Ed. E. Wolf), North-Holland, Amsterdam, pp. 203–260.

Duncan, D. B. and Wattis, J. A. D. (1992). Approximations of solitary waves on lattices using weak and variational formulations, *Chaos, Solitons and*

Fractals, **2**, pp. 505–518.

Dunford, N. and Schwartz, J. T. (1988a). *Linear Operators, Part I. General Theory*, Wiley, New York.

Dunford, N. and Schwartz, J. T. (1988b). *Linear Operators, Part II. Spectral Theory. Selfadjoint Operators in Hilbert Space*, Wiley, New York.

Edmuns, D. E. and Evans, W. D. (1987). *Differential Operators*, Oxford Univ. Press.

Eilbeck, J. C. (1991). Numerical studies of solitons on lattices, in "Nonlinear Coherent Structures in Physics and Biology", *Lect. Notes Phys.*, **393**, Springer, Berlin, pp. 143–150.

Eilbeck, J. C. and Flesch, R. (1990). Calculation of families of solitary waves on discrete lattices, *Phys. Lett. A*, **149**, pp. 200–202.

Ekeland, I. (1990). *Convexity Methods in Hamiltinian Mechanics*, Springer, Berlin.

Evans, L. and Gariepy, R.E. (1992). *Theory and Fine Properties of Functions*, CRC Press, Boca Raton.

Filip, A.-M. and Venakides, S. (1999). Existence and modulation of traveling waves in particle chains, *Commun. Pure Appl. Math.*, **52**, pp. 693–735.

Fermi, E., Pasta, J. and Ulam, S. (1955). Studies of nonlinear problems, *Los Alamos Sci. Lab. Rept.*, **LA-1940**; Reprinted in *Lect. Appl. Math.*, **15** (1974), pp. 143–156.

Flach, S. and Willis, C. R. (1998). Discrete breathers, *Phys. Repts*, **295**, pp. 181–264.

Flytzanis, N., Pnevmatikos, S. and Peyrard M. (1989). Discrete lattice solitons: properties and stability, *J. Phys. A*, **22**, pp. 783–801.

Friesecke, G. and Matthies, K. (2002). Atomic scale localization of high-energy solitary waves on lattices, *Prysica D*, **171**, pp. 211–220.

Friesecke, G. and Pego, R. L. (1999). Solitary waves on FPU lattices, I. Qualitative properties, renormalization and continuum limit, *Nonlinearity*, **12**, pp. 1601–1627.

Friesecke G. and Pego R. L. (2002). Solitary waves on FPU lattices, II. Linear implies nonlinear stability, *Nonlinearity*, **15**, pp. 1343–1359.

Friesecke G. and Pego R. L. (2004a). Solitary waves on FPU lattices, III. Howland-type Floquet Theory, *Nonlinearity*, **17**, pp. 207–227.

Friesecke, G. and Pego R. L. (2004b). Solitary waves on FPU lattices, IV. Proof of stability at low energy, *Nonlinearity*, **17**, pp. 229–251.

Friesecke, G. and Wattis, J. A. D. (1994). Existence theorem for solitary waves on lattices, *Commun. Math. Phys.*, **161**, pp. 391–418.

Gel'fond, A. O. (1971). *Calculus of Finite Differences*, Hindustan Publ., Delhi.

Georgieva, A., Kriecherbauer, T. and Venakides S. (1999). Wave propagation and resonances in a one-dimensional nonlinear discrete medium, *SIAM J. Appl. Math.*, **60**, pp. 272–294.

Georgieva, A., Kriecherbauer, T. and Venakides, S. (2000). 1:2 resonance mediated second harmonic generation in a 1-D nonlinear discrete periodic medium, *SIAM J. Appl. Math.*, **61**, pp. 1802–1815.

Glazman, I. M. (1966). *Direct Methods of Qualitative Spectral Analysis of Singular*

Differential Operators, Daniel Davey & Co., New York.

Hartman, P. (2002). *Ordinary Differential Equations*, SIAM, Philadelphia, PA.

Hennig, D. and Tsironis, G. P. (1999). Wave transmission in nonlinear lattices, *Physics Repts*, **309**, pp. 333-432.

Hochstrasser, D., Mertens, F. G. and Buttner, H. (1989). An iterative method for the calculation narrow solitary excitations on atonic chains, *Physica D*, **35**, pp. 259-266.

Howland, J. S. (1974). Stationary scattering theory for time-dependent Hamiltonians, *Math. Ann.*, **207**, pp. 315-335.

Howland, J. S. (1979). Scattering theory for Hamiltonians periodic in time, *Indiana Univ. Math. J.*, **28**, pp. 471-494.

Iooss G. (2000). Traveling waves in the Fermi-Pasta-Ulam lattice, *Nonlinearity*, **13**, pp. 849-866.

Iooss, G. and Kirschgässner, K. (2000). Traveling waves in a chain of coupled nonlinear oscillators, *Commun. Math. Phys.*, **211**, pp. 439-464.

James, G. (2001). Existence of breathers on FPU lattices, *C. R. Acad. Sci. Paris*, **332**, pp. 581-586.

James, G. (2003). Center manifold reduction for quasilinear discrete systems, *J. Nonlinear Sci.*, **13**, pp. 27-63.

James, G. and Noble, P. (2004). Breathers on diatomic Fermi-Pasta-Ulam lattices, *Physica D*, **196**, pp. 124-171.

Kevrekidis, P. G. and Weinstein, M. I. (2003). Breathers on a background: periodic and quasiperiodic solutions of extended discrete nonlinear wave systems, *Math. Comp. Simul.*, **62**, pp. 65-78.

Kryszewski, W. and Szulkin, A. (1998). Generalized linking theorem with an application to semilinear Schrödinger equation, *Adv. Differ. Equat.*, **3**, pp. 441-472.

Kuchment, P. (1993). *Floquet Theory for Partial Differential Equations*, Birkhäuser, Basel.

Lanczos, C. (1961). *Linear Differential Operators*, Van Nostrand, London – New York.

Levitan, B. M. and Sargsyan, I. S. (1975). *Introduction to Spectral Theory: Self-adjoint Ordinary Differential Operators*, Amer. Math. Soc., Providence, R. I.

Levitan, B. M. and Sargsyan, I. S. (1991). *Sturm-Liouville and Dirac Operators*, Kluwer Acad. Publ., Dordrecht.

Li, Y. and Wang, Z.-Q. (2001). Gluing approximate solutions of minimum type on the Nehari manifold, USA-Chili Workshop on Nonlinear Analisis, *Electr. J. Differ. Eqns.*, Conf. **06**, pp. 215-223.

Lions, J. L. (1969). *Quelques méthodes de résolution des problèmes aux limites non lineaires*, Dunod, Paris.

Lions, J.-L. and Magenes, E. (1972). *Non-Homogeneous Boundary Value Problems and Applications, I*, Springer, New York.

Lions, P.-L. (1984). The concentration-compactness method in the calculus of variations. The locally compact case, I, II, *Ann. Inst. H. Poincaré, Anal. Non Lin.*, **1**, pp. 109-145, pp. 223-283.

Livi, R., Spicci, M, and MacKay, R. (1997). Breathers in a diatonic FPU chain, *Nonlinearity*, **10**, pp. 1421–1434.

MacKay, R. S. and Aubry, S. (1994) Proof of existence of breathers for time-reversible or Hamiltonian networks of weakly coupled oscillators, *Nonlinearity*, **7**, pp. 1623–1643.

Marsden, J. and Ratiu, T. S. (1989). *Introduction to Mechanics and Symmetry*, Springer, New York.

Mawhin, J. and Willem, M. (1989). *Critical Point Theory and Hamiltonian Systems*, Springer, New York.

Morgante, A. M., Johansson, M., Kopidakis, G. and Aubry, S. (2002). Standing waves instability in a chain of nonlinear coupled oscillators, *Physica D*, **162**, pp. 53–94.

Nehari, Z. (1960). On a class of nonlinear second-order differential equations, *Trans. Amer. Math. Soc.*, **95**, pp. 101–123.

Nehari, Z. (1961). Characteristic values associated with a class of non-linear second-order differential equations, *Acta Math.*, **105**, pp. 141–175.

Nirenberg, L. (2001). *Topics in Nonlinear Functional Analysis. (Revised reprint of the 1974 original)*, Amer. Math. Soc., Providence, R. I.

Omana, W. and Willem, M. (1992). Homoclinic orbits for a class of Hamiltinian systems, *Differ. Integr. Equat.*, **5**, pp. 1115–1120.

Pankov, A. (2000). *Introduction to Spectral Theory of Schrödinger Operators*, Lecture Notes, Giessen University.

Pankov, A. (2004). Periodic Nonlinear Schrödinger Equation with an Application to Photonic Crystals, *Acta Appl. Math.*, submitted (preprint version available at arXiv, math.AP/0404450).

Pankov, A. and Pflüger, K. (1999). Periodic and solitary travelling wave solutions for the generalized Kadomtsev-Petviashvili equation, *Math. Meth. Appl. Sci.*, **22**, pp. 733–752.

Pankov, A. and Pflüger, K. (2000a). On ground travelling waves for the generalized Kadomtsev-Petviashvili equation, *Math. Phys., Anal., Geom.*, **3**, pp. 593–609.

Pankov, A. and Pflüger, (2000b). Travelling waves in lattice dynamical systems, *Math. Meth. Appl. Sci.*, **23**, pp. 1223–1235.

Pankov, A. and Zakharchenko, N. (2001). On some discrete variational problems, *Acta Appl. Math.*, **65**, pp. 295–303.

Parton, D. N., Rich, R. and Vischer, W. M. (1967). Lattice termal conductivity in disordered harmonic and anharmonic crystal models, *Phys. Rev.*, **160**, pp. 706–711.

Peyrard, M., Pnevmatikos, S. and Flytzanis, N. (1986). Discreteness effects on non-topological kink soliton dynamics in nonlinear lattices, *Physica D*, **19**, pp. 268–281.

Poggi, P. and Ruffo, S. (1997). Exact solutions in the FPU oscillator chain, *Physica D*, **103**, pp. 251–272.

Rabinowitz, P. H. (1986). *Minimax Methods in Critical Point Theory with Applications to Differential Equations*, AMS Reg. Conf. Ser. Math., **65**, Amer. Math. Soc., Providence, R. I.

Rabiniwitz, P. H. (1990). Homoclinic orbits for a class of Hamiltonian systems, *Proc. Roy. Soc. Edinburgh*, **114**, pp. 33–38.

Rabinowitz, P. H. (1991). A note on a semilinear elliptic equation on $\mathbb{R}^n$, *Nonlinear Analysis (a tribute in honour of G. Prodi)*, Quad. Sc. Norm. Sup. Pisa, pp. 307–318.

Rabinowitz, P. H. (1993). Multibump solutions of a semilinear elliptic PDE on $\mathbb{R}^n$, *Degenerate diffusions (Minneapolis, MN, 1991)*, pp. 175–185, Springer, New York.

Rabinowitz, P. H. (1996). Multibump solutions of differential equations: an overview, *Chinese J. Math.*, **24**, pp. 1–36.

Rabinowitz, P. H. and Stredulinsky, E. (2003). Mixed states for an Allen-Cahn type equation, *Comm. Pure Appl. Math.*, **56**, pp. 1078–1134.

Rabinowitz, P. H. and Tanaka, K. (1991). Some results on connecting orbits for a class of Hamiltonian systems, *Math. Z.*, **206**, pp. 472–499.

Reed, M. (1976) *Abstract Non-Linear Wave Equation*, Lecture Notes Math., **507**, Springer, Berlin.

Reed, M. and Simon, B. (1975). *Methods of Modern Mathematical Physics. II. Fourier Analysis, Self-adjointess*, Acad. Press, New York – London.

Reed, M. and Simon, B. (1978). *Methods of Modern Mathematical Physics. IV. Analysis of Operators*, Acad. Press, New York – London.

Reed, M. and Simon, B. (1979). *Methods of Modern Mathematical Physics. III. Scattering Theory*, Acad. Press, New York – London.

Reed, M. and Simon, B. (1980). *Methods of Modern Mathematical Physics. I. Functional Analysis*, 2nd Ed., Acad. Press, New York.

Rosenau P. (1989). Dynamics of nonlinear mass string chains near the continuum limit, *Phys. Lett. A*, **118**, pp. 222–227.

Ruf, B. and Srikanth, P. N. (1994). On periodic motions of lattices of Toda type via critical point theory, *Arch. Ration. Mech. Anal.*, **126**, pp. 369–385.

Samarskii, A. A. (2001). *The Theory of Difference Schemes*, Marsel Dekker, New York.

Sattinger, D. (1968). On global solutions of nonlinear hyperbolic equaions, *Arch. Rat. Mech. Anal.*, **30**, pp. 148–172.

Schechter, M. (1981). *Operator Methods in Quantum Mechanics*, North Holland, New York.

Schechter, M. (1999). *Linking Methods in Critical Point Theory*, Birkhäuser, Boston, MA.

Smets, D. (1998). Traveling waves for an infinite lattice: multibump type solutions, *Topol. Meth. Nonlin. Anal.*, **12**, pp. 79–90.

Smets, D. and Willem, M. (1997). Solitary waves with prescribed speed on infinite lattices, *J. Funct. Anal.*, **149**, pp. 266–275.

Srikanth, P. N. (1998). On periodic motions of two-dinensional lattices, *Pitman Res. Notes Math. Ser.*, **377**, Longman, Harlow, pp. 118–122

Strauss, W. A. (1989). *Nonlinear wave equations*, Amer. Math. Soc., Providrnce, R. I.

Struwe, M. (2000). *Variational Methods*, 3rd Ed., Springer, Berlin.

Szulkin, A. and Zou, W. (2001). Homoclinic orbits for asymptotically linear

Hamiltonian systems, *J. Funct. Anal.*, **187**, pp. 25–41.

Tarallo, M. and Terracini, S. (1995). On the existence of periodic and solitary traveling waves in some nonlinear lattices, *Dynam. Syst. Appl.*, **4**, pp. 429–458.

Teschl, G. (2000). *Jacobi Operators and Completely Integrable Nonlinear Lattices*, Amer. Math. Soc., Providence, R. I.

Toda, M. (1989). *Theory of Nonlinear Lattices*, Springer, Berlin.

Treschev, D. (2004). Travelling waves in FPU lattices, *Discr. Cont. Dynam. Syst. A*, **11**, pp. 867–880.

Valkering, T. P. (1978). Periodic permanent waves in an anharmonic chain with nearest neighbour interaction, *J. Phys. A*, **11**, pp. 1885–1897.

Van Moerbeke, P. (1976). The specrum of Jacobi matrices, *Invent. Math.*, **37**, pp. 45–81.

Wattis, J. A. D. (1993a). Approximations to solitary waves on lattices, II: Quasicontinuum methods for fast and slow waves, *J. Phys. A: Math. Gen*, **26**, pp. 1193–1209.

Wattis, J. A. D. (1993b). Solitary waves on a two-dimentional lattice, *Phys. Scr.*, **5**, pp. 238–242.

Wattis, J. A. D. (1996). Approximations to solitary waves on lattices, III: the monoatonic lattice with second-neighbour interactions, *J. Phys. A: Math. Gen.*, **29**, pp. 8139–8157.

Weinstein, M. I. (1999). Exitation thresholds for nonlinear localized modes on lattices, *Nonlinearity*, **12**, pp. 673–691.

Willem, M. (1996). *Minimax Methods*, Birkhäuser, Boston.

Zeidler, E. (1985). *Nonlinear Functional Analysis and its Applications. III. Variational Methods and Optimization*, Springer, New York.

Zeidler, E. (1986). *Nonlinear Functional Analysis and its Applications. I. Fixed-Point Theorems*, Springer, New York.

Zeidler, E. (1988). *Nonlinear Functional Analysis and its Applications. IV. Applications to Mathematical Physics*, Springer, New York.

Zeidler, E. (1990a). *Nonlinear Functional Analysis and its Applications. II/A. Linear Monotone Operators*, Springer, New York.

Zeidler, E. (1990b). *Nonlinear Functional Analysis and its Applications. II/B. Nonlinear Monotone Operators*, Springer, New York.

Zeidler, E. (1995a). *Applied Functional Analysis. Applications to Mathematical Physics*, Springer, New York.

Zeidler, E. (1995b). *Applied Functional Analysis. Main Principles and their Applications*, Springer, New York.

Index

Abel's summation by parts formula, 186
average kinetic energy, 123
average potential energy, 123

band function, 17
Bloch condition, 16
Bloch eigenfunction, 17
Bloch wave, 17

chain of coupled nonlinear oscillators, 18
compression wave, 80
critical point, 180
critical value, 180

diatomic lattice, 8
difference operation, 185
difference operation in divergence form, 186
discrete φ^4-equation, 23
discrete breather, 29
discrete Laplacian, 7, 186
discrete nonlinear Klein-Gordon equation, 22
discrete nonlinear wave equation, 22
discrete wave equation, 7
dispersion relation, 17
DNKG equation, 22
DNW equation, 22

energy inequality, 133

expansion wave, 80

Floquet condition, 16
Floquet transform, 16
FPU α-model, 7
FPU β-model, 7
Fréchet derivative, 179
Frenkel-Kontorova model, 22
functional of class C^1, 179
functional of class C^2, 180

gap breather, 76
gap solution, 76
Gateaux derivative, 179
Gelfand transform, 16
geometrically distinct solutions, 63
gradient, 179

harmonic lattice, 12
homoclinic, 74

lattice with impurities, 8
left difference, 185
left shift, 185
Lennard-Jones potential, 8
linking geometry, 182
linking theorems, 181

monoatomic lattice, 8
mountain pass geometry, 181
mountain pass theorem, 180
mountain pass value, 181

multiatomic lattice, 8
multibump solution, 62

Nehari functional, 108
Nehari manifold, 108

Palais-Smale condition, 180
periodic ground wave, 106
periodic travelling wave, 78
plane wave, 18
profile, 77
profile function, 77
profile function for relative
  displacements, 78

quasimomentum, 16

relative displacement, 5
right difference, 185
right shift, 185

second Fréchet derivative, 180
second Gateaux derivative, 179
Sobolev embedding theorem, 173
Sobolev spaces, 172
solitary ground wave, 111
solitary travelling wave, 78
spectral band, 17
spectral gap, 17
speed of sound, 80
speed of wave, 77
subadditivity inequality, 132
subsonic periodic travelling wave, 88

Toda lattice, 8
travelling transition layer, 167
travelling wave, 77

weak derivative, 172

COMPARATIVE & VETERINARY MEDICINE

A GUIDE TO THE
RESOURCE LITERATURE

Compiled by
ANN E. KERKER and
HENRY T. MURPHY

The University of Wisconsin Press

*Publication of this book was supported in part
by NIH grant LM 00695 from the National Library of Medicine*

Published 1973
The University of Wisconsin Press
Box 1379, Madison, Wisconsin 53701
The University of Wisconsin Press, Ltd.
70 Great Russell Street, London

Copyright © 1973
The Regents of the University of Wisconsin System
All rights reserved

First Printing

Printed in the United States of America
George Banta Company, Inc., Menasha, Wisconsin

ISBN 0-299-06330-5; LC 72-7989

CONTENTS

Preface, xi
Abbreviations, xv

PART I MATERIALS OF GENERAL INTEREST   1

1 Indexes and Abstracts for the Periodical Literature, 3
   Retrospective Periodicals, 4
   Current Periodicals, 7
      Primary Interest, 7
      Cognate Interest, 10
   Current Awareness Services, 17

2 Bibliographies of Periodicals, Conferences and Congresses,
   Translations, and Special Subjects, 19
      Periodicals, 19
      Conferences and Congresses, 21
         Forthcoming Meetings, 21
         Proceedings Available, 21
         Retrospective Listings, 21
         Papers Presented at Meetings, 22
      Translations, 22
      Specialized Subject Bibliographies, 22

3 Periodicals, 27
      Primary Veterinary Medical Interest, 27
      Cognate Disciplines, 30
      School, Regional, State, and Commercial Publications, 37
      Additional Veterinary Medical Titles, 38
      Periodically Occurring Congresses and Conferences, 39

4 *Review Serials and Serial Monographs*, 40
   Review Serials, 41
   Serial Monographs, 45

5 *Reference Works*, 58
   Guides to the Literature, 59
   Dictionaries and Encyclopedias, 59
      English Language, General Biomedical, 59
      Foreign Language, 60
      Abbreviations and Synonyms, 61
      Nomenclature, 61
      Special Subject Dictionaries, 62
      Medical Terminology and Word Structure, 63
      General Encyclopedic Works, 63
   Biographical Tools and Directories, 63
   Histories and Source Books, 65
   Miscellany, 66
      Libraries and/or Information Centers and Their Services, 66
      Research Methodology, 68
      Research Support, 68
      Medical Writing, Speaking, Illustration, and Audiovisual
         Materials, 69
      Industrial Information, 70

6 *Handbooks and Manuals of Laboratory Technique*, 71
   General Biomedical Interest, 71
   Clinical Laboratory Methods, 72
   Microscopy and Microphotography, 74
   Microbiological and Parasitological Cultures, Techniques,
      and Classification, 75
   Histological and Histochemical Technique, Stains and
      Staining, 77
   Cell and Tissue Culture, 78
   Dissection, Autopsy, Necropsy, and Specimen Preparation, 79
   Instrumentation and Techniques, 80
      General, 80
      Electrocardiology, 82
      Electroencephalography, 82
      Electromyography and Electroneurography, 83
      Ultrasonics, 83

PART II SPECIFIC DISCIPLINES   85

7 *Anatomy*, 87
   General and Specific, 87
   Comparative, 88
   Developmental, Including Teratology and Aging, 91
   Histological and Cytological, 94

Ultrastructural, 97
Systematic, 98
   Musculoskeletal, 98
   Nervous and Sensory, 99
   Integumentary, 102
   Cardiovascular, 103
   Respiratory, 103
   Gastrointestinal, 103
   Genitourinary, 103
   Endocrine, 104
   Hematopoietic, 104

8 *Physiology*, 105
   General and Specific, 105
   Comparative, 107
   Developmental, Including Aging, 108
   Systematic, 109
      Musculoskeletal, 109
      Nervous and Sensory, 109
      Integumentary, 111
      Cardiovascular, 111
      Respiratory, 111
      Gastrointestinal, 111
      Genitourinary, 112
      Endocrine, 112
      Hematopoietic, 114
   Reproductive, 115

9 *Neurology*, 117
   General, 117
   Clinical, 117
   Behavior, 117

10 *Biochemistry*, 121
   General, 121
   Physiological Chemistry, Including Metabolism, 122
   Neurochemistry, 124
   Histochemistry, 125
   Immunochemistry, 126

11 *Biophysics and Related Disciplines*, 127
   Biophysics, 127
   Molecular Biology, 127
   Bioengineering, 128
   Biomathematics, 129

12 *Pharmacology, Toxicology, and Therapeutics*, 130
   Pharmacology, 130
   Toxicology, 132

Poisonous Plants and Animals, 133
Therapeutics, 133
Drug Testing, 135
Psychopharmacology, 135

13 *Microbiology*, 136
General, 136
Medical Bacteriology, 137
Virology, 138
Mycology, 141
Infectious Diseases, 141
General, 141
Bacterial, 142
Viral, 142
Mycotic, 143
Zoonotic, 143
Epidemiology and Public Health, 144

14 *Parasitology and Entomology*, 145
Parasitology, 145
Parasitic Diseases, 147
Medical Entomology, 148

15 *Immunology*, 149
Principles, 149
Mechanisms, 150
Clinical Applications, 151
Experimental, 152

16 *Genetics*, 152
Principles, 152
Comparative, 153
Clinical Applications, 153

17 *Internal Medicine*, 154
General, 154
Diagnosis, 154
General, 154
Physical, 155
Laboratory, 155
Systematic Diseases, 157
Musculoskeletal, 157
Nervous and Sensory, 157
Integumentary, 158
Cardiovascular, 159
Respiratory, 159
Gastrointestinal, 160
Genitourinary, 160
Endocrine, 161
Hematopoietic, 162

18 *Pathology, Including Neoplasia,* 163
    General Pathology, 163
    General Neoplasia, 164
    Histopathology, 167
    Clinical Pathology, 167
    Immunopathology, 167
    Systematic Pathology and Neoplasia, 168
        Musculoskeletal, 168
        Nervous and Sensory, 168
        Integumentary, 169
        Cardiovascular, 169
        Respiratory, 169
        Gastrointestinal, 170
        Genitourinary, 170
        Endocrine, 170
        Hematopoietic, 171

19 *Surgery and Obstetrics,* 171
    General, 171
    Technique and Materials, 172
    Anesthesia, 172
    Special Surgery, 173
    Wound Healing, 174

20 *Radiology and Radiation Biology,* 175
    Principles, 175
    Clinical Applications, 175
    Radioisotopes and Radiotherapy, 176

PART III VETERINARY MEDICINE   179

21 *Animal Management,* 181
    Clinical Medicine and Surgery, 181
    Animal Nutrition, 182
    Animal Production and Management, 183
    Veterinary Medical Practice, 184

22 *Cattle,* 185
    Bovine Medicine and Surgery, 185
    Production, Nutrition, and Management, 186

23 *Horses,* 187
    Equine Medicine and Surgery, 187
    Production, Management, and Horsemanship, 187

24 *Sheep and Goats,* 189
    Ovine Medicine and Surgery, 189
    Production and Management, 189

25 *Swine*, 190
    Porcine Medicine and Surgery, 190
    Production, Nutrition, and Management, 190

26 *Dogs*, 191
    Canine Medicine and Surgery, 191
    Breeding, Nutrition, and Kennel Management, 192

27 *Cats*, 193
    Feline Medicine and Surgery, 193
    Breeding, Nutrition, and Management, 194

28 *Furbearing Animals*, 195
    Medical Aspects, 195
    Breeding, Nutrition, and Management, 195

29 *Birds and Poultry*, 196
    Avian Medicine, 196
    Breeding, Care, and Management, 197

30 *Fish and Aquatic Mammals*, 198
    Medical and Research Aspects, 198
    Biology and Management, 199

31 *Reptiles and Amphibians*, 200
    Medical and Research Aspects, 200
    Biology and Classification, 200

32 *Wildlife and Zoo Animals*, 201
    Medical Aspects, 201
    Natural Habitat and Conservation, 201
    Zoo Management, 202

33 *Zoology*, 202
    Vertebrate Biology, 202
    Classification, 203

34 *Companion Animals*, 203

35 *Education and Training, Veterinary Medicine and Allied Health*, 204

    PART IV   LABORATORY ANIMALS   207

36 *Experimental Animals (Vertebrates): General Applications*, 209
    Breeding, Management, Nutrition, and Care, 209
    Anatomy and Physiology, 210
    Surgery, Anesthesia, Radiology, and Special Techniques, 211

Diseases and Therapeutics, 211
Animal Models for Research, 212
Vivesection, 214
Legal Aspects, Regulations, and Standards, 214
Miscellaneous, 214

37 *Primates*, 216
Anatomy, Physiology, and Genetics, 216
Behavior, Ecology, and Evolution, 217
Diseases, 218
Models for Research, 218
Miscellaneous, 218

38 *Rodents*, 219
Anatomy and Genetics, 219
Behavior and Ecology, 220
Medicine and Surgery, 220
Models for Research, 220
Miscellaneous, 221

39 *Other Laboratory Animals (Vertebrates)*, 222
Anatomy and Physiology, 222
Surgery, 222
Models for Research, 222
Miscellaneous, 223

40 *Invertebrates*, 224
Anatomy, Physiology, and Classification, 224
Biomedical Relationships, 224
Miscellaneous, 224

Author Index, 229
Subject Index Key, 266
Subject Index, 268

PREFACE

After the publication of the authors' *Biological and biomedical resource literature* by Purdue University in 1968, a need was evidenced for a more specialized compilation with emphasis on the role of the animal in the life sciences, including both veterinary medicine and human medical laboratory research. The purpose of this bibliography is to fill that need by presenting a selective guide to the recent literature for research workers and practitioners in comparative and veterinary medicine and in related biomedical disciplines which utilize animals as subjects. The bibliography has been compiled with the veterinarian, the research scientist, and the graduate student in mind. It is hoped that the librarian and the information specialist will find it useful also, not only as a subject guide to the resource literature but as a selection medium for acquisitions in the fields covered.

Veterinarians in recent years have progressed from the traditional large animal practice (cows and horses), small animal practice (dogs and cats), and routine regulatory public health activities to a broader responsibility for a wider diversity of species and involvement in medical research activities affecting animals and man alike. The space age put chimpanzees in orbit and brought primate research to its present prominence. Behavioral research has utilized insects, rodents, dolphins, and gorillas, as well as many other vertebrate and invertebrate models. With the need for the veterinarian to become familiar with a greater number of species and with the researcher requiring detailed specific information concerning these often unfamiliar animals, an extensive literature in the veterinary and comparative medical field has developed.

Because physiological processes, bacterial invasions, immunological responses, etc., are essentially the same in all species, many of the publications useful to the veterinarian are also useful to the physician or biomedical researcher. Human-oriented medical research often makes use of a variety of animal models; as a result, the medical librarian finds it necessary to have ready access to the nonhuman animal literature. One of the goals of this publication is to help to fill this need by serving as a reference guide to the literature that is germane.

With rapid changes occurring in the biomedical and related fields, the literature relevant to these changes plays a vital role for the scientist. He must know what is currently being done by others in his area of interest, but he must also know what has already been done, if he is to establish a solid foundation for adding new knowledge. Publications in these fields are dynamic in nature. There is a high rate of productivity and the publications undergo frequent changes. It is important, therefore, that the nature of this literature be learned by the student and the scientist if he is to be at ease with the vast output of the myriad of publishers. *Comparative and veterinary medicine: a guide to the resource literature* attempts to provide direction not only to the literature itself but to the indexes, abstracting journals, and reference tools which lead one to the published works within a specific subject area.

Emphasis is on recent materials, with lesser attention given to the older or classical works. English-language publications, especially U.S. publications, dominate the bibliography. Elementary textbooks and popular works have usually been excluded.

In general, the entries used are those established by the National Library of Medicine in its *Current catalog* or by the Library of Congress in its *National union catalog*. An attempt has been made to indicate second authors, editors, compilers, and in some instances translators. The Subject Index is based on the terms used in the National Library of Medicine, *Medical subject headings*.

The arrangement places material of general interest in Part I, Chapters 1–6, which discusses indexing and abstracting journals, bibliographies, serials, reference works, handbooks, and manuals. Part II, Chapters 7–20, presents material in the traditional subject disciplines of the biomedical sciences, such as anatomy, physiology, and biophysics. Part III, Chapters 21–35, covers veterinary medicine under the various species of animals normally encountered in practice. Part IV, Chapters 36–40, covers laboratory animals, including invertebrates as well as the species that are more traditional to biomedical research.

Many persons and organizations have been involved directly or indirectly in the preparation and production of this work. The authors acknowledge their assistance and support with deep gratitude.

The National Library of Medicine provided financial support through its publications grant program. Without this support and encouragement, the publication would not have been produced.

Many individuals associated with Purdue University provided much valuable assistance, advice, and service. Persons who offered encouragement, support, or subject expertise include Dr. R. L. Davis, Purdue Research Foundation; Dean J. J. Stockton, Dean R. L. Morter, Dr. T. Burnstein, Dr. G. L. Coppoc, Dr. S. M. Gaafar, Dr. R. W. Hughes, and Dr. D. C. Van Sickle, School of Veterinary Science and Medicine; Director J. M. Dagnese and Assistant Director K. Dowden, Libraries/Audiovisual Center; and Dr. J. L. Albright, School of Agriculture.

The artwork incorporated in the book jacket and promotional brochure was designed by Mr. N. N. Harris, scientific illustrator, Veterinary Medical Illustration and Communications, Purdue.

Many persons helped in the active preparation of the manuscript. Particular gratitude is expressed to Miss Alice Dunlap, American Hospital Association, for her assistance and expertise in the preparation of the Subject Index and to Miss Ruth Maxwell for typing it. Persons who were engaged in preliminary bibliographic searching, checking, typing, or other necessary activities include Mrs. Barbara Giger, Mrs. Gloria Van Order, Mrs. Pamela Kemp, Mrs. Betty Thomas, and Mr. George Steedly. Those who assisted in proofreading the completed manuscript include Miss Margaret Sullivan, Miss Katherine Markee, Miss Elizabeth Smith, and Mrs. Nancy Baumrucker. We are grateful to Mrs. Rosemary Bohannan for her careful work in the composition of the book.

A number of persons in other institutions offered subject expertise or made their library collections available. Special mention may be made of the cooperation and assistance extended by Miss Isabel McDonald, Oregon Regional Primate Research Center, and Mr. Marjan Merala, College of Veterinary Medicine, University of California-Davis.

The staff of the Veterinary Medical Library, Purdue, not only extended direct service in many duties related to the publication, but by assuming an inordinate amount of the daily library routine provided release time for the senior author to devote the time needed for publication activities. We wish to thank particularly Mrs. Nancy Weisiger, Mrs. Rhonda Greer, Mrs. Pamela De Bonte, Mrs. Ruth Wilson, and Miss Pamela Highlen.

The authors alone accept responsibility for errors and omissions and for the selection of material. They hope that this guide will be helpful to the many persons who will have occasion to use it and that it will serve the function for which it has been designed. They will welcome suggestions for improvement in organization or for additional appropriate material.

Ann E. Kerker
Henry T. Murphy

ABBREVIATIONS

AAMC	Association of American Medical Colleges
ACCESS	Automatic Computer-Controlled Electronic Scanning System
ACLAM	American College of Laboratory Animal Medicine
ACS	American Chemical Society
AEC	Atomic Energy Commission
AMA	American Medical Association
ASCA	Automatic Subject Citation Alert
ATA	American Translators Association
AVMA	American Veterinary Medical Association
BIOSIS	Biological Sciences Information Service
CAS	Chemical Abstracts Service
CATLINE	Cataloging On-Line
CBAC	Chemical-Biological Activities
CBE	Council of Biology Editors
CFSTI	Clearinghouse for Federal Scientific and Technical Information
CLASS	Current Literature Alerting Search Service
CSIRO	Commonwealth Scientific and Industrial Research Organisation
DHEW	Department of Health, Education and Welfare
ETC	European Translations Centre
FAO	Food and Agricultural Organization of the United Nations
FASEB	Federation of American Societies for Experimental Biology
FDA	Food and Drug Administration
FRAME	Fund for the Replacement of Animals in Medical Experiments
GPO	Government Printing Office
HEW	Health, Education and Welfare
IAEC	International Atomic Energy Commission
IAT	Institute for Applied Technology (National Bureau of Standards)
ILAR	Institute of Laboratory Animal Resources
IM	Index Medicus

ISA	Instrument Society of America
ISI	Institute for Scientific Information
IUPHAR	International Union of Pharmacology
KWIC	Keyword in Context
MEDLARS	Medical Literature Analysis and Retrieval System
MEDLINE	MEDLARS On-Line
MESH	Medical Subject Headings
MLA	Medical Library Association
NAL	National Agricultural Library
NAS	National Academy of Sciences
NFSAIS	National Federation of Science Abstracting and Indexing Services
NIH	National Institutes of Health
NLL	National Lending Library for Science and Technology
NLM	National Library of Medicine
NMAC	National Medical Audiovisual Center
NRC	National Research Council
NSF	National Science Foundation
NUC	National Union Catalog
SERLINE	Serials On-Line
SCI	Science Citation Index
SLA	Special Libraries Association
UCLA	University of California at Los Angeles
UFAW	Universities Federation for Animal Welfare
IUA	Union of International Associations
UICC	Union Internationale Contre le Cancer
UNESCO	United Nations Educational, Scientific and Cultural Organization
WHO	World Health Organization

PART I
Materials of General Interest

Chapters 1—6 represent library materials of general interest to practitioners, educators, students, research investigators, and special librarians in the biomedical sciences regardless of specific discipline. Some of these materials are general in scope, including the physical sciences, social sciences, and humanities, but the majority are more specifically slanted toward biomedical interests.

An annotated list of indexes and abstract serials attempts to explain specific coverage and scope, gives indexing and other information pertinent to their effective use, and supplies publishing information for their acquisition. Sources of bibliographic information for periodicals, conferences, proceedings, and translations are given, and an extensive listing of specialized subject bibliographies is provided.

A list of serials that are useful in comparative and veterinary medicine is included. A chapter on review serials and serial monographs draws attention to many titles that may be used advantageously by the student, academician, practitioner, or research scientist who is seeking an overview of the literature of a specific subject.

The chapters covering the "ready reference tools," such as dictionaries, encyclopedias, biographical tools, and guides to the literature, may be equally useful to the science librarian, student, educator, and research investigator. Handbooks, manuals, and compilations of laboratory techniques are arranged by specific interest group. It is hoped that the scope and the varying levels of the materials listed will be suitable for technical assistants, graduate students, and primary investigators alike. A few source books, histories, and classical treatises indicate a small part of the store of knowledge upon which present medical research is based. A section of miscellany includes notations to specific information services and centers and guides to special library collections; sources of information on research and its support; a number of guides to medical writing and speaking, illustration, and audiovisual materials; and a few sources for industrial information.

1

INDEXES AND ABSTRACTS FOR THE PERIODICAL LITERATURE

To make an effective and comprehensive survey of the periodical literature in any area of science, it is necessary to become familiar with the materials that provide the key to the literature. The literature requirements for research and teaching in the biomedical disciplines, including comparative and veterinary medicine, are multidisciplinary in character, touching on not only the basic and applied biomedical sciences, but also agriculture, chemistry, pharmacy, physics, engineering, and more recently the space sciences. Therefore, familiarity with a wide variety of bibliographic aids is essential for the research worker and the educator in order to carry out an adequate review of the literature pertinent to a given problem. For the practitioner in the health sciences to keep abreast of new developments in his own and ancillary fields, it is equally important that he be knowledgeable in the use of guides to the periodical literature.

Serially issued indexes and abstracts are among the most useful of the library aids in carrying out a literature search. Current awareness services bear increasing importance to research by focusing attention on work in progress. Not all of these materials are designed to accomplish the same result; therefore, the user can save valuable time by becoming familiar with the salient features of each before a search is started.

Prior to World War I, the indexes tended to be general in character because of the relatively small volume of literature and the few scientific societies or other agencies capable of assuming the responsibility of compiling such guides. Since that time, the vast expansion of the literature and the increase in the number of interested organizations have led to more specialized coverage of the literature.

Abstract journals in general did not appear as early as those providing indexing only; however, a few are noted that will prove helpful for searches of the very early literature. Extensive holdings of abstract journals can influence materially the effectiveness of a library collection for the research worker in the medical sciences. The vastness of the literature pertinent to any given area, as well as the professional demands on the researcher's time, necessitates the use of abstracts to determine which articles are of sufficient interest to him and warrant more complete study. Since only the largest research libraries have extensive holdings of the older literature and the more obscure publications, particularly those from foreign countries, abstracts have particular value to the smaller library by furnishing keys to this segment of the literature, thereby opening vast stores of knowledge to the user. The extension of interdisciplinary interests makes necessary the use of abstracts to keep abreast of research and professional developments in fields ancillary to those under consideration. To prevent repetition of work and duplication of effort, it is essential that adequate literature searches be carried out. For these searches, indexes and abstracts are vital.

The relatively recent use of computerized systems for storage and retrieval of information has provided remarkable access to vast stores of knowledge through the production of hand-tailored bibliographic searches and of continuing and recurring bibliographic services. Specific mention may be made of the U.S. National Library of Medicine's (NLM) Medical Literature Analysis and Retrieval System (MEDLARS), which provides a most comprehensive mechanism for bibliographic control of the vast world literature of the health sciences. Specific searches tailored to specific needs have been available to practitioners and scientists since 1964 through retrieval from the extensive data base of MEDLARS. A more recent development (1972) provides on-line searches (MEDLINE) from a smaller, but in most instances a more practical, data base. Many recurring bibliographies on specific subjects have been the result of cooperative arrangements between the National Library of Medicine and other governmental or private agencies.

The titles that follow are noted for their usefulness in retrospective searches. Some have ceased publication; others, which are being published currently, have very long histories of publication and therefore are useful both retrospectively and currently.

British abstracts
Royal Society of London, *Catalogue of scientific papers*

Chemical abstracts
Index-catalogue of medical and veterinary zoology

Index-catalogue of the library of the Surgeon General's Office, United States Army (Armed Forces Medical Library), authors and subjects. 1800-1961.
Index medicus (original series)
International catalogue of scientific literature
Institut Pasteur, *Bulletin*
Jahresbericht Veterinaer-medizin

Physiological abstracts
Quarterly cumulative index to current medical literature
Tropical veterinary bulletin
Zentralblatt für Bakteriologie, Parasitenkunde, Infektionskrankheiten und Hygiene
Zoological record

Some titles that are no longer in publication but will, in general, be found useful for the immediate past are:

Abstracts of Japanese medicine
Abstracts of Soviet medicine
Abstracts of world medicine
Carcinogenesis abstracts

Current list of medical literature
Current tissue culture literature
Quarterly cumulative index medicus

For broad coverage of the recent literature in the basic medical, biological, and agricultural sciences, these titles are suggested:

Bibliography of agriculture
Biological abstracts
Bioresearch index
Chemical abstracts
Current contents: agriculture, biology and environmental sciences
Current contents: clinical practice
Current contents: life sciences

Excerpta medica (40 subject sections)
Index medicus (new series)
Index veterinarius
International abstracts of the biological sciences
Science citation index
Veterinary bulletin

The reader is referred to the annotated list that follows. The list includes many titles of indexes and abstracts, in both general and specialized areas, that have extensive application to comparative and veterinary medicine.

The compilers have attempted to arrange in the first section those indexes and abstracts noted for their usefulness retrospectively. A second section listing current materials is divided into those that are of primary and general interest and those of cognate interest, most of which cover specialized areas. In the final section, a few current awareness services are noted.

RETROSPECTIVE PERIODICALS

0001. Abstracts of world medicine. v. 1-45. London, British Medical Association, 1947-1971.

A monthly abstract journal which covered the more important articles appearing in some 1,600 medical periodicals from many countries. Emphasis was on articles dealing with medicine and its special branches, including those aspects of surgery which were of particular concern to the physician. Published in two volumes a year, with author and subject indexes in each volume.

0002. Arthritis and rheumatic diseases abstracts. 6 v. Bethesda, Md., National Institute of Arthritis and Metabolic Diseases, 1964-1970. (Prepared for the Institute by *Excerpta medica*.)

An abstract journal that was published monthly. It includes articles and abstracts from the world biomedical literature covering the fields of arthritis and rheumatology. Arrangement was by broad subject areas. Each issue had author and subject indexes which were cumulated annually.

0003. Birth defects. v. 1-7. New York, The National Foundation, 1964-1970.

This abstract journal has ceased temporarily. During years of publication more than 2,600 journals of U.S. and foreign origin were searched for pertinent articles. The abstracts were prepared by the Medical Documentation Service, College of Physicians of Philadelphia. It was published monthly. Its usefulness was limited somewhat by a lack of indexes and a table of contents as well as its coverage of human birth defects only.

However, it did serve as an awareness tool for the specialist.

0004. British abstracts. London, Bureau of Abstracts, 1926-1953.

From 1926-1937 called *British chemical abstracts* and from 1938-1944 *British chemical and physiological abstracts*, which also included material formerly in *Physiological abstracts* (see entry 0015). Continued the abstracts previously published from 1871 in the Chemical Society *Journal* and 1878 in the Society of Chemical Industry, London, *Journal*. From 1944 issued in three parts: Part A—Pure chemistry, Sect. 1, 2, and 3; Part B—Applied chemistry, Sect. 1, 2, and 3; and Part C—Analysis and apparatus. Ceased publication in Dec. 1953. Replaced or superseded by the following: Part A—Pure chemistry, Sect. 1 and 2 replaced in part by Chemical Society, London, *Current chemical papers*, 1954- , which is a world list of new papers in pure chemistry arranged in 13 classes and issued monthly; Part A—Sect. 3 replaced by *International abstracts of biological sciences*, v. 5-, 1957-, v. 1-4 as *British abstracts of medical sciences*, 1954-1956.

0005. Current list of medical literature. v. 1-36. Washington, D.C., U.S. Army Medical Library, 1941-1959.

Volumes 1-18, 1941-1949, issued as a weekly current awareness tool; v. 19-36, published monthly. There is a "Register of Articles" arranged by title of periodical, with the pertinent articles numbered. 1,300 serial titles current at that time (45% from North America) were searched. Includes biographical material, significant obituaries, scientific reports presented at congresses, and reports of councils of the American Medical Association, in addition to periodical articles. Does not index reprints, abstracts, book reviews, and notices of an organizational nature. Has good foreign literature coverage. Detailed indexes in each issue, with annual author and subject indexes, facilitate use. Special features: "Register of Articles" is followed by "Medical Project Reports," listed by sponsor. After Jan. 1957 has a section "Recent U.S. Publications," useful for government documents of a medical nature.

0006. Current tissue culture literature. v. 1-5. New York, October House, 1961-1965.

A quarterly publication designed to serve as a key to the world's periodical literature in this specialized field. Citations are arranged by broad subject. Each issue contains author and detailed subject indexes. Coverage of the literature is continued by *Index of tissue culture, 1966-1969* (entry 0011) and by *Index of tissue culture, 1970-* (entry 0066).

0007. Index-catalogue of medical and veterinary zoology. (Authors A-Z.) By C. W. Stiles and A. Hassall. Washington, D.C., U.S. Bureau of Animal Industry, 1902-1912. (Bulletin no. 39, v. 1-5.)

An author index of the literature of medical and veterinary zoology up to 1912. The original plan to publish three series including author, subject, and key catalogs never completed. The U.S. Bureau of Animal Industry Bulletins 140, 142, 148, 150, 155, and 159 include some useful key catalogs. U.S. Hygienic Laboratory Bulletins 37 (1908) *Trematoda and trematode diseases*, 85 (1912) *Cestoda and cestodaria*, and 114 (1920) *Roundworms* give a general and species approach. A revision of this index catalog is described in entry 0008.

0008. Index-catalogue of medical and veterinary zoology. v. 1- U.S. Bureau of Animal Industry. Zoological Division. Washington, D.C., 1932-1952. Suppl. 1953-

A revision and continuation of the *Index-catalogue of medical and veterinary zoology—authors* (entry 0007). This revision, published during 1932-1952, consists of 18 parts. Beginning in 1953 Suppl. 1-6 were published to cover the backlog. Since 1957 supplements covering authors A-Z have been issued on an annual basis. The value has been somewhat restricted by the lack of a subject approach, but beginning with Suppl. 15 the parasite-subject catalogs have been issued. These are arranged by parasite name. In using them, it is necessary to consult the corresponding author catalog for complete bibliographic information. Currently issued by the National Animal Parasite Laboratory of the U.S. Dept. of Agriculture's Agricultural Research Service.

0009. Index-catalogue of the library of the Surgeon General's Office. 61 v. Bethesda, Md., National Library of Medicine, 1880-1961.

Ser. 1, Z-A, 1880-1895 (16 v.)
Ser. 2, A-Z, 1896-1915 (21 v.)
Ser. 3, A-Z, 1918-1932 (10 v.)
Ser. 4, A-Mn, 1936-1955 (11 v.)
Ser. 5, A-Z, 1959-1961 (3 v.)
Dictionary catalog of one of the largest

medical libraries in the world. Useful for its retrospective coverage of the biomedical literature of the latter part of the 19th century and early part of the 20th century. Includes references to books, pamphlets, and (in smaller type) a large number of periodical articles. Good for old theses. Especially good for medical and scientific biography. Books listed under author and subject; periodicals under subject only. Publication of this famous set now abandoned. Last published volume of the 4th series, v. 11 (Mh-Mn). NLM issued Ser. 5 in 3 volumes, a supplementary series, greatly modified in form and content. NLM *Current catalog* continues coverage for books and *Index medicus* continues coverage for periodicals.

0010. Index medicus. 1st series, v. 1-21, 1879-1899. 2d series, v. 1-18, 1903-1920. 3d series, v. 1-6, 1921-1927. Washington, D.C., Carnegie Institution of Washington.

A classified list with annual author and subject indexes. Covers periodical articles in all principle languages as well as books, pamphlets, and theses. Some early literature of veterinary interest is indexed.

0011. Index of tissue culture, 1966-1969. Ed. by Helen E. Cesvet. New York, October House, 1970.

For coverage of the earlier literature see entries 0006, 0180, and 0181.

0012. International catalogue of scientific literature. 1901-1914. London, Harrison, 1902-1919.

Annual bibliography of books and periodicals in 17 subject divisions of science. Divided by sections with letters designating each section; e.g. N—Zoology, O—Human anatomy, Q—Physiology, R—Bacteriology. International in scope and includes papers from independent works as well as serials. Most important bibliography covering all science issued for the period 1901-1914. Ceased with v. 14, 1914, although some volumes were not actually published until 1920. Schedules and indexes in four languages. Has author and subject catalog (index).

0013. Jahresbericht Veterinärmedizin. Berlin, 1928-1943.

A German abstract serial formerly published as *Jahresbericht über die Leistungen auf dem Gebiete der Veterinärmedizin*, Berlin, 1881-1927.

0014. Multiple sclerosis abstracts. v. 1-15. New York, National Multiple Sclerosis Society, 1956-1970.

Abstracts of studies on demyelinating diseases in man and animals. Arranged in two broad groups: A—containing abstracts for the clinical neurologist designed for rapid, general information, and B—including abstracts of papers concerning the basic sciences of interest in multiple sclerosis. It was prepared in cooperation with the Excerpta Medica Foundation and issued approximately semimonthly. There were annual author and subject indexes.

0015. Physiological abstracts. v. 1-22. London, Physiological Society, 1916-1937.

Combined with *British chemical abstracts* to form *British chemical and physiological abstracts*, which was published from 1938-1944. From 1945-1953 abstracts of physiological interest were published in *British abstracts*; from 1954-1956 in v. 1-4 of *British abstracts of medical sciences*. Title changed with v. 5, 1957, to *International abstracts of biological sciences*. Through v. 9, 1924-1925, issued in cooperation with the American Physiological Society.

0016. Quarterly cumulative index medicus. v. 1-60. Chicago, American Medical Association, 1927-1956.

An author and subject index published by the American Medical Association through 1956. The Army Medical Library sponsored its publication with the AMA from 1927-1931. Covers over 1,200 periodicals in many languages. Has a separate section for books by author and subject. Good source for all medical references, including biography, for the years covered.

0017. Quarterly cumulative index to current medical literature. v. 1-12. Chicago, American Medical Association, 1916-1926.

An author and subject index to 300 clinical medical journals and society transactions, including considerable foreign language material. Includes medical books and government documents on medical subjects. Merged into *Quarterly cumulative index medicus* (entry 0016).

0018. Royal Society of London. Catalogue of scientific papers. v. 1-19 (in 4 series), 1800-1900. Cambridge, England, The Univ. Press, 1914-1925.

A monumental author index for the 19th

century, covering 1,555 periodicals and transactions in various languages. Includes publications of most of the European academies and learned societies. Papers published as separate monographs not included. Good source for dates of scientific papers. Covers both physical and biological sciences. Continued in part by *International catalogue of scientific literature* (entry 0012).

0019. Royal Society of London. Catalogue of scientific papers. Subject-index. 1800-1900. Cambridge, England, The Univ. Press, 1908-1914. v. 1—Pure mathematics. v. 2—Mechanics. v. 3—Physics.

No additional volumes were published, therefore the subject index is of very limited value. Continued by *International catalogue of scientific literature* (entry 0012).

0020. Survey of pathology in medicine and surgery. v. 1-5. Baltimore, Williams & Wilkins, 1964-1970.

A bimonthly publication that presented abstracts, condensations, digests, and reviews of articles selected from medical scientific periodicals arranged by organic system or subject discipline. Subject matter of the articles covered not limited strictly to pathology, but covers other basic scientific information having a relationship to pathology. Abstracts are arranged by broad subject. No indexes available for v. 1-2. With v. 3 there is a subject index, which cumulates with each successive number.

0021. Tissue culture bibliography. v. 1-10. Bethesda, Md., Microbiological Associates, 1960-1970.

A monthly index, arranged by author, with a numerical notation added to indicate area or areas of interest covered, as defined by a key list in the index. Sixty-five journals were surveyed for articles in this limited subject area. During the years of publication it was most useful as a current awareness tool. Its use retrospectively is limited by the lack of cumulated indexes.

0022. Tropical veterinary bulletin. London, v. 1-18, 1912-1930.

A quarterly publication, containing abstracts of periodical articles and reviews of books dealing with tropical diseases of veterinary interest, arranged by broad subject. Table of contents in each issue.

Replaced by *Veterinary bulletin* (entry 0039).

CURRENT PERIODICALS

Primary Interest

0023. Accumulative veterinary index. v. 1- Arvada, Colo., Index Incorporated, 1960-

A bibliography of references to periodical literature of clinical veterinary medical interest. The index covers only ten periodicals; however these are titles most commonly subscribed to by the practicing veterinarian, which renders the tool useful in this context. The limited coverage makes it less useful for research purposes. Quarterly issues are cumulated annually. Quinquennial volumes covering 1960-1965 and 1966-1970 have been published.

0024. Animal breeding abstracts. v. 1- Farnham Royal, Bucks, England, Commonwealth Agricultural Bureaux, 1933-

A quarterly periodical containing abstracts, arranged in broad subject areas, of articles from worldwide sources concerned with various aspects of animal breeding, including all farm animals, poultry, furbearing animals, laboratory mammals, and other animals of economic importance (camel, dog, etc.). Each issue contains book reviews and one review article on a topic of animal breeding interest. There is an author index and a listing of papers from relevant meetings in each issue. Cumulated author and detailed subject index published annually.

0025. Animaux de laboratoire: revue bibliographique. v. 1- Paris, Centre National de la Recherche Scientifique, 1964-

A monthly index, in which one section is arranged by species of laboratory animals and gives complete citations. An added section which covers techniques and specific disciplines includes only reference numbers referring back to specific citations in the first section. Published in French, but titles are given in the original language. Worldwide coverage with many citations to articles in English. It is most useful as a current awareness tool. Retrospective searching is difficult because of lack of indexes.

0026. Bibliography of medical reviews. v. 1- Bethesda, Md., 1955-

A monthly index of references to review

articles of biomedical interest compiled by the National Library of Medicine. The references appear in *Index medicus*, but inclusion in the Bibliography indicates that the article cited represents a critical review of the subject indicated. Citations arranged by subject and author, following the format and subject headings used in *Index medicus*. Starting with v. 13, 1968, an annual cumulation will be found in *Cumulated index medicus* only. Earlier volumes appeared as separate annual cumulations. Volume 6 represents a cumulation of citations from v. 1-6, 1955-1960.

0027. Bibliography of reproduction. v. 1- Cambridge, England, Reproduction Research Information Service, 1963-

A systematic monthly title list of current research papers and books dealing with the mechanisms of reproduction in vertebrates, including man. These are taken from the world literature of human, comparative, and veterinary medicine, biology, and agriculture. Over 600 titles included in each issue, arranged in 33 subject sections, with author and animal indexes. A subject index completes each semiannual volume. Starting in 1965 the coverage of the zoological literature, especially that dealing with fishes, amphibia, and reptiles, was increased. January, April, and July issues list future meetings concerned with aspects of reproduction.

0028. Bibliography on exotic animal diseases. v. 1- Greenport, N.Y., Plum Island Animal Disease Laboratory, 1961-

A monthly listing of references to animal diseases not ordinarily found in the U.S., such as scrapie, rinderpest, fowl plague, foot-and-mouth disease, and many others. Compiled primarily for use by researchers at the Plum Island Animal Disease Laboratory; however, the list is very useful to all who may encounter or investigate these diseases. References arranged in alphabetic order by disease and author within each disease category. A number of bibliographies on specific diseases have been prepared by the librarians at Plum Island Animal Disease Laboratory, several of which are cited in entry 0186.

0029. Biological abstracts. v. 1- Philadelphia, BIOSIS, 1926-

A semimonthly abstracting journal produced by Biosciences Information Service (BIOSIS) covering the world's bioscience research reports from 100 countries in nearly 7,600 serial publications. Over 140,000 abstracts are to be included in 1972. Abstracts are arranged in 623 subject sections. Each issue contains the following, in order of appearance: table of contents, including an alphabetized outline of subject categories; list of new books and periodicals received (including authors' names, title of publication, publisher, and price); approximately 6,000 abstracts arranged according to the alphabetical subject categories and accompanied by full bibliographic citations; and four indexes—author, biosystematic, cross, and subject (BASIC) color coded and keyed to the abstracts. A computer-produced index is published annually within one month of the close of the volume. Preceded by *Abstracts of bacteriology*, v. 1-9, 1917-1925, and *Botanical abstracts*, v. 1-15, 1918-1925.

0030. Bioresearch index. v. 1- Philadelphia, BIOSIS, 1967-

A computer-produced monthly publication prepared by Biosciences Information Service (BIOSIS) to supplement *Biological abstracts*. It provides access to more than 100,000 research papers annually, in addition to those reported in *Biological abstracts*. Citations only are furnished; no abstracts. Each issue composed of six parts: a permuted title index, giving a subject approach through key words; a bibliographic section, arranged by journal title and issue; an author index; the cross index; the subject index (BASIC); and the biosystematic index.

0031. Chirurgia veterinaria. v. 1- Berlin, Paul Parey Verlag, 1964-

A quarterly journal that publishes abstracts in German and English of periodical articles, dissertations, reviews, and other reports of original studies concerned with veterinary surgery or of significance to veterinary medicine. Occasionally reviews of books or other monographic materials will be included. The basic arrangement is by broad subject, with annual author and subject indexes.

0032. Excerpta medica. v. 1- Amsterdam, Excerpta Medica Foundation, 1947-

A monthly abstract service for the medical periodical literature, worldwide in coverage and language but with abstracts in English. Published in 40 sections (1972), each representing a specialty subject area in the medical sciences. Each section contains monthly author and subject indexes. Both indexes cumulated annually. Sections have

been divided or new sections started at various times. The initial date for v. 1 of each section is noted below:
- Sect. 1 Anatomy, anthropology, embryology, and histology. 1947.
- Sect. 2 Physiology. 1948.
- Sect. 3 Endocrinology. 1947.
- Sect. 4 Microbiology, bacteriology, virology, mycology, and parasitology. 1948.
- Sect. 5 General pathology and pathological anatomy. 1948.
- Sect. 6 Internal medicine. 1947.
- Sect. 7 Pediatrics. 1947.
- Sect. 8 Neurology and neurosurgery. 1948.
- Sect. 9 Surgery. 1947.
- Sect. 10 Obstetrics and gynecology. 1948.
- Sect. 11 Otorhinolaryngology. 1948.
- Sect. 12 Ophthalmology. 1947.
- Sect. 13 Dermatology and venereology. 1947.
- Sect. 14 Radiology. 1947.
- Sect. 15 Chest diseases, thoracic surgery, and tuberculosis. 1948.
- Sect. 16 Cancer. 1953.
- Sect. 17 Public health, social medicine, and hygiene. 1955.
- Sect. 18 Cardiovascular diseases and cardiovascular surgery. 1957.
- Sect. 19 Rehabilitation and physical medicine. 1958.
- Sect. 20 Gerontology and geriatrics. 1958.
- Sect. 21 Developmental biology and teratology. 1961.
- Sect. 22 Human genetics. 1963.
- Sect. 23 Nuclear medicine. 1964.
- Sect. 24 Anesthesiology. 1966.
- Sect. 25 Hematology. 1967.
- Sect. 26 Immunology, serology, and transplantation. 1967.
- Sect. 27 Biophysics, bioengineering, and medical instrumentation. 1967.
- Sect. 28 Urology and nephrology. 1967.
- Sect. 29 Biochemistry. 1948.
- Sect. 30 Pharmacology and toxicology. 1948.
- Sect. 31 Arthritis and rheumatism. 1965.
- Sect. 32 Psychiatry. 1948.
- Sect. 33 Orthopedic surgery. 1956.
- Sect. 34 Plastic surgery. 1970.
- Sect. 35 Occupational health and industrial medicine. 1971.
- Sect. 36 Health economics. 1971.
- Sect. 47 Virology. 1971.
- Sect. 48 Gastroenterology. 1971.
- Sect. 60 Epilepsy. 1971.
- Sect. unnumbered Environmental health and pollution control. 1971.

The service also has published from time to time a number of special abstract journals devoted to specific disease entities and other selected topics. Many of them were supported by interested scientific organizations or industry. The service published:
- *Abstracts of Japanese medicine*, v. 1-2, 1960-1962.
- *Abstracts of Soviet medicine*. Part A—Basic medical sciences, v. 1-4, 1957-1960. Part B—Clinical medicine, v. 1-4, 1957-1960.

Now these abstracts are included in the appropriate section of *Excerpta medica*. Also, some special single-volume works of reference are published.

0033. Horseman's abstracts. v. 1- Goleta, Calif., Leisure Abstracts, 1969-

Includes book reviews, abstracts of periodical articles, bibliographies, news notes, and a cumulated index.

0034. Index medicus. v. 1- Bethesda, Md., National Library of Medicine, 1960-

The outstanding current English-language indexing tool for the medical sciences, particularly valuable for its extensive coverage of the basic biomedical sciences and comparative medicine. The coverage of clinical veterinary medicine is quite adequate and is expanding with the addition of new periodicals. Monthly issues arranged in four sections, including subject and author sections for *Bibliography of medical reviews* as well as for *Index medicus* In the subject section all titles are in English, bracketed if in translation; an abbreviation denotes the original language of the article. In the author section the titles are in the original language. The monthly issues note National Library of Medicine publications, as well as literature searches produced by MEDLARS, which are available for single copy distribution.

Annually, the issues are cumulated into *Cumulative index medicus*, which includes separate volumes for author and subject sections as well as for *Bibliography of medical reviews*.

Index medicus was preceded by entries 0005, 0010, and 0016.

0035. Index veterinarius. v. 1- Farnham Royal, Bucks, England, Commonwealth Agricultural Bureaux, 1933-

Formerly a quarterly publication; published monthly since Jan. 1972. With v. 39 (4), 1971, separate subject and author sections were initiated; previously author and subject entries were interfiled. Inclusion of titles in the original language, which was suspended in 1962, has been resumed under the author entry, but the translated title is used under the subject entries. There is a separate section, "Veterinary subject headings." The cross-references previously printed have been omitted because they are now available in the list of headings. To search the Index effectively the user should first select appropriate key words from the subject headings. 828 serial publications are regularly searched for the items included. Serial titles are now given in full rather than abbreviated. The Index covers the veterinary literature of the world.

0036. International Veterinary Reference Service. 12 v. Santa Barbara, Calif., American Veterinary Publications, 1966- (Formerly *Modern veterinary reference series*.)

Represents a bound edition of the looseleaf monthly releases of *MVP: International veterinary reference service* (formerly *MVP: Reference and data library*) reorganized into books and chapters. The books are arranged by species and each includes a subject index. The bound volumes are published at five-year intervals and each incorporates the monthly releases during this time span. The volumes published in the first series are:

Book 1 Progress in equine practice. v. 1.
Book 2 Progress in feline practice; including caged birds and exotic animals. v. 1.
Book 3 Progress in swine practice. v. 1.
Book 4 Progress in canine practice. v. 1.
Book 5 Progress in canine practice. v. 2.
Book 6 Progress in canine practice. v. 3.
Book 7 Progress in veterinary practice; questions and answers.
Book 8 Progress in cattle and sheep practice. v. 1.
Book 9 Progress in cattle and sheep practice. v. 2.
Book 10 Progress in cattle and sheep practice. v. 3.
Book 11 Progress in equine practice. v. 2.
Book 12 Progress in feline practice; including caged birds and exotic animals. v. 2.

The service is continued by entry 0037.

0037. Modern veterinary practice, International Veterinary Reference Service. v. 1- Wheaton, Ill., American Veterinary Publications, 1920-

This veterinary reference series represents a monthly loose-leaf abstract service that is published in two sections: large animal and small animal. Through 1967 it was called *Modern veterinary practice: reference and data library*. Within each section the abstracts are arranged by species and broad subject. Approximately 300 biomedical periodicals of veterinary and related interest, English language and foreign, are searched. All abstracts are in English. There are monthly subject indexes which are cumulated quarterly, annually, and quinquennially. Every five years the abstracts are reorganized into books and chapters, which are described in entry 0036.

0038. Review of medical and veterinary mycology. v. 1- Kew, Surrey, England, Commonwealth Mycological Institute, 1943-

A quarterly abstract journal prepared by the Commonwealth Mycological Institute, covering the world's literature on mycoses of man and animals, including fish and large invertebrates. Approximately 250 periodicals are included as well as reports of meetings and short reviews of appropriate books. Arranged by broad subject, with an author index in each issue. Cumulated author and subject indexes are prepared for each triennial volume.

0039. Veterinary bulletin. v. 1- Weybridge, Surrey, England, Commonwealth Bureau of Animal Health, Commonwealth Agricultural Bureaux, 1931-

The outstanding abstract journal covering the research and clinical literature of veterinary medicine. It is issued monthly and includes abstracts arranged by broad subject as well as an author index. A detailed subject index is compiled annually. Most of the monthly issues contain a separate review article in addition to the abstracts.

Cognate Interest

0040. Abstracts of mycology. v. 1- Philadelphia, BIOSIS, 1967-

A monthly abstract journal prepared by Biosciences Information Services, with subject, author, biosystematic, and cross-indexes which are cumulated annually.

0041. Abstracts on Hygiene. v. 1- London,

Bureau of Hygiene and Tropical Diseases, 1926-

A monthly abstract journal arranged in nine main sections dealing with various aspects of public health, such as sanitation, communicable diseases, mycology, and occupational hygiene. Book reviews are included and usually a review article, as well as citations to the periodical literature. Author and source indexes are included in each issue and cumulated annually. Several specific subject indexes are published annually. It is published as a complement to *Tropical diseases bulletin* (entry 0083).

0042. Anesthesiology bibliography. v. 1- American Society of Anesthesiologists. Wood Library, Museum of Anesthesiology. Park Ridge, Ill., 1968-

Published quarterly with the aid of the National Library of Medicine and MEDLARS, it covers English-language articles from approximately 2,300 medical journals. It is divided into a subject index and an author index.

0043. Aquatic biology abstracts. v. 1- New York, CCM Information Corp., 1969-

A monthly abstracting service which selectively covers 2,750 international journals for material of marine biological interest, including sea life, water microbiology, etc.

0044. Artificial kidney bibliography. v. 1- Bethesda, Md., National Institute of Arthritis, Metabolism and Digestive Diseases, 1967- (Subscriptions handled by Supt. of Documents, U.S. Govt. Printing Office, Washington, D.C.)

Published quarterly with the cooperation of NLM and includes appropriate citations from the MEDLARS data base. Citations included in any one issue represent material from the corresponding three monthly issues of *Index medicus*. There are author and subject sections. Included are citations to the literature dealing with kidney failure as well as to the use of animals as models in transplantation experiments.

0045. Bibliographia neuroendrocrinologica. v. 1- Bronx, N.Y., Albert Einstein College of Medicine, Dept. of Anatomy, 1964-

Abstracts of periodical literature issued quarterly; arranged alphabetically by title. Has an editorial section with short reviews of new books and notices of the publication of papers presented at symposia and meetings. Subject index in each issue. Many citations refer to the use of mammals, vertebrates, amphibians, etc., as models in neuroendocrine research.

0046. Bibliography of agriculture. v. 1- New York, CCM Information Corp., 1942-

A monthly computer-produced index to the literature of agriculture and allied sciences received in the National Agricultural Library (NAL). Since 1970 the data has been provided by NAL on magnetic tapes, but the index has been processed and published by CCM. Volumes 1-33, 1942-1969 were published by NAL. A substantial amount of veterinary medicine is included. The citations in the main entry section are arranged by broad subject. There are monthly author and detailed subject indexes. Cumulated individual author index, a corporate author index, and a detailed subject index were issued in December each year prior to 1970. For 1970 individual author and subject indexes have been cumulated. Prior to 1970 this tool included citations to U.S. and state government publications of agricultural interest in separate sections. From Jan. 1970 to the present time (mid-1972) only a few of these have been included; however, it has been announced that coverage will be increased in future issues.

0047. Biological and agricultural index. v. 1- Bronx, N.Y., H. W. Wilson Co., 1916-

An alphabetical subject index to nearly 200 English-language periodicals in the biological, agricultural, and environmental sciences. Coverage is very broad, with the most frequently cited journals in these fields selected by subscriber vote for inclusion. Coverage ranges from agricultural chemistry through zoology, including food science, marine biology, and veterinary medicine among its broad spectrum of subjects. Prior to 1964 title was *Agricultural index* with more emphasis on agriculture than on biology. Included bulletins and publications of the U.S. Department of Agriculture, state agricultural experiment stations and extension services. Very detailed indexing. Published monthly except August. Monthly issues superseded by annual cumulated volume.

0048. Bulletin signalétique. v. 1- Paris, Centre National de la Recherche Scientifique, 1940-

Over 40 separate sections covering all scientific and technical fields published monthly. In each issue short abstracts are arranged under broad subject categories with an author index. Titles of articles appear in French translation as well as in the original language. Many sections of biomedical interest, e.g. Sect. 320, Biochimie, Biophysique; Sect. 330, Sciences Pharmacologiques, Toxicologie; Sect. 340, Microbiologie, Virologie, Immunologie, etc. Worldwide coverage. Annual author and subject indexes.

0049. Cancer chemotherapy abstracts. v. 1- Washington, D.C., Information Resources Press, 1960-

The service covers significant articles in the world's literature dealing with cancer chemotherapy, including the treatment of animal neoplasms through the use of animals as models for the treatment of experimental tumors. Volumes 1-12, 1960-1972, were published bimonthly by Scientific Literature Corp. through a cooperative arrangement between the National Cancer Institute (NCI), NLM, and the College of Physicians Library (Philadelphia) and distributed by Supt. of Documents, U.S. Govt. Printing Office, Washington, D.C. Starting with v. 13, June 1972, Herner Information Services is preparing it for NCI. Distribution will be monthly and there will be some revision in subject sections.

0050. Carcinogenesis abstracts. v. 1-3, 7. Bethesda, Md., National Cancer Institute, 1963-1965, 1969-

Publication of this monthly abstract journal resumed with v. 7, no. 1. At a later date the National Cancer Institute will attempt to provide coverage of the carcinogenesis literature between 1965 and 1969. Current issues normally contain about 200 abstracts and some additional citations to current and significant articles describing research at NCI, other government agencies, and private institutions. The literature abstracted covers all areas of carcinogenesis, including neoplasia in animals and their use as models.

0051. Cerebrovascular bibliography. v. 1- Bethesda, Md., Joint Council Subcommittee on Cerebrovascular Disease, National Institute of Neurological Diseases and Blindness, and the National Heart Institute, 1961-

A bibliography prepared with the assistance of the National Library of Medicine presenting a compendium of the cerebrovascular and selected neurological, vascular, and hematological listings under 60 relevant headings in *Index medicus*. Machine produced and issued quarterly. Each issue contains a listing of the search terms used as well as additional headings that may provide useful entry points into the bibliography. There is a bibliography section, review section, author index, and subject index. Many citations refer to articles dealing with animals as models for cerebrovascular disorders.

0052. Chemical abstracts. v. 1- Columbus, Ohio, American Chemical Society, 1970-

Published weekly, one of the best and most comprehensive abstract journals covering the world's literature of chemistry and allied fields. Particularly valuable for biomedical research through its extensive coverage of biochemistry, microbiology, nutrition, and pharmacology. In 1973 over 409,000 documents will be referenced in CA from about 12,000 periodicals of more than 100 countries as well as patents issued by 26 countries. Generally abstracts appear within three to three and one-half months after publication of the original paper in the periodical or patent literature. Over a two-week period the abstracts appear in 80 subject categories in five sections with cross-references between sections. Biochemistry and organic chemistry sections are published one week, with macromolecular, applied chemistry and chemical engineering, and physical and analytical chemistry sections published in alternate weeks. Two volumes are issued annually.

Each issue contains keyword, author, numerical patent, and patent concordance. CA issued author and formula indexes and numerical patent list through 1959. In 1960, changed to semiannual indexes that appear about four months after completion of the period covered. Decennial indexes were published for 1907-1916, 1917-1926, 1927-1936, 1937-1946, and 1947-1956. A sixth collective index, covering the five-year period 1957-1961, was published late in 1963. The seventh collective index covers 1962-1966 and includes indexes by subject, ring systems, author, formula, and numerical patent index and patent concordance—a total of 24 books. The eighth covers the period 1967-1971 and distribution started early in 1972. It includes subject, author, numerical patent, registry number, hetero-atom-in-context and formula indexes as well as a patent concordance, an index of ring systems, and an index guide.

The total regular-issue content of CA is available in five individually packaged section groupings, including Sect. 1—Biochemistry Sections (CABS) with 114,000 abstracts. Each issue of a section grouping includes the corresponding issue keyword subject indexes. The author index, numerical patent index, and patent concordance are not included. Section groupings, including 26 issues, may be subscribed to individually but do not include volume indexes.

0053. Clinical lab digest. v. 1- Columbus, Ohio, American Chemical Society, 1972-

A new monthly publication which will feature abstracts of articles of particular interest to professionals whose work involves clinical laboratories. Sources of the articles to be abstracted include all ACS publications and selected domestic and foreign journals.

0054. Current bibliography of epidemiology. v. 1- Bethesda, Md., National Library of Medicine, 1969- (Subscriptions handled by Supt. of Documents, U.S. Govt. Printing Office, Washington, D.C.) Volumes 1-3 published by the American Public Health Association.

A monthly bibliography of periodical articles concerning the etiology, epidemiology, prevention, and control of diseases that is cosponsored by the American Public Health Association and prepared from the MEDLARS tapes for *Index medicus* of the corresponding date. The subject bibliography is arranged in two parts: Selected Subject Headings; Diseases, Organisms, Vaccines. There is an author index.

0055. Dairy science abstracts. v. 1- Farnham Royal, Bucks, England, Commonwealth Agricultural Bureaux, 1939-

A monthly compilation, prepared by the Commonwealth Bureau of Dairy Science and Technology, of abstracts from periodicals, reports, and books of the world's literature, covering research and new developments in dairy science. The production and processing of milk and milk products, including husbandry, engineering, economics, physiology, chemistry, microbiology, etc., are included. Most issues contain a comprehensive review article. Author and keyword indexes appear monthly and are cumulated annually.

0056. Diabetes literature index. v. 1- Bethesda, Md., National Institute of Arthritis, Metabolism and Digestive Diseases, 1966- (Subscriptions handled by Supt. of Documents, U.S. Govt. Printing Office, Washington, D.C.)

A monthly index to the world's literature of diabetes and related areas. Prepared by the Institute in cooperation with the Univ. of Minnesota and the Univ. of Rochester, using digital information generated from original input to the MEDLARS system. Arrangement of citations is by hierarchical subject index, keyword-in-title index and author index. There are some citations to articles concerned with animals as models for diabetes research.

0057. Dissertation abstracts international. v. 1- Ann Arbor, Mich., University Microfilms, 1938-

A monthly compilation of abstracts of doctoral dissertations of more than 140 cooperating universities, with a broad subject arrangement. There is a monthly author index, and a keyword index has been included in each issue starting with v. 30, no. 1, May 1969. Each index is cumulated annually. Cooperating institutions are listed inside front cover. In July, 1966, *Dissertation abstracts* divided into two sections: Sect. A, Humanities; and Sect. B, Sciences and Engineering (two separate volumes). A complete listing of dissertations by all U.S. and Canadian students who were granted doctoral degrees will be found in *Index to American doctoral dissertations*.

0058. Electroencephalography and clinical neurophysiology. Index to current literature. Prepared by the Brain Information Service of the University of California at Los Angeles. v. 1- Amsterdam, Elsevier, 1967-

The index is published quarterly as a part of the periodical *Electroencephalography and clinical neurophysiology*, but may be purchased separately. As complete coverage as possible is given to the current literature dealing with electroencephalography. Citations are drawn from a list of regularly scanned journals in the field, as well as from other biomedical journals, proceedings of symposia and conferences, and multiple-author monographs. Citations arranged alphabetically by author within broad subject categories. A list of journals cited is supplied in each issue.

0059. Endocrinology index. v. 1- Bethesda, Md., National Institute of Arthritis and Metabolic Diseases, 1968- (Subscriptions handled by Supt. of Documents, U.S. Govt. Printing Office, Washington, D.C.)

Published monthly with citations from the

data base of NLM's MEDLARS system. Time and journal coverage is the same as for the corresponding issue of *Index medicus*. Includes: table of contents; subject and review sections which give complete citations; a method section and an author section which contains citations and subject headings; a subject index and an author index. The method section includes new methods or techniques, modifications of existing methods and comparisons. There is also a section of *Current catalog* selections noting monographic materials. Many citations refer to endocrine problems in animals or to the use of animal models in endocrine research.

0060. Fibrinolysis, thrombolysis, and blood clotting, a bibliography. v. 1- Bethesda, Md., National Heart and Lung Institute, 1965-

A monthly bibliography designed to promote understanding and progress in the fields of thrombolysis and blood clotting. Citations are processed by the MEDLARS system through cooperation with NLM. There are separate subject, review, and author sections, along with subject and author indexes in each issue. Many references are of veterinary interest as well as biomedical research interest with animals used as models. The subject index provides access to citations under all added subject headings from the data file. Bibliography is cumulated annually.

0061. Gastroenterology abstracts and citations. Washington, D.C., U.S. Public Health Service, 1966-

A monthly publication, one of several continuing bibliographies as a by-product of MEDLARS, under the sponsorship of the National Institute of Arthritis and Metabolic Diseases. Abstracts prepared for about one-third of the citations. Arrangement of citations and abstracts is by broad subject. Has author and subject indexes. Yearly cumulative subject and author indexes are published.

0062. Genetics abstracts. v. 1- London, Information Retrieval Ltd., 1968-

Each issue of this monthly journal contains about 1,000 abstracts arranged under 55 subject headings and subheadings, including many aspects of human, animal, plant, and bacterial genetics; as well as immuno-, behavioral, ecological, molecular, and statistical genetics. There are many cross-references from one heading to appropriate articles in other sections. There is an author index. Also included is a section citing books, and one giving notification of published proceedings of relevant meetings.

0063. Helminthological abstracts. Ser. A: Animal and human helminthology. v. 1- Farnham Royal, Bucks, England, Commonwealth Agricultural Bureaux, 1932-

A quarterly journal prepared by the Commonwealth Bureau of Helminthology, containing abstracts of the world's periodical literature on helminths and their vectors, especially in relation to veterinary, medical, agricultural, and fishery science. A small section on nonperiodical literature notes new books and contains abstracts of theses and papers. Each issue contains an author index; starting with v. 30, 1961, each issue has a subject index. Indexes cumulated annually. With v. 39, 1970, HA divided into two series, with Ser. B covering plant nematology.

0064. Index of dermatology. v. 3, no. 1- Bethesda, Md., American Academy of Dermatology and others, 1971- (v. 1, no. 1- v. 2, no. 12, 1969-1970, have the title *Index of investigative dermatopathology and dermatology* and were published by Universities Associated for Research and Education in Pathology.)

The index is a monthly guide to the periodical literature in clinical and investigational dermatology as well as dermatopathology and is divided into four sections: review, general, bioscience, and author. The format, typographical style, and hierarchical organization of subject headings are identical to those used by NLM in publishing *Index medicus*. Citations are computer-selected from the current month's total input of the MEDLARS data base.

0065. Index of rheumatology. v. 1- New York, American Rheumatism Association, Arthritis Foundation, 1965-

A monthly computer-produced index to the periodical literature of interest in the study of rheumatology. Prepared by NLM as a by-product of the MEDLARS program. Citations arranged by subject and author. No cumulations. Articles of veterinary interest are included.

0066. Index of tissue culture, 1970- Ed. by Helen E. Cesvet. New York, October House, 1971-

For coverage of the earlier literature see entries 0006, 0011, and 0181.

0067. Institut Pasteur. Bulletin. v. 1- Paris, Masson & Cie, 1903-

Issued monthly. Contains abstracts, classified by subject, covering many phases of microbiology and related areas of general biology, medical bacteriology, physiology, neoplasia, biochemistry, and hematology. Especially good for foreign literature. Includes references, abstracts, and book reviews.

0068. International abstracts of biological sciences. London, Robert Maxwell Publications, 1957- (v. 1-4 as *British abstracts of medical science*, 1954-1956).

A comprehensive survey of world literature comprising abstracts of significant papers in experimental biology covering many fields, including anatomy, animal behavior, biochemistry, biophysics, cytology, experimental zoology, experimental pathology, genetics, immunology, microbiology, pharmacology and physiology from the fundamental point of view. Clinical medicine is not included. Each issue is arranged by broad subject and includes an author index. Annual cumulated author and subject indexes are published.

0069. Leukemia abstracts. v. 1- Chicago, Research Information Service. John Crerar Library, 1953-

A monthly compilation of abstracts of articles dealing with leukemia, particularly from a research point of view. There are many abstracts of veterinary interest. Annual cumulated author index, a keyword and subject index.

0070. Medical electronics and communications abstracts. v. 1- London, Multiscience Publishing, 1966-

A bimonthly abstracting tool, arranged by broad subject, covering measurement and recording, diagnostics and prosthetics, communication, and circuits. Abstracts are short; proceedings of pertinent meetings as well as reviews and bibliographies are noted. There are no indexes for individual numbers, but annual author and subject indexes are provided. It is useful for articles citing species of laboratory animals used as models.

0071. Microbiology abstracts. v. 1- London, Information Retrieval, 1965-

A monthly publication arranged in two sections: A—Industrial microbiology; and B—General microbiology and bacteriology. International in scope and extensive in its coverage of 2,750 journals. (Formerly *Industrial microbiology abstracts*.)

0072. Muscular dystrophy abstracts. v. 1- New York, Muscular Dystrophy Associations of America, 1957-

A serial publishing abstracts from the world's current literature of muscular dystrophy, related disorders, and those areas of basic science pertinent to them. Prepared in cooperation with *Excerpta medica* and published semimonthly, with annual author and subject indexes. Citations to articles of veterinary interest and research involving animal models.

0073. Nuclear science abstracts. v. 1- U.S. Atomic Energy Commission, Oak Ridge, Tenn., 1948-

A semimonthly publication providing coverage of the world's nuclear science literature. It includes scientific and technical reports of the AEC and its contractors, other U.S. Government agencies, and other governments, universities, and research institutions. It also covers books, conference proceedings, patents, and the international journal literature. Abstracts in each semimonthly issue arranged by broad subject, e.g. chemistry, environmental and earth sciences, isotopes and radiation source technology. Particularly useful in biomedical research for its section on life sciences. Each issue has four indexes: subject, personal author, corporate author, and report number. Quarterly and annual cumulations of indexes issued. Author and subject indexes also in five-year cumulations. Cumulative report number indexes issued covering 1947-1961, 1962-1966, and 1967-1971.

0074. Nutrition abstracts and reviews. v. 1- Bucksburn, Aberdeen, Scotland, Commonwealth Bureau of Animal Nutrition, 1931-

Published quarterly. Worldwide in coverage and useful for both human and animal nutrition. Most issues contain a review article covering in some detail the literature and advances in a particular phase of nutrition. Abstracts, book reviews, and citations to various society proceedings comprise the remainder of each issue and are arranged under broad headings. Each issue contains an author index. There are cumulated annual author and subject indexes.

0075. Parkinson's disease and related disorders. v. 1- Bethesda, Md., National Institute of Neurological Diseases and Stroke, 1963-

A monthly index of relevant citations retrieved from the data base of NLM's MEDLARS system. Contains a list of subject headings used, an author section, and a subject section. "Core" documents primarily clinical in approach or which relate research to a specific disorder are included. For a cumulated bibliography prepared by the same agency, see entry 0183.

0076. Pesticides documentation bulletin. v. 1-5. Washington, D.C., National Agricultural Library, 1965-1969.

Began as a computer-produced keyword index to the literature on diseases, insects, parasites, weeds, etc., affecting animals, man, and plants. Covered the multidisciplinary literature in the pest control and related fields. Special emphasis on biological, toxicological, physiological, biochemical, pathological, biophysical, and epidemiological aspects of pests and their control by chemical and nonchemical methods. Published biweekly. From 1968 format changed from keyword approach to a categorized bibliography, with citations arranged alphabetically by author under eleven broad subjects, e.g. entomology, residues, toxicology. Detailed subject index for each volume. Superseded by entry 0077.

0077. Pesticides documentation bulletin. v. 1- Silver Spring, Md., National Information Services, 1970-

This monthly publication starting with v. 1, July 31, 1970, continues the title of entry 0076 and has essentially the same scope. Author, organizational, and subject indexes are included.

0078. Review of allergy: selected abstracts of allergology. v. 1- St. Paul, Minn., 1947-

A monthly publication in which some original articles appear in addition to comprehensive abstracts of current literature, book reviews, news of interest to allergologists, and a classified current bibliography on allergy and applied immunology. Arrangement is by broad subject as noted in the table of contents for each issue. There is an annual index with author and title entries interfiled. (Formerly *Review of allergy and applied immunology.*)

0079. Science citation index. v. 1- Philadelphia, Institute for Scientific Information, 1961-

A computer-produced index that approaches the literature from a specific key article to which later papers on the same subject refer. It provides a system whereby it is possible to tell what additional research has been done since a specific article was published. It is unique in providing this ability to move forward in the literature. Over 2,200 source journals are included in the index which is multidisciplinary in coverage to enable the searcher to obtain information across disciplines. SCI is published quarterly with annual cumulations. The two main sections are the Citation Index and the Source Index. The Citation Index is an alphabetical listing by cited authors and within this listing chronologically by cited year. Each citation gives the author's name, publication title, volume and page, and the date. Under the author appears the source article citing his paper. The Source Index is an author index to the current literature and includes coauthors, article title, journal title, volume, page, and year. SCI enables the searcher to pinpoint specific information through authors rather than through a complex nomenclature structure. Once the technique of its use is mastered, it can be very rewarding to the research worker in any of the biomedical or physical sciences.

0080. Selected references on environmental quality as it relates to health. v. 1- Bethesda, Md., National Library of Medicine, 1970- (Subscriptions handled by Supt. of Documents, U.S. Govt. Printing Office, Washington, D.C.)

A monthly bibliography, produced as a by-product of the MEDLARS system, which provides subject and author sections noting citations to journal articles covering aspects of environmental pollution that concern health. Citations and articles on pollutants will be found under general rather than highly specific chemical terms. The bibliography is cumulated annually.

0081. Tissue culture abstracts. v. 1- Grand Island, N.Y., Grand Island Biological Co., 1964-

A bimonthly publication furnishing short abstracts of literature dealing with tissue culture. Arrangement is random and there are no indexes, but it serves as a current awareness service for workers in the field.

0082. Toxicity bibliography. v. 1- Bethesda, Md., National Library of Medicine, 1968- (Subscriptions handled by Supt. of Documents, U.S. Govt. Printing Office, Washington, D.C.)

A quarterly publication covering the adverse and toxic effects of drugs and chemicals reported in the approximately 2,300 journals indexed for *Index medicus* and available from the MEDLARS tapes, from which the references are drawn selectively. One section, "Drugs and Chemicals," provides author and subject indexes as well as citations under specific drugs and chemicals. A second section, "Adverse Reactions to Drugs and Chemicals," has citations arranged in broad subject categories, including the various body systems, immunologic reactions, and diseases of animals.

0083. Tropical diseases bulletin. v. 1- London, Bureau of Hygiene and Tropical Diseases, 1912-

A monthly abstracting journal arranged in 22 subject classes dealing with all aspects of tropical diseases. Book reviews, reports, and surveys are included as well as the periodical literature. There are monthly author and source indexes, and annual author and title indexes. It is published as a complement to *Abstracts on hygiene* (entry 0041), which was titled *Bulletin of hygiene* up to v. 43, 1968.

0084. Virology abstracts. v. 1- London, Information Retrieval Ltd., 1967-

A monthly publication including abstracts of current diseases, diagnosis, and techniques for human and animal viruses. There are also book reviews and notification of publication of proceedings of relevant meetings. Indexes included in each issue are by virus name, patentee and assignee, and author. The author index is cumulated annually.

0085. Zentralblatt fuer Bakteriologie, Parasitenkunde, Infektionskrankheiten und Hygiene. Bd. 1-16, Stuttgart, Germany, 1887-1894. Abteilung 1. Medizinisch-hygienische, Bakteriologie, Virusforschung, und tierische Parasitologie. v. 17-30, 1895-1901. Replaced by two publications:
    Originale, v. 31, 1902- Section consists of original papers.
    Referate, v. 31, 1902- Section of abstracts arranged by classes.

Extensive coverage in each section, especially for European literature. Both include original research. The subject areas covered are pure and applied microbiology, including virology, helminthology, protozoology, infectious diseases, chemotherapy, disinfection, and agricultural aspects of bacteriology. Both sections are issued twice a month.

0086. Zoological record. v. 1- London, Zoological Society of London, 1864-

This outstanding zoological bibliography has been published continuously since 1864. Aims at making available a complete list of articles and books of zoological interest published in any one year. Particularly useful for taxonomic references. Coverage is worldwide. Published annually in 20 sections, 18 of which record the year's literature relating to a phylum or class of the animal kingdom (e.g. Aves, Pisces, Mammalia). Sect. 1 is devoted to comprehensive zoology, which includes those works dealing with more than one branch of zoology. Sect. 20 lists the new generic and subgeneric names recorded in *Zoological record* for that year. Each section is divided into three parts:
    1. Author index. Arranged alphabetically by author, with complete bibliographic citation.
    2. Subject index. Arranged under about 200 subjects, such as behavior, distribution, ecology, fertility, reproduction, etc.
    3. Systematic index. Arrangement varies in different sections, e.g. Insecta (Sect. 13) in alphabetical order by genera, and Aves (Sect. 18) in the arrangement used in Peters' *Checklist of birds of the world*.

Full bibliographic citation given only under the author index. Must refer from subject and systematic indexes to author index for complete citation.

CURRENT AWARENESS SERVICES

0087. Biomedical research in progress. v. 1- New York, Walker-Carrollton Associates, 1970-

A service available in microfilm or hard copy that provides information retrieval for ongoing biomedical research projects through cooperation with the Science Information Exchange of the Smithsonian Institution. There are detailed subject indexes, a research organization index, and an investigator index.

0088. Chemical-biological activities (CBAC). v. 1- Columbus, Ohio, American Chemical Society, 1965-

A biweekly computer-readable file of abstracts that reports the current scientific and technical literature related to the field of biochemistry. The literature covered includes journals, patents, reports and conference and symposia proceedings that report the interactions of chemical substances with biological systems in-vivo and in-vitro. Searchable information includes the names of authors, primary bibliographic citations, words in titles and full abstracts, molecular formulas, names of substances, CA-section numbers, and registry numbers. Abstracts appearing in CBAC are those published in Sect. 1-5 of *Chemical abstracts*. About 34,000 abstracts will be included in CBAC in 1973.

0089. Current contents: agriculture, biology and environmental sciences. v. 1- Philadelphia, Institute for Scientific Information, 1970- (Formerly *Current contents: agricultural, food and veterinary sciences*.)

An up-to-date and comprehensive survey of agricultural, food, and veterinary sciences through the photoduplication of tables of contents of more than 800 journals published worldwide and reporting research in these subject disciplines. The publishers changed the emphasis somewhat in late 1972 by added coverage of biology and ecology journals, at which time the journal title was changed to reflect this interest in environmental sciences. There are usually 12 or 13 subject sections in each issue, as well as an author index and address directory and a listing of the journals indexed that week. Complete journal coverage is listed approximately each six months.

0090. Current contents: clinical practice. v. 1- Philadelphia, Institute for Scientific Information, 1973-

This service, which will be instituted in 1973, will cover 700 of the world's most important clinical journals oriented to the practice of human medicine.

0091. Current contents: life sciences. v. 1- Philadelphia, Institute for Scientific Information, 1958-

An up-to-date and comprehensive weekly survey of chemical, pharmaco-medical, and life sciences publications through the photoduplication of the tables of contents appearing in more than one thousand foreign and domestic journals. Usually each issue is divided into eight sections: biochemical, pharmaco, animal and plant science, experimental, behavioral, multi, chemical, and clinical. Each issue contains an alphabetical listing of the journals being indexed that week and an author address directory. Every six months a complete list of journals indexed appears. A weekly subject index compiled from words in titles may be obtained as a separate publication or bound into each weekly issue of CC/LS.

0092. Current contents/life sciences: weekly subject index. no. 1- Philadelphia, Institute for Scientific Information, 1972-

0093. Index of current research on pigs. v. 1- Shinfield, Berks., England, Agricultural Research Council, Technical Committee on Pig Research, 1955-

An annual compilation of swine research throughout the world. Notes work in progress, as well as publications which appeared during the current year. Arrangement is geographical by country and alphabetical by location within a geographical unit. Institutions covered include agricultural experiment stations, universities, institutes, and other centers of research. Each volume contains a subject index.

0094. Unlisted drugs. v. 1- Chatham, N.J., 1948-

A monthly journal that describes and identifies new drug compounds and products that have not yet been listed in standard compendia. Some pertinent book reviews and other data are included. Members of the Pharmaceutical Section, Special Libraries Association (SLA) participate in the collection of data, and from v. 1-18, 1948-1966, published the compilation. Drugs are arranged in two sections by number and name. Information includes composition, manufacturer, actions, and references. There are semiannual indexes with various cumulations available. The service is available also in card format.

2
BIBLIOGRAPHIES OF PERIODICALS
CONFERENCES AND CONGRESSES, TRANSLATIONS
AND SPECIAL SUBJECTS

For purposes of journal selection, collection building, or as vehicles for presentation or publication of papers, the bibliographies of periodicals noted below will serve as guides to titles within the periodical literature. Some of these guides are very comprehensive in coverage, while others are quite specialized. The list, while selected, attempts to call attention to the major bibliographies covering English-language serial publications and the most broadly used foreign periodicals. Also noted are several sources of information relative to forthcoming regional, national, and international meetings, conferences, and congresses. Additional titles give information about published proceedings of meetings, while other titles merely note locations and dates of previously held congresses and meetings. A selected list of translation sources includes the major agencies that maintain files of translations or information about their availability. In the final section, many selected bibliographies on specific subjects are listed. For additional bibliographies, the reader may be referred to several sources, among which may be mentioned: the subheading "Bibliography" under specific subject headings in the National Library of Medicine *Current catalog* and in the *National union catalog*; the lists of NLM literature searches in each monthly issue of *Index medicus* and of new searches announced in NLM *News*, single copies of which are available on request; and the *Bibliography of medical bibliographies*, published by NLM.

PERIODICALS

0095. Benton, Mildred E., and C. W. Shilling. Japanese biomedical serials: their identification and an analysis of their contents. Washington, D.C., Biological Sciences Communication Project, 1965.

0096. Biosciences Information Service. BIOSIS. List of serials with title abbreviations. Philadelphia. Revised annually.

0097. Boalch, D. E., ed. Current agricultural serials: a world list of serials in agriculture and related subjects (excluding forestry and fisheries) current in 1964. 2 v. Oxford, International Association of Agricultural Librarians and Documentalists, 1965. (Kept up to date by section "New agricultural serials" in the IAALD *Quarterly bulletin*.)

0098. British union catalogue of periodicals. New periodical titles. (Incorporating *World list of scientific periodicals*.) London, Butterworth, 1964- (Quarterly, with annual cumulations.)

0099. Chemical Abstracts Service. ACCESS: key to the source literature of the chemical sciences. Columbus, Ohio, American Chemical Society, 1969. (Continued by quarterly supplements through no. 3, 1970. Continued by *Chemical Abstracts Service source index quarterly*, no. 4, 1970- . See entry 0993.)

0100. Collison, R. L. Abstracts and abstracting services. Santa Barbara, Calif., ABC-Clio, 1971.

0101. Commonwealth Bureau of Animal Health. List of journals scanned in the production of *Veterinary bulletin* and *Index veterinarius*. New Haw, Weybridge, Surrey, England, 1972. (Formerly *List of publications searched by the Commonwealth Bureau of Animal Health*.)

0102. Fowler, Maureen J. Guides to scientific periodicals: annotated bibliography. London, The Library Association, 1966.

0103. Hammack, Gloria M. The serial literature of entomology: a descriptive study. College Park, Md., Entomological Society of America, 1970.

0104. Institut national de recherches vétérinaires. Bibliotheque. Catalogue des periodiques conserves par la bibliotheque de l'Institut national de recherches vétérinaires. Brussels, Belgium, Georges Stiers, 1970.

0105. Irregular serials and annuals: an international directory. A classified guide to current foreign and domestic serials, excepting periodicals issued more frequently than once a year. New York, Bowker, 1967. (2d ed. in press 1972).

0106. National Federation of Science Abstracting and Indexing Services. Washington, D.C., and Philadelphia.
    A guide to U.S. indexing and abstracting services in science and technology. 1960. (Report no. 101.)
    A guide to the world's abstracting and indexing services in science and technology. 1963. (Report no. 102.)
    A list of science serials covered by members of the N.F.S.A.I.S. 2 v. 1962.
    Proceedings of annual meeting. 12 v. 1958-1969.
    Digest of annual meeting. 1970- (Continues the Federation's proceedings of annual meeting.)
    Technical report no. 1- Oct. 1969-

0107. Periodicals relevant to microbiology and immunology. A world list—1968. Ed. by Gösta Tunevall. New York, Wiley, 1969.

0108. Plum Island Animal Disease Laboratory. List of serials. Comp. by Robert Uskavitch. Greenport, L.I., N.Y., 1972.

0109. Serials in microform/1972. Ann Arbor, Mich., University Microfilms, 1972. (Kept up to date by University Microfilms, *Serials bulletin*, a bimonthly publication. As of July 1972 a quarterly fiche up-date service is also available.)

0110. Standard periodical directory. 3d ed. New York, Oxbridge, 1970.
    ... Suppl. ed. by Leon Garry. 1972.

0111. Ulrich's international periodicals directory: a classified guide to current periodicals, foreign and domestic. 14th ed. 2 v. New York, Bowker, 1971/72.

0112. Union catalog of medical periodicals. 2d ed. v. 1- New York, Medical Library Center of New York, 1967- . Phase 1—Listing medical and paramedical periodicals in existence in 1950 and new titles published since, with holdings of 83 libraries in the New York metropolitan area, as of Dec. 31, 1967. Phase 2—Medical and paramedical periodicals in existence prior to 1950; including also nonsubstantive items later than 1950 that were excluded from the companion volume, UCMP/I. Holdings of 83 libraries in the New York metropolitan area are listed, as of Sept. 1, 1968.
    ... Rev. June 1970.
    ... Suppl., 1971-

0113. Union list of serials in libraries of the United States and Canada. 5 v. 3d ed. Bronx, N.Y., Wilson, 1965.

0114. U.S. Library of Congress. New serial titles: classed subject arrangement. Washington, D.C., 1955- (Monthly, no cumulations.)

0115. U.S. National Agricultural Library. Serial publications indexed in the bibliography of agriculture. rev. ed. Washington, D.C., 1968. (Library list no. 75.)

0116. U.S. National Library of Medicine. Biomedical serials, 1950-1960: a selective list of serials in the NLM. Washington, D.C., 1962.

0117. ——. Index of NLM serial titles: a keyword listing of serial titles currently received by the National Library of Medicine. Bethesda, Md., 1972.

0118. ——. List of journals indexed in *Index medicus*. Washington, D.C., 1960- (Annual; published also as Part 2 of Jan. issue of *Index medicus*.)

0119. Visual Science Information Center. Vision union list of serials. 2d ed. Berkeley, Univ. of California, 1970.

0120. Vital notes on medical periodicals. v. 1- Chicago, Medical Library Association, 1952- (Five-year cumulated index, 1952-1957.)

0121. World list of scientific periodicals published in the years 1900-1960. 3 v. 4th ed. London, Butterworth, 1963-1965.

0122. World medical periodicals. 3d ed. Comp. by C. H. A. Fleurent. Ed. by H. A. Clegg. New York, World Medical Association, 1961.
    ... Suppl., 1968. Comp. and ed. by Martin Ware.

CONFERENCES AND CONGRESSES

Forthcoming Meetings

0123. Calendar of international congresses of medical sciences. no. 1- Paris, Council for International Organizations of Medical Sciences, 1952-

0124. Calendar of regional congresses of medical sciences. no. 5- Paris, Council for International Organizations of Medical Sciences, 1961-

0125. International associations: monthly review of international organizations and meetings. Brussels, Union of International Associations, 1949- Index v. 1-10, 1949-1958.

0126. Scientific meetings: describing future meetings of technical, scientific, medical and management organizations and universities. New York, Special Libraries Association, 1956-
 Arranged in five parts. The parts consist of an alphabetical directory of sponsoring organizations, a listing of meetings, a subject index of key words, a geographical index, and a listing of short courses. (Quarterly.)

0127. World meetings, outside U.S.A. and Canada: a two-year registry of future medical, scientific, and technical meetings. New York, CCM Information Corp., 1968-
 Contains five indexes: by date, key word, location, deadline for papers, and sponsor. (Quarterly.)

0128. World meetings, United States and Canada: a two-year registry of future medical, scientific, and technical meetings. (Formerly *Technical meetings index.*) New York, CCM Information Corp., 1963-
 Contains five indexes: by date, key word, location, deadline for papers, and sponsor. (Quarterly.)

Note. Many journals include sections announcing forthcoming scientific meetings: e.g. American Medical Association *Journal*; *American Journal of Veterinary Research*; American Veterinary Medical Association, *Journal*; *Bio-science*; *Nature*; *Science*, etc.

Proceedings Available

0129. Chemical Abstracts Service. ACCESS: key to the source literature of the chemical sciences. Columbus, Ohio, American Chemical Society, 1969.
 Lists proceedings, transactions and other publications of associations. (Continued by quarterly supplements through no. 3, 1970. Continued by *Chemical Abstracts Service source index quarterly*, no. 4, 1970- . See entry 0993.)

0130. Directory of published proceedings. Series SEMT-Science/Engineering/Medicine/Technology. Harrison, N.Y., Interdok, 1967-
 Includes preprints as well as published proceedings. (Monthly Sept.-June, cumulated annually.)

0131. Irregular serials and annuals: an international directory. A classified guide to current foreign and domestic serials, excepting periodicals issued more frequently than once a year. New York, Bowker, 1967.
 Includes a section "Title grouping of international meetings' publications" and within subject sections notes specific information relative to each publication. (2d ed. in press 1972.)

0132. Proceedings in print. Mattapan, Mass., 1964- (From 1964–April 1966, published by Aerospace Division, Special Libraries Association.)
 Contains Part 1, "Current entries," listing proceedings published within last two years; and Part 2, "Retrospective entries," for those published earlier. (Bimonthly, with a cumulative index in the last issue of each annual volume.)

Retrospective Listings

0133. Calendar of international and regional congresses of medical sciences: regional calendar. no. 1-4. Paris, Council for International Organizations of Medical Sciences, 1957-1960. (For continued coverage see entries 0123 and 0124.)

0134. Council for International Organizations of Medical Sciences. Bibliography of international congresses of medical sciences. Springfield, Ill., Thomas, 1958.

0135. U.S. National Library of Medicine. Congresses: tentative chronological and bibliographical reference list of national and international meetings of physicians, scientists and experts. Washington, D.C., 1938. (Index-catalogue of the library of the Surgeon General's Office. Ser. 4, v. 3.) ...First additions (Ser. 4, v. 4), 1939.

0136. World list of future international meetings. Part 1—Science, technology, agriculture, and medicine. Part 2—Social, cultural, commercial, and humanistic. Washington, D.C., Library of Congress, International Organizations Section, 1959-1969. (Ceased publication in May 1969. May be useful in retrospective searching about meetings that were held during the time span covered.)

Papers Presented at Meetings

0137. Current index to conference papers: science and technology. v. 1- New York, CCM Information Corp., 1969- (Continues *Current index to conference papers in life sciences*; *Current index to conference papers in chemistry*; and *Current index to conference papers in engineering*.)

TRANSLATIONS

0138. ATA professional services directory. 2d ed. New York, American Translators Association, 1969.
An index of professional translators arranged by broad subject disciplines, e.g. bacteriology, surgery, etc., and by source language. Also has geographical index arranged by zip code number.

0139. Great Britain Dept. of Scientific and Industrial Research. no. 1-117. London, 1949-1958.
Translated contents lists of Russian periodicals; with list of recent accessions of Russian scientific and technical books and parts of serial publications available in the British Museum. (Continued by *NLL Translations bulletin*, entry 0142.)

0140. Groot-de-Rook, A.S. de. Translations journals. Delft, European Translations Centre, 1971.
List of periodicals translated cover to cover, abstracted publications and periodicals containing selected articles.

0141. Kaiser, Frances E. Translators and translations: services and sources in science and technology. 2d ed. New York, Special Libraries Association, 1965.

0142. NLL Translations bulletin. v. 1- London, National Lending Library for Science and Technology, 1959-
Title varies: v. 1-2, issued as *LLU Translations bulletin*, by Great Britain Dept. of Scientific and Industrial Research, Lending Library Unit. Gives abstracts of some articles from the USSR and from China, and lists translations available for purchase or loan from the NLL. (Monthly.)

0143. Translations register-index and list of translations notified to E.T.C. (Combined into one publication Jan. 1972.) Chicago, National Translation Center, John Crerar Library, 1967-
Includes new accessions of the National Translation Center and also items listed by the National Technical Information Service in *U.S. Government research and development reports*. The National Translation Center is a depository for unpublished translations in the natural, physical, and social sciences. (Quarterly cumulative indexes.)

0144. World index of scientific translations. Delft, European Translations Centre, 1967-
A monthly list of translations of serial articles, patents, and standards relating to science and technology. It is essentially a finding list for translations from non-Western languages into Western languages. (Quarterly cumulative indexes.)

SPECIALIZED SUBJECT BIBLIOGRAPHIES

0145. Adams, C. E., and Mary Abbott. Bibliography on recovery and transfer of mammalian eggs and ovarian transplantation: literature, 1887-1971. Cambridge, England, 1971. (Reproduction Research Information Service. Bibliography. no. 45.)

0146. American Association for Laboratory Animal Science. Biological Safety Committee. Laboratory infections bibliography. Joliet, Ill., 1966.

0147. Animal phonation. 1963-1967. Baltimore, Information Center for Hearing, Speech, and Disorders of Human Communication, 1968. (26 refs.)

0148. Basmajian, J. V. Muscles alive;

their functions revealed by electromyography. 2d ed. Baltimore, Williams & Wilkins, 1967. (Bibliography, 1868-1966, p. 369-410. 608 refs.)

0149. Beary, E. G. Laboratory animals; their care and use in research: a bibliography. Natick, Mass., U.S. Army Natick Laboratory, 1968. (U.S. Army Natick Laboratories. Bibliographic Series 68-1.)

0150. Behavioral dimorphisms in vertebrates and primates. 1966-1969. Chicago, John Crerar Library, 1970. (Bibliography no. 45. 8 refs.)

0151. Besterman, Theodore. Biological sciences: a bibliography of bibliographies. Totowa, N.J., Rowman & Littlefield, 1971.

0152. Bibliography of medical bibliographies. Bethesda, Md., National Library of Medicine. Pilot issue 1- 1970-

0153. Bibliography on scrapie. Comp. by C. J. Gibbs et al. Bethesda, National Institute of Neurological Diseases and Stroke, 1969.

0154. Bickford, R. G., et al. KWIC index of EEG literature. New York, American Elsevier, 1965.

0155. Boreva, L. I., and E. M. Panoya, comps. Bibliography of bioastronautics. Washington, D.C., Joint Publications Research Service, 1966. (Technical translation: 66-35-086.)

0156. Cass, J. S., ed. Laboratory animals: an annotated bibliography of informational resources covering medicine—science (including husbandry)—technology. New York, Hafner, 1971. This updates entry 0165.

0156a. The chimpanzee: a topical bibliography. 2d ed. Ed. by F. H. Rohless, Jr. Manhattan, Kansas, Institute for Environmental Research, Kansas State University, 1972.

0157. Drug effects on memory and learning in animals. 1966-1969. Los Angeles, UCLA Brain Information Service, 1969. (175 refs.)

0158. Dunn, A. M. Veterinary helminthology. Philadelphia, Lea & Febiger, 1969. (Bibliography, 1924-1967, p. 291-294. 179 refs.)

0159. Electrophysiology of the developing mammalian brain. 1958-1968. New York, Parkinson Information Center, 1968. (93 refs.)

0160. Fenner, Frank. The biology of animal viruses. 2 v. New York, Academic Press, 1968. v. 1—Molecular and cellular biology. v. 2—The pathogenesis and ecology of viral infections. (Bibliography, 1896-1968, p. 401-474. Approx. 1400 refs.)

0161. Foreign compound metabolism in mammals: a review of the literature published between 1960 and 1969. v. 1- Reported by D. E. Hathaway. London, The Chemical Society, 1970-

0162. Fraenkel-Conrat, Heinz. The chemistry and biology of viruses. New York, Academic Press, 1969. (Bibliography, 1798-1969, p. 231-263. 579 refs.)

0163. Garrison, F. H. A medical bibliography (Garrison and Morton): an annotated checklist of texts illustrating the history of medicine. By L. T. Morton. 3d ed. Philadelphia, Lippincott, 1970.

0163a. The gerbil: an annotated bibliography. Comp. by Victor Schwentker. West Brookfield, Massachusetts, Tumblebrook Farm, Inc., 1972.

0164. Gray, Annie P. Mammalian hybrids, a checklist with bibliography. Farnham Royal, Slough, England, Commonwealth Agricultural Bureaux, 1971. (Commonwealth Bureau of Animal Breeding and Genetics. Technical communication no. 10.)

0165. A guide to production, care and use of laboratory animals: an annotated bibliography. Federation proceedings 19(4) Part 3; 22(2) Part 3. Washington, D.C., Federation of American Societies for Experimental Biology, 1960-1963. (Updated by entry 0156.)

0166. Haywood, B. J. Thin layer chromatography: an annotated bibliography, 1964-1968. Ann Arbor, Mich., Ann Arbor Science Publishers, 1968.

0167. Hinde, R. A. Animal behaviour, a synthesis of ethology and comparative psychology. 2d ed. New York, McGraw-Hill, 1970. (Bibliography, 1911-1969, p. 691-824. Approx. 2730 refs.)

0168. Hoffman, R. A., et al. The golden

hamster: its biology and use in medical research, with a master bibliography by Hulda Magalhaes and including a stereotaxic atlas of the brain of the golden hamster by Karl M. Knigge and Shirley A. Joseph. Ames, Iowa State Univ. Press, 1968. (Bibliography, p. 321-542.)

0169. Hoijer, Dorothy T. A bibliographic guide to neuroenzyme literature. New York, Plenum Press, 1969.

0170. Jay, P. C., ed. Primates: studies in adaptation and variability. New York, Holt, Rinehart & Winston, 1968. (Bibliography, 1901-1967, p. 504-519. 347 refs.)

0171. Jones, E. G., ed. A bibliography of the dog: books published in the English language, 1570-1965. London, The Library Association, 1971.

0172. Love, R. M. The chemical biology of fishes. New York, Academic Press, 1970. (Bibliography, 1894-1968, p. 451-541. 1407 refs.)

0173. McLean, F. C. Bone: fundamentals of the physiology of the skeletal tissue. 3d ed. Chicago, Univ. of Chicago Press, 1968. (Bibliography, 1736-1968, p. 269-299. 485 refs.)

0174. Mason, M. M. Bibliography of the dog. Ames, Iowa State University Press, 1959.

0175. Maupin, Bernard. Blood platelets in man and animals. 2 v. New York, Pergamon Press, 1969. (Bibliography, 1938-1965. Approx. 7500 refs.)

0176. Memory and learning—animal studies. Bulletin. no. 1- Los Angeles. Brain Research Institute and Biomedical Library, 1971- (A continuing bibliography.)

0177. Merck & Co. Coccidiosis: annotated bibliography. Rahway, N.J., 1970.

0178. Mouse hemoglobin. 1960-1968. Knoxville, Univ. of Tennessee, Mooney Memorial Library, 1968. (340 refs.)

0179. Murray, Margaret R., and Gertrude Kopech. A bibliography of the research in tissue culture, 1884-1950. New York, Academic Press, 1953.

Supplementary author list, 1950- (This entry is continued by entry 0180.)

0180. ——, and ——. Tissue culture bibliography, 1950-1960. New York, Columbia Univ. Press, 1965. (In microfilm copy, 11 reels.)

This entry is continued by *Current tissue culture literature* (entry 0006).

0181. Mycoplasmataceae: a bibliography and index 1852-1970. Ed. by C. H. Domermuth and J. G. Rittenhouse. Blacksburg, Va., Virginia Polytechnic Institute and State Univ., 1971. (Research Division Bulletin 61.)

0182. New York Society of Electron Microscopists. The international bibliography of electron microscopy. v. 1, 1950-1955. v. 2, 1956-1961. New York, 1962.

0183. Parkinson's disease and related disorders. Cumulative bibliography. 1800-1970. 3 v. Bethesda, Md., National Institute of Neurological Diseases and Stroke, 1971. (Sold in 3 v.-set by Supt. of Documents, U.S. Govt. Printing Office, Washington, D.C.) v. 1—Citations. v. 2—Author index. v. 3—Subject index.

MESH subject headings are used. 4,500 current serials and over 100 current bibliographic services and numerous books, monographs, and retrospective bibliographies were searched. Subject descriptors are assigned to each citation. These terms are in MESH and those compiled by the Parkinson Information Center. For current coverage see entry 0075.

0184. Pekas, J. C., and L. K. Bustad. A selected list of references on swine in biomedical research. Richland, Wash., Pacific Northwest Laboratory, 1965.

0185. Petchesky, Rosalind P. Issues in biological engineering: a review of recent literature. New York, Columbia Univ., Institute for the Study of Science in Human Affairs, 1969. (ISHA Bulletin no. 7. 71 annotated refs.)

0186. Plum Island Animal Disease Laboratory. A number of comprehensive bibliographies on specific exotic animal diseases have been compiled by the library staff of the U.S. Agricultural Research Service, Plum Island Animal Disease Laboratory, Greenport, New York. Included are:

Foot-and-mouth disease; a guide to reference sources: comprehensive works—review articles—bibliographies. By Bela Balassa. 1969.

Vesicular stomatitis, 1826-1963. By Bela Balassa. 1964.
...Suppl. no. 1, 1964-1967. By Bela Balassa. 1968.
...Suppl. no. 2, 1968-Sept. 1971. By Robert Uskavitch. 1971.
Vesicular exanthema of swine, 1933-1963. By Bela Balassa. 1963.
...Suppl. no. 1, 1964/June 1971. By Robert Uskavitch.
Borna disease, 1926-1971. By Robert Uskavitch. 1971.
Ephemeral fever, 1878-July 1971. By Robert Uskavitch. 1971.
African swine fever, 1921-1965. By Bela Balassa. 1965.
...Suppl. no. 1, 1966-June 1967. By Bela Balassa.
...Suppl. no. 2, July 1967-July 1971. By Robert Uskavitch.
African horse sickness, 1806-June, 1967. By Bela Balassa. 1967.
...Suppl. no. 1, July 1967-July 1971. By Robert Uskavitch.
...Suppl. no. 2, Aug. 1971-July 1972. By Robert Uskavitch.
Contagious bovine pleuropneumonia and mycoplasma mycoides var. mycoides, a bibliography and index 1852-1968. By M. Shifrine, H. Neimark, and B. Balassa.
...Appendix, 1969-April 1970.
...Suppl. no. 1, May 1970-April 1971. By Robert Uskavitch.
...Suppl. no. 2, May 1971-April 1972. By Robert Uskavitch.

0187. Radioautography in animal brain research. 1958-1968. New York, Parkinson Information Center, 1968. (221 refs.) 1969. (18 additional refs.)

0188. Reproduction Research Information Service. Interference with implantation: bibliography, 1962-1969. New York, International Publications Service, 1962-1969.

0189. Research using transplanted tumors of laboratory animals: cross-referenced bibliography. Ed. by D. C. Roberts. v. 1- London, Imperial Cancer Research Fund, 1964- (Annual.)

0190. Rich, S. T. Selected bibliography on gerbillinae. Los Angeles, Div. of Laboratory Animal Medicine, Center for the Health Sciences, Univ. of California, 1968.
...Addendum no. 1, April 1, 1971.
...Addendum no. 2, 1972.

0191. Rosenfeld, Gastão, and Eva M. A. Kelen. Bibliography of animal venoms, envenomations, and treatments, period 1500-1968. São Paulo, Instituto Butantan, 1969.

0192. Russell, F. E., and R. S. Scharffenberg. Bibliography of snake venoms and venomous snakes. West Covina, Calif., Bibliographic Associates, 1964.

0193. Sanborn, W. R., comp. Immunofluorescence, an annotated bibliography. 7 parts. Ft. Detrick, Md., U.S. Army Biological Laboratory, 1965. Part 1—Bacterial studies. Part 2—Viral studies. Part 3—Studies of fungi, metazoa, protozoa, and rickettsiae. Part 4—Studies of animal physiology. Part 5—Diagnostic applications and review articles. Part 6—Technical procedures. Suppl.—Author and subject indexes. (Its Misc. Publ. no. 3.)

0194. Schmidt, J. A. Cellular biology of vertebrate regeneration and repair. Chicago, Univ. of Chicago Press, 1968. (Bibliography, 1891-1967, p. 321-414. Approx. 1300 refs.)

0195. Schwabe, C. W. Veterinary medicine and human health. 2d ed. Baltimore, Williams & Wilkins, 1969. (Bibliography, p. 635-657. Approx. 1600 refs.)

0196. Searching the neoplastic literature: a guide to selected reference tools and other publications. 1950-1967. By Virginia MacDonald. Bethesda, Md., National Library of Medicine, 1968. (217 refs.)

0197. Selected abstracts on animal models for biomedical research. Comp. and ed. by C. B. Frank and Marilyn J. Anderson. Washington, D.C., Institute of Laboratory Animal Resources, National Research Council, National Academy of Sciences, 1971.

0198. Selye, Hans. Hormones and resistance. 2 parts. Heidelberg, Springer-Verlag, 1971. (Bibliography, p. 865-1051.)

0199. Skerman, V. B. D. Abstracts of microbiological methods. New York, Wiley, 1969. (Approx. 3400 annotated refs.)

0200. Southwest Foundation for Research and Education. The baboon: an annotated bibliography 1607-1964. San Antonio, Texas, 1964- (For further coverage see entry 4482.)

0201. The toxicity of herbicides to mammals, aquatic life, soil microorganisms, beneficial insects, and cultivated plants, 1950-1965: a list of selected references. By

Patricia A. Condon. Washington, D.C., U.S. National Agricultural Library, 1968. (Its Library List, no. 87. 1695 refs.)

0202. U.S. Armed Forces Institute of Pathology. International Reference Center for Comparative Oncology. Bibliography of comparative oncology. Comp. by F. M. Garner and R. J. Brown. Washington, D.C., 1967.

0203. U.S. Armed Forces Medical Library. The structure, composition and growth of bone, 1930-1953. A bibliography by Marjory C. Spencer and Katherine Uhler. U.S. Armed Forces Medical Library, Reference Division, Washington, D.C., 1955.

0204. U.S. Army Biological Warfare Laboratories. Ft. Detrick, Frederick, Md. Microbiological safety bibliography. By G. B. Phillips. Springfield, Va., CFSTI, 1965. (Its Misc. Publ. no. 6.)

0205. U.S. Defense Documentation Center. Venezuelan equine encephalomyelitis: a DDC bibliography. v. 1- Alexandria, Va., 1971- (A continuing bibliography.)

0206. U.S. Dept. of Agriculture. Animal and Plant Health Inspection Service. Veterinary Service. Listing of "91" series publications. Hyattsville, Md., 1972.

0207. U.S. Dept. of the Interior. Fish and Wildlife Service. Bureau of Sport Fisheries. Literature survey on general and comparative enzyme biochemistry of birds. Washington, D.C., U.S.G.P.O., 1970. (Special scientific report. Wildlife no. 143.)

0208. U.S. National Cancer Institute. Cancer and virus: a guide and annotated bibliography to monographs, reviews, symposia, and survey articles with emphasis on human neoplasm. 1950-1963. Ed. by Elizabeth Koenig. Washington, D.C., 1966. (U.S. Public Health Service Publ. no. 1424.)

0209. U.S. National Library of Medicine literature searches. Bethesda, Md., 1967-
Because they are considered to be of broad interest to the biomedical community, these reprints of literature searches prepared for individual health professionals as a by-product of the MEDLARS system are made available for single copy distribution by the National Library of Medicine.
Noted below are samples of those in the series which are germane to this publication. A complete listing of available literature searches will be found in each current issue of *Index medicus.*

Rabies in bats. Jan. 1964—July 1967. 24 refs.

15-67 Amino acid transport in non-human mammals. Mid-1963—Sept. 1967. 131 refs.

68-2 Neoplasm models. Mid-1963—Aug. 1967. 79 refs.

68-3 Tumors of cold-blooded vertebrates. Mid-1963—May 1967. 123 refs.

68-17 Veterinary dentistry. Jan. 1964—July 1968. 69 refs.

68-31 Models of neural activity. Jan. 1964—May 1968. 279 refs.

9-69 Sarcoidosis. Jan. 1966—Dec. 1968. 546 refs.

70-12 Risk of infection in the microbiology laboratory. Jan. 1967—Dec. 1969. 143 refs.

70-14 Dopa in neuromuscular disorders. Jan. 1967—March 1970. 98 refs.

70-30 Automation or use of computers in laboratory diagnosis. Jan. 1968—April 1970. 419 refs.

70-43 Prostaglandins and smooth muscles. Jan. 1967—July 1970. 147 refs.

70-9 Programmed instruction in medicine and allied professions. Jan. 1967—Dec. 1969. 108 refs.

71-5 Behavioral responses to animal interactions. Jan. 1968—March 1971. 180 refs.

71-7 Rabies in wildlife. Jan. 1968—March 1971. 99 refs.

71-24 Venezuelan equine encephalitis virus. Jan. 1969—Oct. 1971. 85 refs.

72-2 Programmed instruction in medicine and allied health professions. Jan. 1969—Feb. 1972. 129 refs.

72-9 Electrical safety and hazards in the hospital. Jan. 1969—March 1972. 102 refs.

72-15 Malnutrition and mental development in man. Jan. 1969—June 1972. 56 refs.

72-24 Care and maintenance of laboratory animals. Jan. 1969—June 1972. 299 refs.

72-25 Physical management of rhematoid arthritis. Jan. 1969—July 1972. 54 English-language refs.

0210. University of Maine at Orono. Dept. of Animal and Veterinary Sciences. A bibliography of avian mycosis. 3d ed. By E. S. Barden et al. Orono, University of Maine, 1971.

0211. Use of dopa in animal experiments.

1957-1968. New York, Parkinson Information Center, 1968. (133 refs.) 1969. (13 additional refs.)

0212. Vollman, R. Fifty years of research on mammalian reproduction. A bibliography of the scientific publications of C. G. Hartman. Washington, D.C., 1965. (U.S. Public Health Service publ. no. 1281.)

0213. Walter, Pat L., ed. A KWIC index to EEG and allied literature, 1966-1969. New York, Elsevier, 1970. (*Electroencephalography and clinical neurophysiology.* Suppl. 29.)

0214. Warren, K. S., and V. A. Newill.

Schistosomiasis: a bibliography of the world's literature from 1852 to 1962. 2 v. Cleveland, Western Reserve Univ. Press, 1967. (10,286 refs.)

0215. West, Billy. Bibliography of literature on preservation of microorganisms, blood, tissues and vaccines with emphasis on freezing and freeze-drying. Atlanta, National Communicable Disease Center, 1968. (762 refs.)

0216. Wineburgh, Margaret, ed. A KWIC index to EEG and allied literature, 1964-1966. New York, Elsevier, 1971. (*Electroencephalography and clinical neurophysiology.* Suppl. 30.)

3

PERIODICALS

The periodical literature and conference proceedings are among the most important primary sources of published information in the biomedical sciences. The results of research, as well as reports of work in progress, are usually published first in journals, proceedings or transactions of learned societies, or other serially issued publications. The broad spectrum of medical research, its interdisciplinary nature as well as its international character, has contributed to the very rapid and extensive expansion of the pertinent periodical literature in recent years, particularly during the period after World War II.

The aim of this chapter is to present a broad but selective cross section of the biomedical periodical literature and conference proceedings bearing special pertinence to comparative and veterinary medicine. The chapter is arranged under five headings: Primary Veterinary Medical Interest; Cognate Disciplines; School, Regional, State, and Commercial Publications; Additional Veterinary Medical Titles; and Periodically Occurring Congresses and Conferences.

For more comprehensive listings of periodicals, see the *World list of scientific periodicals*, *Ulrich's international periodicals directory*, and *Union list of serials in libraries of the U.S. and Canada*, or the more specialized lists, such as the *List of journals indexed in Index Medicus*. Additional titles of published bibliographies of periodicals will be found in Chapter 2, entries 0095-0122.

PRIMARY VETERINARY MEDICAL INTEREST

0217. Académie vétérinaire de France. Bulletin. v. 1- Paris, Vigot Freres Editeurs, 1928-

0218. Acta medica veterinaria. v. 1- Naples, Facolta' di Medicina Veterinaria, 1955-

0219. Acta veterinaria. (Academiae Scientiarum Hungaricae.) v. 1- Budapest, Akademiai Kiado, 1951-

0220. Acta veterinaria scandinavica. v. 1- Copenhagen, Den danske Drylaegforening, 1959-

0221. American Animal Hospital Association. Journal. (Animal hospital.) v. 1- Elkhart, Ind., 1965-

0222. American journal of veterinary research. v. 1- Chicago, American Veterinary Medical Association, 1940-

0223. American Veterinary Medical Association. Journal. v. 1- Chicago, 1869-

Periodicals

0224. American Veterinary Radiology Society. Journal. v. 1- Elkhart, Ind., 1960-

0225. Animal behaviour. v. 1- London, Bailliere, Tindall & Cassell, 1953-

0226. Animal blood groups and biochemical genetics. v. 1- Wageningen, Netherlands, Centre for Agricultural Publishing and Documentation, 1970-

0227. Animal health. v. 1- London, Animal Health Trust, 1964-

0228. Archiv fuer experimentelle Veterinaermedizin. v. 1- Leipzig, Hirzel Verlag, 1950-

0229. Archiv fuer Tierernaehrung. v. 1- Berlin, Akademie-Verlag, 1950-

0230. Australian veterinary journal. v. 1- Sydney, Australian Veterinary Association, 1927-

0231. Avian diseases. v. 1- College Station, Texas, American Association of Avian Pathologists, 1957-

0232. Behaviour: an international journal of comparative ethology. v. 1- Leiden, Netherlands, E. J. Brill, 1947-

0233. British veterinary journal. v. 1- London, Bailliere, Tindall & Cassell, 1880-

0234. Canadian veterinary journal. (Revue vétérinaire canadienne.) v. 1- Guelph, Canadian Veterinary Medical Association, 1960-

0235. Comparative biochemistry and physiology. v. 1- New York, Pergamon Press, 1960- In 1971 divided into A—Comparative physiology, and B—Comparative biochemistry.

0236. Cornell veterinarian. v. 1- Ithaca, New York State Veterinary College, 1911-

0237. Deutsche tieraerztliche Wochenschrift. v. 1- Hannover, Schaper, 1893-

0238. Equine veterinary journal. v. 1- London, British Equine Veterinary Association, 1968-

0239. Federal veterinarian. v. 1- Washington, D.C., National Association of Federal Veterinarians, 1943-

Primary Veterinary Medical

0240. Feline practice: the journal of feline medicine and surgery for the practitioner. v. 1- Santa Barbara, Calif., 1972-

0241. General and comparative endocrinology. v. 1- New York, Academic Press, 1961-

0242. Institute of Animal Technicians. Journal. v. 1- Aberdeen, England, Animal Dept., Medical School, 1950-

0243. Irish veterinary journal. v. 1- Dublin, Irish Veterinary Association, 1946-

0244. The Japanese journal of veterinary research. v. 1- Sapporo, Japan, Hokkaido University, Faculty of Veterinary Medicine, 1951?-

0245. Journal of animal science. v. 1- Albany, American Society of Animal Science, 1942-

0246. Journal of comparative laboratory medicine. v. 1-3. Fort Collins, Colo., Veterinary Research Publications, 1967-1969. (Formerly *American journal of veterinary clinical pathology.*)

0247. Journal of comparative neurology. v. 1- Philadelphia, Wistar Institute of Anatomy and Biology, 1891-

0248. Journal of comparative pathology. v. 1- Liverpool, Liverpool Univ. Press, 1888-

0249. Journal of small animal practice. v. 1- Oxford, Pergamon Press, 1960-

0250. Journal of zoo animal medicine. v. 1- Washington, D.C., American Association of Zoo Veterinarians, National Zoological Park, 1970-

0251. Laboratory animal science. v. 1- Joliet, Ill., American Association for Laboratory Animal Science, 1950-

0252. Modern veterinary practice. v. 1- Wheaton, Ill., American Veterinary Publications, 1920-

0253. Monatshefte fuer Veterinaermedizin. v. 1- Leipzig, Fischer Verlag, 1946-

0254. National Institute of Animal Health quarterly. v. 1- Tokyo, 1961-

0255. Netherlands journal of veterinary science. v. 1- Utrecht, Royal Netherlands Veterinary Association, 1968- (English edition of selected articles published in *Tijdschrift voor diergeneeskunde.*)

0256. New Zealand veterinary journal. v. 1- Wellington, New Zealand Veterinary Association, 1952-

0257. Nordisk veterinaermedicin. v. 1- Copenhagen, 1949- (Issued by the veterinary medical associations of Denmark, Finland, Norway, and Sweden.)

0258. Onderstepoort journal of veterinary research. v. 1- Pretoria, South Africa, Department of Agricultural Technical Services, 1933-

0259. Philippine journal of veterinary medicine. v. 1- Quezon City, University of the Philippines, College of Veterinary Medicine, 1962-

0260. Recueil de médecine vétérinaire d'Alfort. v. 1- Paris, Vigot Freres Editeurs, 1924-

0261. Refuah veterinarith. v. 1- Bet Dagan, Israel, Israel Veterinary Medical Association, 1944-

0262. Research in veterinary science. v. 1- Oxford, Blackwell, 1960-

0263. Royal Army Veterinary Corps. Journal. v. 1- London, Royal Army Veterinary Corps School and Stores, 1929-

0264. Schweizer Archiv fuer Tierheilkunde. v. 1- Zurich, Druck, 1859-

0265. Society for Experimental Biology and Medicine. Proceedings. v. 1- New York, 1903-

0266. South Africa Veterinary Medical Association. Journal. v. 1- Pretoria, 1929-

0267. Tijdschrift voor diergeneeskunde. v. 1- Utrecht, Maatschappij voor Diergeneeskunde, 1862-

0268. Tropical animal health and production. v. 1- Edinburgh, Livingstone, 1969-

0269. U.S. Agricultural Research Service. Animal Health Division. Disease control services activities. Quarterly report. Washington, D.C., 1968-

0270. U.S. Dept. of Agriculture. Research Service. Veterinary biologics division notice. v. 1- Hyattsville, Md., 1965-

0271. The veterinarian: an international journal. v. 1-6. New York, Pergamon Press, 1963-1969.

0272. Veterinariya. v. 1- Moscow, Mezhdunarod-nayakniga, 1924-

0273. Veterinary economics: the veterinarian's business magazine. v. 1- Cleveland, United Publishing, 1960-

0274. Veterinary medicine/Small animal clinician. v. 1- Kansas City, Mo., Veterinary Medicine Publishing Co., 1905-

0275. Veterinary pathology. v. 1- New York, Karger, 1964- (Formerly *Pathologia veterinaria.*)

0276. Veterinary record. v. 1- London, British Veterinary Association, 1888-

0277. Wiener tieraerztliche Monatsschrift. v. 1- Vienna, Verlag Brueder Hollinek, 1914-

0278. Wildlife disease. v. 1- Ames, Iowa, Wildlife Disease Association, 1959- (Microcards.)

0279. Wildlife Disease Association. Journal. v. 1- Ames, Iowa, 1965- (Formerly Wildlife Disease Association. *Bulletin.*)

0280. World's poultry science journal. v. 1- London, World's Poultry Science Association, 1945-

0281. Zeitschrift fuer Tierphysiologie, Tierernaehrung und Futtermittelkunde. v. 1- Hamburg, Paul Parey, 1938-

0282. Zeitschrift fuer Tierpsychologie. v. 1- Berlin, Paul Parey, 1937-

0283. Zeitschrift fuer Versuchstierkunde. v. 1- Jena, Fischer Verlag, 1961-

0284. Zentralblatt fuer Veterinaermedizin: Reihe A, B, and C. v. 1- Berlin, Paul Parey, 1953-

0285. Zoonoses research. v. 1- New York, Lyceum Press, 1960- (Irregular.)

0286. U.S. Center for Disease Control. (Formerly Communicable Disease Center.) Atlanta.

Many continuing reports are issued by the above agency with annual or more current summaries. These include:
Zoonoses Surveillance Reports for Brucellosis, Leptospirosis, Listeriosis, Malaria, Psittacosis, Rabies, Trichinosis.
Disease Surveillance Reports for Encephalitis, Hepatitis, Malaria, Neurotropic viral diseases, Rubella, Salmonella, Shigella.

COGNATE DISCIPLINES

0287. Acta allergologica. v. 1- Copenhagen, Munksgaard, 1948-

0288. Acta anatomica. v. 1- Basel, Karger, 1945-

0289. Acta endocrinologica. v. 1- Copenhagen, Periodica, 1948-

0290. Acta haematologica. v. 1- Basel, Karger, 1948-

0291. Acta microbiologica. v. 1- Budapest, Akademiai Kiado, 1953-

0292. Acta microbiologica polonica. v. 1- Warsaw, Polskie Towarzystuo Mickrobiologow, 1952-

0293. Acta morphologica neerlandoscandinavica. v. 1- Lisse, Netherlands, Swets & Zeitlinger, 1956-

0294. Acta neuropathologica. v. 1- Berlin, Springer-Verlag, 1961-

0295. Acta pathologica et microbiologica scandinavica. v. 1- Copenhagen, Munksgaard, 1924-

0296. Acta pathologica japonica. v. 1- Tokyo, Japanese Pathological Society, 1950-

0297. Acta pharmacologica et toxicologica. v. 1- Copenhagen, Munksgaard, 1944-

0298. Acta physiologica. v. 1- Budapest, Akademiai Kiado, 1950-

0299. Acta physiologica scandinavica. v. 1- Stockholm, Karolinska Institute, 1940-

0300. Acta tropica. v. 1- Basel, Verlag fuer Recht und Gesellschaft A. G., 1944-

0301. Acta virologica. v. 1- Prague, Czechoslovak Academy of Sciences, 1957-

0302. American heart journal. v. 1- St. Louis, Mosby, 1925-

0303. American journal of anatomy. v. 1- Philadelphia, Wistar Institute of Anatomy and Biology, 1901-

0304. American journal of clinical pathology. v. 1- Baltimore, Williams & Wilkins, 1931-

0305. American journal of epidemiology. v. 1- Baltimore, Johns Hopkins University, School of Hygiene and Public Health, 1921- (Formerly *American journal of hygiene.*)

0306. American journal of obstetrics and gynecology. v. 1- St. Louis, Mosby, 1920-

0307. American journal of pathology. v. 1- New York, Harper & Row, 1925-

0308. American journal of physiology. v. 1- Bethesda, Md., American Physiological Society, 1898-

0309. American journal of public health and the nation's health. v. 1- New York, American Public Health Association, 1911-

0310. American journal of roentgenology, radium therapy and nuclear medicine. v. 1- Springfield, Ill., Thomas, 1906-

0311. American journal of surgery. v. 1- New York, Donnelley, 1891-

0312. American journal of the medical sciences. v. 1- Thorofare, N.J., Charles B. Slack, Inc., 1820-

0313. American journal of tropical medicine and hygiene. v. 1- Lawrence, Kansas, Allen Press, 1921-

0314. American Medical Association. Journal. v. 1- Chicago, 1848-

0315. American review of respiratory disease. v. 1- New York, National Tuberculosis Association, 1917-

0316. Anatomical record. v. 1- Philadelphia, Wistar Institute of Anatomy and Biology, 1906-

0317. Anesthesia and analgesia: current

researches. v. 1- Cleveland, International Anesthesia Research Society, 1921-

0318. Animal production. v. 1- Edinburgh, Longman Group Ltd., 1959-

0319. Annales medicinae experimentalis et biologiae fenniae. v. 1- Helsinki, Finnish Medical Society Duodecim, 1947-

0320. Annals of applied biology. v. 1- London, Cambridge Univ. Press, 1914-

0321. Annals of tropical medicine and parasitology. v. 1- Liverpool, Liverpool Univ. Press, 1907-

0322. Antonie van Leeuwenhoek. Journal of microbiology and serology. v. 1- Amsterdam, Swets & Zeitlinger, 1935-

0323. Applied microbiology. v. 1- Baltimore, American Society for Microbiology, 1953-

0324. Archiv fuer die gesamte Virusforschung. v. 1- New York, Springer-Verlag, 1939-

0325. Archives of pathology. v. 1- Chicago, American Medical Association, 1926-

0326. Archives of surgery. v. 1- Chicago, American Medical Association, 1920-

0327. Arthritis and rheumatism. v. 1- New York, Harper & Row, 1934-

0328. Australian journal of experimental biology and medical science. v. 1- Adelaide, Univ. of Adelaide, 1924-

0329. Behavior biology. v. 1- New York, Academic Press, 1968- (Continues *Communications in behavioral biology.* Part A.)

0330. Biken's journal. v. 1- Osaka, Japan, Research Institute for Microbial Diseases, 1958-

0331. Biochemical journal. v. 1- London, Biochemical Society, 1906-

0332. Biochemistry. v. 1- Washington, D.C., American Chemical Society, 1964-

0333. The biochemistry of disease. v. 1- New York, Dekker, 1971-

0334. Biological structure and function. v. 1- Cambridge, England, The Univ. Press, 1972-

0335. Biology of reproduction. v. 1- New York, Academic Press, 1969-

0336. Bio-medical engineering. v. 1- London, United Trade Press, 1965- (English translation of *Meditsinskaya tekhuika.*)

0337. Biotechnology and bioengineering. v. 1- New York, Interscience, 1959-

0338. Blood: journal of hematology. v. 1- New York, Grune & Stratton, 1946-

0339. Blut. v. 1- Munich, Lehmanns Verlag, 1954-

0340. Brain research. v. 1- Amsterdam, Elsevier, 1966-

0341. British journal of anesthesia. v. 1- Cheshire, England, John Sherrat & Son Ltd., 1922-

0342. British journal of cancer. v. 1- London, H. K. Lewis, 1947-

0342a. British journal of experimental pathology. v. 1- London, H. K. Lewis, 1920-

0343. British journal of haematology. v. 1- Oxford, Blackwell, 1955-

0344. British journal of nutrition. v. 1- London, Cambridge Univ. Press, 1947-

0345. British journal of surgery. v. 1- Baltimore, Williams & Wilkins, 1913-

0346. British medical bulletin. v. 1- London, Medical Dept., British Council, 1943-

0347. British medical journal. v. 1- London, British Medical Association, 1832-

0348. British poultry science. v. 1- Edinburgh, Longman, 1960-

0349. Bulletin of entomological research. v. 1- London, Commonwealth Institute of Entomology, 1910-

0350. Bulletin of epizootic diseases of Africa. v. 1- London, Organization of African Unity, 1953-

0351. Calcified tissue research. v. 1- Berlin, Springer-Verlag, 1966-

0352. Canadian journal of animal science. v. 1- Ottawa, Agricultural Institute of Canada, 1921-

0353. Canadian journal of biochemistry. Ottawa, National Research Council of Canada, 1929-

0354. Canadian journal of comparative medicine. v. 1- Ottawa, Canadian Veterinary Medical Association, 1937-

0355. Canadian journal of microbiology. v. 1- Ottawa, National Research Council of Canada, 1954-

0356. Canadian journal of physiology and pharmacology. v. 1- Ottawa, National Research Council of Canada, 1964-

0357. Canadian journal of public health. v. 1- Toronto, Canadian Public Health Association, 1909-

0358. Canadian journal of zoology. v. 1- Ottawa, National Research Council of Canada, 1929-

0359. Cancer. v. 1- Philadelphia, Lippincott, 1948-

0360. Cancer research. v. 1- Baltimore, Williams & Wilkins, 1941-

0361. Cardiovascular research. v. 1- Boston, British Medical Journal, 1967-

0362. Cellular immunology. v. 1- New York, Academic Press, 1970-

0363. Circulation research. v. 1- New York, American Heart Association, 1953-

0364. Clinical and experimental immunology. v. 1- Oxford, Blackwell, 1966-

0365. Clinical toxicology. v. 1- New York, Marcel Dekker, 1968-

0366. Comparative biochemistry and physiology. v. 1- New York, Pergamon Press, 1959-

0367. Conditional reflex: a Pavlovian journal of research and therapy. v. 1- Philadelphia, Lippincott, 1966-

0368. Connective tissue research. v. 1- New York, Gordon & Breach, 1972-

0369. Cryobiology. v. 1- Washington, D.C., Georgetown Univ., 1964-

0370. Current topics in surgical research. v. 1- New York, Academic Press, 1969-

0371. Cytogenetics. v. 1- Basel, S. Karger, 1962-

0372. Dairy herd management. v. 1- Minneapolis, Miller Publishing Co., 1965-

0373. Developmental biology. v. 1- New York, Academic Press, 1959-

0374. Electroencephalography and clinical neurophysiology. v. 1- Amsterdam, Elsevier, 1949-

0375. Endocrinology. v. 1- Philadelphia, Lippincott, 1917-

0376. European journal of immunology. v. 1- New York, Academic Press, 1971-

0377. Experimental and molecular pathology. v. 1- New York, Academic Press, 1962-

0378. Experimental brain research. v. 1- Berlin, Springer-Verlag, 1966-

0379. Experimental cell research. v. 1- New York, Academic Press, 1950-

0380. Experimental eye research. v. 1- New York, Academic Press, 1961-

0381. Experimental gerontology. v. 1- New York, Pergamon Press, 1964-

0382. Experimental hematology. v. 22- Houston, International Society for Experimental Hematology, 1972- (Continues entry 0656, with modification in style and content.)

0383. Experimental medicine and surgery. v. 1- New York, Brooklyn Medical Press, 1943-

0384. Experimental neurology. v. 1- New York, Academic Press, 1959-

0385. Experimental parasitology. v. 1- New York, Academic Press, 1951-

0386. FDA papers. (Food and Drug Administration.) v. 1- Washington, D.C., U.S. Govt. Printing Office, 1967-

0387. Federation of American Societies for Experimental Biology. Federation proceedings. v. 1- Bethesda, Md., 1942-

0388. Feedstuffs: the weekly newspaper for agribusiness. v. 1- Minneapolis, Miller Publishing Co., 1929-

0389. Fertility and sterility. v. 1- Baltimore, Williams & Wilkins, 1949-

0390. Folia primatologica. v. 1- Basel, Karger, 1963-

0391. Gann: Japanese journal of cancer research. v. 1- Tokyo, Japan Publications Trading Co., 1907-

0392. Gastroenterology. v. 1- Baltimore, Williams & Wilkins, 1943-

0393. Gut. (British Society of Gastroenterology.) v. 1- London, British Medical Association, 1960-

0394. Haematologia. v. 1- Budapest, Akademiai Kiado, 1961- (Formerly *Haematologia Hungarica*.)

0395. Helminthological Society of Washington. Proceedings. v. 1- Lawrence, Kan., 1934-

0396. Heredity. v. 1- Edinburgh, Longman Group, 1947-

0397. Histochemical journal. v. 1- London, Chapman & Hall, 1968-

0398. Histochemie. v. 1- Berlin, Springer-Verlag, 1958- (Formerly *Zeitschrift fuer Zellforschung und mikroskopische Anatomie; Abteilung Histochemie*.)

0399. Hoard's dairyman. v. 1- Fort Atkinson, Wis., W. D. Hoard & Sons, 1885-

0400. Hormones; European review of endocrinology. v. 1- New York, Karger, 1970-

0401. Hormones and behavior. v. 1- New York, Academic Press, 1969-

0402. Immunochemistry. v. 1- New York, Pergamon Press, 1964-

0403. Immunology. v. 1- Oxford, Blackwell, 1958-

0404. In vitro. v. 6- Rockville, Md., Tissue Culture Association, 1970- (v. 1-5 issued annually as *In vitro* in a series representing the proceedings of the Annual Symposium sponsored by the Tissue Culture Association, entry 0785.)

0405. Infection and immunity. v. 1- Washington, D.C., American Society for Microbiology, 1970-

0406. Institut Pasteur. Annales. v. 1- Paris, Masson & Cie, 1887-

0407. International archives of allergy and applied immunology. v. 1- Basel, Karger, 1949-

0408. International journal of applied radiation and isotopes. v. 1- New York, Pergamon Press, 1956-

0409. International journal of cancer. v. 1- Geneva, International Union Against Cancer, 1966- (Formerly *Acta: unio internationalis contra cancrum*. Louvain.)

0410. International journal of fertility. v. 1- Baltimore, Waverly Press, 1956-

0411. International journal of radiation biology and related studies in physics, chemistry, and medicine. v. 1- London, Taylor & Francis, 1959-

0412. Investigative ophthalmology. v. 1- St. Louis, Mosby, 1962-

0413. Investigative radiology: clinical and laboratory studies in diagnosis. v. 1- Philadelphia, Lippincott, 1966-

0414. Investigative urology. v. 1- Baltimore, Williams & Wilkins, 1963-

0415. Israel journal of medical sciences. v. 1- Jerusalem, 1965- (Incorporating *Israel journal of experimental medicine* and *Israel medical journal*.)

0416. Israel journal of zoology. v. 1- Jerusalem, Weizmann Science Press of Israel, 1951-

0417. Japanese journal of experimental medicine. v. 1- Tokyo, Institute of Medical Science, 1922-

0418. Japanese journal of medical science

and biology. v. 1- Tokyo, National Institute of Health, 1952-

0419. Japanese journal of microbiology. v. 1- Tokyo, Igaku Shoin, 1957-

0420. Journal of anatomy. v. 1- London, Cambridge Univ. Press, 1966-

0421. Journal of applied bacteriology. v. 1- New York, Academic Press, 1938-

0422. Journal of applied physiology. v. 1- Bethesda, American Physiological Society, 1948-

0423. Journal of bacteriology. v. 1- Washington, D.C., American Society for Microbiology, 1916-

0424. Journal of biological chemistry. v. 1- Baltimore, American Society of Biological Chemists, 1905-

0425. Journal of biomechanics. v. 1- New York, Pergamon Press, 1968-

0426. Journal of biomedical materials research. v. 1- New York, Wiley, 1967-

0427. Journal of bone and joint surgery. v. 1-29. London, British Orthopaedic Association, 1930-1947.

0428. Journal of bone and joint surgery. American volume. v. 30- Boston, American Orthopaedic Association, 1948-

0429. Journal of bone and joint surgery. British volume. v. 30- London, British Orthopaedic Association, 1948-

0430. Journal of cell biology. v. 1- New York, Rockefeller Univ. Press, 1955-

0431. Journal of clinical endocrinology and metabolism. v. 1- Philadelphia, Lippincott, 1941-

0432. Journal of clinical investigation. v. 1- Boston, American Society for Clinical Investigation, 1924-

0433. Journal of clinical pathology. v. 1- London, British Medical Association, 1947-

0434. Journal of comparative and physiological psychology. v. 1- Washington, D.C. American Psychological Association, 1908-

0435. Journal of dairy science. v. 1- Champaign, Ill., American Dairy Science Association, 1917-

0436. Journal of economic entomology. v. 1- College Park, Md., Entomological Society of America, 1908-

0437. Journal of electrocardiology. v. 1- Kettering, Ohio, Research in Electrocardiology, 1968-

0438. Journal of embryology and experimental morphology. v. 1- London, Cambridge Univ. Press, 1953-

0439. Journal of endocrinology. v. 1- London, Society for Endocrinology, 1939-

0440. Journal of evolutionary biochemistry and physiology. v. 5- New York, Consultants Bureau, 1969- (English translation of *Zhurnal Evolyutsionnoi Biokhimi i Fiziologii*.)

0441. Journal of experimental biology. v. 1- London, Cambridge Univ. Press, 1923-

0442. Journal of experimental medicine. v. 1- New York, Rockefeller Univ. Press, 1896-

0443. Journal of general microbiology. v. 1- London, Cambridge Univ. Press, 1947-

0444. Journal of general physiology. v. 1- Baltimore, Rockefeller Univ. Press, 1918-

0445. Journal of general virology. v. 1- London, Cambridge Univ. Press, 1967-

0446. Journal of gerontology. v. 1- Washington, D.C., Gerontological Society, 1946-

0447. Journal of helminthology. v. 1- London, London School of Hygiene and Tropical Medicine, 1923-

0448. Journal of heredity. v. 1- Washington, D.C., American Genetic Association, 1910-

0449. Journal of histochemistry and cytochemistry. v. 1- Baltimore, Williams & Wilkins, 1953-

0450. Journal of hygiene. v. 1- Cambridge, England, The Univ. Press, 1901-

0451. Journal of immunological methods.

v. 1- Amsterdam, North-Holland Publishing Co., 1971-

0452. Journal of immunology. v. 1- Baltimore, Williams & Wilkins, 1916-

0453. Journal of infectious diseases. v. 1- Chicago, Univ. of Chicago Press, 1904-

0454. Journal of invertebrate pathology. v. 1- New York, Academic Press, 1959-

0455. Journal of investigative dermatology. v. 1- Baltimore, Williams & Wilkins, 1938-

0456. Journal of laboratory and clinical medicine. v. 1- St. Louis, Mosby, 1915-

0457. Journal of mammalogy. v. 1- Lawrence, Kan., Allen Press, 1919-

0458. Journal of medical education. v. 1- Chicago, Association of American Medical Colleges, 1926-

0459. Journal of medical entomology. v. 1- Honolulu, Bishop Museum Press, 1964-

0460. Journal of medical microbiology. v. 1- Edinburgh, Livingstone, 1968-

0461. Journal of medical primatology. v. 1- New York, Karger, 1972-

0462. Journal of medicine: experimental and clinical. v. 1- Basel, Karger, 1970- (Formerly *Medicina experimentalis* and *Medicina et Pharmacologica experimentalis.* Sect. A.)

0463. Journal of membrane biology. v. 1- New York, Springer-Verlag, 1969-

0464. Journal of microscopy. v. 1- Oxford, Blackwell, 1878- (Formerly Royal Microscopical Society. *Journal.*)

0465. Journal of morphology. v. 1- Philadelphia, Wistar Institute of Anatomy and Biology, 1887-

0466. Journal of neuropathology and experimental neurology. V. 1- Baltimore, 1942-

0467. Journal of neurosurgery. v. 1- Chicago, American Association of the Neurological Surgeons, 1944-

0468. Journal of nuclear medicine. v. 1- Chicago, Society of Nuclear Medicine, 1960-

0469. Journal of nutrition. v. 1- Bethesda, Md., American Institute of Nutrition, 1928-

0470. Journal of parasitology. v. 1- Lawrence, Kan., American Society of Parasitology, 1914-

0471. Journal of pathology. v. 1- Edinburgh, Longman Group, 1892- (Formerly *Journal of pathology and bacteriology.*)

0472. Journal of pharmaceutical sciences. v. 1- Washington, D.C., American Pharmaceutical Association, 1912- (Formerly *Journal of the American Pharmaceutical Association.*)

0473. Journal of pharmacology and experimental therapeutics. v. 1- Baltimore, Williams & Wilkins, 1909-

0474. Journal of physiology. v. 1- London, Cambridge Univ. Press, 1878-

0475. Journal of protozoology. v. 1- New York, Society of Protozoologists, 1954-

0476. Journal of reproduction and fertility. v. 1- Oxford, Blackwell, 1960-

0477. Journal of surgical research. v. 1- New York, Academic Press, 1961-

0478. Journal of the science of food and agriculture. v. 1- London, Society of Chemical Industry, 1950-

0479. Journal of trauma. v. 1- Baltimore, Williams & Wilkins, 1961-

0480. Journal of ultrastructure research. v. 1- New York, Academic Press, 1957-

0481. Journal of urology. v. 1- Baltimore, Williams & Wilkins, 1917-

0482. Journal of virology. v. 1- Bethesda, Md., American Society for Microbiology, 1967-

0483. Laboratory investigation. v. 1- Baltimore, Williams & Wilkins, 1952-

0484. Lancet. v. 1- London, Lancet, 1823- (English edition.)

0485. Lancet. v. 1- Boston, Little, Brown, 1966- (North American edition.)

0486. Life sciences. (In two parts: Part 1—Physiology & pharmacology; Part 2—Biochemistry, general and molecular biology.) v. 1- New York, Pergamon Press, 1962-

0487. Lymphology. v. 1- Stuttgart, Germany, Georg Thieme Verlag, 1968-

0488. Mammalian chromosomes newsletter. Houston, Univ. of Texas, M. D. Anderson Hospital and Tumor Institute. (Available to qualified research personnel.)

0489. Medical journal of Australia. v. 1- Sydney, Australasian Medical Publishing Co., 1914-

0490. Medical laboratory technology. v. 9- London, Academic Press, 1951- (Continues *Laboratory journal* and *Journal of medical laboratory technology*.)

0491. Metabolism: clinical and experimental. v. 1- New York, Stratton, 1952-

0492. Microvascular research. v. 1- New York, Academic Press, 1968-

0493. Nature: new biology. Wednesday edition. v. 229- London, Macmillan, 1971- (v. 1-228, 1869-1970, as *Nature*.)

0494. Neurobiology; biochemistry and morphology. v. 1- Copenhagen, Munksgaard, 1971-

0495. Neuroendocrinology. v. 1- Basel, Karger, 1965-

0496. Neurology. v. 1- Minneapolis, American Academy of Neurology, 1951-

0497. New York Academy of Sciences. Annals. v. 1- New York, New York Academy of Sciences, 1877-

0498. New Zealand medical journal. v. 1- Dunedin, New Zealand, Otago Daily Times, 1885-

0499. Oncology: journal of clinical and experimental cancer research. v. 1- Basel, Karger, 1948- (Formerly *Oncologia*.)

0500. Parasitology. v. 1- New York, Cambridge Univ. Press, 1908-

0501. Perspectives in biology and medicine. v. 1- Chicago, Univ. of Chicago Press, 1957-

0502. Pfluegers Archiv: European journal of physiology. v. 1- Berlin, Springer-Verlag, 1868- (Formerly *Pfluegers Archiv fuer die gesamte Physiologie des Menschen und der Tiere*.)

0503. Physics in medicine and biology. v. 1- London, Institute of Physics, 1956-

0504. Physiology and behavior. v. 1- Oxford, Pergamon Press, 1966-

0505. Plastic and reconstructive surgery. v. 1- Baltimore, Williams & Wilkins, 1946-

0506. Postgraduate medicine. v. 1- Minneapolis, McGraw-Hill, 1925-

0507. Poultry science. v. 1- College Station, Tex., Poultry Science Association, 1908-

0508. Primate behavior: developments in field and laboratory research. Ed. by L. A. Rosenblum. v. 1- New York, Academic Press, 1970-

0509. Public health reports. v. 1- Washington, D.C., U.S. Public Health Service, 1878-

0510. Quarterly journal of experimental physiology and cognate medical sciences. v. 1- Edinburgh, Livingstone, 1908-

0511. Radiation research. v. 1- New York, Academic Press, 1954-

0512. Radiology. v. 1- Syracuse, N.Y., Radiological Society of North America, 1915-

0513. Respiration physiology. v. 1- Amsterdam, North-Holland Publishing Co., 1965-

0514. Reticuloendothelial Society. Journal. v. 1- New York, 1964-

0515. Royal Society of Tropical Medicine and Hygiene. Transactions. v. 1- London, Manson House, 1907-

0516. Scandinavian journal of clinical and laboratory investigation. v. 1- Copenhagen, Universitetsforlaget, 1948-

0517. Scandinavian journal of immunology. v. 1- Oslo, Universitetsforlaget, 1972-

0518. Scandinavian journal of infectious diseases. v. 1- Stockholm, Almqvist & Wiksell, 1969-

0519. Science. v. 1- Washington, D.C., American Association for the Advancement of Science, 1880-

0520. South African journal of medical sciences. v. 1- Johannesburg, Witwatersrand Univ. Press, 1935-

0521. Stain technology. v. 1- Baltimore, Williams & Wilkins, 1925-

0522. Surgery. v. 1- St. Louis, Mosby, 1937-

0523. Surgery, gynecology & obstetrics. v. 1- Chicago, Franklin H. Martin Memorial Foundation, 1905-

0524. Teratology: journal of abnormal development. v. 1- Philadelphia, Wistar Institute Press, 1968-

0525. Texas reports on biology and medicine. v. 1- Galveston, Univ. of Texas Medical Branch, 1943-

0526. Tissue and cell. v. 1- Edinburgh, Oliver & Boyd, 1969-

0527. Tissue antigens. v. 1- Copenhagen, Munksgaard, 1971-

0528. Toxicology and applied pharmacology. v. 1- New York, Academic Press, 1959-

0529. Toxicon. v. 1- New York, Pergamon Press, 1962-

0530. Transplantation. v. 1- Baltimore, Williams & Wilkins, 1963- (Formerly *Transplantation bulletin*. Great Falls, Mont.)

0531. Transplantation proceedings. v. 1- New York, Transplantation Society, 1969-

0532. U.S. National Cancer Institute. Journal. v. 1- Washington, D.C., Dept. of Health, Education & Welfare, 1940-

0533. Virology. v. 1- New York, Academic Press, 1955-

0534. WHO chronicle. v. 1- Geneva, World Health Organization, 1947-

0535. Zeitschrift fuer Zellforschung und mikroskopische Anatomie. v. 1- New York, Springer-Verlag, 1924-

SCHOOL, REGIONAL, STATE, AND COMMERCIAL PUBLICATIONS

0536. Auburn veterinarian. v. 1- Auburn, Ala., Student Chapter of the American Veterinary Medical Association, 1945-

0537. California veterinarian. v. 1- Oakland, California Veterinary Medical Association, 1947-

0538. Fort Dodge biochemic review. v. 1- Fort Dodge, Iowa, Fort Dodge Laboratories, 1930-

0539. Georgia veterinarian. v. 1- Athens, Georgia Veterinary Medical Association, 1948-

0540. Illinois veterinarian. v. 1- Urbana, Univ. of Illinois College of Veterinary Medicine, 1958-

0541. Iowa State University veterinarian. v. 1- Ames, Iowa State University, 1938- (Triennial. Formerly *Iowa State College veterinarian*.)

0542. Iowa veterinarian. v. 1- Des Moines, Iowa Veterinary Medical Association, 1920-

0543. Kansas veterinarian. v. 1- Manhattan, Ag Press, 1957- (Quarterly of the Kansas Veterinary Medical Association.)

0544. Maryland veterinarian. v. 1- Baltimore, Ray Thompson, 1959-

0545. Minnesota veterinarian. v. 1- St. Paul, College of Veterinary Medicine, Univ. of Minnesota, 1961- (Approximately semiannual.)

0546. Missouri veterinarian. v. 1- Columbia, Missouri Student Chapter of the American Veterinary Medical Association, 1951-

0547. Norden news. v. 1- Lincoln, Neb. Norden Laboratories, 1927-

0548. North Carolina veterinarian. v. 1-

Smithfield, North Carolina Veterinary Medical Association, 1954-

0549. Oklahoma veterinarian. v. 1- Stillwater, Heritage Press, 1949-

0550. Pennsylvania veterinarian. v. 1- Camp Hill, Pennsylvania Veterinary Medical Association, 1959-

0551. Practicing veterinarian. v. 1- Washington Crossing, N.J., Pitman-Moore, 1930- (Formerly *Allied veterinarian*.)

0552. Pulse. (Southern California Veterinary Medical Association.) v. 1- Pico Rivera, 1959-

0553. Southern veterinarian. v. 1- Springville, Ala., Southern Veterinarian, Inc., 1964-

0554. Southwestern veterinarian. v. 1- College Station, Tex., American Veterinary Medical Association, 1946-

0555. Texas veterinary medical journal. v. 1- Austin, Texas Veterinary Medical Association, 1938-

0556. Veterinary news. v. 1- Utica, New York State Veterinary Medical Society, 1937-

0557. Veterinary scope. v. 1- Kalamazoo, Mich., Upjohn Co., 1954-

0558. Virginia veterinarian. v. 1- Richmond, Cavalier Press, 1958-

ADDITIONAL VETERINARY MEDICAL TITLES

0559. Acta veterinaria, Brno. v. 1- Brno, Central Library of the University School of Veterinary Medicine, 1922- (Formerly *Acta universitatis agriculturae. Facultas veterinaria: Rada B.*)

0560. Acta veterinaria japonica. v. 1- Tokyo, Dai-Nihon Printing, 1956-

0561. Ankara üniversite. Veteriner fakültesi dergisi. v. 1- Ankara, 1954-

0562. Annales de médecine vétérinaire. v. 1- Brussels, Corps Enseignant de la faculté de Médecine Vétérinaire de l'Université de Liege, 1849-

0563. Archiva veterinaria. v. 1- Bucharest, Institutul de Cercetari si Biopreparate Veterinare "Pasteur," 1965-

0564. Archivio veterinario italiano. v. 1- Milan, Consiglio Nazionalle delle Ricerche, 1950-

0565. Berliner und Muenchener tieraerztliche Wochenschrift. Berlin and Munich veterinarian's weekly. v. 1- Berlin, Paul Parey, 1888-

0566. Ceylon veterinary journal. v. 1- Colombo, Ceylon Veterinary Association, 1953-

0567. Clinica veterinaria. v. 1- Milan, Istituto Sieroterapico Milanese Serafino Belfanti, 1878-

0568. Hellenic veterinary medicine. v. 1- Thessaloniki, Greece, Nicolas Aspiotis, 1958-

0568a. ILAR news. v. 10- Washington, D.C., Institute of Laboratory Animal Resources, 1966-

0569. Indian journal of veterinary science and animal husbandry. v. 1- New Delhi, Indian Council of Agricultural Research, 1931-

0570. Indian veterinary journal. v. 1- Madras, Indian Veterinary Association, 1924-

0571. The Japanese journal of veterinary science. v. 1- Tokyo, Japanese Society of Veterinary Science, 1939- (Most articles in Japanese.)

0572. Kajian veterinaire. (Journal of the Association of Veterinary Surgeons.) v. 1- Singapore, Malaysia, 1967-

0573. Kleintier-Praxis: Archiv fuer klein Haus- und Nutztiere sowie Laboratoriums- und Zoo-Tiere. v. 1- Hannover-Waldhausen, M. H. Schaper, 1956-

0574. Medycyna weterynaryjna. (Polish Society of Veterinary Sciences.) v. 1- Lublin, Poland, Panstwowe Wydawnictwo Relnicze i Lesne, 1945-

0575. Nippon Juishikai Zasshi. (Journal of the Japan Veterinary Medical Association.) v. 1- Tokyo, 1948-

0576. Office international des épizooties. Bulletin. v. 1- Paris, 1931-

0577. Revue d'élevage et de médecine vétérinaire des pays tropicaux. v. 1- Paris, Vigot Freres, 1947-

0578. Revue de médecine vétérinaire. v. 1- Toulouse, Ecoles Nationales Vétérinaires de Lyon et de Toulouse, 1900-

0579. Tieraerztliche Umschau. v. 1- Konstanz, Terra Verlag, 1948-

0580. Veterinaria italiana. v. 1- Teramo, Italy, Istituto Sperimentale Zooprofilattico, 1954-

0581. Veterinary practice. v. 1- Stoneleigh, Ewell, Surrey, England, N.C. Magazines Ltd., 1966-

0582. Zooprofilassi. v. 1- Rome, Istituti Zooprofilattici Sperimentali, 1946-

0583. Zootechnia; acta societatis internationalis veterinariorum zootechnicorum. v. 1- Madrid, Garsi, S.L., 1946-

0584. Zootecnica e veterinaria. v. 1- Milan, 1946-

0585. Zuchthygiene. v. 1- Berlin, Paul Parey, 1966-

PERIODICALLY OCCURRING CONGRESSES AND CONFERENCES

0586. American Animal Hospital Association. v. 1- Scientific presentations. Elkhart, Ind., 1934- (Annual.)

0587. American Association of Bovine Practitioners. Annual convention. 1st- Proceedings. Stillwater, Okla., 1968-

0588. American Association of Equine Practitioners. Annual convention. 1st- Proceedings. Golden, Colo., 1956- (Proceedings of earlier meetings were not published separately.)

0589. American College of Veterinary Ophthalmologists. Annual meeting. 1st- Proceedings. Chicago, 1970-

0590. Biomedical sciences instrumentation symposia. v. 1- Proceedings. Pittsburgh, Instrument Society of America, 1963- (Annual.)

0591. Conference and workshop on histocompatibility testing. 1st- Baltimore, Williams & Wilkins, 1964- (Approximately biennial.)

0592. Conference on biology of hard tissue. 1st- Proceedings. 1965- (Annual; publishers vary.)

0593. Conference on fetal homeostasis. v. 1- Proceedings. New York, Appleton-Century-Crofts, 1966- (Approximately annual.)

0594. Inter-American meeting on foot-and-mouth disease and zoonoses control. 1st- Washington, D.C., Pan American Health Organization, 1968- (Annual.)

0595. International congress for microbiology. 1st- Proceedings. 1930- (Congresses held irregularly at 3-5 year intervals; publishers vary.)

0596. International congress of immunology. 1st- Washington, D.C., 1971-

0597. International congress of neuropathology. 1st- Proceedings. Paris, Masson et Cie, Editeurs, 1952- (Approximately triennial.)

0598. International congress on animal reproduction and artificial insemination. 1st- Proceedings. 1948- (Quadrennial; publishers vary.)

0599. International Society of Haematology. v. 1- Proceedings of the International Congress of Haematology. Cambridge, England, 1948- (Biennial; publishers vary.)

0600. International symposium on vectorcardiography. v. 1- Proceedings. Philadelphia, Lippincott, 1965- (Semiannual.)

0601. Leucocyte culture conference. 1st- 1965- (Approximately annual; publishers vary.)

0602. Società italiana delle scienze veterinarie. Atti. v. 1- Faenza, Italy, 1947- (Annual.)

0603. Technical conference on artificial insemination and reproduction. 1st- Proceedings. Columbia, Mo., National Association of Animal Breeders, 1966?- (Biennial.)

0604. United States Animal Health Association. Annual meeting. 1st- Proceedings. Richmond, Va., 1897- (Continues United States Livestock Sanitary Association. Proceedings.)

0604a. World Association of Veterinary Food Hygienists. 1st- Proceedings. 1956- (Formerly International Association of Veterinary Food Hygienists. Approximately triennial; publishers vary.)

0605. World congress on fertility and sterility. 1st- Proceedings. 1953- (Approximately biennial; publishers vary.)

0606. World veterinary congress. 1st- Proceedings. 1893- (Approximately quadrennial; publishers vary.)

0607. World's poultry congress. 1st- Proceedings. 1921- (Congresses held irregularly at 3-5 year intervals; publishers vary.)

Note: For additional conferences and congresses see the section on "Serial Monographs" (entries 0731-0801) in Chap. 4.

4

REVIEW SERIALS AND SERIAL MONOGRAPHS

Review serials serve as secondary information sources by presenting articles that review, synthesize, survey, and in some cases evaluate the "state of the art" of a unique and timely segment of a specialized subject area over a designated span of time. This is in contrast to primary source journals, which ordinarily report original research. In many instances, review serials are published annually under the direction of an editorial board composed of specialists in the subject under consideration. Although the serial may be published annually, individual articles will usually survey the literature covering the work and advances of many years in a narrow subject field. In general, the articles are written by specialists upon invitation of the editor, or represent papers presented at a symposium, conference, or other meeting of experts. Review articles are usually characterized by comprehensive, in-depth, definitive, and balanced coverage of the subject matter, and include extensive bibliographies of relevant articles published over a specified span of years. This type of serial is of particular value to clinicians who will find current information of value in a practice situation, as well as to individuals who wish an overview of a subject ancillary to their own specific area of competence. Researchers may need a review of the literature preparatory to the initiation of a new investigation, and students who are preparing for comprehensive examinations will find them valuable.

Review articles of biomedical interest are sometimes found in journals of broader scope than the review serials, such as *American journal of the medical sciences*, or in more specialized periodicals, such as *Journal of infectious diseases*, along with articles reporting original research. In some instances, review articles may be issued as journal supplements or special issues, examples of which are the supplements to *Acta physiologica scandinavica* and special issues of *Postgraduate medicine*. On occasion they may be found in appropriate index and abstract journals. For example, each issue of the various abstract journals published by the Commonwealth Agricultural Bureaux usually includes one extensive review article. For a key to review articles of biomedical interest that appear in general periodicals as well as in review serials, the reader is referred to *Bibliography of medical reviews* (see entry 0026). A very useful guide for bibliographic information about review serials as well as selection of them is *Irregular serials and annuals: an international directory* (described in entry 0105).

Serial monographs are publications which have distinctive titles and usually can stand by themselves as individual books but are numbered as parts of series. They are published as a result of their being symposia, colloquia, proceedings, collected papers, lecture series, congress or conference reports, publisher's series, or reports of sponsored research projects. The frequency of publication is often irregular. Usually these monographs include some previously unpublished material as well as secondary source materials.

In some libraries, individual serial monographs may not be brought to the attention of the reader if only the title of the series is indicated in the catalog and the separate titles are not analyzed. If the library analyzes each monograph in a series, it may be found under author, title, and subject, in the same manner as any book.

Some of the series included in the present work have been in publication for many years, while a number are of more recent origin. In contrast to multivolume sets of books, the completion of which is planned to fall within a given span of time or a given number of volumes, the publication of a series of monographs, like periodicals, is planned to continue indefinitely.

A selected list of review serials and a number of series with some titles of individual monographs that are of direct or related interest to the researcher, clinician, or educator whose interest is in comparative and veterinary medicine are noted below.

REVIEW SERIALS

0608. Advances in applied microbiology. v. 1- New York, Academic Press, 1959-

0609. Advances in biological and medical physics. v. 1- New York, Academic Press, 1948-

0610. Advances in biomedical engineering. v. 1- New York, Academic Press, 1971-

0611. Advances in biomedical engineering and medical physics. v. 1- New York, Wiley, 1968-

0612. Advances in cell and molecular biology. v. 1- New York, Academic Press, 1971-

0613. Advances in clinical chemistry. v. 1- New York, Academic Press, 1958-

0614. Advances in comparative physiology and biochemistry. v. 1- New York, Academic Press, 1962-

0615. Advances in enzyme regulation. New York, Pergamon Press, 1963-

0616. Advances in genetics. v. 1- New York, Academic Press, 1947-

0617. Advances in gerontological research. v. 1- New York, Academic Press, 1964-

0618. Advances in marine biology. v. 1- New York, Academic Press, 1963-

0619. Advances in metabolic disorders. v. 1- New York, Academic Press, 1964-

0620. Advances in microbial physiology. v. 1- New York, Academic Press, 1967-

0621. Advances in microcirculation. v. 1- New York, Karger, 1968-

0622. Advances in morphogenesis. v. 1- New York, Academic Press, 1961-

0623. Advances in nephrology from the Necker Hospital. v. 1- Chicago, Year Book Medical Publishers, 1971-

0624. Advances in oral biology. v. 1- New York, Academic Press, 1964-

0625. Advances in parasitology. v. 1- New York, Academic Press, 1963-

0626. Advances in pharmacology and chemotherapy. v. 1- New York, Academic Press, 1962-

0627. Advances in reproductive physiology. v. 1- New York, Academic Press, 1966-

0628. Advances in small animal practice. v. 1-4. Oxford, Pergamon Press, 1959-1962.

0629. Advances in steroid biochemistry and pharmacology. v. 1- New York, Academic Press, 1970-

0630. Advances in teratology. v. 1- New York, Academic Press, 1966-

0631. Advances in the study of behavior. v. 1- New York, Academic Press, 1965-

0632. Advances in veterinary science and omparative medicine. v. 1- New York, cademic Press, 1953-

0633. Advances in virus research. v. 1- New York, Academic Press, 1953-

0634. Animal behaviour monographs. v. 1- London, Bailliere, Tindall & Cassell, 1968-

0635. Annual review of biochemistry. v. 1- Palo Alto, Calif., Annual Reviews, 1932-

0636. Annual review of biophysics and bioengineering. v. 1- Palo Alto, Calif., Annual Reviews, 1972-

0637. Annual review of medicine. v. 1- Palo Alto, Calif., Annual Reviews, 1950-

0638. Annual review of pharmacology. v. 1- Palo Alto, Calif., Annual Reviews, 1961-

0639. Annual review of physiology. v. 1- Palo Alto, Calif., Annual Reviews, 1939-

0640. Bacteriological reviews. v. 1- Baltimore, American Society for Microbiology, 1937-

0641. Biochemical Society, London. Symposia. v. 1- London, Cambridge Univ. Press, 1942-

0642. Biomembranes. v. 1- New York, Plenum Press, 1971-

0643. Ciba Foundation Study Group. v. 1- London, 1959- (distributed in U.S. by Little, Brown of Boston).

0644. Clinical anesthesia. v. 1- Philadelphia, Davis, 1963-

0645. Clinical obstetrics and gynecology. v. 1- New York, Harper & Row, 1958-

0646. Clinical orthopaedics and related research. no. 1- Philadelphia, Lippincott, 1953-

0647. Contemporary topics in immunobiology. v. 1- New York, Plenum Press, 1972-

0648. Contributions to sensory physiology. v. 1- New York, Academic Press, 1965-

0649. Current topics in cellular regulation. v. 1- New York, Academic Press, 1969-

0650. Current topics in comparative pathobiology. v. 1- New York, Academic Press, 1971-

0651. Current topics in developmental biology. v. 1- New York, Academic Press, 1966-

0652. Current topics in experimental endocrinology. v. 1- New York, Academic Press, 1971-

0653. Current topics in microbiology and immunology. Ergebnisse der Mikrobiologie und Immunitätsforschung. v. 1- Berlin, Springer-Verlag, 1967-

0654. Current topics in radiation research. v. 1- New York, Wiley, 1965-

0655. Essays in toxicology. v. 1- New York, Academic Press, 1969-

0656. Experimental hematology. v. 1-21. Oak Ridge, Tenn., Oak Ridge National Laboratory, 1957-1971. (Continues *Fundamental and clinical aspects of radiation protection and recovery.*)

0656a. Experiments in physiology and biochemistry. v. 1- New York, Academic Press, 1968-

0657. Food and Agriculture Organization of the United Nations. Animal health yearbook. Rome, 1957-

0658. Frontiers in neuroendocrinology. v. 1- New York, Oxford Univ. Press, 1969-

0659. Gaines veterinary symposium. 1st- Papers presented. New York, Gaines Dog Research Center, 1951-

0660. Hematologic reviews. v. 1- New York, Marcel Dekker, 1968-

0661. Institute of Biology. Symposia. 1st- 1952- (Publishers vary.)

0662. International convocation on immunology. 1st- New York, Karger, 1968- (Biennial.)

0663. International review of connective tissue research. v. 1- New York, Academic Press, 1963-

0664. International review of cytology. v. 1- New York, Academic Press, 1952-

Suppl. 1. Microbodies and related particles.

Suppl. 2. Cellular mechanisms of chromosome distribution.

0665. International review of experimental pathology. v. 1- New York, Academic Press, 1962-

0666. International review of general and experimental zoology. v. 1- New York, Academic Press, 1964-

0667. International review of neurobiology. v. 1- New York, Academic Press, 1959-

0668. International review of tropical medicine. v. 1- New York, Academic Press, 1961-

0669. International Society for Cell Biology. Symposia. v. 1- New York, Academic Press, 1962-

0670. International zoo yearbook. v. 1- London, Zoological Society of London, 1959-

0671. Kyosoba Hoken Kenkyusho. Hokoku. Experimental reports of equine health laboratory. v. 1- Tokyo, 1963-

0672. Major problems in clinical surgery. v. 1- Philadelphia, Saunders, 1964-

0673. Medical clinics of North America. v. 1- Philadelphia, Saunders, 1917-

0674. Methods in cell physiology. v. 1- New York, Academic Press, 1964-

0675. Methods in medical research. v. 1- Chicago, Year Book Medical Publishers, 1948-

0676. Methods in virology. v. 1- New York, Academic Press, 1967-

0677. Methods of biochemical analysis. v. 1- New York, Wiley, 1954-

0678. Modern treatment. v. 1- New York, Harper & Row, 1964-

0679. Modern trends in immunology. v. 1- Washington, D.C., Butterworth, 1963-

0680. Modern trends in medical virology. v. 1- New York, Appleton-Century-Crofts, 1967-

0681. Modern trends in pathology. v. 1- New York, Academic Press, 1959-

0682. Modern trends in pharmacology and therapeutics. v. 1- New York, Appleton-Century-Crofts, 1967-

0683. Modern trends in surgery. v. 1- New York, Appleton-Century-Crofts, 1962-

0684. Modern trends in toxicology. v. 1- New York, Appleton-Century-Crofts, 1968-

0685. Modern trends in vascular surgery. v. 1- New York, Appleton-Century-Crofts, 1970-

0686. Neurosciences research. v. 1- New York, Academic Press, 1968-

0687. Nutrition reviews. v. 1- New York, Nutrition Foundation, 1942-

0688. Pathobiology annual. v. 1- New York, Appleton-Century-Crofts, 1971-

0689. Pathology annual. v. 1- New York, Appleton-Century-Crofts, 1966-

0690. Pharmacological reviews. v. 1- Baltimore, Williams & Wilkins, 1949-

0691. Primate behavior: developments in field and laboratory research. v. 1- New York, Academic Press, 1970-

0692. Progress in allergy. v. 2- Basel, Karger, 1949-

0693. Progress in experimental tumor research. v. 1- Basel, Karger, 1960-

0694. Progress in histochemistry and cytochemistry. v. 1- Stuttgart, Gustav Fischer Verlag, 1970-

0695. Progress in immunobiological standardization. v. 1- Basel, Karger, 1962-

0696. Progress in liver diseases. v. 1- New York, Grune & Stratton, 1961-

0697. Progress in medical laboratory technique. v. 1- New York, Appleton-Century-Crofts, 1962-

0698. Progress in medical virology. v. 1- New York, S. Karger, 1958-

0699. Progress in neurology and psychiatry. v. 1- New York, Grune & Stratton, 1946-

0700. Progress in neuropathology. v. 1- New York, Grune & Stratton, 1971-

0701. Progress in nucleic acid research and molecular biology. v. 1- New York, Academic Press, 1963-

0702. Progress in surgery. v. 1- Basel, Karger, 1961-

0703. Progress of allergology. v. 1- Baltimore, University Park Press, 1970-

0704. Radiologic clinics of North America. v. 1- Philadelphia, Saunders, 1963-

0705. Recent progress in hormone research. v. 1- New York, Academic Press, 1947-

0706. Recent results in cancer research. v. 1- Berlin, Springer-Verlag, 1965-

0707. Research on steroids. v. 1- Amsterdam, North-Holland, 1964- (International Study Group for Steroid Hormones, proceedings of the biennial meeting.)

0708. Series haematologica. v. 1-10. Copenhagen, Munksgaard, 1965. (Suppl. to *Scandinavian journal of haematology*, and selected papers from 10th Congress of the International Society of Haematology, Stockholm, 1964.)

0709. Series haematologica. v. 1- Baltimore, Williams & Wilkins, 1968- (Continues entry 0708.)

0710. Society for Developmental Biology. Symposium. v. 1- New York, Academic Press, 1939-

0711. Society for Endocrinology. Memoirs. v. 1- Cambridge, England, The Univ. Press, 1953-

0712. Standard methods of clinical chemistry. v. 1- New York, Academic Press, 1953-

0713. Surgical clinics of North America. v. 1- Philadelphia, Saunders, 1921-

0714. Symposium on fundamental cancer research. v. 1- Annual. Houston, Univ. of Texas, M. D. Anderson Hospital and Tumor Institute, 1947- (Publishers vary.)

0715. Veterinary annual. v. 1- Bristol, John Wright & Sons, 1959-

0716. The veterinary clinics of North America. v. 1- Philadelphia, Saunders, 1971-

0717. Vitamins and hormones: advances in research and applications. v. 1- New York, Academic Press, 1943-

0718. World Health Organization. Veterinary Public Health Unit. World survey of rabies. v. 1- Geneva, 1961-

0719. World review of nutrition and dietetics. v. 1- New York, S. Karger, 1959-

0720. Year book of anesthesia. Chicago, Year Book Medical Publishers, 1963-

0721. Year book of cardiovascular medicine and surgery. Chicago, Year Book Medical Publishers, 1961. (Formerly *Year book of cardiovascular and renal diseases*.)

0722. Year book of dermatology. Chicago, Year Book Medical Publishers, 1933- (Formerly *Year book of dermatology and syphilology*.)

0723. Year book of drug therapy. Chicago, Year Book Medical Publishers, 1900- (Formerly *Year book of general therapeutics*.)

0724. Year book of endocrinology. Chicago, Year Book Medical Publishers, 1946- (Supersedes in part the *Year book of endocrinology, metabolism and nutrition*.)

0725. Year book of general surgery. Chicago, Year Book Medical Publishers, 1933-

0726. Year book of nuclear medicine. Chicago, Year Book Medical Publishers, 1966-

0727. Year book of opthalmology. Chicago, Year Book Medical Publishers, 1957/58- (Supersedes in part *Year book of the eye, ear, nose and throat* and *Year book of the ear, nose, throat and maxillofacial surgery*.)

0728. Year book of orthopedics, traumatic and plastic surgery. Chicago, Year Book Medical Publishers, 1947- (Formerly *Year book of orthopedics and traumatic surgery*.)

0729. Year book of pathology and clinical pathology. Chicago, Year Book Medical Publishers, 1947- (Formerly *Year book of pathology and immunology*.)

0730. Year book of urology. Chicago, Year Book Medical Publishers, 1933-

SERIAL MONOGRAPHS

0731. Advances in biology of skin. v. 1- Ed. by W. Montagna. New York, Appleton-Century-Crofts, 1960-
- v. 1 Cutaneous innervation. 1960.
- v. 2 Blood vessels and circulation. 1961.
- v. 3 Eccrine sweat glands and eccrine sweating. 1962.
- v. 4 The sebaceous glands. 1963.
- v. 5 Wound healing. 1964.
- v. 6 Aging. 1965.
- v. 7 Carcinogenesis. 1966.
- v. 8 The pigmentary system. 1967.
- v. 9 Hair growth. 1969.
- v. 10 The dermis. 1970.
- v. 11 Immunology and the skin. 1971.

0732. Advances in experimental medicine and biology. v. 1- New York, Plenum Press, 1967-
- v. 3 Germ-free biology: experimental and clinical aspects. Ed. by E. A. Mirand and N. Back. 1969.
- v. 5 Lymphatic tissue and germinal centers in immune response. Ed. by L. Fiore-Donati and M. G. Hanna, Jr. 1969.
- v. 6 Red cell metabolism and function. Ed. by G. J. Brewer. 1970.
- v. 9 Shock: biochemical, pharmocological, and clinical aspects. Ed. by A. Bertelli and N. Back. 1970.
- v. 12 Morphological and functional aspects of immunity. Ed. by K. Lindahl-Kiessling, G. Alm, and M. G. Hanna, Jr. 1971.
- v. 13 Chemistry and brain development. Ed. by R. Paoletti and A. N. Davison. 1971.
- v. 15 The reticuloendothelial system and immune phenomena. Ed. by N. R. Di Luzio and K. Flemming. 1971.
- v. 22 Comparative pathophysiology of circulatory disturbances. Ed. by C. M. Bloor. 1972.
- v. 23 The fundamental mechanisms of shock. Ed. by L. B. Hinshaw and Barbara G. Cox. 1972.
- v. 26 Pharmacological control of lipid metabolism. Ed. by W. L. Holmes et al. 1972.
- v. 27 Drugs and fetal development. Ed. by M. A. Klingberg et al. 1972.

(The above noted monographs are a sample of those contained in the series and are germane to this publication. A complete listing may be found in the publisher's catalog.)

0733. Animal reproduction symposium. Proceedings of biennial symposia. 1953-
- 1st Conference on female reproduction in farm animals. 1953. Iowa State College Journal of Science 28(1): 1-138.
- 2d Reproduction and infertility. 1955. Michigan State Univ. Centennial Symposium Publication.
- 3d Reproduction and infertility. 1958. Ed. by F. X. Gassner. New York, Pergamon Press.
- 4th The effect of germ cell damage on animal reproduction. 1960. J. Dairy Science 43; Suppl. Ed. by N. L. Van Demark, symposium chairman.
- 5th Biennial symposium on animal reproduction. 1961. Univ. of Tennessee, Knoxville. (Proceedings not published.)
- 6th Gonadotropins: their chemical and biological properties and secretory control. 1964. Ed. by H. H. Cole. San Francisco, Freeman.
- 7th Environmental influences on reproductive processes. 1965. Ed. by William Hansel and R. H. Dutt. J. Animal Science 25; Suppl. 1966. Published by the American Society of Animal Science.

0734. Bibliotheca haematologica. (Suppl. to *Acta haematologica*.) v. 1- Basel, Karger, 1955- (The series includes *Ergebnisse der Bluttransfusionsforschung*.)
- fasc. 18 Erythrocytometric methods and their standardization. 1964.
- fasc. 29, pt. 1 Blood genetics and physiology. Haemolytic disease of the newborn. 1966.
- fasc. 29, pt. 2 Autoimmunity. Immunoglobulins. Cellular immunology. 1966.
- fasc. 29, pt. 3 Problems in blood transfusion. Blood and bone marrow preservation. 1966.
- fasc. 29, pt. 4 Haemoglobin. Haemostasis. Miscellaneous. 1966.
- fasc. 30 Leukemia in animals and man. 1967.
- fasc. 36 Comparative leukemia research. 1970.
- fasc. 38, pt. 1 Immunohematology—immunology, transplantation problems, leukemia, coagulation. 1969.

(The above noted fascicles contain proceedings of meetings or papers that are germane to this publication. A complete listing may be found in the publisher's catalog.)

0735. Bibliotheca microbiologica. v. 1- Basel, Karger, 1960-
- v. 1 Pranatale infektionen. Wiener Colloquium, 1959. Ed. by A. Grumbach. 1960.
- v. 2 Die enteralen Staphylokkeninfektionen des Kindes. Ed. by M. Kienitz. 1962.
- v. 3 Brucella phages, properties and application. Ed. by Józef Parnas. 1963.
- v. 4 Oecology and adaptation of microorganisms. Swiss Society of Microbiology Symposium. 1963. Ed. by A. Grumbach. 1964.
- v. 5 Susceptibility to pyelonephritis after treatment with some therapeutic agents: a screening study in mice. Ed. by P. Toivanen, Auli Toivanen, and K. Lauslahti. 1966.
- v. 6 Systematik der Streptomyceten unter besonderer Bërucksichtigung der von ihnen gebildeten Antibiotica. Ed. by Ralf Hütter. 1967.
- v. 7 Das Versuchstier. Ed. by A. Grumbach. 1969.
- v. 8 Immunological methods in brucellosis research. Part 1—In vitro procedures. Ed. by A. Olitzki. 1970.
- v. 9 Immunological methods in brucellosis research. Part 2—In vivo procedures. Ed. by A. Olitzki. 1970.
- v. 10 Enteric fevers causing organisms and host's reactions. Ed. by A. Olitzki. 1972.

0736. Bibliotheca "Nutritio et Dieta." v. 1- Basel, Karger, 1960-
- no. 6 Group of European nutritionists. Symposium. 2d. Oslo, 1963. Proceedings. Nutrition and cardiovascular diseases. Ed. by J. C. Somogyi. 1964.
- no. 9 Group of European nutritionists. Symposium. 4th. Hameenlinna, Finland, 1965. Proceedings. Minor constituents in foods. Ed. by J. C. Somogyi. 1967.
- no. 10 Group of European nutritionists. Symposium. 5th. Jouy-en-Josas, France, 1966. Proceedings. Antibiotics in agriculture. Ed. by J. C. Somogyi. 1968.
- no. 13 Nutritional aspects of the development of bone and connective tissue. Ed. by J. C. Somogyi. 1969.

(The above noted monographs are a sample of those contained in the series and are germane to this publication. A complete listing may be found in the publisher's catalog.)

0737. Bibliotheca primatologica. v. 1- Basel, Karger, 1962-
- v. 9 Circadian rhythms in nonhuman primates. Ed. by F. H. Rohles, Jr. 1969.
- v. 10 Taxonomy and evolution of the monkeys of Celebes. By Jack Fooden. 1969.
- v. 11 The behavioral repertoire of the stumptail macaque: a descriptive and comparative study. By Mireille Bertrand. 1969.
- v. 12 Baboon ecology. African field research. By Stuart Altmann and Jeanne Altmann. 1970.
- v. 13 Comparative ecology of Gorilla gorilla (Savage and Wyman) and Pan troglodytes (Blumenbach) in Rio Muni, West Africa. By Clyde Jones and J. S. Pi. 1971.
- v. 14 Functional myology of the hip and thigh of cebid monkeys and its implications for the evolution of erect posture. By J. T. Stern. 1971.

(The above noted monographs are a sample of those contained in the series and are germane to this publication. A complete listing may be found in the publisher's catalog.)

0738. Biology colloquium. Proceedings. v. 1- Corvallis, Oregon State Univ., 1940-
- v. 22 Physiology of reproduction. Ed. by F. L. Hisaw. 1963.
- v. 26 Host-parasite relationships. Ed. by J. E. McCauley. 1965.
- v. 27 Animal orientation and navigation. Ed. by R. M. Storm. 1966.
- v. 29 Biochemical coevolution. Ed. by K. L. Chambers. 1968.
- v. 30 Biological ultrastructure: the origin of cell organelles. Ed. by Patricia J. Harris. 1969.

(The above noted monographs are a sample of those contained in the series and are germane to this publication. A complete listing may be found in the publisher's catalog.)

0739. British Society for Parasitology. Symposium. 1st- Philadelphia, Davis, 1963-
- v. 1 Techniques in parasitology. Ed. by Angela R. Taylor. 1963.
- v. 2 Host-parasite relationships in in-

vertebrate hosts. Ed. by Angela R. Taylor. 1964.
v. 3 Evolution of parasites. Ed. by Angela R. Taylor. 1965.
v. 4 Pathology of parasitic diseases. Ed. by Angela R. Taylor. 1966.
v. 5 Problems of in vitro culture. Ed. by Angela R. Taylor. 1967.
v. 6 Immunity to parasites. Ed. by Angela R. Taylor. 1968.
v. 7 Nippostrongylus and toxoplasma. Ed. by Angela R. Taylor. 1969.
v. 8 Aspects of fish parasitology. Ed. by Angela R. Taylor and R. Muller. 1970.
v. 9 Isolation and maintenance of parasites in vitro. Ed. by Angela R. Taylor and R. Muller. 1971.
v. 10 Functional aspects of parasitic surfaces. Ed. by Angela R. Taylor and R. Muller. 1972.

0740. Cold Spring Harbor symposia on quantitative biology. Cold Spring Harbor, N.Y., Cold Spring Laboratory on Quantitative Biology, 1933-
v. 27 Basic mechanisms in animal virus biology. 1962.
v. 30 Sensory receptors. 1965.
v. 32 Antibodies. 1967.
v. 35 Transcription of genetic material. 1970.
v. 36 Structure and function of proteins at the third-dimensional level. 1972.
(The above noted monographs are a sample of those contained in the series and are germane to this publication. A complete listing may be found in the publisher's catalog.)

0741. Columbia University. Biological series. New York, Columbia Univ. Press, 1894-
v. 19 Evolution above the species level. By B. Rensch. 1960.
v. 20 Principles of animal taxonomy. By G. G. Simpson. 1961.
v. 21 New patterns in genetics and development. By C. H. Waddington. 1962.
v. 22 The hemoglobins in genetics and evolution. By V. M. Ingram. 1963.
v. 23 Neuroendocrinology. By Ernst and Berta Scharrer. 1963.
v. 24 Virus-induced enzymes. By S. S. Cohen. 1968.
(The above noted monographs are a sample of those contained in the series and are germane to this publication. A complete listing may be found in the publisher's catalog.)

0742. Commonwealth Bureau of Animal Health. Review series. Weybridge, England, 1937-
v. 3 Photosensitization in diseases of domestic animals. By N. T. Clare. 1952.
v. 4 Neoplasms of the domesticated mammals. A guide to the literature. By E. Cotchin. 1956.
v. 6 Fungal diseases of animals. By G. C. Ainsworth and P. K. C. Austwick. 1959.
v. 7 Escherichia coli in domestic animals and poultry. By W. J. Sojka. 1965.
v. 8 Pig trypanosomiasis in tropical Africa. By L. E. Stephen. 1966.
v. 9 Prenatal survival in pigs. By A. E. Wrathall. 1971.
(The above noted monographs are a sample of those contained in the series and are germane to this publication. A complete listing may be found in the publisher's catalog.)

0743. Commonwealth Bureau of Animal Nutrition. Technical communications. v. 1- Farnham Royal, Bucks, England, 1948-
no. 20 Diet in relation in reproduction and the viability of the young. (Part 2. Sheep: World survey of reproduction and review of feeding experiments.) By W. Thomson. 1959.
no. 21 Diet in relation to reproduction and the viability of the young. (Part 3. Pigs.) By D. L. Duncan and G. A. Lodge. 1960.
no. 22 The nutrition of the young pig. By I. A. M. Lucas and G. A. Lodge. 1961.
no. 23 Feeding of fur-bearing animals. By F. C. Aitken. 1963.
no. 25 Vitamins in feeds for livestock. By F. C. Aitken and R. G. Hamkin. 1970.
(The above noted monographs are a sample of those contained in the series and are germane to this publication. A complete listing may be found in the publisher's catalog.)

0744. Commonwealth Bureau of Helminthology. Technical communications. St. Albans, England, 1962-
no. 34 The biology of cestode life-cycles. By J. D. Smyth. 1963.
no. 35 Nematode parasite population in

sheep and on pasture. By H. D. Crofton. 1963.
  no. 37 The helminth parasites of the herring gull (Larus agentatus Pontopp). By W. Threlfall. 1966.
  no. 38 Checklist of the helminth parasites of African mammals. By M. C. Round. 1968.
(The above noted monographs are a sample of those contained in the series and are germane to this publication. A complete listing may be found in the publisher's catalog.)

0745. Conference on experimental medicine and surgery in primates. 1st- 1967-
  1st New York, 1967. Experimental medicine and surgery in primates. Ed. by E. I. Goldsmith and J. Moor-Jankowski. New York, New York Academy of Sciences. Annals. v. 162, art. 1.
  2d New York, 1969. Medical primatology. 1970. Ed. by E. I. Goldsmith and J. Moor-Jankowski. Basel, Karger, 1971.

0746. Developmental immunology workshop. 1st- Sanibel Island, Fla. Proceedings. 1965-
  1st 1965. Phylogeny of immunity. Ed. by R. T. Smith, P. A. Miescher, and R. A. Good. Gainesville, Univ. of Florida Press, 1965.
  2d 1966. Ontogeny of immunity. Ed. by R. T. Smith, R. A. Good, and P. A. Miescher. Gainesville, Univ. of Florida Press, 1967.
  3d 1967. Immunologic deficiency diseases of man. Ed. by D. Bergsman et al. New York, National Foundation Press, 1968.
  4th 1968. Cellular recognition. Ed. by R. T. Smith and R. A. Good. New York, Appleton-Century-Crofts, 1969.

0747. European Association for Animal Production. 1st- Proceedings. 1958-
  1st Copenhagen, 1958.
  2d Wageningen, Netherlands, 1961.
  3d Troon, Scotland, 1964. Energy metabolism. Ed. by K. L. Blaxter. New York, Academic Press, 1965.
  4th Warsaw, 1967. Energy metabolism of farm animals. Ed. by K. L. Blaxter et al. Newcastle-upon-Tyne, England, Oriel Press, 1969.

0748. European conference on animal blood groups and biochemical polymorphism. (The 9th conference is the first published proceeding under the sponsorship of European Society for Animal Blood Group Research. Earlier conferences [1st-8th] were held in various European locations, but published proceedings are not available.)
  9th Prague, 1964. Blood groups of animals. Ed. by Josef Matousek. The Hague, Junk, 1965.
  10th Paris, 1966. Polymorphismes biochimiques des animaux. Paris, Institut National de la Recherche Agronomique, 1967.
  11th Warsaw, 1968. Proceedings. The Hague, Junk, 1970.
  12th Budapest, 1970. Proceedings. The Hague, Junk, 1972.

0749. European Society for the Study of Drug Toxicity. Proceedings of the meetings. Amsterdam, Excerpta Medica Foundation, 1963-
Proceedings of the meetings are published in Excerpta Medica international congress series. The series number follows each title.
  v. 1 Paris, 1963. Effects of drugs on the foetus. no. 64.
  v. 3 Lausanne, 1964. Evaluation of the potential carcinogenic action of a drug. no. 75.
  v. 5 Bad Homburg, 1965. Advances in toxicological methodology. no. 90.
  v. 8 Prague, 1966. Neurotoxicity of drugs. no. 118.
  v. 10 Oxford, England, 1968. Sensitizations to drugs. no. 181.
  v. 12 Uppsala, 1970. The correlation of adverse effects in man with observations in animals. no. 220.
(The above noted monographs are a sample of those contained in the series and are germane to this publication. A complete listing may be found in the publisher's catalog.)

0750. European symposium on calcified tissues. 1st- Proceedings. 1963-
  1st Oxford, 1963.
  2d Liege, 1964.
  3d Davos, Switzerland, 1965. New York, Springer-Verlag, 1966.
  4th Leyden and Noordwijk-aan-zee, 1966. Ed. by P. J. Gaillard et al. Amsterdam, Excerpta Medica Foundation, 1966. (Excerpta Medica international congress series. no. 120.)
  5th Bordeaux, 1967. Ed. by G. Milhaud et al. Paris, Societe d'edition d'enseignement superieur, 1968.
  6th Lund, Sweden, 1968. Ed. by J. F.

Dymling and G. C. H. Bauer. New York, Springer-Verlag, 1968. (Calcified tissue research. v. 2. Suppl. 1968.)

7th Montecatini Terme, Italy, 1970. Ed. by R. Ampino, A. Ascenzi, and B. de Bernard. New York, Springer-Verlag, 1970. (Calcified tissue research. v. 4. Suppl. 1970.)

8th Jerusalem, 1971. Ed. by J. Menczel and A. Harell. New York, Academic Press, 1971.

0751. Excerpta Medica. International congress series. no. 1- New York, Excerpta Medica Foundation, 1952-

no. 152 Experimental surgery. Abstracts of papers from the 2d Congress of European Society for Experimental Surgery. Louvain, 1967. Ed. by J. J. Haxhe, P. J. Kestens, and R. Vaughan-Jones. 1967.

no. 157 Endocrinology. Abstracts of brief communications from the 3d International congress of endocrinology. Mexico, 1968. Ed. by Carlos Gual. 1968.

no. 158 Growth hormone. Proceedings of the 1st International symposium on growth hormone. Milan, 1967. Ed. by A. Pecile and E. E. Muller. 1968.

no. 188 Inflammation, biochemistry and drug interaction. Proceedings of the international symposium on inflammation, biochemistry and drug interaction. Como, Italy, 1968. Ed. by A. Bertelli and J. C. Houck. 1969.

no. 218 Regeneration of striated muscle, and myogenesis. Proceedings of the international conference convened by Muscular Dystrophy Associations of America. New York, 1969. Ed. by Alexander Mauro et al. 1970.

no. 220 The correlation of adverse effects in man with observations in animals. Proceedings of an international symposium of the European Society for the Study of Drug Toxicity. Uppsala, Sweden, 1970. Ed. by S. B. De C. Baker. 1971.

no. 237 International congress on muscle diseases. 2d. Perth, Australia, 1971. Abstracts of papers presented. 1971.

0752. Federation of European Biochemical Societies. Proceedings of the meetings. New York, Academic Press, 1966-

3d Warsaw, 1966. The biochemistry of blood platelets. Ed. by E. Kowalski and S. Niewiarowski. Biochemistry of mitochondria. Ed. by E. C. Slater, Z. Kaniuga, and L. Wojtczak.

4th Oslo, 1967. The biochemistry of virus replication. Ed. by S. G. Laland and L. O. Frøholm. Structure and function of the endoplasmic reticulum in animal cells. Ed. by F. C. Gran.

5th Prague, 1968. Enzymes and isoenzymes: structure, properties, and function. Ed. by D. Shugar.

(The above noted monographs are a sample of those presented at the meetings.)

0753. Food and Agricultural Organization of the United Nations. Agricultural study. no. 1- Rome, 1948-

no. 33 Trichomonas foetus infection of cattle. By J. A. Laing. 1957.

no. 51 Vibrio fetus infection of cattle. rev. ed. By J. A. Laing. 1960.

no. 85 Joint FAO/WHO Expert Committee on Brucellosis. 1970.

no. 86 Contagious bovine pleuropneumonia. By J. R. Hudson. 1971.

(The above noted monographs are a sample of those contained in the series and are germane to this publication. A complete listing may be found in the publisher's catalog.)

0754. Gann monographs. v. 1- Tokyo, Japanese Cancer Association, 1966- (Consists of proceedings of international conferences and symposia.)

v. 1 Biological and biochemical evaluation of malignancy in experimental hepatomas. Ed. by Tomizo Yoshida. 1966.

v. 5 Experimental animals in cancer research. Ed. by Tomizo Yoshida, 1968.

v. 7 Virus and human cancer. Ed. by the Japanese Cancer Association. 1969.

v. 8 Experimental carcinoma of the glandular stomach. Ed. by H. P. Morris and Tomizo Yoshida. 1969.

v. 10 Nasopharyngeal carcinoma and related topics. Ed. by the Japanese Cancer Association. 1970.

v. 12 Experimental leukemogenesis. Ed. by Tadashi Yamamoto and Haruo Sugano. 1972.

(The above noted monographs are a sample of those contained in the series and are germane to this publication. A complete listing may be found in the publisher's catalog.)

0755. Inter-American meeting on foot-and-mouth disease and zoonoses control. 1st-

Washington, D.C., Pan American Health
Organization, 1968-
   1st Washington, D.C., 1968. Papers.
(Pan American Sanitary Bureau. Sci.
publ. no. 172.)
   2d Rio de Janeiro, 1969. Papers. (Pan
American Sanitary Bureau. Sci. publ.
no. 196.)
   3d Buenos Aires, 1970. Papers. (Pan
American Sanitary Bureau. Sci. publ.
no. 218.)

0756. International Academy of Pathology.
Monographs in pathology. v. 1- Baltimore,
Williams & Wilkins, 1960-
   v. 1 The lymphocyte and lymphocytic
tissue. Ed. by J. W. Rebuck and R. E.
Stowell. 1960.
   v. 2 The adrenal cortex. Ed. by H. D.
Moon and R. E. Stowell. 1961.
   v. 3 The ovary. Ed. by H. C. Grady and
D. E. Smith. 1962.
   v. 4 The peripheral blood vessels. Ed.
by J. L. Orbison and D. E. Smith.
1963.
   v. 5 The thyroid. Ed. by J. B. Hazard
and D. E. Smith. 1964.
   v. 6 The kidney. Ed. by F. K. Mostofi
and D. E. Smith. 1966.
   v. 7 The connective tissue. Ed. by
B. M. Wagner and D. E. Smith. 1967.
   v. 8 The lung. Ed. by A. A. Liebow and
D. E. Smith. 1968.
   v. 9 The central nervous system. Ed.
by O. T. Bailey and D. E. Smith. 1968.
   v. 10 The skin. Ed. by E. B. Helwig and
F. K. Mostofi. 1971.
   v. 11 The platelet. Ed. by K. M. Brinkhous and N. F. Rodman, Jr. 1971.

0757. International Committee on Laboratory Animals. Symposium. 1st- Proceedings. 1958-
   1st Paris, 1958.
   2d Liblice and Smolenice, 1961. The
problems of laboratory animal disease.
Ed. by R. J. C. Harris. New York,
Academic Press, 1962.
   3d Dublin, 1965. Husbandry of laboratory animals. Ed. by M. L. Conalty.
New York, Academic Press, 1967.
   4th Washington, D.C., 1969. Defining
the laboratory animal. Washington,
D.C., National Academy of Sciences,
1971.

0758. International conference on equine
infectious diseases. 1st- Proceedings.
1966-
   1st Stresa, Italy, 1966. Proceedings.

Ed. by J. T. Bryans. Lexington, Ky.,
The Grayson Foundation, 1966.
   2d Paris, 1969. Proceedings. Ed. by
J. T. Bryans and H. Gerber. New York,
Karger, 1970.

0759. International congress of chemotherapy. Proceedings. 1st- 1959-
   1st Geneva, 1959. (Antibiotica et
chemotherapia, v. 8?)
   2d Naples, 1961. 6 pts. (Part 1 also as
Antibiotica et chemotherapia, v. 11;
Part 5 also as Chemotherapia, v. 6.)
   3d Stuttgart, 1963. 2 v. Stuttgart,
Thieme, 1964.
   5th Vienna, 1967. 4 v in 6. Verlag der
Wiener Medizinischen Akademie,
1967.
   6th Tokyo, 1969. Progress in antimicrobial and anticancer chemotherapy. 2
v. Tokyo, Univ. of Tokyo Press, 1970.

0760. International congress of primatology. 1st- Proceedings. 1966-
   1st Frankfurt a.M., 1966. Neue Ergebnisse der Primatologia (Progress in
primatology). Stuttgart, Fischer, 1967.
   2d Atlanta, 1968. 3 v. New York, Karger,
1969. v. 1—Behavior. Ed. by C. R.
Carpenter. v. 2—Recent advances in
primatology. Ed. by H. O. Hofer.
v. 3—Neurology, physiology and infectious diseases. Ed. by H. O. Hofer.

0761. International histological classification of tumours (including slides and text).
v. 1- Geneva, World Health Organization,
1967-
   no. 1 Histological typing of lung tumours.
By Leiv Kreyberg. 1967.
   no. 2 Histological typing of breat tumours. By R. W. Scarff. 1968.
   no. 3 Histological typing of soft tissue
tumours. By F. M. Enzinger. 1969.
   no. 4 Histological typing of oral and
oropharyngeal tumours. By P. N. Wahi.
1971.
   no. 5 Histological typing of odontogenic
tumours, jaw cysts and allied lesions.
By J. J. Pindborg. 1971.

0762. International symposium on immunopathology. 1st- Proceedings. Ed. by P. A.
Miescher and P. Grabar. Basel, Schwabe,
1959-
   1st Basel/Seelisberg, 1958.
   2d Brook Lodge, Mich., 1961. Mechanism
of cell and tissue damage produced by
immune reactions.
   3d La Jolla, Calif., 1963.

4th Monte Carlo, 1965. Immunopathology of malignancy.
5th Punta Ala, Italy, 1967.
6th Grindelwald, Switzerland, 1970.

0763. International symposium on medical and applied virology. 1st- Proceedings. 1964-
  1st Boca Raton, Florida, 1964. Applied virology. Ed. by Murray Sanders and E. H. Lennette. Sheboygan, Wis., Ellis Corporation, 1965.
  2d Ft. Lauderdale, Florida, 1966. Medical and applied virology. Ed. by Murray Sanders and E. H. Lennette. St. Louis, Green, 1968.
  3d Ft. Lauderdale, Florida, 1969-1970. Viruses affecting man and animals. Ed. by Murray Sanders and Morris Schaeffer. St. Louis, Green, 1971.

0764. International symposium on stereoencephalotomy. 1st Basel, Karger, 1961-
  1st Philadelphia, 1961.
  2d Copenhagen and Vienna, 1965. Advances in stereoencephalotomy. 2 v. Ed. by E. A. Spiegel and H. T. Wycis. Part 1—Methodology and extrapyramidal system. (Reprint from *Confinia neurologica*. v. 26, no. 3-5, 1965.) Part 2—Pain, convulsive disorders, behavioral and other effects of stereoencephalotomy. (Reprint from *Confinia neurologica*. v. 27, no. 1-3, 1966.)
  3d Madrid, 1967. Advances in stereoencephalotomy. Part 3—Diskinesias. Sensory, emotional and mental aspects. Methods and various stimulation effects. Ed. by E. A. Spiegel and H. T. Wycis. (Reprint from *Confinia neurologica*. v. 29, no. 2-5, 1967.)

0765. International symposium on the physiology of digestion in the ruminant. 1st- Proceedings. 1960-
  1st Nottingham, England, 1960. Digestive physiology and nutrition of the ruminant. Ed. by D. Lewis. London, Butterworth, 1961. (Nottingham Easter School in Agricultural Science. 7th.)
  2d Ames, Iowa, 1964. Physiology of digestion in the ruminant. Ed. by R. W. Dougherty. Washington, D.C., Butterworth, 1965.
  3d Cambridge, England, 1969. Physiology of digestion and metabolism in the ruminant. Ed. by A. T. Phillipson. Newcastle-upon-Tyne, England, Oriel Press, 1970.

0766. International symposium on the response of the nervous system to ionizing radiation. 1st- Proceedings. 1960-
  1st Chicago, 1960. Proceedings. Ed. by T. J. Haley and R. S. Snider. New York, Academic Press, 1962.
  2d Los Angeles, 1963. Proceedings. Ed. by T. J. Haley and R. S. Snider. Boston, Little, Brown, 1964.

0767. International Union Against Cancer. UICC monograph series. v. 1-2. Copenhagen, Munksgaard, 1967. v. 3- Berlin, Springer-Verlag, 1967-
  v. 2 Specific tumor antigens. Ed. by R. J. C. Harris. 1967.
  v. 4 Cancer detection. 1966.
  v. 6 Mechanisms of invasion in cancer. Ed. by Pierre Denoix. 1967.
  v. 7 Potential carcinogenic hazards from drugs: evaluation of risks. Ed. by René Truhaut. 1967.
  v. 8 Treatment of Burkitt's tumour. Ed. by J. H. Burchenal and D. P. Burkitt. 1967.
  v. 9-10 International cancer congress. 9th. Tokyo, 1966. Proceedings. 2 v. 1967.
  v. 12 Thyroid cancer. Ed. by C. E. Hedinger. 1969.
(The above noted monographs are a sample of those contained in the series and are germane to this publication. A complete listing may be found in the publisher's catalog.)

0768. Journal of reproduction and fertility. Supplement series. no. 3- Oxford, Blackwell, 1968-
  no. 6 Biology of reproduction in mammals. Ed. by J. S. Perry and I. W. Rowlands. 1969. (The Nairobi symposium of the Society for the Study of Fertility.)
  no. 7 Intersexuality. Ed. by J. S. Perry. 1969. (3d symposium of the Society for the Study of Fertility.)
  no. 11 Reproductive behaviour. Ed. by J. S. Perry. 1970. (6th symposium of the Society for the Study of Fertility.)
  no. 13 Spermatogenesis and sperm maturation. Ed. by J. S. Perry. 1971. (8th symposium of the Society for the Study of Fertility.)
  no. 14 Experiments on mammalian eggs and embryos. Ed. by J. S. Perry. 1971. (9th symposium of the Society for the Study of Fertility.)
  no. 15 Genetics and reproduction. Ed. by J. S. Perry. 1972. (10th symposium of the Society for the Study of Fertility.)

no. 16 Control of parturition. Ed. by
J. S. Perry. 1972. (11th symposium
of the Society for the Study of Fertility.)
(The above noted monographs are a sample
of those contained in the series and are
germane to this publication. A complete
listing may be found in the publisher's
catalog.)

0769. Methods and achievements in experimental pathology. v. 1- Ed. by E.
Bajusz and G. Jasmin. Basel, Karger, 1966-
- v. 1 An introduction to experimental
  pathology. 1966.
- v. 2 Investigative techniques. 1967.
- v. 3 Ultrastructural, histopathologic
  and chemical approaches. 1967.
- v. 4 Examples of descriptive and functional morphology. 1969.
- v. 5 Functional morphology of the heart.
  W. Hort and G. Baroldi, consulting
  eds. 1971.

0770. Monographs in allergy. v. 1- New
York, Karger, 1966-
- v. 1 Immunotolerance to simple chemicals. By A. L. De Weck and J. R.
  Frey. 1966.
- v. 2 Tissue mast cells in immune response. By R. Keller. 1966.
- v. 3 Mechanism and role of immunological tolerance. By I. Hraba. 1968.
- v. 4 Hypersensitivity diseases of the
  lungs due to fungi and organic dusts.
  By J. Pepys. 1969.
- v. 5 Immunology of nematode infections.
  Trichinosis in guinea-pigs as a model.
  By David Catty. 1969.
- v. 6 The complement system. By D. R.
  Schultz. 1971.
- v. 7 The chemistry of atopic allergens.
  By L. Berrens. 1971.

0771. Monographs in virology. v. 1-
Basel, Karger, 1967-
- v. 1 Rhinoviruses. By Dorothy Hamre.
  1968.
- v. 2 Enzyme induction by viruses. By
  Saul Kit and Del Rose Dubbs. 1969.
- v. 3 Persistent and slow virus infections.
  By John Hotchin. 1971.
- v. 4 Viral structural components as im-
  munogens of prophylactic value. By
  A. R. Neurath and B. A. Rubin. 1971.
- v. 5 Classification and nomenclature of
  viruses. By Peter Wildy. 1971.

0772. Monographs on endocrinology. v. 1-
New York, Springer-Verlag, 1967-
- v. 1 Sex chromosomes and sex-linked
  genes. By S. Ohno. 1967.
- v. 2 Gas phase chromatography of steroids. By K. B. Eik-Nes and E. C.
  Horning. 1968.
- v. 3 Hypothalamic control of lactation.
  By F. G. Sulman. 1970.
- v. 4 Steroid-protein interactions. By
  U. Westphal. 1971.
- v. 5 Regulation of aldosterone biosynthesis. By J. Müller. 1971.
- v. 6 Immunopathology of insulin. By K.
  Federlin. 1971.
- v. 7 Prostaglandins. By E. W. Horton.
  1972.

0773. Nottingham Easter School in Agricultural Science. Proceedings. London, Butterworth, 1955-
- 7th 1960. Digestive physiology and nutrition of the ruminant. Ed. by D.
  Lewis.
- 8th 1961. Nutrition of pigs and poultry.
  Ed. by J. T. Morgan.
- 9th 1962. Antibiotics in agriculture. Ed.
  by Malcolm Woodbine.
- 13th 1967. Reproduction in the female
  mammal. Ed. by G. E. Lamming and
  E. C. Amoroso.
- 14th 1968. Growth and development of
  mammals. Ed. by G. E. Lodge and
  G. E. Lamming.
- 16th 1970. Proteins as human food. Ed.
  by R. A. Lawrie.
- 17th 1971. Lactation. Ed. by I. R. Falconer.
- 18th 1971. Pig production. Ed. by D. J.
  A. Cole.

(The above noted monographs are a sample
of those contained in the series and are
germane to this publication. A complete
listing may be found in the publisher's catalog.)

0774. Perspectives in virology. Gustav
Stern symposium. v. 1- Ed. by Morris Pollard. 1959-
- v. 1-4 1959-1965. Publishers vary.
- v. 5 Virus-directed host response. New
  York, Academic Press, 1967.
- v. 6 Virus-induced immunopathology.
  New York, Academic Press, 1968.
- v. 7 From molecules to man. New York,
  Academic Press, 1971.

0775. Physiological Society of London.
Monographs. v. 1- London, Arnold, 1954?-
- v. 18 The climatic physiology of the pig.
  By L. E. Mount. 1968.
- v. 20 Mechanisms of urine concentration

and dilution in mammals. By S. E. Dicker. 1970.
v. 22 Biogenesis and physiology of histamine. By A. T. Cowie and J. S. Tindal. 1971.
v. 23 Mammalian muscle receptors and their central actions. By P. B. C. Matthews. 1972.

(The above noted monographs are a sample of those contained in the series and are germane to this publication. A complete listing may be found in the publisher's catalog.)

0776. Primates in medicine. v. 1- New York, Karger, 1968-
  v. 1 1st Holloman symposium on primate immunology and molecular genetics. Proceedings. Ed. by C. H. Kratochvil. 1968.
  v. 2-3 Using primates in medical research. Part 1—Husbandry and technology. Ed. by W. I. B. Beveridge. 1969. Part 2—Recent comparative research. Ed. by W. I. B. Beveridge. 1969.
  v. 4 Chimpanzee: central nervous system and behavior; a review. Ed. by H. H. Reynolds. 1969.
  v. 5 Conservation of nonhuman primates in 1970. Ed. by Barbara Harrisson. 1971.
  v. 6 Chimpanzee: immunological specificities of blood. Ed. by C. H. Kratochvil. 1972.

0777. Progress in brain research. v. 1- New York, Elsevier, 1963-
  v. 26 Developmental neurology. 1967.
  v. 29 Brain barrier systems. Ed. by Abel Lajtha and D. H. Ford. 1968.
  v. 30 Cerebral circulation. Ed. by W. Luyendijk. 1968.
  v. 34 Histochemistry of nervous transmission. Ed. by Olavi Eränkö. 1971.
  v. 35 Cerebral blood flow. Ed. by J. S. Meyer. 1972.

(The above noted monographs are a sample of those contained in the series and are germane to this publication. A complete listing may be found in the publisher's catalog.)

0778. Progress in experimental tumor research. v. 1- Ed. by F. Homburger. New York, Karger, 1960-
  v. 13 Immunological aspects of neoplasia. Ed. by R. S. Schwartz. 1970.
  v. 14 Inhibition of carcinogenesis. Ed. by B. A. Rubin and B. L. van Duuren. 1971.
  v. 16 Pathology of the Syrian hamster. Ed. by F. Homburger. 1971.
  v. 17 Symposium on brain tumor. Ed. by W. G. Bingham. 1972.
  v. 18 Oncogenic adenoviruses. Ed. by L. Merkow and M. Slifkin. 1972.

(The above noted monographs are a sample of those contained in the series and are germane to this publication. A complete listing may be found in the publisher's catalog.)

0779. Queensland University. Faculty of Veterinary Science. Papers. v. 1- St. Lucia, Queensland Univ. Press, 1961-1966.
  no. 1 An environmental lipofuscin pigmentation of livers. By Hans Winter. 1961.
  no. 2 A general theory of social organization and behaviour. By G. McBride. 1964.
  no. 3 The bone marrow cells of sheep. By Hans Winter. 1965.
  no. 4 Studies on the direct visualization of the bovine ovaries through a retained cannula in the paralumbar fossa. By A. A. Baker. 1966.

0780. Recent results in cancer research.
v. 1- New York, Springer-Verlag, 1965-
  v. 21 Scientific basis of cancer chemotherapy. Ed. by Georges Mathé. 1969.
  v. 22 Tumor specific transplantation antigen. Ed. by P. Koldovsky. 1969. Biology of amphibian tumors. Ed. by Merle Mizell. 1969. (Special suppl.)
  v. 28 Antitumor and antiviral substances. Ed. by E. S. Meek. 1970.
  v. 29 Aseptic environment and cancer treatment. Ed. by Georges Mathé. 1970.
  v. 30 Advances in the treatment of acute (blastic) leukemias. Ed. by Georges Mathé. 1970.
  v. 36 Current concepts in the management of leukemia and lymphoma. Ed. by J. E. Utmann et al. 1971.
  v. 37 Endolymphatic radiotherapy in malignant lymphomas. Ed. by S. Chiappa et al. 1971.

(The above noted monographs are a sample of those contained in the series and are germane to this publication. A complete listing may be found in the publisher's catalog.)

0781. Society for General Microbiology. Symposia. v. 1- Cambridge, The Univ. Press, 1949-
  no. 17 Airborne microbes. Ed. by P. H. Gregory and J. L. Monteith. 1967.

no. 18 The molecular biology of viruses. Ed. by L. V. Crawford and M. G. P. Stoker. 1968.
no. 21 Microbes and biological productivity. Ed. by D. E. Hughes and A. H. Rose. 1971.
no. 22 Microbial pathogenicity in man and animals. Ed. by Harry Smith and J. H. Pearce. 1972.
(The above noted monographs are a sample of those contained in the series and are germane to this publication. A complete listing may be found in the publisher's catalog.)

0782. Symposia series in immunobiological standardization. v. 1- Ed. by R. H. Regamey et al. New York, Karger, 1966-
   v. 1 International symposium on rabies. 1966.
   v. 2 International symposium on neurovirulence of viral vaccines. 1967.
   v. 3 International symposium on biotechnical developments in bacterial vaccine production. 1965.
   v. 4 International symposium on immunological methods of biological standardization. 1967.
   v. 5 International symposium on laboratory animals. 1966.
   v. 6 International symposium on adjuvants of immunity. 1967.
   v. 7 International symposium on combined vaccines. 1967.
   v. 8 International symposium on foot-and-mouth disease. Variants and immunity. 1968.
   v. 9 International symposium on pseudotuberculosis. 1968.
   v. 10 International symposium on biological assay methods. 1969.
   v. 11 International symposium on rubella vaccines. 1969.
   v. 12 International symposium on brucellosis. Standardization and control of vaccines and reagents. 1970.
   v. 13 International symposium on pertussis. 1970.
   v. 14 International symposium on interferon and interferon inducers. 1970.
   v. 15 International symposium on enterobacterial vaccines. 1970.
   v. 16 International symposium on antilymphocyte serum. 1971.
   v. 17 International symposium on B C G vaccine. 1971.

0783. Symposium on veterinary medical education. 1st- Proceedings. Chicago, American Veterinary Medical Association, 1965-
   1st Colloquium on veterinary education in radiology, radiobiology, and radioisotope utilization. (Am. J. Vet. Res. 26(111) pt. 2:381.)
   2d Symposium on clinical education: professional, graduate and continuing. (Am. J. Vet. Res. 26(115) pt. 2:1497.)
   3d Symposium on preprofessional veterinary education—student selection and curriculum. Michigan State Univ., 1966.
   4th Symposium on graduate education in veterinary medicine. Texas A&M Univ., 1969.

0784. Texas University. Annual clinical conference on cancer. Papers. 1st- Chicago, Year Book Medical Publishers, 1956-
   10th Cancer of the gastrointestinal tract. 1967.
   11th Cancer of the uterus and ovary. 1969.
   12th Neoplasia in childhood. 1969.
   13th Breast cancer: early and late. 1970.
   14th Leukemia-lymphoma. 1970.
(The above noted monographs are a sample of those contained in the series and are germane to this publication. A complete listing may be found in the publisher's catalog.)

0785. Tissue Culture Association. Annual symposium. Proceedings. v. 1-5. Baltimore, Williams & Wilkins, 1965-1970.
   1965 Miami. The chromosome: structural and functional aspects. Ed. by C. J. Dawe. 1967. (*In vitro.* v. 1.)
   1966 San Francisco. Phenotypic expression: immunological, biochemical, and morphological. 1967. (*In vitro.* v. 2.)
   1967 Philadelphia. Differentiation and defense mechanisms in lower organisms. Ed. by M. M. Sigel. 1968. (*In vitro.* v. 3.)
   1968. San Juan. Hemic cells in vitro. Ed. by Patricia Farnes. 1969. (*In vitro.* v. 4.)
   Advances in tissue culture. Ed. by Charity Waymouth. Baltimore, Williams & Wilkins, 1970. (Published as *In vitro*; however, it is a separate review publication and does not represent the proceedings of a symposium.)
Papers presented at later symposia may be found in the bimonthly journal *In vitro* (entry 0404).

0786. Transplantation reviews. Ed. by

Göran Möller. v. 1- Baltimore, Williams & Wilkins, 1969-
v. 1 Antigen sensitive cells, their source and differentiation. 1969.
v. 2 Liver transplantation. Ed. by T. E. Starzl. 1969.
v. 3 Strong and weak histocompatibility antigens. Ed. by W. H. Heldemann. 1970.
v. 4 Human transplantation antigens. Ed. by F. Kissmeyer-Nielsen and E. Thorsby. 1970.
v. 5 Antigen-binding lymphocyte receptors. Ed. by O. Makelä. 1970.
v. 6 Surface antigens on nucleated cells. 1971.
v. 8 Immunological tolerance: effect of antigen on different cell populations. Ed. by W. O. Weigle et al. 1972.
v. 9 Lymphoid cell replacement therapy. Ed. by D. W. van Bekkum. Use and abuse of hemopoietic cell grafts in immune deficiency diseases. Ed. by D. W. van Bekkum. Adoptive immunotherapy in malignant disease. Ed. by R. B. Thompson and G. Mathé. 1972.
v. 10 Antigen recognition in cell-mediated immunity. Ed. by A. L. de Weck et al. 1972.

0787. A treatise of skin. v. 1- New York, Wiley-Interscience, 1971-
v. 1 Elden, H. R. Biophysical properties of skin. 1971.

0788. U.C.L.A. forum in medical sciences. v. 1- Berkeley, Univ. of California Press, 1962-
v. 1 Brain function: Cortical excitability and steady potentials; relations of basic research to space biology. Ed. by Mary A. B. Brazier. 1963.
v. 2 Brain function: RNA and brain function; memory and learning. Ed. by Mary A. B. Brazier. 1965.
v. 3 Brain and behavior: The brain and gonadal function. Ed. by R. A. Gorski and R. E. Whalen. 1966.
v. 4 Brain function: Speech, language and communication. Ed. by E. C. Carterette. 1966.
v. 5 Gastrin: Proceedings of a conference. Ed. by M. I. Grossman. 1966.
v. 6 Brain function: Brain function and learning. Ed. by D. B. Lindsley and A. A. Lumsdaine. 1967.
v. 7 Brain function: Aggression and defense, neural mechanisms and social patterns. Ed. by C. D. Clemente and D. B. Lindsley. 1967.

v. 8 The retina: Morphology, function, and clinical characteristics. Ed. by B. R. Straatsma et al. 1969.
v. 9 Image processing in biological science. Ed. by D. M. Ramsey. 1969.
v. 10 Pathophysiology of congenital heart disease. Ed. by F. H. Adams, J. C. Swan, and V. E. Hall. 1970.
v. 11 The interneuron. Ed. by Mary A. B. Brazier. 1969.
v. 12 The history of medical education. Ed. by C. D. O'Malley. 1970.
v. 13 Cardiovascular beta adrenergic responses. Ed. by A. A. Kattus, G. Ross, and V. E. Hall. 1970.
v. 14 Cellular aspects of neural growth and differentiation. Ed. by D. C. Pease. 1971.
v. 15 Steroid hormones and brain function. Ed. by C. H. Sawyer and R. A. Gorski. 1971.

0789. U.S. Dept. of the Interior. Fish and Wildlife Service. Bureau of Sport Fisheries. and Wildlife. Washington, D.C. Special scientific report—Wildlife.
no. 143 Literature survey on general and comparative enzyme biochemistry of birds. By H. P. Pan. 1970.
no. 152 Comparative dietary toxicities of pesticides to birds. By R. G. Heath et al. 1972.

0790. U.S. National Cancer Institute. Monographs. v. 1- Bethesda, Md., 1959-
v. 22 Conference on murine leukemia. Ed. by M. A. Rich and J. B. Moloney. 1966.
v. 23 International symposium on the nucleolus: its structure and function. Ed. by W. S. Vincent and O. L. Miller, Jr. 1966.
v. 26 Second decennial review conference on cell tissue and organ culture. 1967.
v. 29 Cell cultures for virus vaccine production. Ed. by D. J. Merchant. 1968.
v. 32 Comparative morphology of hematopoietic neoplasms. Ed. by C. H. Lingeman and F. M. Garner. 1969.
(The above noted monographs are a sample of those contained in the series and are germane to this publication. A complete listing may be found in the publisher's catalog.)

0791. U.S. National Research Council. Committee on Animal Nutrition. Nutrient requirements of domestic animals. no. 1- Washington, D.C.

no. 1 Poultry. rev. 1971. Publ. no. 1345.
no. 2 Swine. rev. 1968. Publ. no. 1599.
no. 3 Dairy cattle. rev. 1971. Publ. no. 1349.
no. 4 Beef cattle. rev. 1970. Publ. no. 1137.
no. 5 Sheep. rev. 1968. Publ. no. 1693.
no. 6 Horses. rev. 1968. Publ. no. 1401.
no. 7 Mink and foxes. rev. 1968. Publ. no. 1676.
no. 8 Dogs. rev. 1962. Publ. no. 989.
no. 9 Rabbits. rev. 1966. Publ. no. 1194.
no. 10 Laboratory animals. rev. 1972. Publ. no. 990.

0792. ———. Institute of Laboratory Animal Resources. Laboratory animals. v. 1- Washington, D.C., National Academy of Sciences, 1961-
I Guide for shipments of small laboratory animals. 1961. (NRC publ. no. 846.)
II Animals for research. 8th rev. ed. 1971. (NRC publ. no. 1678.)
III Recommended minimum standards for the shipment of laboratory primates. 1963. (NRC publ. no. 971.)
IV Graduate education in laboratory animal medicine, proceedings of a workshop, 1964. 1965. (NRC publ. no. 1284.)
V Workshop in animal technician training, 1964. 1965. (NRC publ. no. 1285.)

0793. ———. ———. Committee on Standards. Standards for the breeding, care and management of laboratory animals. v. 1- Washington, D.C., National Research Council, 1960-
Hamsters, 1960. Mice, 1962. Rats, 1962. Cats, 1964. Dogs, 1964. Guinea pigs, 1964. Rabbits, 1965. Chickens, 1966. Nonhuman primates, 1968. Coturnix, 1969. Rodents, 1969. Gnotobiotes, 1970. Swine, 1971.

0794. The veterinary clinics of North America. v. 1- Philadelphia, Saunders, 1971-
v. 1, no. 1 Symposium on physical diagnosis in small animals. Ed. by R. W. Kirk. 1971.
v. 1, no. 2 Symposium on feline medicine. Ed. by R. L. Stansbury. 1971.
v. 1, no. 3 Symposium on orthopedic surgery in small animals. Ed. by D. L. Piermattei. 1971.
v. 2, no. 1 Symposium on gastrointestinal medicine and surgery. Ed. by A. Palminteri. 1972.
v. 2, no. 2 Symposium on emergencies in veterinary practice. Ed. by S. I. Bistner. 1972.
v. 2, no. 3 Symposium on practice management and hospital design. Ed. by Stanton Williamson. 1972.

0795. Virology monographs. v. 1- New York, Springer-Verlag, 1968-
v. 1 Echoviruses. By H. A. Wenner and A. M. Behbehani. Reoviruses. By L. Rosen. 1968.
v. 2 The simian viruses. By R. N. Hull. Rhinoviruses. By D. A. J. Tyrrell. 1968.
v. 3 Cytomegaloviruses. By J. B. Hanshaw. Rinderpest virus. By W. Plowright. Lumpy skin disease virus. By K. E. Weiss. 1968.
v. 4 The influenza viruses. By L. Hoyle. 1968.
v. 5 Herpes simplex and pseudorabies viruses. By A. S. Kaplan. 1969.
v. 6 Interferon. By J. Vilcek. 1969.
v. 7 Polyoma virus. By B. E. Eddy. Rubella virus. By E. Norrby. 1969.
v. 8 Spontaneous and virus induced transformation in cell culture. By J. Pontén. 1971.
v. 9 African swine fever virus. By W. R. Hess. Bluetongue virus. By P. G. Howell. 1971.
v. 10 Lymphocytic choriomeningitis virus. By F. Lehmann-Grube. 1971.
v. 11 Canine distemper virus. By M. J. G. Appel and J. H. Gillespie. 1972.

0796. Wenner-Gren Center Foundation. International symposium series. v. 1- New York, Pergamon Press, 1962-
v. 6 Comparative leukaemia research. Ed. by G. Winqvist. 1966.
v. 12 Physiology and patho-physiology of plasma protein metabolism. Ed. by G. Birke. 1969.
v. 16 The structure and metabolism of the pancreatic islets. Ed. by I. B. Täljedal. 1970.
v. 17 Human anti-human gammaglobulins; their specificity and function. Ed. by R. Grubb and G. Samuelsson. 1971.
(The above noted monographs are a sample of those contained in the series and are germane to this publication. A complete listing may be found in the publisher's catalog.)

0797. Wistar Institute of Anatomy and Bi-

ology. Symposium monograph. v. 1- Philadelphia, Wistar Institute Press, 1964-
- v. 6 Lipid metabolism in tissue culture cells. Ed. by G. H. Rothblat and David Kritchevsky. 1966.
- v. 7 Growth regulating substances for animal cells in culture. Ed. by Vittorio Defendi and Michael Stoker. 1967.
- v. 8 Biological properties of the mammalian surface membrane. Ed. by L. A. Manson. 1968.
- v. 9 Heterospecific genome interactions. Ed. by Vittorio Defendi. 1969.

(The above noted monographs are a sample of those contained in the series and are germane to this publication. A complete listing may be found in the publisher's catalog.)

0798. World Association for Buiatrics. International conference on cattle diseases. Reports.
- 1st Hanover, 1960.
- 2d Vienna, 1962.
- 3d Copenhagen, 1964. *Nordisk Veterinaermedicin*, 1964, 16. Suppl. 1.
- 4th Zurich, 1966. Art. Institut Orell Füssli AG, Zurich, 1967.
- 5th Opatija, Yugoslavia, 1968. Institut Za Patologiju I Terapiju Domacih Zivotinja, 1968.
- 6th Philadelphia, 1970. Heritage Press, Stillwater, Okla., 1970.

0799. World Association for the Advancement of Veterinary Parasitology. International conference. 1st- 1963-
- 1st Hanover, 1963. Symposium on the evaluation of anthelminthics. Proceedings. Ed. by E. J. L. Soulsby. New York, Merck, 1964.
- 2d Univ. of Pennsylvania, 1965. Biology of parasites; emphasis on veterinary parasites. Ed. by E. J. L. Soulsby. New York, Academic Press, 1966.
- 3d Lyons, 1967. The reaction of the host to parasitism. Ed. by E. J. L. Soulsby. Lyons, School of Veterinary Medicine, 1968.
- 4th Glasgow, Scotland, 1969. Pathology of parasitic diseases. Ed. by S. M. Gaafar et al. Lafayette, Ind., Purdue Univ. Studies, 1971.

0800. World Health Organization. Monograph series. v. 1- Geneva, World Health Organization, 1951-
- no. 23 Laboratory techniques in rabies. 2d ed. Ed. by Pascu Atanasiu. 1966.
- no. 50 Snail control in the prevention of bilharziasis. Ed. by J. M. Watson. 1965.
- no. 55 Laboratory techniques in brucellosis. By G. G. Alton and Lois M. Jones. 1967.
- no. 57 Interactions of nutrition and infection. By N. S. Scrimshaw, C. E. Taylor, and J. E. Gordon. 1968.

(The above noted monographs are a sample of those contained in the series and are germane to this publication. A complete listing may be found in the publisher's catalog.)

0801. Zoological Society of London. Symposia. v. 1- New York, Academic Press, 1960-
- v. 15 Comparative biology of reproduction in mammals. Ed. by I. W. Rolands, 1966.
- v. 17 Some recent developments in comparative medicine. Ed. by R. N. Fiennes, 1966.
- v. 21 Comparative nutrition of wild animals. Ed. by M. A. Crawford. 1968.
- v. 24 Diseases in free-living wild animals. Ed. by A. McDiarmid. 1969.
- v. 27 The haemostatic mechanism in man and other animals. Ed. by R. G. Macfarlane. 1970.
- v. 30 Diseases of fish. Ed. by L. E. Mawdesley-Thomas. 1971.

(The above noted monographs are a sample of those contained in the series and are germane to this publication. A complete listing may be found in the publisher's catalog.)

5
REFERENCE WORKS

Reference books have been defined by Constance M. Winchell as "those which are meant to be consulted or referred to for some definite piece of information" as opposed to those to be read through for pleasure or information. The line of demarcation between the two is often not a sharp one. For example, an anthology of poetry would ordinarily be read for enjoyment, but if it is well indexed it could become a useful reference book in a library. Similarly, the Nobel lectures in physiology and medicine will be read by many scientists for pleasure, but may serve also as excellent reference sources for the research person who is interested in comparative and veterinary medicine.

The purpose of this chapter is to bring together those works which the biomedical or veterinary research worker will find useful for locating specific bits of information on a variety of subjects relating to his work. Such a compilation as this cannot be all-inclusive. Therefore, it may be helpful to consult the bibliographies of special subject fields noted in Chapter 2.

This chapter is arranged under five headings: Guides to the Literature; Dictionaries and Encyclopedias; Biographical Tools and Directories; Histories and Source Books; and Miscellany.

Guides to the Literature. In addition to the indexes and abstract journals noted in Chapter 1 and to the bibliographies of periodicals, conferences and congresses, translations, and specialized subjects in Chapter 2, there are many guides of either general or specific nature that note the existing literature, particularly monographs, of interest to those engaged in biomedical or veterinary research. They can be especially useful to the librarian as selection aids in building or strengthening a collection, and to the student or research worker who wishes to become familiar with the pertinent publications in his own field or in fields ancillary to his own discipline. A typical guide to the literature within a specific field will note related histories, biographic and bibliographic tools, specialized dictionaries, directories, handbooks, manuals, systems of taxonomy or classification, and, in some instances, sources of trade and industrial information. Some of the compilations noted here are primarily listings of the literature, while others discuss the use of this material as well.

Dictionaries and Encyclopedias. Dictionaries listed here include those that define or identify terminology or give synonymous terms in the same language and those that note equivalent terms in other languages. Both are useful tools for the student and the research worker. Although general unabridged dictionaries are not included, their usefulness for scientific as well as general terminology should not be overlooked.

Abbreviations, synonyms, and word structure all pose problems to the research worker in any science, but in the biomedical field these may be particularly difficult, especially for someone engaged in interdisciplinary work. Some of the publications useful in understanding and resolving these problems have been listed in this section.

The encyclopedic works listed represent both general and specific subject treatises that serve to give an overview of various topics within the biomedical disciplines.

Biographical Tools and Directories. Biographical tools and directories of the "Who's Who" type are among the useful ready reference materials serving the scientist who needs to maintain contact with colleagues and research activities both at home and abroad. They serve as valuable resources for the research or academic administrator by furnishing him with pertinent information about possible candidates for positions, committee assignments, or participants in symposia and meetings, and by giving him information about areas of support and sources of specialized tools, materials, and services. In general, materials noted here are useful for current information about scientists and their organizations.

Histories and Source Books. In order to provide a base for understanding current research problems, it is highly desirable that the scientist or student consult those publications which relate the historical development of the biomedical sciences, particularly those areas pertinent to his work. A number of historical treatises are noted here. A few of the biographic works include sketches of the individuals who laid the early framework for various areas of biomedicine. Some source books and classical treatises are included.

Miscellany. The biomedical scientist as well as the veterinary medical clinician will use a wide variety of reference tools in his various academic, research, or practice activities. Most of these can readily be categorized by subject or format, but there always remains a group which falls outside the available groupings. These have been assembled here and include material on: Libraries and/or Information Centers and Their Services; Research Methodology; Research Support; Medical Writing, Speaking, Illustration, and Audiovisual Materials; and Industrial Information.

Several sources of specialized information services of both governmental and private origin are described briefly. Various research methodologies and experimental design are included in several of the treatises noted, while others list or describe funding agencies which have supplied financial support for biomedical research. Several guides to useful techniques in medical writing, speaking, and illustration are noted, as well as some listings of pertinent audiovisual materials. A few sources of industrial information are given that may be useful in the procurement of equipment and drugs and the handling of industrial materials.

GUIDES TO THE LITERATURE

0802. Blake, J. B., and Charles Roos, eds. Medical reference works, 1679-1966: a selected bibliography. Chicago, Medical Library Association, 1967.
... Suppl. 1, 1970. (Medical Library Association publ. no. 3.)

0803. Bottle, R. T., and H. V. Wyatt. The use of biological literature. 2d ed. Hamden, Conn., Shoe String Press, 1971.

0804. Bowker's medical books in print. New York, R. R. Bowker, 1972.
Classified by subject with author and title indexes. The available books in medicine, dentistry, nursing, and veterinary medicine in 5,000 subject categories.

0805. Ebert, Myrl. An introduction to the literature of the medical sciences. 3d ed. Chapel Hill, N.C., Univ. of North Carolina Press, 1970.

0806. Guide to microforms in print. Washington, D.C., Microcard Editions. (Revised annually.)

0807. Jenkins, Frances B. Science reference sources. 5th ed. Cambridge, Mass., M.I.T. Press, 1969.

0808. Kerker, Ann E., and H. T. Murphy. Biological and biomedical resource literature. Lafayette, Ind., Purdue Univ., 1968.

0809. Mellon, M. G. Chemical publications. 4th ed. New York, McGraw-Hill, 1965.

0810. Smith, R. C., and W. M. Reid. Guide to the literature of the life sciences. 8th ed. Minneapolis, Burgess, 1972. (Formerly *Guide to the literature of the zoological sciences.*)

0811. U.S. National Library of Medicine. Current catalog. Bethesda, Md., 1966–

0812. Vaillancourt, Pauline M. Bibliographic control of the literature of oncology, 1800-1960. Metuchen, N.J., Scarecrow Press, 1969.

0813. Walford, A. J., ed. Guide to reference material. 2d ed. v. 1: Science and technology. London, The Library Association, 1966–

0814. Winchell, Constance M. Guide to reference books. 8th ed. Chicago, American Library Association, 1967.
... Suppl. 1, 1968.
... Suppl. 2, 1970.

DICTIONARIES AND ENCYCLOPEDIAS

English Language, General Biomedical

0815. Black's veterinary dictionary. 10th ed. By W. C. Miller and G. P. West. London, Black, 1972.

0816. The British medical dictionary. rev. ed. Ed. by A. S. MacNalty. London, Caxton, 1963.

0817. Burton, Maurice. Systematic dictionary of mammals of the world. New York, Crowell, 1962.

0818. Dorland, W. A. Dorland's illustrated medical dictionary. 24th ed. Philadelphia, Saunders, 1965.

0819. Durham, R. H. Encyclopedia of medical syndromes. New York, Harper & Row, 1960.

0820. Henderson, Isabella F., and W. D. Henderson. A dictionary of biological terms: pronunciation, derivation, and definition of terms in biology, botany, zoology, anatomy, cytology, genetics, embryology, physiology. 8th ed. Ed. by J. H. Kenneth. Princeton, N.J., Van Nostrand, 1963.

0821. Hoerr, N. L., and A. Osol, eds. Blakiston's Gould medical dictionary. 3d ed. New York, McGraw-Hill, 1972.

0822. Schmidt, J. E. Paramedical dictionary. Springfield, Ill., Thomas, 1969.

0823. ———. Reversicon: a medical word finder. Springfield, Ill., Thomas, 1958.

0824. Stedman, T. L. Stedman's medical dictionary: a vocabulary of medicine and its allied sciences, with pronunciations and derivations. 22d ed. Baltimore, Williams & Wilkins, 1972.

0825. Taber, C. W. Cyclopedic medical dictionary. 11th ed. Philadelphia, Davis, 1969.

0826. Thomson, W. A., ed. Black's medical dictionary. 29th ed. Scranton, Pa., Harper & Row, 1971.

Foreign Language

0827. Carpovich, E. A. Russian-English biological and medical dictionary. 2d ed. New York, Technical Dictionaries, 1960.

0828. Clairville, A. L. Dictionnaire polyglotte des termes medicaux. rev. ed. Paris, Heinman, 1953.

0829. Condoyannis, G. E. Scientific Russian. New York, Wiley, 1959.

0830. De Vries, Louis. French-English science dictionary. 3d ed. New York, McGraw-Hill, 1962.

0831. ———. German-English science dictionary. 3d ed. New York, McGraw-Hill, 1959.

0832. Dumbleton, C. W. Russian-English biological dictionary. Edinburgh, Oliver & Boyd, 1964.

0833. Elsevier's lexicon of parasites and diseases in livestock, including parasites and diseases of all farm and domestic animals, free-living wild fauna, fishes, honeybee and silkworm, and parasites of products of animal origin. Latin, English, French, German, Italian, and Spanish. Comp. by Manuel Merino-Rodriguez. Amsterdam, Elsevier, 1964.

0834. Elsevier's medical dictionary in five languages: English/American, French, Italian, Spanish and German. Comp. and arr. on an English alphabetical base. By A. Sliosberg. Amsterdam, Elsevier, 1964.

0835. Finch, B., ed. Multilingual guide for medical personnel. Flushing, N.Y., Medical Examination Publishing, 1967.

0836. Goldberg, Morris. English-Spanish chemical and medical dictionary. New York, McGraw-Hill, 1947.

0837. ———. Spanish-English chemical and medical dictionary. New York, McGraw-Hill, 1952.

0838. Hirschhorn, H. H. Spanish-English and English-Spanish medical guide. New York, Regents, 1969.

0839. Hitti, Y. K. Hitti's English-Arabic medical dictionary. Syracuse, N.Y., Syracuse Univ. Press, 1967.

0840. Jablonski, Stanley. Russian-English medical dictionary. Ed. by B. S. Levine. New York, Academic Press, 1958.

0841. Karpovich, E. A. Russian-English biological and medical dictionary. New York, Technical Dictionaries, 1958.

0842. Kusama, Yoshio, ed. Ei-Wa igaku sho-jiten. [Kaitei zoho han] English-Japanese medical dictionary. rev. and enl. ed. Tokyo, Kanehara, 1965.

0843. Lee-Delisle, Dora. English-Hungarian, Hungarian-English medical dictionary. New York, Saphrograph, n.d.

0844. Lejeune, Fritz. German-English, English-German dictionary for physicians. 2d ed. 2 v. Stuttgart, Thieme Intercontinental Medical Book Corp., 1968-1969.

0845. Lépine, Pierre. Dictionnaire français-anglais, anglais-français des termes médicaux et biologiques. Paris, Flammarion, 1952.

0846. Manuila, A., et al., eds. Dictionnaire français de médecine et de biologie. 4 v. New York, Stechert-Hafner, 1970-1972.

0847. Meyboom, F. Quaestionarium Medicum. (Polyglot.) New York, Elsevier, 1961.

0848. Nobel, A. Medical dictionary. 5th rev. and enl. ed. (English, German, French.) New York, Springer-Verlag, 1969.

0849. Rokitskiĭ, P. F., ed. Anglo-russkiĭ biologicheskiĭ slovar'. Moscow, Glavnaya Redaktsiya Inostrannykh Nauchno-Tekhnicheskikh Slovarei Fizmatgiza, 1963.

0850. Ruiz Torres, F. Diccionario inglés-español y español-inglés de medicina. 3d ed. Madrid, Editorial Alhambra, 1965.

0851. Unseld, D. W. Medical dictionary of the English and German languages. 6th ed., rev. and enl. Stuttgart, Wissenschaftliche Verlagsgesellschaft, 1971.

0852. Veillon, Emmanuel. Medical dictionary. Medizinisches Wörterbuch. Dictionnaire médical. Dicionario médico. 5th ed. Ed. by Albert Nobel. Berlin, Springer-Verlag, 1970.

0853. Villemin, M. Dictionnaire des termes vétérinaires et zootechniques. Paris, Vigot Freres Editeurs, 1963.

Abbreviations and Synonyms

0854. Acronyms and initialisms dictionary: a guide to alphabetic designations, contractions, acronyms, initialisms, and similar condensed appellations. 2d ed. Ed. by R. C. Thomas, J. M. Ethridge, and F. G. Ruffner, Jr., Detroit, Gale Research Co., 1965.
...Suppl. 1968-1969.

0855. Crowe, Barry. Concise dictionary of Soviet terminology, institutions and abbreviations. 1st ed. Oxford, Pergamon Press, 1969.

0856. Gardner, William. Chemical synonyms and trade names: a dictionary and commercial handbook. 7th ed., rev. and enl. by E. I. Cooke. Cleveland, CRC Press, 1971.

0857. Haynes, William. Chemical trade names and commercial synonyms: a dictionary of American usage. 2d ed., rev. and enl. Princeton, N.J., Van Nostrand, 1959.

0858. Jablonski, Stanley. Illustrated dictionary of eponymic diseases and their synonyms. Philadelphia, Saunders, 1969.

0859. Kerr, A. H. Medical hieroglyphs, abbreviations and symbols. Chicago, Clissold Books, 1970.

0860. Marler, E. E. J., comp. Pharmacological and chemical synonyms. 5th ed. New York, Excerpta Medica, 1972.

0861. Reverse acronyms and initialisms dictionary. A companion volume to *Acronyms and initialisms dictionary*, with terms arranged alphabetically by meaning. Detroit, Gale Research Co., 1972.

0862. Royal Society of Medicine. Units, symbols, and abbreviations: a guide for biological and medical editors and authors. Ed. by George Ellis. 1st ed. London, 1971.

0863. Steen, E. B. Dictionary of abbreviations in medicine and related sciences. 3d ed. Philadelphia, Davis, 1971.

0864. Union of International Associations. International initialese: guide to initials. Brussels, 1963. (U.I.A. publ. no. 182.)

Nomenclature

0865. American Medical Association. Current medical information and terminology. 4th ed. Ed. by B. L. Gordon. Chicago, 1971. (Formerly *Current medical terminology*.)

0866. ———. Current procedural terminology. Ed. by B. L. Gordon. 1st- Chicago, 1966-

0867. American Veterinary Medical Association. Special Committee on Nomenclature of Diseases. A basis for nomenclature of animal diseases, topographic classifications, and etiologic categories. Chicago, 1956.

0868. College of American Pathologists. Committee on Nomenclature and Classification of Disease. Systematized nomenclature of pathology. Chicago, 1965.

0869. International Anatomical Nomenclature Committee. Nomina anatomica. 3d ed. Amsterdam, Excerpta Medica, 1966.

0870. ———. Subcommittee on Embryology. Nomina embryologica. June 1970 revision. Bethesda, Md., Federation of American Societies for Experimental Biology, 1970.

0871. International Union Against Cancer. Committee on Tumor Nomenclature. Illustrated tumor nomenclature. 2d ed. New York, Springer-Verlag, 1969.

0872. Mack, Roy. Veterinary subject headings for use in *Index veterinarius* and *Veterinary bulletin*. Farnham Royal, England, Commonwealth Agricultural Bureaux, 1972.

0873. Steffanides, G. F. The scientist's thesaurus. 3d ed. Boston, Best Printers, 1963.

0874. U.S. National Cancer Institute. Epizootiology Section. Standard nomenclature of veterinary diseases and operations. 1st ed., rev. Bethesda, Md., National Cancer Institute, 1966. (U.S. Public Health Service publ. no. 1466.)
...1971 coding suppl. Ed. by W. A. Priester. Bethesda, Md., 1971.

0875. U.S. Public Health Service. Medical and health related services thesaurus. rev. ed. Washington, D.C., 1970. (USPHS publ. no. 1031.)

0876. Zuckerman, Solly. The nomenclature of primates commonly used in laboratory work. New Haven, Conn., Yale Univ. School of Medicine, 1934.

Special Subject Dictionaries

0877. Ainsworth, G. C. Ainsworth & Bisby's Dictionary of the fungi. 6th ed. Kew, England, Commonwealth Mycological Institute, 1971.

0878. Bander, E. J., and Jeffrey Wallach. Medical legal dictionary. Dobbs Ferry, N.Y., Oceana, 1970.

0879. Dobson, Jessie. Anatomical eponyms. London, Livingstone, 1962.

0880. Donáth, Tibor. Anatomical dictionary with nomenclatures and explanatory notes. New York, Pergamon Press, 1969.

0881. Drazil, J. V. Dictionary of quantities and units. Cleveland, CRC Press, 1971.

0882. Etter, L. E. Glossary of words and phrases used in radiology, nuclear medicine and ultrasound. 2d ed. Springfield, Ill., Thomas, 1970.

0883. Gluckstein, Fritz. Veterinary medicine: selected glossary and indexing instructions. Bethesda, Md., National Library of Medicine, 1969.

0884. Gray, Peter. Dictionary of the biological sciences. New York, Reinhold, 1967.

0885. Halliday, W. J. Glossary of immunological terms. London, Butterworth, 1971.

0886. Hawley, G. G. Condensed chemical dictionary. 8th ed. Van Nostrand Reinhold, 1971.

0887. Herbert, W. J., and P. C. Wilkinson, eds. Dictionary of immunology. Philadelphia, Davis, 1971.

0888. Hughes, E. C., ed. Obstetric-gynecologic terminology with section on neonatology and glossary of congenital anomalies. Philadelphia, Davis, 1972.

0889. King, R. C. A dictionary of genetics. 2d ed. New York, Oxford Univ. Press, 1972.

0890. Kingzett, C. T. Kingzett's chemical encyclopedia. 8th ed. Princeton, N.J., Van Nostrand, 1967.

0891. Leader, R. W. Dictionary of comparative pathology and experimental biology. Philadelphia, Saunders, 1971.

0892. Leider, Morris. Dictionary of dermatological words, terms and phrases. New York, McGraw-Hill, 1968.

0893. Magalini, Sergio. Dictionary of medical syndromes. Philadelphia, Lippincott, 1971.

0894. Mason, I. L. A world dictionary of livestock breeds, types, and varieties. 2d ed. Farnham Royal, Bucks, England, Commonwealth Agricultural Bureaux, 1969. (Commonwealth Bureau of Animal Breeding and Genetics. Technical communication no. 8.)

0895. The Merck index: an encyclopedia of chemicals and drugs. 8th ed. Rahway, N.J., Merck, 1968.

0896. Peters, J. A. Dictionary of herpetology. New York, Hafner, 1964.

0897. Snell, W. H., and Esther A. Dick. A glossary of mycology. rev. ed. Cambridge, Harvard Univ. Press, 1971.

0898. Winburne, J. N. A dictionary of agricultural and allied terminology. East Lansing, Michigan State Univ. Press, 1962.

0899. Young, Clara G. Medical specialty terminology. v. 1- St. Louis, Mosby, 1971- v. 1: Pathology, clinical cytology, and clinical pathology.

Medical Terminology and Word Structure

0900. Ayers, D. M. Bioscientific terminology: words from Latin and Greek terms. Tucson, Ariz., Univ. of Arizona Press, 1972.

0901. Bolander, D. O., et al. Instant medical spelling dictionary. Mundelein, Ill., Career Institute, 1970.

0902. Frenay, Agnes C. Understanding medical terminology. rev. 4th ed. St. Louis, Catholic Hospital Association, 1971.

0903. Gross, V. E. Mastering medical terminology. North Hollywood, Calif., Halls of Ivy, 1971. (Suppl. 1. Ten study lessons for mastering medical terminology.)

0904. Guide to the indexes of Biological abstracts and Bioresearch index. Philadelphia, Biological Sciences Information Service, 1972.

0905. Hadley, Anne. Pathology and autopsy lessons for the medical transcriber. Philadelphia, Lippincott, 1971.

0906. Harned, Jessie M. Medical terminology made easy. 2d ed. Berwyn, Ill., Physicians Record, 1968.

0907. Lee, R. V., and Doris J. Hofer. How to divide medical words. Carbondale, Ill., Southern Illinois Univ. Press, 1972.

0908. Nybakken, O. E. Greek and Latin in scientific terminology. Ames, Iowa State College Press, 1959.

0909. Roberts, Ffrangcon. Medical terms: their origin and construction. 5th ed. London, Heinemann Medical, 1971.

0910. Schmidt, J. E. English word power for physicians and other professionals: a vigorous and cultured vocabulary. Springfield, Ill., Thomas, 1971.

0911. ———. Structural units of medical and biological terms. Springfield, Ill., Thomas, 1969.

0912. Wain, Harry. Story behind the word: some interesting origins of medical terms. Springfield, Ill., Thomas, 1958.

General Encyclopedic Works

0913. Encyclopedia of the life sciences. v. 1-8. Garden City, N.Y., Doubleday, 1965-1966.

0914. Gray, Peter. The encyclopedia of the biological sciences. 2d ed. New York, Van Nostrand Reinhold, 1970.

0915. International encyclopedia of veterinary medicine. v. 1-5. London, Sweet & Maxwell, 1966.

0916. Shilling, C. W. Atomic energy encyclopedia in the life sciences. Philadelphia, Saunders, 1964.

0917. Van Nostrand's scientific encyclopedia. 4th ed. Princeton, N.J., Van Nostrand, 1968.

0918. Veterinary encyclopedia: diagnosis and treatment. 4 v. Ed. by Kjeld Wamberg. Copenhagen, Medical Book Co., 1969.

0919. Williams, R. J., and E. M. Lansford, Jr. The encyclopedia of biochemistry. New York, Reinhold, 1967.

BIOGRAPHICAL TOOLS AND DIRECTORIES

0920. AMA physician reference listing. Clifton, N.J., Fisher-Stevens, 1972.

0921. American Association of Equine Practitioners. Directory. Golden, Colo., 1969.

0922. American Chemical Society. Committee on Professional Training. Directory of graduate research: faculties, publications, and doctoral theses in departments or divisions of chemistry, chemical engineering, biochemistry, and pharmaceutical and/or medicinal chemistry at universities in the U.S. and Canada. Washington, D.C., 1971.

0923. American Medical Association Directory. 25th ed. 3 parts. Chicago, American Medical Association, 1970.

0924. American men and women of science: the physical and biological sciences. 12th ed. Ed. by Jacques Cattell Press. v. 1– New York, Bowker, 1971–

0925. American Veterinary Medical Association. Directory. Chicago. (Biennial.)

0926. Association of American Medical Colleges. Directory of American medical education. Washington, D.C. (Annual.)

0927. Casey, Donn, ed. World directory of research workers in vertebrate reproduction. Cambridge, England, Reproductive Research Information Service, 1967.

0928. Directory of bioscience departments in the United States and Canada. New York, Reinhold, 1967.

0929. Directory of health and allied sciences libraries and information sources in Illinois, Indiana, Iowa, Minnesota, North Dakota and Wisconsin. Comp. and ed. by Sara L. Moreland. Chicago, Midwest Regional Medical Library, The John Crerar Library, 1971.

0930. Directory of medical specialists holding certification by American specialty boards. 15th ed., 1972-1973. v. 1-2. Chicago, Marquis, 1972. (Revised biennially.)

0931. Directory of special libraries and information centers. 2d ed. Ed. by A. T. Kruzas. Detroit, Gale Research Co., 1968.

0932. Encyclopedia of associations. 6th ed. 3 v. Detroit, Gale Research Co., 1970. v. 1: National associations of the United States. v. 2: Geographic-executive index. v. 3: New associations and projects.

0933. European research index: a guide to European research including medicine, agriculture, and engineering. 2d ed. 2 v. Comp. by C. H. Williams. Guernsey, Channel Islands, Francis Hodgson, 1969.

0934. Federation of American Societies for Experimental Biology. Directory of membership. Bethesda, Md., FASEB, Office of Scientific Personnel. (Published annually.)

0935. Foundation directory. 3d ed. Ed. by Marianna O. Lewis. New York, Russell Sage Foundation, 1967.

0936. Goodman, N. M. International health organizations and their work. 2d ed. Baltimore, Williams & Wilkins, 1971.

0937. Industrial research laboratories of the U.S. 13th ed. Ed. by Jacques Cattell Press. New York, Bowker, 1970.

0938. International directory of genetic services. Comp. by H. T. Lynch. 3d ed. New York, National Foundation–March of Dimes, 1971.

0939. International handbook of universities and other institutions of higher education. 4th ed. Ed. by H. M. R. Keyes. Washington, D.C., American Council on Education, 1968.

0940. Ireland, Norma O. Index to scientists of the world. Boston, Faxon, 1962.

0941. ISI's Who is publishing in science. Philadelphia, Institute for Scientific Information. (Annual.)

0942. Manual and directory of animal diagnostic laboratories in the United States, 1967. Washington, D.C., U.S. Agricultural Research Service, 1967.

0943. The medical directory. 2 v. London, Churchill. (Annual.)

0944. Medical Library Association. Directory. Chicago. (Biennial.)

0945. National faculty directory. 1971. 2 v. Detroit, Gale Research Co., 1971.

0946. Research centers directory, 1972-73. 4th ed. Detroit, Gale Research Co. (Quarterly suppl. New research centers. 1972-)

0947. Scientific and technical societies of the United States. 8th ed. Washington, D.C., National Research Council, 1968.

0948. Scientific research in British universities and colleges. Annual. London, Dept. of Education and Science, and the British Council, 1963/1964. (Formerly *Scientific research in Britain*.)

0949. Scientific, technical and health personnel in the federal government, 1969. Washington, D.C., National Science Foundation, 1971. (NSF-70-44.)

0950. U.S. Dept. of Agriculture. Professional workers in state agricultural experiment stations and other cooperating state institutions. 1971-1972. Washington, D.C., 1971. (Agriculture handbook no. 305.)

0951. U.S. National Institutes of Health. Scientific directory and annual bibliography. Bethesda, Md., National Institutes of Health. (Annual.)

0952. U.S. Public Health Service. Directory of biomedical institutions in the U.S.S.R. Washington, D.C., 1965. (Publ. no. 1354.)

0953. Who's who of British scientists, 1969/1970. London, Longman, 1970- (Formerly *Directory of British scientists*.)

0954. World directory of medical schools. Geneva, World Health Organization, 1963.

0955. World directory of veterinary schools. Geneva, World Health Organization, 1963.

0956. The world of learning, 1970-71. 24th ed. London, Europa Publications, 1971.

0957. World who's who in science: a biographical dictionary of notable scientists from antiquity to the present. Ed. by A. G. Debus. Chicago, Marquis, 1968.

HISTORIES AND SOURCE BOOKS

0958. Ackerknecht, E. W. History and geography of the most important diseases. New York, Hafner, 1965.

0959. Bernard, Claude. Claude Bernard and the experimental method in medicine. By J. M. D. Olmsted and E. H. Olmsted. New York, Abelard-Schuman, 1952. (Life science library, no. 23.)

0960. ———. An introduction to the study of experimental medicine. New York, Dover, 1957. (Reprint of 1927 English translation.)

0961. Bierer, B. W. A short history of veterinary medicine in America. East Lansing, Michigan State Univ. Press, 1955.

0962. Chauveau, A. The comparative anatomy of the domesticated animals. 2d ed. New York, Jenkins, 1885. (Translated and edited by George Fleming.)

0963. Claude Bernard and experimental medicine. Collected papers from a symposium commemorating the centenary of the publication of *An introduction to the study of experimental medicine* and the first English translation of Claude Bernard's *Cahier rouge*. Ed. by Francisco Grande. Cambridge, Mass., Schenkman, 1967.

0964. Cole, F. J. A history of comparative anatomy, from Aristotle to the eighteenth century. London, Macmillan, 1944.

0965. Cooke, Ian, and Mack J. Lipkin, comps. and eds. Cellular neurophysiology: a source book. New York, Holt, Rinehart & Winston, 1972.

0966. Galen. Galen on the usefulness of the parts of the body. 2 v. Ithaca, N.Y. Cornell Univ. Press, 1968.

0967. Garrison, F. H. A medical bibliography (Garrison and Morton): an annotated check-list of texts illustrating the history of medicine. By L. T. Morton. 3d ed. Philadelphia, Lippincott, 1970.

0968. Harvey, William. Exercitatio anatomica de motu cordis et sanguinis in animalibus. 5th ed. Springfield, Ill., Thomas, 1970. (English translation.)

0969. Hill, A. V. Trails and trials in physiology: a bibliography, 1909-1964; with reviews of certain topics and methods and a reconnaissance for future research. Baltimore, Williams & Wilkins, 1966.

0970. Kelley, E. C. Encyclopedia of medical sources. Baltimore, Williams & Wilkins, 1948.

0971. Liebig, Justus. Animal chemistry: or, organic chemistry in its application to physiology and pathology. New York, Johnson

Reprint, 1964. (A facsimile of the Cambridge edition. 1842.)

0972. Merillat, L. A., and D. M. Campbell. Veterinary military history of the United States: with a brief record of the development of veterinary education, practice, organization, and legislation. Kansas City, Mo., Haver-Glover Laboratories, 1935.

0973. Morgagni, G. B. The seats and causes of diseases investigated by anatomy. 3 v. New York, Hafner, 1960. (Facsimile reprint of 1769 edition.)

0974. Needham, Dorothy M. Machina carnis: the biochemistry of muscular contraction in its historical development. New York, Cambridge Univ. Press, 1971.

0975. Neveu-Lemaire, Maurice. Traité de protozoologie médicale et vétérinaire. Paris, Vigot Freres, 1943.

0976. Nobelstiftelsen, Stockholm. Physiology or medicine. 3 v. New York, American Elsevier, 1964-1967. (Nobel lectures, including presentation speeches and laureates' biographies, 1901-1962.)

0977. Papez, J. W. Comparative neurology: a manual and text for the study of the nervous system of vertebrates. New York, Crowell, 1929.

0978. Pavlov, Ivan. Pavlov, a biography. By B. P. Babkin. Chicago, Univ. of Chicago Press, 1949.

0979. Pugh, L. P. From farriery to veterinary medicine, 1785-1795. West Orange, N.J., Saifer, 1970.

0980. Schmidt, J. E. Medical discoveries, who and when. Springfield, Ill., Thomas, 1955.

0981. Smithcors, J. F. The American veterinary profession, its background and development. Ames, Ia., Iowa State Univ. Press, 1963.

0982. ———. Evolution of the veterinary art: a narrative account to 1850. Kansas City, Mo., Veterinary Medical Pub., 1957.

0983. Smythe, R. H. Healers on horseback. Springfield, Ill., Thomas, 1963.

0984. Stevenson, L. G. Nobel prize winners in medicine and physiology, 1901-1950. New York, Henry Schuman, 1953.

0985. Tyson, Edward. Edward Tyson, MD, FRS, 1650-1708, and the rise of human and comparative anatomy in England. By M. F. A. Montagu. Philadelphia, American Philosophical Society, 1943.

0986. U.S. National Research Council. Committee on Animal Health. Historical survey on animal-disease morbidity. Washington, D.C., National Academy of Sciences, 1966.

0987. Whitteridge, Gweneth. William Harvey and the circulation of the blood. New York, American Elsevier, 1971.

0988. World veterinary congress. 17th. Hannover, 1963. Centenary brochure. Wiesbaden, Deutschen Tierarzteschaft, 1963.

MISCELLANY

Libraries and/or Information Centers and Their Services

0989. Animal Models and Genetic Stocks Information Exchange Program. Washington, D.C., Institute of Laboratory Animal Resources.
 An information exchange service started in 1969 and maintained by the National Research Council, Institute of Laboratory Animal Resources, which serves as a central agency to collect, maintain, and disseminate information on vertebrate models and genetic stocks used in biomedical research. The service provides information about locations and sources of supply for particular strains of animals, names of persons who may serve as consultants as well as key references describing animal models.

0990. Ash, Lee, and D. Lorenz. Subject collections: a guide to special book collections in libraries. 3d ed. New York, Bowker, 1967.

0991. Automatic Subject Citation Alert (ASCA). Philadelphia, Institute for Scientific Information.
 This current awareness information system is based on a predetermined profile constructed by an information specialist and the user. ASCA provides a weekly customized computer report of the pertinent scientific

and technical literature specific to the user's interest.

ASCA Topics is an information retrieval system based on a standard profile for each of many specific research areas. This system provides a print-out weekly of citations appropriate to the topic.

Additional information may be obtained from the Institute for Scientific Information.

0992. BIOSIS Current Literature Alerting Search Service (CLASS). Philadelphia. Biosciences Information Service of Biological Abstracts.

This selective dissemination and current awareness service provides to the subscriber printouts of complete references from *BA previews* (a magnetic tape compilation of prepublication material from *Biological abstracts* and *Bioresearch index*) based upon a profile that has been developed by "search strategists" and the user. Additional information regarding subscription cost, profile development, etc., may be obtained from Biosciences Information Service.

0993. Chemical Abstracts Service source index. v. 1- Columbus, Ohio, American Chemical Society, 1970-

A reference work that provides bibliographic descriptions and library holdings information for 25,000 journals, patents, congress and symposium proceedings, as well as monographs cited in *Chemical abstracts* and other bibliographic works. Approximately 30,000 entries representing holdings of 325 major resource libraries in the U.S. and 72 in 27 other countries are included. The entries include complete titles of the cited publications, each abbreviated according to American National Standards Institute recommendations. Each primary entry also includes CODEN identification and the publication title cataloged according to American Library Association cataloging rules. Library identification is by the Library of Congress National Union Catalog (NUC) symbols. A quarterly service provides updating information.

The *CAS source index* is available in printed and computer-readable format. Additional information regarding cost and other services may be obtained from Chemical Abstracts Service.

0994. Cheshier, R. G. Principles of medical librarianship: the environment affecting health sciences libraries. Cleveland, Case Western Reserve Univ., 1970.

0995. Directory of special libraries and information centers. 2d ed. Detroit, Gale Research Center, 1968.

0996. International Cell Research Organization. Directory of cell research laboratories. Paris, UNESCO, 1969.

0997. Medical Library Association. Handbook of medical library practice. Ed. by Gertrude L. Annan and Jacqueline W. Felter. Baltimore, Waverly, 1970.

0998. Simonton, Wesley, and Charlene Mason, eds. Information retrieval with special reference to the biomedical sciences. Minneapolis, Nolte Center for Continuing Ed., Univ. of Minnesota, 1966.

0999. U.S. National Cancer Institute. Viral Oncology Program. Catalog of resources and services. Bethesda, Md., 1971.

1000. U.S. National Library of Medicine. MEDLARS Search Service. Bethesda, Md., 1964-

The Medical Literature Analysis and Retrieval System (MEDLARS) is a computer-based system instituted by the National Library of Medicine to provide rapid access to its vast store of biomedical bibliographic information. This information is supplied to the health sciences community in several forms.

The most widely known and used by-products of this system are *Index medicus* (entry 0034) and *Current catalog* (entry 0811). However, many other services are available. MEDLARS demand literature searches have provided hand-tailored bibliographies covering the world's biomedical literature in a narrow subject area limited to the specific need of a health professional. MEDLARS searches have been replaced to a large extent by MEDline searches, 1972-, which uses a smaller data base from which an on-line computer search can be effected more rapidly. In many instances, the demand literature searches are reprinted for general distribution. Several of these are cited in entry 0183.

Many continuing bibliographies in limited subject areas are prepared from the data base and printed as separate publications. Examples of these are *Endocrinology index* (entry 0059) and *Diabetes literature index* (entry 0056), as well as several others cited in Chapter 1.

In 1973 it is planned to have available SERline and CATline services which will provide serials and cataloging information respectively.

1001. U.S. National Referral Center for Science and Technology. A directory of information resources in the United States: federal government. (With a supplement of government-sponsored information resources.) Washington, D.C., 1967.

1002. ———. A directory of information resources in the United States: general toxicology. Washington, D.C., 1969.

1003. ———. A directory of information resources in the United States: physical sciences, biological sciences, engineering. Washington, D.C., 1965.

1004. World Health Organization. The work of WHO virus reference centers and the services they provide. Geneva, 1968.

Research Methodology

1005. Animal models for biomedical research: proceedings of a symposium sponsored by the Institute of Laboratory Animal Resources and the American College of Laboratory Animal Medicine. v. 1- Washington, D.C., National Academy of Sciences, 1968-
    I—Its publication no. 1594. II—Its publication no. 1736. III—Its publication no. 1854. Revised approximately annually.

1006. Barzun, Jacques, and H. F. Graff. The modern research. rev. ed. New York, Harcourt, Brace & World, 1970.

1007. Brookhaven National Laboratory. Lectures in science: vistas in research. v. 1- New York, Gordon & Breach, 1969-

1008. Cross, Wilbur, and Susan Graves. The new age of medical discovery. New York, Hawthorn Books, 1972.

1009. Experiments in physiology and biochemistry. v. 1- New York, Academic Press, 1968-

1010. Fisher, R. A. The design of experiments. 8th ed. New York, Hafner, 1966.

1011. Gay, W. I., ed. Methods of animal experimentation. 3 v. New York, Academic Press, 1965-1968.

1012. Holman, H. H. Biological research method, a practical guide. 2d ed. New York, Hafner, 1969.

1013. Little, L. M. An introduction to the experimental method, for students of biology and the health sciences. Minneapolis, Burgess, 1961.

1014. Lloyd, L. E. Techniques for efficient research. New York, Chemical Publishing Co., 1966.

1015. Reflections on research and the future of medicine: a symposium and other addresses. Ed. by C. E. Lyght. New York, McGraw-Hill, 1967.

1016. Rosen, Robert. Optimality principles of biology. New York, Plenum Press, 1967.

1017. Schlieper, Carl, ed. Research methods in marine biology. Seattle, Univ. of Washington Press, 1972.

1018. Selected abstracts on animal models for biomedical research. Comp. and ed. by C. B. Frank and Marilyn J. Anderson. Washington, D.C., Institute of Laboratory Animal Resources, National Research Council, National Academy of Sciences, 1971.

1019. Selye, Hans. In vivo: the case for supramolecular biology. New York, Liveright, 1967.

1020. Symposium on cross-disciplinary sciences in biomedical research. Johannesburg, 1969. Johannesburg, South Africa Council for Scientific and Industrial Research, 1969.

1021. U.S. National Research Council. Committee on Physiological Effects of Environmental Factors on Animals. A guide to environmental research on animals. Washington, D.C., National Academy of Sciences, 1971.

1022. Winer, B. J. Statistical principles in experimental design. 2d ed. New York, McGraw-Hill, 1971.

Research Support

1023. Annual register of grant support. Ed. by Alvin Renetzky and Jean L. Aroeste. Los Angeles, Academic Media, 1967-

Miscellany   69   Reference Works

1024. Current financial aids for graduate students. Peoria, Ill., College Opportunities Unlimited, Inc., 1966.

1025. Dermer, Joseph, ed. Where America's large foundations make their grants. New York, Public Service Materials Center, 1971.

1026. Education Services Press. Guide to support programs for education. St. Paul, Education Services Press, 1966.

1027. Foundation Center. The foundation grants index, 1970-1971: a two-year cumulative listing of foundation grants. Ed. by Lee Noe. New York, Columbia Univ. Press, 1972.

1028. Foundation directory. 3d ed. Ed. by Marianna O. Lewis. New York, Russell Sage Foundation, 1967.

1029. The grants register. Postgraduate awards for the English-speaking world. Chicago, St. James Press, 1969/70.

1030. Guide to grants, loans, and other types of government assistance available to students and educational institutions. Washington, D.C., Public Affairs Press, 1967.

1031. Searles, Aysel. Guide to financial aids for students in arts and sciences for graduate and professional study. New York, Arco, 1971.

1032. U.S. National Cancer Institute. Active research grants. Bethesda, Md., 1965-

1033. U.S. National Institutes of Health. Public health service grants and awards. 2 parts. Washington, D.C., Govt. Printing Office, 1970-71.
  Part 1—Research grants, research contracts. Part 2—Training, construction, medical libraries, and summary tables (including research).

1034. ———. Division of Research Grants. A guide to grant and award programs of the National Institutes of Health. rev. ed. Bethesda, Md., 1970. (U.S. Public Health Service publ. no. 1067.)

1035. ———. ———. Training grant programs of the Public Health Service. Bethesda, Md., Career Development Review Branch, 1966.

(U.S. Public Health Service publ. no. 1467.)

1036. U.S. National Science Foundation. Division of Science Resources Studies. An analysis of federal R & D funding by budget function, 1960-1972. Washington, D.C., 1971.

Medical Writing, Speaking, Illustration, and Audiovisual Materials

1037. Audio-visual equipment directory. 15th ed. Fairfax, National Audio-visual Association, 1969.

1038. Blaker, A. A. Photography for scientific publication: a handbook. San Francisco, Freeman, 1965.

1039. Burton, A. L., ed. Cinematographic techniques in biology and medicine. New York, Academic Press, 1971.

1040. Council of Biology Editors. Committee on Form and Style. CBE style manual. 3d ed. Washington, D.C., American Institute of Biological Sciences, 1972.

1041. ———. Committee on Graduate Training in Scientific Writing. Scientific writing for graduate students. Ed. by F. P. Woodford. New York, Rockefeller Univ. Press, 1968.

1042. Currie, D. J., and Arthur Smialowski. Photographic illustration for medical writing. Springfield, Ill., Thomas, 1962.

1043. Eddy, Samuel, C. P. Oliver, and J. P. Turner. Atlas of drawings for vertebrate anatomy. 3d ed. New York, Wiley, 1964.

1044. Fishbein, Morris. Medical writing: the technic and the art. 4th ed. Springfield, Ill., Thomas, 1972.

1045. Garn, S. M. Writing the biomedical research paper. Springfield, Ill., Thomas, 1970.

1046. Hadley, Anne. 101 lessons in X-ray terminology for the medical transcriber. Philadelphia, Lippincott, 1972.

1047. Hawkins, C. F. Speaking and writing in medicine: the art of communication. Springfield, Ill., Thomas, 1967.

1048. King, L. S., and C. G. Roland. Scientific writing. Chicago, American Medical Association, 1968.

1049. Knight, C. R. Animal drawing: anatomy and action for artists. (Animal anatomy and psychology for artists and laymen.) New York, Dover, 1959.

1050. Knudsen, Jens W. Biological techniques, collecting, preserving and illustrating plants and animals. New York, Harper & Row, 1966.

1051. Manko, H. H. Effective technical speeches and sessions: a guide for speakers and program chairmen. New York, McGraw-Hill, 1969.

1052. Moser, R. H., and Erwin Di Cyan. Purpose of medical writing: how medical communications may be improved. Springfield, Ill., Thomas, 1970.

1053. Netter, F. H. The Ciba collection of medical illustrations. v. 1- New York, Ciba Pharmaceutical Co., 1953-
   v. 1: Nervous system. v. 2: Reproductive system. v. 3: Digestive system. v. 4: Endocrine system. v. 5: The heart.

1054. Richer, Paul. Artistic anatomy. New York, Watson-Guptill, 1971.

1055. Roland, C. G., ed. Good scientific writing: an anthology. Chicago, American Medical Association, 1971.

1056. Thorne, Charles. Better medical writing. New York, Grune & Stratton, 1971.

1057. U.S. Federal Advisory Council on Medical Training Aids. Film reference guide for medicine and allied sciences. Rev. ed. Washington, D.C., U.S. Public Health Service, 1970. (Publ. no. 487. Approximately 2,000 films.)

1058. U.S. National Audiovisual Center. A list of U.S. government medical and dental 8mm films for sale by the National Audiovisual Center. Washington, D.C., 1970.

1059. U.S. National Library of Medicine. Current catalog.
   Beginning with v. 6, no. 7 (July 1971) an Audiovisual Section will be found in the NLM *Current catalog*; it contains citations to various types of audiovisual materials which are distributed by the National Medical Audiovisual Center, Atlanta, Ga.

1060. Welsh, J. J. The speech writing guide: professional techniques for regular and occasional speakers. New York, Wiley, 1968.

1061. World Veterinary Association. World catalogue of veterinary films and films of veterinary interest. 2d ed. Utrecht, 1966. (1st ed. has the title *List of veterinary films and films of veterinary interest*.)

1062. Zollinger, R. M. A practical outline for preparing medical talks and papers. New York, Macmillan, 1961.

Industrial Information

1063. American druggist blue book. v. 1- New York, American Druggist, 1962-

1064. Arthur H. Thomas Company. Scientific apparatus and reagents, selected for academic, industrial and medico-biological laboratories. Philadelphia, 1971.

1065. Audio-visual equipment directory. 15th ed. Fairfax, National Audio-visual Association, 1969.

1066. Drug topics red book. v. 1- New York, Topics, 1962-

1067. Dummer, G. W. A., ed. Medical electronic equipment 1969-1970. 4 v. New York, Pergamon Press, 1969-1970.
   v. 1: Clinical, diagnostic and therapeutic equipment. v. 2: Monitoring, recording and computing equipment. v. 3: Laboratory systems and equipment (analytical). v. 4: Laboratory systems and equipment (general).

1068. Establishments holding U.S. veterinary licenses to produce biological products. 2 v. Hyattsville, Md., Veterinary Biologics Division, U.S. Agricultural Research Service, 1967-1968. (Continued by entry 1073.)

1069. Gleason, M. N. Clinical toxicology of commercial products. 3d ed. Baltimore, Williams & Wilkins, 1969.

1070. Sax, N. I. Dangerous properties of industrial materials. 3d ed. New York, Reinhold, 1968.

1071. Smith, D. A. Medical electronics equipment handbook. Indianapolis, W. H. Sams, 1962.

General Biomedical Interest

1072. Thomas' register of American manufacturers. New York, Thomas. (Annual.)

1073. Veterinary biological products: licensees, permitees. Hyattsville, Md., Veterinary Biologics Division, U.S. Agricultural Research Service, 1970- (Continues entry 1068.)

Handbooks of Laboratory Technique

1074. Zimmerman, O. T., and I. Lavine. Industrial Research Service's handbook of material trade names. Dover, N.H., Industrial Research Service, 1953.
... Suppl. 1, 1956.
... Suppl. 2, 1957.
... Suppl. 3, 1960.
... Suppl. 4, 1965.

6

HANDBOOKS AND MANUALS OF LABORATORY TECHNIQUE

In the first two sections of this chapter, the compilers have attempted to note general interest handbooks and tables and techniques manuals that are useful in many areas of the biomedical sciences. A selected number are designed primarily for use in other disciplines, such as chemistry or physics, but they are applicable to study and research in the biomedical sciences as well. Some are quite specialized; however, they find application across the broad spectrum of interests in comparative and veterinary medical research and practice. Many compilations of clinical laboratory methods are included.

Other sections list manuals designed to cover special techniques in microscopy and microphotography as well as various staining, culture, and clinical techniques. Guides to dissection, autopsy, necropsy, and specimen preparation are noted in another section. A final section lists many handbooks that cover the use of general and specialized instruments in medical practice, teaching, and research.

GENERAL BIOMEDICAL INTEREST

1075. Altman, P. L., ed. Blood and other body fluids. Washington, D.C., Federation of American Societies for Experimental Biology, 1961.

1076. ——. Environmental biology. Bethesda, Md., Federation of American Societies for Experimental Biology, 1966.

1077. ——. Respiration and circulation. Bethesda, Md., Federation of American Societies for Experimental Biology, 1971.

1078. Altman, P. L., and Dorothy S. Dittmer, eds. Biology data book. 2d ed. Bethesda, Md., Federation of American Societies for Experimental Biology, 1972. (An earlier ed. was titled *Handbook of biological data*, ed. by W. S. Spector, 1956.)

1079. —— and ——. Growth, including reproduction and morphological development. Washington, D.C., Federation of American Societies for Experimental Biology, 1962.

1080. Association of Official Agricultural Chemists. Official methods of analysis. 11th ed. Washington, D.C., 1970.

1081. Chemical Rubber Co. Handbook of chemistry and physics. 52d ed. Cleveland, 1971.

1082. Damm, H. C. The handbook of biochemistry and biophysics. Cleveland, World, 1966.

1083. Frankel, Joseph, and N. E. Williams. Laboratory manual in principles of animal biology. Minneapolis, Burgess, 1967.

1084. Gray, C. H. Laboratory handbook of toxic agents. 2d ed. Englewood, N.J., Franklin, 1968.

1085. Long, Cyril. Biochemists' handbook. Princeton, N.J., Van Nostrand, 1961.

1086. Sober, H. A., ed. Handbook of biochemistry: selected data for molecular biology. Cleveland, Chemical Rubber Co., 1968.

1087. Spector, W. S. Handbook of biological data. Philadelphia, Saunders, 1956.

CLINICAL LABORATORY METHODS

1088. Archer, R. K. Haematological techniques for use on animals. Oxford, England, Blackwell, 1965.

1089. Arquembourg, P. C., et al. Primer of immunoelectrophoresis. Ann Arbor, Mich., Ann Arbor Science Publishers, 1970.

1090. Assendelft, O. W. van. Spectrophotometry of haemoglobin derivatives. Springfield, Ill., Thomas, 1970.

1091. Barnett, R. N. Clinical laboratory statistics. Boston, Little, Brown, 1971.

1092. Benjamin, Maxine M. Outline of veterinary clinical pathology. 2d ed. Ames, Iowa State Univ. Press, 1965.

1093. Bier, Milan. Electrophoresis. 2 v. New York, Academic Press, 1959-1967.

1094. British Society for Parasitology. Application of new techniques to parasitology. Ed. by Angela E. R. Taylor. Philadelphia, 1963.

1095. Campbell, D. H. Methods in immunology: a laboratory text for instruction and research. 2d ed. New York, W. A. Benjamin, 1970.

1096. Cartwright, G. E. Diagnostic laboratory hematology. 4th ed., rev. and enl. New York, Grune & Stratton, 1968.

1097. Cawley, L. P. Electrophoresis and immunoelectrophoresis. Boston, Little, Brown, 1969.

1098. Cherim, S. M. Chemistry for laboratory technicians. Philadelphia, Saunders, 1971.

1099. Clinical diagnosis by laboratory methods. 14th ed. Ed. by I. Davidsohn. Philadelphia, Saunders, 1969.

1100. Coles, E. H. Veterinary clinical pathology. Philadelphia, Saunders, 1967.

1101. Cutts, J. H. Cell separation: methods in hematology. New York, Academic Press, 1970.

1102. Dacie, J. V., and S. M. Lewis. Practical haematology. 4th ed. London, Churchill, 1968.

1103. Damm, H. C. The handbook of biochemistry and biophysics. Cleveland, World, 1966.

1104. ———, ed. Methods and references in biochemistry and biophysics. Cleveland, World, 1966.

1105. ———. Practical manual for clinical laboratory procedures. Cleveland, Chemical Rubber Co., 1965.

1106. Escamilla, R. F. Laboratory tests in the diagnosis and investigation of endocrine function. 2d ed. Philadelphia, Davis, 1971.

1107. Faulkner, W. R., ed. Manual of clinical laboratory procedures. Cleveland, Chemical Rubber Co., 1970.

1108. Gradwohl, R. B. H. Gradwohl's clinical laboratory methods and diagnosis: a textbook on laboratory procedures and their interpretation. 7th ed. 2 v. Ed. by Sam Frankel et al. St. Louis, Mosby, 1970.

1109. Heftman, Erich, ed. Chromatography: absorption, partition, ion exchange, electrochromatography, column, slab, paper, gas. 2d ed. New York, Reinhold, 1967.

1110. Henry, R. J. Clinical chemistry: Principles and technics. New York, Harper & Row, 1964.

1111. Issitt, P. D. Applied blood group serology. Oxnard, Calif., Spectra Biologicals, 1970.

1112. Jonxis, J. H. P., and T. H. J. Huisman. A laboratory manual on abnormal haemoglobins. 2d ed. Oxford, Blackwell, 1968.

1113. Kamath, S. H. Clinical biochemistry for medical technologists. Boston, Little, Brown, 1972.

1114. Kaneko, J. J., and C. E. Cornelius, eds. Clinical biochemistry of domestic animals. 2d ed. New York, Academic Press, 1970.

1115. Kosakai, Nozomu, ed. Illustrated laboratory techniques. Flushing, N.Y., Medical Examination Publishing, 1969.

1116. Krieg, A. F., et al. Clinical laboratory computerization. Baltimore, Univ. Park Press, 1971.

1117. Kubica, G. P. Laboratory methods for

clinical and public health mycobacteriology. Atlanta, National Communicable Disease Center, 1967. (U.S.P.H.S. publ. no. 1547.)

1118. Kwapinski, J. B. Methods of serological research. New York, Wiley, 1965.

1119. Lamela, Alberto. Introduction to medical laboratory methods. New York, Harper & Row, 1971.

1120. Levinson, S. A., and R. P. MacFate. Clinical laboratory diagnosis. 7th ed. Philadelphia, Lea & Febiger, 1969.

1121. Linné, J. J. Basic laboratory techniques for the medical laboratory technician. New York, McGraw-Hill, 1970.

1122. Lynch, M. J. Medical laboratory technology and clinical pathology. 2d ed. Philadelphia, Saunders, 1969.

1123. McDonald, G. A., T. C. Dodds, and Bruce Cruickshank. Atlas of haematology. 3d ed. Baltimore, Williams & Wilkins, 1970.

1124. MacFate, R. P. Introduction to the clinical laboratory. 3d ed. Chicago, Year Book Medical Publishers, 1972.

1125. Mark, D. D., and Arthur Zimmer. Atlas of clinical laboratory procedures. v. 1- New York, McGraw-Hill, 1967-

1126. Mattenheimer, Hermann. Micromethods for the clinical and biochemical laboratory. 1st American ed. Ann Arbor, Mich., Ann Arbor Science Publishers, 1970. (Revision and translation of 2d ed. of *Mikromethoden für das klinisch-chemische und biochemische Laboratorium.*)

1127. Narin, R. C. Fluorescent protein tracing. 3d ed. Baltimore, Williams & Wilkins, 1969.

1128. Natelson, Samuel. Techniques of clinical chemistry. 3d ed. Springfield, Ill., Thomas, 1971.

1129. Nuzzolo, Lucio. Serological diagnostix. Springfield, Ill., Thomas, 1966.

1130. O'Brien, Donough, et al. Laboratory manual of pediatric, micro- and ultramicrobiochemical techniques. 4th ed. New York, Harper & Row, 1968.

1131. Oppenheim, I. A. Textbook for laboratory assistants. St. Louis, Mosby, 1972.

1132. Osbaldiston, G. W., and E. C. Stowe. Laboratory procedures in clinical veterinary bacteriology. Baltimore, Univ. Park Press, 1972.

1133. Ouchterlony, Örjan. Handbook of immunodiffusion and immunoelectrophoresis. Rev. ed. Ann Arbor, Mich., Ann Arbor Science Publishers, 1968.

1134. Padmore, G. R. A. Elementary calculations in clinical chemistry. Baltimore, Williams & Wilkins, 1972.

1135. Pinkava, J. Handbook of laboratory unit-preparations. New York, Gordon & Breach, 1970.

1136. Ring, A. M. Laboratory correlations manual. Springfield, Ill., Thomas, 1969.

1137. Simmons, Arthur. Technical hematology. Philadelphia, Lippincott, 1968.

1138. Slonim, N. B. Cardiopulmonary laboratory basic methods and calculations: a manual of cardiopulmonary technology. Springfield, Ill., Thomas, 1967.

1139. Smith, Ivor, ed. Chromatographic and electrophoretic techniques. 3d ed. v. 1- New York, Wiley, 1969- v. 1: Chromatography.

1140. Spencer, E. S. Hand atlas of the urinary sediment: bright-field, phase-contrast, and polarized-light. Baltimore, Univ. Park Press, 1971.

1141. Standard methods of clinical chemistry. v. 1- New York, Academic Press, 1953-

1142. Umbreit, W. W., R. H. Burris, and J. F. Stauffer. Manometric and biochemical techniques: a manual describing methods applicable to the study of tissue metabolism. 5th ed. Minneapolis, Burgess, 1972.

1143. U.S. National Institute of General Medical Sciences. Automation in the medical laboratory. Sciences-Review Committee. The mechanization, automation, and increased effectiveness of the clinical laboratory: a status report. Bethesda, Md., 1971. (DHEW publ. no. [NIH] 72-145.)

1144. Whitby, L. E. H., and C. J. C. Britton. Disorders of the blood: diagnosis, pathology, treatment, technique. 10th ed. New York, Grune & Stratton, 1969.

1145. White, Wilma L. Chemistry for medical technologists. 3d ed. St. Louis, Mosby, 1970.

1146. Wiener, A. S. Advances in blood grouping. 3 v. New York, Grune & Stratton, 1961-1970.

1147. Williams, M. Ruth. Introduction to the profession of medical technology. Philadelphia, Lea & Febiger, 1971.

1148. Wintrobe, M. W. Clinical hematology. 6th ed. Philadelphia, Lea & Febiger, 1967.

1149. Zweig, Gunter, and J. R. Whitaker. Paper chromatography and electrophoresis. v. 1- New York, Academic Press, 1967-

MICROSCOPY AND MICROPHOTOGRAPHY

1150. Allen, R. M. Photomicrography. 2d ed. Princeton, N.J., Van Nostrand, 1958.

1151. Braley, A. E., et al. Stereoscopic atlas of slit-lamp biomicroscopy. 2 v. St. Louis, Mosby, 1970.

1152. Causey, Gilbert. Electron microscopy: a textbook for students of medicine and biology. Baltimore, Williams & Wilkins, 1962.

1153. Clark, G. L. The encyclopedia of microscopy. New York, Reinhold, 1961.

1154. Dawes, C. J. Biological techniques in electron microscopy. New York, Barnes & Noble, 1971.

1155. Freeman, J. A. Cellular fine structure, an introductory student text and atlas. New York, McGraw-Hill, 1964.

1156. Fujita, Tsueno, Junichi Tokunaga, and Hajime Inoue. Atlas of scanning electron microscopy in medicine. New York, Elsevier, 1971.

1157. Hayat, M. A. Basic electron microscopy techniques. New York, Van Nostrand Reinhold, 1972.

1158. Kay, D. E. Techniques for electron microscopy. 2d ed. Philadelphia, Davis, 1965.

1159. Loveland, R. P. Photomicrography: a comprehensive treatise. 2 v. New York, Wiley, 1970.

1160. McClung, C. E. Handbook of microscopial technique for workers in animal and plant tissues. 3d ed., rev. and enl. New York, Hafner, 1961.

1161. Meek, G. A. Practical electron microscopy for biologists. New York, Wiley, 1970.

1162. Mercer, E. H. Electron microscopy: a handbook for biologists. 2d ed. Oxford, Blackwell, 1966.

1163. New York Society of Electron Microscopists. The international bibliography of electron microscopy. v. 1, 1950-1955. v. 2, 1956-1961. New York, 1962.

1164. Pantin, C. F. A. Notes on microscopical technique for zoologists. New York, Cambridge Univ. Press, 1960.

1165. Parsons, D. F. Some biological techniques in electron microscopy. New York, Academic Press, 1970.

1166. Ross, K. F. Phase contrast and interference microscopy for cell biologists. New York, St. Martin's Press, 1967.

1167. Sandborn, E. B. Cell and tissues by light and electron microscopy. 2 v. New York, Academic Press, 1970.

1168. Sjöstrand, F. S. Electron microscopy of cells and tissues. v. 1- New York, Academic Press, 1967- v. 1: Instrumentation and techniques.

1169. Stevens, G. W. W. Microphotography: photography at extreme resolution. 2d ed., rev. and expanded. New York, Wiley, 1968.

1170. Szabó, D. Medical colour photomicrography. Budapest, Akademiai Kiado, 1967.

1171. Tolansky, S. Interference microscopy for the biologist. Springfield, Ill., Thomas, 1968.

1172. Toner, P. G., and Katharine E. Carr.

Cell structure: an introduction to biological electron microscopy. 2d ed. Edinburgh, Livingstone, 1971.

1173. Traber, Y. A., ed. The microscope as a camera. New York, Amphoto, 1971.

1174. Wischnitzer, Saul. Introduction to electron microscopy. 2d ed. New York, Pergamon Press, 1970.

MICROBIOLOGICAL AND PARASITOLOGICAL CULTURES, TECHNIQUES, AND CLASSIFICATION

1175. Alton, G. G., and Lois M. Jones. Laboratory techniques in brucellosis. World Health Organization, 1967. (WHO monograph no. 55.)

1176. American Public Health Association. Committee on Evaluation and Standards. Coordinating Committee on Laboratory Methods. Subcommittee on Methods for the Microbiological Examination of Foods. Recommended methods for the microbiological examination of foods. 2d ed. Ed. by J. M. Sharf. New York, 1966.

1177. ———. Recommended procedures for the bacteriological examination of sea water and shellfish. 3d ed. New York, 1962.

1178. ———. Standard methods for the examination of dairy products, microbiological, bioassay and chemical. 12th ed. New York, 1967.

1179. ———. Standard methods for the examination of water and wastewater. 13th ed. Washington, D.C., 1971.

1180. American Type Culture Collection. Catalogue of strains. 10th ed. Rockville, Md., 1972.

1181. Bailey, W. R., and E. G. Scott. Diagnostic microbiology: a textbook for the isolation and identification of pathogenic microorganisms. 3d ed. St. Louis, Mosby, 1970.

1182. Behbehani, A. M. Laboratory diagnosis of viral, bedsonial and rickettsial diseases: handbook for laboratory workers. Springfield, Ill., Thomas, 1971.

1183. Blair, J. E., ed. Manual of clinical microbiology. Bethesda, Md., American Society for Microbiology, 1970.

1184. Branson, Dorothy. Methods in clinical bacteriology: a manual of tests and procedures. Springfield, Ill., Thomas, 1972.

1185. Busby, D. W. G., W. House, and J. R. MacDonald. Virological technique. Boston, Little, Brown, 1964.

1186. Carter, G. R. Diagnostic procedures in veterinary bacteriology and mycology. Springfield, Ill., Thomas, 1967.

1187. Central Veterinary Laboratory. Parasitology Dept. Manual of veterinary parasitological laboratory techniques. London, H. M. Stationery Office, 1971.

1188. Cherry, W. B., Morris Goldman, and T. R. Carsky. Fluorescent antibody techniques in the diagnosis of communicable diseases. Atlanta, Communicable Disease Center, 1960.

1189. Chowdhuri, A. N. Rai, and A. K. Thomas. Rabies: general considerations and laboratory procedures. New Delhi, Indian Council of Medical Research, 1967. (Special report series, no. 58.)

1190. Collins, C. H. Microbiological methods. 2d ed. New York, Plenum Press, 1967.

1191. Conant, N. F., et al. Manual of clinical mycology. 3d ed. Philadelphia, Saunders, 1971.

1192. Cumming, Hamish. Virology—tissue culture. London, Butterworth, 1970.

1193. Cunningham, C. H. A laboratory guide in virology. 6th ed. Minneapolis, Burgess, 1966.

1194. Edwards, P. R. Identification of enterobacteriaceae. 3d ed. Minneapolis, Burgess, 1972.

1195. Finstein, M. S. Pollution microbiology: a laboratory manual. New York, Dekker, 1972.

1196. Gavan, T. L., H. W. McFadden, Jr., and Esther L. Cheatle. Antimicrobial susceptibility testing. Chicago, American Society of Clinical Pathologists, Commission on Continuing Education, 1971.

1197. Gibbs, B. M., ed. Identification methods for microbiologists. 2 v. New York, Academic Press, 1966-1968.

1198. Goldman, Morris. Fluorescent antibody methods. New York, Academic Press, 1968.

1199. Graber, C. D. Rapid diagnostic methods in medical microbiology. Baltimore, Williams & Wilkins, 1970.

1200. Great Britain Ministry of Agriculture, Fisheries and Food. Manual of veterinary parasitological laboratory techniques. London, H. M. Stationery Office, 1971. (Its Technical bulletin no. 18.)

1201. Grist, N. R. Diagnostic methods in clinical virology. Philadelphia, Davis, 1966.

1202. Haley, Leanor D. Diagnostic medical mycology. New York, Appleton-Century-Crofts, 1964.

1203. Harris, R. J. C. Techniques in experimental virology. New York, Academic Press, 1964.

1204. Hazen, Elizabeth L., et al. Laboratory identification of pathogenic fungi simplified. 3d ed. Springfield, Ill., Thomas, 1970.

1205. Hoskins, J. M. Virological procedures. New York, Appleton-Century-Crofts, 1967.

1206. Hsiung, G. D. Diagnostic virology. New Haven, Yale Univ. Press, 1964.

1207. Index Bergeyana: an annotated alphabetic listing of names of the taxa of the bacteria. Ed. by R. E. Buchanan. Baltimore, Williams & Wilkins, 1966.

1208. International Symposium on Continuous Cultivation of Microorganisms. Proceedings. 2d. Prague, 1962. Ed. by I. Malek et al. New York, Academic Press, 1964.

1209. Jeffrey, H. C., and R. M. Leach. Atlas of medical helminthology and protozoology. Baltimore, Williams & Wilkins, 1966.

1210. Kawamura, Akiyoshi. Fluorescent antibody techniques and their applications. Tokyo, Univ. of Tokyo Press, 1969.

1211. Lees, R. Laboratory handbook of methods of food analysis. 2d ed. Cleveland, CRC Press, 1971.

1212. Lennette, E. H., and Nathalie J. Schmidt, eds. Diagnostic procedures for viral and rickettsial infections. 4th ed. Washington, D.C., American Public Health Association, 1969.

1213. Macy, R. W., and A. K. Berntzen. Laboratory guide to parasitology: with introduction to experimental methods. Springfield, Ill., Thomas, 1971.

1214. Manclark, C. R., M. J. Pickett, and H. B. Moore. Laboratory manual for medical bacteriology. 5th ed. New York, Appleton-Century-Crofts, 1972.

1215. Merchant, I. A., et al. Laboratory manual for veterinary bacteriology. Minneapolis, Burgess, 1961.

1216. Meyer, M. C., and L. R. Penner. Laboratory essentials of animal parasitology. Dubuque, Iowa, Brown, 1962.

1217. Miller, J. N., ed. Spirochetes in body fluids and tissues: manual of investigative methods. Springfield, Ill., Thomas, 1971.

1218. National Mastitis Council. Research Committee. Microbiological procedures for the diagnosis of bovine mastitis. Washington, D.C., National Mastitis Council, 1969.

1219. Noble, E. R., and G. A. Noble. Animal parasitology: laboratory manual. Philadelphia, Lea & Febiger, 1962.

1220. Norris, J. R. Methods in microbiology. 3 v. New York, Academic Press, 1969-1970.

1221. Prévot, A. R. Manual for the classification and determination of the anaerobic bacteria. Philadelphia, Lea & Febiger, 1966.

1222. Skerman, V. B. D. A guide to the identification of the genera of bacteria, with methods and digests of generic characteristics. 2d ed. Baltimore, Williams & Wilkins, 1967.

1223. Society of American Bacteriologists. Bergey's manual of determinative bacteriology. 7th ed. By R. S. Breed et al. Baltimore, Williams & Wilkins, 1957.

1224. Sykes, George. Disinfection and sterilization. 2d ed. Philadelphia, Lippincott, 1965.

1225. U.S. Agricultural Research Service. Animal Health Division. Recommended procedure for the isolation of salmonella organisms from animal feeds and feed ingredients. Washington, D.C., Agricultural Research Service, 1971.

1226. U.S. Communicable Disease Center, Atlanta. Laboratory manual for medical mycology. By Liberjo Ajello. Washington, D.C., U.S. Govt. Printing Office, 1963. (U.S. Public Health Service publ. no. 994.)

1227. ———. Leptospirosis: epidemiology, clinical manifestations in man and animals and methods in laboratory diagnosis. By Mildred M. Galton et al. Atlanta, 1962. (U.S. Public Health Service publ. no. 951.)

1228. U.S. Dept. of the Army. Laboratory procedures in virology. Washington, D.C., Dept. of the Army, 1964. (Its Technical manual no. TM8-227-7.)

1229. U.S. National Research Council. Food Protection Committee. Reference methods for the microbiological examination of foods. Washington, D.C., National Academy of Sciences, 1971.

1230. U.S. Public Health Service. Procedures for the isolation and identification of mycobacteria. By Annie L. Vestal. Washington, D.C., U.S. Dept. of HEW, 1969. (U.S.P.H.S. publ. no. 1995.)

1231. Whitlock, J. H. Diagnosis of veterinary parasitisms. Philadelphia, Lea & Febiger, 1960.

HISTOLOGICAL AND HISTOCHEMICAL TECHNIQUE, STAINS AND STAINING

1232. American Registry of Pathology. Manual of histologic staining methods of the Armed Forces Institute of Pathology. 3d ed. Ed. by L. G. Luna. New York, McGraw-Hill, 1968.

1233. Baker, J. R. Cytological technique: the principle underlying routine methods. 5th ed. New York, Wiley, 1966.

1234. Bancroft, J. D. An introduction to histochemical technique. London, Butterworth, 1967.

1235. Barka, Tibor, and P. J. Anderson. Histochemistry: theory, practice and bibliography. New York, Harper, 1963.

1236. Bowling, Mary C. Histopathology laboratory procedures. Bethesda, Md., U.S. National Cancer Institute, 1967. (U.S.P.H.S. publ. no. 1595.)

1237. Brain, E. B. The preparation of decalcified sections. Springfield, Ill., Thomas, 1966.

1238. Carleton, H. M. Carleton's histological technique. Rev. and rewritten by R. A. B. Drury and E. A. Wallington. 4th ed. New York, Oxford Univ. Press, 1967.

1239. Chayen, Joseph, et al. Guide to practical histochemistry. Philadelphia, Lippincott, 1969.

1240. Clark, George, and Margaret P. Clark. A primer in neurological staining procedures. Springfield, Ill., Thomas, 1971.

1241. Clayden, E. C. Practical section cutting and staining. 5th ed. Baltimore, Williams & Wilkins, 1971.

1242. Conn, H. J. Staining procedures used by the Biological Stain Commission. 2d ed. Baltimore, Williams & Wilkins, 1960.

1243. Culling, C. F. A. Handbook of histopathological technique. 2d ed. Washington, D.C., Butterworth, 1963.

1244. Disbrey, Brenda D., and J. H. Rack. Histological laboratory methods. Edinburgh, Livingstone, 1970.

1245. Galigher, A. E., and E. N. Kozloff. Essentials of practical microtechnique. 2d ed. Philadelphia, Lea & Febiger, 1971.

1246. Glick, David. Quantitative chemical techniques of histo- and cytochemistry. v. 1- New York, Wiley, 1961-

1247. Gurr, Edward. Encyclopedia of microscopic stains. Baltimore, Williams & Wilkins, 1960.

1248. Gurr, G. T. Biological staining methods. 7th ed. London, George T. Gurr Ltd., 1963.

1249. Guyer, M. F. Animal micrology: practical exercises in zoological microtechniques. 5th rev. ed. Chicago, Univ. of Chicago Press, 1953.

1250. Handbuch der Histochemie. Hrsg. von Walther Graumann and Karlheinz Neumann. v. 1- Stuttgart, Gustav Fischer Verlag, 1958-

1251. Koss, L. G. Diagnostic cytology and its histopathologic bases. 2d ed. Philadelphia, Lippincott, 1968.

1252. Lillie, R. D. H. J. Conn's biological stains: a handbook on the nature and uses of dyes employed in the biological laboratory. 8th ed. Baltimore, Williams & Wilkins, 1969.

1253. ——. Histopathologic technic and practical histochemistry. 3d ed. New York, McGraw-Hill, 1965.

1254. Nieburgs, H. E. Diagnostic cell pathology in tissues and smears. New York, Grune & Stratton, 1967.

1255. Palocsay, G. P. Review of histology technique. Berkeley, Berkeley Scientific, 1969.

1256. Pease, D. C. Histological techniques for electron microscopy. 2d ed. New York, Academic Press, 1964.

1257. Preece, Ann. A manual for histologic technicians. 3d ed. Boston, Little, Brown, 1972.

1258. Putt, F. A. Manual of histopathological staining methods. New York, Wiley-Interscience, 1972.

1259. Scanga, F. Atlas of electron microscopy. New York, American Elsevier, 1964.

1260. Stiles, K. A. Handbook of histology. 5th ed. New York, McGraw-Hill, 1968.

1261. Thompson, S. W. Selected histochemical and histopathological methods. Springfield, Ill., Thomas, 1966.

1262. U.S. Armed Forces Institute of Pathology. Manual of histologic and special staining techniques. 2d ed. New York, McGraw-Hill, 1960.

1263. Wismar, Beth L. A visual approach to histology. Philadelphia, Davis, 1970.

1264. Zugibe, F. T. Diagnostic histochemistry. St. Louis, Mosby, 1970.

CELL AND TISSUE CULTURE

1265. Ambrose, E. J. The cancer cell in vitro. New York, Appleton-Century-Crofts, 1967.

1266. Cumming, Hamish. Virology—tissue culture. London, Butterworth, 1970.

1267. Elkind, M. M., and G. F. Whitmore. The radiobiology of cultured mammalian cells. New York, Gordon & Breach, 1967.

1268. Humason, Gretchen L. Animal tissue techniques. 2d ed. San Francisco, Freeman, 1967.

1269. International conference of tissue culture in cancer research. Tokyo, 1966. Cancer cells in culture. Ed. by Hajim Katsuta. State College, Pa., Univ. Park Press, 1968.

1270. International Society of Blood Transfusion. Tissue typing: a symposium organized by the International Society of Blood Transfusion at the Xth Congress of the European Society of Haematology. Strasbourg, 1965. Ed. by Jean Dausset. Basel, Karger, 1966. (*Vox sanguinis*, v. 11, no. 3.)

1271. Malinin, Theodore. Processing and storage of viable human tissues. Washington, D.C., 1966. (U.S. Public Health Service publ. no. 1442.)

1272. Merchant, D. J., R. H. Kahn, and W. H. Murphy, Jr. Handbook of cell and organ culture. 2d ed., 3d enl. printing. Minneapolis, Burgess, 1967.

1273. New, D. A. Techniques for culture of vertebrate embryos. New York, Academic Press, 1966.

1274. Nieburgs, H. E. Diagnostic cell pathology in tissue and smears. New York, Grune & Stratton, 1967.

1275. Panel on current problems of bone-marrow cell transplantation with special emphasis on conservation and culture. Moscow, 1968. Vienna, International Atomic Energy Agency, 1969. (Panel proceedings series. STI/PUB/219.)

1276. Parker, R. C. Methods of tissue culture. 3d ed. New York, Harper & Row, 1961.

1277. Paul, J. R. Cell and tissue culture. 4th ed. Baltimore, Williams & Wilkins, 1970.

1278. Rose, G. G. Atlas of vertebrate cells in tissue culture. New York, Academic Press, 1970.

1279. Rothblat, G. H., ed. Growth, nutrition, and metabolism of cells in culture. v. 1- New York, Academic Press, 1972-

1280. Sandritter, Walter. Color atlas and textbook of tissue cellular pathology. 3d ed. Chicago, Year Book Medical Publishers, 1969.

1281. Schindler, R. Die tierische Zelle in Zellkultur. Berlin, Springer, 1965. (*Recent results in cancer research*, v. 1.)

1282. Symposium on aging in cell and tissue culture. Ziukovy, Czechoslovakia, 1969. Aging in cell and tissue culture. Ed. by Emma Holečková and Vincent Cristofalo. New York, Plenum Press, 1970.

1283. Symposium on growth control in cell cultures. London, 1970. Ciba Foundation symposium: growth control in cell cultures. Ed. by G. E. W. Wolstenholme and Julie Knight. Baltimore, Williams & Wilkins, 1971.

1284. Tedeschi, C. G., ed. Neuropathology: methods and diagnosis. Boston, Little, Brown, 1970.

1285. Thomas, J. A., ed. Organ culture. New York, Academic Press, 1970.

1286. Tissue Culture Association. Annual symposium. Proceedings. v. 1-5. Baltimore, Williams & Wilkins, 1965-1970. (For specific titles of volumes see entry 0785.)

1287. U.S. National Institute of Allergy and Infectious Diseases. Transplantation Immunology Branch. Catalog of tissue typing antisera. Ed. by D. B. Hare et al. Bethesda, Md., 1972. (DHEW publ. no. 72-97.)

1288. ———. Manual of tissue typing techniques. Ed. by D. L. Brand and J. G. Ray, Jr. Bethesda, Md., 1970.

1289. Wasley, G. D., and J. W. May. Animal cell culture methods. Oxford, Blackwell, 1970.

1290. Waymouth, Charity, ed. Advances in tissue culture. Baltimore, Williams & Wilkins, 1970.

1291. White, D. O., ed. A postgraduate course in cell culture. Parkville, Australia, Cell Culture Society of Victoria, 1963.

1292. White, P. R. Cultivation of animal and plant cells. 2d ed. New York, Ronald Press, 1963.

1293. Willmer, E. N., ed. Cells and tissues in culture: methods, biology and physiology. 3 v. New York, Academic Press, 1965-1966.

1294. Wilt, F. H. Methods in developmental biology. New York, Crowell, 1967.

1295. Woodliff, H. J. Blood and bone marrow cell culture. Philadelphia, Lippincott, 1964.

DISSECTION, AUTOPSY, NECROPSY, AND SPECIMEN PREPARATION

1296. Earle, K. M. Examination of the brain (necropsy technique). Washington, D.C., American Registry of Pathology, 1968.

1297. Food and Agriculture Organization of the United Nations. Animal Health Branch. Necropsy procedures and the submission of laboratory specimens. Comp. by G. R. Carter. Rome, 1962. (FAO monograph no. 4.)

1298. Grant, J. C. B. Grant's dissector. 6th ed. Baltimore, Williams & Wilkins, 1967.

1299. Grollman, Sigmund. Laboratory manual of mammalian anatomy and physiology. 2d ed. New York, Macmillan, 1969.

1300. Habel, R. E. Guide to the dissection of domestic ruminants. 2d ed. Ithaca, N.Y., published by the author, 1964 (distributed by Edwards Bros., Ann Arbor, Mich.

1301. Harrison, B. M. Dissection of the cat (and comparisons with man): a laboratory manual on Felis domestica. 6th ed. St. Louis, Mosby, 1970.

1302. Hildebrand, Milton. Anatomical preparation. Berkeley, Univ. of California Press, 1968.

1303. Jones, T. C., and C. A. Gleiser, eds. Veterinary necropsy procedures. Philadelphia, Lippincott, 1954.

1304. Kalbus, Barbara H., and K. G. Neal. Dissection guide for the cat. Minneapolis, Burgess, in press 1972.

1305. Kent, G. C. Anatomy of the vertebrates: a lab manual. St. Louis, Mosby, 1967.

1306. Lee, Henry, and Kris Neville. Handbook of biomedical plastics. Pasadena, Calif., Pasadena Technology Press, 1971.

1307. Miller, M. E., and Alexander DeLahunta. Guide to the dissection of the dog. Philadelphia, Saunders, 1971.

1308. Norman, J. C., ed. Organ perfusion and preservation. New York, Appleton-Century-Crofts, 1968.

1309. Rezek, P. R., and Max Millard. Autopsy pathology, a guide for pathologists and clinicians. Springfield, Ill., Thomas, 1963.

1310. Rooney, J. R. Autopsy of the horse: technique and interpretation. Baltimore, Williams & Wilkins, 1970.

1311. ———, W. O. Sack, and R. E. Habel. Guide to the dissection of the horse. Ithaca, N.Y., W. O. Sack, 1967 (distributed by Edwards Bros., Ann Arbor, Mich.).

1312. Ross, C. F. Ross's post-mortem appearances. 6th ed. London, Oxford Univ. Press, 1963.

1313. Saphir, Otto. Autopsy diagnosis and technic. 4th ed. New York, Hoeber, 1961.

1314. Söderström, Nils. Fine-needle aspiration biopsy. New York, Grune & Stratton, 1966.

1315. Thorpe, D. R. Fetal pig: a dissection guide in color. Palo Alto, Calif., National Press Books, 1970.

1316. Tompsett, D. H. Anatomical techniques. 2d ed. Baltimore, Williams & Wilkins, 1970.

1317. U.S. Armed Forces Institute of Pathology. Medical Museum. Laboratory. Manual of macropathological techniques. Washington, D.C., 1957.

1318. Walker, W. F., Jr. Vertebrate dissection. 4th ed. Philadelphia, Saunders, 1970.

1319. Whitehouse, R. H., and A. J. Grove. The dissection of the rabbit, with an appendix on the rat. 6th ed. London, University Tutorial Press, 1967.

1320. Winter, Hans. Post mortem examination of ruminants. St. Lucia, Brisbane, Univ. of Queensland Press, 1966.

1321. Wischnitzer, Saul. Atlas and dissection guide for comparative anatomy. San Francisco, Freeman, 1967.

1322. Woodburne, R. T. A guide to dissection in gross anatomy. 3d ed. New York, Oxford Univ. Press, 1971.

INSTRUMENTATION AND TECHNIQUES

General

1323. Ackermann, P. G. Electronic instrumentation in the clinical laboratory. Boston, Little, Brown, 1972.

1324. Arthur H. Thomas Company. Scientific apparatus and reagents, selected for academic, industrial and medico-biological laboratories. Philadelphia, 1971.

1325. Bartholomew, Davis. Electrical measurements and instrumentation. Boston, Allyn & Bacon, 1963.

1326. Becker, Walter, et al. Atlas of otorhinolaryngology and bronchoesophagology. English ed. Philadelphia, Saunders, 1969.

1327. Bender, G. T. Chemical instrumentation: a laboratory manual based on clinical chemistry. Philadelphia, Saunders, 1972.

1328. Bourne, G. H. In vivo techniques in histology. Baltimore, Williams & Wilkins, 1967.

1329. Brown, C. C. Instrumentation with semiconductors for medical researchers. Springfield, Ill., Thomas, 1964.

1330. Bůres, Jan, Mojmír Petráň, and Jozef Zachar. Electrophysiological methods in biological research. 3d ed. New York, Academic Press, 1967.

1331. D'Amour, F. E. Manual for laboratory work in mammalian physiology. 3d ed. Chicago, Univ. of Chicago Press, 1965.

1332. Dewhurst, D. J. Physical instrumentation in medicine and biology. New York, Pergamon Press, 1966.

1333. Donaldson, P. E. K. Electronic apparatus for biological research. New York, Academic Press, 1958.

1334. Dummer, G. W. A., ed. Medical electronic equipment, 1969-1970. 4 v. (For specific titles see entry 1067.)

1335. Geddes, L. A., and L. E. Baker. Principles of applied biomedical instrumentation. New York, Wiley, 1968.

1336. Goldman, Leon. Biomedical aspects of the laser. New York, Springer-Verlag, 1967.

1337. Hayt, W. H. Engineering circuit analysis. Ed. by J. E. Kemmerly. New York, McGraw-Hill, 1962.

1338. Hine, G. J., ed. Instrumentation in nuclear medicine. v. 1- New York, Academic Press, 1967-

1339. International symposium on freeze-drying. 2d. Recent research in freezing and drying. Oxford, Blackwell, 1958. (Report of the first symposium issued as symposia no. 2 of the Institute of Biology with title: Freezing and drying.)

1340. ISA Symposium on instrumentation methods for the further development of predictive medicine. Los Angeles, 1965. Ed. by T. B. Weber and J. Poyer. Pittsburg, Instrument Society of America, 1966.

1341. Lee, L. W. Elementary principles of laboratory instruments. rev. ed. 2d ed. St. Louis, Mosby, 1970.

1342. Lenihan, J. M. Instrumentation in medicine. Baltimore, Williams & Wilkins, 1968.

1343. Levedahl, B. H., A. A. Barber, and Alan Grinnell. Laboratory experiments in physiology. 8th ed. St. Louis, Mosby, 1971.

1344. Mackay, R. S. Bio-medical telemetry: sensing and transmitting biological information from animals and man. 2d ed. New York, Wiley, 1970.

1345. Malmstadt, H. V., and C. G. Enke. Electronics for scientists: principles and experiments for those who use instruments. New York, Benjamin, 1962.

1346. Martin, A. E. Infra-red instrumentation and techniques. New York, American Elsevier, 1966.

1347. Meloan, C. E. Instrumental analysis using physical properties. Philadelphia, Lea & Febiger, 1967. (Medical technology series. v. 2.)

1348. ———. Instrumental analysis using spectroscopy. Philadelphia, Lea & Febiger, 1968. (Medical technology series. v. 1.)

1349. National biomedical sciences instrumentation symposium. 1st. Biomedical sciences instrumentation: proceedings. Pittsburgh, Instrument Society of America, 1963 (distributed by Plenum Press, New York).

1350. Pollister, A. W., ed. Physical techniques in biological research. 2d ed. v. 1- New York, Academic Press, 1966-
v. 1: Optical techniques. v. 2-A: Physical chemical techniques. v. 2-B: Physical chemical techniques. v. 3-A: Cells and tissues. v. 3-B: Autoradiography at the cellular level. v. 3-C: Cells and tissues. v. 4: Special methods. v. 5: Electrophysiological methods, I. v. 6: Electrophysiological methods, II.

1351. Price, L. W. Electronic laboratory techniques. Baltimore, Williams & Wilkins, 1969.

1352. Rudin, S. C., et al. Bioinstrumentation experiments in physiology. Millis, Mass., Harvard Apparatus Foundation, 1971.

1353. Schindler, Rudolf. Gastroscopy: the

endoscopic study of gastric pathology. 2d ed. New York, Hafner, 1966.

1354. Smith, D. A. Medical electronics equipment handbook. Indianapolis, W. H. Sams, 1962.

1355. Symposium on medical applications of plastics. Toronto, 1970. Medical applications of plastics. Ed. by H. P. Gregor. New York, Interscience, 1971. (*Journal of biomedical materials research*, v. 5, no. 2.)

1356. Udenfriend, Sidney. Fluorescence assay in biology and medicine. 2 v. New York, Academic Press, 1962-1969.

1357. Umeda, Noritsugu. Diagnosis by gastrophotography. Philadelphia, Saunders, 1971.

1358. Whitfield, I. C. Manual of experimental electrophysiology. New York, Pergamon Press, 1964.

1359. Willard, H. H., Lynne L. Merritt, and J. A. Dean. Instrumental methods of analysis. 4th ed. Princeton, N.J., Van Nostrand, 1965.

1360. Windhager, E. E. Micropuncture techniques and nephron function. New York, Appleton-Century-Crofts, 1968.

1361. Winstead, Martha. Instrument check systems. Philadelphia, Lea & Febiger, 1971.

1362. Zucker, M. H. Electronic circuits for the behavioral and biomedical sciences: a reference book of useful solid state circuits. San Francisco, Freeman, 1969.

Electrocardiology

1363. Bernreiter, Michael. Electrocardiography. 2d ed. Philadelphia, Lippincott, 1963.

1364. Blake, T. M. An introduction to electrocardiography. New York, Appleton-Century-Crofts, 1964.

1365. Burch, G. E., and Travis Winsor. A primer of electrocardiography. 6th ed. Philadelphia, Lea & Febiger, 1972.

1366. Dimond, E. G. Electrocardiography and vectorcardiography. 4th ed. Boston, Little, Brown, 1967.

1367. Goldberger, Emanuel. How to interpret electrocardiograms in terms of vectors: a practical manual. Springfield, Ill., Thomas, 1968.

1368. Goldman, M. J. Principles of clinical electrocardiography. 7th ed. Los Altos, Calif., Lange Medical Publications, 1970.

1369. Grant, R. P. Grant's clinical electrocardiography: the spatial vector approach. 2d ed. Rev. by J. R. Beckwith. New York, McGraw-Hill, 1970.

1370. International symposium on vectorcardiography. v. 1- Proceedings. Philadelphia, Lippincott, 1965- (Semiannual.)

1371. Marriott, H. J. L. Practical electrocardiography. 4th ed. Baltimore, Williams & Wilkins, 1968.

1372. Winsor, Travis. Primer of vectorcardiography. Philadelphia, Lea & Febiger, 1972.

Electroencephalography

1373. Cooper, R., J. W. Osselton, and J. C. Shaw. EEG technology. London, Butterworth, 1969.

1374. Gibbs, F. A. Medical electroencephalography. Reading, Mass., Addison-Wesley, 1967.

1375. ———, and Erna L. Gibbs. Atlas of electroencephalography. 2d ed. 3 v. Reading, Mass., Addison-Wesley, 1950-1964.

1376. Kiloh, L. G., A. J. McComas, and J. W. Osselton. Clinical electroencephalography. 3d ed. New York, Appleton-Century-Crofts, 1972.

1377. Klemm, W. R. Animal electroencephalography. New York, Academic Press, 1969.

1378. Kooi, K. A. Fundamentals of electroencephalography. New York, Harper & Row, 1971.

1379. Pampiglione, Giuseppe. Development of cerebral function in the dog. London, Butterworth, 1963.

1380. Walter, Pat L., ed. A KWIC index to EEG and allied literature, 1966-1969. New York, Elsevier, 1970. (*Electroencephalography and clinical neurophysiology*. Suppl. 29.)

1381. Wineburgh, Margaret, ed. A KWIC index to EEG and allied literature, 1964-1966. New York, Elsevier, 1971. (*Electroencephalography and clinical neurophysiology*. Suppl. 30.)

Electromyography and Electroneurography

1382. Basmajian, J. V. Muscles alive: their functions revealed by electromyography. 2d ed. Baltimore, Williams & Wilkins, 1967.

1383. Buchthal, Fritz. An introduction to electromyography. Copenhagen, Gyldendal, 1957.

1384. Camougis, George. Nerves, muscles, and electricity: an introductory manual of electrophysiology. New York, Appleton-Century-Crofts, 1970.

1385. Cohen, H. L., and Joel Brumlik. Manual of electronneuromyography. New York, Harper & Row, 1968.

1386. Goodgold, Joseph, and Arthur Eberstein. Electrodiagnosis of neuromuscular diseases. Baltimore, Williams & Wilkins, 1972.

1387. Kendall, H. O., Florence P. Kendall, and Gladys E. Wadsworth. Muscles, testing and function. 2d ed. Baltimore, Williams & Wilkins, 1971.

1388. Lenman, J. A. R. Clinical electromyography. Philadelphia, Lippincott, 1970.

1389. Licht, S. H., ed. Electrodiagnosis and electromyography. 3d ed. New Haven, Licht, 1971.

1390. Marinacci, A. A. Applied electromyography. Philadelphia, Lea & Febiger, 1968.

1391. Smorto, M. P., and J. V. Basmajian. Clinical electroneurography: an introduction to nerve conduction tests. Baltimore, Williams & Wilkins, 1972.

Ultrasonics

1392. Brown, B., and D. Gordon, eds. Ultrasonic techniques in biology and medicine. Springfield, Ill., Thomas, 1967.

1393. Él'piner, I. E. Ultrasound: physical, chemical and biological effects. New York, Consultants Bureau, 1964.

1394. Goldberg, R. E., and L. K. Sarin, eds. Ultrasonics in opthalmology, diagnostic and therapeutic applications. Philadelphia, Saunders, 1967.

1395. Grossman, C. C. Diagnostic ultrasound: proceedings of the first international conference, Univ. of Pittsburgh, 1965. New York, Plenum Press, 1966.

1396. Kobayashi, Mitsunao. Atlas of ultrasonography in obstetrics and gynecology. New York, Appleton-Century-Crofts, 1972.

1397. Rand, Elias, ed. Recent advances in diagnostic ultrasound. Springfield, Ill., Thomas, 1971.

1398. Symposium on ultrasound in biology and medicine. 1962. Ultrasonic energy: biological investigations and medical applications. Ed. by Elizabeth Kelly. Urbana, Univ. of Illinois Press, 1965.

1399. Wells, P. N. T. Physical principles of ultrasonic diagnosis. New York, Academic Press, 1969.

PART II
Specific Disciplines

Chapters 7—20 represent selected lists of library materials arranged according to specific subject interest, with a strong emphasis on the comparative aspects between man and other vertebrates in each discipline. While a few citations in any given chapter or section may be multidisciplinary in character, in general the literature cited within a unit is most pertinent to the practitioner, educator, student, or research investigator who is interested in that particular biomedical discipline. A given work may be cited in more than one chapter or section if the subject content is appropriate. The arrangement follows quite historical or traditional subject categories within the basic and clinical medical sciences because of the broad spectrum of interests that the compilers anticipate will be reflected by the audience using the bibliography.

In some instances the materials cited within a given chapter are separated into only a few subheadings, especially if the body of literature germane to this publication is relatively small. In other instances the broad subject is divided into several categories, e.g., a body system breakdown, as will be noted in the chapters covering anatomy, physiology, pathology, and internal medicine. Further, because anatomy and physiology are disciplines which have large aggregates of comparative and developmental literature, separate sections are provided for materials of traditional comparative interest and those concerned with developmental processes. Materials included in other chapters representing basic or clinical sciences may be comparative in nature, or may be geared more specifically to either human or animal medicine; however, in most instances the principles involved are applicable equally to all vertebrate species.

The biomedical or behavioral scientist whose work involves animals as models for human medical or behavioral research, or as primary research subjects, will find very useful the citations to clinical materials noted in those chapters devoted to internal medicine, surgery, and radiology, and to the basic sciences, including microbiology, genetics, neurology, and others that fall within the parameters established for this work.

7

ANATOMY

GENERAL AND SPECIFIC

1400. Basmajian, J. V. Primary anatomy. 6th ed. Baltimore, Williams & Wilkins, 1970.

1401. Beck, Ernest. The horse, in anatomical transparencies, with additional detailed color illustrations. Grafton, Wis., Animal Technical Publications, 1971.

1402. Becker, R. F. The anatomical basis of medical practice. Baltimore, Williams & Wilkins, 1971.

1403. Bensley, B. A. Practical anatomy of the rabbit: an elementary textbook in mammalian anatomy. 8th ed. Ed. by E. H. Craigie. Toronto, Univ. of Toronto Press, 1948.

1404. Booth, E. S. Laboratory anatomy of the cat. 4th ed. Rev. by R. B. Chiasson. Dubuque, Ia., Brown, 1967.

1405. Bradley, O. C. Topographical anatomy of the dog. 6th ed. New York, Macmillan, 1959.

1406. Brantigan, O. C. Clinical anatomy. New York, McGraw-Hill, 1963.

1407. Chiasson, R. B. Laboratory anatomy of the white rat. 2d ed. Dubuque, Ia., Brown, 1969.

1408. Cook, Margaret J. The anatomy of the laboratory mouse. New York, Academic Press, 1965.

1409. Craigie, E. H. A laboratory guide to the anatomy of the rabbit. 2d ed. Toronto, Univ. of Toronto Press, 1966.

1410. Crouch, J. E. Text-atlas of cat anatomy. Philadelphia, Lea & Febiger, 1969.

1411. Dyce, K. M., and C. J. G. Wensing. Essentials of bovine anatomy. Utrecht, Oosthoek, 1971.

1412. Ede, D. A. Bird structure, an approach through evolution development and function in the fowl. London, Hutchinson Educational, 1964.

1413. Field, H. E. An atlas of cat anatomy. 2d ed. Chicago, Univ. of Chicago Press, 1969.

1414. Fitzgerald, T. C. The Coturnix quail: anatomy and histology. Ames, Iowa State Univ. Press, 1969.

1415. Gardner, E. D., D. J. Gray, and R. O'Rahilly. Anatomy: a regional study of human structure. 3d ed. Philadelphia, Saunders, 1969.

1416. Getty, Robert. Atlas for applied veterinary anatomy. 2d ed. Ames, Iowa State Univ. Press, 1964.

1417. Gilbert, S. G. Pictorial anatomy of the cat. Seattle, Univ. of Washington Press, 1968.

1418. Grant, J. C. B. An atlas of anatomy. 6th ed. Baltimore, Williams & Wilkins, 1972.

1419. Gray, Henry. Anatomy of the human body. 28th ed. Ed. by C.M. Gross. Philadelphia, Lea & Febiger, 1966.

1420. Greene, Eunice C. Anatomy of the rat. New York, Hafner, 1959.

1421. Hartman, C. G., and W. L. Straus, Jr., eds. The anatomy of the rhesus monkey (Macaca mulatta). By T. H. Bast et al. New York, Hafner, 1933. (Reprinted with corrections 1961, 1965.)

1422. Hill, W. C. Primates: comparative anatomy and taxonomy. v. 1- New York, Wiley, 1953-
v. 1: Strepsirhini. v. 2: Haplorhini; Tarsioidea. v. 3: Pithecoidea; Platyrrhini. v. 4: Part A—Cebidae. v. 5: Part B—Cebidae. v. 6: Catarrhini; Cercopithecoidea; Cercopithecinae. v. 7: Cynopithecinae; Papio, Mandrillus, Theropithecus.

1423. Hollinshead, W. H. Anatomy for surgeons. 2d ed. 3 v. New York, Harper & Row, 1968-1971.
v. 1: The head and neck. v. 2: The thorax, abdomen and pelvis. v. 3: The back and limbs.

1424. Hugh, A. E., and J. N. Glanville. Pro-

Anatomy

grammed primers of anatomy and physiological functions. New York, Appleton-Century-Crofts, 1971.

1425. International Anatomical Nomenclature Committee. Nomina anatomica. 3d ed. Amsterdam, Excerpta Medica, 1966.

1426. International Committee on Veterinary Anatomical Nomenclature. Nomina anatomica veterinaria. Vienna, 1968.

1427. Kiss, F., and J. Szentagothai. Atlas of human anatomy. 3 v. New York, Pergamon Press, 1964.

1428. McLaughlin, C. A. Laboratory anatomy of the rabbit. Dubuque, Ia., Brown, 1970.

1429. McLeod, W. M. Bovine anatomy. 2d ed. Rev. by D. M. Trotter and J. W. Lumb. Minneapolis, Burgess, 1958.

1430. May, N. D. S. The anatomy of the sheep. 3d ed. Queensland, Univ. of Queensland Press, 1970.

1431. Mayer, Edmund. Introduction to dynamic morphology. New York, Academic Press, 1963.

1432. Miller, M. E., G. C. Christensen, and H. E. Evans. Anatomy of the dog. Philadelphia, Saunders, 1965. (2d ed. in preparation 1972.)

1433. Morgagni, G. B. The seats and causes of diseases investigated by anatomy. 3 v. New York, Hafner, 1960. (Facsimile reprint of 1769 edition.)

1434. Morris, Henry. Human anatomy: a complete systematic treatise. 12th ed. Ed. by B. J. Anson. New York, McGraw-Hill, 1966.

1435. Netter, F. H. The Ciba collection of medical illustrations. v. 1- New York, Ciba Pharmaceutical Co., 1953-
 v. 1: Nervous system. v. 2: Reproductive system. v. 3: Digestive system. v. 4: Endocrine system. v. 5: The heart.

1436. Pernkopf, Eduard. Atlas of topographical and applied human anatomy. 2 v. Ed. by Helmut Ferner. Philadelphia, Saunders, 1964.

1437. Reighard, J. E., and H. S. Jennings.

Comparative

Anatomy of the cat. 3d ed. New York, Holt, Rinehart & Winston, 1963.

1438. Robinson, M. C. Laboratory anatomy of the domestic chicken. Ed. by E. S. Booth and R. B. Chiasson. Dubuque, Ia., Brown, 1970.

1439. Romer, A. S. The vertebrate body. 4th ed. Philadelphia, Saunders, 1970.

1440. Schebitz, H., and H. Wilkens. Atlas of radiographic anatomy of dog and horse. Berlin, Paul Parey, 1968.

1441. Sealander, J. A., and C. E. Hoffman. Laboratory manual of elementary mammalian anatomy, with emphasis on the rat. 2d ed. Minneapolis, Burgess, 1967.

1442. Way, R. F., and D. G. Lee. The anatomy of the horse. Philadelphia, Lippincott, 1965.

1443. Yapp, W. B. Vertebrates: their structure and life. New York, Oxford Univ. Press, 1965.

COMPARATIVE

1444. Alexander, R. M. Size and shape. London, Arnold, 1971. (Institute of Biology. Studies in biology, no. 29.)

1445. Anderson, W. D. Thorax of sheep and man. Minneapolis, Dillon Press, 1971.

1446. Ariens-Kappers, C. U., et al. Comparative anatomy of the nervous system of vertebrates, including man. 3 v. 2d ed. New York, Hafner, 1960.

1447. Baer, J. G. Comparative anatomy of vertebrates. Washington, D.C., Butterworth, 1964.

1448. Ballard, W. W. Comparative anatomy and embryology. New York, Ronald Press, 1964.

1449. Barone, R. Anatomie comparée des mammifères domestiques. 2 v. Lyon, Laboratoire d'anatomie, École National vétérinaire, 1966-1968.
 v. 1: Ostéologie. v. 2: Arthrologie et myologie.

1450. Beklemishev, W. N. Principles of comparative anatomy of invertebrates. 2 v.

Chicago, Univ. of Chicago Press, 1970. (Translation of the 3d (1964) edition of *Osnovy sravnitel'noĭ anatomii bespozvonochnykh*.)

1451. Chauveau, A. The comparative anatomy of the domesticated animals. 2d ed. New York, Jenkins, 1885. (Translated and edited by George Fleming.)

1452. Colbert, E. H. Evolution of the vertebrates: a history of the backboned animals through time. 2d ed. New York, Wiley, 1969.

1453. Cole, F. J. A history of comparative anatomy, from Aristotle to the eighteenth century. London, Macmillan, 1944.

1454. Dobberstein, J. C. Lehrbuch der vergleichenden Anatomie der Haustiere. 2., verb. und erweiterte Aufl. v. 1- Leipzig, Hirzel, 1961-

1455. Eaton, T. H., Jr. Comparative anatomy of the vertebrates. 2d ed. New York, Harper & Row, 1960.

1456. Eddy, Samuel, C. P. Oliver, and J. P. Turner. Atlas of drawings for vertebrate anatomy. 3d ed. New York, Wiley, 1964.

1457. Ellenberger, Wilhelm. Handbuch der vergleichenden Anatomie der Haustiere. 2 v. Berlin, Springer, 1943.

1458. Foust, H. L., and R. Getty. Atlas and dissection guide for the study of the anatomy of domestic animals. 3d ed. Ames, Iowa State Univ. Press, 1956.

1459. Frandson, R. D. Anatomy and physiology of farm animals. Philadelphia, Lea & Febiger, 1965.

1460. Freedman, Russell, and J. E. Morriss. The brains of animals and man. New York, Holiday House, 1972.

1461. Gans, Carl, and J. F. Storr. Comparative anatomy atlas. New York, Academic Press, 1962.

1462. Goin, C. J., and Olive B. Goin. Comparative vertebrate anatomy. New York, Barnes & Noble, 1965.

1463. Graham-Jones, Oliver, ed. Aspects of comparative ophthalmology. New York, Pergamon Press, 1966.

1464. Griffin, D. R. Animal structure and function. 2d ed. New York, Holt, Rinehart & Winston, 1970.

1465. Grollman, Sigmund. Laboratory manual of mammalian anatomy and physiology. 2d ed. New York, Macmillan, 1969.

1466. Habel, R. E. Applied anatomy: a laboratory guide for veterinary students. 5th ed. Ithaca, N.Y., the author, 1965.

1467. Harrison, B. M. Dissection of the cat (and comparisons with man): a laboratory manual on Felis domestica. 6th ed. St. Louis, Mosby, 1970.

1468. ———. Manual of comparative anatomy: a general laboratory guide. 3d ed. St. Louis, Mosby, 1970.

1469. Hayward, C. L. A laboratory study of the vertebrates. 2d ed. Minneapolis, Burgess, 1967.

1470. Huettner, A. F. Fundamentals of comparative embryology of the vertebrates. rev. ed. New York, Macmillan, 1949.

1471. Igarashi, Shiro, and Toshiro Kamiya. Atlas of the vertebrate brain: morphological evolution from cyclostomes to mammals. Baltimore, University Park Press, 1972.

1472. Jollie, M. T. Chordate morphology. New York, Reinhold, 1962.

1473. Kent, G. C., Jr. Anatomy of the vertebrates: a lab manual. St. Louis, Mosby, 1967.

1474. ———. Comparative anatomy of the vertebrates. 2d ed. St. Louis, Mosby, 1969.

1475. Koldovsky, Otakar. Development of the functions of the small intestine in mammals and man. White Plains, N.Y., Phiebig, 1969.

1476. Larsell, Olof. The comparative anatomy and histology of the cerebellum. Ed. by Jan Jansen. v. 1- Minneapolis, Univ. of Minnesota Press, 1967-
v. 1: The cerebellum from myxinoids through birds. v. 2: The cerebellum from monotremes through apes.

1477. Leach, W. J. Functional anatomy: mammalian and comparative. 3d ed. New York, McGraw-Hill, 1961.

1478. Leghissa, Silvano. Anatomia comparata dei vertebrati. Bologna, L. Tinarelli, 1960.

1479. McFadyean, J. Osteology and arthrology of the domesticated animals. 4th ed. Ed. by H. V. Hughes and J. W. Dransfield. London, Bailliere, Tindall & Cox, 1953.

1480. Montagna, William. Comparative anatomy. New York, Wiley, 1959.

1481. ——, and Walter Kenworthy. A laboratory manual of comparative anatomy. 2d ed. New York, Wiley, 1963.

1482. Mossman, H. W., and K. L. Duke. Comparative morphology of the mammalian ovary. Madison, Univ. of Wisconsin Press, 1973.

1483. Newman, H. H. The phylum Chordata: biology of vertebrates and their kin. New York, Macmillan, 1939.

1484. Newth, D. R. Animal growth and development. Baltimore, Williams & Wilkins, 1970. (Institute of Biology. Studies in biology, no. 24.)

1485. Nickel, Richard. Lehrbuch der Anatomie der Haustiere. v. 1- Berlin, Parey, 1954-
  v. 1: Bewegungsapparat. v. 2: Eingeweide.

1486. Nieberle, Karl, and Paul Cohrs. Textbook of the special pathological anatomy of domestic animals. New York, Pergamon Press, 1967.

1487. Patt, D. I., and Gail R. Patt. Comparative vertebrate histology. New York, Harper & Row, 1969.

1488. Piermattei, D. L., and R. G. Greeley. An atlas of surgical approaches to the bones of the dog and cat. Philadelphia, Saunders, 1966.

1489. Popesko, Peter. Atlas of topographic anatomy of the domestic animal. 3 v. Philadelphia, Saunders, 1971. (English translation by Robert Getty.)

1490. Prince, J. H. Anatomy and histology of the eye and orbit in domestic animals. Springfield, Ill., Thomas, 1960.

1491. ——. Comparative anatomy of the eye. Springfield, Ill., Thomas, 1956.

1492. Saunders, J. T., and S. M. Manton. A manual of practical vertebrate morphology. 4th ed. Oxford, Clarendon Press, 1969.

1493. Schmid, Elisabeth. Atlas of animal bones for prehistorians, archaeologists, and quaternary geologists. New York, Elsevier, 1972. (Knochenatlas für Prähistoriker, Archäologen, und Quartärgeologen.)

1494. Schmidt, J. A. Cellular biology of vertebrate regeneration and repair. Chicago, Univ. of Chicago Press, 1968.

1495. Schwarze, E. Kompendium der Veterinär-Anatomie. v. 1- Jena, Fischer, 1960-
  v. 1: Einführung in die Veterinär-Anatomie der Bewegungsapparat. v. 2: Das Eingeweidesystem. v. 3: Kreisläufapparat äussere Haut. v. 4: Nervensystem. Sinnesorgane. v. 5: Anatomie des Hausgeflügels. v. 6: Embryologie.

1496. Sisson, S. The anatomy of the domestic animals. Rev. by J. D. Grossman. 4th ed., rev. Philadelphia, Saunders, 1956.

1497. Stokoe, W. M. A guide to comparative veterinary anatomy. Baltimore, Williams & Wilkins, 1967.

1498. Symposium on body composition in animals and man. University of Missouri, 1967. Proceedings. Washington, D.C., National Academy of Sciences, 1968. (Its publication no. 1598.)

1499. Taylor, J. A. Regional and applied anatomy of the domestic animals. v. 1- Philadelphia, Lippincott, 1955-

1500. Tyson, Edward. Edward Tyson, MD, FRS, 1650-1708, and the rise of human and comparative anatomy in England. By M. F. A. Montagu. Philadelphia, American Philosophical Society, 1943.

1501. Verhaart, W. J. C. Comparative anatomical aspects of the mammalian brain stem and the cord. Assen, Netherlands, Van Gorcum, 1970. (Studies in neuro-anatomy, 9.)

1502. Walker, W. F., Jr. A study of the cat, with reference to man. 2d ed. Philadelphia, Saunders, 1972.

1503. ——. Vertebrate dissection. 4th ed. Philadelphia, Saunders, 1970.

1504. Waterman, A. J. Chordate structure and function. New York, Macmillan, 1971.

1505. Weichert, C. K. Anatomy of the chordates. 4th ed. New York, McGraw-Hill, 1970.

1506. Wischnitzer, Saul. Atlas and dissection guide for comparative anatomy. San Francisco, Freeman, 1967.

DEVELOPMENTAL, INCLUDING TERATOLOGY AND AGING

1507. Altman, P. L., and Dorothy S. Dittmer, eds. Growth, including reproduction and morphological development. Washington, D.C., Federation of American Societies for Experimental Biology, 1962.

1508. Andrew, Warren. Anatomy of aging in man and animals. New York, Grune & Stratton, 1971.

1509. Arey, L. B. Developmental anatomy: a textbook and laboratory manual of embryology. 7th ed. Philadelphia, Saunders, 1965.

1510. Balinsky, B. I. An introduction to embryology. 2d ed. Philadelphia, Saunders, 1965.

1511. Ballard, W. W. Comparative anatomy and embryology. New York, Ronald Press, 1964.

1512. Bell, Eugene. Molecular and cellular aspects of development. rev. ed. New York, Harper & Row, 1967.

1513. Bellairs, Ruth. Developmental processes in higher vertebrates. Coral Gables, Fla., Univ. of Miami Press, n.d.

1514. Blandau, R. J., ed. Biology of the blastocyst. Chicago, Univ. of Chicago Press, 1971.

1515. Bourne, G. H. Structural aspects of ageing. New York, Hafner, 1961.

1516. Brachet, Jean. The biochemistry of development. New York, Pergamon Press, 1960.

1517. Bullough, W. S. The evolution of differentiation. London, Academic Press, 1967.

1518. Cameron, I. L., and J. D. Thrasher, eds. Cellular and molecular renewal in the mammalian body. New York, Academic Press, 1971.

1519. Cohen, J. Living embryos. New York, Pergamon Press, 1963.

1520. Comfort, Alexander. Ageing, the biology of senescence. rev. ed. New York, Holt, Rinehart & Winston, 1964.

1521. Daniel, J. C., Jr., ed. Methods in mammalian embryology. San Francisco, Freeman, 1971.

1522. De Beer, G. R. The development of the vertebrate skull. Oxford, Clarendon Press, 1971.

1523. DeHaan, R. L., and Heinrich Ursprung. Organogenesis. New York, Holt, Rinehart & Winston, 1965.

1524. Detwiler, S. R. Neuroembryology: an experimental study. New York, Hafner, 1964. (Reprint of 1936 edition.)

1525. Deuchar, Elizabeth M. Biochemical aspects of amphibian development. New York, Wiley, 1966.

1526. Eakin, R. M. Vertebrate embryology: a laboratory manual. 2d ed. Berkeley, Univ. of California Press, 1971.

1527. Ebert, J. D. Interacting systems in development. New York, Holt, Rinehart & Winston, 1965.

1528. Giroud, Antoine. The nutrition of the embryo. Springfield, Ill., Thomas, 1970.

1529. Gray, Annie P. Mammalian hybrids, a checklist with bibliography. Farnham Royal, Slough, England, Commonwealth Agricultural Bureaux, 1971. (Commonwealth Bureau of Animal Breeding and Genetics. Technical communication no. 10.)

1530. Hadek, Robert. Mammalian fertilization. An atlas of ultrastructure. New York, Academic Press, 1969.

1531. Hamburger, Viktor. A manual of experimental embryology. rev. ed. Chicago, Univ. of Chicago Press, 1960.

1532. Harris, W. H., and R. P. Heaney. Skeletal renewal and metabolic bone disease. Boston, Little, Brown, 1969.

1533. Harrison, B. M. Embryology of the chick and pig. rev. ed. Dubuque, Ia., William C. Brown, 1971.

1534. Harrison, R. G. Organization and development of the embryo. New Haven, Yale Univ. Press, 1969.

1535. Hasegawa, Toshio. Trophoblastic neoplasia: its basic and clinical aspects. Baltimore, Williams & Wilkins, 1971.

1536. Hay, E. D., and J. P. Revel. Fine structure of the developing avian cornea. New York, Karger, 1969. (Monographs in developmental biology, v. 1.)

1537. Hendrickx, A. G. Embryology of the baboon. Chicago, Univ. of Chicago Press, 1971.

1538. Hsu, T. C., and K. Benirschke. Atlas of mammalian chromosomes. v. 1-5. New York, Springer-Verlag, 1967-1971.

1539. International workshop in teratology. 2d. Kyoto, 1968. Methods for teratological studies in experimental animals and man. Flushing, N.Y., Medical Examination Publishing, 1969.

1540. Jacobson, Marcus. Developmental neurobiology. New York, Holt, Rinehart & Winston, 1970.

1541. Johnstone, P. N. Studies on the physiological anatomy of the embryonic heart. Springfield, Ill., Thomas, 1971.

1542. Kalter, Harold. Teratology of the central nervous system: induced and spontaneous malformations of laboratory, agricultural, and domestic animals. Chicago, Univ. of Chicago Press, 1968.

1543. Kohn, R. R. Principles of mammalian aging. Englewood Cliffs, N.J., Prentice-Hall, 1971.

1544. Korenchevsky, V. Physiological and pathological ageing. Ed. by G. H. Bourne. New York, Hafner, 1961.

1545. The laboratory animal in gerontological research: proceedings of a symposium. St. Petersburg, Fla., 1967. Washington, D.C., National Academy of Sciences, 1968. (Its publication no. 1591.)

1546. Langman, Jan. Medical embryology: human development normal and abnormal. 2d ed. Baltimore, Williams & Wilkins, 1969.

1547. Lentz, T. L. Primitive nervous systems. New Haven, Yale Univ. Press, 1968.

1548. Lopashov, G. V., and O. G. Stroeva. Development of the eye: experimental studies. New York, Davey, 1964.

1549. Malformations congénitales des mammifères. Ed. by H. Tuchmann-Duplessis. Paris, Masson, 1971.

1550. Mammalian oogenesis. Papers by Hannah Peters et al. New York, MSS Information Corp., 1972.

1551. Manner, H. W. Elements of comparative vertebrate embryology. New York, Macmillan, 1964.

1552. Marrable, A. W. The embryonic pig: a chronological account. London, Pitman Medical, 1971.

1552a. Mathews, W. W. Atlas of descriptive embryology. New York, Macmillan, 1972.

1553. Metz, C. B., and Alberto Monroy, eds. Fertilization: comparative morphology, biochemistry and immunology. 2 v. New York, Academic Press, 1967-1969.

1554. Mills, Harlan. Pig manual: photographed dissection of the fetal pig. Dubuque, Ia., Brown, 1963.

1555. Nelsen, O. E. Comparative embryology of the vertebrates. New York, McGraw-Hill, 1953.

1556. Nishimura, Hideo, and J. R. Miller, eds. Methods for teratological studies in experimental animals and man. Flushing, N.Y., Medical Examination Publishing, 1969.

1557. Odlaug, T. O. Laboratory anatomy of the fetal pig. 4th ed. Ed. by E. S. Booth and R. B. Chiasson. Dubuque, Ia., Brown, 1969.

1558. Patten, B. M. Early embryology of the chick. 5th ed. New York, McGraw-Hill, 1971.

1559. ——. Embryology of the pig. 3d ed. New York, McGraw-Hill, 1948.

1560. ——. Foundations of embryology. 2d ed. New York, McGraw-Hill, 1964.

1561. ——. Human embryology. 3d ed. New York, McGraw-Hill, 1968.

1562. Polezhaev, L. V. Loss and restoration of regenerative capacity in tissues and organs of animals. Cambridge, Mass., Harvard Univ. Press, 1972. (Technical translation 70-50103.)

1563. Rafferty, K. A. Methods in experimental embryology of the mouse. Baltimore, Johns Hopkins Press, 1970.

1564. Ramón y Cajal, Santiago. Studies on vertebrate neurogenesis. Springfield, Ill., Thomas, 1960.

1565. Raven, C. P. Morphogenesis. 2d ed. New York, Pergamon Press, 1966.

1566. ——. Oögenesis: the storage of developmental information. New York, Pergamon Press, 1961.

1567. Regulatory mechanisms for protein synthesis in mammalian cells. Ed. by A. San Pietro. New York, Academic Press, 1968.

1568. Romanoff, A. L. The avian embryo: structural and functional development. New York, Macmillan, 1960.

1569. ——, and A. J. Romanoff. The avian egg. New York, Wiley, 1949.

1570. ——, and ——. Biochemistry of the avian embryo. New York, Wiley-Interscience, 1967.

1571. Rose, S. M. Regeneration: key to understanding normal and abnormal growth and development. New York, Appleton-Century-Crofts, 1970.

1572. Rugh, Roberts. Experimental embryology: techniques and procedures. 3d ed. Minneapolis, Burgess, 1962.

1573. ——. Laboratory manual of vertebrate embryology. 5th ed. Minneapolis, Burgess, 1961.

1574. ——. Mouse: its reproduction and development. Minneapolis, Burgess, 1968.

1575. ——. Vertebrate embryology: the dynamics of development. New York, Harcourt, Brace & World, 1964.

1576. Saunders, J. W., Jr. Animal morphogenesis. New York, Macmillan, 1968.

1577. ——. Patterns and principles of animal development. New York, Macmillan, 1970.

1578. Schering symposium on intrinsic and extrinsic factors in early mammalian development. Venice, 1970. [Proceedings.] Ed. by Gerhard Raspé. New York, Pergamon Press, 1970. (*Advances in the biosciences*, 6.)

1579. Shelesnyak, M. C., ed. Ovum implantation. New York, Gordon & Breach, 1969.

1580. Spemann, Hans. Embryonic development and induction. New York, Hafner, 1962.

1581. Sussman, Maurice. Growth and development. 2d ed. Englewood Cliffs, N.J., Prentice-Hall, 1964.

1582. Teir, Harald, and Tapio Rytömaa. Control of cellular growth in adult organisms. New York, Academic Press, 1967. (Sigrid Jusélius symposium. Proceedings.)

1583. Thorpe, D. R. Fetal pig: a dissection guide in color. Palo Alto, Calif., National Press Books, 1970.

1584. Tissue interactions during organogenesis. Ed. by Etienne Wolff. New York, Gordon & Breach, 1970. (Documents on biology, v. 1.)

1585. Torrey, T. W. Morphogenesis of the vertebrates. 3d ed. New York, Wiley, 1971.

1586. Villee, C. A. The placenta and fetal membranes. Baltimore, Williams & Wilkins, 1960.

1587. Vogel, F., and G. Röhrborn. Chemical mutagenesis in mammals and man. New York, Springer-Verlag, 1970.

1588. Walford, R. L. Immunologic theory of aging. Baltimore, Williams & Wilkins, 1969.

1589. Watterson, R. L. Laboratory studies of chick, pig and frog embryos. 2d ed. Minneapolis, Burgess, 1970.

1590. Weber, Rudolph. The biochemistry of animal development. 2 v. New York, Academic Press, 1965-1967.
    v. 1: Descriptive biochemistry of animal development. v. 2: Biochemical control mechanisms and adaptations in development.

1591. Weiss, P. A. Principles of development: a text in experimental embryology. New York, Hafner, 1969. (Reprint of 1939 edition.)

1592. Willier, B. H., and Jane M. Oppenheimer. Foundations of experimental embryology. Englewood Cliffs, N.J., Prentice-Hall, 1964.

1593. Willis, R. A. The borderland of embryology and pathology. 2d ed. Washington, D.C., Butterworth, 1962.

1594. Wilt, F. H. Methods in developmental biology. New York, Crowell, 1967.

1595. Workshop in teratology. 1st. 1964. Teratology, principles and techniques. Ed. by J. G. Wilson. Chicago, Univ. of Chicago Press, 1964.

1596. Zamboni, Luciano. Fine morphology of mammalian fertilization. New York, Harper & Row, 1971.

HISTOLOGICAL AND CYTOLOGICAL

1597. Adam, W. S., et al. Microscopic anatomy of the dog: a photographic atlas. Springfield, Ill., Thomas, 1970.

1598. Afzelius, Bjorn. Anatomy of the cell. Chicago, Univ. of Chicago Press, 1966.

1599. Andrew, Warren. Microfabric of man: a textbook of histology. Chicago, Year Book Medical Publishers, 1966.

1600. ———. Textbook of comparative histology. New York, Oxford Univ. Press, 1959.

1601. Andrews, G. S. Exfoliative cytology. Springfield, Ill., Thomas, 1971.

1602. Arey, L. B. Human histology: a textbook in outline form. 3d ed. Philadelphia, Saunders, 1968.

1603. Bailey, F. R. Textbook of histology. 16th ed. By W. M. Copenhaver, R. P. Bunge, and Mary B. Bunge. Baltimore, Williams & Wilkins, 1971.

1604. Balazs, E. A., ed. Chemistry and molecular biology of the intercellular matrix. 3 v. New York, Academic Press, 1970.
    v. 1: Collagen, basal laminae elastin.
v. 2: Glycosaminoglycans and proteoglycans.
v. 3: Structural organization and function of the matrix.

1605. Banks, W. J., Jr. Atlas of veterinary histology. Baltimore, Williams & Wilkins, in press 1972.

1606. Bevelander, Gerrit. Essentials of histology. 6th ed. St. Louis, Mosby, 1970.

1607. ———. Outline of histology. 7th ed. St. Louis, Mosby, 1971.

1608. Bloom, B. R., and P. R. Glade, eds. In vitro methods in cell-mediated immunity. New York, Academic Press, 1971.

1609. Bloom, William, and D. W. Fawcett. A textbook of histology. 9th ed. Philadelphia, Saunders, 1968.

1610. Bourne, G. H. An introduction to functional histology. 2d ed. Boston, Little, Brown, 1960.

1611. ———. "In vivo" techniques in histology. Baltimore, Williams & Wilkins, 1967.

1612. Brachet, Jean, and A. E. Mirsky, eds. The cell: biochemistry, physiology, morphology. 6 v. New York, Academic Press, 1959-1964.

1613. Brook Lodge conference on inflammation. 1967. Chemical biology of inflammation. Ed. by B. K. Forscher. New York, Pergamon Press, 1968.

1614. Brown, W. V. Textbook of cytogenetics. St. Louis, Mosby, 1972.

1615. ———, and E. M. Bertke. Textbook of cytology. St. Louis, Mosby, 1969.

1616. Burnet, F. M. Cellular immunology. 2 v. New York, Cambridge Univ. Press, 1969. (Book 1 also issued under the title *Self and not self.*)

1617. Busch, Harris. The nucleolus. New York, Academic Press, 1970.

1618. Calhoun, Mary L. Microscopic anatomy of the digestive system of the chicken. Ames, Iowa State Univ. Press, 1954.

1619. Cameron, I. L. Cellular and molecular renewal in the mammalian body. New York, Academic Press, 1971.

1620. ———. Developmental aspects of the cell cycle. New York, Academic Press, 1971.

1621. Campbell, P. N. The structure and function of animal cell components: an introductory text. New York, Pergamon Press, 1966.

1622. Ciba Foundation. The frozen cell, a Ciba Foundation symposium. Ed. by G. E. W. Wolstenholme. Edinburgh, Churchill, 1970.

1623. Clark, W. E. L. The tissues of the body. 6th ed. Oxford, Clarendon Press, 1971.

1624. Conference on comparative mammalian cytogenetics. Dartmouth Medical School, 1968. Proceedings. Ed. by K. Benirschke. New York, Springer-Verlag, 1969.

1625. Cowdry, E. V. Special cytology: the form and functions of the cell in health and disease; a textbook for students of biology and medicine. 2d ed. 3 v. New York, Hafner, 1963.

1626. ———, and J. C. A. Finerty. A textbook of histology. 5th ed. Philadelphia, Lea & Febiger, 1960.

1627. Dellmann, Horst-Dieter. Veterinary histology. Philadelphia, Lea & Febiger, 1971.

1628. Fine, B. S. Ocular histology: a text and atlas. New York, Harper & Row, 1972.

1629. Fiore, M. S. H. di. An atlas of human histology. 3d ed. Philadelphia, Lea & Febiger, 1967.

1630. Foust, H. L. Veterinary histology and embryology (lecture outlines and laboratory guide). 4th ed. Minneapolis, Burgess, 1958.

1631. Garber, E. D. Cytogenetics: an introduction. New York, McGraw-Hill, 1972.

1632. Giese, A. C. Cell physiology. 3d ed. Philadelphia, Saunders, 1968.

1633. Hall, M. C. The locomotor system: functional histology. Springfield, Ill., Thomas, 1965.

1634. Ham, A. W. Histology. 6th ed. Philadelphia, Lippincott, 1969.

1635. Herrath, Ernst von. Atlas of histology: normal microscopic anatomy of man. New York, Hafner, 1966. (First English edition, based upon the second German edition.)

1636. Holmes, R. L. Living tissues: an introduction to functional histology. New York, Pergamon Press, 1965.

1637. Hurry, S. W. The microstructure of cells. Boston, Houghton Mifflin, 1964.

1638. International conference on the morphological precursors of cancer. University of Perugia, 1961. Morphological precursors of cancer. Ed. by Lucio Severi. Perugia, Div. of Cancer Research, 1962.

1639. International Society of Blood Transfusion. Tissue typing: a symposium organized by the International Society of Blood Transfusion at the 10th Congress of the European Society of Haematology. Strasbourg, 1965. Ed. by Jean Dausset. Basel, Karger, 1966. (*Vox sanguinis*, v. 11, no. 3.)

1640. International symposium on cytoecology. Leningrad, 1963. The cell and environmental temperature; proceedings. Ed. by A. S. Troshin. New York, Pergamon Press, 1967.

1641. Kavanau, J. L. Structure and function in biological membranes. 2 v. San Francisco, Holden-Day, 1965.

1642. Kedrevskiĭ, B. V. The cytology of the protein synthesis in an animal cell. New York, Gordon & Breach, 1965.

1643. Kihlman, B. A. Actions of chemicals on dividing cells. Englewood Cliffs, N.J., Prentice-Hall, 1966.

1644. Koss, L. G. Diagnostic cytology and its histopathologic bases. 2d ed. Philadelphia, Lippincott, 1968.

1645. Laborit, Henri. Stress and cellular function. Philadelphia, Lippincott, 1959.

1646. Langley, L. L. Cell function. 2d ed. New York, Reinhold, 1968.

1647. Lawrence, H. S., and Maurice Landy, eds. Mediators of cellular immunity. New York, Academic Press, 1969.

1648. Leeson, T. S., and C. R. Leeson. Histology. 2d ed. Philadelphia, Saunders, 1970.

1649. Levine, Laurence, ed. The cell in mitosis. New York, Academic Press, 1963.

1650. Loewy, A. G., and Philip Siekevitz. Cell structure and function. 2d ed. New York, Holt, Rinehart & Winston, 1969.

1651. Mazia, Daniel, and Albert Tyler. General physiology of cell specialization. New York, McGraw-Hill, 1963.

1652. Patt, D. I., and Gail R. Patt. Comparative vertebrate histology. New York, Harper & Row, 1969.

1653. Penfield, Wilder. Cytology and cellular pathology of the nervous system. 3 v. New York, Hafner, 1965. (Reprint of 1932 edition.)

1654. Priest, J. H. Cytogenetics. Philadelphia, Lea & Febiger, 1969.

1655. Reith, E. J., and M. H. Ross. Atlas of descriptive histology. 2d ed. New York, Harper & Row, 1970.

1656. Robertis, E. D. P. de, W. W. Nowinski, and F. A. Saez. Cell biology. 5th ed. Philadelphia, Saunders, 1970.

1657. Rose, G. G. Atlas of vertebrate cells in tissue culture. New York, Academic Press, 1970.

1658. Rothblat, G. H., ed. Growth, nutrition, and metabolism of cells in culture. v. 1- New York, Academic Press, 1972-

1659. Sandborn, E. B. Cell and tissues by light and electron microscopy. 2 v. New York, Academic Press, 1970.

1660. San Pietro, A. G., ed. Regulatory mechanisms for protein synthesis in mammalian cells. New York, Academic Press, 1968.

1661. Schmid, F., and J. Stein. Cell research and cellular therapy. Thun, Switzerland, Ott, 1967. (Translation of *Zellforschung und Zellulartherapie*.)

1662. Spriggs, A. I., and M. M. Boddington. The cytology of effusions. 2d ed. New York, Grune & Stratton, 1968.

1663. Stiles, K. A. Handbook of histology. 5th ed. New York, McGraw-Hill, 1968.

1664. Swanson, C. P. The cell. 3d ed. Englewood Cliffs, Prentice-Hall, 1969.

1665. Symposium on rhythmic research. Wiesbaden, 1965. The cellular aspects of biorhythms. Ed. by H. von Mayersbach. New York, Springer-Verlag, 1967.

1666. Taylor, R. G. W. Practical cytology. New York, Academic Press, 1967.

1667. Thompson, S. W., E. D. Jenkins, and Mary A. Fox. A histopathologic and histologic coding system for tissue collections. Springfield, Ill., Thomas, 1961.

1668. Trautmann, Alfred, and Josef Fiebiger. Fundamentals of the histology of domestic animals. London, Bailliere, Tindall & Cox, 1952.

1669. Tritsch, G. L., ed. Axenic mammalian cell reactions. New York, Dekker, 1969.

1670. Valdés-Dapena, Marie A. An atlas of fetal and neonatal histology. Philadelphia, Lippincott, 1957.

1671. Wainio, W. W. The mammalian mitochondrial respiratory chain. New York, Academic Press, 1970.

1672. Williamson, M. E. Histologic patterns in tumor pathology. New York, Hoeber, 1968.

1673. Wilson, G. B. Cell division and the mitotic cycle. New York, Reinhold, 1966.

1674. Windle, W. F. Textbook of histology. 4th ed. New York, McGraw-Hill, 1969.

1675. Wismar, Beth L. A visual approach to histology. Philadelphia, Davis, 1970.

ULTRASTRUCTURAL

1676. Wynn, R. M. Cellular biology of the uterus. New York, Appleton-Century-Crofts, 1967.

1677. Austin, C. R. Ultrastructure of fertilization. New York, Holt, Rinehart & Winston, 1968.

1678. Babel, Jean. Ultrastructure of the peripheral nervous system and sense organs. St. Louis, Mosby, 1970.

1679. Bjorkman, Nils. Atlas of placental fine structure. Baltimore, Williams & Wilkins, 1971.

1680. Breathnach, A. S. An atlas of the ultrastructure of human skin: development, differentiation, and post-natal features. Baltimore, Williams & Wilkins, 1971.

1681. Dalton, A. J., and F. Haguenau, eds. Ultrastructure in biological systems. 4 v. New York, Academic Press, 1962-1968.
    v. 1: Tumors induced by viruses. v. 2: Ultrastructure of the kidney. v. 3: The nucleus. v. 4: The membranes.

1682. Dodge, J. D. Atlas of biological ultrastructure. New York, American Elsevier, 1968.

1683. Engström, Arne. Engström-Finean biological ultrastructure. 2d ed. Ed. by J. B. Finean. New York, Academic Press, 1967.

1684. Fawcett, D. W. An atlas of fine structure: the cell, its organelles and inclusions. Philadelphia, Saunders, 1966.

1685. Freeman, J. A. Cellular fine structure: an introductory student text and atlas. New York, McGraw-Hill, 1964.

1686. Hay, E. D., and J. P. Revel. Fine structure of the developing avian cornea. New York, Karger, 1969. (Monographs in developmental biology, v. 1.)

1687. Hirohata, Kazushi, and Kazuo Morimoto. Ultrastructure of bone and joint diseases. 1st ed. New York, Grune & Stratton, 1971.

1688. Huhn, Dieter, and Walther Stich, eds. Fine structure of blood and bone marrow. New York, Hafner, 1969.

1689. Kurtz, S. M. Electron microscopic anatomy. New York, Academic Press, 1964.

1690. Laguens, R. P., and V. R. Miatello. Atlas: ultrastructure renal. Kidney fine structure. Ultrastructre du rena. Buenos Aires, Inter-Medica, 1964.

1691. Lentz, T. L. Cell fine structure: an atlas of drawings of whole-cell structure. Philadelphia, Saunders, 1971.

1692. Lupulescu, A. Ultrastructure of the thyroid gland. Baltimore, Williams & Wilkins, 1968.

1693. Mori, Yoshitaka, and Karl Lennert. Electron microscopic atlas of lymph node cytology and pathology. New York, Springer-Verlag, 1969.

1694. Pease, D. C. Histological techniques for electron microscopy. 2d ed. New York, Academic Press, 1964.

1695. Peters, Alan. Fine structure of the nervous system: the cells and their processes. New York, Harper & Row, 1970.

1696. Poon, T. P., Asao Hirano, and H. M. Zimmerman. Electron microscopic atlas of brain tumors. New York, Grune & Stratton, 1971.

1697. Porter, K. R., and Mary A. Bonneville. Fine structure of cells and tissues. 3d ed. Philadelphia, Lea & Febiger, 1968.

1698. Rhodin, J. A. An atlas of ultrastructure. Philadelphia, Saunders, 1963.

1699. Scanga, F. Atlas of electron microscopy. New York, American Elsevier, 1964.

1700. Sjöstrand, F. S. Electron microscopy of cell and tissues. New York, Academic Press, 1967- v. 1: Instrumentation and techniques.

1701. Threadgold, L. T. Ultrastructure of the animal cell. New York, Pergamon Press, 1967.

1702. Toner, P. G. The digestive system: an ultrastructural atlas and review. New York, Appleton-Century-Crofts, 1971.

1703. Zamboni, Luciano. Fine morphology of mammalian fertilization. New York, Harper & Row, 1971.

1704. Zelickson, A. S. Ultrastructure of normal and abnormal skin. Philadelphia, Lea & Febiger, 1967.

SYSTEMATIC

Musculoskeletal

1705. Alexander, R. M. Animal mechanics. Seattle, Univ. of Washington Press, 1969.

1706. Andrew, B. L. Control and innervation of skeletal muscle. Edinburgh, Livingstone, 1966.

1707. Banga, Ilona. Structure and function of elastin and collagen. Budapest, Akademiai Kiado, 1966.

1708. Basmajian, J. V. Muscles alive: their functions revealed by electromyography. 2d ed. Baltimore, Williams & Wilkins, 1967.

1709. Bendall, J. R. Muscles, molecules and movement. New York, American Elsevier, 1969.

1710. Bernshtein, N. A. The co-ordination and regulation of movements. New York, Pergamon Press, 1967. (1st English edition.)

1711. Bourne, G. H. Structure and function of muscle. 3 v. New York, Academic Press, 1960.
  v. 1: Structure. v. 2: Biochemistry and physiology. v. 3: Pharmacology and disease.

1712. Brookes, Murray. The blood supply of bone: an approach to bone biology. New York, Appleton-Century-Crofts, 1971.

1713. Bulbring, Edith, et al., eds. Smooth muscle. Baltimore, Williams & Wilkins, 1970.

1714. Burkhardt, Rolf. Bone marrow and bone tissue: color atlas of clinical histopathology. New York, Springer-Verlag, 1971. (Translated by W. J. Hirsch.)

1715. Ernst, Jenö. Biophysics of the striated muscle. 2d ed. Budapest, Akademiai Kiado, 1963.

1716. Evans, F. G. Studies on the anatomy and function of bones and joints. New York, Springer-Verlag, 1966.

1717. European symposium on calcified tissues. Proceedings. 1st- 1963- (For additional coverage see entry 0750.)

1718. Fêng, Yuan-chêng, ed. Biomechanics. New York, American Society of Mechanical Engineers, 1966.

1719. Fourman, Paul, and Pierre Royer. Calcium metabolism and the bone. 2d ed. Oxford, Blackwell, 1968.

1720. Frankel, V. H., and A. H. Burstein. Orthopaedic biomechanics: the application of engineering to the musculoskeletal system. Philadelphia, Lea & Febiger, 1970.

1721. Frost, H. M. Bone dynamics. Boston, Little, Brown, 1964.

1722. ———. An introduction to biomechanics. Springfield, Ill., Thomas, 1967.

1723. George, J. C., and A. J. Berger. Avian myology. New York, Academic Press, 1966.

1724. Gray, James. Animal locomotion. New York, Norton, 1968.

1725. Greenblatt, G. M. Cat musculature. Chicago, Univ. of Chicago Press, 1954.

1726. Hancox, N. M. Biology of bone. Cambridge, England, The Univ. Press, 1972.

1727. Harvey, E. B., H. E. Kaiser, and L. E. Rosenberg. Atlas of the domestic turkey (Meleagris gallopavo): myology and osteology. Washington, D.C., U.S. Atomic Energy Commission, 1968.

1728. Howell, A. B. Speed in animals: their specialization for running and leaping. New York, Hafner, 1965. (Reprint of 1944 edition.)

1729. Laki, Koloman. Contractile proteins and muscle. New York, Dekker, 1971.

1730. Matzen, P. F., and H. K. Fleissner. Orthopedic roentgen atlas. New York, Grune & Stratton, 1970. (English edition arranged by L. S. Michaelis.)

1731. Muybridge, Eadweard. Animal locomotion: an electrophotographic investigation of consecutive phases of animal movements, 1872-1885. Philadelphia, Lippincott, 1887.

1732. ———. Animals in motion. Ed. by Lewis S. Brown. New York, Dover, 1957.

1733. Needham, Dorothy M. Machina carnis: the biochemistry of muscular contraction in its historical development. New York, Cambridge Univ. Press, 1971.

1734. Rang, Mercer, ed. The growth plate and its disorders. Baltimore, Williams & Wilkins, 1969.

1735. Rock Island Arsenal biomechanics symposium. 1st. Augustana College, 1967. Biomechanics. Ed. by David Bootzin and H. C. Muffley. New York, Plenum Press, 1969.

1736. Silverstein, Alvin, and Virginia B. Silverstein. The skeletal system: frameworks for life. Englewood Cliffs, N.J., Prentice-Hall, 1972.

1737. Smythe, R. H. The dog: structure and movement. New York, Arco, 1970.

1738. Somogyi, J. C., ed. Nutritional aspects of the development of bone and connective tissue. Basel, Karger, 1969. (Bibliotheca "Nutritio et Dieta," no. 13.)

1739. Strömberg, Berndt. The normal and diseased superficial flexor tendon in race horses. 1971. (*Acta radiologica*. Suppl. 305.)

1740. Symposium on the foundations and objectives of biomechanics. La Jolla, Calif., 1970. Biomechanics: its foundations and objectives. Ed. by Y. C. Fung, N. Perrone, and M. Anliker. Englewood Cliffs, N.J., Prentice-Hall, 1972.

1741. Thomson, R. G. Vertebral body osteophytes in bulls. Ed. by C. N. Barron and P. Cohrs. White Plains, N.Y., Phiebig, 1969.

1742. Tricker, R. A. R., and B. J. K. Tricker. The science of movement. New York, American Elsevier, 1967.

1743. U.S. Armed Forces Medical Library. The structure, composition and growth of bone, 1930-1953. (For further coverage see entry 0203.)

1744. Way, R. F. The anatomy of the bovine foot: a pictorial approach. Philadelphia, Univ. of Pennsylvania Press, 1954.

1745. Weitbrecht, Josias. Syndesmology or a description of the ligaments of the human body. Philadelphia, Saunders, 1969.

1746. Wilkie, D. R. Muscle. New York, St. Martin's Press, 1968. (Institute of Biology. Studies in biology, no. 11.)

1747. Zachar, Jozef. Electrogenesis and contractility in skeletal muscle cells. Baltimore, University Park Press, 1971.

Nervous and Sensory

1748. Adrianov, O. S., and T. A. Mering. Atlas of the canine brain. Ed. by E. F. Domino. Ann Arbor, Mich., Edwards Bros., 1964. (Translated by E. Ignatieff.)

1749. Albe-Fessard, D., et al. Atlas stéréotaxique du diencéphale du rat blanc. Paris, Centre National de la Recherche Scientifique, 1971.

1750. American Animal Hospital Association. Committee on Ophthalmology. Canine and feline ocular fundus: an atlas. Elkhart, Ind., American Animal Hospital Association, 1965?

1751. Austin, George. The spinal cord: basic aspects and surgical considerations. 2d ed. Springfield, Ill., Thomas, 1972.

1752. Barnett, K. C. Variations of the normal ocular fundus of the dog. Elkhart, Ind., American Animal Hospital Association, 1972.

1753. Berman, A. L. The brain stem of the cat: a cytoarchitectonic atlas with stereotaxic coordinates. Madison, Univ. of Wisconsin Press, 1968.

1754. Bossy, Jean. Atlas of neuroanatomy and special sense organs. Philadelphia, Saunders, 1970.

1755. Bourne, G. H. Structure and function of nervous tissue. v. 1- New York, Academic Press, 1968-
 v. 1: Structure I. v. 2: Structure II and physiology. v. 3: Biochemistry and disease.

1756. Bowsher, David. Introduction to the anatomy and physiology of the nervous system. 2d ed. Oxford, Blackwell Scientific Pub., 1970.

1757. Brauer, Kurt. Katalog der Säugetiergehirne. Catalogue of mammalian brains. Jena, Fischer-Verlag, 1970.

1758. Brodal, A. Neurological anatomy in relation to clinical medicine. 2d ed. New York, Oxford Univ. Press, 1969.

1759. Bullock, T. H., and G. A. Horridge. Structure and function in the nervous systems of invertebrates. 2 v. San Francisco, Freeman, 1965.

1760. Ciba Foundation. Taste and smell in vertebrates: a Ciba Foundation symposium. London, Churchill, 1970.

1761. Ciba Foundation symposium. Hearing mechanisms in vertebrates. Baltimore, Williams & Wilkins, in press 1972.

1762. Davis, Ross, and R. D. Huffman. Stereotaxic atlas of the brain of the baboon (Papio). Published for the Southwest Foundation for Research and Education, San Antonio. Austin, Univ. of Texas Press, 1968.

1763. Davison, A. N., and Alan Peters. Myelination. Springfield, Ill., Thomas, 1970.

1764. Davson, Hugh, ed. The eye. 2d ed. 4 v. New York, Academic Press, 1969-1972.
v. 1: Vegetative physiology and biochemistry. v. 2A-B: The visual process. v. 3: Muscular mechanisms. v. 4: Visual optics and optical space sense.

1765. DeLucchi, M. R., et al. A stereotaxic atlas of the chimpanzee brain (Pan satyrus). Berkeley, Univ. of California Press, 1965.

1766. Dua-Sharma, Sushil, K. N. Sharma, and H. L. Jacobs. The canine brain in stereotaxic coordinates: full sections in frontal, sagittal, and horizontal planes. Cambridge, Mass., MIT Press, 1970.

1767. Dubrul, E. L. Evolution of the speech apparatus. Springfield, Ill., Thomas, 1958.

1768. Duke-Elder, W. S., ed. System of ophthalmology. v. 1- St. Louis, Mosby, 1958-
v. 1: The eye in evolution. v. 2: The anatomy of the visual system. v. 3: Normal and abnormal development. v. 4: The physiology of the eye and vision. v. 5: Ophthalmic optics and refraction. v. 6: Ocular motility and strabismus. v. 7: The foundation of ophthalmology. v. 8: Diseases of the outer eye. v. 9: Disease of the uveal tract. v. 10: Diseases of the retina. v. 11: Diseases of the lens and virteous; glaucoma and hypotony. v. 12: Neuro-ophthalmology. (In preparation as of 1972 are v. 13: The ocular adnexa; v. 14: Injuries; and v. 15: Index of general and systemic ophthalmology.)

1769. Elliott, H. C. Textbook of neuroanatomy. 2d ed. Philadelphia, Lippincott, 1969.

1770. Emmers, Raimond, and Konrad Akert. A sterotaxic atlas of the brain of the squirrel monkey (Saimiri sciureus). Madison, Univ. of Wisconsin Press, 1963.

1771. Everett, N. B. Functional neuroanatomy, including an atlas of the brain stem. 6th ed. Philadelphia, Lea & Febiger, 1971.

1772. Fine, B. S. Ocular histology: a text and atlas. New York, Harper & Row, 1972.

1773. Gatz, A. J. Manter's essentials of clinical neuroanatomy and neurophysiology. 4th ed. Philadelphia, Davis, 1970.

1774. Guyton, A. C. Structure and function of the nervous system. Philadelphia, Saunders, 1972.

1775. Herrick, C. J. Brains of rats and men: a survey of the origin and biological significance of the cerebral cortex. New York, Hafner, 1963. (Reprint of 1926 edition.)

1776. Horridge, G. A. Interneurons: their origin, action, specificity, growth and plasticity. San Francisco, Freeman, 1968.

1777. House, E. L., and Ben Pansky. A functional approach to neuroanatomy. 2d ed. New York, McGraw-Hill, 1967.

1778. International symposium on the skin senses. 1st. Florida State University, March, 1966. The skin senses. Ed. by D. R. Kenshalo. Springfield, Ill., Thomas, 1968.

1779. Jacobson, Marcus. Developmental neurobiology. New York, Holt, Rinehart & Winston, 1970.

1780. Jenkins, T. W. Functional mammalian neuroanatomy with emphasis on dog and cat, including an atlas of dog central nervous system. Philadelphia, Lea & Febiger, 1972.

1781. Kalter, Harold. Teratology of the central nervous system: induced and spontaneous malformations of laboratory, agricultural, and domestic animals. Chicago, Univ. of Chicago Press, 1968.

1782. Karten, H. J., and William Hodos. A stereotaxic atlas of the brain of the pigeon, Columba livia. Baltimore, Johns Hopkins Press, 1967.

1783. Kemali, M., and V. Braitenberg. Atlas of the frog's brain. New York, Springer-Verlag, 1969.

1784. König, J. F. R., and R. A. Klippel. The rat brain: a stereotaxic atlas of the forebrain and lower parts of the brain stem. Baltimore, Williams & Wilkins, 1963.

1785 Kovac, Werner, and Helmut Denk. Der Hirnstamm der Maus. Wien, Springer-Verlag, 1968.

1786. Kuhlenbeck, Hartwig. The central nervous system of vertebrates: a general survey of its comparative anatomy with an introduction to the pertinent fundamental biologic and logical concepts. v. 1- New York, Academic Press, 1967-
  v. 1: Propaedeutics to comparative neurology. v. 2: Invertebrates and the origin of vertebrates. v. 3: Part 1—Structural elements; biology of nervous tissue.

1787. Larsell, Olof. The comparative anatomy and histology of the cerebellum. Ed. by Jan Jansen. v. 1- Minneapolis, Univ. of Minnesota Press, 1967-
  v. 1: The cerebellum from myxinoids through birds. v. 2: The cerebellum from monotremes through apes.

1788. Lentz, T. L. Primitive nervous systems. New Haven, Yale Univ. Press, 1968.

1789. Lin, K'o-shêng. A stereotaxic atlas of the dog's brain. By R. K. S. Lim, Channao Liu, and R. L. Moffitt. Springfield, Ill., Thomas, 1960.

1790. Luparello, T. J. Stereotaxic atlas of the forebrain of the guinea pig. Baltimore, Williams & Wilkins, 1967.

1791. Madigan, J. C. Cerebellum of the rhesus monkey. Baltimore, University Park Press, 1971.

1792. Manocha, S. L., T. R. Shantha, and G. H. Bourne. A stereotaxic atlas of the brain of the cebus monkey. Oxford, Clarendon Press, 1968.

1793. Matzke, H. A. Synopsis of neuroanatomy. 2d ed. New York, Oxford Univ. Press, 1972.

1794. Nakai, Junnosuke, ed. Morphology of neuroglia. Springfield, Ill., Thomas, 1963.

1795. Nauta, W. J. H., and S. O. E. Ebbesson, eds. Contemporary research methods in neuroanatomy. New York, Springer-Verlag, 1970.

1796. Noback, C. R., and William Montagna, eds. The primate brain. New York, Appleton-Century-Crofts, 1970. (*Advances in primatology*, v. 1.)

1797. Pearson, Ronald. The avian brain. New York, Academic Press, 1972.

1798. Pellegrino, Louis J., and Anna J. Cushman. A stereotaxic atlas of the rat brain. New York, Appleton-Century-Crofts, 1967.

1799. Peters, Alan. Fine structure of the nervous system: the cells and their processes. New York, Harper & Row, 1970.

1800. Pick, Joseph. The autonomic nervous system: morphological, comparative, clinical and surgical aspects. Philadelphia, Lippincott, 1970.

1801. Polyak, S. L. The vertebrate visual system. Ed. by W. H. Klüver. Chicago, Univ. of Chicago Press, 1957.

1802. Prince, J. H. Anatomy and histology of the eye and orbit in domestic animals. Springfield, Ill., Thomas, 1960.

1803. ———. Comparative anatomy of the eye. Springfield, Ill., Thomas, 1956.

1804. Progress in brain research. v. 1- New York, Elsevier, 1963- (For specific titles see entry 0777.)

1805. Ramón y Cajal, Santiago. The structure of the retina. Springfield, Ill., Thomas, 1972. (Comp. and tr. by Sylvia A. Thorpe and Mitchell Glickstein. Incorporates translation of the author's *La rétine des vertébrés*, published in 1892; *Die Retina der Wirbelthiere*, published in 1894 (translated by R. Greeff); and *La rétine des vertebrés*, published in 1933.)

1806. Rodahl, Kåre. Nerve as a tissue: proceedings of a conference held at the Lankenau Hospital, 1964. New York, Harper & Row, 1966.

1807. Schadé, J. P. The peripheral nervous system. New York, American Elsevier, 1966.

1808. School on neural networks. Ravello, 1967. Neural networks: proceedings. Ed. by E. R. Cainiello. New York, Springer-Verlag, 1968.

1809. Shantha, T. R., S. L. Manocha, and G. H. Bourne. A stereotaxic atlas of the Java monkey brain (Macaca irus). New York, Karger, 1968.

1810. Sherwood, Nancy M., and P. S. Timiras. A stereotaxic atlas of the developing rat brain. Berkeley, Univ. of California Press, 1970.

1811. Shkol'nik-IAarros, Ekaterina G. Neurons and interneuronal connections of the central visual system. New York, Plenum Press, 1971.

1812. Sidman, R. L., J. B. Angevine, Jr., and Elizabeth T. Pierce. Atlas of the mouse brain and spinal cord. Cambridge, Mass., Harvard Univ. Press, 1971.

1813. Singer, Marcus. The brain of the dog in section. Philadelphia, Saunders, 1962.

1814. Smythe, R. H. Animal vision: what animals see. Springfield, Ill., Thomas, 1961.

1815. Snider, R. S., and J. C. Lee. A stereotaxic atlas of the monkey brain. Chicago, Univ. of Chicago Press, 1961.

1816. ———, and W. T. Niemer. A stereotaxic atlas of the cat brain. Chicago, Univ. of Chicago Press, 1961.

1817. Truex, R. C., and M. B. Carpenter. Strong and Elwyn's human neuroanatomy. 6th ed. Baltimore, Williams & Wilkins, 1969.

1818. Van den Akker, L. M. An anatomical outline of the spinal cord of the pigeon. Philadelphia, Davis, 1970.

1819. Wolf, J. K. The classical brain stem syndromes: translations of the original papers with notes on the evolution of chemical neuroanatomy. Springfield, Ill., Thomas, 1971.

1820. Wolff, Eugene. Anatomy of the eye and orbit. 6th ed. Rev. by R. J. Last. Philadelphia, Saunders, 1968.

1821. Yoshikawa, Tetsuo. Atlas of the brains of domestic animals. University Park, Pennsylvania State Univ. Press, 1967.

1822. Young, J. Z. A model of the brain. New York, Oxford Univ. Press, 1964.

1823. Zeman, Wolfgang, and J. R. M. Innes. Craigie's neuroanatomy of the rat. New York, Academic Press, 1963.

Integumentary

1824. Advances in biology of skin. v. 1- Ed. by W. Montagna. New York, Appleton-Century-Crofts, 1960- (For specific titles of volumes see entry 0731.)

1825. Allen, A. C. The skin. 2d ed. New York, Grune & Stratton, 1967.

1826. Champion, R. H., ed. An introduction to the biology of the skin. Philadelphia, Davis, 1970.

1827. Ebling, F. J., et al., eds. The mammalian epidermis and its derivatives. New York, Academic Press, 1964. (Zoological Society of London. Symposia. no. 12.)

1828. Epstein, Ervin, ed. Skin surgery. 3d ed. Springfield, Ill., Thomas, 1970.

1829. Miles, A. E. W. Structural and chemical organization of teeth. 2 v. New York, Academic Press, 1967.

1830. Montagna, William, and W. C. Lobitz, eds. The epidermis. New York, Academic Press, 1964.

1831. A treatise of skin. v. 1- New York, Wiley-Interscience, 1971- (For specific titles see entry 0787.)

1832. Zelickson, A. S. Ultrastructure of normal and abnormal skin. Philadelphia, Lea & Febiger, 1967.

Cardiovascular

1833. Cardiology: an encyclopedia of the cardiovascular system. v. 1- (loose-leaf). Ed. by A. A. Luisada. New York, McGraw-Hill, 1959- (Kept up to date by revision sheets called supplements.)
    v. 1: Normal heart and vessels. v. 2: Methods. v. 3: Clinical cardiology. v. 4: Clinical cardiology-therapy. v. 5: Related specialty fields. Suppl. no. 1.

1834. Chiodi, Valentino. Conducting system of the vertebrate heart. Italy, Edizioni Calderini, 1967.

1835. Dintenfass, Leopold. Blood microrheology-viscosity factors in blood flow, ischaemia and thrombosis. New York, Appleton-Century-Crofts, 1971.

1836. Krogh, August. The anatomy and physiology of capillaries. rev. and enl. New York, Hafner, 1959.

1837. The microcirculation: a symposium. Comp. and ed. by W. L. Winters, Jr., and A. N. Brest. Springfield, Ill., Thomas, 1969.

1838. Rushmer, R. F. Cardiovascular dynamics. 3d ed. Philadelphia, Saunders, 1970.

Respiratory

1839. Anderson, W. D. Thorax of sheep and man: an anatomy atlas. Minneapolis, Dillon Press, 1972.

1839a. Fraser, R. G., and J. A. P. Paré. Structure and function of the lung, with emphasis on roentgenology. Philadelphia, Saunders, 1971.

1840. Nagaishi, Chuzo. Functional anatomy and histology of the lung. Baltimore, University Park Press, 1972.

Gastrointestinal

1841. Benzie, David, and A. T. Phillipson. The alimentary tract of the ruminant. London, Oliver & Boyd, 1957.

1842. Calhoun, Mary L. Microscopic anatomy of the digestive system of the chicken. Ames, Iowa State Univ. Press, 1954.

1843. DiDio, L. J. A., and M. C. Anderson. The "sphincters" of the digestive system: anatomical, functional and surgical considerations. Baltimore, Williams & Wilkins, 1968.

1844. Dowdy, G. S., Jr. The biliary tract. Philadelphia, Lea & Febiger, 1969.

1845. Elias, Hans. Morphology of the liver. New York, Academic Press, 1969.

1846. Morton, John. Guts: the form and function of the digestive system. New York, St. Martin's Press, 1967. (Institute of Biology. Studies in biology, no. 7.)

1847. Rouiller, Charles, ed. The liver: morphology, biochemistry, physiology. 2 v. New York, Academic Press, 1963-1964.

1848. Toner, P. G. The digestive system: an ultrastructural atlas and review. New York, Appleton-Century-Crofts, 1971.

Genitourinary

1849. Bjorkman, Nils. Atlas of placental fine structure. Baltimore, Williams & Wilkins, 1971.

1850. De Wardener, H. E. The kidney. 3d ed. Boston, Little, Brown, 1967.

1851. Graves, F. T. Arterial anatomy of the kidney: the basis of surgical technique. Baltimore, Williams & Wilkins, 1971.

1852. Hutch, J. A. Anatomy and physiology of the bladder, trigone, and urethra. New York, Appleton-Century-Crofts, 1972.

1853. Johnson, A. D., ed. The testis. 3 v. New York, Academic Press, 1970.
    v. 1: Development, anatomy, and physiology. v. 2: Biochemistry. v. 3: Influencing factors.

1854. The kidney: morphology, biochemistry, physiology. 4 v. Ed. by Charles Rouiller and A. F. Muller. New York, Academic Press, 1969-1971.

1855. Laguens, R. P., and V. R. Miatello. Atlas: ultrastructure renal. Kidney fine structure. Ultrastructre du rena. Buenos Aires, Inter-Medica, 1964.

1856. Mc Kerns, K. W. The gonads. New York, Appleton-Century-Crofts, 1969.

1857. The mammalian oviduct: comparative biology and methodology. Ed. by E. S. E. Hafez and R. J. Blandau. Chicago, Univ. of Chicago Press, 1969.

1858. Masui, Kiyoshi. Sex determination and sexual differentiation in the fowl. Ames, Iowa State Univ. Press, 1967.

1859. Simkiss, K. Calcium in reproductive physiology. New York, Reinhold, 1967.

1860. Wynn, R. M. Cellular biology of the uterus. New York, Appleton-Century-Crofts, 1967.

Endocrine

1861. Eisenstein, A. B. The adrenal cortex. Boston, Little, Brown, 1967.

1862. Harris, G. W., and B. T. Donovan. The pituitary gland. 3 v. Berkeley, Univ. of California Press, 1966.

1863. The hypothalamus. Ed. by W. Haymaker, E. Anderson, and W. J. H. Nauta. Springfield, Ill., Thomas, 1969.

1864. Jones, I. C. The adrenal cortex. New York, Cambridge Univ. Press, 1957.

1865. Lupulescu, A. Ultrastructure of the thyroid gland. Baltimore, Williams & Wilkins, 1968.

1866. Mühlen, Klaus aus der. The hypothalamus of the guinea pig: a topographic survey of its nuclear regions. New York, Karger, 1966.

1867. Pitt-Rivers, Rosalind, and W. R. Trotter, eds. The thyroid gland. 2 v. Washington, D.C., Butterworth, 1964.

1868. Purpura, D. P., and M. D. Yahr. The thalamus. New York, Columbia Univ. Press, 1966.

1869. Schalm, O. W. A syllabus on the bovine mammary glands in health and disease. Davis, Univ. of California, Dept. of Clinical Pathology, School of Veterinary Medicine, 1960.

1870. Wurtman, R. J. The pineal. New York, Academic Press, 1968.

Hematopoietic

1871. Abramson, D. I. Blood vessels and lymphatics. New York, Academic Press, 1962.

1872. American National Red Cross scientific symposium. 2d. Washington, D.C., 1969. Red cell membrane: structure and function. Ed. by G. Jamieson and T. J. Greenwalt. Philadelphia, Lippincott, 1969.

1873. Andrew, Warren. Comparative hematology. New York, Grune & Stratton, 1965.

1874. Bibliotheca haematologica. (Suppl. to *Acta haematologica*.) v. 1- Basel, Karger, 1955- (For specific titles see entry 0734.)

1875. Bishop, C. W., and D. M. Surgenor, eds. The red blood cell: a comprehensive treatise. New York, Academic Press, 1964.

1876. Branemark, Per-Ingvar. Intravascular anatomy of blood cells. Basel, Karger, 1971.

1877. Burkhardt, Rolf. Bone marrow and bone tissue: color atlas of clinical histopathology. New York, Springer-Verlag, 1971. (Translated by W. J. Hirsch.)

1878. Custer, P. L. An atlas of the blood and bone marrow. Philadelphia, Saunders, in press 1972.

1879. Elves, M. W. The lymphocytes. 2d ed. London, Lloyd-Luke, 1972.

1880. Fernex, Michel. The mast-cell system. Baltimore, Williams & Wilkins, 1968.

1881. Harris, J. W., and R. W. Kellermeyer. The red cell; production, metabolism, destruction: normal and abnormal. Cambridge, Mass., Harvard Univ. Press, 1970.

1882. Hayhoe, F. G. J., and R. J. Flemans. An atlas of haematological cytology. New York, Wiley, 1970.

1883. Holmes, W. L., ed. Blood cells as a tissue. New York, Plenum Press, 1970.

1884. Huhn, Dieter, and Walther Stich, eds. Fine structure of blood and bone marrow. New York, Hafner, 1969.

1885. Huisman, T. H. J., and W. A. Schroeder. New aspects of the structure, function, and synthesis of hemoglobins. Cleveland, CRC Press, 1971.

1886. Huser, H. J. Atlas of comparative primate hematology. New York, Academic Press, 1970.

1887. International congress of lymphology. 2d. Miami, 1968. Progress in lymphology: selected papers. Ed. by Manuel Viamonte et al. Stuttgart, Georg Thieme, 1970.

1888. International Society of Hemorheology. Theoretical and clinical hemorheology. New York, Springer-Verlag, 1971. (Proceedings of its 2d international conference, Heidelberg, 1969.)

1889. International symposium on lymphology. Zurich, 1966. Progress in lymphology. Ed. by A. Rüttimann. Stuttgart, Thieme, 1967.

1890. Kampmeier, O. F. Evolution and comparative morphology of the lymphatic system. Springfield, Ill., Thomas, 1969.

1891. Laki, Koloman. Fibrinogen. New York, Dekker, 1968.

1892. Lucas, A. M., and Casimir Jamroz. Atlas of avian hematology. Washington, D.C., U.S. Dept. of Agriculture, 1961. (Agriculture monograph no. 25.)

1893. Maupin, Bernard. Blood platelets in man and animals. 2 v. New York, Pergamon Press, 1969.

1894. Miale, J. B. Laboratory medicine: hematology. 4th ed. St. Louis, Mosby, 1972.

1895. Quick, A. J. Hemorrhagic diseases and thrombosis. 2d ed. rev. Philadelphia, Lea & Febiger, 1966.

1896. Red cell membrane. Ed. by R. I. Weed et al. New York, Grune & Stratton, 1971. (Seminars in hematology.)

1897. Schalm, O. W. Veterinary hematology. 2d ed. Philadelphia, Lea & Febiger, 1965.

1898. Schermer, Siegmund. Blood morphology of laboratory animals. 3d ed. Philadelphia. Davis, 1967.

1899. Selye, Hans. The mast cells. Washington, D.C., Butterworth, 1965.

1900. Stuart, A. E. The reticuloendothelial system. London, Livingstone, 1970.

1901. Tocantins, L. M., and L. A. Kazal, eds. Blood coagulation, hemorrhage and thrombosis: methods of study. 2d rev. ed. New York, Grune & Stratton, 1964.

1902. Wiener, A. S. Advances in blood grouping. 3 v. New York, Grune & Stratton, 1961-1970.

1903. Wintrobe, M. M. Clinical hematology. 6th ed. Philadelphia, Lea & Febiger, 1967.

1904. Yoffey, J. M. Bone marrow reactions. Baltimore, Williams & Wilkins, 1966.

1905. ———, ed. The lymphocyte in immunology and haemopoiesis: the symposium held at Bristol, 1966. London, Edward Arnold, 1967.

1906. Yoffey, J. M., and F. C. Courtice. Lymphatics, lymph, and the lymphomyeloid complex. New York, Academic Press, 1970.

8

PHYSIOLOGY

GENERAL AND SPECIFIC

1907. Altman, P. L., ed. Respiration and circulation. Bethesda, Md., Federation of American Societies for Experimental Biology, 1971.

1908. Andrew, B. L., ed. Experimental physiology. 8th ed. Edinburgh, Livingstone, 1969.

1909. Beament, J. W. L., and J. E. Treherne, eds. Insects and physiology. New York, American Elsevier, 1968.

1910. Bell, D. J., and B. M. Freeman, eds. Physiology and biochemistry of the domestic fowl. 3 v. New York, Academic Press, 1971.

1911. Bellville, J. W. Techniques in clinical physiology: a survey of measurements in anesthesiology. New York, Macmillan, 1969.

1912. Best, C. H., and N. B. Taylor. The physiological basis of medical practice: a text in applied physiology. 8th ed. Baltimore, Williams & Wilkins, 1966.

1913. Breazile, J. E. Textbook of veterinary physiology. Philadelphia, Lea & Febiger, 1971.

1914. Burke, Shirley R. The composition and function of body fluids. St. Louis, Mosby, 1972.

1915. Circadian rhythms in nonhuman primates. Ed. by F. H. Rohles. New York, Karger, 1969.

1916. D'Amour, F. E., et al. Manual for laboratory work in mammalian physiology. 3d ed. Chicago, Univ. of Chicago Press, 1965.

1917. Danielli, J. F. Cell physiology and pharmacology. New York, Hafner, 1968.

1918. Davson, Hugh. A textbook of general physiology. 4th ed. Baltimore, Williams & Wilkins, 1970.

1919. Dukes, H. H. Dukes' physiology of domestic animals. 8th ed. Ed. by M. J. Swenson. Ithaca, N.Y., Cornell Univ. Press, 1970.

1920. Dunn, Arnold. Experimental animal physiology: experiments in cellular and general physiology. New York, Holt, Rinehart & Winston, 1969.

1921. Folk, G. E., Jr. Introduction to environmental physiology: environmental extremes and mammalian survival. 2d ed. Philadelphia, Lea & Febiger, in press 1972.

1922. Frandson, R. D. Anatomy and physiology of farm animals. Philadelphia, Lea & Febiger, 1965.

1923. Ganong, W. F. Review of medical physiology. 5th ed. Los Altos, Calif., Lange Medical Publications, 1971.

1924. Gordon, M. S. Animal function: principles and adaptations. New York, Macmillan, 1968.

1925. Guyton, A. C. Basic human physiology: normal function and mechanisms of disease. Philadelphia, Saunders, 1971.

1926. ———. Function of the human body. 3d ed. Philadelphia, Saunders, 1969.

1927. ———. Textbook of medical physiology. 4th ed. Philadelphia, Saunders, 1971.

1928. Handbook of physiology: a critical, comprehensive presentation of physiological knowledge and concepts. v. 1- Washington, D.C., American Physiological Society, 1959-
  Sect. 1—v. 1-3: Neurophysiology.
  Sect. 2—v. 1-3: Circulation.
  Sect. 3—v. 1-2: Respiration.
  Sect. 4—Adaptation to the environment.
  Sect. 5—Adipose tissue.
  Sect. 6—Alimentary canal. v. 1: The control of food and water intake. v. 2: Secretion. v. 3: Intestinal absorption. v. 4: Motility. v. 5: Bile, digestion; ruminal physiology.
  Sect. 7—v. 1: Endocrinology; endocrine pancreas.

1929. Hoar, W. S., and D. J. Randall, eds. Fish physiology. 6 v. New York, Academic Press, 1969-1972.
  v. 1: Excretion, ionic regulation and metabolism. v. 2: The endocrine system. v. 3: Reproduction and growth; bioluminescence pigments and poisons. v. 4: The nervous system, circulation and respiration. v. 5: In press 1972. v. 6: Environmental relations and behavior.

1930. Hugh, A. E., and J. N. Glanville. Programmed primers of anatomy and physiological functions. New York, Appleton-Century-Crofts, 1971.

1931. Langley, L. L. Review of physiology. 3d ed. New York, McGraw-Hill, 1971.

1932. Larimer, James. An introduction to animal physiology. Dubuque, Ia., Brown, 1968.

1933. Levedahl, B. H., A. A. Barber, and Alan Grinnel. Laboratory experiments in physiology. 8th ed. St. Louis, Mosby, 1971.

1934. Lockwood, A. P. M. Animal body

fluids and their regulation. Cambridge, Harvard Univ. Press, 1964.

1935. Love, R. M. The chemical biology of fishes. New York, Academic Press, 1970.

1936. McNaught, Ann B., and Robin Callander. Illustrated physiology. 2d ed. Baltimore, Williams & Wilkins, 1970.

1937. Marshall, A. J. Biology and comparative physiology of birds. 2 v. New York, Academic Press, 1960-1961.

1938. Marshall, P. T., and G. M. Hughes. The physiology of mammals and other vertebrates: a textbook for schools and colleges. New York, Cambridge Univ. Press, 1965.

1939. Mayer, W. V., and R. G. Van Gelder, eds. Physiological mammology. 2 v. New York, Academic Press, 1963-1965.

1940. Mazia, Daniel, and Albert Tyler. General physiology of cell specialization. New York, McGraw-Hill, 1963.

1941. Mendelsohn, E. I. Heat and life: the development of the theory of animal heat. Cambridge, Mass., Harvard Univ. Press, 1964.

1942. Milhorn, H. T. The application of control theory to physiological systems. Philadelphia, Saunders, 1966.

1943. Olsen, Gjerding. Laboratory manual of vertebrate physiology. Dubuque, Ia., Brown, 1966.

1944. Pathophysiology: altered regulatory mechanisms in diseases. Ed. by E. D. Frohlich. Philadelphia, Lippincott, 1972.

1945. Phillips, R. W., and W. J. Tietz. Physiological basis of veterinary practice. Baltimore, Williams & Wilkins, in press 1972.

1946. Physiological Society of London. v. 1- London, Arnold, 1954?- (For specific titles see entry 0775.)

1947. Prosser, C. L., and F. A. Brown, Jr. Comparative animal physiology. 2d ed. Philadelphia, Saunders, 1961.

1948. Rose, A. H. Thermobiology. New York, Academic Press, 1967.

1949. Rothstein, Howard. General physiology: the cellular and molecular basis. Waltham, Mass., Xerox College Pub., 1971.

1950. Ruch, T. C., and H. D. Patton. Physiology and biophysics. 19th ed. Philadelphia, Saunders, 1965.

1951. Rudin, S. C., et al. Bioinstrumentation experiments in physiology. Millis, Mass., Harvard Apparatus Foundation, 1971.

1952. Selkurt, E. E. Physiology. 3d ed. Boston, Little, Brown, 1971.

1953. Snively, W. D., Jr., and Donna R. Beshear. Textbook of pathophysiology. Philadelphia, Lippincott, 1972.

1954. Sodeman, W. A., and W. A. Sodeman, Jr. Pathologic physiology: mechanisms of disease. 4th ed. Philadelphia, Saunders, 1967.

1955. Starling, E. H., and C. A. L. Evans. Principles of human physiology. 14th ed. Ed. by Hugh Davson and M. Grace Eggleton. Philadelphia, Lea & Febiger, 1968.

1956. Sturkie, P. D. Avian physiology. 2d ed. Ithaca, N.Y., Comstock, 1965.

1957. Symposium on the use of radioisotopes in animal nutrition and physiology. Prague, 1964. Radioisotopes in animal nutrition and physiology. Vienna, International Atomic Energy Agency, 1965.

1958. Tuttle, W. W., and B. A. Schottelius. Textbook of physiology. 16th ed. St. Louis, Mosby, 1969.

1959. Wilson, J. A. Principles of animal physiology. New York, Macmillan, 1972.

1960. Wood, D. W. Principles of animal physiology. New York, American Elsevier, 1970.

1961. Wright, Samson. Samson Wright's applied physiology. 12th ed. Rev. by C. A. Keele and Eric Neil. London, Oxford Univ. Press, 1971.

COMPARATIVE

1962. Barcroft, Joseph. Features in the architecture of physiological function. New York, Hafner, 1972.

1963. Dukes, H. H. Dukes' physiology of domestic animals. 8th ed. Ed. by M. J. Swenson. Ithaca, N.Y., Cornell Univ. Press, 1970.

1964. Dunn, Arnold. Experimental animal physiology: experiments in cellular and general physiology. New York, Holt, Rinehart & Winston, 1969.

1965. Florey, Ernst. An introduction to general and comparative animal physiology. Philadelphia, Saunders, 1966.

1966. Frieden, Earl, and Harry Lipner. Biochemical endocrinology of the vertebrates. Englewood Cliffs, N.J., Prentice-Hall, 1971.

1967. Gordon, M. S. Animal physiology: principles and adaptations. 2d ed. New York, Macmillan, 1972.

1968. Graham-Jones, Oliver, ed. Aspects of comparative ophthalmology. New York, Pergamon Press, 1966.

1969. Griffin, D. R. Animal structure and function. New York, Holt, Rinehart & Winston, 1970.

1970. Hoar, W. S. General and comparative physiology. Englewood Cliffs, N.J., Prentice-Hall, 1966.

1971. Jones, J. D. Comparative physiology of respiration. London, Arnold, 1972.

1972. Kaplan, H. M. The rabbit in experimental physiology. 2d ed. New York, Scholar's Library, 1962.

1973. Koldovsky, Otakar. Development of the functions of the small intestine in mammals and man. White Plains, N.Y., Phiebig, 1969.

1974. McCauley, W. J. Vertebrate physiology. Philadelphia, Saunders, 1971.

1975. Pantelouris, E. M. Introduction to animal physiology and physiological genetics. New York, Pergamon Press, 1967.

1976. Prosser, C. L., and F. A. Brown, Jr. Comparative animal physiology. 2d ed. Philadelphia, Saunders, 1961.

1976a. Riegel, J. A. Comparative physiology of renal excretion. Edinburgh, Oliver & Boyd, 1972.

1977. Scheer, B. T. Comparative physiology. Dubuque, Ia., Brown, 1968.

1978. Western Society of Naturalists. Comparative physiology of carbohydrate metabolism in heterothermic animals. Ed. by A. W. Martin. Seattle, Univ. of Washington Press, 1961.

1979. Whittow, G. C. Comparative physiology of thermoregulation. v. 1- New York, Academic Press, 1970- v. 1: Invertebrates and nonmammalian vertebrates.

1980. Yapp, W. B. An introduction to animal physiology. 3d ed. Oxford, Clarendon Press, 1970.

DEVELOPMENTAL, INCLUDING AGING

1981. Blumenthal, H. T., ed. The regulatory role of the nervous system in aging. New York, Karger, 1970.

1982. Dawes, G. S. Foetal and neonatal physiology: a comparative study of the changes at birth. Chicago, Year Book Medical Publishers, 1968.

1983. Fertilization: comparative morphology, biology and immunology. v. 1- Ed. by C. B. Metz and Alberto Monroy. New York, Academic Press, 1967-

1984. Kohn, R. R. Principles of mammalian aging. Englewood Cliffs, N.J., Prentice-Hall, 1971.

1985. The laboratory animal in gerontological research: proceedings of a symposium. St. Petersburg, Fla., 1967. Washington, D.C., National Academy of Sciences, 1968. (Its publication no. 1591.)

1986. Needham, Joseph. Chemical embryology. 3 v. Cambridge, England, The Univ. Press, 1931. (Reprint by Hafner, New York, 1963.)

1987. Raven, C. P. An outline of developmental physiology. 3d ed. New York, Pergamon Press, 1966.

1988. Stave, Uwe, ed. Physiology of the

perinatal period: functional and biochemical development in mammals. New York, Appleton-Century-Crofts, 1970.

1989. Tolerance, autoimmunity and aging. Compiled by M. M. Sigel and R. A. Good. Springfield, Ill., Thomas, 1972.

SYSTEMATIC

Musculoskeletal

1990. Bourne, G. H., ed. The biochemistry and physiology of bone. 2d ed. New York, Academic Press, 1971.

1991. Camougis, George. Nerves, muscles, and electricity: an introductory manual of electrophysiology. New York, Appleton-Century-Crofts, 1970.

1992. Chvapil, Milos. Physiology of connective tissue. New York, Appleton-Century-Crofts, 1967.

1993. Ebashi, Setsuro. Molecular biology of muscular contraction. New York, American Elsevier, 1965.

1994. Granit, Ragnar. The basis of motor control: integrating the activity of muscles, alpha and gamma motoneurons and their leading control systems. New York, Academic Press, 1970.

1995. Karpovich, P. V., and W. E. Sinning. Physiology of muscular activity. Philadelphia, Saunders, 1971.

1996. Kendall, H. O., Florence P. Kendall, and Gladys E. Wadsworth. Muscles, testing and function. 2d ed. Baltimore, Williams & Wilkins, 1971.

1997. McLean, F. C. Bone: fundamentals of the physiology of the skeletal tissue. 3d ed. Chicago, Univ. of Chicago Press, 1968.

1998. Symposium on myotatic, kinesthetic and vestibular mechanisms. London, 1966. Ed. by A. V. S. deReuck and Julie Knight. Boston, Little, Brown, 1967.

1999. Symposium on the physiology and biochemistry of muscle as a food. 1st- Madison, Univ. of Wisconsin Press, 1966-
    1st University of Wisconsin, 1965. Proceedings. Ed. by E. J. Briskey, R. G. Cassens, and J. C. Trautman. 1966.
    2d University of Wisconsin, 1969. Proceedings. Ed. by E. J. Briskey, R. G. Cassens, and B. B. Marsh. 1970.

2000. Vaughan, Janet M. Physiology of bone. Oxford, Clarendon Press, 1970.

Nervous and Sensory

2001. Abraham, Ambrus. Microscopic innervation of the heart and blood vessels in vertebrates including man. New York, Pergamon Press, 1969. (1st English edition.)

2002. Aidley, D. J. The physiology of excitable cells. New York, Cambridge Univ. Press, 1971.

2003. Adler, F. H. Physiology of the eye: clinical application. 5th ed. St. Louis, Mosby, 1970.

2004. Bajusz, Eörs, ed. Physiology and pathology of adaptation mechanisms: neural-neuroendocrine-humoral. New York, Pergamon Press, 1969.

2005. Bondy, S. C., and F. L. Margolis. Sensory deprivation and brain development: the avian visual system as a model. Jena, Fischer, 1971. (Abhandlungen aus dem Gebiet der Hirnforschung und Verhaltenphysiologie, bd. 4.)

2006. Bowsher, David. Introduction to the anatomy and physiology of the nervous system. 2d ed. Oxford, Blackwell Scientific Pub., 1970.

2007. Brazier, Mary A. B. The electrical activity of the nervous system: a textbook for students. 3d ed. Baltimore, Williams & Wilkins, 1968.

2008. Bullock, T. H., and G. A. Horridge. Structure and function in the nervous systems of invertebrates. 2 v. San Francisco, Freeman, 1965.

2009. Burn, J. H. The autonomic nervous system: for students of physiology and of pharmacology. 4th ed. Oxford, Blackwell, 1971.

2010. Camougis, George. Nerves, muscles, and electricity: an introductory manual of electrophysiology. New York, Appleton-Century-Crofts, 1970.

2011. Campbell, H. J. Correlative physiology of the nervous system. New York, Academic Press, 1965.

2012. Case, James. Sensory mechanisms. New York, Macmillan, 1966.

2013. Ciba Foundation. Taste and smell in vertebrates: a Ciba Foundation symposium. London, Churchill, 1970.

2014. Cooke, Ian, and Mack J. Lipkin, comps. and eds. Cellular neurophysiology: a source book. New York, Holt, Rinehart & Winston, 1972.

2015. Davson, Hugh. Physiology of the cerebrospinal fluid. London, Churchill, 1967.

2016. ———. The physiology of the eye. New York, Academic Press, 1972.

2017. Dethier, V. G. The physiology of insect senses. London, Methuen, 1963.

2018. Duke-Elder, W. S., ed. System of ophthalmology. v. 1- St. Louis, Mosby, 1958- (For titles of specific volumes see entry 1768.)

2019. Eccles, J. C. Physiology of synapses. New York, Academic Press, 1964.

2020. Gatz, A. J. Manter's essentials of clinical neuroanatomy and neurophysiology. 4th ed. Philadelphia, Davis, 1970.

2021. Griffith, J. S. Mathematical neurobiology: an introduction to the mathematics of the nervous system. New York, Academic Press, 1971.

2022. Guyton, A. C. Structure and function of the nervous system. Philadelphia, Saunders, 1972.

2023. Handbook of electroencephalography and clinical neurophysiology. v. 1- Amsterdam, Elsevier, 1971-

2024. Handbook of sensory physiology. v. 1- New York, Springer-Verlag, 1971-

2025. International symposium on the response of the nervous system to ionizing radiation. 1st- Proceedings. 1960- (For additional coverage see entry 0766.)

2026. Ivanov-Smolenskiĭ, A. G., and L. I. Kotliarevskiĭ, eds. Pathophysiology and experimental therapy of disorders in animal higher nervous activity during some intoxications and infections. Moscow, Academy of Sciences of USSR, 1959. (Akademiĭa Nauk SSSR. Works of the Institute of Higher Nervous Activity. Pathophysiological series, v. 6.)

2027. Lofts, Brian. Animal photoperiodism. London, Arnold, 1970. (Institute of Biology. Studies in biology, no. 25.)

2028. Mellon, Deforest, Jr. Physiology of sense organs. San Francisco, Freeman, 1968.

2029. Monnier, Marcel. Functions of the nervous system. v. 1- New York, American Elsevier, 1968-
   v. 1: General physiology, autonomic functions (neurohormonal regulations). v. 2: Motor and psychomotor functions. v. 3: Sensory and perceptual functions. v. 4: Psychic functions. (v. 3-4 in preparation 1972.)

2030. Naumenko, A. I. The physiological mechanisms of cerebral blood circulation. Springfield, Ill., Thomas, 1970.

2031. Ochs, Sidney. Elements of neurophysiology. New York, Wiley, 1965.

2032. Parakkal, P. F., and Nancy J. Alexander. Keratinization: a survey of vertebrate epithelia. New York, Academic Press, 1972.

2033. Shtark, M. B. The brain of hibernating animals. Washington, D.C., National Aeronautics and Space Administration, 1972. (NASA technical translation F619.)

2034. Smythe, R. H. Animal vision: what animals see. Springfield, Ill., Thomas, 1961.

2035. Stebbins, W. C., ed. Animal psychophysics: the design and conduct of sensory experiments. New York, Appleton-Century-Crofts, 1970.

2036. Stevens, C. F. Neurophysiology. New York, Wiley, 1966.

2037. Symposium on myotatic, kinesthetic

and vestibular mechanisms. London, 1966. Ed. by A. V. S. deReuck and Julie Knight. Boston, Little, Brown, 1967.

2038. Tansley, Katharine. Vision in vertebrates. Scranton, Pa., Harper & Row, 1965.

2039. Treherne, J. E., and J. W. L. Beament, eds. The physiology of the insect central nervous system. New York, Academic Press, 1965. (International congress of entomology. 12th. London, 1964. Proceedings.)

Integumentary

2040. Rook, A. J., and G. S. Walton. Comparative physiology and pathology of the skin. Philadelphia, Davis, 1965.

2041. Rothman, Stephen. Physiology and biochemistry of the skin. Chicago, Univ. of Chicago Press, 1954.

2042. Tregear, R. T. Physical functions of the skin. New York, Academic Press, 1966.

Cardiovascular

2043. Altman, P. L., ed. Respiration and circulation. Bethesda, Md., Federation of American Societies for Experimental Biology, 1971.

2044. Baroldi, Giorgio, and Guiseppe Scomazzoni. Coronary circulation in the normal and the pathologic heart. Washington, D.C., Armed Forces Institute of Pathology, 1967.

2045. Berne, R. M. Cardiovascular physiology. 2d ed. St. Louis, Mosby, 1972.

2046. Burton, A. C. Physiology and biophysics of the circulation. Chicago, Year Book Medical Publishers, 1965.

2047. Henry, J. P. The circulation: an integrative physiologic study. Chicago, Year Book Medical Publishers, 1971.

2048. Hirsch, E. F. The innervation of the vertebrate heart. Springfield, Ill., Thomas, 1970.

2049. Robb, Jane S. Comparative basic cardiology. New York, Grune & Stratton, 1965.

Respiratory

2050. Altman, P. L., ed. Respiration and circulation. Bethesda, Md., Federation of American Societies for Experimental Biology, 1971.

2051. Breslav, I. S. Perception of the respiratory medium and gas preference in man and animals. Washington, D.C., National Aeronautics and Space Administration, 1972.

2052. Comroe, J. H., Jr. Physiology of respiration, an introductory text. Chicago, Year Book Medical Publishers, 1965.

2053. Fraser, R. G., and J. A. P. Paré. Structure and function of the lung, with emphasis on roentgenology. Philadelphia, Saunders, 1971.

2054. Jones, J. D. Comparative physiology of respiration. London, Arnold, 1972.

2055. Rossier, P. H., A. A. Buhlmann, and K. Wiesinger. Respiration: physiologic principles and their clinical applications. St. Louis, Mosby, 1960. (1st English edition.)

2056. Steen, J. B. Comparative physiology of the respiratory mechanisms. New York, Academic Press, 1971.

2057. West, J. B. Ventilation/blood flow and gas exchange. 2d ed. Oxford, Blackwell, 1970.

Gastrointestinal

2058. Armstrong, W. M., and A. S. Nunn, Jr., eds. Intestinal transport of electrolytes, amino acids, and sugars. Springfield, Ill., Thomas, 1971.

2059. Benzie, David, and A. T. Phillipson. The alimentary tract of the ruminant. London, Oliver & Boyd, 1957.

2060. Brooks, F. P. Control of gastrointestinal function: an introduction to the physiology of the gastrointestinal tract. New York, Macmillan, 1970.

2061. Church, D. C. Digestive physiology and nutrition of ruminants. 3 v. Corvallis, Oregon State Univ. Book Stores, 1969-1972.

2062. Davenport, H. W. Physiology of the digestive tract: an introductory text. 3d ed. Chicago, Year Book Medical Publishers, 1971.

2063. Hargreaves, Tom. The liver and bile metabolism. New York, Appleton-Century-Crofts, 1968.

2064. International symposium on the physiology of digestion in the ruminant. (For additional coverage see entry 0765.)

2065. Koldovsky, O. Development of the functions of the small intestine in mammals and man. New York, Karger, 1969.

2066. Rouiller, Charles, ed. The liver: morphology, biochemistry, physiology. 2 v. New York, Academic Press, 1963-1964.

2067. Semb, L. S., and J. Myren. The physiology of gastric secretion. Baltimore, Williams & Wilkins, 1968.

2068. Sheehy, T. W., and M. H. Floch. The small intestine: its function and diseases. N.Y., Hoeber Medical Division, Harper & Row, 1964.

2069. Texter, E. C., Jr. Physiology of the gastrointestinal tract. St. Louis, Mosby, 1968.

2070. Ugolev, A. M. Physiology and pathology of membrane digestion. New York, Plenum Press, 1968.

Genitourinary

2071. Golden, Abner, and J. F. Maher. The kidney. Baltimore, Williams & Wilkins, 1971.

2072. Keitzer, W. A., and G. C. Huffman. Urodynamics. Springfield, Ill., Thomas, 1971.

2073. The kidney: morphology, biochemistry, physiology. 4 v. Ed. by Charles Rouiller and A. F. Muller. New York, Academic Press, 1969-1971.

2074. Pitts, R. F. Physiology of the kidney and body fluids: an introductory text. 2d ed. Chicago, Year Book Medical Publishers, 1968.

2075. Reuterskiöld, A. G. The wide ureter: functional studies in children and dogs. Uppsala, Soderstrom & Finn, 1969.

2076. Urodynamics: hydrodynamics of the ureter and renal pelvis. Ed. by Saul Boyarsky et al. New York, Academic Press, 1971.

2077. Windhager, E. E. Micropuncture techniques and nephron function. New York, Appleton-Century-Crofts, 1968.

Endocrine

2078. Adipose tissue: regulation and metabolic functions. Ed. by B. Jeanrenaud et al. New York, Academic Press, 1970. (*Hormone and metabolic research.* Suppl. 2.)

2079. Asboe-Hansen, Gustav. Hormones and connective tissue. Baltimore, Williams & Wilkins, 1966.

2080. Astwood, E. B., ed. Clinical endocrinology. v. 1- New York, Grune & Stratton, 1960-

2081. Bajusz, Eörs. An introduction to clinical neuroendocrinology. Baltimore, Williams & Wilkins, 1967.

2082. Bentley, P. J. Endocrines and osmoregulation: a comparative account of the regulation of water and salt in vertebrates. New York, Springer-Verlag, 1971. (*Zoophysiology and ecology*, v. 1.)

2083. Biochemical actions of hormones. v. 1- Ed. by Gerald Litwack. New York, Academic Press, 1970-

2084. Ciba Foundation symposium. Melbourne, 1965. The thymus: experimental and clinical studies. Ed. by G. E. W. Wolstenholme and Ruth Porter. Boston, Little, Brown, 1966.

2085. Clegg, Catherine, and A. G. Clegg. Hormones, cells and organisms: the role of hormones in mammals. Stanford, Calif., Stanford Univ. Press, 1969.

2086. Danowski, T. S. Clinical endocrinology. 4 v. Baltimore, Williams & Wilkins, 1962.
 v. 1: Pineal, hypothalamus, pituitary and gonads. v. 2: Thyroid. v. 3: Calcium, phosphorus, parathyroids and bone. v. 4: Adrenal cortex and medulla.

2087. Donovan, B. T. Mammalian neuroendocrinology. New York, McGraw-Hill, 1970.

2088. Dorfman, R. I., ed. Methods in hormone research. 2d ed. v. 1- New York, Academic Press, 1968-
 v. 1: Chemical determinations. v. 2: Part A—Bioassay.

2089. Eisenstein, A. B. The adrenal cortex. Boston, Little, Brown, 1967.

2090. Fisher, J. W. Kidney hormones. New York, Academic Press, 1971.

2091. Folley, S. J. The physiology and biochemistry of lactation. Edinburgh, Oliver & Boyd, 1956.

2092. Frieden, Earl, and Harry Lipner. Biochemical endocrinology of the vertebrates. Englewood Cliffs, N.J., Prentice-Hall, 1971.

2093. Frye, B. E. Hormonal control in vertebrates. New York, Macmillan, 1967.

2094. Functions of the adrenal cortex. 2 v. Ed. by K. W. McKerns. New York, Appleton-Century-Crofts, 1968.

2095. Gardiner-Hill, H. Modern trends in endocrinology. 3d series. New York, Appleton-Century-Crofts, 1967.

2096. Goldstein, Gideon. The human thymus. St. Louis, Green, 1969.

2097. Gorbman, Aubrey, and H. A. Bern. A textbook of comparative endocrinology. New York, Wiley, 1962.

2098. Gray, C. H., and A. L. Bacharach, eds. Hormones in blood. 2d ed. 2 v. New York, Academic Press, 1967.

2099. Harris, G. W., and B. T. Donovan. The pituitary gland. 3 v. Berkeley, Univ. of California Press, 1966.

2100. Höhn, E. O. Hormones in man and animals: the endocrine physiology of man and warm-blooded animals. London, English Universities Press, 1966.

2101. Hormones in development. Ed. by Max Hamburgh. New York, Appleton-Century-Crofts, 1971.

2102. Hubinont, P. E., et al. Basic actions of sex steroids on target organs. New York, Karger, 1971. (International seminar on reproductive physiology and sexual endocrinology. 3d. Brussels, 1970. Proceedings.)

2103. International congress on hormonal steroids. 2d. Milan, 1966. Abstract of papers. Ed. by L. Martini, F. Fraschini, and M. Motta. New York, Excerpta Medica, 1966.

2104. International Society of Psychoneuroendocrinology. Proceedings of meeting held at Downstate Medical Center, Brooklyn, N.Y., 1970. Influence of hormones on the nervous system. Ed. by D. H. Ford. Basel, Karger, 1971.

2105. Jenkin, Penelope M. Animal hormones: a comparative survey. v. 1- New York, Pergamon Press, 1962- (International series of monographs on pure and applied biology, zoology division, v. 6.)

2106. Kitay, J. I., and M. D. Altschule. The pineal gland: a review of the physiologic literature. Cambridge, Mass., Harvard Univ. Press, 1954.

2107. Kon, S. K., and A. T. Cowie, eds. Milk: the mammary gland and its secretion. 2 v. New York, Academic Press, 1961.

2108. Lee, Julius, and F. G. W. Knowles. Animal hormones. London, Hutchinson Univ. Library, 1965.

2109. McDonald, L. E. Veterinary endocrinology and reproduction. Philadelphia, Lea & Febiger, 1969.

2110. McKerns, K. W. Steroid hormones and metabolism. New York, Appleton-Century-Crofts, 1969.

2111. Mantegazza, P., and E. W. Horton, eds. Prostaglandins, peptides and amines. New York, Academic Press, 1969.

2112. Martini, Luciano, and W. F. Ganong. Neuroendocrinology. 2 v. New York, Academic Press, 1966-1967.

2112a. Methods in investigative and diagnostic endocrinology. Ed. by S. A. Berson. v. 1- Amsterdam, North-Holland Publishing, 1972-
 v. 1: The thyroid and biogenic amines. By J. E. Rall and I. J. Kopin.

2113. Monographs on endocrinology. v. 1-

2114. Mount, L. E. Climatic physiology of the pig. London, Arnold, 1968. (Physiological Society of London. Monograph no. 18.)

2115. Mrosovsky, Nicholas. Hibernation and the hypothalamus. New York, Appleton-Century-Crofts, 1971.

2116. Neurochemical aspects of hypothalmic function. Ed. by L. Martini and Joseph Meites. New York, Academic Press, 1970.

2117. Paschkis, K. E. Clinical endocrinology. 3d ed. New York, Harper & Row, 1967.

2118. The physiology of lactation. Ed. by A. T. Cowie. London, Arnold, 1971. (Physiological Society of London. Monograph no. 22.)

2119. Pincus, Gregory, and K. V. Thimann, eds. The hormones: physiology, chemistry, and application. v. 1- New York, Academic Press, 1948-

2120. Pitt-Rivers, Rosalind, and W. R. Trotter, eds. The thyroid gland. 2 v. Washington, D.C., Butterworth, 1964.

2121. Progress in comparative endocrinology. v. 1- New York, Academic Press, 1961- (Suppl. to *General and comparative endocrinology*.)

2122. Purpura, D. P., and M. D. Yahr. The thalamus. New York, Columbia Univ. Press, 1966.

2123. Richardson, G. S. Ovarian physiology. Boston, Little, Brown, 1967.

2124. Sawin, C. T. The hormones: endocrine physiology. Boston, Little, Brown, 1969.

2125. Schalm, O. W., E. J. Carroll, and N. C. Jain. Bovine mastitis. Philadelphia, Lea & Febiger, 1971.

2126. Selye, Hans. Hormones and resistance. Parts 1-2. Heidelberg, Springer-Verlag, 1971.

2127. Shearman, R. P. Induction of ovulation. Springfield, Ill., Thomas, 1969.

2128. Study group on hormones and the immune response. London, 1970. Hormones and the immune response. Ed. by G. E. W. Wolstenholme and Julie Knight. London, Churchill, 1970. (Ciba Foundation study group, no. 36.)

2129. Symposium on lactogenesis. University of Pennsylvania, 1968. Lactogenesis: the initiation of milk secretion at parturition. Ed. by M. Reynolds and S. J. Folley. Philadelphia, Univ. of Pennsylvania Press, 1969.

2130. Tepperman, Jay. Metabolic and endocrine physiology. 2d ed. Chicago, Year Book Medical Publishers, 1968.

2131. Tudhope, G. R. The thyroid and the blood. Springfield, Ill., Thomas, 1969.

2132. Turner, C. D. General endocrinology. 5th ed. Philadelphia, Saunders, 1971.

2133. Williams, R. H. Textbook of endocrinology. 4th ed. Philadelphia, Saunders, 1968.

2134. Zarrow, M. X., J. M. Yochim, and J. L. McCarthy. Experimental endocrinology: a sourcebook of basic techniques. New York, Academic Press, 1964.

Hematopoietic

2135. Altman, P. L. Blood and other body fluids. Washington, D.C., Federation of American Societies for Experimental Biology, 1961.

2136. Andrew, Warren. Comparative hematology. New York, Grune & Stratton, 1965.

2137. Applied seminar on the clinical pathology of hemoglobin, its precursors and metabolites. Washington, D.C., 1962. Hemoglobin: its precursors and metabolites. Ed. by F. W. Sunderman and F. W. Sunderman, Jr. Philadelphia, Lippincott, 1964.

2138. Bibliotheca haematologica. (Suppl. to *Acta haematologica*.) v. 1- Basel, Karger, 1955- (For specific titles see entry 0734.)

2139. Bishop, C. W., and D. M. Surgenor, eds. The red blood cell: a comprehensive treatise. New York, Academic Press, 1964.

2140. Dintenfass, Leopold. Blood microrheology: viscosity factors in blood flow, ischaemia, and thrombosis. An introduction to molecular and clinical haemorheology. New York, Appleton-Century-Crofts, 1971.

2141. Faraday Society of London. Colloid and Biophysics Committee. Flow properties of blood and other biological systems. Ed. by A. L. Copley and G. Stainsby. New York, Pergamon Press, 1960.

2142. Haesler, W. E., Jr. Immunohematology. Philadelphia, Lea & Febiger, 1969.

2143. Huisman, T. H. J., and W. A. Schroeder. New aspects of the structure, function, and synthesis of hemoglobins. Cleveland, CRC Press, 1971.

2144. Ling, N. R. Lymphocyte stimulation. New York, Wiley, 1968.

2145. Lucas, A. M., and Casimir Jamroz. Atlas of avian hematology. Washington, D.C., U.S. Dept. of Agriculture, 1961. (Its Agriculture monograph no. 25.)

2146. Marcus, A. J. The physiology of blood platelets. New York, Grune & Stratton, 1965.

2147. Schalm, O. W. Veterinary hematology. 2d ed. Philadelphia, Lea & Febiger, 1965.

2148. Seegers, W. H. Blood clotting enzymology. New York, Academic Press, 1967.

2149. Seitz, J. F. Biochemistry of the cells of blood and bone marrow. Springfield, Ill., Thomas, 1969.

2150. Selye, Hans. The mast cells. Washington, D.C., Butterworth, 1965.

2151. Tocantins, L. M., and L. A. Kazal, eds. Blood coagulation, hemorrhage and thrombosis: methods of study. 2d rev. ed. New York, Grune & Stratton, 1964.

2152. Wiener, A. S. Advances in blood grouping. 3 v. New York, Grune & Stratton, 1961-1970.

2153. Yoffey, J. M. Bone marrow reactions. Baltimore, Williams & Wilkins, 1966.

2154. Zmijewski, C. M., June L. Fletcher, and R. L. St. Pierre. Immunohematology. New York, Appleton-Century-Crofts, 1968.

REPRODUCTIVE

2155. Animal reproduction symposium. Proceedings of biennial symposia. (For specific titles see entry 0733.)

2156. Asdell, S. A. Cattle fertility and sterility. 2d ed. Boston, Little, Brown, 1968.

2157. ———. Patterns of mammalian reproduction. 2d ed. Ithaca, N.Y., Comstock, 1964.

2158. Assali, N. S., ed. Biology of gestation. 2 v. New York, Academic Press, 1968.
  v. 1: The maternal organism. v. 2: The fetus and neonate.

2159. Austin, C. R. Fertilization. Englewood Cliffs, N.J., Prentice-Hall, 1965.

2160. ———. Reproduction in mammals. 5 v. Cambridge, The Univ. Press, 1972.
  v. 1: Germ cells and fertilization. v. 2: Embryonic and fetal development. v. 3: Hormones in reproduction. v. 4: Reproductive patterns. v. 5: Artificial control of reproduction.

2161. ———. Ultrastructure of fertilization. New York, Holt, Rinehart & Winston, 1968.

2162. Bullough, W. S. Vertebrate reproductive cycles. 2d ed. New York, Wiley, 1961.

2163. Casey, Donn, ed. World directory of research workers in vertebrate reproduction. Cambridge, England, Reproduction Research Information Service, 1967.

2164. Cole, H. H., and P. T. Cupps, eds. Reproduction in domestic animals. 2d ed. New York, Academic Press, 1969.

2165. Faulkner, L. C., ed. Abortion diseases of livestock. Springfield, Ill., Thomas, 1968.

2166. Frye, B. E. Hormonal control in vertebrates. New York, Macmillan, 1967.

2167. Gibian, H., and E. J. Plotz, eds. Mammalian reproduction. New York, Springer-Verlag, 1970. (Gesellschaft für biologische Chemie. Colloquim no. 21.)

2168. Hadek, Robert. Mammalian fertilization. An atlas of ultrastructure. New York, Academic Press, 1969.

2169. Hafez, E. S. E., ed. Comparative reproduction of nonhuman primates. Springfield, Ill., Thomas, 1971.

2170. ———. Reproduction in farm animals. 2d ed. Philadelphia, Lea & Febiger, 1968.

2171. ———, and R. J. Blandau, eds. Mammalian oviduct: comparative biology and methodology. Chicago, Univ. of Chicago Press, 1969.

2172. Harmar, Hilary. Dogs and how to breed them. London, Gifford, 1968.

2173. Harrop, A. E. Reproduction in the dog. Baltimore, Williams & Wilkins, 1960.

2174. Harvard University. Control of ovulation. Ed. by C. A. Villee. New York, Pergamon Press, 1961.

2175. Johansson, Ivar, and Jan Rendel. Genetics and animal breeding. San Francisco, Freeman, 1968.

2176. Johnson, A. D., ed. The testis. 3 v. New York, Academic Press, 1970.
v. 1: Development, anatomy, and physiology. v. 2: Biochemistry. v. 3: Influencing factors.

2177. Journal of reproduction and fertility. Supplement series. no. 3- Oxford, Blackwell, 1968- (For specific titles see entry 0768.)

2178. Kiddy, C. A., and H. D. Hafs, eds. Sex ratio at birth—prospects for control: symposium. Champaign, Ill., American Society of Animal Science, 1971.

2179. Laing, J. A., ed. Fertility and infertility in the domestic animals. 2d ed. Baltimore, Williams & Wilkins, 1970.

2180. Marshall, I. H. A. Physiology of reproduction. 3d ed. 3 v. Ed. by A. S. Parkes. Boston, Little, Brown, 1960-1966.

2181. Moghissi, K. S., and E. S. Hafez. Biology of mammalian fertilization and implantation. Springfield, Thomas, 1972.

2182. Morris, Desmond. Patterns of reproductive behaviour. New York, McGraw-Hill, 1971.

2183. Mossman, H. W., and K. L. Duke. Comparative morphology of the mammalian ovary. Madison, Univ. of Wisconsin Press, 1973.

2184. Nalbandov, A. V. Reproductive physiology: comparative reproductive physiology of domestic animals, laboratory animals, and man. 2d ed. San Francisco, Freeman, 1964.

2185. Nishikawa, Y. Studies on reproduction in horses: singularity and artificial control in reproductive phenomena. Tokyo, Japan Racing Association, 1959.

2186. Odell, W. D., and D. L. Moyer. Physiology of reproduction. St. Louis, Mosby, 1971.

2187. Pathophysiology of gestation. v. 1- v. 1: Maternal disorders. Ed. by N. S. Assali. New York, Academic Press, 1972-

2188. Perry, E. J. Artificial insemination of farm animals. 4th ed. New Brunswick, N.J., Rutgers Univ. Press, 1968.

2189. Reproduction research information service. Interference with implantation: bibliography 1962-1969. New York, International Publications Service, 1962-1969.

2190. Salisbury, G. W., and N. L. Van Demark. Physiology of reproduction and artificial insemination of cattle. San Francisco, Freeman, 1961.

2191. Schmidt, G. H. Biology of lactation. San Francisco, Freeman, 1971.

2192. Shelesnyak, M. C., ed. Ovum implantation. New York, Gordon & Breach, 1969.

2193. Simkiss, K. Calcium in reproductive physiology. New York, Reinhold, 1967.

2194. Van Tienhoven, Ari. Reproductive physiology of vertebrates. Philadelphia, Saunders, 1968.

2195. Vollman, R. Fifty years of research on mammalian reproduction. A bibliography of the scientific publications of C. G. Hartman. Washington, D.C., 1965. (U.S. Public Health Service publ. no. 1281.)

2196. Young, W. C., ed. Sex and internal secretions. 3d ed. 2 v. Baltimore, Williams & Wilkins, 1961. (Earlier editions by Edgar Allen.)

2197. Zemjanis, R. Diagnostic and therapeutic techniques in animal reproduction. 2d ed. Baltimore, Williams & Wilkins, 1970.

Note. See also entries 2078—2134.

9
NEUROLOGY

GENERAL

2198. Brazier, Mary A. B. The electrical activity of the nervous system: a textbook for students. 3d ed. Baltimore, Williams & Wilkins, 1968.

2199. Chusid, J. G. Correlative neuroanatomy and functional neurology. 14th ed. Los Altos, Calif., Lange, 1970.

2200. Freedman, Russell, and J. E. Morriss. The brains of animals and man. New York, Holiday House, 1972.

2201. Gardner, E. D. Fundamentals of neurology. 5th ed. Philadelphia, Saunders, 1968.

2202. International research conference. 5th. Lankenau Hospital, Philadelphia, 1966. Mind as a tissue. Ed. by Charles Rupp. New York, Harper & Row, 1968.

2203. Matthews, L. H., and Maxwell Knight. The senses of animals. New York, Philosophical Library, 1963.

2204. Merritt, H. H. A textbook of neurology. 4th ed. Philadelphia, Lea & Febiger, 1967.

2205. Palmer, A. C. Introduction to animal neurology. Philadelphia, Davis, 1965.

2206. Sidman, R. L., M. C. Green, and S. H. Appel. Catalog of the neurological mutants of the mouse. Cambridge, Mass., Harvard Univ. Press, 1965.

2207. Skinner, J. E. Neuroscience: a laboratory manual. Philadelphia, Saunders, 1971.

2208. Stark, Lawrence. Neurological control systems-studies in bioengineering. New York, Plenum Press, 1968.

2209. Walton, J. N. Essentials of neurology. 2d ed. Philadelphia, Lippincott, 1966.

CLINICAL

2210. Brain, W. R. Brain's clinical neurology. 3d ed. New York, Oxford Univ. Press, 1969.

2211. Brock, Samuel, ed. Injuries of the brain and spinal cord and their coverings. 4th ed. New York, Springer, 1960.

2212. Decker, Kurt. Clinical neuroradiology. Ed. and tr. by W. H. Shehadi. New York, McGraw-Hill, 1966.

2213. Elliott, F. A. Clinical neurology. 2d ed. Philadelphia, Saunders, 1971.

2214. Handbook of electroencephalography and clinical neurophysiology. v. 1- Amsterdam, Elsevier, 1971-

2215. Hoerlein, B. F. Canine neurology: diagnosis and treatment. 2d ed. Philadelphia, Saunders, 1971.

2216. International symposium on stereoencephalotomy. 1st- Basel, Karger, 1961- (For additional coverage see entry 0764.)

2217. Kiloh, L. G., A. J. McComas, and J. W. Osselton. Clinical electroencephalography. 3d ed. New York, Appleton-Century-Crofts, 1972.

2218. McGrath, J. T. Neurologic examination of the dog with clinicopathologic observations. 2d rev. ed. Philadelphia, Lea & Febiger, 1960.

2219. Peele, T. L. The neuroanatomic basis for clinical neurology. 2d ed. New York, McGraw-Hill, 1961.

2220. Rooney, J. R. Clinical neurology of the horse. 1st ed. Kennett Square, Pa., KNA Press, 1971.

2221. Sunderland, Sydney. Nerves and nerve injuries. Baltimore, Williams & Wilkins, 1968.

BEHAVIOR

2222. Altman, Joseph. Organic foundations of animal behavior. New York, Holt, Rinehart & Winston, 1966.

2223. Altmann, S. A., ed. Social communication among primates. Chicago, Univ. of Chicago Press, 1967.

2224. Alverdes, Friedrich. Social life in the animal world. New York, Kraus, 1969. (Reprint of 1927 edition.)

2225. Ardrey, Robert. African genesis. New York, Atheneum, 1961.

2226. ——. Social contract. New York, Dell, 1971.

2227. ——. Territorial imperative. New York, Dell, 1971.

2228. Barnett, S. A. Instinct and intelligence: behavior of animals and man. Englewood Cliffs, N.J., Prentice-Hall, 1967.

2229. ——. The rat: a study in behavior. Chicago, Aldine, 1963.

2230. Beritashvili, I. S. Neural mechanisms of higher vertebrate behavior. Boston, Little, Brown, 1965.

2231. ——. Vertebrate memory: characteristics and origin. New York, Plenum Press, 1971.

2232. Bliss, E. L., ed. Roots of behavior: genetics, instinct, and socialization in animal behavior. New York, Harper & Row, 1962.

2233. Bourne, G. H., ed. The chimpanzee. v. 1- Baltimore, University Park Press, 1969-
v. 1: Anatomy, behavior, and diseases of chimpanzees. v. 2: Physiology, behavior, serology, and diseases of chimpanzees. v. 3: Immunology, infections, hormones, anatomy, and behavior. v. 4: Behavior, growth, and pathology of chimpanzees. v. 5: Histology, reproduction, and restraint.

2234. Breland, Keller, and Marian Breland. Animal behavior. New York, Macmillan, 1966.

2235. Burkhardt, Dietrich, et al., eds. Signals in the animal world. New York, McGraw-Hill, 1968.

2236. Burtt, H. E. The psychology of birds: an interpretation of bird behavior. New York, Macmillan, 1967.

2237. Calhoun, J. B. The ecology and sociology of the Norway rat. Washington, D.C., U.S. Public Health Service, 1962. (Its publication no. 1008.)

2238. Carpenter, C. R. Naturalistic behavior of nonhuman primates. University Park, Pennsylvania State Univ. Press, 1964.

2239. Central nervous system and fish behavior. Ed. by David Ingle. Chicago, Univ. of Chicago Press, 1968.

2240. Cohen, J. E. Casual groups of monkeys and men: stochastic models of elemental social systems. Cambridge, Mass., Harvard Univ. Press, 1971.

2240a. Communication by chemical signals. New York, Appleton-Century-Crofts, 1970. (*Advances in chemoreception*, v. 1.)

2241. Comparative psychopathology: animal and human. Ed. by Joseph Zubin and H. F. Hunt. New York, Grune & Stratton, 1967. (American Psychopathological Association. Proceedings. 55th annual meeting, New York, 1965.)

2242. Darwin, Charles. Expression of the emotions in man and animals. Chicago, Univ. of Chicago Press, 1965.

2242a. ——. The voyage of the Beagle. Ed. by Leonard Engel. Garden City, N.Y., 1962?

2243. Davis, D. E. Integral animal behavior. New York, Macmillan, 1966.

2243a. Dembeck, Hermann. Willingly to school: how animals are taught. New York, Taplinger, 1972. (Translated from German by Charles Johnson.)

2244. Denenberg, V. H., ed. Readings in the development of behavior. Stamford, Conn., Sinauer Associates, 1972.

2245. Denny, M. R., and S. C. Ratner. Comparative psychology: research in animal behavior. Homewood, Ill., Dorsey Press, 1970.

2246. Dethier, V. G., and Eliot Stellar. Animal behavior. 3d ed. Englewood Cliffs, N.J., Prentice-Hall, 1970.

2247. Dimond, S. J. Social behavior of animals. London, Batsford, 1970.

2248. Dolhinow, Phyllis, ed. Primate patterns. New York, Holt, Rinehart & Winston, 1972.

2249. Dröscher, V. B. The mysterious senses of animals. New York, Dutton, 1965.

2250. Eibl-Eibesfeldt, Irenaus. Ethology, the biology of behavior. New York, Holt, Rinehart & Winston, 1970. (Translated by Erich Klinghammer.)

2251. Etkin, William. Social behavior from fish to man. Chicago, Univ. of Chicago Press, 1967. (Chap. 1-4 originally appeared in *Social behavior and organization among vertebrates.* Ed. by William Etkin. Chicago, Univ. of Chicago Press, 1964.)

2252. Evans, W. F. Communication in the animal world. New York, Crowell, 1968.

2253. Ewer, R. F. Ethology of mammals. New York, Plenum Press, 1968.

2254. Fox, M. W. Behaviour of wolves, dogs and related canids. London, Jonathan Cape, 1971.

2255. ——. Canine behavior. Springfield, Ill., Thomas, 1965.

2256. ——. Integrative development of brain and behavior in the dog. Chicago, Univ. of Chicago Press, 1971.

2257. ——, ed. Abnormal behavior in animals. Philadelphia, Saunders, 1968.

2258. Fraenkel, G. S., and D. L. Gunn. The orientation of animals: kineses, taxes, and compass reactions. New York, Dover, 1961.

2259. Fraser, A. F. Reproductive behaviour in ungulates. New York, Academic Press, 1968.

2259a. Freeman, B. M., and R. F. Gordon, eds. Aspects of poultry behaviour. Edinburgh, Oliver & Boyd, 1970.

2260. Friedrich, Heinz, ed. Man and animal: studies in behaviour. London, Paladin, 1972. (Translation of *Mensch und Tier: Ausdrucksformen des Lebendigen.*)

2261. Geist, Valerius. Mountain sheep: a study in behavior and evolution. Chicago, Univ. of Chicago Press, 1971.

2262. Gilbert, R. M., and N. S. Sutherland, eds. Animal discrimination learning. New York, Academic Press, 1969.

2263. Grzimek, Bernhard. Grzimek's animal life encyclopedia. v. 1- New York, Van Nostrand-Reinhold, 1972- (To be complete in 13 v.; 3 v. published by mid-1972. Other volumes in preparation.)

2264. Hafez, E. S. E. Adaptation of domestic animals. Philadelphia, Lea & Febiger, 1968.

2265. ——. Behaviour of domestic animals. 2d ed. Baltimore, Williams & Wilkins, 1969.

2265a. ——, and I. A. Dyer. Animal growth and nutrition. Philadelphia, Lea & Febiger, 1969. (See especially Chap. 5, "Social environment and growth," by J. L. Albricht.)

2266. Hainsworth, Marguerite D. Experiments in animal behaviour. Boston, Houghton Mifflin, 1968.

2267. Hediger, Heini. Man and animal in the zoo. New York, Delacorte Press, 1969. (Translated by Gwynne Vevers and Winwood Reade.)

2268. ——. Psychology and behaviour of animals in zoos and circuses. New York, Dover, 1955. (Original title: *Skizzen zu einer Tierpsychologie im Zoo und im Zirkus.* English translation by Geoffrey Sircom.)

2269. ——. Wild animals in captivity. New York, Dover, 1964. (English translation by Geoffrey Sircom.)

2270. Herrick, C. J. Neurological foundations of animal behavior. New York, Hafner, 1962. (Reprint of 1924 edition.)

2270a. Hinde, R. A. Animal behaviour: a synthesis of ethology and comparative psychology. 2d ed. New York, McGraw-Hill, 1970.

2271. Jay, Phyllis C., ed. Primates: studies in adaptation and variability. New York, Holt, Rinehart & Winston, 1968.

2272. Johnsgard, P. A. Animal behavior. 2d ed. Dubuque, Ia., Brown, 1972.

2273. Johnson, R. N. Aggression in man and animals. Philadelphia, Saunders, 1972.

2274. Jolly, Alison. The evolution of primate behavior. New York, Macmillan, 1972.

2275. Klopfer, P. H., and J. P. Hailman. Function and evolution of behavior: an historical sample from the pens of ethologists. Reading, Mass., Addison-Wesley, 1972.

2276. Krushinski, L. V. Animal behavior: its normal and abnormal development. New York, Plenum Press, 1962.

2277. Lockley, R. M. Animal navigation. New York, Hart, 1967.

2278. Lorenz, Konrad Z. King Solomon's ring: new light on animal ways. New York, Crowell, 1952.

2279. ——. Man meets dog. Translated by Marjorie K. Wilson. Baltimore, Penguin, 1970. (Reprint of 1953 edition.)

2280. ——. On aggression. New York, Harcourt, Brace & World, 1966.

2281. ——. Studies in animal and human behaviour. v. 1- Cambridge, Mass., Harvard Univ. Press, 1970-

2282. McFarland, D. J. Feedback mechanisms in animal behavior. New York, Academic Press, 1971.

2283. Maier, N. R. F., and T. C. Schneirla. Principles of animal psychology. New York, Dover, 1964. (Enlarged and corrected republication of the 1935 edition.)

2284. Maier, R. A., and Barbara M. Maier. Comparative animal behavior. Monterey, Calif., Brooks/Cole, 1970.

2285. Marler, P. R., and W. J. Hamilton. Mechanisms of animal behavior. New York, Wiley, 1966.

2286. Mech, L. D. The wolf: ecology and behavior of an endangered species. New York, Natural History Press, 1970.

2287. Moltz, Howard, ed. The ontogeny of vertebrate behavior. New York, Academic Press, 1971.

2288. Morris, Desmond. Human zoo. New York, McGraw-Hill, 1969.

2289. ——. Naked ape. New York, Dell, 1969.

2290. ——. Patterns of reproductive behaviour. New York, McGraw-Hill, 1971.

2291. Pfaffenberger, C. J. The new knowledge of dog behavior. New York, Howell Book House, 1971.

2292. Rheingold, Harriet L., ed. Maternal behavior in mammals. New York, Wiley, 1963.

2293. Sadleir, R. M. Ecology of reproduction in wild and domestic mammals. London, Methuen, 1969. (Distributed in U.S. by Barnes & Noble, New York.)

2294. Saslow, C. Animal psychophysics: an examination of the field. Arlington, Va., U.S. Clearinghouse for Scientific and Technical Information, 1967.

2295. Schrier, A. M., ed. Behavior of nonhuman primates: modern research trends. v. 1-4. New York, Academic Press, 1965-1971.

2296. Schultz, D. P. Sensory restrictions: effects on behavior. New York, Academic Press, 1965.

2297. Scott, J. P., and J. L. Fuller, eds. Genetics and the social behavior of the dog. Chicago, Univ. of Chicago Press, 1965.

2298. Sebeok, T. A., ed. Animal communication: Techniques of study and results of research. Bloomington, Indiana Univ. Press, 1968.

2299. ——, and Alexandra Ramsay, eds. Approaches to animal communication. New York, Humanities Press, 1970.

2300. Sluckin, Wladyslaw. Early learning in man and animal. London, Allen & Unwin, 1970. (Advances in psychology series, no. 2.)

2301. Smith, D. D. Mammalian learning and behavior: a psychoneurological theory. Philadelphia, Saunders, 1965.

2302. Smythe, R. H. Animal habits: the things animals do. Springfield, Ill., Thomas, 1962.

2303. ——. Animal psychology: a book of comparative psychology which discusses the behaviour of animals and man. Springfield, Ill., Thomas, 1961.

2304. Society for Veterinary Ethology. Symposium. Proceedings. 1st. Univ. of Edinburgh, 1966. London, Society of Universities

Federation for Animal Welfare, 1967. (Proceedings of subsequent meetings are published approximately semiannually in *British veterinary journal*.)

2305. Sommer, Robert. Personal space: the behavioral basis of design. Englewood Cliffs, N.J., Prentice-Hall, 1969.

2306. Southwick, C. H. Animal aggression: selected readings. New York, Van Nostrand Reinhold, 1970.

2307. Stebbins, W. C., ed. Animal psychophysics: the design and conduct of sensory experiments. New York, Appleton-Century-Crofts, 1970.

2308. Stevenson, H. W., et al., eds. Early behavior: comparative and developmental approach. New York, Wiley, 1967.

2309. Stokes, A. W., ed. Animal behavior in laboratory and field. San Francisco, Freeman, 1968.

2310. Storm, R. M. Animal orientation and navigation. Corvallis, Oregon State Univ. Press, 1967. (Biology colloquim no. 27.)

2311. Sutherland, N. S., and N. J. Mackintosh. Mechanisms of animal discrimination learning. New York, Academic Press, 1971.

2312. Tavolga, W. N. Principles of animal behavior. New York, Harper & Row, 1969.

2313. Thévenin, René. Animal migration. New York, Walker, 1963.

2314. Thorndike, E. L. Animal intelligence: experimental studies. New York, Hafner, 1965.

2315 Thorpe, W. H. Learning and instinct in animals. rev. and enl. Cambridge, Mass., Harvard Univ. Press, 1963.

2316. Tinbergen, Niko. Animal behavior. New York, Time-Life, 1965.

2317. ———. Curious naturalists. Garden City, N.Y., Natural History Press, 1968.

2318. ———. The study of instinct. Oxford, Clarendon Press, 1969.

2319. Tolman, E. C. Purposive behavior in animals and man. New York, Appleton-Century-Crofts, 1967.

2320. Vernberg, F. J., and Winona B. Vernberg. The animal and the environment. New York, Holt, Rinehart & Winston, 1970.

2321. Völgyesi, F. A., and Gerhard Klumbies. Hypnosis of man and animals, with special reference to the development of the brain in the species and the individual. 2d ed. Baltimore, Williams & Wilkins, 1966.

2321a. Whitehead, G. K. The ancient white cattle of Britain and their descendents. London, Faber & Faber, 1953.

2322. Wynne-Edwards, V. C. Animal dispersion in relation to social behaviour. New York, Hafner, 1962.

2323. Zajonc, R. B. Animal social psychology: a reader of experimental studies. New York, Wiley, 1969.

Note. See also the section on "Psychopharmacology" (entries 2675-2690) in Chap. 12.

10

BIOCHEMISTRY

GENERAL

2324. Baldwin, Ernest. Introduction to comparative biochemistry. 5th ed. New York, Cambridge Univ. Press, 1967.

2325. Campbell, J. W., ed. Comparative biochemistry of nitrogen metabolism. v. 1- New York, Academic Press, 1970- v. 1: The invertebrates. v. 2: The vertebrates.

2326. Cantarow, Abraham, and Bernard Schepartz. Biochemistry. 4th ed. Philadelphia, Saunders, 1967.

2327. Comar, C. L., and Felix Bronner. Mineral metabolism: an advanced treatise. 3 v. in 5. New York, Academic Press, 1960-1969.
  v. 1: Parts A & B—Principles, processes and systems. v. 2: Parts A & B—The elements. v. 3: Calcium physiology.

2328. Damm, H. C. The handbook of biochemistry and biophysics. Cleveland, World, 1966.

2329. ———, ed. Methods and references in biochemistry and biophysics. Cleveland, World, 1966.

2330. Falconer, I. R. Mammalian biochemistry. Baltimore, Williams & Wilkins, 1969.

2331. Florkin, Marcel. Comprehensive biochemistry. v. 1-28 in five sections. New York, American Elsevier, 1962-1971.

2332. Fruton, J. S., and Sofia Simmonds. General biochemistry. 2d ed. New York, Wiley, 1960.

2333. Greenberg, D. M. Metabolic pathways. 3d ed. 4 v. New York, Academic Press, 1967-1970.
  v. 1: Energetics, tricarbosylic acid cycle, and carbohydrates. v. 2: Lipids, steroids, and carotenoids. v. 3: Amino acids and tetrapyrroles. v. 4: Nucleic acids, protein synthesis, and coenzymes.

2334. Harrow, Benjamin. Textbook of biochemistry. 10th ed. Ed. by Abraham Mazur. Philadelphia, Saunders, 1971.

2335. Haywood, B. J. Thin-layer chromatography: an annotated bibliography, 1964-1968. Ann Arbor, Mich., Ann Arbor-Humphrey Science Publishers, 1968.

2336. Karlson, P. Introduction to modern biochemistry. 3d ed. New York, Academic Press, 1968.

2337. Long, Cyril. Biochemists' handbook. Princeton, N.J., Van Nostrand, 1961.

2338. Neil, M. W. Vertebrate biochemistry in preparation for medicine. 2d ed. Philadelphia, Lippincott, 1965.

2339. Orten, J. M. Biochemistry. 8th ed. St. Louis, Mosby, 1970.

2340. Sebrell, W. H., Jr., and R. S. Harris, eds. The vitamins. v. 1- New York, Academic Press, 1967- (v. 6-7 ed. by Paul Gyorgy and W. N. Pearson.)

2341. Sober, H. A., ed. Handbook of biochemistry: selected data for molecular biology. Cleveland, Chemical Rubber Co., 1968.

2342. Thorpe, W. V. Biochemistry for medical students. 9th ed. Baltimore, Williams & Wilkins, 1970.

2343. Tietz, N. W., ed. Fundamentals of clinical chemistry. Philadelphia, Saunders, 1970.

2344. Treatise on collagen. Ed. by G. N. Ramachandran. v. 1- New York, Academic Press, 1967-
  v. 1: Chemistry of collagen. v. 2: Parts A & B—Biology of collagen. v. 3: Chemical pathology of collagen (in preparation 1972).

2345. West, E. S. Textbook of biochemistry. 4th ed. New York, Macmillan, 1965.

2346. White, Abraham. Principles of biochemistry. 4th ed. New York, McGraw-Hill, 1968.

PHYSIOLOGICAL CHEMISTRY, INCLUDING METABOLISM

2347. Aikawa, J. K. The relationship of magnesium to disease in domestic animals and in humans. Springfield, Ill., Thomas, 1971.

2348. Altman, P. L., ed. Metabolism. Bethesda, Md., Federation of American Societies for Experimental Biology, 1968.

2349. Arnow, L. E. Introduction to physiological and pathological chemistry. 8th ed. St. Louis, Mosby, 1972.

2350. Bentley, P. J. Endocrines and osmoregulation: a comparative account of the regulation of water and salt in vertebrates. New York, Springer-Verlag, 1971. (*Zoophysiology and ecology*, v. 1.)

2351. Beutler, Ernest. Red cell metabolism: a manual of biochemical methods. New York, Grune & Stratton, 1971.

2352. Blaxter, K. L. Energy metabolism of ruminants. 2d ed. London, Hutchinson, 1967.

2353. Bloom, F. Blood chemistry of the dog and cat. New York, Gamma Publications, 1960.

2354. Ciba Foundation symposium. Peptide transport in bacteria and mammalian gut. New York, American Elsevier, in press 1972.

2355. Clegg, Catherine, and A. G. Clegg. Hormones, cells and organisms: the role of hormones in mammals. Stanford, Calif., Stanford Univ. Press, 1969.

2356. Eastham, R. D. Biochemical values in clinical medicine. 4th ed. Baltimore, Williams & Wilkins, 1971.

2357. European symposium on calcified tissues. 1st- Proceedings. 1963- (For additional coverage see entry 0750.)

2358. Federation of European Biochemical Societies. Proceedings of the meetings. New York, Academic Press, 1964- (For specific titles see entry 0752.)

2359. Florkin, Marcel, and B. T. Scheer, eds. Chemical zoology. v. 1- New York, Academic Press, 1967-
    v. 1: Protozoa. v. 2: Porifera, coelenterata, and platyhelminthes. v. 3: Echinodermata, nematoda, and acanthocephala. v. 4: Annelida, echiura, and sipuncula. v. 5: Arthropoda, Part A. v. 6: Arthropoda, Part B.

2360. Fourman, Paul, and Pierre Royer. Calcium metabolism and the bone. 2d ed. Oxford, Blackwell, 1968.

2361. Gallagher, C. H. Nutritional factors and enzymological disturbances in animals. Philadelphia, Lippincott, 1964.

2362. Gozsy, Bela, and Laszlo Kato. Balancing mechanisms in acute inflammation. Montreal, 1970. (Institute of Microbiology and Hygiene of Montreal University. Monograph, no. 5.)

2363. Graymore, C. N. Biochemistry of the eye. New York, Academic Press, 1970.

2364. Hall, L. W. Fluid balance in canine surgery. Baltimore, Williams & Wilkins, 1967.

2365. Harper, H. A. Review of physiological chemistry. Los Altos, Calif., Lange Medical Publications, 1969.

2366. Hawk, P. B. Hawk's physiological chemistry. 14th ed. Ed. by B. L. Oser. New York, McGraw-Hill, 1965.

2367. Henry, R. J. Clinical chemistry: principles and technics. New York, Harper & Row, 1964.

2368. Hoffman, W. S. The biochemistry of clinical medicine. 4th ed. Chicago, Year Book Medical Publishers, 1970.

2369. Hsia, D. Yi-Yung. Inborn errors of metabolism. 2d ed. 2 parts. Chicago, Year Book Medical Publishers, 1966.
    Part 1—Clinical aspects. Part 2—Laboratory methods.

2370. International symposium on natural mammalian hibernation. 3d. Toronto, 1965. Mammalian hibernation, III. Ed. by K. C. Fisher et al. Edinburgh, Oliver & Boyd, 1967.

2371. International symposium on the newer trace elements in nutrition. Grand Forks, N.D., 1970. New trace elements in nutrition. Ed. by Walter Mertz and W. E. Cornatzer. New York, Dekker, 1971.

2372. Kaneko, J. J., and C. E. Cornelius, eds. Clinical biochemistry of domestic animals. 2d ed. v. 1- New York, Academic Press, 1970-

2373. Kleiber, Max. The fire of life: an introduction to animal energetics. New York, Wiley, 1961.

2374. Larner, Joseph. Intermediary metabolism and its regulation. Englewood Cliffs, N.J., Prentice-Hall, 1971.

2375. Latner, A. L. Isoenzymes in biology and medicine. New York, Academic Press, 1968.

2376. Lehninger, A. L. Bioenergetics: the molecular basis of biological energy transformations. 2d ed. Menlo Park, Calif., W. A. Benjamin, 1971.

2377. Linford, J. H. An introduction to energetics, with applications to biology. London, Butterworth, 1966.

2378. Manuel, Y., J. P. Revillard, and H. Betuel, eds. Proteins in normal and pathological urine. Baltimore, University Park Press, 1970.

2379. Maxwell, M. H., ed. Clinical disorders of fluid and electrolyte metabolism. 2d ed. New York, McGraw-Hill, 1972.

2380. Miller, A. T. Energy metabolism. Philadelphia, Davis, 1968.

2381. Mitchell, H. H. Comparative nutrition of man and domestic animals. 2 v. New York, Academic Press, 1962-1964.

2382. Munro, H. N., and J. B. Allison, eds. Mammalian protein metabolism. 4 v. New York, Academic Press, 1969.

2383. Panel on the use of nuclear techniques in studies of mineral metabolism and disease in domestic animals. Vienna, 1970. Mineral studies with isotopes in domestic animals: proceedings. Vienna, International Atomic Energy Agency, 1971.

2384. Patton, A. R. Biochemical energetics and kinetics. Philadelphia, Saunders, 1965.

2385. Physiological chemistry series. Ed. by E. J. Masoro. v. 1- Philadelphia, Saunders, 1968-
 v. 1 Physiological chemistry of lipids in mammals. By E. J. Masoro. 1968.
 v. 2 Physiological chemistry of proteins and nucleic acids in mammals. By George Kaldor. 1969.
 v. 3 Energy transformations in mammals: regulatory mechanisms. By F. L. Hoch. 1971.
 v. 4 Acid-base regulation: its physiology and pathophysiology. By E. J. Masoro and P. D. Siegel. 1971.

2386. Physiological Society of London. Monographs. v. 1- London, Arnold, 1954?- (For specific titles see entry 0775.)

2387. Schubert, Maxwell, and David Hamerman. Primer of connective tissue biochemistry. Philadelphia, Lea & Febiger, 1968.

2388. Searcy, R. L. Diagnostic biochemistry. New York, McGraw-Hill, 1969.

2389. Seegers, W. H. Blood clotting enzymology. New York, Academic Press, 1967.

2390. Stanbury, J. B., J. B. Wyngaarden, and D. S. Frederickson, eds. The metabolic basis of inherited disease. 3d ed. New York, McGraw-Hill, 1972.

2391. Symposium on rhythmic research. Wiesbaden, 1965. The cellular aspects of biorhythms. Ed. by H. von Mayersbach. New York, Springer-Verlag, 1967.

2392. Symposium: sulfur in nutrition. Oregon State University, 1969. Ed. by O. H. Muth. Westport, Conn., Avi Publishing, 1970.

2393. Taylor, W. H. Fluid therapy and disorders of electrolyte balance. 2d ed. Philadelphia, Davis, 1970.

2394. Thompson, R. H. S., and E. J. King. Biochemical disorders in human disease. 3d ed. New York, Academic Press, 1970.

2395. Underwood, E. J. Trace elements in human and animal nutrition. 3d ed. New York, Academic Press, 1971.

2396. Weber, Rudolph. The biochemistry of animal development. 2 v. New York, Academic Press, 1965-1967.
 v. 1: Descriptive biochemistry of animal development. v. 2: Biochemical control mechanisms and adaptations in development.

2397. Wills, M. R. The biochemical consequences of chronic renal failure. Baltimore, University Park Press, 1971.

2398. World Association for Animal Production. International Biological Programme. Aberdeen, Scotland. Trace element metabolism in animals: proceedings of WAAP/IBP international symposium. Aberdeen, Scotland, July, 1969. Ed. by C. F. Mills. Edinburgh, Livingstone, 1970.

NEUROCHEMISTRY

2399. Adams, C. W. M. Neurohistochemistry. New York, American Elsevier, 1965.

2400. Arvy, Lucie. Histoenzymology of the endocrine glands. 1st English ed. New York, Pergamon Press, 1971.

2401. Bajusz, Eörs. An introduction to clinical neuroendocrinology. Baltimore, Williams & Wilkins, 1967.

2402. Cumings, J. N., and M. Kremer, eds. Biochemical aspects of neurological disorders, third series. Oxford, Blackwell Scientific Pub., 1968.

2403. Davison, A. N., ed. Applied neurochemistry. Oxford, Blackwell, 1968.

2404. Fried, Rainer, ed. Methods of neurochemistry. v. 1- New York, Dekker, 1971-

2405. Friede, R. L. Topographic brain chemistry. New York, Academic Press, 1966.

2406. Gabe, M. Neurosecretion. New York, Pergamon Press, 1966.

2407. Histochemistry of nervous transmission. Ed. by Olavi Eränkö. New York, Elsevier, 1971. (*Progress in brain research*, v. 34.)

2408. Hoijer, Dorothy J. A bibliographic guide to neuroenzyme literature. New York, Plenum Press, 1969.

2409. International neurochemical conference. Oxford, 1965. Variation in chemical composition of the nervous system, as determined by developmental and genetic factors. Ed. by G. B. Ansell. New York, Pergamon Press, 1966.

2410. International neurochemical symposium. 1st-5th. Proceedings. 1954-1962. Various publishers.

2411. International Society of Psychoneuroendocrinology. Proceedings of meeting held at Downstate Medical Center, Brooklyn, New York, 1970. Influence of hormones on the nervous system. Ed. by D. H. Ford. Basel, Karger, 1971.

2412. Lajtha, Abel, ed. Handbook of neurochemistry. 8 v. New York, Plenum Press, 1969-1972.
v. 1: Chemical architecture of the nervous system. v. 2: Structural neurochemistry. v. 3: Metabolic reactions in the nervous system. v. 4: Control mechanisms in the nervous system. v. 5: Part B—Metabolic turnover in the nervous system. v. 6: Alterations of chemical equilibrium in the nervous system. v. 7: Pathological chemistry of the nervous system.

2413. Lissák, Kálmán, and E. Endroczi. The neuroendocrine control of adaptation. New York, Pergamon Press, 1965.

2414. McIlwain, Henry. Biochemistry and the central nervous system. 3d ed. Boston, Little, Brown, 1966.

2415. Martini, Luciano, and W. F. Ganong. Neuroendocrinology. 2 v. New York, Academic Press, 1966-1967.

2416. Neurochemical aspects of hypothalamic function. Ed. by L. Martini and Joseph Meites. New York, Academic Press, 1970.

2417. Palladin, A. V. Biochemistry of the nervous system. Springfield, Va., U.S. Clearinghouse for Scientific and Technical Information, 1967.

2418. Richter, Derek, ed. Comparative neurochemistry. New York, Macmillan, 1964. (International neurochemical symposium. 5th, 1962.)

2419. Ruščák, Michal, and Dagmar Ruščáková. Metabolism of the nerve tissue in relation to ion movements in vitro and in situ. Baltimore, University Park Press, 1971.

2420. Triggle, D. J. Chemical aspects of the autonomic nervous system. New York, Academic Press, 1965.

HISTOCHEMISTRY

2421. Baillie, A. H., M. M. Ferguson, and D. M. Hart. Developments in steroid histochemistry. New York, Academic Press, 1966.

2422. Barka, Tibor, and P. J. Anderson. Histochemistry: theory, practice and bibliography. New York, Harper, 1963.

2423. Brown, H. D., ed. Chemistry of the cell interface. v. 1- New York, Academic Press, 1971-

2424. Burstone, M. S. Enzyme histochemistry, and its applications in the study of neoplasms. New York, Academic Press, 1962.

2425. Handbuch der Histochemie. Hrsg. von Walther Graumann and Karlheinz Neumann. v. 1- Stuttgart, Gustav Fischer Verlag, 1958-

2426. International congress of histochemistry and cytochemistry. 3d. New York, 1968. Summary reports. New York, Springer-Verlag, 1968.

2427. International Society of Blood Transfusion. Tissue typing: a symposium organized by the International Society of Blood Transfusion at the 10th congress of the European Society of Haematology. Strasbourg, 1965. Ed. by Jean Dausset. Basel, Karger, 1966. (*Vox sanguinis*, v. 11, no. 3.)

2428. Pearse, A. G. E. Histochemistry: theoretical and applied. 3d ed. v. 1- Boston, Little, Brown, 1968-

2429. Raekallio, Jyrki. Enzyme histochemistry of wound healing. Stuttgart, Portland, Fischer, 1970. (*Progress in histochemistry and cytochemistry*, v. 1, no. 2.)

2430. San Pietro, A. G., ed. Regulatory mechanisms for protein synthesis in mammalian cells. New York, Academic Press, 1968.

2431. Seitz, J. F. Biochemistry of the cells of blood and bone marrow. Springfield, Ill., Thomas, 1969.

2432. Thompson, S. W. Selected histochemical and histopathological methods. Springfield, Ill., Thomas, 1966.

2433. Wied, George, ed. Introduction to quantitative cytochemistry. v. 1- New York, Academic Press, 1966-
   v. 1: Introduction to quantitative cytochemistry. v. 2: Introduction to quantitative cytochemistry, II.

2434. Zugibe, F. T. Diagnostic histochemistry. St. Louis, Mosby, 1970.

IMMUNOCHEMISTRY

2435. Angevine, C. D., et al. The immunochemistry and biochemistry of connective tissue and its disease states. New York, Karger, 1970.

2436. Arquembourg, P. C., et al. Primer of immunoelectrophoresis: with interpretation of pathologic human serum patterns. Ann Arbor, Mich., Ann Arbor-Humphrey Science Publishers, 1970.

2437. Barrett, J. T. Textbook of immunology: an introduction to immunochemistry and immunobiology. St. Louis, Mosby, 1970.

2438. Cawley, L. P. Electrophoresis and immunoelectrophoresis. Boston, Little, Brown, 1969.

2439. Conference on biologic aspects and clinical uses of immunoglobulins. Ohio State University, 1969. Immunoglobulins. Ed. by Ezio Merler. Washington, D.C., National Academy of Sciences, 1970.

2440. Crowle, A. J. Immunodiffusion. New York, Academic Press, 1961.

2441. Day, E. D. Advanced immunochemistry. Baltimore, Williams & Wilkins, 1972.

2442. ———. Foundation of immunochemistry. Baltimore, Williams & Wilkins, 1966.

2443. Haurowitz, Felix. Immunochemistry and the biosynthesis of antibodies. New York, Wiley, 1968.

2444. Kabat, E. A. Experimental immunochemistry. 2d ed. Springfield, Ill., Thomas, 1971.

2445. ———. Structural concepts in immunology and immunochemistry. New York, Holt, Rinehart & Winston, 1968.

2446. Kwapinski, J. B. Research in immunochemistry and immunobiology. 4 v. Baltimore, University Park Press, 1971.

2447. Nezlin, R. S. Biochemistry of antibodies. New York, Plenum Press, 1970.

2448. Nowotny, Alois. Basic exercises in immunochemistry. New York, Springer-Verlag, 1969.

2449. Nucleic acids in immunology: proceedings of a symposium held at the Institute of Microbiology of Rutgers, The State University, New Brunswick, N.J., 1967. Ed. by O. J. Plescia and Werner Braun. New York, Springer-Verlag, 1968.

2450. Ouchterlony, Örjan. Handbook of immunodiffusion and immunoelectrophoresis. Ann Arbor, Mich., Ann Arbor-Humphrey Science Publishers, 1968.

2451. Sanborn, W. R., comp. Immunofluorescence, an annotated bibliography. 7 parts. (For specific titles see entry 0193.)

2452. Williams, C. A., and M. W. Chase, eds. Methods in immunology and immunochemistry. v. 1- New York, Academic Press, 1967-
   v. 1: Preparation of antigens and antibodies. v. 2: Physical and chemical methods. v. 3: Reactions of antibodies with soluble antigens.

11
BIOPHYSICS AND RELATED DISCIPLINES

BIOPHYSICS

2453. Burton, A. C. Physiology and biophysics of the circulation. Chicago, Year Book Medical Publishers, 1965.

2454. Casey, E. J. Biophysics: concepts and mechanisms. New York, Reinhold, 1962.

2455. Damm, H. C. The handbook of biochemistry and biophysics. Cleveland, World, 1966.

2456. ———, ed. Methods and references in biochemistry and biophysics. Cleveland, World, 1966.

2457. Ernst, Jenö. Biophysics of the striated muscle. 2d ed. Budapest, Akademiai Kiado, 1963.

2458. Faraday Society of London. Colloid and Biophysics Committee. Flow properties of blood and other biological systems. Ed. by A. L. Copley and G. Stainsby. New York, Pergamon Press, 1960.

2459. Glasser, Otto, ed. Medical physics. 3 v. Chicago, Year Book Medical Publishers, 1944-1960.

2460. Greenwood, Maud E. An illustrated approach to medical physics. 2d ed. Philadelphia, Davis, 1966.

2461. Hope, A. B. Ion transport and membranes: a biophysical outline. Baltimore, University Park Press, 1971.

2462. Hughes, D. O. Physics for chemists and biologists. New York, Chemical Publishing, 1968.

2463. International Society of Hemorheology. Theoretical and clinical hemorheology. New York, Springer-Verlag, 1971. (Proceedings of its 2d international conference, Heidelberg, 1969.)

2464. Klip, Willem. Theoretical foundations of medical physics. 2 v. University, Ala., Univ. of Alabama Press, 1969.
 v. 1: Mathematics for the basic sciences and clinical research. v. 2: An introduction to medical physics.

2465. Minckler, Jeff. Introduction to neuroscience. St. Louis, Mosby, 1972.

2466. Richardson, I. W. Physics for biology and medicine. New York, Wiley, 1972.

2467. Ruch, T. C., and H. D. Patton. Physiology and biophysics. 19th ed. Philadelphia, Saunders, 1965.

2468. Symposium on biophysics and physiology of biological transport. Frascati, Italy. June 15-18, 1965. Proceedings. Ed. by Liana Bolis et al. New York, Springer-Verlag, 1967.

2469. Symposium on comparative bioelectrogenesis. Rio de Janeiro, 1959. Bioelectrogenesis: a comparative survey of its mechanisms with particular emphasis on electric fishes. Ed. by Carlos Chagas and A. P. de Carvalho. New York, Elsevier, 1961.

2470. Whitmore, R. L. Rheology of the circulation. New York, Pergamon Press, 1968.

MOLECULAR BIOLOGY

2471. Allen, J. M. Molecular organization and biological function. New York, Harper & Row, 1967.

2472. Bonner, J. F. The molecular biology of development. New York, Oxford Univ. Press, 1965.

2473. Bradshaw, L. J. Introduction to molecular biological techniques. Englewood Cliffs, N.J., Prentice-Hall, 1966.

2474. Ciba Foundation symposium. Principles of biomolecular organization. London, 1965. Ed. by G. E. W. Wolstenholme and Maeve O'Connor. Boston, Little, Brown, 1966.

2475. Daudel, Pascaline, and Raymond Daudel. Chemical carcinogenesis and molecular biology. New York, Wiley, 1966.

2476. Ebashi, Setsuro. Molecular biology of muscular contraction. New York, American Elsevier, 1965.

2477. Green, D. E., and R. F. Goldberger.

Molecular insights into the living process.
New York, Academic Press, 1967.

2478. International conference on biological membranes. Frascati, Italy. June, 1967. Membrane models and the formation of biological membranes. Ed. by Liana Bolis and B. A. Pethica. New York, Wiley, 1968.

2479. King, D. W. Ultrastructural aspects of disease. New York, Harper & Row, 1966.

2480. Reiner, J. M. The organism as an adaptive control system. Englewood Cliffs, N.J., Prentice-Hall, 1968.

2481. Rhinesmith, H. S. Macromolecules of living systems: structure and chemistry. New York, Reinhold, 1968.

2482. Rosenberg, Eugene. Cell and molecular biology: an appreciation. New York, Holt, Rinehart & Winston, 1971.

2483. Rothfield, L. I., ed. Structure and function of biological membranes. New York, Academic Press, 1971.

2484. Spencer, Frank. Introduction to human and molecular biology. London, Butterworth, 1970.

2485. Szent-Gyorgyi, Albert. The living state, with observations on cancer. New York, Academic Press, 1972.

2486. Watson, J. D. Molecular biology of the gene. 2d ed. New York, Benjamin, 1970.

BIOENGINEERING

2487. Brown, J. H. U., J. E. Jacobs, and Lawrence Stark. Biomedical engineering. Philadelphia, Davis, 1971.

2488. Caceres, C. A. Biomedical telemetry. New York, Academic Press, 1965.

2489. Chemical engineering in medicine and biology. Ed. by D. Hershey. New York, Plenum Press, 1967.

2490. Clynes, Manfred. Biomedical engineering systems. New York, McGraw-Hill, 1970.

2491. Frankel, V. H., and A. H. Burstein. Orthopaedic biomechanics: the application of engineering to the musculoskeletal system. Philadelphia, Lea & Febiger, 1970.

2492. Frost, H. M. Introduction to biomechanics. Springfield, Ill., Thomas, 1967.

2493. Fêng, Yuan-chêng, ed. Biomechanics. New York, American Society of Mechanical Engineers, 1966.

2494. Future goals of engineering in biology and medicine. Ed. by J. F. Dickson III. New York, Academic Press, 1969.

2495. Hill, D. W. Principles of electronics in medical research. New York, Appleton-Century-Crofts, 1965.

2496. International seminar on biomechanics. 2d. Eindhoven, Netherlands, 1969. Biomechanics, II. Ed. by J. Vredenbregt and J. Wartenweiler. Baltimore, University Park Press, 1971.

2497. Mackay, R. S. Bio-medical telemetry: sensing and transmitting biological information from animals and man. 2d ed. New York, Wiley, 1970.

2498. Rushmer, R. F. Medical engineering: projections for health care delivery. New York, Academic Press, 1972.

2499. Rutstein, D. D., and Murray Eden. Engineering and living systems: interfaces and opportunities. Cambridge, Mass., M.I.T. Press, 1970.

2500. Schwan, H. P., ed. Biological engineering. New York, McGraw-Hill, 1969.

2501. Seminar-workshop in biomaterials. Battelle Seattle Research Center, 1969. Biomaterials: bioengineering applied to materials for hard and soft tissue replacement. Ed. by A. L. Bement, Jr. Seattle, Univ. of Washington Press, 1971.

2502. Simpson, D. C. Modern trends in biomechanics. 1st series. New York, Appleton-Century-Crofts, 1970.

2503. Stark, Lawrence. Neurological control systems-studies in bioengineering. New York, Plenum Press, 1968.

2504. Symposium on biomechanics and related bio-engineering topics. Glasgow, 1964. Biomechanics and related bioengineering topics: proceedings. Ed. by R. M. Kenedi. New York, Pergamon Press, 1965.

2505. Symposium on the foundations and objectives of biomechanics. La Jolla, Calif.,

1970. Biomechanics: its foundations and objectives. Ed. by Y. C. Fung, N. Perrone, and M. Anliker. Englewood Cliffs, N.J., Prentice-Hall, 1972.

2506. Tromp, S. W. Medical biometerology: weather, climate, and the living organism. New York, American Elsevier, 1963.

2507. Whitfield, I. C. An introduction to electronics for physiological workers. 2d ed. London, Macmillan, 1960.

2508. Wiener, Norbert. Cybernetics: or control and communication in the animal and the machine. 2d ed. New York, M.I.T. Press, 1961.

2509. Wolff, H. S. Biomedical engineering. New York, McGraw-Hill, 1970.

2510. Workshop on the status of research and training in biomaterials. University of Illinois, 1968. Biomaterials. Ed. by Lawrence Stark. New York, Plenum Press, 1969.

2511. Yamada, Hiroshi. Strength of biological materials. Baltimore, Williams & Wilkins, 1970.

2512. Yanof, H. M. Biomedical electronics. 2d ed. Philadelphia, Davis, 1971.

2513. Zucker, M. H. Electronic circuits for the behavioral and biomedical sciences: a reference book of useful solid state circuits. San Francisco, Freeman, 1969.

BIOMATHEMATICS

2514. Armitage, P. Statistical methods in medical research. New York, Wiley, 1971.

2515. Bailey, N. T. J. The mathematical approach to biology and medicine. New York, Wiley, 1967.

2516. Barnett, R. N. Clinical laboratory statistics. Waltham, Mass., Little, Brown, 1971.

2517. Bishop, O. N. Statistics for biology: a practical guide for the experimental biologist. London, Longmans, 1966.

2518. Blackith, R. E., and R. A. Reyment. Multivariate morphometrics. New York, Academic Press, 1971.

2519. Bliss, C. I. Statistics in biology. v. 1- New York, McGraw-Hill, 1967-

2520. Bourke, G. J., and James McGilvray. Interpretation and uses of medical statistics. Philadelphia, Davis, 1969.

2521. Burdette, W. J., and E. A. Gehan. Planning and analysis of clinical studies. Springfield, Ill., Thomas, 1970.

2522. Crowe, Alan. Mathematics for biologists. New York, Academic Press, 1969.

2523. Fisher, R. A., and Frank Yates. Statistical tables for biological, agricultural, and medical research. 6th ed. rev. New York, Hafner, 1963.

2524. Griffith, J. S. Mathematical neurobiology: an introduction to the mathematics of the nervous system. New York, Academic Press, 1971.

2525. Hill, A. B. Principles of medical statistics. 9th ed. Fair Lawn, N.J., Oxford Univ. Press, 1971.

2526. ———. Statistical methods in clinical and preventive medicine. Fair Lawn, N.J., Oxford Univ. Press, 1962.

2527. Huntsberger, D. V., and P. E. Leaverton. Statistical inference in the biomedical sciences. Boston, Allyn & Bacon, 1970.

2528. Jackson, H. L. Mathematics of radiology and nuclear medicine. St. Louis, Green, 1971.

2529. Ledley, R. S. Use of computers in biology and medicine. New York, McGraw-Hill, 1965.

2530. Lewis, A. E. Biostatistics. New York, Reinhold, 1966.

2531. Mason, E. E., and W. G. Bulgren. Computer applications in medicine. Springfield, Ill., Thomas, 1964.

2532. Oldham, P. D. Measurement in medicine: the interpretation of numerical data. Philadelphia, Lippincott, 1968.

2533. Riggs, D. S. The mathematical approach to physiological problems: a critical primer. Baltimore, Williams & Wilkins, 1963.

2534. Saunders, Leonard, and Robert Fleming. Mathematics and statistics for use in the biological and pharmaceutical sciences. 2d ed., rev. and enl. to incorporate an introduction to computer techniques. London, Pharmaceutical Press, 1971.

2535. Simon, William. Mathematical techniques for physiology and medicine. New York, Academic Press, 1972.

2536. Smith, C. A. B. Biomathematics: the principles of mathematics for students of biological and general science. v. 1- 4th ed. New York, Hafner, 1966-

2537. Snedecor, G. W., and W. G. Cochran. Statistical methods. 6th ed. Ames, Iowa State Univ. Press, 1967.

2538. Sokal, R. R. Biometry: the principles and practice of statistics in biological research. San Francisco, Freeman, 1969.

2539. Stacy, R. W., and B. D. Waxman. Computers in biomedical research. v. 1- New York, Academic Press, 1965-

2540. Stibitz, G. R. Mathematics in medicine and the life sciences. Chicago, Year Book Medical Publishers, 1966.

2541. Vann, Edwin. Fundamentals of biostatistics. Lexington, Mass., Heath, 1972.

12

PHARMACOLOGY, TOXICOLOGY, AND THERAPEUTICS

PHARMACOLOGY

2542. Alexander, Frank. An introduction to veterinary pharmacology. 2d ed. Baltimore, Williams & Wilkins, 1969.

2543. American drug index. Philadelphia, Lippincott, 1972.

2544. American Pharmaceutical Association. The national formulary. 13th ed. Washington, D.C., 1970.

2545. Baslow, M. H. Marine pharmacology. Baltimore, Williams & Wilkins, 1969.

2546. Bertelli, A., and A. P. Monaco, eds. Pharmacological treatment in organ and tissue transplantation. Baltimore, Williams & Wilkins, 1970.

2547. British veterinary codex. 2d ed. London, Pharmaceutical Press, 1965. Suppl., 1970.

2548. Clinical pharmacology: basic principles in therapeutics. Ed. by K. L. Melmon and H. F. Morrelli. New York, Macmillan, 1972.

2549. Cutting, W. C. Handbook of pharmacology: the actions and uses of drugs. 4th ed. New York, Appleton-Century-Crofts, 1969.

2550. Danielli, J. F. Cell physiology and pharmacology. New York, Hafner, 1968.

2551. D'Arcy, P. F., and J. P. Griffin. Iatrogenic diseases. New York, Oxford Univ. Press, 1972.

2552. Dawson, Mary. Cellular pharmacology: the effects of drugs on living vertebrate cells in vitro. Springfield, Ill., Thomas, 1972.

2553. Daykin, P. V. Veterinary applied pharmacology and therapeutics. 2d ed. Ed. by G. C. Brander and D. M. Pugh. Baltimore, Williams & Wilkins, 1971.

2554. Drill's pharmacology in medicine. 4th ed. Ed. by J. R. DiPalma. New York, McGraw-Hill, 1971.

2555. Drug-induced diseases. Ed. by L. Meyler. v. 1- New York, Excerpta Medica, 1965-

2556. Drugs in current use and new drugs. Ed. by Walter Modell. New York, Springer-Verlag, 1972.

2557. Drugs of choice, 1972-1973. Ed.

by Walter Modell. St. Louis, Mosby, 1972.

2558. Gelatt, K. N. Veterinary ophthalmic pharmacology and therapeutics. Bonner Springs, Kan., VM Publ., 1970.

2559. Goldstein, Avram, et al. Principles of drug action: the basis of pharmacology. New York, Hoeber Medical Division, Harper & Row, 1968.

2560. Goodman, L. S., and Alfred Gilman, eds. The pharmacological basis of therapeutics: a textbook of pharmacology, toxicology, and therapeutics for physicians and medical students. 4th ed. New York, Macmillan, 1970.

2561. Goth, Andres. Medical pharmacology: principles and concepts. 6th ed. St. Louis, Mosby, 1972.

2562. Grollman, Arthur, and Evelyn F. Grollman. Pharmacology and therapeutics: a textbook for students and practitioners of medicine and its allied professions. 7th ed. Philadelphia, Lea & Febiger, 1970.

2563. International encyclopedia of pharmacology and therapeutics. Sponsored by the International Union of Pharmacology (IUPHAR). v. 1- Oxford, Pergamon Press, 1966-

2564. Jones, L. M. Veterinary pharmacology and therapeutics. 3d ed. Ames, Iowa State Univ. Press, 1965.

2565. Koppányi, Theodore, and A. G. Karczmar. Experimental pharmacodynamics. 3d ed. Minneapolis, Burgess, 1963.

2566. Krantz, J. S., and C. J. Carr. The pharmacologic principles of medical practice. 8th ed. By D. M. Aviado, M.D. Baltimore, Williams & Wilkins, 1972.

2567. LaDu, B. N., H. G. Mandel, and E. L. Way, eds. Fundamentals of drug metabolism and drug disposition. Baltimore, Williams & Wilkins, 1971.

2568. Martin, E. W., et al. Hazards of medication: a manual on drug interactions, incompatibilities, contraindications, and adverse effects. Philadelphia, Lippincott, 1972.

2569. The Merck index: an encyclopedia of chemicals and drugs. 8th ed. Rahway, N.J., Merck & Co., 1968.

2570. Miya, T. S., et al. Laboratory guide in pharmacology. Minneapolis, Burgess, 1968.

2571. Osol, Arthur, ed. The United States dispensatory and physicians' pharmacology. 26th ed. Philadelphia, Lippincott, 1967.

2572. The pharmacopeia of the United States of America. 18th revision. Bethesda, Md., U.S. Pharmacopeial Convention, 1970.

2573. Root, W. S., ed. Physiological pharmacology. A comprehensive treatise. v. 1- New York, Academic Press, 1963-
 v. 1: The nervous system, Part A—Central nervous system drugs. v. 2: The nervous system, Part B—Central nervous system drugs. v. 3: The nervous system, Part C—Autonomic nervous system drugs. v. 4: The nervous system, Part D—Autonomic nervous system drugs.

2574. Side effects of drugs. Ed. by L. Meyler. v. 1- New York, Excerpta Medica, 1963-

2575. Slack, R. Medical and veterinary chemicals. 2 v. New York, Pergamon Press, 1968.
 v. 1: Part 1—The pharmaceutical industry; organization and methods; Part 2—Reviews of selected types of drugs. v. 2: Part 3—A tabulated classification and chronology of drug categories.

2576. Spector, W. G. The pharmacology of inflammation. New York, Grune & Stratton, 1968.

2577. Symposium on the interaction of drugs and subcellular components in animal cells. Middlesex Hospital Medical School, 1967. Proceedings. Ed. by P. N. Campbell. Boston, Little, Brown, 1968.

2577a. Trease, G. E., and W. C. Evans. Pharmacognosy. 10th ed. Baltimore, Williams & Wilkins, 1972.

2578. Vida, J. A. Androgens and anabolic agents: chemistry and pharmacology. New York, Academic Press, 1969.

2579. Ward, C. O. Experimental methods in pharmacology. Jamaica, N.Y., St. John's Univ. Press, 1969.

TOXICOLOGY

2580. British Crop Protection Council. Pesticide manual: basic information on the chemicals used as active components of pesticides. Ed. by Hubert Martin. 2d ed. Worcester, England, 1971.

2581. Brown, R. L. Pesticides in clinical practice: identification, pharmacology, and therapeutics. Springfield, Ill., Thomas, 1966.

2582. Burchfield, H. P., and D. E. Johnson. Guide to the analysis of pesticide residues. v. 1- Washington, D.C., Govt. Printing Office, 1965- (Loose-leaf.)

2583. Chemical mutagens: principles and methods for their detection. 2 v. Ed. by Alexander Hollaender. New York, Plenum Press, 1971.

2584. Deichmann, W., and H. Gerarde. Toxicology of drugs and chemicals. 4th ed. New York, Academic Press, 1969.

2585. Essays in toxicology. v. 1- New York, Academic Press, 1969-

2586. European Society for the Study of Drug Toxicity. Proceedings. Amsterdam, Excerpta Medica Foundation, 1963- (For specific titles see entry 0749.)

2587. Garner, R. J. Veterinary toxicology. 3d ed. Revised by E. G. C. Clarke and Myra L. Clarke. London, Bailliere, Tindall & Cassell, 1967.

2588. Gleason, M. N. Clinical toxicology of commercial products. 3d ed. Baltimore, Williams & Wilkins, 1969.

2589. Goldblatt, L. A. Aflatoxin: scientific background, control, and implications. New York, Academic Press, 1969. (Food science and technology: a series of monographs. no. 7.)

2590. Gray, C. H. Laboratory handbook of toxic agents. 2d ed. Englewood, N.J., Franklin, 1968.

2591. International symposium on mycotoxins in foodstuffs. Cambridge, Mass., 1964. Mycotoxins in foodstuffs. Ed. by G. N. Wogan. Cambridge, Mass., M.I.T. Press, 1965.

2592. Lillie, R. J. Air pollutants affecting the performance of domestic animals: a literature review. Slightly revised. Washington, D.C., U.S. Agricultural Research Service, 1972.

2593. Polson, C. J. Clinical toxicology. 2d ed. Philadelphia, Lippincott, 1969.

2594. Radeleff, R. D. Veterinary toxicology. 2d ed. Philadelphia, Lea & Febiger, 1970.

2595. Sax, N. I. Dangerous properties of industrial materials. 3d ed. New York, Reinhold, 1968.

2596. Simpson, L. L., ed. Neuropoisons: their pathophysiological actions. v. 1- New York, Plenum Press, 1971-

2597. Society of Chemical Industry. Veterinary pesticides. London, 1969. (Its Monograph no. 33.)

2598. Spector, W. S. Handbook of toxicology. 5 v. Philadelphia, Saunders, 1956-1959.

2599. Sunshine, Irving, ed. Manual of analytical toxicology. Cleveland, Chemical Rubber Co., 1971.

2600. Thienes, C. H. Clinical toxicology. 5th ed. Philadelphia, Lea & Febiger, 1972.

2601. Tucker, R. K., and D. G. Crabtree. Handbook of toxicity of pesticides to wildlife. Washington, D.C., U.S. Govt. Printing Office, 1970. (U.S. Bureau of Sport Fisheries and Wildlife. Resource publication no. 84.)

2602. U.S. National Referral Center for Science and Technology. A directory of information resources in the United States: general toxicology. Washington, D.C., 1969.

2603. Zweig, Gunter. Analytical methods for pesticides, plant growth, regulators, and food additives. v. 1- New York, Academic Press, 1963-
 v. 1: Principles, methods and general applications. v. 2: Insecticides. v. 3: Fungicides, nematocides and soil fumigants, rodenticides, and food and feed additives. v. 4: Herbicides. v. 5: Additional principles and methods of analysis.

POISONOUS PLANTS AND ANIMALS

2604. Bucherl, Wolfgang. Venomous animals and their venoms. v. 1- New York, Academic Press, 1968- v. 1-2: Venomous vertebrates.

2605. Dodge, N. N. Poisonous dwellers of the desert. 12th ed. Globe, Ariz., Southwest Parks and Monuments, 1970.

2606. Evers, R. A., and R. P. Link. Poisonous plants of the Midwest and their effect on livestock. Urbana, Univ. of Illinois Press, 1972.

2607. Gay, C. W. Poisonous range plants. Las Cruces, New Mexico State Univ., 1967. (Cooperative Extension Service circular 391.)

2608. Gillis, W. T. Symptoms of plant poisoning. East Lansing, Michigan State Univ., 1960.

2609. Halstead, B. W. Dangerous marine animals. Cambridge, Md., Cornell Maritime Press, 1959.

2610. ———. Poisonous and venomous marine animals of the world. 3 v. Washington, D.C., U.S. Govt. Printing Office, 1965-1970.
  v. 1: Invertebrates. v. 2-3: Vertebrates.

2611. Hardin, J. W. Human poisoning from native and cultivated plants. Durham, N.C., Duke Univ. Press, 1969.

2612. Harmon, R. W., and C. B. Pollard. Bibliography of animal venoms. Gainesville, Fla., Univ. of Florida Press, 1948.

2613. International symposium on animal and plant toxins. Tel-Aviv, Israel. Toxins of animal and plant origin. Proceedings. 3 v. Ed. by A. De Vries and E. Kochva. New York, Gordon & Breach, 1972.

2614. International symposium on animal toxins. 1st. Atlantic City, 1966. Animal toxins. Ed. by F. E. Russell and P. R. Saunders. New York, Pergamon Press, 1967.

2615. Kingsbury, John. Deadly harvest: a guide to common poisonous plants. New York, Holt, Rinehart & Winston, 1965.

2616. Lampe, K. F., and Rune Fagerström. Plant toxicity and dermatitis. Baltimore, Williams & Wilkins, 1968.

2617. Lodge, R. W. Stock-poisoning plants of western Canada. Ottawa, Canada Dept. of Agriculture, 1968. (Publication no. 1361.)

2618. Minton, S. A. Snake venoms and envenomation. New York, Dekker, 1971.

2619. North, Pamela M. Poisonous plants and fungi in colour. London, Blandford, 1967.

2620. Panel on the radiation sensitivity of toxins and animal poisons. Bangkok, 1969. Vienna, International Atomic Energy Agency, 1970. (Panel proceedings series. STI/PUB/243.)

2621. Rickett, H. W. The new field book of American wild flowers. New York, Putnam, 1963.

2622. Rosenfeld, Gastão, and Eva M. A. Kelen. Bibliography of animal venoms, envenomations, and treatments, period 1500-1968. São Paulo, Instituto Butantan, 1969.

2623. Russell, F. E., and R. S. Scharffenberg. Bibliography of snake venoms and venomous snakes. West Covina, Calif., Bibliographic Associates, 1964.

2624. Schmutz, E. M. Livestock-poisoning plants of Arizona. Tucson, Univ. of Arizona Press, 1968.

2625. Toxic constituents of plant foodstuffs. Ed. by I. E. Liener. New York, Academic Press, 1969.

2626. Twenty-two plants poisonous to livestock in the Western states. Washington, D.C., U.S. Dept. of Agriculture, 1968. (Agriculture information bulletin no. 327.)

2627. U.S. Bureau of Medicine and Surgery. Poisonous snakes of the world: a manual for use by U.S. amphibious forces. 2d ed. Washington, D.C., Govt. Printing Office, 1965.

THERAPEUTICS

2628. The actions of corticosteroids and their application in veterinary medicine: a symposium held at Royal Society of Medicine, London, 1971. Greenford, England, Glaxo Laboratories, 1971.

2629. American druggist blue book. v. 1- New York, American Druggist, 1962-

2630. Barnes, C. D., and L. G. Eltherington. Drug dosage in laboratory animals: a handbook. Berkeley, Univ. of Calif. Press, 1964.

2631. Brion, A. Vade-mecum du vétérinaire: formulaire vétérinaire de pharmacologie, de therapeutique et d'hygiene. Paris, Vigot Freres, 1966.

2632. The chemist's veterinary handbook: a survey of modern methods in veterinary medicine, including diseases and treatment. 12th ed. London, Offices of the Chemist and Druggist, 1964.

2633. Clinical pharmacology: basic principles in therapeutics. Ed. by K. L. Melmon and H. F. Morrelli. New York, Macmillan, 1972.

2634. Current veterinary therapy: small animal practice. v. 1- Ed. by R. W. Kirk. Philadelphia, 1964/1965- (Approximately biennial.)

2635. D'Arcy, P. F., and J. P. Griffin. Iatrogenic diseases. New York, Oxford, Univ. Press, 1972.

2636. Daykin, P. V. Veterinary applied pharmacology and therapeutics. 2d ed. Ed. by G. C. Brander and D. M. Pugh. Baltimore, Williams & Wilkins, 1971.

2637. Drug-induced diseases. Ed. by L. Meyler. v. 1- New York, Excerpta Medica, 1965-

2638. Drug topics red book. v. 1- New York, Topics, 1962-

2639. Drugs in current use and new drugs. New York, Springer, 1970.

2640. Drugs of choice, 1972-1973. Ed. by Walter Modell. St. Louis, Mosby, 1972.

2641. Establishments holding U.S. veterinary licenses to produce biological products. 2 v. Hyattsville, Md., Veterinary Biologics Division, Agricultural Research Service, 1967-1968. (Continued by entry 1073.)

2642. Garrod, L. P. Antibiotic and chemotherapy. 3d ed. London, Churchill Livingstone, 1971.

2643. Gelatt, K. N. Veterinary ophthalmic pharmacology and therapeutics. Bonner Springs, Kan., VM Publ., 1970.

2644. Griffenhagen, G. B., and Linda L. Hawkins. Handbook of non-prescription drugs. 1971 ed. Washington, D.C., American Pharmaceutical Association, 1971.

2645. Haugen, Jostein. Anticoagulant treatment of rabbits. 2 v. Bergen, Norwegian Universities Press, 1967.

2646. Hedgecock, L. W. Antimicrobial agents. Philadelphia, Lea & Febiger, 1968. (Medical technology series, v. 3.)

2647. International congress of chemotherapy. 5th. Vienna, 1967. 4 v. in 6. Vienna, Verlag der Wiener Medizinischen Akademie, 1967. (*Chemotherapy in veterinary medicine*, pp. 653-. For specific titles see entry 0759.)

2648. International symposium on selenium in biomedicine. 1st. Oregon State University, 1966. Proceedings. Ed. by O. H. Muth. Westport, Conn., Avi Publishing, 1967.

2649. Jones, L. M. Veterinary pharmacology and therapeutics. 3d ed. Ames, Iowa State Univ. Press, 1965.

2650. Martin, E. W., et al. Hazards of medication: a manual on drug interactions, incompatibilities, contraindications, and adverse effects. Philadelphia, Lippincott, 1972.

2651. Merck and Company, Inc. Merck manual of diagnosis and therapy. 12th ed. Rahway, N.J., Merck, Sharp & Dohme Research Laboratories, 1972.

2652. The Merck index: an encyclopedia of chemicals and drugs. 8th ed. Rahway, N.J., Merck & Co., 1968.

2653. Merck veterinary manual: a handbook of diagnosis and therapy for the veterinarian. Ed. by O. H. Siegmund et al. Rahway, N.J., Merck & Co., 1967.

2654. Mihich, Enrico, ed. Immunity, cancer and chemotherapy. New York, Academic Press, 1967.

2655. Modern drug encyclopedia and therapeutic index. 11th ed. New York, Donnelley, 1970.

2656. National drug code directory. 3d ed. June, 1971. Washington, D.C., U.S. Public Health Service, 1971.

2657. Parish, H. J. Antisera, toxoids, vaccines and tuberculins in prophylaxis and treatment. 6th ed. Baltimore, Williams & Wilkins, 1962.

2658. Physicians' desk reference to pharmaceutical specialties and biologicals. 24th ed. Oradell, N.J., Medical Economics, 1970.

2659. Rossoff, Irving. Veterinary drugs in current use. 2d ed. New York, Springer, in press 1972.

2660. Seiden, R. Veterinary drugs in current use. New York, Springer, 1960.

2661. Side effects of drugs. Ed. by L. Meyler. v. 1- New York, Excerpta Medica, 1963-

2662. Thomas, Picton. Guide to steroid therapy. Philadelphia, Lippincott, 1968.

2663. U.S. Dept. of Agriculture. Veterinary biologics checklist. Washington, D.C., 1970-71. (Program aid no. 841.)

2664. Veterinarians' blue book. 15th ed. New York, Donnelley, 1967.

2665. Veterinarian's product and therapeutic reference. 1972. A guide to pharmaceuticals and biologicals with a therapeutic index. Ed. by R. M. Grey. Clifton, N.J., Therapeutic Communications, 1972.

2666. Veterinary biological products: licensees, permittees. Hyattsville, Md., Veterinary Biologics Division, Agricultural Research Service, 1970- (Continues entry 1068.)

2667. Williams, F. L. Veterinary formulary. (Loose-leaf.) Pullman, Wash., Lestor, 1958.

2668. World Health Organization. Cumulative list of proposed international nonproprietary names for pharmaceutical substances. Geneva, 1971.

DRUG TESTING

2669. Burger, Alfred, ed. Selected pharmacological testing methods. New York, Dekker, 1968. (Medicinal research: a series of monographs. v. 3.)

2670. Domer, F. R. Animal experiments in pharmacological analysis. Springfield, Ill., Thomas, 1971.

2671. Nodine, J. H., P. E. Siegler, and J. H. Moyer, eds. Animal and clinical pharmacologic techniques in drug evaluation. v. 1- Chicago, Year Book Medical Publishers, 1964-

2672. Symposium on the use of subhuman primates in drug evaluation. San Antonio, 1967. Use of non-human primates in drug evaluation. Ed. by Harold Vagtborg. Published for the Southwest Foundation for Research and Education, San Antonio. Austin, Univ. of Texas Press, 1968.

2673. Universities Federation for Animal Welfare. Symposium. Birmingham, England, 1969. The use of animals in toxicological studies. Potters Bar, England, 1969.

2674. Ward, C. O. Experimental methods in pharmacology. Jamaica, N.Y., St. John's Univ. Press, 1969.

PSYCHOPHARMACOLOGY

2675. Advances in biochemical psychopharmacology. v. 1- New York, Raven Press, 1969-

2676. Ban, T. A. Psychopharmacology. Baltimore, Williams & Wilkins, 1969.

2677. Brucke, F. Th. von, and O. Hornykiewicz. The pharmacology of psychotherapeutic drugs. New York, Springer, 1969. (Formerly *Pharmakologie der Pscyhopharmaka*. Translated and rev. by E. B. Sigg.)

2678. Burger, Alfred, ed. Drugs affecting the central nervous system. New York, Dekker, 1968.

2679. Clark, W. G., et al., eds. Principles of psychopharmacology: a textbook for physicians, medical students and behavioral scientists. New York, Academic Press, 1970.

2680. Cooper, J. R., et al. The biochemical

basis of neuropharmacology. New York, Oxford Univ. Press, 1970.

2681. Crismon, Cathrine. Repeated administration to animals of phenothiazines, barbiturates, central stimulants, and muscle relaxants: a selected annotated bibliography. Menlo Park, Calif., Stanford Research Institute, 1965.

2682. DiMascio, Alberto, and R. I. Shader, eds. Clinical handbook of psychopharmacology. New York, Science House, 1970.

2683. Furchtgott, Ernest, ed. Pharmacological and biophysical agents and behavior. New York, Academic Press, 1971.

2684. Gordon, Maxwell, ed. Psychopharmacological agents. 2 v. New York, Academic Press, 1964-1967.

2685. Joyce, C. R. B., ed. Psychopharmacology, dimensions and perspectives. Philadelphia, Lippincott, 1968.

2686. Lear, Edwin. Chemistry and applied pharmacology of tranquilizers: a primer for students and practitioners. Springfield, Ill., Thomas, 1966.

2687. Rech, R. H., and K. E. Moore, eds. An introduction to psychopharmacology. Amsterdam, North-Holland Publishing, 1971.

2688. Shader, R. I., and Alberto DiMascio. Psychotropic drug side effects: clinical and theoretical perspectives. Baltimore, Williams & Wilkins, 1970.

2689. Steinberg, Hannah. Animal behavior and drug action. A Ciba Foundation symposium, 1963. Boston, Little, Brown, 1964.

2690. Symposium on anticholinergic drugs and brain functions in animals and man. 6th. Washington, D.C., 1966. Anticholinergic drugs and brain functions in animals and man. Ed. by P. B. Bradley and M. Fink. New York, Elsevier, 1968.

13

MICROBIOLOGY

GENERAL

2691. Bibliotheca microbiologica. v. 1- Basel, Karger, 1960- (For specific titles see entry 0735.)

2692. Blair, J. E., E. H. Lennette, and J. P. Truant, eds. Manual of clinical microbiology. Bethesda, Md., American Society for Microbiology, 1970.

2693. Buchner, Paul. Endosymbiosis of animals with plant microorganisms. New York, Wiley, 1965.

2694. Carpenter, P. L. Microbiology. 3d ed. Philadelphia, Saunders, 1972.

2695. Carter, G. R. Outline of veterinary bacteriology and mycology. Columbia, Mo., Lucas Brothers, 1970.

2696. Collins, C. H. Microbiological methods. 2d ed. New York, Plenum Press, 1967.

2697. Davis, B. D., et al. Microbiology. New York, Harper & Row, 1967.

2698. Gibbs, B. M., F. A. Skinner, and D. A. Shapton. Identification methods for microbiologists. 2 v. New York, Academic Press, 1966-1968.

2699. Gillies, R. R., and T. C. Dodds. Bacteriology illustrated. 2d ed. Baltimore, Williams & Wilkins, 1968.

2700. Henry, S. M. Symbiosis. 2 v. New York, Academic Press, 1966-1967.
   v. 1: Associations of microorganisms, plants, and marine organisms. v. 2: Associations of invertebrates, birds, ruminants, and other biota.

2701. Index Bergeyana: an annotated alphabetic listing of names of the taxa of the bacteria. Ed. by R. E. Buchanan. Baltimore, Williams & Wilkins, 1966.

2702. International symposium on continuous cultivation of microorganisms. Proceedings. 2d. Prague, 1962. Ed. by I. Malek et al. New York, Academic Press, 1964.

2703. Kubitschek, H. E. Introduction to research with continuous cultures. Englewood Cliffs, N.J., Prentice-Hall, 1970.

2704. McAllister, H. A. Procedures for the identification of microorganisms from the higher animals. Columbia, Mo., Lucas Brothers, 1970.

2705. Mathews, Christopher. Bacteriophage biochemistry. New York, Van Nostrand Reinhold, 1971.

2706. Merchant, I. A., and R. A. Packer. Veterinary bacteriology and virology. 7th ed. Ames, Iowa State Univ. Press, 1967.

2707. Meynell, G. G., and Elinor Meynell. Theory and practice in experimental bacteriology. 2d ed. New York, Cambridge Univ. Press, 1970.

2708. Prévot, A. R. Manual for the classification and determination of the anaerobic bacteria. Philadelphia, Lea & Febiger, 1966. (1st English edition.)

2709. Salle, A. J. Fundamental principles of bacteriology. 6th ed. New York, McGraw-Hill, 1967.

2710. Skerman, V. B. D. A guide to the identification of the genera of bacteria, with methods and digests of generic characteristics. 2d ed. Baltimore, Williams & Wilkins, 1967.

2711. Slack, J. M., V. F. Gerencser, and S. J. Deal. Experimental microbiology for the health sciences. rev. 3d ed. Minneapolis, Burgess, 1971. (Original title: *Experimental pathogenic microbiology*.)

2712. Society of American Bacteriologists. Bergey's manual of determinative bacteriology. 7th ed. By R. S. Breed, E. G. D. Murray, and N. R. Smith. Baltimore, Williams & Wilkins, 1957.

2713. Stokes, E. Joan. Clinical bacteriology. 3d ed. Baltimore, Williams & Wilkins, 1968.

2714. Topley, W. W. C., G. S. Wilson, and A. A. Miles. Topley and Wilson's principles of bacteriology and immunity. 5th ed. 2 v. Baltimore, Williams & Wilkins, 1964.

2715. U.S. Army Biological Warfare Laboratories. Ft. Detrick, Frederick, Md., Microbiological safety bibliography. By G. B. Phillips. Springfield, Va., CFSTI, 1965. (Its Misc. publ. no. 6.)

2716. Willis, R. A., and A. T. Willis. Principles of pathology and bacteriology. 3d ed. New York, Appleton-Century-Crofts, 1972.

2717. Zinsser, Hans. Zinsser's microbiology. 15th ed. Ed. by W. K. Joklik and D. T. Smith. New York, Appleton-Century-Crofts, 1972.

MEDICAL BACTERIOLOGY

2718. Bailey, W. R., and E. G. Scott. Diagnostic microbiology: a textbook for the isolation and identification of pathogenic microorganisms. 3d ed. St. Louis, Mosby, 1970

2719. Burrows, William. Textbook of microbiology. 19th ed. 2 v. Philadelphia, Saunders, 1968.
    v. 1: The biology of microorganisms.
    v. 2: The pathogenic microorganisms.

2720. Carter, G. R. Diagnostic procedures in veterinary bacteriology and mycology. Springfield, Ill., Thomas, 1967.

2721. Ciba Foundation symposium. Pathogenic mycoplasmas. New York, American Elsevier, in press 1972.

2722. Cowan, S. T., and K. J. Steel. Manual for the identification of medical bacteria. New York, Cambridge Univ. Press, 1965.

2723. Cruickshank, Robert, J. P. Duguid, and R. H. A. Swain, eds. Medical microbiology: a guide to the laboratory diagnosis and control of infection. 11th ed. Baltimore, Williams & Wilkins, 1965.

2724. Eckmann, Leo, ed. Principles on tetanus. Bern, Hans Huber. (International conference on tetanus, Bern, 1966.)

2725. Edwards, P. R., and W. H. Ewing. Identification of enterobacteriaceae. 3d ed. Minneapolis, Burgess, 1972.

2726. Graber, C. D. Rapid diagnostic

methods in medical microbiology. Baltimore, Williams & Wilkins, 1970.

2727. Jawetz, Ernest, J. L. Melnick, and E. A. Adelberg. Review of medical microbiology. 10th ed. Los Altos, Calif., Lange Medical Publications, 1972.

2728. Kauffmann, Fritz. The bacteriology of enterobacteriaceae. Baltimore, Williams & Wilkins, 1966.

2729. Klieneberger-Nobel, Emmy. Pleuropneumonia-like organisms (PPLO): mycoplasmataceae. New York, Academic Press, 1962.

2730. Kubica, G. P., and W. E. Dye. Laboratory methods for clinical and public health: mycobacteriology. Atlanta, National Communicable Disease Center, 1967. (U.S.P.H.S. publ. no. 1547.)

2731. Manclark, C. R., M. J. Pickett, and H. B. Moore. Laboratory manual for medical bacteriology. 5th ed. New York, Appleton-Century-Crofts, 1972.

2732. Mycoplasmataceae: a bibliography and index, 1852-1970. Ed. by C. H. Domermuth and J. G. Rittenhouse. Blacksburg, Va., Virginia Polytechnic Institute and State Univ., 1971. (Research Division bulletin 61.)

2733. Neter, Erwin. Medical microbiology. 6th ed. Philadelphia, Davis, 1969.

2734. Norris, J. R., and D. W. Ribbons, eds. Methods in microbiology. 3 v. New York, Academic Press, 1969-1970.

2735. Osbaldiston, G. W., and E. C. Stowe. Laboratory procedures in clinical veterinary bacteriology. Baltimore, University Park Press, 1972.

2736. Pease, P. E. L-forms, episomes and autoimmune disease. Baltimore, Williams & Wilkins, 1965.

2737. Sharp, J. T., ed. The role of mycoplasmas and L forms of bacteria in disease. Springfield, Ill., Thomas, 1970.

2738. Slack, J. M., V. F. Gerencser, and S. J. Deal. Experimental pathogenic microbiology. 2d ed. Minneapolis, Burgess, 1964.

2739. Smith, Louis D. S., and Lillian V. Holdeman. The pathogenic anaerobic bacteria. Springfield, Ill., Thomas, 1968.

2740. Smith, P. F. The biology of mycoplasmas. New York, Academic Press, 1971.

2741. Society for General Microbiology. v. 1- Cambridge, The Univ. Press, 1949- (For specific titles see entry 0781.)

2742. Storz, Johannes. Chlamydia and chlamydia-induced diseases. Springfield, Ill., Thomas, 1971.

2743. Waterson, A. P. Recent advances in medical microbiology. Boston, Little, Brown, 1967.

2744. Willis, A. T. Anaerobic bacteriology in clinical medicine. 2d ed. Washington, D.C., Butterworth, 1964.

2745. ——. Clostridia of wound infection. New York, Appleton-Century-Crofts, 1969.

VIROLOGY

2746. Andrewes, C. H. The natural history of viruses. New York, Norton, 1967.

2747. ——. Viruses and cancer. London, Weidenfeld & Nicolson, 1970.

2748. ——, and H. G. Pereira. Viruses of vertebrates. 3d ed. Baltimore, Williams & Wilkins, 1972.

2749. Andrews, R. D. Viral interference and interferon. London, Heinemann Medical Books, 1970.

2750. Banatvala, J. E. Current problems in clinical virology. Edinburgh, Churchill Livingstone, 1971.

2751. Burnet, F. M. Principles of animal virology. 2d ed. New York, Academic Press, 1960.

2752. ——, and W. M. Stanley, eds. The viruses; biochemical, biological and biophysical properties. 3 v. New York, Academic Press, 1959.
    v. 1: General virology. v. 2: Plant and animal viruses. v. 3: Animal viruses.

2753. Busby, D. W. G., W. House, and J. R. MacDonald. Virological technique. Boston, Little, Brown, 1964.

2754. Cohen, Arnold. Textbook of medical virology. Oxford, Blackwell, 1970.

2755. Cohen, S. S. Virus-induced enzymes. New York, Columbia Univ. Press, 1968.

2756. Cunningham, C. H. A laboratory guide in virology. 6th ed. Minneapolis, Burgess, 1966.

2757. Dalton, A. J., and Francoise Haguenau. Tumors induced by viruses: ultrastructural studies. New York, Academic Press, 1962.

2758. Dubes, G. R. Methods for transfecting cells with nucleic acids of animal viruses: a review. Basel, Birkhauser, 1971. (*Experientia.* Suppl. 16.)

2759. Fenner, F. J. The biology of animal viruses. 2 v. New York, Academic Press, 1968.
   v. 1: Molecular and cellular biology.
   v. 2: The pathogenesis and ecology of viral infections.

2760. ——, and D. O. White. Medical virology. New York, Academic Press, 1970.

2761. Fraenkel-Conrat, H. L. The chemistry and biology of viruses. New York, Academic Press, 1969.

2762. ——. Design and function at the threshold of life: the viruses. New York, Academic Press, 1962.

2763. ——. Molecular basis of virology. New York, Reinhold, 1968.

2764. Fraser, Dean. Viruses and molecular biology. New York, Macmillan, 1967.

2765. Gajdusek, D. C., et al., eds. Slow, latent, and temperate virus infections. Washington, D.C., National Institute of Neurological Diseases and Blindness, 1965. (U.S. Public Health Service publ. no. 1378.)

2766. Grist, N. R., et al. Diagnostic methods in clinical virology. Philadelphia, Davis, 1966.

2767. Gross, Ludwik. Oncogenic viruses. 2d ed. New York, Pergamon Press, 1970.

2768. Habel, Karl, and N. P. Salzman, ed. Fundamental techniques in virology. New York, Academic Press, 1969.

2769. Harris, R. J. C. Techniques in experimental virology. New York, Academic Press, 1964.

2770. Hoskins, J. M. Virological procedures. New York, Appleton-Century-Crofts, 1967.

2771. Hsiung, G. D., and J. R. Henderson. Diagnostic virology. New Haven, Yale Univ. Press, 1964.

2772. International symposium on interferons. Siena, Italy, 1967. The Interferons. Ed. by Geo Rita. New York, Academic Press, 1968.

2773. International symposium on medical and applied virology. 1st- Proceedings. 1964- (For specific titles see entry 0763.)

2774. International symposium on resistance to infectious disease. Western College of Veterinary Medicine, Saskatoon, 1969. Ed. by R. H. Dunlop and H. W. Moon. Saskatoon, Univ. of Saskatchewan, 1970.

2775. International virology. v. 1- Proceedings of international congress for virology. Basel, Karger, 1969-
   v. 1: Helsinki, 1968; ed. by J. L. Melnick. v. 2: Budapest, 1971.

2776. Juhasz, S. E., and Gordon Plummer, eds. Host-virus relationships in mycobacterium, nocardia and actinomyces. Proceedings of a symposium. Springfield, Ill., Thomas, 1970.

2777. Lennette, E. H., and Nathalie J. Schmidt, eds. Diagnostic procedures for viral and rickettsial infections. 4th ed. Washington, D.C., American Public Health Association, 1969.

2778. Lepetit colloquia on biology and medicine. International Lepetit colloquium. 2d. Paris, 1970. The biology of oncogenic viruses. Ed. by L. G. Silvestri. New York, American Elsevier, 1971.

2779. Levy, H. B. The biochemistry of viruses. New York, Dekker, 1969.

2780. Luria, S. E., and J. E. Darnell. General virology. 2d ed. New York, Wiley, 1967.

2781. Maramorosch, Karl, ed. Viruses, vectors, and vegetation. New York, Wiley, 1969.

2782. ——, and Hilary Koprowsky, eds. Methods in virology. 4 v. New York, Academic Press, 1967-1968.

2783. ——, and Edouard Kurstak, eds. Comparative virology. New York, Academic Press, 1971.

2784. Mascoli, Carmine C., and R. G. Burrell. Experimental virology. 3d ed. Minneapolis, Burgess, 1970.

2785. Monographs in virology. v. 1- Basel, Karger, 1967- (For specific titles see entry 0771.)

2786. Perspectives in virology. Gustav Stern symposium. v. 1- Ed. by Morris Pollard. 1959- (For specific titles see entry 0774.)

2787. Rhodes, A. J., and C. E. van Rooyen. Textbook of virology for students and practitioners of medicine. 5th ed. Baltimore, Williams & Wilkins, 1968.

2788. Röhrer, Heinz. Handbuch der Virusinfektionen bei Tieren. 4 v. Jena, Gustav Fischer Verlag, 1967-1969.

2789. Rosenberg, Nancy, and L. Z. Cooper. Vaccines and viruses. New York, Grosset & Dunlap, 1971.

2790. Smith, K. M. The biology of viruses. New York, Oxford Univ. Press, 1965.

2791. Stent, G. S. Molecular biology of bacterial viruses. San Francisco, Freeman, 1963.

2792. Swain, R. H. A., and T. C. Dodds. Clinical virology. London, Livingstone, 1967.

2793. Symposium on cellular biology of myxovirus infections. London, 1964. Ciba Foundation symposium: cellular biology of myxovirus infections. Ed. by G. E. W. Wolstenholme and Julie Knight. Boston, Little, Brown, 1964.

2794. Symposium on interferon. New York, 1969. Interferon. Sponsored by the New York Heart Association. Boston, Little, Brown, 1970.

2795. Symposium on oncogenicity of virus vaccines. Zagreb, Yugoslavia, 1968. Proceedings. Zagreb, Immunoloske Zavod Rockefellervoa, 1968.

2796. Symposium on strategy of the viral genome. London, 1971. Strategy of the viral genome. A Ciba Foundation symposium. Ed. by G. E. W. Wolstenholme and Maeve O'Connor. Edinburgh, Livingstone, 1971.

2797. Symposium on the molecular biology of viruses. Univ. of Alberta, Canada, 1966. The molecular biology of viruses. Ed. by J. S. Colter and William Paranchych. New York, Academic Press, 1967.

2798. Taylor, R. M. Catalogue of arthropodborne viruses of the world. Washington, D.C., U.S. Public Health Service, 1967. (Publ. no. 1760.)

2799. Thompson, P. D. The virus realm. Philadelphia, Lippincott, 1968.

2800. U.S. Dept. of the Army. Laboratory procedures in virology. Washington, D.C., 1964. (Its Technical manual no. TM8-227-7.)

2801. U.S. Federal Water Quality Administration. Advanced Waste Treatment Research Laboratory. Virology Section. Viruses in waste, renovated, and other waters. Ed. by Gerald Berg and F. Dianne White. Cincinnati, 1970?

2802. U.S. National Cancer Institute. Cancer and virus: a guide and annotated bibliography to monographs, reviews, symposia, and survey articles with emphasis on human neoplasm. 1950-1963. Ed. by Elizabeth Koenig. Washington, D.C., 1966. (U.S. Public Health Service publ. no. 1424.)

2803. Virology monographs. v. 1- New York, Springer-Verlag, 1968- (For specific titles see entry 0795.)

2804. Waterson, A. P. Introduction to animal virology. 2d ed. New York, Cambridge Univ. Press, 1968.

2805. Wilner, B. I. A classification of the major groups of human and other animal viruses. 4th ed. Minneapolis, Burgess, 1969.

2806. World Health Organization. The work of WHO virus reference centers and the services they provide. Geneva, 1968.

2807. Zil'ber, L. A., and G. I. Abelev. Virology and immunology of cancer. New York, Pergamon Press, 1968.

MYCOLOGY

2808. Beneke, E. S., and A. L. Rogers. Medical mycology. 3d ed. Minneapolis, Burgess, 1971.

2809. Carter, G. R. Outline of veterinary bacteriology and mycology. Columbia, Mo., Lucas Brothers, 1970.

2810. Conant, N. F., et al. Manual of clinical mycology. 3d ed. Philadelphia, Saunders, 1971.

2811. Dvořák, Jaroslav, and Miloš Otčenášek. Mycological diagnosis of animal dermatophytoses. Prague, Academia, 1969. (Translated by Eva Kalinova.)

2812. Emmons, C. W., C. H. Binford, and J. P. Utz. Medical mycology. 2d ed. Philadelphia, Lea & Febiger, 1970.

2813. Haley, Leanor D. Diagnostic medical mycology. New York, Appleton-Century-Crofts, 1964.

2814. Hazen, Elizabeth L., and F. C. Reed. Laboratory identification of pathogenic fungi simplified. 3d ed. Springfield, Ill., Thomas, 1970.

2815. International symposium on mycotoxins in foodstuffs. Cambridge, Mass., 1964. Mycotoxins in foodstuffs. Ed. by G. N. Wogan. Cambridge, Mass., M.I.T. Press, 1965.

2816. Jungerman, P. F., and R. M. Schwartzman. Veterinary medical mycology. Philadelphia, Lea & Febiger, 1972.

2817. Medical Research Council. Medical Mycology Committee. Nomenclature of fungi pathogenic to man and animals. 3d ed. London, Her Majesty's Stationery Office, 1967. (Medical Research Council memorandum no. 23.)

2818. Moss, Emma S., and A. L. McQuown. Atlas of medical mycology. 3d ed. Baltimore, Williams & Wilkins, 1969.

2819. Rayner, R. W. A mycological colour chart. Kew, England, Commonwealth Mycological Institute and the British Mycological Society, 1970.

2820. U.S. Communicable Disease Center, Atlanta. Laboratory manual for medical mycology. By Libero Ajello. Washington, D.C., Govt. Printing Office, 1963. (U.S. Public Health Service publ. no. 994.)

2821. Waksman, S. A. The actinomycetes: a summary of current knowledge. New York, Ronald Press, 1967.

INFECTIOUS DISEASES

General

2822. Behbehani, A. M. Human viral, bedsonial and rickettsial diseases: a diagnostic handbook for physicians. Springfield, Ill., Thomas, 1972.

2823. ——. Laboratory diagnosis of viral, bedsonial and rickettsial diseases: handbook for laboratory workers. Springfield, Ill., Thomas, 1971.

2824. Burnet, F. M. Natural history of infectious disease. 3d ed. New York, Cambridge Univ. Press, 1962.

2825. Christie, A. B. Infectious diseases: epidemiology and clinical practice. Edinburgh, Livingstone, 1969.

2826. Cockburn, Aidan. Infectious diseases. Springfield, Ill., Thomas, 1967.

2827. Davis, J. W., et al., eds. Infectious and parasitic diseases of wild birds. Ames, Iowa State Univ. Press, 1971.

2828. Davis, J. W., D. O. Trainer, and L. H. Karstad, eds. Infectious diseases of wild mammals. Ames, Iowa State Univ. Press, 1970.

2829. Food and Agriculture Organization. Emerging diseases of animals. New York, Unipub, 1968. (FAO agricultural studies, no. 61.)

2830. Hagan, W. A., D. W. Bruner, and J. H. Gillespie. Hagan's infectious diseases of domestic animals: with special reference to etiology, diagnosis, and biologic therapy. 5th ed. Ithaca, N.Y., Cornell Univ. Press, 1966.

2831. Henning, M. W. Animal diseases in South Africa: being an account of the infectious diseases of domestic animals. 3d ed. Johannesburg, Central News Agency, 1956.

2832. Hudson, J. R. Contagious bovine pleuropneumonia. Rome, Food and Agricultural Organization of the United Nations, 1971. (FAO agricultural studies, no. 86.)

2833. International conference on equine infectious diseases. (For additional coverage see entry 0758.)

2834. Konferentsiia po prirodnoĭ ochagovosti bolezneĭ i voprosam parazitologii Kazakhstana i respubik Sredneĭ Azii. 4th Alma-Ata, Kazakhstan, 1959. Natural nidality of diseases and questions of parasitology. Ed. by N. D. Levine. Urbana, Univ. of Illinois Press, 1968.

2835. Merchant, I. A., and R. D. Berner. Outline of infectious diseases of domestic animals. 3d ed. Ames, Iowa State Univ. Press, 1964.

2836. Mudd, Stuart. Infectious agents and host reactions. Philadelphia, Saunders, 1970.

2837. Soltys, M. A. Bacteria and fungi pathogenic to man and animals. Baltimore, Williams & Wilkins, 1963.

2838. Top, F. H. Communicable and infectious diseases: diagnosis, prevention, treatment. 6th ed. St. Louis, Mosby, 1968.

2839. Weinman, David, and Miodrag Ristic. Infectious blood diseases of man and animals. 2 v. New York, Academic Press, 1968.

2840. Woodruff, A. W., and S. Bell. A synopsis of infectious and tropical diseases. Baltimore, Williams & Wilkins, 1968.

Bacterial

2841. American Association of Avian Pathologists. Salmonellosis workshop, Las Vegas, 1970. Proceedings. Hyattsville, Md., Animal Health Division, U.S. Agricultural Research Service, 1971. (ARS 91-101.)

2842. Dubos, R. J., and J. G. Hirsch, eds. Bacterial and mycotic infections of man. 4th ed. Philadelphia, Lippincott, 1965.

2843. Fennestad, K. L. Experimental leptospirosis in calves. Copenhagen, Munksgaard, 1963. (Translated by H. Cowen.)

2844. Myers, J. A. Tuberculosis: a half century of study and conquest. St. Louis, Mosby, 1970.

2845. Stableforth, A. W., and I. A. Galloway, eds. Infectious diseases of animals: diseases due to bacteria. 2 v. New York, Plenum Press, 1959.

2846. U.S. Center for Disease Control (formerly Communicable Disease Center). Leptospirosis: epidemiology, clinical manifestations in man and animals and methods in laboratory diagnosis. By Mildred M. Galton et al. Atlanta, 1962. (U.S. Public Health Service publ. no. 951.)

Viral

2847. Adams, J. M. Newer virus diseases. New York, Macmillan, 1960.

2848. Betts, A. O., and C. J. York. Viral and rickettsial infections of animals. v. 1- New York, Academic Press, 1967-

2849. Bisseru, Balideo. Rabies. London, Heinemann Medical Books, 1972.

2850. Chowdhuri, A. N. Rai, and A. K. Thomas. Rabies: general considerations and laboratory procedures. New Delhi, Indian Council of Medical Research, 1967. (Special report series, no. 58.)

2850a. Conference on atypical virus infections. Possible relevance to animal models and rheumatic disease. Proceedings. Atlanta, 1968. Ed. by C. L. Christian et al. New York, Arthritis Foundation, 1968.

2851. Eichwald, Claus, and Horst Pitzschke. Die Tollwut bei Mensch und Tier. (Rabies.) Jena, Fischer, 1966.

2852. Fenner, F. J., and F. N. Ratcliffe. Myxomatosis. New York, Cambridge Univ. Press, 1965.

2853. Fowl Pest Review Panel. Fowl pest: Newcastle disease epidemic, 1970-71; report of the Review Panel. London, H.M. Stationery Office, 1971.

2854. Horsfall, F. L., and Igor Tamm. Viral and rickettsial infections of man. 4th ed. Philadelphia, Lippincott, 1965. (A revision of T. M. Rivers' *Viral and rickettsial infections of man.*)

2855. Institut français de la fievre aphteuse. (Hoof and mouth disease.) Publications 1956-1960. Lyon, Imprimerie Giraud-Rivoire, 1967-

2856. International conference on foot and mouth disease. 1st. New York City, 1969. Proceedings. Ed. by Fred Rapp. New York, Gustav Stern Foundation, 1970.

2857. Juel-Jensen, B. E. Herpes simplex, varicella, and zoster: clinical manifestations and treatment. Philadelphia, Lippincott, 1972.

2858. Machado, M. A., Jr. Industry in crisis: Mexican-United States cooperation in the control of foot-and-mouth disease. Berkeley, Univ. of Calif. Press, 1968.

2859. Marburg virus disease. Ed. by G. A. Martini and R. Siegert. New York, Springer-Verlag, 1971.

2860. National rabies symposium. U.S. National Communicable Disease Center, 1966. Proceedings. Atlanta, National Communicable Disease Center, 1966.

2861. Piazza, Marcello. Experimental viral hepatitis. Springfield, Ill., Thomas, 1969.

2862. Scott, G. R. Diagnosis of rinderpest. Rome, Food and Agricultural Organization of the United Nations, 1967. (FAO agricultural studies, no. 71.)

2863. Stuart-Harris, C. H. Influenza and other virus infections of the respiratory tract. 2d ed. Baltimore, Williams & Wilkins, 1965.

2863a. West, G. P. Rabies: in animals and man. North Pomfret, Vt., David & Charles, 1972.

2864. Whitlock, Ralph. The great cattle plague: an account of the foot-and-mouth epidemic of 1967-1968. London, John Baker, 1968.

2865. Working conference on rabies. Tokyo, 1970. Rabies: proceedings. Ed. by Yasuiti Nagano and F. M. Davenport. Baltimore, University Park Press, 1970.

2866. World's poultry congress. XIII. Kiev, U.S.S.R., 1966. Proceedings. Field experience with the control of Newcastle disease and infectious bronchitis with inactivated vaccines. By P. G. Box. Kiev, U.S.S.R., 1966.

2867. Zuckerman, A. J. Virus diseases of the liver. New York, Appleton-Century-Crofts, 1970.

Mycotic

2868. Ainsworth, G. C., and P. K. C. Austwick. Fungal diseases of animals. Farnham Royal, England, Commonwealth Agricultural Bureaux, 1959. (Commonwealth Bureau of Animal Health reviews, no. 6.)

2869. Bronner, Marcell, and Max Bronner. Actinomycosis. Baltimore, Williams & Wilkins, 1969.

2870. Buechner, H. A., ed. Management of fungus diseases of the lungs. Springfield, Ill., Thomas, 1971.

2871. Dubos, R. J., and J. G. Hirsch, eds. Bacterial and mycotic infections of man. 4th ed. Philadelphia, Lippincott, 1965.

2872. Fetter, B. F., W. S. Hendry, and G. K. Klintworth. Mycoses of the central nervous system. Baltimore, Williams & Wilkins, 1967.

2873. Jellison, W. L. Adiaspiromycosis. Missoula, Mont., Mountain Press, 1969.

2874. National conference on histoplasmosis. 2d. Proceedings. Atlanta, U.S. Center for Disease Control, 1969. Springfield, Ill., Thomas, 1971.

2875. Symposium on Candida infections. London, 1965. Ed. by H. I. Winner and Rosalinde Hurley. Edinburgh, Livingstone, 1966.

2876. Symposium on coccidioidomycosis. 2d. Phoenix, 1965. Papers. Ed. by Libero Ajello. Tucson, Univ. of Arizona Press, 1965.

2877. Vanbreuseghem, R. Mycoses of man and animals. Springfield, Ill., Thomas, 1968. (Translated by J. Wilkinson.)

Zoonotic

2878. Alston, J. M., and Broom, J. C. Leptospirosis in man and animals. London, Livingstone, 1958.

2879. Bisseru, Balideo. Diseases of man acquired from his pets. Philadelphia, Lippincott, 1968.

2880. Constantine, D. G. Rabies transmission by air in bat caves. Altanta, U.S. Communicable Disease Center, 1967.

2881. Faust, E. C., P. C. Beaver, and R. C. Jung. Animal agents and vectors of human diseases. 3d ed. Philadelphia, Lea & Febiger, 1968.

2882. Felsenfeld, Oscar. Borrelia: strains, vectors, human and animal borreliosis. St. Louis, Green, 1971.

2883. Fiennes, Richard. Zoonoses of primates: the epidemiology and ecology of simian diseases in relation to man. Ithaca, N.Y., Cornell Univ. Press, 1967.

2884. Gould, S. E., ed. Trichinosis in man and animals. Springfield, Ill., Thomas, 1970.

2885. Hoeden, Jacob van der, ed. Zoonoses. New York, American Elsevier, 1965.

2886. Inter-American meeting on foot-and-mouth disease and zoonoses control. World Health Organization, 1968- (For additional coverage see entry 0755.)

2887. Machado, M. A. Aftosa: a historical survey of foot-and-mouth disease and inter-American relations. Albany, State Univ. of New York Press, 1969.

2888. Myers, J. A., and J. H. Steele. Bovine tuberculosis control in man and animals. St. Louis, Green, 1969.

2889. Pavlovskiĭ, E. N. Natural nidality of transmissible diseases: with special reference to the landscape epidemiology of zooanthroponoses. Ed. by N. D. Levine. Urbana, Univ. of Illinois Press, 1966. (English translation by F. K. Plous, Jr.)

2890. Pavlovskiy, Y. N. Natural foci of transmissible diseases as related to territorial epidemiology of zooanthroponoses. New York, CCM Information Corp., 1964.

2891. Schwabe, C. W. Diseases in animals and man's well being. Saskatoon, University of Saskatchewan, 1970. (Its University lectures, no. 24.)

2892. ———. Veterinary medicine and human health. 2d ed. Baltimore, Williams & Wilkins, 1969.

2893. Some diseases of animals communicable to man in Britain. Proceedings of a symposium. London, 1966. Ed. by O. Graham-Jones. New York, Pergamon Press, 1968.

2893a. Trends in epidemiology: application to health service, research and training. Ed. by G. T. Stewart. Springfield, Ill., Thomas, 1972.

2894. U.S. Center for Disease Control (formerly Communicable Disease Center). Veterinary Public Health Section. Epidemiological aspects of some of the zoonoses. Atlanta, 1966.

2895. ———. For titles of serially issued reports see entry 0286.

EPIDEMIOLOGY AND PUBLIC HEALTH

2896. Bjornson, B. F., H. D. Pratt, and K. S. Littig. Control of domestic rats and mice. rev. Rockville, Md., U.S. Environmental Control Administration, 1968.

2897. Brandly, P. J., George Migaki, and K. E. Taylor. Meat hygiene. 3d ed. Philadelphia, Lea & Febiger, 1966.

2898. Busvine, J. R. Insects and hygiene: the biology and control of insect pests of medical and domestic importance. 2d ed. Barnes & Noble, 1966.

2899. Code of federal regulations. [Title] 9. Animals and animal products. rev. as of Jan. 1, 1972. Washington, D.C., Office of the Federal Register, 1972.

2900. Cohen, Daniel. Epidemiological methods in veterinary medicine. Baltimore, Williams & Wilkins, in press 1972.

2901. Dykstra, R. R. Animal sanitation and disease control. 6th ed. Danville, Ill., Interstate, 1961.

2902. Felsenfeld, Oscar. The cholera problem. St. Louis, Green, 1967.

2903. Fox, J. P., Carrie E. Hall, and Lila

R. Elveback. Epidemiology. Toronto, Macmillan, 1970.

2904. Frazier, W. C. Food microbiology. 2d ed. New York, McGraw-Hill, 1967.

2905. Grant, Murray. Handbook of preventive medicine and public health. Philadelphia, Lea & Febiger, 1967.

2906. Hanlon, J. J. Principles of public health administration. 5th ed. St. Louis, Mosby, 1969.

2907. Jay, J. M. Modern food microbiology. New York, Van Nostrand, 1970.

2908. Lawrence, C. A., and S. S. Block, eds. Disinfection, sterilization, and preservation. Philadelphia, Lea & Febiger, 1968.

2909. Longree, Karla, and Gertrude G. Blaker. Sanitary techniques in food service. New York, Wiley, 1971.

2910. MacMahon, Brian, and T. F. Pugh. Epidemiology: principles and methods. Boston, Little, Brown, 1970.

2911. Mitchell, Ralph, ed. Water pollution microbiology. New York, Wiley-Interscience, 1972.

2912. Paul, J. R. Clinical epidemiology. rev. ed. Chicago, Univ. of Chicago Press, 1966.

2913. Pederson, C. S. Microbiology of food fermentations. Westport, Conn., Avi Publishing, 1971.

2914. Riemann, Hans, ed. Food-borne infections and intoxications. New York, Academic Press, 1969. (Food science and technology series. Monograph no. 5.)

2915. Thornton, Horace. Textbook of meat inspection. 5th ed. Baltimore, Williams & Wilkins, 1968.

2916. U.S. Agricultural Research Service. Animal Health Division. Recommended procedure for the isolation of salmonella organisms from animal feeds and feed ingredients. Washington, D.C., 1971.

2917. U.S. Dept. of Agriculture. Manual of meat inspection procedures. Washington, D.C., 1971.

2918. U.S. National Research Council. Food Protection Committee. Reference methods for the microbiological examination of foods. Washington, D.C., National Academy of Sciences, 1971.

2919. World Association of Veterinary Food Hygienists. Symposium. 5th. Opatija, 1969. Proceedings. Ed. by M. van Schothorst et al. Belgrade, Savez veterinara i veterinarskih technicara Jugoslavije, 1971.

14

PARASITOLOGY AND ENTOMOLOGY

PARASITOLOGY

2920. Adam, K. M., and V. Zaman. Medical and veterinary protozoology. Baltimore, Williams & Wilkins, 1971.

2921. Baer, J. G. Animal parasites. New York, McGraw-Hill, 1971. (Translated from the French by Kathleen Lyons.)

2922. Beck, J. W., and Elizabeth Barrett-Connor. Medical parasitology. St. Louis, Mosby, 1971.

2923. Belding, D. L. Textbook of parasitology. 3d ed. New York, Appleton-Century-Crofts, 1965.

2924. Benbrook, E. A. Outline of parasites reported for domesticated animals in North America. 6th ed. Ames, Iowa State Univ. Press, 1963.

2925. Boch, Josef, and R. Supperer. Veterinarmedizinische Parasitologie. Berlin, Paul Parey, 1971.

2926. British Society for Parasitology.

Symposium. 1st- Philadelphia, Davis, 1963- (For specific titles see entry 0739.)

2927. Central Veterinary Laboratory. Parasitology Dept. Manual of veterinary parasitological laboratory techniques. London, H.M. Stationery Office, 1971.

2928. Cheng, T. C. The biology of animal parasites. Philadelphia, Saunders, 1964.

2929. Commonwealth Bureau of Helminthology. Technical communications. St. Albans, England, 1962- (For specific titles see entry 0744.)

2930. Dunn, A. M. Veterinary helminthology. Philadelphia, Lea & Febiger, 1969.

2931. Elsevier's lexicon of parasites and diseases in livestock, including parasites and diseases of all farm and domestic animals, free-living wild fauna, fishes, honeybee and silkworm, and parasites of products of animal origin. Latin, English, French, German, Italian, and Spanish. Comp. by Manuel Merino-Rodriguez. Amsterdam, Elsevier, 1964.

2932. Faust, E. C., P. C. Beaver, and R. C. Jung. Animal agents and vectors of human diseases. 3d ed. Philadelphia, Lea & Febiger, 1968.

2933. ——, P. F. Russell, and R. C. Jung. Craig and Faust's clinical parasitology. 8th ed. Philadelphia, Lea & Febiger, 1970.

2934. Garnham, P. C. C. Malaria parasites and other haemosporidia. Oxford, Blackwell, 1966.

2935. ——. Progress in parasitology. London, Athlone Press, 1971.

2936. Georgi, J. R. Parasitology for veterinarians. Philadelphia, Saunders, 1969.

2937. Hoare, C. A. The trypanosomes of mammals. Oxford, Blackwell Scientific Pub., 1972.

2938. Hoffman, G. L. Parasites of North American freshwater fishes. Berkeley, Univ. of California Press, 1967.

2939. Jackson, George, Robert Herman, and Ira Singer, eds. Immunity to parasitic animals. 2 v. New York, Appleton-Century-Crofts, 1968-1970.

2940. Jeffrey, H. C., and R. M. Leach. Atlas of medical helminthology and protozoology. Baltimore, Williams & Wilkins, 1966.

2941. Kheĭsin, E. M. Life cycles of coccodia of domestic animals. Ed. by K. S. Todd, Jr. Baltimore, University Park Press, 1972. (Translated by F. K. Plous, Jr.)

2942. Konferentsiia po prirodnoĭ ochagovosti bolezneĭ i voprosam parazitologii Kazakhstana i respubik Srednei Aziĭ. (For further coverage see entry 2834.)

2943. Krull, W. H. Notes in veterinary parasitology. v. 1- (loose-leaf). Lawrence, Univ. of Kansas Press, 1969-

2944. Lapage, Geoffrey. Veterinary parasitology. 2d ed. Springfield, Ill., Thomas, 1968.

2945. Levine, N. D. Nematode parasites of domestic animals and of man. Minneapolis, Burgess, 1968.

2946. ——. Protozoan parasites of domestic animals and of man. Minneapolis, Burgess, 1961.

2947. ——, and Virginia Ivens, eds. The coccidian parasites (protozoa, sporozoa) of ruminants. Urbana, Univ. of Illinois Press, 1970. (Illinois biological monographs, no. 44.)

2948. Macy, R. W., and A. K. Berntzen. Laboratory guide to parasitology: with introduction to experimental methods. Springfield, Ill., Thomas, 1971.

2949. Markell, E. K., and Marietta Voge. Medical parasitology. 3d ed. Philadelphia, Saunders, 1971.

2950. Meyer, M. C., and L. R. Penner. Laboratory essentials of animal parasitology. Dubuque, Ia., Brown, 1962.

2951. Neveu-Lemaire, Maurice. Traité d'helminthologie médicale et vétérinaire. Paris, Vigot Freres, 1936.

2952. Newell, R. C. Biology of intertidal animals. New York, Elsevier, 1970.

2953. Noble, E. R., and G. A. Noble. Parasitology: the biology of animal parasites. 3d ed. Philadelphia, Lea & Febiger, 1971.

2954. Pellérdy, L. P. Coccidia and coccidiosis. Budapest, Akademiai Kiado, 1965.

2955. Read, C. P. Animal parasitism. Englewood Cliffs, N.J., Prentice-Hall, 1972.

2956. Richardson, U. F., and S. B. Kendall. Veterinary protozoology. 3d rev. ed. Edinburgh. Oliver & Boyd, 1963.

2957. Sloss, Margaret W. Veterinary clinical parasitology. 4th ed. Ames, Iowa State Univ. Press, 1970.

2958. Smyth, J. D. Physiology of cestodes. San Francisco, Freeman, 1969.

2959. Soulsby, E. J. L., ed. Helminths, arthropods and protozoa of domesticated animals. 6th ed. Baltimore, Williams & Wilkins, 1968. (Previous edition by H. O. Mönnig: *Veterinary helminthology and entomology.*)

2960. Sprent, J. F. Parasitism: an introduction to parasitology and immunology for students of biology, veterinary science and medicine. Baltimore, Williams & Wilkins, 1963.

2961. Textbook of veterinary clinical parasitology. v. 1- Philadelphia, Davis, 1965-
v. 1: Helminths. Ed. by E. J. L. Soulsby.

2962. Theodor, Oskar, and M. Costa. Ectoparasites: a survey of the parasites of wild mammals and birds in Israel, Part 1. Jerusalem, Israel Academy of Sciences and Humanities, 1967. (Other volumes in preparation 1972.)

2963. Whitlock, J. H. Diagnosis of veterinary parasitisms. Philadelphia, Lea & Febiger, 1960.

2964. Work, Kresten. Toxoplasmosis, with special reference to transmission and life cycle of Toxoplasma gondii. Copenhagen, Munksgaard, 1971.

2965. World Association for the Advancement of Veterinary Parasitology. International conference. 1st- 1963- (For specific titles see entry 0799.)

2966. Wright, C. A. Flukes and snails. London, Allen & Unwin, 1971.

2967. Yamaguti, Satyu. Synopsis of digenetic trematodes of vertebrates. rev. ed. 2 v. Tokyo, Keigaku Publishing Co., 1971.

PARASITIC DISEASES

2968. Abuladze, K. I. Taeniata of animals and man and diseases caused by them. Jerusalem, Israel Program for Scientific Translations, 1970.

2969. Alicata, J. E. Parasitic infections of man and animals in Hawaii. Honolulu, Hawaii Agricultural Experiment Station, College of Tropical Agriculture, Univ. of Hawaii, 1964. (Technical bulletin no. 61.)

2970. British Veterinary Association. Technical Development Committee. Parasitic diseases of cattle. 3d ed. London, 1969.

2971. Coatney, G. R., et al. The primate malarias. Bethesda, Md., National Institute of Allergy and Infectious Diseases, 1971.

2972. Davis, J. W., ed. Infectious and parasitic diseases of wild birds. Ames, Iowa State Univ. Press, 1971.

2973. ——, and R. C. Anderson, eds. Parasitic diseases of wild mammals. Ames, Iowa State Univ. Press, 1971.

2974. Ershov, V. S. Parasitology and parasitic diseases of livestock. Moscow, State Pub. House for Agricultural Literature, 1956. (Available from the Office of Technical Services, U.S. Dept. of Commerce, Washington, D.C.)

2975. Euzéby, Jacques. Les maladies vermineuses animaux domestiques et leurs incidences sur la pathologie humaine. 1961- Paris, Vigot Freres.
v. 1: Maladies dues aux némathelminthes (2 fascs.). v. 2: Maladies dues aux plathelminthes (3 fascs.).

2976. Galuzo, I. G. Toxoplasmosis of animals. Ed. by P. R. Fitzgerald. Urbana, College of Veterinary Medicine, Univ. of Illinois, 1970. (Translation of *Toxsoplazmoz zhivotnykh.*)

2977. International symposium on canine heartworm disease, 1969. Proceedings. Ed. by R. E. Bradley. Gainesville, Univ. of Florida, 1969.

2978. Marcial-Rojas, R. A., and Esteban Moreno, eds. Pathology of protozoal and helminthic disease, with clinical correlation. Baltimore, Williams & Wilkins, 1971.

2979. Mozgovoĭ, A. A. Ascaridata of animals and man and the diseases caused by them. Jerusalem, Israel Program for Scientific Translations, 1968. (Translation of *Askaridaty zhivothykh i cheloveka i vyzyvaemye imi zabolevaniia.* Essentials of nematodology, v. 2.)

2980. Mulligan, H. W., and W. H. Potts, eds. African trypanosomiases. New York, Wiley, 1970.

2981. Orlov, N. P. Coccidiosis of farm animals. Jerusalem, Israel Program for Scientific Translations, 1970.

2982. Warren, K. S., and V. A. Newill. Schistosomiasis: a bibliography of the world's literature from 1852 to 1962. 2 v. Cleveland, Western Reserve Univ. Press, 1967.

MEDICAL ENTOMOLOGY

2983. Allred, D. M. Medical entomology: laboratory guide. 2d ed. Minneapolis, Burgess, 1963.

2984. Beament, J. W. L., and J. E. Treherne, eds. Insects and physiology. New York, American Elsevier, 1968.

2985. Busvine, J. R. Insects and hygiene: the biology and control of insect pests of medical and domestic importance. 2d ed. New York, Barnes & Noble, 1966.

2986. Gordon, R. M., and M. M. J. Lavoipierre. Entomology for students of medicine. Oxford, Blackwell, 1962.

2987. Horsfall, W. R. Medical entomology: arthropods and human diseases. New York, Ronald Press, 1962.

2988. James, M. T., and R. F. Harwood. Herms' medical entomology. 6th ed. London, Macmillan, 1969.

2989. Leclercq, Marcel. Entomological parasitology: the relations between entomology and the medical sciences. Elmsford, N.Y., Pergamon Press, 1968.

2990. Mattingly, P. F. The biology of mosquito-borne disease. New York, American Elsevier, 1969.

2991. Metcalf, C. L., and W. P. Flint, eds. Destructive and useful insects: their habits and control. 4th ed. New York, McGraw-Hill, 1962.

2992. Roth, L. M., and E. R. Willis. The medical and veterinary importance of cockroaches. Washington, D.C., Smithsonian Institution, 1957.

2993. Roy, D. N., and A. W. A. Brown. Entomology (medical and veterinary), including insecticides and insect and rat control. 3d ed. Bangalore, India, Bangalore Printing & Publishing, 1970.

2994. Travis, B. V., H. H. Lee, and R. M. Labadan. Arthropods of medical importance in America north of Mexico. Natick, Mass., Natick Laboratories, U.S. Army Material Command, 1969. (Other treatises issued by the same laboratory describe arthropods of medical importance in Africa, Asia, European USSR, Australia, and the Pacific Islands, and in Latin America.)

2995. U.S. Naval Medical School. Medical entomology. Comp. by Oscar Taboada. Bethesda, Md., 1967.

15

IMMUNOLOGY

PRINCIPLES

2996. Barrett, J. T. Textbook of immunology: an introduction to immunochemistry and immunobiology. St. Louis, Mosby, 1970.

2997. Bayer symposium. 1st. Grosse Ledder, 1968. Current problems in immunology. Ed. by O. Westphal, H. E. Bock, and E. Grundmann. New York, Springer-Verlag, 1969.

2998. Bellanti, J. A. Immunology. Philadelphia, Saunders, 1971.

2999. Boyd, W. C. Fundamentals of immunology. 4th ed. New York, Wiley, 1967.

3000. Burnet, F. M. Cellular immunology. 2 v. New York, Cambridge Univ. Press, 1969. (Book one also issued under the title *Self and not self*.)

3001. Cruse, J. M. Immunology examination review book. v. 1- Flushing, N.Y., Medical Examination Publishing, 1971-

3002. Developmental immunology workshop. 1st- Sanibel Island, Fla. Proceedings. 1965- (For specific titles see entry 0746.)

3003. Fudenberg, H. H., et al. Basic immunogenetics. Fair Lawn, N.J., Oxford Univ. Press, 1971.

3004. Good, R. A. Immunobiology: current knowledge of basic concepts in immunology and their clinical applications. Stamford, Conn., Sinauer Associates, 1972.

3005. Gray, D. F. Immunology. 2d ed. New York, Elsevier, 1970.

3006. Greenwalt, T. J., ed. Advances in immunogenetics. Philadelphia, Lippincott, 1967.

3007. Haesler, W. E., Jr. Immunohematology. Philadelphia, Lea & Febiger, 1972.

3008. Halliday, W. J. Glossary of immunological terms. London, Butterworth, 1971.

3009. Herbert, W. J., and P. C. Wilkinson.
A dictionary of immunology. Oxford, Blackwell Scientific Pub., 1971.

3010. Hildemann, W. H. Immunogenetics. San Francisco, Holden-Day, 1970.

3011. Humphrey, J. H., and R. G. White. Immunology for students of medicine. 3d ed. Philadelphia, Davis, 1970.

3012. International congress of immunology. 1st- Washington, D.C., 1971-

3013. International symposium on immunopathology. 1st- Proceedings. Ed. by P. Miescher and P. Grabar. Basel, Schwabe, 1959- (For additional coverage see entry 0762.)

3014. Nossal, G. J. V., and G. L. Ada. Antigens, lymphoid cells, and the immune response. New York, Academic Press, 1971.

3015. Sela, Michael, and Moshe Prywes. Topics in basic immunology. New York, Academic Press, 1969.

3016. Symposia series in immunobiological standardization. v. 1- Ed. by R. H. Regamey et al. New York, Karger, 1966- (For specific titles see entry 0782.)

3017. Transplantation reviews. Ed. by Göran Möller. v. 1- Baltimore, Williams & Wilkins, 1969- (For specific titles see entry 0786.)

3018. Turk, J. L. Immunology in clinical medicine. New York, Appleton-Century-Crofts, 1969.

3019. Weir, D. M. Immunology for undergraduates. 2d ed. Baltimore, Williams & Wilkins, 1972.

3020. Weiser, R. S., Q. N. Myrvik, and Nancy N. Pearsall. Fundamentals of immunology for students of medicine and related sciences. Philadelphia, Lea & Febiger, 1969.

3021. Wilson, David. Body and antibody: a report on the new immunology. New York, Knopf, 1972.

MECHANISMS

3022. Abramoff, P., and M. La Via. Biology of the immune response. New York, McGraw-Hill, 1970.

3023. Adinolfi, Matteo. Immunology and development. London, Spastics International Medical, 1969. (Clinics in developmental medicine, no. 34.)

3024. Canadian Society for Immunology. 3d International symposium. Toronto, 1968. Cellular and humoral mechanisms in anaphylaxis and allergy. Ed. by H. Z. Movat. New York, Karger, 1969.

3025. Ciba Foundation symposium. Ontogeny of acquired immunity. New York, American Elsevier, in press 1972.

3026. Cinader, Bernhard, ed. Regulation of the antibody response. 2d ed. Springfield, Ill., Thomas, 1971.

3027. Conference on the secretory immunologic system. Vero Beach, Florida, 1969. Proceedings. Ed. by D. H. Dayton, Jr., et al. Bethesda, Md., National Institute of Child Health and Human Development, 1969.

3028. Developmental aspects of antibody formation and structure. v. 1- Proceedings of symposium. Prague and Slapy, 1969. Ed. by J. Sterzl and I. Ríha. New York, Academic Press, 1970-

3029. The immune response and its suppression. Ed. by E. Sorkin. Basel, Karger, 1969. (*Antibiotica et chemotherapia*, v. 15.)

3030. Immunological tolerance: a reassessment of mechanisms of the immune response. Ed. by Maurice Landy. New York, Academic Press, 1969.

3031. International symposium on blood and tissue antigens. University of Michigan Medical Center, 1969. Blood and tissue antigens: proceedings. Ed. by David Aminoff. New York, Academic Press, 1970.

3032. Kabat, E. A. Structural concepts in immunology and immunochemistry. New York, Holt, Rinehart & Winston, 1968.

3033. Lawrence, H. S., and Maurice Landy, eds. Mediators of cellular immunity. New York, Academic Press, 1969.

3034. Makelä, O., ed. Cell interactions and receptor antibodies in immune responses. New York, Academic Press, 1971. (Sigrid Jusélius symposium. 3d. Proceedings.)

3035. Monographs in allergy. v. 1- New York, Karger, 1966- (For specific titles see entry 0770.)

3036. Nossal, G. J., and G. L. Ada. Antigens, lymphoid cells and the immune response. New York, Academic Press, 1971.

3037. Panel on radiation and the control of immune response. Paris, 1967. Report. Vienna, International Atomic Energy Agency, 1968. (Panel proceedings series. STI/PUB/175.)

3038. Revillard, J. P., ed. Cell-mediated immunity: in vitro correlates. Baltimore, University Park Press, 1971.

3039. Sherman, W. B. Hypersensitivity: mechanisms and management. Philadelphia, Saunders, 1968.

3040. Smith, R. T., P. A. Miescher, and R. A. Good, eds. Ontogeny of immunity. Gainesville, Univ. of Florida Press, 1971.

3041. Studer, A., and H. Cottier, eds. Immune reactions. New York, Springer-Verlag, 1970.

3042. Study group on hormones and the immune response. London, 1970. Hormones and the immune response. Ed. by G. E. W. Wolstenholme and Julie Knight. London, Churchill, 1970. (Ciba Foundation study group, no. 36.)

3043. Taliaferro, W. H., Lucy G. Taliaferro, and B. N. Jaroslow. Radiation and immune mechanisms. New York, Academic Press, 1964.

3044. Tolerance, autoimmunity and aging. Compiled by M. M. Sigel and R. A. Good. Springfield, Ill., Thomas, 1972.

3045. Uhr, J. W., and Maurice Landy, eds. Immunologic intervention. New York, Academic Press, 1971.

3046. U.S. National Institute of Allergy and Infectious Diseases. Selected references on antilymphocyte sera. Bethesda, Md., National Institutes of Health, 1968.

3047. Walford, R. L. Immunologic theory of aging. Baltimore, Williams & Wilkins, 1969.

3048. Warren, Katherine B., ed. Differentiation and immunology. New York, Academic Press, 1968. (International Society for Cell Biology. Symposium. v. 7.)

3049. Weiss, Leon. The cells and tissues of the immune system: structure, functions, interactions. Englewood Cliffs, N.J., Prentice-Hall, 1972.

3050. Yoffey, J. M., ed. The lymphocyte in immunology and haemopoiesis: the symposium held at Bristol, 1966. London, Edward Arnold, 1967.

3051. Zmijewski, C. M., June L. Fletcher, and R. L. St. Pierre. Immunohematology. New York, Appleton-Century-Crofts, 1968.

CLINICAL APPLICATIONS

3052. Anderson, J. R., W. W. Buchanan, and R. B. Goudie. Autoimmunity: clinical and experimental. Springfield, Ill., Thomas, 1967.

3053. Bertelli, A., and A. P. Monaco, eds. Pharmacological treatment in organ and tissue transplantation. Baltimore, Williams & Wilkins, 1970.

3054. Billingham, R. E., and Willys Silvers. The immunobiology of transplantation. Englewood Cliffs, N.J., Prentice-Hall, 1971.

3055. Brambell, F. W. The transmission of passive immunity from mother to young. New York, Elsevier, 1970. (*Frontiers of biology*, v. 18.)

3056. Conference on biologic aspects and clinical uses of immunoglobulins. Ohio State University, 1969. Immunoglobulins. Ed. by Ezio Merler. Washington, D.C., National Academy of Sciences, 1970.

3057. Criep, L. H. Clinical immunology and allergy. 2d ed. New York, Grune & Stratton, 1969.

3058. Cross-reacting antigens and neoantigens: with implications for autoimmunity and cancer immunity. Ed. by J. J. Trentin. Baltimore, Williams & Wilkins, 1967.

3059. Gell, R. G. H., and R. R. A. Coombs. Clinical aspects of immunology. 2d ed. Oxford, Blackwell Scientific Pub., 1968.

3060. Glynn, L. E., and E. J. Holborrow. Autoimmunity and disease. Philadelphia, David, 1965.

3061. Hamburger, Jean, et al. Renal transplantation: theory and practice. Baltimore, Williams & Wilkins, 1972.

3062. Histocompatibility workshop on primates. 1st. Rijswijk, 1971. Transplantation genetics of primates. Ed. by Hans Balner and Jon J. van Rood. New York, Grune & Stratton, 1972. (Issued also as *Transplantation proceedings*, v. 4, no. 1, with title "Proceedings of a workshop and symposium on transplantation genetics of primates.")

3063. Immunoglobulins: biologic aspects and clinical uses. Washington, D.C., National Academy of Sciences, 1971.

3064. Immunology and connective tissue disorders. New York, Grune & Stratton, 1971. (*Seminars in arthritis and rheumatism*, v. 1, no. 2.)

3065. Immunology of the liver. Ed. by Martin Smith and Roger Williams. Philadelphia, Davis, 1971. (Proceedings of an international meeting, London, 1970.)

3066. International symposium on infections and immunosuppression in subhuman primates. Rijswijk, Netherlands, Dec. 17-19, 1969. Proceedings. Ed. by H. Balner and W. I. B. Beveridge. Copenhagen, Munksgaard, 1970.

3067. Jackson, G. J., Robert Herman, and Ira Singer, eds. Immunity to parasitic animals. 2 v. New York, Appleton-Century-Crofts, 1968-1970.

3068. Magnusson, Bertil, and A. M. Kligman. Allergic contact dermatitis in the guinea pig: identification of contact allergens. Springfield, Ill., Thomas, 1970.

3069. Montagna, William, and R. E. Billingham, eds. Immunology and the skin. New York, Appleton-Century-Crofts, 1971.

3070. Movat, H. Z., ed. Inflammation, immunity, and hypersensitivity. 1st ed. New York, Harper & Row, 1971.

3071. Mudd, Stuart. Infectious agents and host reactions. Philadelphia, Saunders, 1970.

3072. Natural and acquired immunological unresponsiveness. Ed. by W. O. Weigle. Cleveland, World, 1968. (Monographs in microbiology, no. 3.)

3073. Samter, Max, and H. L. Alexander, eds. Immunological diseases. 2d ed. Waltham, Mass., Little, Brown, 1970.

3074. Smith, R. T., and Maurice Landy, eds. Immune surveillance. New York, Academic Press, 1971.

3075. Transplantation reviews. Ed. by Göran Möller. v. 1- Baltimore, Williams & Wilkins, 1969- (For specific titles see entry 0786.)

EXPERIMENTAL

3076. Ackroyd, J. F. Immunological methods. Philadelphia, Davis, 1964.

3077. Bloom, B. R., and P. R. Glade, eds. In vitro methods in cell-mediated immunity. New York, Academic Press, 1971.

3078. Burrell, R. G., and Carmine C. Mascoli. Experimental immunology. 3d ed. Minneapolis, Burgess, 1971.

3079. Campbell, D. H. Methods in immunology: a laboratory text for instruction and research. 2d ed. New York, W. A. Benjamin, 1970.

3080. Carpenter, P. L. Immunology and serology. 2d ed. Philadelphia, Saunders, 1965.

3081. Kwapinski, J. B. Methods of serological research. New York, Wiley, 1965.

3082. Research in immunochemistry and immunobiology. v. 1- Baltimore, University Park Press, 1972-

3083. Symposia series in immunobiological standardization. v. 1- Ed. by R. H. Regamey et al. New York, Karger, 1966- (For specific titles see entry 0782.)

3084. U.S. National Institute of Allergy and Infectious Diseases. Transplantation Immunology Branch. Manual of tissue typing techniques. Ed. by D. L. Brand and J. G. Ray, Jr. Bethesda, Md., 1970.

3085. Waksman, B. H. Atlas of experimental immunology and immunopathology. New Haven, Yale Univ. Press, 1970.

3086. Weir, D. M. Handbook of experimental immunology. Philadelphia, Davis, 1967.

3087. Williams, C. A., and M. W. Chase, eds. Methods in immunology and immunochemistry. v. 1- (For specific titles see entry 2452.)

Note. For additional entries see also the section on "Immunochemistry" (entries 2435-2452) in Chap. 10, and the section on "Immunopathology" (entries 3477-3480) in Chap. 18.

16

GENETICS

PRINCIPLES

3088. Beadle, G. W. Genetics and modern biology. Philadelphia, American Philosophical Society, 1963.

3089. Brown, W. V. Textbook of cytogenetics. St. Louis, Mosby, 1972.

3090. Buettner-Janusch, John, ed. Evolutionary and genetic biology of primates. 2 v. New York, Academic Press, 1963-1964.

3091. Conference on genetics. 2d. Princeton, 1960. Mutations. Ed. by W. J. Schull. Ann Arbor, Univ. of Michigan Press, 1962.

3092. Fudenberg, H. H., et al. Basic immunogenetics. Fair Lawn, N.J., Oxford Univ. Press, 1971.

3093. Greenwalt, T. J., ed. Advances in

immunogenetics. Philadelphia, Lippincott, 1967.

3094. Hildemann, W. H. Immunogenetics. San Francisco, Holden-Day, 1970.

3095. International directory of genetic services. Comp. by H. T. Lynch. 3d ed. New York, National Foundation-March of Dimes, 1971.

3096. Mittwoch, Ursula. Sex chromosomes. New York, Academic Press, 1967.

3097. Pirchner, Franz. Population genetics in animal breeding. San Francisco, Freeman, 1969.

3098. Priest, J. H. Cytogenetics. Philadelphia, Lea & Febiger, 1969.

3099. Roberts, J. A. F. Introduction to medical genetics. 5th ed. New York, Oxford Univ. Press, 1970.

3100. Sinnott, E. W., L. C. Dunn, and T. Dobzhansky. Principles of genetics. 5th ed. New York, McGraw-Hill, 1968.

3101. Taylor, J. H., ed. Molecular genetics. 2 v. New York, Academic Press, 1963-1968.

3102. White, M. J. D. The chromosomes. 5th ed. New York, Barnes & Noble, 1961.

3103. Wilkie, David. The cytoplasm in heredity. New York, Wiley, 1964.

COMPARATIVE

3104. Chiarelli, A. B., ed. Comparative genetics in monkeys, apes and man. New York, Academic Press, 1971. (Proceedings of a symposium on comparative genetics in primates and human heredity, Erice, Sicily, 1970.)

3105. Conference on comparative mammalian cytogenetics. Dartmouth Medical School, 1968. Proceedings. Ed. by K. Benirschke. New York, Springer-Verlag, 1969.

3106. Hsu, T. C., and K. Benirschke. An atlas of mammalian chromosomes. v. 1- New York, Springer-Verlag, 1967-

3107. Lush, I. E. The biochemical genetics of vertebrates except man. Philadelphia, Saunders, 1966.

3108. Robinson, Roy. Gene mapping in laboratory mammals. 2 v. New York, Plenum Press, 1971-1972.

3109. Searle, A. G. Comparative genetics of coat colour in mammals. New York, Academic Press, 1968.

CLINICAL APPLICATIONS

3110. Asdell, S. A. Dog breeding: reproduction and genetics. Boston, Little, Brown, 1966.

3111. Benirschke, Kurt, and T. C. Hsu, eds. Chromosome atlas: fish, amphibians, reptiles and birds. New York, Springer-Verlag, 1971- (Loose-leaf.)

3112. Burdette, W. J., ed. Methodology in mammalian genetics. San Francisco, Holden-Day, 1963.

3113. Burns, Marca, and Margaret N. Fraser. Genetics of the dog: the basis of successful breeding. 2d ed. Edinburgh, Oliver & Boyd, 1966.

3114. Genetics in laboratory animal medicine: proceedings of a symposium. Boston, 1968. Washington, D.C., National Academy of Sciences, 1969. (Its publication no. 1724.)

3115. Hsia, D. Yi-Yung. Inborn errors of metabolism. 2d ed. 2 parts. Chicago, Year Book Medical Publishers, 1966.
  Part 1—Clinical aspects. Part 2—Laboratory methods.

3116. Hutt, F. B. Animal genetics. New York, Ronald Press, 1964.

3117. ———. Genetic resistance to disease in domestic animals. Ithaca, N.Y., Comstock, 1958.

3118. Johansson, Ivar, and Jan Rendel. Genetics and animal breeding. San Francisco, Freeman, 1968.

3119. Jones, W. E., and Ralph Bogart. Genetics of the horse. Ann Arbor, Mich., Edwards, 1971.

3120. Lasley, J. F. Genetics of livestock improvement. 2d ed. Englewood Cliffs, N.J., Prentice-Hall, 1972.

3121. Robinson, Roy. Genetics for cat breeders. New York, Pergamon Press, 1971.

3122. ———. Genetics of the Norway rat. New York, Pergamon Press, 1965.

3123. Scott, J. P., and J. L. Fuller, eds. Genetics and the social behavior of the dog. Chicago, Univ. of Chicago Press, 1965.

3124. Valontine, G. H. The chromosome disorders: an introduction for clinicians. 2d ed. Philadelphia, Lippincott, 1969.

3125. Winge, Øjvind. Inheritance in dogs: with special reference to hunting breeds. Ithaca, N.Y., Comstock, 1950. (Translated from Danish by Catherine Roberts.)

17

INTERNAL MEDICINE

GENERAL

3126. Adams, A. R. D., and B. G. Maegraith. Clinical tropical diseases. 5th ed. Oxford, Blackwell Scientific Pub., 1971.

3127. Blum, H. F. Photodynamic action and diseases caused by light. New York, Hafner, 1964.

3128. Bone, J. F., ed. Equine medicine and surgery, a text and reference work. 2d ed. Wheaton, Ill., American Veterinary Publications, 1972.

3129. Bovine medicine and surgery. Ed. by W. J. Gibbons, E. J. Catcott, and J. F. Smithcors. Wheaton, Ill., American Veterinary Publications, 1970.

3130. Catcott, E. J. Feline medicine and surgery, a text and reference work. Wheaton, Ill., American Veterinary Publications, 1964.

3131. ———, ed. Canine medicine: first Catcott edition. Santa Barbara, American Veterinary Publications, 1968. (Previous editions edited by H. P. Hoskins.)

3132. Cecil, R. L. Cecil-Loeb textbook of medicine. 13th ed. Ed. by P. B. Beeson and Walsh McDermott. Philadelphia, Saunders, 1971.

3133. The chemist's veterinary handbook: a survey of modern methods in veterinary medicine, including diseases and treatment. 12th ed. London, Offices of the Chemist and Druggist, 1964.

3134. Harrison, T. R. Principles of internal medicine. 6th ed. 2 v. New York, McGraw-Hill, 1970.

3135. Harvey, A. M., et al. The principles and practice of medicine. 18th ed. New York, Appleton-Century-Crofts, 1972.

3136. McCombs, R. P. Fundamentals of internal medicine: a physiologic and clinical approach to disease. 4th ed. Chicago, Year Book Medical Publishers, 1971.

3137. Manson-Bahr, P. H., and Charles Wilcocks, eds. Manson's tropical diseases: a manual of the diseases of warm climates. 16th ed. Baltimore, Williams & Wilkins, 1966.

3138. Talso, P. J., and A. P. Remenchik, eds. Internal medicine, based on mechanisms of disease. St. Louis, Mosby, 1968.

3139. Thal, A. P., et al. Shock: a physiologic basis for treatment. Chicago, Year Book Medical Publishers, 1971.

3140. Weil, M. H., and Herbert Shubin. Diagnosis and treatment of shock. Baltimore, Williams & Wilkins, 1967.

3141. Wright, F. J., and Alexander Biggam. Tropical diseases. 2d ed. London, Livingstone, 1967.

DIAGNOSIS

General

3142. Berrier, H. H. Diagnostic aids in the practice of veterinary medicine. 3d ed. St. Louis, Alban Professional Books, 1968.

3143. Boddie, G. F. Diagnostic methods in veterinary medicine. 6th ed. Philadelphia, Lippincott, 1969.

3144. Gibbons, W. J. Clinical diagnosis of diseases of large animals. Philadelphia, Lea & Febiger, 1966.

3145. Harvey, A. M., and James Bordley. Differential diagnosis. 2d ed. Philadelphia, Saunders, 1970.

3146. Kelly, W. R. Veterinary clinical diagnosis. Baltimore, Williams & Wilkins, 1967.

3147. Kirk, H. Index of diagnosis (clinical and radiological) for the canine and feline surgeon. 4th ed. London, Bailliere, Tindall & Cox, 1953.

3148. Kirk, R. W., and S. I. Bistner. Handbook of veterinary procedures and emergency treatment. Philadelphia, Saunders, 1969.

3149. Merck and Company, Inc. Merck manual of diagnosis and therapy. 12th ed. Rahway, N.J., Merck, Sharp & Dohme Research Laboratories, 1972.

3150. Merck veterinary manual: a handbook of diagnosis and therapy for the veterinarian. Ed. by O. H. Siegmund et al. Rahway, N.J., Merck & Co., 1967.

3151. Wirth, David. Veterinary clinical diagnosis. London, Bailliere, Tindall & Cox, 1956. (1st English edition.)

Physical

3152. Burch, G. E. A primer of venous pressure. Springfield, Ill., Thomas, 1972.

3153. Chamberlain, E. N., and Colin Ogilvie. Symptoms and signs in clinical medicine: an introduction to medical diagnosis. 8th ed. Bristol, Wright, 1967.

3154. Friedberg, C. K., ed. Physical diagnosis in cardiovascular disease. New York, Grune & Stratton, 1969.

3155. Gardner, A. F. Pathology of oral manifestations of systematic diseases. New York, Hafner, 1972.

3156. Geddes, L. A. The direct and indirect measurement of blood pressure. Chicago, Year Book Medical Publishers, 1970.

3157. Hyman, H. T. Differential diagnosis: an integrated handbook. Philadelphia, Lippincott, 1965.

3158. Klostermann, G. F., et al. Color atlas of external manifestations of disease. Ed. by Marcel Tuchmann. New York, McGraw-Hill, 1966.

3159. Leopold, S. S. Principles and methods of physical diagnosis. By H. U. Hopkins. 3d ed. Philadelphia, Saunders, 1965.

3160. Licht, S. H., ed. Electrodiagnosis and electromyography. 3d ed. New Haven, Licht, 1971.

3161. MacBryde, C. M., and R. S. Blacklow, eds. Signs and symptoms: applied pathologic physiology and clinical interpretation. 5th ed. Philadelphia, Lippincott, 1970.

3162. Major, R. H., and R. T. Manning. Major's physical diagnosis. 7th ed. Philadelphia, Saunders, 1968.

3163. Slonim, N. B., Bettye P. Bell, and Sherryl E. Christensen. Cardiopulmonary laboratory basic methods and calculations: a manual of cardiopulmonary technology. Springfield, Ill., Thomas, 1967.

3164. Swansburg, R. C. The measurement of vital signs. New York, Putnam, 1970.

Laboratory

3165. American Veterinary Medical Association. Radiologic diagnosis. Chicago, 1962.

3166. Applied seminar on the laboratory diagnosis of endocrine disorders. Washington, D.C., 1969. Laboratory diagnosis of endocrine diseases. Ed. by F. W. Sunderman and F. W. Sunderman, Jr., St. Louis, Green, 1971.

3167. Cartwright, G. E. Diagnostic laboratory hematology. 4th ed. New York, Grune & Stratton, 1968.

3168. Cohen, H. L., and Joel Brumlik. A manual of electroneuromyography. New York, Hoeber, 1968.

3169. Coles, E. H. Veterinary clinical pathology. Philadelphia, Saunders, 1967.

3170. Collins, R. D. Illustrated manual of laboratory diagnosis: indications and interpretations. Philadelphia, Lippincott, 1968.

3171. Damm, H. C., and J. W. King, eds. Handbook of clinical laboratory data. Comp. by Carolyn Thomas. Cleveland, Chemical Rubber Co., 1965.

3172. Davidsohn, Israel, and J. B. Henry, eds. Todd-Sanford clinical diagnosis by laboratory methods. 14th ed. Philadelphia, Saunders, 1969.

3173. De Gruchy, G. C. Clinical haematology in medical practice. 3d ed. Philadelphia, Davis, 1970.

3174. Eastham, R. D. A laboratory guide to clinical diagnosis. 2d ed. Baltimore, Williams & Wilkins, 1970.

3175. Escamilla, R. F., ed. Laboratory tests in diagnosis and investigation of endocrine function. 2d ed. Philadelphia, Davis, 1971.

3176. Gibbs, F. A., and Erna L. Gibbs. Medical electroencephalography. Reading, Mass., Addison-Wesley, 1967.

3177. Goldman, M. J. Principles of clinical electrocardiography. 7th ed. Los Altos, Calif., Lange Medical Publications, 1970.

3178. Goodale, R. H., and Frances K. Widmann. Clinical interpretation of laboratory tests. 6th ed. Philadelphia, Davis, 1969.

3179. Goodgold, Joseph, and Arthur Eberstein. Electrodiagnosis of neuromuscular diseases. Baltimore, Williams & Wilkins, 1972.

3180. Hoffman, W. S. The biochemistry of clinical medicine. 4th ed. Chicago, Year Book Medical Publishers, 1970.

3181. Holt, S. H. Laboratory aids in diagnosis. Baltimore, Williams & Wilkins, 1971.

3182. Kaneko, J. J., and C. E. Cornelius, eds. Clinical biochemistry of domestic animals. 2d ed. New York, Academic Press, 1971.

3183. Klemm, W. R. Animal electroencephalography. New York, Academic Press, 1969.

3184. Koepke, J. A. Guide to clinical laboratory diagnosis. New York, Appleton-Century-Crofts, 1969.

3185. Lenman, J. A. R., and A. E. Ritchie. Clinical electromyography. Philadelphia, Lippincott, 1970.

3186. Levinson, S. A., and R. P. MacFate. Clinical laboratory diagnosis. 7th ed. Philadelphia, Lea & Febiger, 1969.

3187. McDonald, G. A., T. C. Dodds, and Bruce Cruickshank. Atlas of haematology. 3d ed. Baltimore, Williams & Wilkins, 1970.

3188. Maizels, M., T. A. J. Prankerd, and J. D. M. Richards. Haematology in diagnosis and treatment. Baltimore, Williams & Wilkins, 1968.

3189. Medway, William, J. E. Prier, and J. S. Wilkinson. A textbook of veterinary clinical pathology. Baltimore, Williams & Wilkins, 1969.

3190. Miale, J. B. Laboratory medicine: hematology. 4th ed. St. Louis, Mosby, 1972.

3191. Nieburgs, H. E. Diagnostic cell pathology in tissues and smears. New York, Grune & Stratton, 1967.

3192. Ravel, Richard. Clinical laboratory medicine: application of laboratory data. Chicago, Year Book Medical Publishers, 1969.

3193. Sandritter, Walter. Color atlas and textbook of tissue cellular pathology. 3d ed. Chicago, Year Book Medical Publishers, 1969. (Translated and edited by W. B. Wartman.)

3194. Scheuer, P. J. Liver biopsy interpretation. Baltimore, Williams & Wilkins, 1968.

3195. Searcy, R. L. Diagnostic biochemistry. New York, McGraw-Hill, 1969.

3196. Umeda, Noritsugu. Diagnosis by gastrophotography. Philadelphia, Saunders, 1971.

3197. Wells, P. N. T. Physical principles of ultrasonic diagnosis. New York, Academic Press, 1969.

SYSTEMATIC DISEASES

Musculoskeletal

3198. Adams, O. R. Lameness in horses. 2d ed. Philadelphia, Lea & Febiger, 1966.

3199. Barry, H. C. Paget's disease of bone. Edinburgh, Livingstone, 1969.

3200. Bennett, J. C. Vistas in connective tissue diseases. Springfield, Ill., Thomas, 1968.

3201. Bethlem, J. Muscle pathology: introduction and atlas. New York, American Elsevier, 1970.

3202. Cyriax, J. H. Textbook of orthopaedic medicine. 5th ed. Baltimore, Williams & Wilkins, 1969.

3203. Gardner, D. L. Pathology of the connective tissue diseases. Baltimore, Williams & Wilkins, 1966.

3204. Goodgold, Joseph, and Arthur Eberstein. Electrodiagnosis of neuromuscular diseases. Baltimore, Williams & Wilkins, 1972.

3205. Harris, W. H., and R. P. Heaney. Skeletal renewal and metabolic bone disease. Boston, Little, Brown, 1970.

3206. Hickman, John. Veterinary orthopaedics. Edinburgh, Oliver & Boyd, 1964.

3207. Hollander, J. L., et al. Arthritis and allied conditions. 8th ed. Philadelphia, Lea & Febiger, 1972.

3208. Jaffe, H. L. Metabolic, degenerative, and inflammatory diseases of bones and joints. Philadelphia, Lea & Febiger, 1972.

3209. Lenman, J. A. R., and A. E. Ritchie. Clinical electromyography. Philadelphia, Lippincott, 1970.

3210. Licht, S. H., ed. Electrodiagnosis and electromyography. 3d ed. New Haven, Licht, 1971.

3211. Lichtenstein, Louis. Diseases of bone and joints. St. Louis, Mosby, 1970.

3212. Pettit, G. D., ed. Intervertebral disc protrusion in the dog. New York, Appleton-Century-Crofts, 1966.

3213. Salter, R. B. Textbook of disorders and injuries of the musculoskeletal system. Baltimore, Williams & Wilkins, 1970.

3214. Smillie, I. S. Osteochondritis dissecans; loose bodies in joints: etiology, pathology, treatment. Edinburgh, Livingstone, 1960.

3215. Sokoloff, Leon. The biology of degenerative joint disease. Chicago, University of Chicago Press, 1969.

3216. Strömberg, Berndt. The normal and diseased superficial flexor tendon in race horses. 1971. (*Acta radiologica.* Suppl. 305.)

3217. Symposium on equine bone and joint diseases. Cornell University, 1967. Proceedings. Ithaca, N.Y., 1968. (*Cornell veterinarian*, v. 58. Suppl.)

3218. Telford, I. R., and Larus Einarson. Experimental muscular dystrophies in animals: a comparative study. Springfield, Ill., Thomas, 1971.

Nervous and Sensory

3219. American Society of Veterinary Ophthalmology. Diseases of the canine and feline conjunctiva and cornea. Elkhart, Ind., American Animal Hospital Association, 1970.

3220. Brain, W. R., and J. N. Walton. Diseases of the nervous system. 7th ed. London, Oxford Univ. Press, 1969.

3221. Cohen, H. L., and Joel Brumlik. A manual of electroneuromyography. New York, Hoeber, 1968.

3222. Donaldson, D. D. Atlas of external diseases of the eye. v. 1- St. Louis, Mosby, 1966-
  v. 1: Congenital anomalies and systemic diseases. v. 2: Orbit, lacrimal apparatus, eyelids, and conjuctiva. v. 3: Cornea and sclera.

3223. Duke-Elder, W. S., ed. System of ophthalmology. v. 1- St. Louis, Mosby, 1958- (For titles of specific volumes see entry 1768.)

3224. Gelatt, K. N. Diagnostic procedures of comparative opthalmology. Elkhart,

Ind., American Animal Hospital Association, 1969.

3225. Gibbs, F. A., and Erna L. Gibbs. Medical electroencephalography. Reading, Mass., Addison-Wesley, 1967.

3226. Goldberg, R. E., and L. K. Sarin, eds. Ultrasonics in ophthalmology, diagnostic and therapeutic applications. Philadelphia, Saunders, 1967.

3227. International congress of ultrasonography in ophthalmology. 4th. Philadelphia, 1968. Ophthalmic ultrasound. Ed. by K. A. Gitter et al. St. Louis, Mosby, 1969.

3228. Jensen, H. E. Stereoscopic atlas of clinical ophthalmology of domestic animals. St. Louis, Mosby, 1971.

3229. Klemm, W. R. Animal electroencephalography. New York, Academic Press, 1969.

3230. Klostermann, G. F., et al. Color atlas of external manifestations of disease. Ed. by Marcel Tuchmann. New York, McGraw-Hill, 1966.

3231. McGrath, J. T. Neurologic examination of the dog, with clinicopathologic observations. 2d rev. ed. Philadelphia, Lea & Febiger, 1960.

3232. Magrane, W. G. Canine ophthalmology. 2d ed. Philadelphia, Lea & Febiger, 1971.

3233. Maumenee, A. E., and A. M. Silverstein. Immunopathology of uveitis. Baltimore, Williams & Wilkins, 1964.

3234. May, C. H. May and Worth's manual of diseases of the eye. 13th ed. Ed. by T. K. Lyle, A. G. Cross, and C. A. G. Cook. Philadelphia, Davis, 1968.

3235. Sloan symposium on uveitis. 4th. Baltimore, 1967. Clinical methods in uveitis. Ed. by S. B. Aronson et al. St. Louis, Mosby, 1968.

3236. Smorto, M. P., and J. V. Basmajian. Clinical electroneurography: an introduction to nerve conduction tests. Baltimore, Williams & Wilkins, 1972.

3237. Startup, F. G. Diseases of the canine eye. Baltimore, Williams & Wilkins, 1969.

3238. Sunderland, Sydney. Nerves and nerve injuries. Baltimore, Williams & Wilkins, 1968.

3239. Vinken, P. J., and G. W. Bruyn. Diseases of nerves. New York, American Elsevier, 1970.

3240. Virus diseases and the nervous system. Ed. by C. W. M. Whitty, J. T. Hughes, and F. O. MacCallum. Oxford, Blackwell, 1969.

3241. Walshe, F. M. R. Diseases of the nervous system. 10th ed. Baltimore, Williams & Wilkins, 1963.

Integumentary

3242. Domonkos, Anthony. Andrew's diseases of the skin: clinical dermatology. 6th ed. Philadelphia, Saunders, 1971.

3243. Fitzpatrick, T. B., et al., eds. Dermatology in general medicine. New York, McGraw-Hill, 1971.

3244. Graciansky, P. de, and M. Boulle. Color atlas of dermatology. 5 v. Chicago, Year Book Medical Publishers, 1965.

3245. Hashimoto, Ken, and W. F. Lever. Appendage tumors of the skin. Springfield, Ill., Thomas, 1968.

3246. Hunziker, Nicole. Experimental studies on guinea pig's eczema. New York, Springer-Verlag, 1969.

3247. Maddin, Stuart, and T. H. Brown, eds. Current dermatologic management. St. Louis, Mosby, 1970.

3248. Montgomery, Hamilton. Dermatopathology. New York, Hoeber Medical Division, Harper & Row, 1967.

3249. Muller, G. H. Canine skin lesions. Chicago, American Animal Hospital Association, 1968.

3250. ———, and R. W. Kirk. Small animal dermatology. Philadelphia, Saunders, 1969.

3251. Percival, G. H., G. L. Montgomery, and T. C. Dodds. Atlas of histopathology of the skin. 2d ed. Baltimore, Williams & Wilkins, 1962.

3252. Pillsbury, D. M., W. B. Shelley, and A. M. Kligman. Dermatology. Philadelphia, Saunders, 1957.

3253. ———, M. B. Sulzberger, and C. S. Livingood. Manual of dermatology. Philadelphia, Saunders, 1971.

3254. Pinkus, Hermann, and A. H. Mehregan. A guide to dermatohistopathology. New York, Appleton-Century-Crofts, 1969.

3255. Rook, A. J., D. S. Wilkinson, and F. J. G. Ebling, eds. Textbook of dermatology. 2 v. Philadelphia, Davis, 1968.

3256. Shelley, W. B. Consultations in dermatology. Philadelphia, Saunders, 1972.

Cardiovascular

3257. Allen, Edgar van Nuys. Peripheral vascular diseases. 4th ed. Ed. by J. F. Fairbairn II, J. L. Juergens, and J. A. Spittell, Jr. Philadelphia, Saunders, 1972.

3258. American Animal Hospital Association. A manual of clinical cardiology. Elkhart, Ind., 1972.

3259. Burch, G. E., and Travis Winsor. A primer of electrocardiography. 6th ed. Philadelphia, Lea & Febiger, 1972.

3260. Byron, F. B. The hypertensive vascular crisis: an experimental study. New York, Grune & Stratton, 1969.

3261. Cardiology: an encyclopedia of the cardio-vascular system. v. 1- (looseleaf). Ed. by A. A. Luisada. New York, McGraw-Hill, 1959- (For titles of specific volumes see entry 1833.)

3262. Comparative atherosclerosis. Ed. by J. C. Roberts and Reuben Straus. New York, Harper & Row, 1965.

3263. Ettinger, S. J., and P. F. Suter. Canine cardiology. Philadelphia, Saunders, 1970.

3264. Friedberg, C. K., ed. Physical diagnosis in cardiovascular disease. New York, Grune & Stratton, 1969.

3265. Goldman, M. J. Principles of clinical electrocardiography. 7th ed. Los Altos, Calif., Lange Medical Publications, 1970.

3266. Group of European nutritionists. Symposium. 2d. Oslo, 1963. Proceedings. Nutrition and cardiovascular diseases. Ed. by J. C. Somogyi. Basel, Karger, 1964. (Bibliotheca "Nutritio et Dieta," no. 6.)

3267. Progress in cardiology. v. 1- Ed. by P. N. Yu and J. F. Goodwin. Philadelphia, Lea & Febiger, 1972-

3268. Selye, Hans. Experimental cardiovascular diseases. 2 v. New York, Springer-Verlag, 1970.

Respiratory

3269. Aviado, D. M. The lung circulation. 2 v. Oxford, Pergamon Press, 1965.
  v. 1: Physiology and pharmacology.
  v. 2: Pathologic physiology and therapy of diseases.

3270. Bates, D. V., R. V. Christie, and P. T. Macklem. Respiratory function in disease: an introduction to the integrated study of the lung. 2d ed. Philadelphia, Saunders, 1971.

3271. Becker, Walter, et al. Atlas of otorhinolaryngology and bronchoesophagology. English ed. Philadelphia, Saunders, 1969.

3272. Buechner, H. A., ed. Management of fungus diseases of the lungs. Springfield, Ill., Thomas, 1971.

3273. Cole, R. B. Essentials of respiratory disease. Philadelphia, Lippincott, 1972.

3274. Crofton, J. W., and Andrew Douglas. Respiratory diseases. Philadelphia, Davis, 1969.

3275. Equen, Murdock. Magnetic removal of foreign bodies: the use of the Alnico magnet in the recovery of foreign bodies from the air passages, the esophagus, stomach and duodenum. Springfield, Ill., Thomas, 1957.

3276. Filley, G. F., et al. Pulmonary insufficiency and respiratory failure. Philadelphia, Lea & Febiger, 1967.

3277. Luisada, A. A. Pulmonary edema in man and animals. St. Louis, Green, 1970.

3278. Moore, F. D., et al. Post-traumatic pulmonary insufficiency: pathophysiology of respiratory failure and principles of respiratory care after surgical operations, trauma, hemorrhage, burns, and shock. Philadelphia, Saunders, 1969.

3279. Pace, W. R. Pulmonary physiology in clinical practice. 2d ed. Philadelphia, Davis, 1970.

3280. Rubin, E. H., and S. S. Siegelman. The lungs in systemic diseases. Springfield, Ill., Thomas, 1969.

3281. Spencer, Herbert. Pathology of the lung, excluding pulmonary tuberculosis. 2d ed. New York, Pergamon Press, 1968.

Gastrointestinal

3282. Benzie, David, and A. T. Phillipson. The alimentary tract of the ruminant. London, Oliver & Boyd, 1957.

3283. Bockus, H. L. R. Gastroenterology. 2d ed. 3 v. Philadelphia, Saunders, 1963-1965.

3284. Boley, S. J., S. S. Schwartz, and L. F. Williams, Jr., eds. Vascular disorders of the intestine. New York, Appleton-Century-Crofts, 1971.

3285. Brooks, F. P. Control of gastrointestinal function: an introduction to the physiology of the gastrointestinal tract. New York, Macmillan, 1970.

3286. Burket, L. W. Oral medicine: diagnosis and treatment. 6th ed. Philadelphia, Lippincott, 1971.

3287. Di Dio, L. J. A., and M. C. Anderson. The "sphincters" of the digestive system: anatomical, functional and surgical considerations. Baltimore, Williams & Wilkins, 1968.

3288. Dowdy, G. S., Jr. The biliary tract. Philadelphia, Lea & Febiger, 1969.

3289. Equen, Murdock. Magnetic removal of foreign bodies: the use of the Alnico magnet in the recovery of foreign bodies from the air passages, the esophagus, stomach and duodenum. Springfield, Ill., Thomas, 1957.

3290. Foulk, W. T., ed. Diseases of the liver. New York, McGraw-Hill, 1968.

3291. Immunology of the liver. Ed. by Martin Smith and Roger Williams. Philadelphia, Davis, 1971.

3292. Palmer, E. D. Practical points in gastroenterology. Flushing, N.Y., Medical Examination Publishing, 1971.

3293. Paton, Alexander. Liver disease. London, Heinemann, 1969.

3294. Paulson, Moses, ed. Gastroenterologic medicine. Philadelphia, Lea & Febiger, 1969.

3295. Piazza, Marcello. Experimental viral hepatitis. Springfield, Ill., Thomas, 1969.

3296. Scheuer, P. J. Liver biopsy interpretation. Baltimore, Williams & Wilkins, 1968.

3297. Schiff, Leon. Diseases of the liver. 3d ed. Philadelphia, Lippincott, 1969.

3298. Schindler, Rudolf. Gastroscopy: the endoscopic study of gastric pathology. 2d ed. New York, Hafner, 1966.

3299. Sheehy, T. W., and M. H. Floch. The small intestine: its function and diseases. New York, Hoeber Medical Division, Harper & Row, 1964.

3300. Sherlock, Sheila. Diseases of the liver and biliary system. 4th ed. Philadelphia, Davis, 1968.

3301. Umeda, Noritsugu. Diagnosis by gastrophotography. Philadelphia, Saunders, 1971.

3302. Woodruff, A. W., ed. Alimentary and haematological aspects of tropical disease. London, Arnold, 1970.

3303. Zuckerman, A. J. Virus diseases of the liver. New York, Appleton-Century-Crofts, 1970.

Genitourinary

3304. Arthur, G. H. Wright's veterinary obstetrics: including diseases of reproduction. 3d ed. Baltimore, Williams & Wilkins, 1964.

3305. Berlyne, G. M. A course in renal diseases. 2d ed. Philadelphia, Davis, 1968.

3306. Campbell, M. F., and J. H. Harrison. Urology. 3d ed. 3 v. Philadelphia, Saunders, 1970.

3307. De Wardener, H. E. The kidney. 3d ed. Boston, Little, Brown, 1967.

3308. Gompel, Claude, and S. G. Silverberg. Pathology in gynecology and obstetrics. Philadelphia, Lippincott, 1969.

3309. Osborne, C. A., D. G. Low, and D. R. Finco. Canine and feline urology. Philadelphia, Saunders, 1972.

3310. Roberts, S. J. Veterinary obstetrics and genital diseases. (Theriogenology.) 2d ed. Ithaca, N.Y., 1971 (distributed by Edwards Bros., Ann Arbor, Mich.).

3311. Sophian, John. Pregnancy nephropathy. v. 1-2. New York, Appleton-Century-Crofts, 1972.

3312. Stamey, T. A. Urinary infections. Baltimore, Williams & Wilkins, 1972.

3312a. Williams, G. A colour atlas of renal diseases. Fort Myers, Fla., Wolfe, 1972.

3313. Williams, W. L. The diseases of the genital organs of domestic animals. 3d ed. Worcester, Mass., Ethel Williams Plimpton, 1950.

3314. Wills, M. R. The biochemical consequences of chronic renal failure. Baltimore, University Park Press, 1971.

Endocrine

3315. Applied seminar on the laboratory diagnoses of endocrine disorders. Washington, D.C., 1969. Laboratory diagnosis of endocrine diseases. Ed. by F. W. Sunderman and F. W. Sunderman, Jr. St. Louis, Green, 1971.

3316. Bloodworth, J. M. B. Endocrine pathology. Baltimore, Williams & Wilkins, 1968.

3317. Conference on parturient paresis in dairy animals. University of Illinois, 1968. Parturient hypocalcemia. Ed. by J. J. B. Anderson. New York, Academic Press, 1970.

3318. Danowski, T. S. Outline of endocrine gland syndromes. 2d ed. Baltimore, Williams & Wilkins, 1968.

3319. Ellenberg, Max, and Harold Rifkin, eds. Diabetes mellitus: theory and practice. New York, McGraw-Hill, 1970.

3320. Escamilla, R. F., ed. Laboratory tests in diagnosis and investigation of endocrine function. 2d ed. Philadelphia, Davis, 1971.

3321. Hans Selye conference. Mont Tremblant, Quebec, 1968. Endocrine aspects of disease processes. Ed. by Gaëtan Jasmin. St. Louis, Green, 1968.

3322. Heidrich, H. J., and W. Renk. Diseases of the mammary glands of domestic animals. Philadelphia, Saunders, 1967.

3323. Joslin, E. P. Diabetes mellitus. Ed. by Alexander Marble et al. 11th ed. Philadelphia, Lea & Febiger, 1971.

3324. Methods in investigative and diagnostic endocrinology. Ed. by S. A. Berson. v. 1- Amsterdam, North-Holland Publishing, 1972-
   v. 1: The thyroid and biogenic amines. By J. E. Rall and I. J. Kopin.

3325. Oakley, W. G., D. A. Pyke, and K. W. Taylor. Clinical diabetes and its biochemical basis. Oxford, Blackwell, 1968.

3326. Schalm, O. W. A syllabus on the bovine mammary glands in health and disease. Davis, Univ. of California, Dept. of Clinical Pathology, School of Veterinary Medicine, 1960.

3327. Symington, Thomas. Functional pathology of the human adrenal gland. Baltimore, Williams & Wilkins, 1969.

3328. Warren, Shields, et al. The pathology of diabetes mellitus. 4th ed. Philadelphia, Lea & Febiger, 1966.

Hematopoietic

3329. Atamer, M. A. Blood diseases. New York, Grune & Stratton, 1963.

3330. Bishop, C. W., and D. M. Surgenor, eds. The red blood cell: a comprehensive treatise. New York, Academic Press, 1964.

3331. Cartwright, G. E. Diagnostic laboratory hematology. 4th ed., rev. & enl. New York, Grune & Stratton, 1968.

3332. Dacie, J. V., and S. M. Lewis. Practical haematology. 4th ed. London, Churchill, 1968.

3333. Dameshek, William, and R. M. Dutcher, eds. Perspectives in leukemia. New York, Grune & Stratton, 1968.

3334. ——, and Frederick Gunz. Leukemia. 2d ed. New York, Grune & Stratton, 1964.

3335. De Gruchy, G. C. Clinical haematology in medical practice. 3d ed. Philadelphia, Davis, 1970.

3336. Dougherty, W. M. Introduction to hematology. St. Louis, Mosby, 1971.

3337. Harris, J. W., and R. W. Kellermeyer. The red cell: production, metabolism, destruction; normal and abnormal. Cambridge, Mass., Harvard Univ. Press, 1970.

3338. Hematology. Ed. by W. J. Williams et al. New York, McGraw-Hill, 1972.

3339. Hutchison, H. E. An introduction to the haemoglobinopathies and the methods used for their recognition. London, Arnold, 1967.

3340. International conference on leukemia-lymphoma. Proceedings. Ed. by C. J. D. Zarafonetis. Philadelphia, Lea & Febiger, 1968.

3341. Laufman, Harold, and R. B. Erichson. Hematologic problems in surgery. Philadelphia, Saunders, 1970.

3342. Leavell, B. S., and O. A. Thorup, Jr. Fundamentals of clinical hematology. 3d ed. Philadelphia, Saunders, 1971.

3343. Ljungqvist, Ulf. Platelet response to acute haemorrhage in the dog. Stockholm, Almqvist & Wiksell, 1970. (*Acta chirurgica scandinavica*. Suppl. 411.)

3344. McDonald, G. A., T. C. Dodds, and Bruce Cruickshank. Atlas of haematology. 3d ed. Baltimore, Williams & Wilkins, 1970.

3345. Maizels, M., T. A. J. Prankerd, and J. D. M. Richards. Haematology in diagnosis and treatment. Baltimore, Williams & Wilkins, 1968.

3346. Mengel, C. E., Emil Frei III, and Ralph Nachman. Hematology: principles and practice. Chicago, Year Book Medical Publishers, 1972.

3347. Miale, J. B. Laboratory medicine: hematology. 4th ed. St. Louis, Mosby, 1972.

3348. Myeloproliferative disorders of animals and man. Ed. by W. J. Clarke, E. B. Howard, and P. L. Hackett. Washington, D.C., U.S. Atomic Energy Commission, 1970. (U.S. Atomic Energy Commission. Annual Hanford biology symposium. 8th. Richland, Washington, 1968. Also A.E.C. symposium series, no. 19.)

3349. Mecherles, T. F., D. M. Allen, and H. E. Finkel. Clinical disorders of hemoglobin structure and synthesis. New York, Appleton-Century-Crofts, 1969.

3350. Platt, W. R. Color atlas and textbook of hematology. Philadelphia, Lippincott, 1969.

3351. Progress in hemostasis and thrombosis. v. 1- New York, Grune & Stratton, 1972.

3352. Quick, A. J. Bleeding problems in clinical medicine. Philadelphia, Saunders, 1970.

3353. Rapaport, S. I. Introduction to hematology. 1st ed. New York, Harper & Row, 1971.

3354. Schalm, O. W. Veterinary hematology. 2d ed. Philadelphia, Lea & Febiger, 1965.

3355. Weinman, David, and Miodrag Ristic. Infectious blood diseases of man and animals. 2 v. New York, Academic Press, 1968.

3356. Whitby, L. E. H., and C. J. C. Britton. Disorders of the blood: diagnosis, pathology, treatment, technique. 10th ed. New York, Grune & Stratton, 1969.

3357. Wintrobe, M. M. Clinical hematology. 6th ed. Philadelphia, Lea & Febiger, 1967.

3358. Woodruff, A. W., ed. Alimentary and haematological aspects of tropical disease. London, Arnold, 1970.

3359. Zmijewski, C. M., June L. Fletcher, and R. L. St. Pierre. Immunohematology. New York, Appleton-Century-Crofts, 1968.

Note. For additional entries see the section on "Systematic Pathology and Neoplasia: Hematopoietic" (entries 3556-3564) in Chap. 18.

18

PATHOLOGY, INCLUDING NEOPLASIA

GENERAL PATHOLOGY

3360. Aggarwala, A. C., and R. M. Sharma. Laboratory manual of practical veterinary pathology. New York, Asia Publishing, 1962.

3361. Anderson, W. A. D. Pathology. 6th ed. 2 v. St. Louis, Mosby, 1971.

3362. ——, and T. M. Scotti. Synopsis of pathology. 8th ed. St. Louis, Mosby, 1972.

3363. Boyd, William. An introduction to the study of disease. 6th ed. Philadelphia, Lea & Febiger, 1971.

3364. ——. Textbook of pathology. 8th ed. Philadelphia, Lea & Febiger, 1970.

3365. Color atlas of pathology. Prepared under the auspices of the U.S. Naval Medical School. v. 1- Philadelphia, Lippincott, 1950-

3366. Crowley, L. V. Introductory concepts in pathology: a manual for students in the health professions. Chicago, Year Book Medical Publishers, 1972.

3367. Fazzini, Eugene, et al. A manual for surgical pathologists. Springfield, Ill., Thomas, 1972.

3368. Florey, H. W., ed. General pathology. 4th ed. Philadelphia, Saunders, 1970.

3369. Frei, Walter, ed. Algemeine Pathologie für Tierärzte und Studierende der Tiermedizin. 6, neubearb. Aufl., hrsg. von H. Stünzl und E. Weiss unter Mitwirkung von K. Kammrich et al. Berlin, Parey, 1972.

3370. Frohlich, E. D., ed. Pathophysiology: altered regulatory mechanisms in disease. Philadelphia, Lippincott, 1972.

3371. Gardner, A. F. Paramedical pathology: fundamentals of pathology for the allied medical occupations. Springfield, Ill., Thomas, 1972.

3372. Gresham, G. A. Color atlas of general pathology. Chicago, Year Book Medical Publishers, 1971.

3373. Hutyra, Ference, Joseph Marek, and Rudolph Manninger. Special pathology and therapeutics of the diseases of domestic animals. 5th English ed. Ed. by J. R. Grieg, J. R. Mohler, and Adolf Eichhorn. Chicago, Eger, 1946.

3374. International Academy of Pathology. Monographs in pathology. v. 1- Baltimore, Williams & Wilkins, 1960- (For specific titles of volumes see entry 0756.)

3375. Jennings, A. R. Animal pathology. London, Bailliere, Tindall & Cassell, 1970.

3376. Joest, Ernst. Handbuch der speziellen pathologischen Anatomie der Hausteire. 3d Aufl. Bd. 1- hrsg. von J. Dobberstein et al. Berlin, Parey, 1963-

3377. Jubb, K. V. F., and P. C. Kennedy. Pathology of domestic animals. 2d ed. 2 v. New York, Academic Press, 1970.

3378. Lapin, B. A., and L. A. Yakovleva. Comparative pathology in monkeys. Springfield, Ill., Thomas, 1963.

3379. La Via, Mariano F., and R. B. Hill. Principles of pathobiology. New York, Oxford Univ. Press, 1971.

3380. Methods and achievements in experimental pathology. v. 1- Ed. by E. Bajusz and G. Jasmin. Basel, Karger, 1966- (For specific titles see entry 0769.)

3381. Minckler, Jeff, H. B. Anstall, and T. M. Minckler, eds. Pathobiology: an introduction. St. Louis, Mosby, 1971.

3382. Muir, Robert. Textbook of pathology. 9th ed. Rev. by D. F. Cappell and J. R. Anderson. London, Arnold, 1971.

3383. Nieberle, Karl, and Paul Cohrs. Textbook of the special pathological anatomy of domestic animals. New York, Pergamon Press, 1967.

3384. Pathology of simian primates. Ed. by R. N. Fiennes. v. 1- Basel, Karger, 1972-
Part I—General pathology. Part II—Infectious and parasitic diseases.

3385. Pathophysiology: altered regulatory mechanisms in disease. Ed. by E. D. Frohlich. Philadelphia, Lippincott, 1972.

3386. Peery, T. M., and F. N. Miller, Jr. Pathology: a dynamic introduction to medicine and surgery. 2d ed. Boston, Little, Brown, 1971.

3387. Robbins, S. L. Pathology. 3d ed. 2 v. Philadelphia, Saunders, 1967.

3388. Runnells, R. A., W. S. Monlux, and A. W. Monlux. Principles of veterinary pathology. 7th ed. Ames, Iowa State Univ. Press, 1965.

3389. Saphir, Otto, ed. A text on systemic pathology. 2 v. New York, Grune & Stratton, 1958-1959.

3390. Scadding, J. G. Sarcoidosis. New York, Barnes & Noble, 1967.

3391. Selye, Hans. Calciphylaxis. Chicago, Univ. of Chicago Press, 1962.

3392. Sisson, J. A. The bare facts of general pathology. Philadelphia, Lippincott, 1971.

3393. ———. The bare facts of systemic pathology. Philadelphia, Lippincott, 1972.

3394. Smith, H. A., and T. C. Jones. Veterinary pathology. 4th ed. Philadelphia, Lea & Febiger, 1972.

3395. Snively, W. D., Jr., and Donna R. Beshear. Textbook of pathophysiology. Philadelphia, Lippincott, 1972.

3396. Sodeman, W. A., and W. A. Sodeman, Jr. Pathologic physiology: mechanisms of disease. 4th ed. Philadelphia, Saunders, 1967.

3397. Walter, J. B., and M. S. Israel. General pathology. 3d ed. London, Churchill, 1970.

3398. Willis, R. A., and A. T. Willis. Principles of pathology and bacteriology. 3d ed. New York, Appleton-Century-Crofts, 1972.

3399. Wolman, Moshe. Pigments in pathology. New York, Academic Press, 1969.

3400. Wright, G. P., and W. St. C. Symmers, eds. Systemic pathology. 2 v. New York, American Elsevier, 1967.

3401. Zweifach, B. W., Lester Grant, and R. T. McCluskey. The inflammatory process. New York, Academic Press, 1965.

GENERAL NEOPLASIA

3402. Andrewes, C. H. Viruses and cancer. London, Weidenfeld & Nicolson, 1970.

3403. Andrews, G. S. Exfoliative cytology. Springfield, Ill., Thomas, 1971.

3404. Berenblum, Isaac. Cancer research today. New York, Pergamon Press, 1967.

3405. Burdette, W. J., ed. Viruses inducing cancer: implications for therapy. Salt Lake City, Univ. of Utah Press, 1966.

3406. Burstone, M. S. Enzyme histochemistry, and its applications in the study of neoplasms. New York, Academic Press, 1962.

3407. Busch, Harris. An introduction to the

biochemistry of the cancer cell. New York, Academic Press, 1962.

3408. ——, ed. Methods in cancer research. 6 v. New York, Academic Press, 1967-1971.

3409. Campbell, J. G. Tumours of the fowl. Philadelphia, Lippincott, 1969.

3410. Cotchin, E. Neoplasms of the domesticated mammals: a review. Farnham Royal, England, Commonwealth Agricultural Bureaux, 1956.

3411. Dalton, A. J., and Francoise Haguenau. Tumors induced by viruses: ultrastructural studies. New York, Academic Press, 1962.

3412. Daudel, Pascaline, and Raymond Daudel. Chemical carcinogenesis and molecular biology. New York, Wiley, 1966.

3413. Evans, R. W. Histological appearances of tumours. 2d ed. Baltimore, Williams & Wilkins, 1966.

3414. Everson, T. C., and W. H. Cole. Spontaneous regression of cancer. Philadelphia, Saunders, 1966.

3415. Foulds, Leslie. Neoplastic development. v. 1- New York, Academic Press, 1969-

3416. Gann monographs. v. 1- Tokyo, Japanese Cancer Association, 1966- (Consists of proceedings of international conferences and symposia. For specific titles see entry 0754.)

3417. Gauze, G. F. Microbial models of cancer cells. New York, Wiley, 1966.

3418. Graham, Ruth M. The cytologic diagnosis of cancer. 2d ed. Philadelphia, Saunders, 1963.

3419. Green, H. N., et al. An immunological approach to cancer. London, Butterworth, 1967.

3420. Gross, Ludwik. Oncogenic viruses. 2d ed. New York, Pergamon Press, 1970.

3421. Harris, J. E., and J. G. Sinkovics. The immunology of malignant disease. St. Louis, Mosby, 1970.

3422. Huxley, J. S. Biological aspects of cancer. New York, Harcourt, Brace & World, 1958.

3423. International cancer congress. 10th. Houston, 1970. Oncology. 5 v. Ed. by R. L. Clark et al. Chicago, Year Book Medical Publishers, 1970.
v. 1-A: Cellular and molecular mechanisms of carcinogenesis. v. 1-B: Regulation of gene expression. v. 2: Experimental cancer therapy. v. 3: Diagnosis and management of cancer: general considerations. v. 4: Diagnosis and management of cancer: specific sites. v. 5-A: Environmental causes. v. 5-B: Epidemiology and demography. v. 5-C: Cancer education. (A supplementary volume contains abstracts of proffered papers that are not included in the volumes noted above.)

3424. International conference of tissue culture in cancer research. Tokyo, 1966. Cancer cells in culture. Ed. by Hajim Katsuta. State College, Pa., Univ. Park Press, 1968.

3425. International conference on the morphological precursors of cancer. University of Perugia, 1961. Morphological precursors of cancer. Ed. by Lucio Severi. Perugia, Division of Cancer Research, 1962.

3426. International histological classification of tumours (including slides and text). v. 1- Geneva, World Health Organization, 1967- (For specific titles see entry 0761.)

3427. International Union Against Cancer. UICC monograph series. v. 1-2—Copenhagen, Munksgaard, 1967. v. 3—Berlin, Springer-Verlag, 1967- (For specific titles see entry 0767.)

3428. Lepetit colloquia on biology and medicine. International Lepetit colloquium. 2d. Paris, 1970. The biology of oncogenic viruses. Ed. by L. G. Silvestri. New York, American Elsevier, 1971.

3429. Mihich, Enrico, ed. Immunity, cancer and chemotherapy. New York, Academic Press, 1967.

3430. Moulton, J. E. Tumors in domestic animals. Berkeley, Univ. of California Press, 1961.

3431. Naib, Z. M. Exfoliative cytopathology. Boston, Little, Brown, 1970.

3432. Nieburgs, H. E. Diagnostic cell pathology in tissues and smears. New York, Grune & Stratton, 1967.

3433. Progress in experimental tumor research. v. 1- Ed. by F. Homburger. New York, Karger, 1960- (For specific titles see entry 0778.)

3433a. Raven, R. W., ed. Cancer. 9 v. London, Butterworth, 1957-1963.
v. 1: Part 1—Research into causation. v. 2: Part 2—Pathology of malignant tumours. v. 3: Part 3—Additional pathological aspects; Part 4—Geography of cancer; Part 5—Occupational cancer; Part 6—Cancer education; Part 7—Cancer detection. v. 4: Part 8—Clinical aspects. v. 5: Part 9—Radiotherapy. v. 6: Part 10—Chemotherapy; Part 11—Public health and nursing aspects of cancer. v. 7: Indices. v. 8: Progress volume 1960. v. 9: Progress volume 1963.

3434. Recent advances in human tumor virology and immunology: proceedings of the 1st International symposium of the Princess Takamatsu Cancer Research Fund. Ed. by Waro Nakahara et al. Baltimore, University Park Press, 1971.

3435. Recent results in cancer research. v. 1- New York, Springer-Verlag, 1965- (For specific titles see entry 0780.)

3436. Reid, Eric. Biochemical approaches to cancer. New York, Pergamon Press, 1965.

3437. Research using transplanted tumors of laboratory animals: cross-referenced bibliography. Ed. by D. C. Roberts. v. 1- London, Imperial Cancer Research Fund, 1964- (Issued annually.)

3438. Snapper, I., and A. Kahn. Myelomatosis: fundamentals and clinical features. Baltimore, University Park Press, 1971.

3439. Stewart, H. L., et al. Transplantable and transmissible tumors of animals. Washington, D.C., U.S. Armed Forces Institute of Pathology, 1959. (*Atlas of tumor pathology*, Sect. XII, fasc. 40.)

3440. Symposium on oncogenicity of virus vaccines. Zagreb, Yugoslavia, 1968. Proceedings. Zagreb, Immunoloske Zavod Rockefellervoa, 1968.

3441. Takahashi, Masayoshi. Color atlas of cancer cytology. Philadelphia, Lippincott, 1971.

3442. U.S. Armed Forces Institute of Pathology. International Reference Center for Comparative Oncology. Bibliography of comparative oncology. Comp. by F. M. Garner and R. J. Brown. Washington, D.C., 1967.

3443. U.S. National Cancer Institute. Cancer and virus: a guide and annotated bibliography to monographs, reviews, symposia, and survey articles with emphasis on human neoplasm. 1950-1963. Ed. by Elizabeth Koenig. Washington, D.C., 1966. (U.S. Public Health Service publ. no. 1424.)

3444. ———. Monographs. v. 1- Bethesda, Md., 1959- (For specific titles see entry 0790.)

3445. U.S. National Research Council. Committee on Pathology. Atlas of tumor pathology. 1st series. Washington, D.C., 1949-1968. (40 fascicles published in ten sections covering all body systems. Superseded by entry 3446.)

3446. Universities Associated for Research and Education in Pathology. Atlas of tumor pathology. 2d series. Washington, D.C., U.S. Armed Forces Institute of Pathology, 1967- (Continues entry 3445.)

3447. Vaillancourt, Pauline M. Bibliographic control of the literature of oncology, 1800-1960. Metuchen, N.J., Scarecrow Press, 1969.

3448. Waldenström, Jan. Diagnosis and treatment of multiple myeloma. New York, Grune & Stratton, 1970.

3449. Weiss, D. W., ed. Immunological parameters of host-tumor relationships. New York, Academic Press, 1971.

3450. Williamson, M. E. Histologic patterns in tumor pathology. New York, Hoeber, 1968.

3451. Willis, R. A. Pathology of tumours. 4th ed. New York, Appleton-Century-Crofts, 1967.

3452. Zil'ber, L. A., and G. I. Abelev. Virology and immunology of cancer. New York, Pergamon Press, 1968.

HISTOPATHOLOGY

3453. Bolande, R. P. Cellular aspects of developmental pathology. Philadelphia, Lea & Febiger, 1967.

3454. Curran, R. C. Color atlas of histopathology. New York, Oxford Univ. Press, 1966.

3455. Hughes, Helena E., and T. C. Dodds. Handbook of diagnostic cytology. Baltimore, Williams & Wilkins, 1968.

3456. Koss, L. G. Diagnostic cytology and its histopathologic bases. 2d ed. Philadelphia, Lippincott, 1968.

3457. Putt, F. A. Manual of histopathological staining methods. New York, Wiley-Interscience, 1972.

3458. Sandritter, Walter. Color atlas and textbook of tissue cellular pathology. 3d ed. Chicago, Year Book Medical Publishers, 1969. (Translated and edited by W. B. Wartman.)

3459. ——, and Julius Schorn. Histopathologie: Lehrbuch und Atlas für Studierende und Arzte. Stuttgart, Schattauer-Verlag, 1965.

CLINICAL PATHOLOGY

3460. Applied seminar on the clinical pathology of hemoglobin, its precursors and metabolites. Washington, D.C., 1962. Hemoglobin: its precursors and metabolites. Ed. by F. W. Sunderman and F. W. Sunderman, Jr. Philadelphia, Lippincott, 1964.

3461. Arnow, L. E. Introduction to physiological and pathological chemistry. 8th ed. St. Louis, Mosby, 1972.

3462. Arquembourg, P. C., J. E. Salvaggio, and J. N. Bickers. Primer of immunoelectrophoresis, with interpretation of pathologic human serum patterns. White Plains, N.Y., Phiebig, 1970.

3463. Benjamin, Maxine M. Outline of veterinary clinical pathology. 2d ed. Ames, Iowa State Univ. Press, 1965.

3464. Bologna, Camillo V. Understanding laboratory medicine. St. Louis, Mosby, 1971.

3465. Coles, E. H. Veterinary clinical pathology. Philadelphia, Saunders, 1967.

3466. Doxey, D. L. Veterinary clinical pathology. Baltimore, Williams & Wilkins, 1971.

3467. Dyke, S. C. Recent advances in clinical pathology. 5th series. Boston, Little, Brown, 1968.

3468. Gray, C. H. Clinical chemical pathology. 6th ed. Baltimore, Williams & Wilkins, 1971.

3469. Kaneko, J. J., and C. E. Cornelius, eds. Clinical biochemistry of domestic animals. 2d ed. New York, Academic Press, 1970-

3470. Koepke, J. A. Guide to clinical laboratory diagnosis. New York, Appleton-Century-Crofts, 1969.

3471. McAllister, R. A. Theory of chemical pathology technique. New York, Appleton-Century-Crofts, 1967.

3472. Manuel, Y., J. P. Revillard, and H. Betuel. Proteins in normal and pathological urine. Baltimore, University Park Press, 1970.

3473. Medway, William, J. E. Prier, and J. S. Wilkinson. A textbook of veterinary clinical pathology. Baltimore, Williams & Wilkins, 1969.

3474. Rezek, P. R., and Max Millard. Autopsy pathology, a guide for pathologists and clinicians. Springfield, Ill., Thomas, 1963.

3475. Textbook of clinical pathology. 8th ed. Ed. by S. E. Miller and J. M. Weller. Baltimore, Williams & Wilkins, 1971.

3476. Wells, B. B., and J. A. Halstead. Clinical pathology: interpretation and application. 4th ed. Philadelphia, Saunders, 1967.

IMMUNOPATHOLOGY

3477. International symposium on immunopathology. 1st- Proceedings. Ed. by P. A. Miescher and P. Grabar. Basel, Schwabe, 1959- (For additional coverage see entry 0762.)

3478. Maumenee, A. E., and A. M. Silverstein. Immunopathology of uveitis. Baltimore, Williams & Wilkins, 1964.

3479. Miescher, P. A., and H. J. Mueller-Eberhard, eds. Textbook of immunopathology. 2 v. New York, Grune & Stratton, 1968-1969.

3480. Waksman, B. H. Atlas of experimental immunology and immunopathology. New Haven, Yale Univ. Press, 1970.

SYSTEMATIC PATHOLOGY AND NEOPLASIA

Musculoskeletal

3481. Angevine, C. D., et al. The immunochemistry and biochemistry of connective tissue and its disease states. New York, Karger, 1970.

3482. Commissie voor Beentumoren. Radiological atlas of bone tumours. v. 1- Baltimore, Williams & Wilkins, 1966-

3483. Dahlin, D. C. Bone tumors: general aspects and data on 3,987 cases. 2d ed. Springfield, Ill., Thomas, 1967.

3484. Enneking, W. F. Principles of musculoskeletal pathology. Gainesville, Fla., Storter, 1970.

3485. Graham, W. D. Bone tumours. London, Butterworth, 1966.

3486. Jacobson, S. A. The comparative pathology of the tumors of bone. Springfield, Ill., Thomas, 1971.

3487. Lichtenstein, Louis. Bone tumors. 4th ed. St. Louis, Mosby, 1972.

3488. Owen, L. N. Bone tumours in man and animals. London, Butterworth, 1969.

3489. Rubin, Philip. Dynamic classification of bone dysplasias. Chicago, Year Book Medical Publishers, 1964.

Nervous and Sensory

3490. Adams, R. D., and R. L. Sidman. Introduction to neuropathology. New York, McGraw-Hill, 1968.

3491. Blackwood, William, T. C. Dodds, and J. C. Sommerville. Atlas of neuropathology. 2d ed. Baltimore, Williams & Wilkins, 1964.

3492. Dublin, W. B. Fundamentals of neuropathology. 2d ed. Springfield, Ill., Thomas, 1967.

3493. Greenfield, J. G. Neuropathology. 2d ed. Ed. by W. Blackwood et al. Baltimore, Williams & Wilkins, 1963.

3494. Innes, J. R. M., and L. Z. Saunders. Comparative neuropathology. New York, Academic Press, 1962.

3495. Jane, J. A., and David Yashon. Cytology of tumors affecting the nervous system. Springfield, Ill., Thomas, 1969.

3496. Minckler, Jeff, ed. Pathology of the nervous system. v. 1- New York, McGraw-Hill, 1968-

3497. Neuropoisons: their pathophysiological actions. v. 1- Ed. by L. L. Simpson. New York, Plenum Press, 1971-

3498. Penfield, Wilder. Cytology and cellular pathology of the nervous system. 3 v. New York, Hafner, 1965. (Reprint of 1932 edition.)

3499. Poon, T. P., Asao Hirano, and H. M. Zimmerman. Electron microscopic atlas of brain tumors. New York, Grune & Stratton, 1971.

3500. Rio-Hortega, Pio del. The microscopic anatomy of tumors of the central and peripheral nervous system. Springfield, Ill., Thomas, 1962. (Translated by Anselmo Pineda, G. V. Russell, and K. M. Earle.)

3501. Russell, Dorothy S., and L. J. Rubinstein. Pathology of tumours of the nervous system. 3d ed. Baltimore, Williams & Wilkins, 1969.

3502. Slager, Ursula E. T. Basic neuropathology. Baltimore, Williams & Wilkins, 1970.

3503. Tedeschi, C. G., ed. Neuropathology: methods and diagnosis. Boston, Little, Brown, 1970.

3504. Zacks, S. I. Atlas of neuropathology. New York, Harper & Row, 1971.

3505. Zülch, K. J. Brain tumors: biology and pathology. 2d ed. New York, Springer, 1965.

Integumentary

3506. Colyer, Frank. Variations and diseases of the teeth of animals. London, John Bale, Sons, 1936.

3507. Hashimoto, Ken, and W. F. Lever. Appendage tumors of the skin. Springfield, Ill., Thomas, 1968.

3508. Lever, W. F. Histopathology of the skin. 4th ed. Philadelphia, Lippincott, 1967.

3509. Montgomery, Hamilton. Dermatopathology. New York, Hoeber Medical Division, Harper & Row, 1967.

3510. Percival, G. H., G. L. Montgomery, and T. C. Dodds. Atlas of histopathology of the skin. 2d ed. Baltimore, Williams & Wilkins, 1962.

3511. Schwartzman, R. M., and Frank Král. Atlas of canine and feline dermatoses. Philadelphia, Lea & Febiger, 1967.

3512. ———, and Milton Orkin. Comparative study of skin disease of dog and man. Springfield, Ill., Thomas, 1962.

Cardiovascular

3513. Adams, F. H., H. J. C. Swan, and V. E. Hall. Pathophysiology of congenital heart disease. Berkeley, Univ. of California Press, 1970. (UCLA forum in medical sciences, no. 10.)

3514. Bajusz, Eörs, ed. Electrolytes and cardiovascular diseases: physiology, pathology, therapy. 2 v. Baltimore, Williams & Wilkins, 1965-1966.

3515. Baroldi, Giorgio, and Guiseppe Scomazzoni. Cornary circulation in the normal and pathologic heart. Washington, D.C., U.S. Armed Forces Institute of Pathology, 1967.

3516. Davies, M. J. Pathology of conducting tissue of the heart. London, Butterworth, 1971.

3517. Edwards, J. E., et al. Congenital heart disease: correlation of pathologic anatomy and angiocardiography. 2 v. Philadelphia, Saunders, 1965.

3518. Friedberg, C. K. Diseases of the heart. 3d ed. 2 v. Philadelphia, Saunders, 1966.

3519. Gould, S. E. Pathology of the heart and blood vessels. 3d ed. Springfield, Ill., Thomas, 1968.

3520. Hudson, R. E. B. Cardiovascular pathology. 3 v. Baltimore, Williams & Wilkins, 1965-1970.

3521. Lannigan, Robert. Cardiac pathology. London, Butterworth, 1966.

3522. New York Heart Association. Diseases of the heart and blood vessels: nomenclature and criteria for diagnosis. 6th ed. Boston, Little, Brown, 1964.

3523. Selye. Hans. Experimental cardiovascular diseases. 2 v. New York, Springer-Verlag, 1970.

Respiratory

3524. Bates, D. V., R. V. Christie, and P. T. Macklem. Respiratory function in disease: an introduction to the integrated study of the lung. 2d ed. Philadelphia, Saunders, 1971.

3525. Conference on the morphology of experimental respiratory carcinogenesis. Oak Ridge National Laboratory, Gatlinburg, Tenn., 1970. Proceedings. Ed. by P. Nettesheim et al. Springfield, Va., U.S. Dept. of Commerce, 1970. (AEC symposium series, 21.)

3526. International conference on lung tumours in animals. University of Perugia, 1965. Lung tumours in animals. Ed. by Lucio Severi. Perugia, Division of Cancer Research, 1966.

3527. Spencer, Herbert. Pathology of the lung, excluding pulmonary tuberculosis. 2d ed. New York, Pergamon Press, 1968.

3528. Tsuboi, Eitaka. Atlas of transbronchial biopsy: early diagnosis of peripheral pulmonary carcinomas. 1st ed. Baltimore, Williams & Wilkins, 1970.

3529. Vascular diseases of the lung. Comp. and ed. by H. A. Lyons. Springfield, Ill., Thomas, 1967.

Gastrointestinal

3530. Gardner, A. F. Pathology of oral manifestations of systematic diseases. New York, Hafner, 1972.

3531. Gibbs, D. D. Exfoliative cytology of the stomach. London, Butterworth, 1968.

3532. Henning, Norbert, and Siegfried Witte. Atlas of gastrointestinal cytodiagnosis. 2d rev. and enl. ed. Stuttgart, Thieme, 1970.

3533. Lowe, W. C. Neoplasms of the gastrointestinal tract. Flushing, N.Y., Medical Examination Publishing, 1972.

3534. McNeer, Gordon, and G. T. Pack. Neoplasms of the stomach. Philadelphia, Lippincott, 1967.

3535. Valdes-Dapena, A. M., and G. N. Stein. Morphologic pathology of the alimentary canal: gross, radiographic and microscopic. Philadelphia, Saunders, 1970.

3536. Wepler, Wilhelm, and Egmont Wildhirt. Clinical histopathology of the liver: an atlas. New York, Grune & Stratton, 1972. (American edition arranged by Leo Lowbeer.)

Genitourinary

3537. Bennington, J. L., and R. M. Kradjian. Renal carcinoma. Philadelphia, Saunders, 1967.

3538. Caldwell, W. L. Cancer of the urinary bladder. St. Louis, Green, 1970.

3539. Dallenbach-Hellweg, G. Histopathology of the endometrium. New York, Springer-Verlag, 1972. (English translation by F. D. Dallenbach.)

3540. Emmett, J. L., and D. M. Witten. Clinical urography: an atlas and textbook of roentgenologic diagnosis. 3d ed. Philadelphia, Saunders, 1971.

3541. Golden, Abner, and J. F. Mather. The kidney: structure and function in disease. Baltimore, Williams & Wilkins, 1971.

3542. Hasegawa, Toshio. Trophoblastic neoplasia: its basic and clinical aspects. Baltimore, Williams & Wilkins, 1971.

3543. Kirkman, Hadley. Estrogen-induced tumors of the kidney in the Syrian hamster. Bethesda, Md., U.S. Govt. Printing Office, 1959. (U.S. National Cancer Institute monograph 1.)

3544. Lang, E. K. Roentgenographic diagnosis of bladder tumors. Springfield, Ill., Thomas, 1968.

3545. ———. The roentgenographic diagnosis of renal mass lesions. St. Louis, Green, 1971.

3546. Maltry, Emile, ed. Benign and malignant tumors of the urinary bladder. Flushing, N.Y., Medical Examination Publishing, 1971.

3547. Patten, S. F. Diagnostic cytology of the uterine cervix. New York, Karger, 1969. (Monographs in clinical cytology, v. 3.)

3548. Reagan, J. W., and A. B. P. Ng. The cells of uterine adenocarcinoma. Basel, Karger, 1965.

3549. Sarma, K. P. Tumours of the urinary bladder. New York, Appleton-Century-Crofts, 1969.

3550. Scott, J. W. Stereocolposcopic atlas of the uterine cervix. Kendall, Fla., Zephyr Publishers, 1971.

Endocrine

3551. Bloodworth, J. M. B. Endocrine pathology. Baltimore, Williams & Wilkins, 1968.

3552. Hale, J. F., and E. F. Scowen. Thymic tumours: their association with myasthenia gravis and their treatment by radiotherapy. London, Lloyd-Luke, 1967.

3553. Simionescu, Nicoale. The histogenesis of thyroid cancer. London, Heinemann, 1970.

3554. Smithers, D. W., ed. Tumours of the thyroid gland. Edinburgh, Livingstone, 1970. (Monographs on neoplastic disease at various sites, v. 6.)

3555. Warner, Nancy E. Basic endocrine pathology. Chicago, Year Book Medical Publishers, 1971.

Hematopoietic

3556. Amromin, G. D. Pathology of leukemia. New York, Harper & Row, 1968.

3557. Burkhardt, Rolf. Bone marrow and bone tissue: color atlas of clinical histopathology. New York, Springer-Verlag, 1971. (Translated by W. J. Hirsch.)

3558. Dameshek, William, and R. M. Dutcher, eds. Perspectives in leukemia. New York, Grune & Stratton, 1968.

3559. ———, and Frederick Gunz. Leukemia. 2d ed. New York, Grune & Stratton, 1964.

3560. International conference on leukemia-lymphoma. Ann Arbor, Michigan, 1967. Proceedings. Ed. by C. J. D. Zarafonetis. Philadelphia, Lea & Febiger, 1968.

3561. Israels, M. C. G. Atlas of bone-marrow pathology. 4th ed. New York, Grune & Stratton, 1971.

3562. Mackenzie, D. H. The differential diagnosis of fibroblastic disorders. Oxford, Blackwell, 1970.

3563. Pegg, D. E. Bone marrow transplantation. London, Lloyd-Luke, 1966.

3564. Rich, M. A. Experimental leukemia. New York, Appleton-Century-Crofts, 1968.

19

SURGERY AND OBSTETRICS

GENERAL

3565. Annis, J. R., and A. R. Allen. An atlas of canine surgery: basic surgical procedures with emphasis on the gastrointestinal and urogenital systems. v. 1- Philadelphia, Lea & Febiger, 1967-

3566. Arthur, G. H. Wright's veterinary obstetrics. 4th ed. Baltimore, Williams & Wilkins, in press 1972.

3567. Berge, Ewald, and Melchior Westhues. Veterinary operative surgery. Copenhagen, Medical Book Co., 1965. (1st English edition.)

3568. Bone, J. F., ed. Equine medicine and surgery, a text and reference work. 2d ed. Wheaton, Ill., American Veterinary Publications, 1972.

3569. Bovine medicine and surgery. Ed. by W. J. Gibbons, E. J. Catcott, and J. F. Smithcors. Wheaton, Ill., American Veterinary Publications, 1970.

3570. Brockis, J. Gwynne. The scientific fundamentals of surgery. New York, Appleton-Century-Crofts, 1972.

3571. Canine surgery. Ed. by James Archibald. Santa Barbara, American Veterinary Publications, 1965. (Earlier editions by H. P. Hoskins, Karl Mayer, J. V. Lacroix.)

3572. Catcott, E. J., ed. Feline medicine and surgery, a text and a reference work. Wheaton, Ill., American Veterinary Publications, 1964.

3573. Cole, W. H. Cole and Zollinger textbook of surgery. 9th ed. New York, Appleton-Century-Crofts, 1970.

3574. Fazzini, Eugene, et al. A manual for surgical pathologists. Springfield, Ill., Thomas, 1972.

3575. Frank, E. R. Veterinary surgery. 7th ed. Minneapolis, Burgess, 1964.

3576. Hall, L. W. Fluid balance in canine surgery: an introduction. Baltimore, Williams & Wilkins, 1967.

3577. Hobday, F. T. G. Surgical diseases of the dog and cat. 6th ed. Ed. by James McCunn. London, Bailliere, Tindall & Cox, 1953.

3578. Laufman, Harold, and R. B. Erichson. Hematologic problems in surgery. Philadelphia, Saunders, 1970.

3579. Leonard, E. P. Fundamentals of small animal surgery. Philadelphia, Saunders, 1968.

3580. Marks, Charles. Applied surgical anatomy. Springfield, Ill., Thomas, 1972.

3581. Oehme, F. W., and J. E. Prier. Textbook of large animal surgery. Baltimore, Williams & Wilkins, in press 1972.

3582. Ormrod, A. N. Surgery of the dog and cat, practical guide. London, Bailliere, Tindall & Cassell, 1966.

3583. Rob, Charles, and Rodney Smith, eds. Operative surgery. 2d ed. 14 v. Philadelphia, Lippincott, 1968-1971.

3584. Roberts, S. J. Veterinary obstetrics and genital diseases. (Theriogenology.) 2d ed. Ithaca, N.Y., 1971 (distributed by Edwards Bros., Ann Arbor, Mich.).

3585. Smythe, R. H. Clinical veterinary surgery. 2 v. London, Lockwood, 1959-1962.

TECHNIQUE AND MATERIALS

3586. Bierschwal, C. J., and C. H. deBois. The technique of fetotomy in large animals: a manual. Bonner Springs, Kan., VM Publ., 1972.

3586a. Bloch, Bernard, and G. W. Hastings. Plastics materials in surgery. 2d ed. Springfield, Ill., Thomas, 1972.

3587. Charnley, John. Acrylic cement in orthopaedic surgery. Baltimore, Williams & Wilkins, 1970.

3588. Cryogenics in surgery. Ed. by Hans von Leden. New York, Medical Examination Publishing, 1971.

3589. Cryosurgery. Ed. by R. W. Rand, A. P. Rinfret, and Hans von Leden. Springfield, Ill., Thomas, 1968.

3590. Ethicon, Inc. Manual of operative procedure and surgical knots. 9th ed. Sommerville, N.J., 1970.

3591. Graves, F. T. Arterial anatomy of the kidney: the basis of surgical technique.
Baltimore, Williams & Wilkins, 1971.

3592. International symposium on stereoencephalotomy. 1st- Basel, Karger, 1961- (For additional coverage see entry 0764.)

3593. Leahy, J. R., and Pat Barrow. Restraint of animals. 2d ed. Ithaca, N.Y., Cornell Campus Store, 1954.

3594. Lee, Henry, and Kris Neville. Handbook of biomedical plastics. Pasadena, Calif., Pasadena Technology Press, 1971.

3595. Matsumoto, Teruo. Tissue adhesives in surgery. New York, Medical Examination Publishing, 1972.

3596. Müller, M. E., M. Allgöwer, and H. Willenegger. Manual of internal fixation: technique recommended by the AO-Group. New York, Springer-Verlag, 1970.

3597. Nealon, T. F. Fundamental skills in surgery. 2d ed. Philadelphia, Saunders, 1971.

3598. Söderström, Nils. Fine-needle aspiration biopsy. New York, Grune & Stratton, 1966.

3598a. Sykes, George. Disinfection and sterilization. 2d ed. Philadelphia, Lippincott, 1965.

3599. Symposium on medical applications of plastics. Toronto, 1970. Medical applications of plastics. Ed. by H. P. Gregor. New York, Interscience, 1971. (*Journal of biomedical materials research*, v. 5, no. 2.)

3600. Symposium on plastics in surgical implants. Indianapolis, 1964. Plastics in surgical implants. Philadelphia, American Society for Testing and Materials, 1965. (ASTM special technical publication no. 386.)

3601. Thorek, Philip. Atlas of surgical techniques. Philadelphia, Lippincott, 1970.

3602. Zacarian, S. A. Cryosurgery of skin cancer and cryogenic techniques in dermatology. Springfield, Ill., Thomas, 1969.

ANESTHESIA

3603. Bellville, J. W., and C. S. Weaver, eds. Techniques in clinical physiology, a survey of measurements in anesthesiology. New York, Macmillan, 1969.

3604. Croft, Phyllis G. An introduction to

Special Surgery 173 Surgery and Obstetrics

the anaesthesia of laboratory animals. 2d ed. London, UFAW (Universities Federation for Animal Welfare), 1964.

3605. De Jong, R. H. Physiology and pharmacology of local anesthesia. Springfield, Ill., Thomas, 1970.

3606. Gray, T. C., and F. J. Nunn, eds. General anesthesia. 3d ed. New York, Appleton-Century-Crofts, 1971.

3607. Hill, D. W. Electronic measurement techniques in anaesthesia and surgery. New York, Appleton-Century-Crofts, 1970.

3608. Nunn, J. F. Applied respiratory physiology with special reference to anaesthesia. New York, Appleton-Century-Crofts, 1969.

3609. Quimby, C. W. Anesthesiology: a manual of concept and management. New York, Appleton-Century-Crofts, 1972.

3610. Soma, L. R., ed. Textbook of veterinary anesthesia. Baltimore, Williams & Wilkins, 1971.

3611. Westhues, Melchior, and Rudolf Fritsch. Animal anaesthesia. v. 1- Philadelphia, Lippincott, 1964-
   v. 1: Local anaesthesia. v. 2: General anaesthesia.

3612. Wright, J. G. Veterinary anaesthesia and analgesia. 7th ed. Ed. by L. W. Hall. London, Bailliere, Tindall, 1971.

SPECIAL SURGERY

3613. Anderson, J. M., ed. The biology and surgery of tissue transplantation. Philadelphia, Davis, 1970.

3614. Austin, George. The spinal cord: basic aspects and surgical considerations. 2d ed. Springfield, Ill., Thomas, 1972.

3615. Bertelli, A., and A. P. Monaco, eds. Pharmacological treatment in organ and tissue transplantation. Baltimore, Williams & Wilkins, 1970.

3616. Billingham, R. E., and Willys Silvers. The immunobiology of transplantation. Englewood Cliffs, N.J., Prentice-Hall, 1971.

3617. Dausset, Jean, et al., eds. Advances in transplantation. Proceedings of the first international congress of the Transplantation Society. Paris, 1967. Baltimore, Williams & Wilkins, 1968.

3618. Demikhov, V. P. Experimental transplantation of vital organs. New York, Consultants Bureau, 1962. (Translated by Basil Haigh.)

3619. Dodson, A. I. Urological surgery. 4th ed. St. Louis, Mosby, 1970.

3620. Epstein, Ervin, ed. Skin surgery. 3d ed. Springfield, Ill., Thomas, 1970.

3621. Hamburger, Jean, et al. Renal transplantation: theory and practice. Baltimore, Williams & Wilkins, 1972.

3622. Hickman, John. Veterinary orthopaedics. Edinburgh, Oliver & Boyd, 1964.

3623. International corneo-plastic conference. 2d. Royal College of Surgeons of England, 1967. Corneo-plastic surgery: proceedings. Ed. by P. V. Rycroft. New York, Pergamon Press, 1969.

3624. Largiadèr, Felix. Organ transplantation. New York, Intercontinental Medical Book, 1970.

3625. Leonard, E. P. Orthopedic surgery of the dog and cat. 2d ed. Philadelphia, Saunders, 1971.

3626. Mathé, Georges, Jean-Louis Amiel, and Leon Schwarzenberg. Bone marrow transplantation and leucocyte transfusion. Springfield, Ill., Thomas, 1971.

3627. Micklem, H. S., and J. F. Loutit. Tissue grafting and radiation. New York, Academic Press, 1966.

3628. O'Neal, L. W. Surgery of the adrenal glands. St. Louis, Mosby, 1968.

3629. Peer, L. A. Transplantation of tissue. v. 1- Baltimore, Williams & Wilkins, 1955-
   v. 1: Cartilage, bone, fascia, tendon, and muscle. v. 2: Cornea, fat, nerves, teeth, blood vessels, endocrine glands, organs, peritoneum, cancer cells.

3630. Pegg, D. E. Bone marrow transplantation. London, Lloyd-Luke, 1966.

3631. Piermattei, D. L., ed. Symposium on orthopedic surgery in small animals.

Philadelphia, Saunders, 1971. (Veterinary clinics of North America, v. 1, no. 3.)

3632. Piermattei, D. L., and R. G. Greeley. An atlas of surgical approaches to the bones of the dog and cat. Philadelphia, Saunders, 1966.

3633. Russell, P. S., and A. P. Monaco. The biology of tissue transplantation. Boston, Little, Brown, 1965.

3634. Starzl, T. E. Experience in hepatic transplantation. Philadelphia, Saunders, 1969.

3635. Strande, Anders. Repair of the ruptured cranial cruciate ligament in the dog; an experimental and clinical study (a few experiments in cats included). Baltimore, Williams & Wilkins, 1968.

3636. Symposium on cross-species transplantation. Ed. by Keith Reemtsma. New York, Stratton, 1970. (Transplantation proceedings, v. 2, no. 4.)

3637. Symposium on gastrointestinal medicine and surgery. Ed. by Anthony Palminteri. Philadelphia, Saunders, 1972. (Veterinary clinics of North America, v. 2, no. 1.)

3638. Te Linde, R. W., and R. F. Mattingly. Operative gynecology. 4th ed. Philadelphia, Lippincott, 1970.

3639. Turek, S. L. Orthopaedics: principles and their application. 2d ed. Philadelphia, Lippincott, 1967.

3640. Walker, D. F., and J. T. Vaughan. Urogenital surgery of the bovine and equine. Philadelphia, Lea & Febiger, in press 1972.

WOUND HEALING

3641. Brook Lodge conference on inflammation. 1967. Chemical biology of inflammation. Ed. by B. K. Forscher. New York, Pergamon Press, 1968.

3642. Bucher, Nancy, and R. A. Malt. Regeneration of liver and kidney. Boston, Little, Brown, 1971.

3642a. Carlson, B. M. The regeneration of minced muscles. New York, Karger, 1972.

(Monographs in developmental biology, v. 4.)

3643. Lister centenary scientific meeting, Glasgow, 1965. Wound healing: a symposium based upon the Lister centenary scientific meeting, held in Glasgow, September 1965. Ed. by Sir Charles Illingworth. London, Churchill, 1966.

3643a. Longacre, J. J. Scar tissue: its use and abuse; the surgical correction of deformation due to hypertrophic scar and the prevention of its formation. Springfield, Ill., Thomas, 1972.

3644. McMinn, R. H., and J. J. Pritchard. Tissue repair. New York, Academic Press, 1969.

3645. Maibach, H. I., and D. T. Rovee. Epidermal wound healing. Chicago, Year Book Medical Publishers, 1972.

3646. Montagna, William, ed. Wound healing. New York, Macmillan, 1964. (*Advances in biology of skin*, v. 5.)

3647. Raekallio, Jyrki. Enzyme histochemistry of wound healing. Stuttgart, Portland, Fischer, 1970. (*Progress in histochemistry and cytochemistry*, v. 1, no. 2.)

3648. Repair and regeneration, the scientific basis for surgical practice. Ed. by J. E. Dunphy and Walton Van Winkle, Jr. New York, McGraw-Hill, 1969.

3649. Said, A. H. Contribution to the study of bone healing with reference to the action of certain therapeutic agents, especially vitamin K. Ghent, State Univ. of Ghent, School of Veterinary Medicine, Clinic for Large Animals, 1960.

3650. Schmidt, A. J. Cellular biology of vertebrate regeneration and repair. Chicago, Univ. of Chicago Press, 1968.

3651. U.S. National Research Council. Committee on Trauma. Wound healing. Proceedings of a workshop conducted by the Committee on Trauma. December 5-8, 1963. Ed. by Stanley M. Levenson et al. Washington, D.C., 1966.

3652. Willis, A. T. Clostridia of wound infection. New York, Appleton-Century-Crofts, 1969.

20
RADIOLOGY AND RADIATION BIOLOGY

PRINCIPLES

3653. Christensen, E. E., T. S. Curry, and James Nunnally. An introduction to the physics of diagnostic radiology. Philadelphia, Lea & Febiger, 1972.

3654. Elkind, M. M., and G. F. Whitmore. The radiobiology of cultured mammalian cells. New York, Gordon & Breach, 1967.

3655. Goodwin, P. N., Edith H. Quimby, and R. H. Morgan. Physical foundations of radiology. 4th ed. New York, Harper & Row, 1970.

3656. International symposium on the response of the nervous system to ionizing radiation. 1st- Proceedings. 1960- (For additional coverage see entry 0766.)

3657. Johns, H. E., and J. R. Cunningham. The physics of radiology. 3d ed. Springfield, Ill., Thomas, 1969.

3658. Kemp, L. A. W., and R. Oliver. Basic physics in radiology. 2d ed. Oxford, Blackwell, 1970.

3659. Micklem, H. S., and J. F. Loutit. Tissue grafting and radiation. New York, Academic Press, 1966.

3660. Oliver, Raymond. Radiation physics in radiology. Philadelphia, Davis, 1966.

3661. Panel on radiation and the control of immune response. Paris, 1967. Report. Vienna, International Atomic Energy Agency, 1968. (Panel proceedings series. STI/PUB/175.)

3662. Panel on radiation sterilization of biological tissues for transplantation. Budapest, 1969. Sterilization and preservation of biological tissues by ionizing radiation. Vienna, International Atomic Energy Agency, 1970. (Panel proceedings series.)

3663. Panel on the effects of various types of ionizing radiations from different sources on haematopoietic tissue. Vienna, 1966. Vienna, International Atomic Energy Agency, 1967. (Panel proceedings series. STI/PUB/134.)

3664. Panel on the radiation sensitivity of toxins and animal poisons. Bangkok, 1969. Vienna, International Atomic Energy Agency, 1970. (Panel proceedings series. STI/PUB/243.)

3665. Petrov, I. R., ed. Influence of microwave radiation on the organism of man and animals. Washington, D.C., National Aeronautics and Space Administration, 1972. (NASA technical translation F708.)

3666. Pizzarello, D. J., and R. L. Witcofski. Medical radiation biology. Philadelphia, Lea & Febiger, 1972.

3667. Radiation biology of the fetal and juvenile mammal. Ed. by M. R. Sikov and D. D. Mahlum. Washington, D.C., U.S. Atomic Energy Commission, 1969. (U.S. Atomic Energy Commission. Annual Hanford biology symposium. 9th. Richland, Washington, 1969. Also A.E.C. symposium series, no. 17.)

3668. Taliaferro, W. H., Lucy G. Taliaferro, and B. N. Jaroslow. Radiation and immune mechanisms. New York, Academic Press, 1964.

3669. Thomson, J. F. Radiation protection in mammals. New York, Reinhold, 1962.

3670. U.S. National Council on Radiation Protection and Measurements. Radiation protection in veterinary medicine. Washington, D.C., NCRP Publications, 1970. (Its NCRP report no. 36.)

CLINICAL APPLICATIONS

3671. American Veterinary Medical Association. Radiologic diagnosis. Prepared with the assistance of W. H. Riser. Chicago, 1962.

3672. Bogardus, C. R. Clinical applications of physics of radiology and nuclear medicine. St. Louis, Green, 1969.

3673. Carlson, W. D. Veterinary radiology. 2d ed. Philadelphia, Lea & Febiger, 1967.

3674. Clifford, Margaret A., and Ann E. Drummond. Radiologic techniques related to pathology. Baltimore, Williams & Wilkins, 1968.

3675. Commissie voor Beentumoren. Radiological atlas of bone tumours. v. 1- Baltimore, Williams & Wilkins, 1966.

3676. Committee for the Preparation of a Technical Guide for Roentgen Techniques in Laboratory Animals. Roentgen techniques in laboratory animals, radiography of the dog and other experimental animals. Ed. by Benjamin Felson. Philadelphia, Saunders, 1968.

3677. Decker, Kurt. Clinical neuroradiology. Ed. and tr. by W. H. Shehadi. New York, McGraw-Hill, 1966.

3678. Douglas, S. W., and H. D. Williamson. Principles of veterinary radiography. 2d ed. Baltimore, Williams & Wilkins, 1972.

3679. ———, and ———. Veterinary radiological interpretation. Philadelphia, Lea & Febiger, 1970.

3680. Fraser, R. G., and J. A. P. Paré. Structure and function of the lung, with emphasis on roentgenology. Philadelphia, Saunders, 1971.

3681. Hadley, Anne. 101 lessons in x-ray terminology for the medical transcriber. Philadelphia, Lippincott, 1972.

3682. Lang, E. K. Roentgenographic diagnosis of bladder tumors. Springfield, Ill., Thomas, 1968.

3683. ———. The roentgenographic diagnosis of renal mass lesions. St. Louis, Green, 1971.

3684. Lodwick, G. S. The bones and joints. Chicago, Year Book Medical Publishers, 1971. (*Atlas of tumor radiology*, v. 2.)

3685. Matzen, P. F., and H. K. Fleissner. Orthopedic roentgen atlas. New York, Grune & Stratton, 1970. (English edition arranged by L. S. Michaelis.)

3686. Morgan, J. P. Radiology in veterinary orthopedics. Philadelphia, Lea & Febiger, 1972.

3687. Murray, R. O., and H. G. Jacobson. The radiology of skeletal disorders: exercises in diagnosis. Baltimore, Williams & Wilkins, 1971.

3688. Olsson, Sten-Erik. Radiological diagnosis in canine and feline emergencies: an atlas of thoracic and abdominal changes. Philadelphia, Lea & Febiger, 1971.

3689. Rubin, Philip, and G. W. Casarett. Clinical radiation pathology. 2 v. Philadelphia, Saunders, 1968.

3690. Schebitz, H., and H. Wilkens. Atlas of radiographic anatomy of dog and horse. Berlin, Parey, 1968.

3691. Schinz, H. R., et al. Roentgen diagnosis. 2d ed. 5 v. Arr. and ed. by L. G. Rigler. New York, Grune & Stratton, 1967-1970.
  v. 1: General principles and methods. v. 2: Skeleton and soft tissues. v. 3: Spinal column, skull, nervous system, eyes, temporal bone, paranasal sinuses, teeth and jaws, tropical diseases. v. 4: Thorax. v. 5: Abdomen.

3692. Schreiber, M. H. Indications and alternatives in x-ray diagnosis. Springfield, Ill., Thomas, 1970.

3693. Shanks, S. C., and Peter Kerley, eds. A textbook of x-ray diagnosis. 4th ed. 6 v. Philadelphia, Saunders, 1969-1970.
  v. 1: Central nervous system; The teeth and jaws; The eye; The paranasal sinuses; The ear and temporal bone. v. 2: Cardiovascular system. v. 3: Respiratory system. v. 4: Alimentary tract; Biliary tract. v. 5: Obstetrics; Gynaecology; Sonar in obstetrics and gynaecology; The abdomen; The urinary tract. v. 6: Bones, joints and soft tissues.

3694. Smith, R. N. Radiography for the veterinary surgeon. Baltimore, Williams & Wilkins, 1960.

RADIOISOTOPES AND RADIOTHERAPY

3695. Arena, Victor. Ionizing radiation and life: an introduction to radiation biology and biological radiotracer methods. St. Louis, Mosby, 1971.

3696. Behrens, C. F., and E. R. King. Atomic medicine. 4th ed. Baltimore, Williams & Wilkins, 1964.

3697. Beierwaltes, W. H., J. W. Keyes, Jr., and J. E. Carey. Manual of nuclear medicine procedures. Cleveland, CRC Press, 1971. (Portions originally published by the Univ. of Michigan as *Nuclear medicine clinical techniques*.)

3698. Blahd, W. H., ed. Nuclear medicine. 2d ed. New York, McGraw-Hill, 1971.

3699. Chiappa, S., R. Musumeci, and C. Uslenghi, eds. Endolymphatic radiotherapy in malignant lymphomas. New York, Springer-Verlag, 1971. (*Recent results in cancer research*, v. 37.)

3700. DeLand, F. H., and H. N. Wagner. Atlas of nuclear medicine. v. 1- Philadelphia, Saunders, 1969-
   v. 1: Brain. v. 2: Lung and heart.

3701. Hale, J. F., and E. F. Scowen. Thymic tumours: their association with myasthenia gravis and their treatment by radiotherapy. London, Lloyd-Luke, 1967.

3702. Hine, G. J., ed. Instrumentation in nuclear medicine. v. 1- New York, Academic Press, 1967-

3703. International Atomic Energy Agency. Laboratory training manual on the use of isotopes and radiation in animal research. Vienna, 1966. (Its Technical report series, no. 60.)

3704. International nuclear medicine symposium. Imperial College of Science and Technology, London, September, 1967. Radioactive isotopes in the localization of tumours. Ed. by V. R. McCready et al. New York, Grune & Stratton, 1969.

3705. Jackson, H. L. Mathematics of radiology and nuclear medicine. St. Louis, Green, 1971.

3706. Panel on the use of isotopes in studies of mineral metabolism and disease in domestic animals. Vienna, 1968. Trace mineral studies with isotopes in domestic animals. Vienna, International Atomic Energy Agency, 1969.

3707. Panel on the use of nuclear techniques in studies of mineral metabolism and disease in domestic animals. Vienna, 1970. Mineral studies with isotopes in domestic animals: proceedings. Vienna, International Atomic Energy Agency, 1971.

3708. Quimby, Edith, ed. Radioactive nuclides in medicine and biology. 3d ed. v. 1- Philadelphia, Lea & Febiger, 1968-
   v. 1: Basic physics and instrumentation. v. 2: Medicine.

3709. Spiers, F. W. Radioisotopes in the human body—physical and biological aspects. New York, Academic Press, 1968.

3710. Symposium on the use of radioisotopes in animal nutrition and physiology. Prague, 1964. Radioisotopes in animal nutrition and physiology. Vienna, International Atomic Energy Agency, 1965.

3711. Wolf, George. Isotopes in biology. New York, Academic Press, 1964.

PART III
Veterinary Medicine

The materials in Chapters 21-35 are those that have been determined to be of the greatest interest and usefulness to the veterinarian in a clinical setting. However, the species approach will prove helpful also to biomedical research personnel who are interested in a special animal species to be used as a model for a human disease process, surgical technique, or therapeutic agent. Many of the entries have been cited in Part II under appropriate subject disciplines, but are included here for the general usefulness of an extensive listing by species. The investigator is referred to Part IV for additional titles on the use of a given species as a model for research.

Materials of broad clinical veterinary medical interest that are not germane to one species alone are included in a chapter on animal management. Here one will find citations to works covering general veterinary medicine and surgery, animal nutrition, and animal production and management, as well as materials geared specifically to the operation of an animal practice.

Within each chapter devoted to a species or a general class of animals, the materials are separated into two or more sections. The first section includes works that are of interest to the veterinarian in his medical practice or to biomedical research personnel. Subsequent sections cite materials that may be useful not only to the veterinarian but also to breeders, managers, zoo personnel, and others who are concerned with the nonmedical aspects of animal nutrition, production, care, conservation, and behavior, or with the classification of a given species.

A chapter on zoology is intended to be representative only and is not all-inclusive for titles in vertebrate biology and classification. Because of the limitation in scope of the work under consideration, no attempt has been made to cover invertebrate zoology with the exception of those titles in Part IV, Chapter 40, which are germane to the study and use of invertebrates as models for research.

Some titles representative of the role of animals in man's social or recreational environment, as well as those dealing with his dependence on animals for work or service, are included in the chapter on companion animals.

Materials concerned with education and training for medicine and allied health disciplines will be found in a separate chapter. Some of the titles listed are intended primarily for the education of physicians but are applicable to veterinary medical education as well. Others deal more specifically with allied health professions, and a few are applicable to the training of support staff. The educator may wish to consult appropriate indexes to the periodical literature for additional material.

21
ANIMAL MANAGEMENT

CLINICAL MEDICINE AND SURGERY

3712. Berrier, H. H. Diagnostic aids in the practice of veterinary medicine. 3d ed. St. Louis, Alban Professional Books, 1968.

3713. Blood, D. C., and J. A. Henderson. Veterinary medicine. 3d ed. Baltimore, Williams & Wilkins, 1968.

3714. Boddie, G. F. Diagnostic methods in veterinary medicine. 6th ed. Philadelphia, Lippincott, 1969.

3715. The chemist's veterinary handbook: a survey of modern methods in veterinary medicine, including diseases and treatment. 12th ed. London, Offices of the Chemist and Druggist, 1964.

3716. Galloway, J. H. Farm animal health and disease control. Philadelphia, Lea & Febiger, 1972.

3717. Gibbons, W. J. Clinical diagnosis of diseases of large animals. Philadelphia, Lea & Febiger, 1966.

3718. Heidrich, J. F., and W. Renk. Diseases of the mammary glands of domestic animals. Philadelphia, Saunders, 1967.

3719. Henderson, G. N., and John Stratton. Bailliere's veterinary handbook. 9th ed. Baltimore, Williams & Wilkins, 1963.

3720. Herbert, W. J. Veterinary immunology. Philadelphia, Davis, 1970.

3721. Hickman, John. Veterinary orthopaedics. Edinburgh, Oliver & Boyd, 1964.

3722. Hungerford, T. G. Diseases of livestock. 7th rev. ed. Sydney, Angus & Robertson, 1970.

3723. International symposium on health aspects of the international movement of animals. San Antonio, 1968. Papers presented. Washington, D.C., Pan American Health Organization, 1969. (Scientific publication no. 182.)

3724. Joest, Ernst. Handbuch der speziellen pathologischen Anatomie der Hausteire. 3d Aufl. Bd. 1- hrsg. von J. Dobberstein et al. Berlin, Parey, 1963.

3725. Kirk, R. W., and S. I. Bistner. Handbook of veterinary procedures and emergency treatment. Philadelphia, Saunders, 1969.

3726. Merchant, I. A., and R. D. Barner. Outline of infectious diseases of domestic animals. 3d ed. Ames, Iowa State Univ. Press, 1964.

3727. Merck and Company, Inc. Merck manual of diagnosis and therapy. 12th ed. Rahway, N.J., Merck, Sharp & Dohme Research Laboratories, 1972.

3728. Merck veterinary manual: a handbook of diagnosis and therapy for the veterinarian. Ed. by O. H. Siegmund et al. Rahway, N.J., Merck & Co., 1967.

3729. Mönnig, H. O., and F. J. Veldman. Handbook on stock diseases. 5th rev. ed. Johannesburg, Nasionale Boekhandel Beperk, 1970.

3730. Moulton, J. E. Tumors in domestic animals. Berkeley, Univ. of California Press, 1961.

3731. Muller, G. H., and R. W. Kirk. Small animal dermatology. Philadelphia, Saunders, 1969.

3732. Oehme, F. W., and J. E. Prier. Textbook of large animal surgery. Baltimore, Williams & Wilkins, in press 1972.

3733. Orlov, N. P. Coccidiosis of farm animals. Jerusalem, Israel Program for Scientific Translations, 1970.

3734. Palmer, A. C. Introduction to animal neurology. Philadelphia, Davis, 1965.

3735. Phillips, R. W., and W. J. Tietz. Physiological basis of veterinary practice. Baltimore, Williams & Wilkins, in press 1972.

3736. Popesko, Peter. Atlas of topographical anatomy of the domestic animals. v. 1-3 in 1 v. Philadelphia, Saunders, 1971. (English translation by Robert Getty.)

3737. Schwabe, C. W. Veterinary medicine and human health. 2d ed. Baltimore, Williams & Wilkins, 1969.

3738. Seddon, H. R. Diseases of domestic animals in Australia. 2d ed. Rev. by H. E. Albiston. v. 1- Canberra, Australia, Dept. of Health, 1965-
Part 1—Helminth infestations. Part 4—Protozoan and virus diseases. Part 5—v. 1: Bacterial diseases; v. 2: Bacterial diseases.

3739. Trinca, J. S. CSL veterinary handbook. Parkville, Australia, Commonwealth Serum Laboratories, 1968.

3740. Udall, D. H. The practice of veterinary medicine. 6th rev. ed. Ithaca, N.Y., the author, 1954.

3741. The veterinary clinics of North America. v. 1- Philadelphia, Saunders, 1971- (For specific titles see entry 0794.)

3742. Veterinary Research Laboratory. Onderstepoort, South Africa. Emerging diseases of animals. Rome, Food and Agricultural Organization of the United Nations, 1963. (FAO agricultural studies, no. 61.)

3743. White, E. G., and F. T. W. Jordan. Veterinary preventive medicine. Baltimore, Williams & Wilkins, 1963.

3744. Wright, J. G. Wright's veterinary anesthesia and analgesia. 7th ed. Ed. by L. W. Hall. London, Bailliere, Tindall & Cassens, 1971.

ANIMAL NUTRITION

3745. Abrams, J. T. Animal nutrition and veterinary dietetics. 4th ed. Baltimore, Williams & Wilkins, 1962.

3746. ——, ed. Recent advances in animal nutrition. Boston, Little, Brown, 1966.

3747. Aikawa, J. K. The relationship of magnesium to disease in domestic animals and in humans. Springfield, Ill., Thomas, 1971.

3748. Church, D. C. Digestive physiology and nutrition of ruminants. 3 v. Corvallis, Oregon State Univ. Book Stores, 1969-1972.

3749. Commonwealth Bureau of Animal Nutrition. Technical communications. v. 1- Farnham Royal, Bucks, England, 1948- (For specific titles see entry 0743.)

3750. Crampton, E. W., and L. W. Harris. Applied animal nutrition. 2d ed. San Francisco, Freeman, 1969.

3751. Cuthbertson, D. P., ed. Nutrition of animals of agricultural importance. 2 parts. New York, Pergamon Press, 1969. (*International encyclopaedia of food and nutrition*, v. 17.)
Part 1—The science of nutrition of farm livestock. Part 2—Assessment of and factors affecting requirements of farm livestock.

3752. Dent, J. B., and H. Casey. Linear programming and animal nutrition. Philadelphia, Lippincott, 1967.

3753. De Rochemont, Richard. The pets' cookbook: a layman's comprehensive guide to the feeding of dogs, cats, birds, fish, and odd animals around the house. 1st ed. New York, Knopf, 1964.

3754. Foreign compound metabolism in mammals: a review of the literature published between 1960 and 1969. v. 1- Reported by D. E. Hathaway. London, The Chemical Society, 1970-

3755. Gutcho, M. Animal feeds, 1970. Park Ridge, N.J., Noyes Data Corp., 1970.

3756. Hafez, E. S. E., and I. A. Dyer. Animal growth and nutrition. Philadelphia, Lea & Febiger, 1969.

3757. Harvey, D. G. Tables of the amino acids in foods and feedingstuffs. 2d ed. Farnham Royal, Bucks, England, Commonwealth Agricultural Bureaux, 1970.

3758. Jennings, J. B. Feeding, digestion, and assimilation in animals. 1st ed. Oxford, Pergamon Press, 1965.

3759. King, J. O. L. Veterinary dietetics: a manual of nutrition in relation to disease in animals. Baltimore, Williams & Wilkins, 1961.

3760. Lenkeit, Walter, Knut Breirem, and Edgar Crasemann. Handbuch der Tierernährung. v. 1- Hamburg, Verlag Paul Parey, 1969- v. 1: Allgemeine Grundlagen.

3761. McDonald, Peter, R. A. Edwards, and J. F. D. Greenhalgh. Animal nutrition. Edinburgh, Oliver & Boyd, 1966.

3762. Maplesden, D. C. Practical nutrition and feeding of domestic animals. Galt, Canada, the author, 1962.

3763. Maynard, L. A., and J. K. Loosli. Animal nutrition. 6th ed. New York, McGraw-Hill, 1969.

3764. Mitchell, H. H. Comparative nutrition of man and domestic animals. 2 v. New York, Academic Press, 1962-1964.

3765. Morris, M. L. Nutrition and diet in small animal medicine. Denver, Mark Morris Associates, 1960.

3766. Morrison, F. B., Elsie B. Morrison, and S. H. Morrison. Feeds and feeding: a handbook for the student and stockman. 22d ed. Ithaca, N.Y., Morrison Publ. Co., 1956.

3767. Perry, T. W. Feed formulations handbook. Danville, Ill., Interstate, 1966.

3768. Symposium on the physiology and biochemistry of muscle as a food. 1st- Madison, University of Wisconsin Press, 1966- (For additional coverage see entry 1999.)

3769. Symposium: sulfur in nutrition. Oregon State Univ., 1969. Ed. by O. H. Muth. Westport, Conn., Avi Publishing, 1970.

3770. Underwood, E. J. Trace elements in human and animal nutrition. 3d ed. New York, Academic Press, 1971.

3771. U.S. National Research Council. Committee on Animal Nutrition. Nutrient requirements of domestic animals. (For additional coverage see entry 0791.)

3772. The use of drugs in animal feeds: proceedings of a symposium. Washington, 1969. Washington, D.C., National Academy of Sciences, 1969. (Its publication no. 1679.)

3773. World congress on animal feeding. Madrid, 1966. Proceedings. 3 v. Madrid, International Veterinary Federation on Zootechnics, 1966.
 v. 1: General reports. v. 2: Free communications. v. 3: Chronicle and conclusions. (In Spanish, German, French, and English.)

3774. Zweig, Gunter, ed. Analytical methods for pesticides, plant growth regulators, and food additives. 6 v. New York, Academic Press, 1963-1972.
 v. 1: Principles, methods and general applications. v. 2: Insecticides. v. 3: Fungicides, nematocides and soil fumigants, rodenticides and food and feed additives. v. 4: Herbicides. v. 5: Additional principles and methods of analysis. v. 6: Gas chromatographic analysis.

ANIMAL PRODUCTION AND MANAGEMENT

3775. American Society of Animal Production. Techniques and procedures in animal production research. Beltsville, Md., 1959.

3776. Cole, H. H., and P. T. Cupps, eds. Reproduction in domestic animals. 2d ed. New York, Academic Press, 1969.

3777. Ensminger, M. E. Animal science. 6th ed. Danville, Ill., Interstate, 1969.

3777a. Epstein, H. The origin of the domestic animals of Africa. Rev. in collaboration with I. L. Mason. New York, Africana Publishing, 1971.

3778. Fertility and infertility in the domestic animals. 2d ed. Ed. by J. A. Laing. Baltimore, Williams & Wilkins, 1970.

3779. Frandson, R. D. Anatomy and physiology of farm animals. Philadelphia, Lea & Febiger, 1965.

3780. Hafez, E. S. E., ed. The behaviour of domestic animals. 2d ed. Baltimore, Williams & Wilkins, 1969.

3781. ———. Reproduction in farm animals. 2d ed. Philadelphia, Lea & Febiger, 1968.

3782. Instruments de médecine vétérinaire (Firm). Straw method of freezing: technique for preparing, preserving and utilizing semen, as practised by Artificial Insemination Centre, l'Aigle. L'Aigle, France, 1969.

3782a. International congress on animal reproduction and artificial insemination. 1st- Proceedings. 1948- (Quadrennial; publishers vary.)

3783. Lasley, J. F., ed. Genetics of livestock improvement. 2d ed. Englewood Cliffs, N.J., Prentice-Hall, 1972.

3783a. Maule, J. P., ed. The semen of animals and artificial insemination. Farnham Royal, Bucks, England, Commonwealth Agricultural Bureaux, 1962. (Commonwealth Bureau of Animal Breeding and Genetics. Technical communication no. 15.)

3784. Milovanov, V. K., ed. Artificial insemination of livestock in the U.S.S.R.: a visual aid album. Jerusalem, Israel Program for Scientific Translations, 1964.

3784a. Parker, W. H. Health and disease in farm animals for those concerned with animal husbandry. New York, Pergamon Press, 1970.

3785. Perry, E. J. Artificial insemination of farm animals. 4th ed. New Brunswick, N.J., Rutgers Univ. Press, 1968.

3786. Pirchner, Franz. Population genetics in animal breeding. San Francisco, Freeman, 1969.

3787. Rice, V. A., et al. Breeding and improvement of farm animals. 6th ed. New York, McGraw-Hill, 1967.

3788. Universities Federation for Animal Welfare. The UFAW handbook on the care and management of farm animals. Edinburgh, Churchill Livingstone, 1971.

3789. World conference on animal production. 1st- 1963-
   1st Rome, 1963. Papers, 3 v. Rome, European Association for Animal Production, 1963.
   2d Univ. of Maryland, College Park, Md., 1968. Papers. Urbana, Ill., American Dairy Science Association, 1969.

3790. Zemjanis, R. Diagnostic and therapeutic techniques in animal reproduction. 2d ed. Baltimore, Williams & Wilkins, 1970.

VETERINARY MEDICAL PRACTICE

3791. American Animal Hospital Association. Manual of standards for animal hospitals prescribed for AAHA member hospitals. Elkhart, Ind., 1971.

3792. American Society of Planning Officials. Zoning for animal hospitals. Prepared by J. H. Pickford. Chicago, 1964.

3793. Bierschwal, C. J., and C. H. deBois. The technique of fetotomy in large animals: a manual. Bonner Springs, Kan., VM Publ., 1972.

3794. British Small Animal Veterinary Association. Jones's animal nursing. 2d ed. Ed. by R. S. Pinniger. Oxford, Pergamon Press, 1972. (Earlier edition by B. V. Jones.)

3795. Catcott, E. J., ed. Animal hospital technology: a text for veterinary aides. Wheaton, Ill., American Veterinary Publishers, 1971.

3796. Hannah, H. W., and D. F. Storm. Law for the veterinarian and livestock owner. 2d ed. Danville, Ill., Interstate, 1965.

3797. Heath, J. S., ed. Aids to veterinary nursing. Baltimore, Williams & Wilkins, 1970.

3798. Judy, W. L. Kennel building and plans. 7th ed. Chicago, Judy Publ. Co., 1968.

3799. Leahy, J. R., and Pat Barrow. Restraint of animals. 2d ed. Ithaca, N.Y., Cornell Campus Store, 1954.

3800. Miller, W. C., and G. P. West. Encyclopedia of animal care. 9th ed. Baltimore, Williams & Wilkins, 1970.

3801. Sainsbury, D. B. W., and Peter Sainsbury. Animal health and housing. Baltimore, Williams & Wilkins, 1967.

3802. Sarner, Harvey. The business management of a small animal practice. Philadelphia, Saunders, 1967.

3803. Soave, O. A. An introduction to veterinary law. Philadelphia, Saunders, 1962.

3804. Sumner, K. C. First aid for small animals. Bristol, John Wright, 1965.

3805. U.S. Dept. of Agriculture. Veterinary biologics checklist. Care and use of veterinary biologics: veterinarian's checklist. Washington, D.C., 1970-71. (Program aid no. 841.)

3806. U.S. National Council on Radiation Protection and Measurements. Radiation protection in veterinary medicine. Washington, D.C., NCRP Publications, 1970. (Its NCRP report no. 36.)

3807. U.S. National Research Council. Committee on Veterinary Medical Research and Education. New horizons for veterinary medicine. Washington, D.C., National Academy of Sciences, 1972.

3808. The veterinary technician. Washington, D.C., U.S. Dept. of the Air Force, 1968. (Air Force manual no. 163-2.)

3809. White, E. G., and F. T. W. Jordan. Veterinary preventive medicine. Baltimore, Williams & Wilkins, 1963.

22

CATTLE

BOVINE MEDICINE AND SURGERY

3810. Baker, L. A. Bovine health programming. Cleveland, United Publ. Corp., 1968.

3811. Belschner, H. G. Cattle diseases. Sidney, Angus & Robertson, 1967.

3812. Blaxter, K. L. Energy metabolism of ruminants. 2d ed. London, Hutchinson, 1967.

3813. Bovine medicine and surgery. Ed. by W. J. Gibbons, E. J. Catcott, and J. F. Smithcors. Wheaton, Ill., American Veterinary Publications, 1970.

3814. British Veterinary Association. Technical Development Committee. Parasitic diseases of cattle. 3d ed. London, 1969.

3815. Conference on parturient paresis in dairy animals. University of Illinois, 1968. Parturient hypocalcemia. Ed. by J. J. B. Anderson. New York, Academic Press, 1970.

3816. The control of bovine mastitis: papers given at a meeting organised by the British Cattle Veterinary Association and the Agricultural Development Association. Reading University, 1971. Ed. by F. H. Dodd and E. R. Jackson. Shinfield, England, National Institute for Research in Dairying, 1971.

3817. Dyce, K. M., and C. J. G. Wensing. Essentials of bovine anatomy. Utrecht, Oosthoek, 1971.

3818. Faulkner, L. C., ed. Abortion diseases of livestock. Springfield, Ill., Thomas, 1968.

3819. Fennestad, K. L. Experimental leptospirosis in calves. Copenhagen, Munksgaard, 1963. (Translated from Danish by H. Cowan.)

3819a. Greenough, P. R., F. J. MacCallum, and A. D. Weaver. Lameness in cattle. Edinburgh, Oliver & Boyd, 1972.

3820. Habel, R. E. Guide to the dissection of domestic ruminants. 2d ed. Ithaca, N.Y., published by the author, 1964 (distributed by Edwards Bros., Ann Arbor, Mich.).

3821. Hudson, J. R. Contagious bovine pleuropneumonia. Rome, Food and Agricultural Organization of the United Nations, 1971. (FAO agricultural studies, no. 86.)

3822. International Dairy Federation. A monograph on bovine mastitis. Brussels, Secrétariat général, 1971.

3823. International symposium on the physiology of digestion in the ruminant. (For additional coverage see entry 0765.)

3824. Jensen, Rue, and D. R. Mackey. Diseases of feedlot cattle. 2d ed. Philadelphia, Lea & Febiger, 1971.

3825. Joint FAO/WHO Expert Committee on Brucellosis. Rome, Food and Agricultural Organization of the United Nations, 1970. (FAO agricultural studies, no. 85.)

3826. Laing, J. A. Trichomonas foetus infection of cattle. Rome, Food and Agricultural Organization of the United Nations, 1957. (FAO agricultural studies, no. 33.)

3827. ———. Vibrio fetus infection of cattle. rev. ed. Rome, Food and Agricultural Organization of the United Nations, 1960. (FAO agricultural studies, no. 51.)

3828. Levine, N. D., and Virginia Ivens, eds. The coccidian parasites (protozoa, sporozoa) of ruminants. Urbana, Univ. of Illinois Press, 1970. (Illinois biological monographs, no. 44.)

3829. McLeod, W. M. Bovine anatomy. 2d ed. Rev. by D. M. Trotter and J. W. Lumb. Minneapolis, Burgess, 1958.

3830. Myers, J. A., and J. H. Steele. Bovine tuberculosis control in man and animals. St. Louis, Green, 1969.

3831. National Mastitis Council. Research Committee. Microbiological procedures for the diagnosis of bovine mastitis. Washington, D.C., 1969.

3832. Schalm, O. W., E. J. Carroll, and N. C. Jain. Bovine mastitis. Philadelphia, Lea & Febiger, 1971.

3833. Scott, G. R. Diagnosis of rinderpest. Rome, Food and Agricultural Organization of the United Nations, 1967. (FAO agricultural studies, no. 71.)

3834. Swett, W. W., and C. A. Mathews. Studies on comparative anatomy of goats and dairy cows. Hyattsville, Md., U.S. Agricultural Research Service, 1967.

3835. Thomson, R. G. Vertebral body osteophytes in bulls. Ed. by C. N. Barron and P. Cohrs. White Plains, N.Y., Phiebig, 1969.

3836. Walker, D. F., and J. T. Vaughan. Urogenital surgery of the bovine and equine. Philadelphia, Lea & Febiger, in press 1972.

3837. Way, R. F. The anatomy of the bovine foot: a pictorial approach. Philadelphia, Univ. of Pennsylvania Press, 1954.

3838. Whitlock, Ralph. The great cattle plague: an account of the foot-and-mouth epidemic of 1967-1968. London, John Baker, 1968.

3839. Winter, Hans. Post mortem examination of ruminants. St. Lucia, Brisbane, Univ. of Queensland Press, 1966.

3840. World Association for Buiatrics. International conference on cattle diseases. Reports. 1st- 1960- (For additional coverage see entry 0798.)

PRODUCTION, NUTRITION, AND MANAGEMENT

3841. Asdell, S. A. Cattle fertility and sterility. 2d ed. Boston, Little, Brown, 1968.

3842. Bailey, J. W. Veterinary handbook for cattlemen. 4th ed. New York, Springer, 1972.

3842a. Dyer, I. A., and C. C. O'Mary, eds. The feedlot. Philadelphia, Lea & Febiger, 1972.

3843. Ensminger, M. E. Beef cattle science. 4th ed. Danville, Ill., Interstate, 1968.

3844. Foley, R. C., et al. Dairy cattle: principles, practices, problems, profits. Philadelphia, Lea & Febiger, 1972.

3845. Gray, J. R. Ranch economics. Ames, Iowa State Univ. Press, 1968.

3846. Loosli, J. K., and I. W. McDonald. Nonprotein nitrogen in the nutrition of ruminants. Rome, Food and Agricultural Organization of the United Nations, 1968. (FAO agricultural studies, no. 75.)

3847. Neumann, A. L., and R. R. Snapp. Beef cattle. 6th ed. New York, Wiley, 1969.

3848. Rouse, J. E. World cattle. 2 v. Norman, Univ. of Oklahoma Press, 1970.

3849. Roy, J. H. B. The calf: nutrition and health. 3d ed. University Park, Pennsylvania State Univ. Press, 1970.

3850. Salisbury, G. W., and N. L. Van Demark. Physiology of reproduction and artificial insemination of cattle. San Francisco, Freeman, 1961.

3851. U.S. National Research Council. Subcommittee on Prenatal and Postnatal Mortality. Prenatal and postnatal mortality in cattle: a report. Washington, D.C., 1968. (Its publication no. 1685.)

23

HORSES

EQUINE MEDICINE AND SURGERY

3852. Beck, Ernest. The horse, in anatomical transparencies, with additional detailed color illustrations. Grafton, Wis., Animal Technical Publications, 1971.

3853. Belschner, H. G. Horse diseases. London, Angus & Robertson, 1969.

3854. Bone, J. F., ed. Equine medicine and surgery: a text and reference work. 2d ed. Wheaton, Ill., American Veterinary Publications, 1972.

3855. Equine infectious anemia progress report. Hyattsville, Md., Agricultural Research Service, Animal Health Division, 1967?- (Reports issued periodically.)

3856. Equine piroplasmosis progress report. Hyattsville, Md., Agricultural Research Service, Animal Health Division, 1966- (Reports issued periodically.)

3857. Fleming, George. Roaring in horses. (Laryngismus paralyticus.) London, Bailliere, Tindall & Cox, 1889.

3858. International conference on equine infectious diseases. (For additional coverage see entry 0758.)

3859. Jones, W. E., and Ralph Bogart. Genetics of the horse. Ann Arbor, Mich., Edwards, 1971.

3860. Nishikawa, Y. Studies on reproduction in horses: singularity and artificial control in reproductive phenomena. Tokyo, Japan Racing Association, 1959.

3861. Rooney, J. R. Autopsy of the horse: technique and interpretation. Baltimore, Williams & Wilkins, 1970.

3862. ———. Biomechanics of lameness in horses. Baltimore, Williams & Wilkins, 1969.

3863. ———. Clinical neurology of the horse. 1st ed. Kennett Square, Pa., KNA Press, 1971.

3864. ———, W. O. Sack, and R. E. Habel. Guide to the dissection of the horse. Ithaca, N.Y., W. O. Sack, 1967 (distributed by Edwards Bros., Ann Arbor, Mich.).

3865. Schebitz, H., and H. Wilkens. Atlas of radiographic anatomy of dog and horse. Berlin, Parey, 1968.

3866. Strömberg, Berndt. The normal and diseased superficial flexor tendon in race horses. 1971. (*Acta radiologica.* Suppl. 305.)

3867. Symposium on equine bone and joint diseases. Cornell University, 1967. Proceedings. Ithaca, N.Y., 1968. (*Cornell veterinarian*, v. 58. Suppl.)

3868. Walker, D. F., and J. T. Vaughan. Urogenital surgery of the bovine and equine. Philadelphia, Lea & Febiger, in press 1972.

3869. Way, R. F., and D. G. Lee. The anatomy of the horse. Philadelphia, Lippincott, 1965.

PRODUCTION, MANAGEMENT, AND HORSEMANSHIP

3870. American Association of Equine Practitioners. Official guide for determining the age of the horse. Fort Dodge, Iowa, Fort Dodge Laboratories, 1966.

3871. Andrist, Friedrich. Mares, foals and foaling: a handbook for the small breeder. London, Allen, 1962. (Translated by Anthony Dent.)

3872. Ensminger, M. E. Horses and horsemanship. 4th ed. Danville, Ill., Interstate, 1969.

3873. Goodall, Daphne M. Horses of the world. New York, Macmillan, 1965.

3874. Greeley, R. G. The art and science of horseshoeing. Philadelphia, Lippincott, 1970.

3875. Harrison, J. C. Care and training of

of the trotter and pacer. Columbus, Ohio, United States Trotting Association, 1968.

3876. Hayes, M. H. Points of the horse: a treatise on the conformation, movements, breeds and evolution of the horse. 7th ed. New York, Arco, 1969.

3877. ——. Stable management and exercise: a book for horse owners and students. 6th ed. Rev. by Sir Andrew Horsbrugh-Porter. New York, Arco, 1969.

3878. ——. Veterinary notes for horse owners. 16th ed. Rev. by S. Paul. London, Stanley Paul, 1968.

3879. Isenbart, Hans-Heinrich. The kingdom of the horse. English version ed. by Barbara Rey. New York, Time-Life Books, 1969. (Translated from German by F. A. Bauchwitz.)

3880. Johnson, Patricia H. Meet the horse. New York, Grosset & Dunlap, 1967.

3881. Kays, D. J. The horse: judging, breeding, feeding, management, selling. New York, Rinehart, 1953.

3882. Lungwitz, Anton. A text-book of horseshoeing for horseshoers and veterinarians. Philadelphia, Lippincott, 1904. (Translated from the 10th German edition by J. W. Adams.)

3883. McGee, W. R. Veterinary notebook: an elementary guide for the practical horseman. Lexington, Ky., The Bloodhorse, 1958.

3884. ——. Veterinary notes for the standardbred breeder. Lexington, Ky., Hagyard-Davidson-McGee, n.d.

3885. Mahaffey, L. W. "Stud and Stable" veterinary handbook, 1962-1969. London, Stud and Stable, 1969.

3886. Manwill, M. C. How to shoe a horse. South Brunswick, N.J., Barnes, 1968.

3887. Osborne, W. D. The quarter horse. New York, Grosset & Dunlap, 1967.

3888. Patten, J. W. The light horse breeds: their origin, characteristics, and principal uses. New York, Barnes, 1960.

3889. Ransom, J. H. History of American saddle horses. Lexington, Ky., Ransom, 1962.

3890. Saunders, G. C. Your horse: his selection, stabling, and care. 2d ed. Princeton, N.J., Van Nostrand, 1966.

3891. Savitt, Sam. America's horses. Garden City, N.Y., Doubleday, 1966.

3892. Self, Margaret C. Horses of today: Arabian, thoroughbred, saddle horse, standardbred, western, pony. 1st ed. New York, Duell, Sloan & Pearce, 1964.

3893. Serth, G. W. The horse owner's guide to common ailments. London, Nelson, 1971.

3894. Smythe, R. H. The horse: structure and movement. London, Allen, 1967.

3895. ——. Horses in action. Springfield, Ill., Thomas, 1964.

3896. ——. The mind of the horse. Brattleboro, Vt., Greene Press, 1965.

3897. Strong, C. L. Horses' injuries: commonsense therapy of muscles and joints for the layman. London, Faber & Faber, 1967.

3898. Stubbs, George. The anatomy of the horse: the original 1766 edition and illustrations, with a modern veterinary paraphrase by James McCunn and C. W. Ottaway. London, Allen, 1965.

3899. Summerhays, R. S. Encyclopedia for horsemen. 4th ed. London, Warne, 1966.

3900. Wiseman, R. F. The complete horseshoeing guide. Norman, Okla., Univ. of Oklahoma Press, 1968.

24
SHEEP AND GOATS

OVINE MEDICINE AND SURGERY

3901. Anderson, W. D. Thorax of sheep and man: an anatomy atlas. Minneapolis, Dillon Press, 1972.

3902. Belschner, H. G. Sheep management and diseases. 8th ed. Sydney, Angus & Robertson, 1965.

3903. Blaxter, K. L. Energy metabolism of ruminants. 2d rev. ed. Springfield, Ill., Thomas, 1967.

3904. Cole, V. G. Diseases of sheep. London, Angus & Robertson, 1966.

3905. Conference on parturient paresis in dairy animals. University of Illinois, 1968. Parturient hypocalcemia. Ed. by J. J. B. Anderson. New York, Academic Press, 1970.

3906. Fraser, Allan, and J. T. Stamp. Sheep husbandry and diseases. 5th ed. London, Lockwood, 1968.

3907. Mackenzie, David. Goat husbandry. Levittown, N.Y., Transatlantic Arts, 1957.

3908. May, N. D. S. The anatomy of the sheep. 3d ed. St. Lucia, Australia, Univ. of Queensland Press, 1970.

3909. Newsom, I. E. Sheep diseases. 3d ed. Ed. by Hadleigh Marsh. Baltimore, Williams & Wilkins, 1965.

3910. Robinson, T. J., ed. Control of the ovarian cycle in the sheep. Sydney, Sydney Univ. Press, 1967.

3911. Rowlands, W. T. Diseases of sheep. 2d ed. London, Great Britain Ministry of Agriculture, Fisheries and Food, 1965. (Bulletin no. 170.)

3912. Swett, W. W., and C. A. Matthews. Studies on comparative anatomy of goats and dairy cows. Hyattsville, Md., U.S. Agricultural Research Service, 1967.

3913. U.S. Chemical Research and Development laboratories. Pathologic findings in the laboratory goat. 3 parts in 1 v. By F. W. Light, Jr. Edgewood Arsenal, Md., 1965. (CRDL special publication, 2-62, 2-67. Part 3 by C. E. Hopkins and F. W. Light, Jr.)

3914. Winter, H. Bone marrow cells of sheep. Portland, Oregon, International Scholarly Book Services, 1965. (Published by Univ. of Queensland Press.)

3915. Wright, S. E. Detoxication mechanisms in the sheep. Portland, Oregon, International Scholarly Book Services, n.d. (Published by Univ. of Queensland Press.)

PRODUCTION AND MANAGEMENT

3916. Dolling, C. H. Breeding merinos. Mystic, Conn., Verry, 1970.

3917. Ensminger, M. E. Sheep and wool science. 4th ed. Danville, Ill., Interstate, 1970.

3918. French, M. H. Observations on the goat. Rome, Food and Agricultural Organization of the United Nations, 1970. (FAO agricultural studies, no. 80.)

3919. Gray, J. R. Ranch economics. Ames, Iowa State Univ. Press, 1968.

3920. Kammlade, W. G. Sheep science. Ed. by R. W. Gregory. Philadelphia, Lippincott, 1947.

3921. Spedding, C. R. W. Sheep production and grazing management. 2d ed. London, Bailliere, Tindall & Cassell, 1970.

3922. Stevens, P. G. Sheep. 2 parts. 2d ed. San Francisco, Tri-Ocean Books, 1967. Part 1—Sheep husbandry. Part 2—Sheep farming development and sheep breeds in New Zealand.

3923. Turner, Helen, and Sydney Young. Quantitative genetics in sheep breeding. Ithaca, N.Y., Cornell Univ. Press, 1969.

25
SWINE

*PORCINE MEDICINE
AND SURGERY*

3924. Agricultural Research Council. (Great Britain.) Index on current research on pigs. v. 1- Shinfield, Reading, England, National Institute for Research in Dairying, 1954-

3925. Anthony, D. J., and E. F. Lewis. Diseases of the pig: a handbook of the diseases of the pig, with an introduction to its husbandry. 5th ed. Baltimore, Williams & Wilkins, 1961.

3926. Belschner, H. G. Pig diseases. Sydney, Angus & Robertson, 1967.

3927. British Veterinary Association. Technical Development Committee. The husbandry and disease of pigs. 4th ed. London, British Veterinary Association, 1967.

3928. Dunne, H. W., ed. Diseases of swine. 3d ed. Ames, Iowa State Univ. Press, 1970.

3929. Gilbert, S. G. Pictorial anatomy of the fetal pig. 2d ed. Seattle, Univ. of Washington Press, 1966.

3930. Kalbus, Barbara H., and K. G. Neal. Dissection guide for the fetal pig. Minneapolis, Burgess, 1971.

3931. Leone, C. A., and P. W. Ogilvie. Fetal pig manual. 2d ed. Minneapolis, Burgess, 1963.

3932. Marrable, A. W. The embryonic pig: a chronological account. London, Pitman Medical, 1971.

3933. Mount, L. E. The climatic physiology of the pig. London, Arnold, 1968. (Physiological Society, London. Monograph no. 18.)

3934. ———, and D. L. Ingram. The pig as a laboratory animal. New York, Academic Press, 1971.

3935. Pekas, J. C., and L. K. Bustad. A selected list of references on swine in biomedical research. Richland, Wash., Pacific Northwest Laboratory, 1965.

3936. Swine in biomedical research: proceedings of a symposium at the Pacific Northwest Laboratory, Richland, Wash., 1965. Ed. by L. K. Bustad, R. O. McClellan, and M. P. Burns. Richland, Wash., Biology Dept., Battelle Memorial Institute, Pacific Northwest Laboratory, 1966.

3937. Tumbleson, M. E. A selected list of references on the use of swine in biomedical research. (Loose-leaf.) Columbia, Univ. of Missouri, 1971.

*PRODUCTION, NUTRITION,
AND MANAGEMENT*

3938. Baker, J. K., and E. M. Juergenson. Approved practices in swine production. 5th ed. Danville, Ill., Interstate, 1971.

3939. Bundy, C. E., and R. V. Diggins. Swine production. 3d ed. Englewood Cliffs, N.J., Prentice-Hall, 1970.

3940. Cole, D. J. A., ed. Pig production. Proceedings of the 18th Easter School in Agricultural Science. University of Nottingham, 1971. London, Butterworth, 1972.

3941. Duncan, D. L., and G. A. Lodge. Diet in relation to reproduction and the viability of the young. Part 3—Pigs. Aberdeen, Scotland, Commonwealth Bureau of Animal Nutrition, 1960. (Its Technical communication no. 21.)

3942. Ensminger, M. E. Swine science. 4th ed. Danville, Ill., Interstate, 1970.

3943. Krider, J. L., and W. E. Carroll. Swine production. 4th ed. New York, McGraw-Hill, 1971.

3944. Lucas, I. A. M., and G. A. Lodge. The nutrition of the young pig: a review. Aberdeen, Scotland, Commonwealth Bureau of Animal Nutrition, 1961. (Its Technical communication no. 22.)

26

DOGS

CANINE MEDICINE AND SURGERY

3945. Adam, W. S., et al. Microscopic anatomy of the dog: a photographic atlas. Springfield, Ill., Thomas, 1970.

3946. Adrianov, O. S., and T. A. Mering. Atlas of the canine brain. Ed. by E. F. Domino. Ann Arbor, Mich., Edwards Bros., 1964. (Translated by E. Ignatieff.)

3947. American Animal Hospital Association. Committee on Ophthalmology. Canine and feline ocular fundus: an atlas. Elkhart, Ind., American Animal Hospital Association, 1965?

3948. American Society of Veterinary Ophthalmology. Diseases of the canine and feline conjunctiva and cornea. Elkhart, Ind., American Animal Hospital Association, 1970.

3949. Andersen, A. C., and Loraine S. Good, eds. The beagle as an experimental dog. Ames, Iowa State Univ. Press, 1970.

3950. Annis, J. R., and A. R. Allen. An atlas of canine surgery: basic surgical procedures with emphasis on the gastrointestinal and urogenital systems. v. 1- Philadelphia, Lea & Febiger, 1967-

3951. Barnett, K. C. Variations of the normal ocular fundus of the dog. Elkhart, Ind., American Animal Hospital Association, 1972.

3952. Bradley, O. C. Topographical anatomy of the dog. 6th ed. New York, Macmillan, 1959.

3953. Canine surgery. Ed. by James Archibald. Santa Barbara, American Veterinary Publications, 1965. (Earlier editions by H. P. Hoskins, Karl Mayer, J. V. Lacroix.)

3954. Catcott, E. J., ed. Canine medicine: first Catcott edition. Santa Barbara, American Veterinary Publications, 1968. (Previous editions edited by H. P. Hoskins.)

3955. Dua-Sharma, Sushil, K. N. Sharma, and H. L. Jacobs. The canine brain in stereotaxic coordinates: full sections in frontal, sagittal, and horizontal planes. Cambridge, Mass., M.I.T. Press, 1970.

3956. Ettinger, S. J., and P. F. Suter. Canine cardiology. Philadelphia, Saunders, 1970.

3957. Fox, M. W. Canine pediatrics, development, neonatal and congenital disease. Springfield, Ill., Thomas, 1966.

3958. Hall, L. W. Fluid balance in canine surgery. Baltimore, Williams & Wilkins, 1967.

3959. Hoerlein, B. F. Canine neurology: diagnosis and treatment. 2d ed. Philadelphia, Saunders, 1971.

3960. International symposium on canine heartworm disease, 1969. Proceedings. Ed. by R. E. Bradley. Gainesville, Univ. of Florida, 1969.

3961. Jenkins, T. W. Functional mammalian neuroanatomy with emphasis on dog and cat, including an atlas of dog central nervous system. Philadelphia, Lea & Febiger, 1972.

3962. Leonard, E. P. Fundamentals of small animal surgery. Philadelphia, Saunders, 1968.

3963. ———. Orthopedic surgery of the dog and cat. 2d ed. Philadelphia, Saunders, 1971.

3964. Lin, K'o-shêng. A stereotaxic atlas of the dog's brain. By R. K. S. Lim, Chan-nao Liu, and R. L. Moffitt. Springfield, Ill., Thomas, 1960.

3965. Ljungqvist, Ulf. Platelet response to acute haemorrhage in the dog. Stockholm, Almqvist & Wiksell, 1970. (*Acta chirurgica scandinavica.* Suppl. 411.)

3966. McGrath, J. T. Neurologic examination of the dog, with clinicopathologic observations. 2d rev. ed. Philadelphia, Lea & Febiger, 1960.

3967. ———. Spinal dysraphism in the dog,

with comments on syringomyelia. New York, Karger, 1965. (*Pathologia veterinaria*, v. 2. Suppl.)

3968. Magrane, W. G. Canine ophthalmology. 2d ed. Philadelphia, Lea & Febiger, 1971.

3969. Miller, M. E., G. C. Christensen, and H. E. Evans. Anatomy of the dog. Philadelphia, Saunders, 1965. (2d ed. in preparation 1972.)

3970. ——, and Alexander DeLahunta. Guide to the dissection of the dog. Philadelphia, Saunders, 1971.

3971. Muller, G. H. Canine skin lesions. Chicago, American Animal Hospital Association, 1968.

3972. Olsson, Sten-Erik. The radiological diagnosis in canine and feline emergencies: an atlas of thoracic and abdominal changes. Philadelphia, Lea & Febiger, 1971.

3973. Ormrod, A. N. Surgery of the dog and cat: a practical guide. London, Bailliere, Tindall & Cassell, 1966.

3974. Osborne, C. A., D. G. Low, and D. R. Finco. Canine and feline urology. Philadelphia, Saunders, in press 1972.

3975. Pettit, G. D., ed. Intervertebral disc protrusion in the dog. New York, Appleton-Century-Crofts, 1966.

3976. Piermattei, D. L., and R. G. Greeley. An atlas of surgical approaches to the bones of the dog and cat. Philadelphia, Saunders, 1966.

3977. Schebitz, Horst, and H. Wilkens. Atlas of radiographic anatomy of dog and horse. Berlin, Parey, 1968.

3978. Schwartzman, R. M., and Frank Král. Atlas of canine and feline dermatoses. Philadelphia, Lea & Febiger, 1967.

3979. ——, and Milton Orkin. Comparative study of skin diseases of dog and man. Springfield, Ill., Thomas, 1962.

3980. Singer, Marcus. The brain of the dog in section. Philadelphia, Saunders, 1962.

3981. Smythe, R. H. The dog: structure and movement. New York, Arco, 1970.

3982. Startup, F. G. Diseases of the canine eye. Baltimore, Williams & Wilkins, 1969.

3983. Strande, Anders. Repair of the ruptured cranial cruciate ligament in the dog: an experimental and clinical study (a few experiments in cats included). Baltimore, Williams & Wilkins, 1968.

BREEDING, NUTRITION, AND KENNEL MANAGEMENT

3984. American Kennel Club. The complete dog book: the histories and standards of breeds admitted to AKC registration, and the feeding, training, care, breeding, and health of pure-bred dogs. rev. ed. Garden City, N.Y., Doubleday, 1968.

3985. Asdell, S. A. Dog breeding: reproduction and genetics. Boston, Little, Brown, 1966.

3986. Bairacli-Levy, Juliette de. The complete herbal book for the dog: a complete handbook of natural care and rearing. new and rev. ed. London, Faber & Faber, 1971.

3987. Burns, Marca, and Margaret N. Fraser. Genetics of the dog: the basis of successful breeding. 2d ed. Edinburgh, Oliver & Boyd, 1966.

3988. Dangerfield, Stanley, and Elsworth Howell, eds. International encyclopedia of dogs. New York, McGraw-Hill, 1971.

3989. Davis, H. P. The new dog encyclopedia. Harrisburg, Pa., Stackpole Books, 1970.

3990. Fox, M. W. Behaviour of wolves, dogs and related canids. London, Jonathan Cape, 1971.

3991. ——. Canine behavior. Springfield, Ill., Thomas, 1965.

3992. ——. Integrative development of brain and behavior in the dog. Chicago, Univ. of Chicago Press, 1971.

3993. Gaines veterinary symposium. The newer knowledge about dogs. Papers presented. White Plains, N.Y., Gaines Dog Research Center, 1951- (Annual.)

3994. Graham-Jones, Oliver, ed. Canine and feline nutritional requirements: proceedings of a symposium organized by the British Small Animal Veterinary Association. London, 1964. Oxford, Pergamon Press, 1965.

3995. Hamilton, Ferelith, ed. The world encyclopedia of dogs. Assoc. ed. in America: A. F. Jones. 1st American ed. New York, World, 1971.

3996. Harmar, Hilary. Dogs and how to breed them. London, Gifford, 1968.

3997. Harrop, A. E. Reproduction in the dog. Baltimore, Williams & Wilkins, 1960.

3998. Hart, A. H. Dog owner's encyclopedia of veterinary medicine. Jersey City, N.J., T.F.H. Publications, 1971.

3999. Jones, A. F., and John Rendel. The treasury of dogs. New York, Golden Press, 1964.

4000. Jones, E. G., ed. A bibliography of the dog: books published in the English language 1570-1965. London, The Library Association, 1971.

4001. Scott, J. P., and J. L. Fuller, eds. Genetics and the social behavior of the dog. Chicago, Univ. of Chicago Press, 1965.

4002. Vine, Louis L. Dogs are my patients. 2d ed. New York, Exposition Press, 1971.

4003. ——. Your dog: his health and happiness. New York, Exposition Press, 1971.

4004. Winge, Øjvind. Inheritance in dogs; with special reference to hunting breeds. Ithaca, N.Y., Comstock, 1950. (Translated from Danish by Catherine Roberts.)

27

CATS

FELINE MEDICINE AND SURGERY

4005. American Animal Hospital Association. Committee on Ophthalmology. Canine and feline ocular fundus: an atlas. Elkhart, Ind., American Animal Hospital Association, 1965?

4006. American Society of Veterinary Ophthalmology. Diseases of the canine and feline conjunctiva and cornea. Elkhart, Ind., American Animal Hospital Association, 1970.

4007. Berman, A. L. The brain stem of the cat: a cytoarchitectonic atlas with stereotaxic coordinates. Madison, Univ. of Wisconsin Press, 1968.

4008. Booth, E. S. Laboratory anatomy of the cat. 4th ed. Rev. by R. B. Chiasson. Dubuque, Ia., Brown, 1967.

4009. Busch, H. F. M. An anatomical analysis of the white matter in the brain stem of the cat. Assen, Netherlands, Van Gorcum, 1961.

4010. Catcott, E. J. Feline medicine and surgery: a text and reference work. Wheaton, Ill., American Veterinary Publications, 1964.

4011. Crouch, J. E. Text-atlas of cat anatomy. Philadelphia, Lea & Febiger, 1969.

4012. Field, H. E., and Mary E. Taylor. An atlas of cat anatomy. 2d ed. Chicago, Univ. of Chicago Press, 1969.

4013. Gilbert, S. G. Pictorial anatomy of the cat. Seattle, Univ. of Washington Press, 1968.

4014. Greenblatt, G. M. Cat musculature. Chicago, Univ. of Chicago Press, 1954.

4015. Harrison, B. M. Dissection of the cat (and comparisons with man): a laboratory manual on Felis domestica. 6th ed. St. Louis, Mosby, 1970.

4016. Jenkins, T. W. Functional mammalian neuroanatomy, with emphasis on dog and cat, including an atlas of dog central nervous system. Philadelphia, Lea & Febiger, 1972.

4017. Joshua, J. O. The clinical aspects of some diseases of cats. Philadelphia, Lippincott, 1965.

4018. Kalbus, Barbara H., and K. G. Neal. Dissection guide for the cat. Minneapolis, Burgess, in press 1972.

4019. Leonard, E. P. Fundamentals of small animal surgery. Philadelphia, Saunders, 1968.

4020. ———. Orthopedic surgery of the dog and cat. 2d ed. Philadelphia, Saunders, 1971.

4021. Olsson, Sten-Erik. Radiological diagnosis in canine and feline emergencies: an atlas of thoracic and abdominal changes. Philadelphia, Lea & Febiger, 1971.

4022. Ormrod, A. N. Surgery of the dog and cat: a practical guide. London, Bailliere, Tindall & Cassell, 1966.

4023. Osborne, C. A., D. G. Low, and D. R. Finco. Canine and feline urology. Philadelphia, Saunders, in press 1972.

4024. Piermattei, D. L., and R. G. Greeley. An atlas of surgical approaches to the bones of the dog and cat. Philadelphia, Saunders, 1966.

4025. Reighard, J. E., and H. S. Jennings. Anatomy of the cat. 3d ed. New York, Holt, Rinehart & Winston, 1963.

4026. Schwartzman, R. M., and Frank Král. Atlas of canine and feline dermatoses. Philadelphia, Lea & Febiger, 1967.

4027. Snider, R. W., and W. T. Niemer. A stereotaxic atlas of the cat brain. Chicago, Univ. of Chicago Press, 1961.

4028. Strande, Anders. Repair of the ruptured cranial cruciate ligament in the dog: an experimental and clinical study (a few experiments in cats included). Baltimore, Williams & Wilkins, 1968.

4029. Symposium on feline medicine. Ed. by R. L. Stansbury. Philadelphia, Saunders, 1971. (Veterinary clinics of North America, v. 1, no. 2.)

4030. Ulmer, M. J., R. E. Haupt, and E. A. Hicks. Anatomy of the cat: atlas and dissection guide. New York, Harper & Row, 1971.

4031. Walker, W. F., Jr. A study of the cat, with reference to man. 2d ed. Philadelphia, Saunders, 1972.

4032. Wilkinson, G. T. Diseases of the cat. Oxford, Pergamon Press, 1966.

BREEDING, NUTRITION, AND MANAGEMENT

4033. Amberson, Rosanne. Raising your cat: a complete illustrated guide. New York, Crown, 1969.

4034. The cat compendium. 1st ed. Comp. by Ann Currah. New York, Meredith Press, 1969.

4035. Denis, Armand. Cats of the world. Boston, Houghton Mifflin, 1964.

4036. Gardner, M. A. The secret of cooking for cats. Garden City, N.Y., Doubleday, 1965.

4037. Graham-Jones, Oliver, ed. Canine and feline nutritional requirements: proceedings of a symposium organized by the British Small Animal Veterinary Association. London, 1964. Oxford, Pergamon Press, 1965.

4038. Greer, Milan. The fabulous feline; or, Dogs are passé. New York, Dial, 1961.

4039. Henderson, G. N., and N. St. C. Mead. Cats: an intelligent owners' guide. London, Faber & Faber, 1966.

4040. Jude, A. C. Cat genetics. Fond du Lac, Wis., All-Pets Books, 1955.

4041. Leyhausen, P. Cat behavior. New York, Van Nostrand Reinhold, in press 1972.

4042. McCoy, J. J. Complete book of cat health and care. New York, Putnam, 1968.

4043. Manton, S. M. Colorpoint, longhair and Himalayan cats: genetics, breeding and care of these and other pedigree cats. New York, Crown, 1971.

4044. Necker, Claire. The natural history of cats. South Brunswick, N.J., A. S. Barnes, 1970.

4045. Robinson, Roy. Genetics for cat breeders. New York, Pergamon Press, 1971.

4046. Smith, R. C. The complete cat book. New York, Walker, 1963.

4047. Vesey-FitzGerald, B. S. The cat owner's encyclopedia. London, Pelham Books, 1963.

4048. Wilson, Kit. Kit Wilson's cat encyclopedia. Kingswood, England, A. G. Elliot, 1951.

28

FURBEARING ANIMALS

MEDICAL ASPECTS

4049. Bensley, B. A. Practical anatomy of the rabbit: an elementary textbook in mammalian anatomy. 8th ed. Ed. by E. H. Craigie. Toronto, Univ. of Toronto Press, 1948.

4049a. Craigie, E. H. A laboratory guide to the anatomy of the rabbit. 2d ed. Toronto, Univ. of Toronto Press, 1966.

4050. Gorham, J. R., et al. Minks: diseases and parasites. Washington, D.C., Agricultural Research Service, 1965. (U.S. Dept. of Agriculture. Agriculture handbook no. 175.)

4051. Haugen, Jostein. Anticoagulant treatment of rabbits. 2 v. Bergen, Norwegian Universities Press, 1967.

4052. Kaplan, H. M. The rabbit in experimental physiology. 2d ed. New York, Scholar's Library, 1962.

4052a. Klingener, David. Laboratory anatomy of the mink. Dubuque, Ia., Brown, 1970.

4053. McLaughlin, C. A. Laboratory anatomy of the rabbit. Dubuque, Ia., Brown, 1972.

4053a. Oswaldo-Cruz, E., and C. E. Rocha-Miranda. Brain of the opossum in stereotaxic coordinates. Baltimore, Williams & Wilkins, 1968.

4054. Prince, J. H. The rabbit in eye research. Springfield, Ill., Thomas, 1964.

4055. Whitehouse, R. H., and A. J. Grove. The dissection of the rabbit, with an appendix on the rat. 6th ed. London, University Tutorial Press, 1967.

BREEDING, NUTRITION, AND MANAGEMENT

4056. Aitken, F. C. Feeding of fur bearing animals. Aberdeen, Scotland, Commonwealth Bureau of Animal Nutrition, 1963. (Its Technical communication no. 23.)

4057. Blount, W. P. Rabbits' ailments: a short treatise on the domestic rabbit in health and disease. rev. 1957 ed. Bradford, England, "Fur and Feather," 1957.

4057a. Caras, R. A. North American mammals: furbearing animals of the United States and Canada. 1st ed. New York, Meredith Press, 1967.

4058. Kennedy, A. H. The mink in health and disease. Toronto, Fur Trade Journal of Canada, 1951.

4059. Leonard, A. H. Modern mink management. St. Louis, Ralston Purina Co., 1966.

4060. Leoschke, W. L. Feeding and nutrition of mink. Basel, Hoffmann-La Roche, 1970?

4061. Naether, C. A. The book of the domestic rabbit. New York, McKay, 1967.

4062. Templeton, G. S. Domestic rabbit production. 4th ed. Danville, Ill., Interstate, 1968.

4063. Wells, T. A. G. The rabbit: a practical guide. 2d ed. London, Heinemann Educational Books, 1964.

Note. For additional entries see Chap. 39, "Other Laboratory Animals (Vertebrates)" (entries 4530-4574).

BIRDS AND POULTRY

AVIAN MEDICINE

4064. American Association of Avian Pathologists. Salmonellosis workshop, Las Vegas, 1970. Proceedings. Hyattsville, Md., Animal Health Division, U.S. Agricultural Research Service, 1971. (ARS 91-101.)

4065. Avian biology. v. 1- Ed. by D. S. Farner and J. R. King. New York, Academic Press, 1971-

4066. Bell, D. J., and B. M. Freeman, eds. Physiology and biochemistry of the domestic fowl. 3 v. New York, Academic Press, 1971.

4067. Benirschke, Kurt, and T. C. Hsu, eds. Chromosome atlas: fish, amphibians, reptiles and birds. New York, Springer-Verlag, 1971- (Loose-leaf.)

4068. Bondy, S. C., and F. L. Margolis. Sensory deprivation and brain development: the avian visual system as a model. Jena, Fischer, 1971. (*Abhandlungen aus dem Gebiet der Hirnforschung und Verhaltenphysiologie*, bd. 4.)

4069. Calhoun, Mary L. Microscopic anatomy of the digestive system of the chicken. Ames, Iowa State Univ. Press, 1954.

4070. Campbell, J. G. Tumours of the fowl. Philadelphia, Lippincott, 1969.

4071. Chiasson, R. B. Laboratory anatomy of the pigeon. 2d ed. Dubuque, Ia., Brown, 1972.

4072. Coffin, D. L. Angell Memorial parakeet and parrot book. Boston, Angell Memorial Animal Hospital, 1953.

4073. Davis, J. W., et al., eds. Infectious and parasitic diseases of wild birds. Ames, Iowa State Univ. Press, 1971.

4074. Diseases of poultry. Ed. by M. S. Hofstad et al. 6th ed. Ames, Iowa State Univ. Press, 1972. (5th ed. edited by H. E. Biester and L. H. Schwarte.)

4075. Duncker, Hans-Rainer. The lung air sac system of birds: a contribution to the functional anatomy of the respiratory apparatus. New York, Springer-Verlag, 1971.

4076. Ede, D. A. Bird structure: an approach through evolution, development, and function in the fowl. London, Hutchinson Educational, 1964.

4077. Fitzgerald, T. C. The coturnix quail: anatomy and histology. Ames, Iowa State Univ. Press, 1969.

4078. Fowl Pest Review Panel. Fowl pest: Newcastle disease epidemic, 1970-71; report of the Review Panel, presented to Parliament by the Secretary of State for Scotland and the Minister of Agriculture, Fisheries and Food, by command of Her Majesty, October 1971. London, H.M. Stationery Office, 1971.

4079. Fox, Herbert. Disease in captive wild mammals and birds: incidence, descriptions, comparisons. Philadelphia, Lippincott, 1923.

4080. George, J. C., and A. J. Berger. Avian myology. New York, Academic Press, 1966.

4080a. Great Britain Ministry of Agriculture, Fisheries and Food. Avian embryo development. London, H.M. Stationery Office, 1972. (Its Technical bulletin no. 23.)

4081. Harvey, E. B., H. E. Kaiser, and L. E. Rosenberg. An atlas of the domestic turkey (Meleagris gallopavo): myology and osteology. Washington, D.C., U.S. Atomic Energy Commission, 1968.

4082. Hay, E. D., and J. P. Revel. Fine structure of the developing avian cornea. New York, Karger, 1969. (Monographs in developmental biology, v. 1.)

4083. Horton-Smith, C., and E. C. Amoroso, eds. Physiology of the domestic fowl. Edinburgh, Oliver & Boyd, 1966.

4084. Hungerford, T. G. Diseases of poultry, including cage birds and pigeons. 4th ed. Sydney, Angus & Robertson, 1969.

4085. Infectious and parasitic diseases of wild birds. Ed. by J. W. Davis et al. Ames, Iowa State Univ. Press, 1971.

4086. Karten, H. J., and William Hodos. A stereotaxic atlas of the brain of the pigeon, Columba livia. Baltimore, Johns Hopkins Press, 1967.

4087. Lucas, A. M., and Casimir Jamroz. Atlas of avian hematology. Washington, D.C., U.S. Dept. of Agriculture, 1961. (Its Agriculture monograph no. 25.)

4088. Marshall, A. J., ed. Biology and comparative physiology of birds. 2 v. New York, Academic Press, 1960-1961.

4089. Masui, Kiyoshi. Sex determination and sexual differentiation in the fowl. Ames, Iowa State Univ. Press, 1967.

4090. Patten, B. M. Early embryology of the chick. 5th ed. New York, McGraw-Hill, 1971.

4091. Pearson, Ronald. The avian brain. New York, Academic Press, 1972.

4092. Petrak, Margaret L. Diseases of cage and aviary birds. Philadelphia, Lea & Febiger, 1969.

4093. Poultry disease and world economy. Ed. by R. F. Gordon and B. M. Freeman. Edinburgh, British Poultry Science, 1971.

4094. Robinson, M. C. Laboratory anatomy of the domestic chicken. Ed. by E. S. Booth and R. B. Chiasson. Dubuque, Ia., Brown, 1970.

4095. Romanoff, A. L. The avian embryo: structural and functional development. New York, Macmillan, 1960.

4096. ———, and A. J. Romanoff. Biochemistry of the avian embryo. New York, Wiley-Interscience, 1967.

4097. Seneviratna, P. Diseases of poultry, including cage birds. 2d ed. Bristol, John Wright & Sons, 1969.

4098. Sturkie, P. D. Avian physiology.

2d ed. Ithaca, N.Y., Comstock, 1965.

4099. Symposium on avian leukosis. Sofia, 1970. Avian leukosis and Marek's disease: reports and communications. Sofia, Publishing House of the Bulgarian Academy of Sciences, 1972.

4100. U.S. Center for Disease Control (formerly Communicable Disease Center). Poultry diseases in public health: review for epidemiologists. By Mildred M. Galton and Paul Arnstein. Atlanta, 1960. (U.S. Public Health Service publ. no. 767.)

4101. U.S. National Research Council. Subcommittee on Avian Diseases. Methods for examining poultry biologics and for identifying and quantifying avian pathogens. Washington, D.C., National Academy of Sciences, 1971. (Supersedes NAS-NRC publication no. 1038, 1963.)

4102. University of Maine at Orono. Dept. of Animal and Veterinary Sciences. A bibliography of avian mycosis. 3d ed. By E. S. Barden et al. Orono, University of Maine, 1971.

4103. Van den Akker, L. M. An anatomical outline of the spinal cord of the pigeon. Philadelphia, Davis, 1970.

BREEDING, CARE, AND MANAGEMENT

4104. Bennett, R. B. Care and breeding of budgerigars, canaries and foreign finches. New York, Arco, 1961.

4105. Card, Leslie E., and M. C. Nesheim. Poultry production. 11th ed. Philadelphia, Lea & Febiger, in preparation 1972.

4106. Ensminger, M. E. Poultry science. Danville, Ill., Interstate, 1971.

4107. Lanyon, W. E. Biology of birds. Garden City, N.Y., American Museum of Natural History Press, 1963.

4108. Levi, W. M. The pigeon. 2d ed. Sumter, S.C., Levi Publ., 1963.

4109. Merck and Company, Inc. The Merck poultry serviceman's manual. 2d ed. Rahway, N.J., Merck Chemical Division, 1967.

4110. Schaible, P. J. Poultry: feeds and nutrition. Westport, Conn., Avi Publishing, 1970.

4111. Scott, M. L., M. C. Nesheim, and R. J. Young. Nutrition of the chicken. Ithaca, N.Y., Scott, 1969.

4112. Titus, H. W., and J. C. Fritz. The scientific feeding of chickens. 5th ed. Danville, Ill., Interstate, 1971.

4113. Welty, J. C. The life of birds. Philadelphia, Saunders, 1962.

30

FISH AND AQUATIC MAMMALS

MEDICAL AND RESEARCH ASPECTS

4114. Alexander, R. M. Functional design in fishes. New York, Hillary House, 1967.

4115. American Fisheries Society. Symposium Committee on Fish Diseases. A symposium on disease of fishes and shellfishes. Ed. by S. F. Snieszko, Washington, D.C., 1970.

4116. Benirschke, Kurt, and T. C. Hsu, eds. Chromosome atlas: fish, amphibians, reptiles and birds. New York, Springer-Verlag, 1971- (Loose-leaf.)

4117. Brown, Margaret E., ed. Physiology of fishes. 2 v. Academic Press, 1957.
    v. 1: Metabolism. v. 2: Behavior.

4118. Central nervous system and fish behavior. Ed. by David Ingle. Chicago, Univ. of Chicago Press, 1968.

4119. Dogel', V. A., G. K. Petrushevskiĭ, and Y. I. Polyanski, eds. Parasitology of fishes. Edinburgh, Oliver & Boyd, 1961. (*Osnovnye problemy parazitlogii ryb*, translated by Z. Kabata.)

4120. Duijn, C. van. Diseases of fishes. 2d ed. Springfield, Ill., Thomas, 1967.

4121. Geisler, Rolf. Aquarium fish diseases. Neptune City, N.J., T.F.H. Publications, 1963.

4122. Halstead, B. W. Poisonous and venomous marine animals of the world. 3 v. Washington, D.C., U.S. Govt. Printing Office, 1965-1970.
    v. 1: Invertebrates. v. 2-3: Vertebrates.

4123. Hoar, W. S., and D. J. Randall, eds. Fish physiology. v. 1- (For specific titles see entry 1929.)

4124. Hoffman, G. L. Parasites of North American freshwater fishes. Berkeley, Univ. of California Press, 1967.

4125. International symposium on cetacean research. 1st. Washington, D.C., 1963. Whales, dolphins, and porpoises. Ed. by K. S. Norris. Berkeley, Univ. of California Press, 1966.

4126. Kleerekoper, Herman. Olfaction in fishes. Bloomington, Indiana Univ. Press, 1969.

4127. Krogh, A. Osmotic regulation in aquatic animals. New York, Dover, 1965.

4128. Lewis, W. M. Maintaining fishes for experimental and instructional purposes. Carbondale, Southern Illinois Univ. Press, 1963.

4129. Love, R. M. The chemical biology of fishes. New York, Academic Press, 1970.

4130. Nikolskiĭ, G. V. The ecology of fishes. New York, Academic Press, 1963. (Translated by L. Birkett.)

4131. Reichenbach-Klinke, H. H., and E. Elkan. The principal diseases of lower vertebrates. New York, Academic Press, 1965.

4132. Ridgway, S. H., ed. Mammals of the sea: biology and medicine. Springfield, Ill., Thomas, 1972.

4133. Satchell, G. H. Circulation in fishes.

New York, Cambridge Univ. Press, 1971. (Cambridge monographs in experimental biology, no. 18.)

4134. Schlieper, Carl, ed. Research methods in marine biology. Seattle, Univ. of Washington Press, 1972.

4135. Sindermann, C. J. Principal diseases of marine fish and shellfish. New York, Academic Press, 1970.

4136. Snieszko, S., et al. Fish diseases: bacteria. Neptune City, N.J., T.F.H. Publications, 1971.

4137. Symposium on comparative bioelectrogenesis. Rio de Janeiro, 1959. Bioelectrogenesis: a comparative survey of its mechanisms with particular emphasis on electric fishes. Ed. by Carlos Chagas and A. P. de Carvalho. New York, Elsevier, 1961.

4138. Symposium on diseases and husbandry of aquatic mammals. 2d. Florida Atlantic University, 1968. Proceedings. Ed. by D. K. Caldwell and Melba C. Caldwell. St. Augustine, Fla., Marineland Research Laboratory, 1968.

4139. Symposium on marine bio-acoustics. 2d. American Museum of Natural History, 1966. Marine bio-acoustics. Proceedings. Ed. by W. N. Tavolga. 1st ed. New York, Symposium Publications Division, Pergamon Press, 1967.

4140. Symposium on the use of fish as an experimental animal in basic research. Vermillion, S.D., 1968. Fish in research. Ed. by O. W. Neuhaus and J. E. Halver. New York, Academic Press, 1969.

4141. Wood, E. J. F. Marine microbial ecology. New York, Van Nostrand, Reinhold, 1965.

4142. Woods Hole, Mass., Marine Biological Laboratory. Methods for obtaining and handling marine eggs and embryos. 2d ed. Ed. by D. P. Costello and Catherine Henley. Woods Hole, Mass., 1971.

BIOLOGY AND MANAGEMENT

4143. Andersen, H. T. The biology of marine mammals. New York, Academic Press, 1969.

4144. Axelrod, H. R., and C. W. Emmens. Exotic marine fishes. Jersey City, N.J., T.F.H. Publications, 1969. (Loose-leaf.)

4145. Friedrich, Hermann. Marine biology: an introduction to its problems and results. Seattle, Univ. of Washington Press, 1969. (Translated by Gwynne Vevers.)

4146. Gotto, R. V. Marine animals: partnerships and other associations. New York, American Elsevier, 1969.

4147. Halstead, B. W. Dangerous marine animals. Cambridge, Md., Cornell Maritime Press, 1959.

4148. Harrison, R. J., and Judith E. King. Marine mammals. New York, Hillary House, 1970.

4149. Jordan, D. S. The genera of fishes, and a classification of fishes. Stanford, Calif., Stanford Univ. Press, 1963.

4150. Kleynenberg, S. Y., ed. Morphological features of aquatic mammals. New York, CCM Information Corp., 1964.

4151. Nicol, J. A. C. The biology of marine animals. 2d ed. London, Pitman, 1967.

31
REPTILES AND AMPHIBIANS

*MEDICAL AND RESEARCH
ASPECTS*

4152. Benirschke, Kurt, and T. C. Hsu, eds. Chromosome atlas: fish, amphibians, reptiles and birds. New York, Springer-Verlag, 1971- (Loose-leaf.)

4153. Brown, Donald, et al. Molecular biology of amphibian development. New York, MSS Information, 1972.

4154. Deuchar, Elizabeth M. Biochemical aspects of amphibian development. New York, Wiley, 1966.

4155. Kemali, M., and V. Braitenberg. Atlas of the frog's brain. New York, Springer-Verlag, 1969.

4156. Langebartel, D. A. The hyoid and its associated muscles in snakes. Urbana, Univ. of Illinois Press, 1968.

4157. Minton, S. A., Jr., and Madge R. Minton. Venomous reptiles. New York, Scribner, 1969.

4158. Reichenbach-Klinke, H. H., and E. Elkan. The principal diseases of lower vertebrates. New York, Academic Press, 1965.

4159. Romer, A. S. Osteology of the reptiles. Chicago, Univ. of Chicago Press, 1956.

4160. Russell, F. E., and R. S. Scharffenberg. Bibliography of snake venoms and venomous snakes. West Covina, Calif., Bibliographic Associates, 1964.

4161. Symposium—biology of amphibian tumors. New Orleans, 1968. Biology of amphibian tumors. Ed. by Merle Mizell. New York, Springer-Verlag, 1969.

4162. Underhill, R. A. Laboratory anatomy of the frog. 2d ed. Dubuque, Ia., Brown, 1969.

4163. U.S. Bureau of Medicine and Surgery. Poisonous snakes of the world: a manual for use by U.S. amphibious forces. 2d ed. Washington, D.C., U.S. Govt. Printing Office, 1965.

*BIOLOGY AND
CLASSIFICATION*

4164. Bellairs, A. A. The life of reptiles. 2 v. London, Weidenfeld, 1969.

4165. ——, and Richard Carrington. The world of reptiles. London, Chatto & Windus, 1966.

4166. Cloudsley-Thompson, J. L. The temperature and water relations of reptiles. Watford, England, Merrow, 1971.

4167. Ditmars, R. L. Reptiles of the world. rev. ed. New York, Macmillan, 1966. (c1933.)

4168. Gans, Carl, ed. Biology of the reptilia. v. 1- New York, Academic Press, 1969-
   v. 1: Morphology A. v. 2: Morphology B. v. 3: Morphology C.

4169. Goin, C. J., and Olive B. Goin. Introduction to herpetology. 2d ed. San Francisco, Freeman, 1971.

4170. Leviton, A. E. Reptiles and amphibians of North America. New York, Doubleday, 1971.

4171. Mertens, Robert. The world of amphibians and reptiles. New York, McGraw-Hill, 1960. (*La vie des amphibiens et reptiles*, translated by H. W. Parker.)

4172. U.S. National Museum. Catalogue of neotropical squamata. 2 v. By J. A. Peters and Braulio Orejas-Miranda. Washington, D.C., Smithsonian Institution Press, 1970.
   v. 1: Snakes. v. 2: Lizards and amphibians. (Its Bulletin no. 297.)

4173. Wright, A. H., and Anna A. Wright. Handbook of snakes of the United States and Canada. 2 v. Ithaca, N.Y., Cornell Univ. Press, 1957.

32
WILDLIFE AND ZOO ANIMALS

MEDICAL ASPECTS

4174. Davis, J. W., L. H. Karstad, and D. O. Trainer. Infectious diseases of wild mammals. Ames, Iowa State Univ. Press, 1970.

4175. ——, and R. C. Anderson, eds. Parasitic diseases of wild mammals. Ames, Iowa State Univ. Press, 1971.

4176. Fox, Herbert. Disease in captive wild mammals and birds: incidence, descriptions, comparisons. Philadelphia, Lippincott, 1923.

4177. Fyvie, Audrey. Manual of common parasites, diseases and anomalies of wildlife in Ontario. 2d ed. Toronto, Dept. of Lands and Forests, 1969.

4178. International symposium on diseases in zoo animals. 4th. Copenhagen, 1962. Proceedings. Copenhagen, Verhandlungen, 1962. (*Nordisk veterinaermedicin*, bd. 14. Suppl. 1.) Other symposia were held in Berlin, Cologne, and Amsterdam, but proceedings have not been determined as being published.

4179. Mackenzie, P. Z., and R. M. Simpson. The African veterinary handbook. 5th ed. New York, Pitman, 1971.

NATURAL HABITAT AND CONSERVATION

4180. Andrewartha, H. G. Introduction to the study of animal populations. 2d ed. London, Methuen, 1970.

4181. Caras, R. A. Dangerous to man: wild animals, a definitive study of their reputed dangers to man. 1st ed. Philadelphia, Chilton Books, 1964.

4182. ——. Vanishing wildlife. Richmond, Va., Westover, 1970.

4183. Crowe, P. K. World wildlife: the last stand. New York, Scribner's 1970.

4184. Fisher, James, et al. Wildlife in danger. New York, Viking Press, 1969.

4185. Fox, M. W. Behaviour of wolves, dogs and related canids. London, Jonathan Cape, 1971.

4186. Guggisberg, C. A. Man and wildlife. New York, Arco, 1970.

4187. Harris, W. C. Portraits of the game and wild animals of southern Africa. Mystic, Conn., Verry, 1969.

4188. Harthoorn, A. M. The flying syringe: ten years of immobilising wild animals in Africa. London, Bles, 1970.

4189. Hornaday, W. T. Our vanishing wildlife: its extermination and preservation. New York, Arno Press, 1970. (Reprint of 1913 edition.)

4190. ——. Thirty years war for wild life. New York, Arno Press, 1970.

4191. Mech, L. D. The wolf: the ecology and behavior of an endangered species. New York, Natural History Press, 1970.

4192. Murphy, Robert. Wild sanctuaries: our national wildlife refuges, a heritage restored. New York, Dutton, 1968.

4193. Sadleir, R. M. The ecology of reproduction in wild and domestic mammals. London, Methuen, 1969 (distributed in U.S. by Barnes & Noble, New York).

4194. Schaller, G. B. The mountain gorilla: ecology and behavior. Chicago, Univ. of Chicago Press, 1963.

4195. Simon, Noel, and Paul Géroudet. Last survivors: the natural history of animals in danger of extinction. New York, World, 1970.

4196. Stonehouse, Bernard. Animals of the Arctic: the ecology of the far North. 1st American ed. New York, Holt, Rinehart & Winston, 1971.

4197. Street, Phillip. Wildlife preservation. Chicago, Regnery, 1971.

4198. Trippensee, R. E. Wildlife manage-

ment. 2 v. New York, McGraw-Hill, 1948-1953.
v. 1: Upland game and general principles. v. 2: Furbearers, waterfowl, and fish.

4199. Tucker, R. K., and D. G. Crabtree. Handbook of toxicity of pesticides to wildlife. Washington, D.C., U.S. Govt. Printing Office, 1970. (U.S. Bureau of Sport Fisheries and Wildlife resource publication no. 84.)

4200. U.S. Dept. of the Interior. Fish and Wildlife Service. Bureau of Sport Fisheries and Wildlife. Washington, D.C. Special scientific report—Wildlife. (For specific titles see entry 0789.)

4201. ———. Wildlife research: problems, programs, progress. Washington, D.C., U.S. Govt. Printing Office. (U.S. Bureau of Sport Fisheries and Wildlife resource publications; issued approximately annually.)

4202. U.S. National Academy of Sciences. Committee on Agricultural Land Use and Wildlife Resources. Land use and wildlife resources. Washington, D.C., 1970.

4203. Van Dersal, W. R. Wildlife for America: the story of wildlife conservation. rev. ed. New York, Walck, 1970. (Previous editions by E. H. Graham and W. R. Van Dersal.)

4204. Wild animals of North America. 5th ed. Ed. by A. R. Kellogg et al. Washington, D.C., National Geographic Society, 1971. (Reprint of 1960 edition.)

4205. Ziswiler, Vinzenz. Extinct and vanishing animals: a biology of extinction and survival. Rev. English ed. by Fred Bunnell and Pille Bunnell. New York, Springer-Verlag, 1967.

ZOO MANAGEMENT

4206. Crandall, L. S. The management of wild mammals in captivity. Chicago, Univ. of Chicago Press, 1964.

4207. Fisher, James. Zoos of the world: the story of animals in captivity. Garden City, N.Y., Natural History Press, 1967.

4208. Gersh, Harry. The animals next door: a guide to zoos and aquariums of the Americas. New York, Fleet Academic Editions, 1971.

4209. Hediger, Heini. Man and animal in the zoo: zoo biology. New York, Delacorte Press, 1969. (Translated by Gwynne Vevers and Winwood Reade.)

4210. ———. Wild animals in captivity. New York, Dover, 1964. (Translated by Geoffrey Sircom.)

4211. International zoo yearbook. v. 1- London, Zoological Society of London, 1959-

4212. Street, Philip. Animals in captivity. London, Faber & Faber, 1965.

33

ZOOLOGY

VERTEBRATE BIOLOGY

4213. Fraser, A. F. Reproductive behaviour in ungulates. New York, Academic Press, 1968.

4214. Golley, F. B., and H. K. Buechner. A practical guide to the study of the productivity of large herbivores. Philadelphia, Davis, 1968.

4215. Hayward, C. L. A laboratory study of the vertebrates. 2d ed. Minneapolis, Burgess, 1967.

4216. Jollie, M. T. Chordate morphology. New York, Reinhold, 1962.

4217. Orr, R. T. Vertebrate biology. 3d ed. Philadelphia, Saunders, 1971.

4218. Vandel, Albert. Biospeleology: the biology of cavernicolous animals. New

York, Pergamon Press, 1965. (Translated by B. E. Freeman.)

4219. Van Gelder, R. G. Biology of mammals. New York, Scribner's, 1969.

4220. Verts, B. J. The biology of the striped skunk. Urbana, Univ. of Illinois Press, 1967.

4221. Villee, C. A., W. F. Walker, Jr., and F. E. Smith. General zoology. 3d ed. Philadelphia, Saunders, 1968.

CLASSIFICATION

4222. Clark, R. B., and A. L. Panchen. Synopsis of animal classification. London, Chapman & Hall, 1971 (distributed in U.S. by Barnes & Noble, New York).

4223. Hall, E. R., and K. R. Kelson. The mammals of North America. 2 v. New York, Ronald Press, 1959.

4224. Mammalian species. no. 1- New York, American Society of Mammalogists, 1970-

4225. Simpson, G. G. The principles of classification and a classification of mammals. New York, American Museum of Natural History, 1945.

4226. Walker, E. P. Studying our fellow mammals. 2d ed. New York, Animal Welfare Institute, 1967.

4227. ——, et al. Mammals of the world. 2 v. 2d rev. ed. by J. L. Paradiso. Baltimore, Johns Hopkins Press, 1968.

34

COMPANION ANIMALS

4228. Benson, Wes. Unusual small animals as pets. 1st ed. New Augusta, Ind., Editors & Engineers, 1966.

4229. Chrystie, Frances N. Pets: a complete handbook on the care, understanding, and appreciation of all kinds of animal pets. new rev. ed. Boston, Little, Brown, 1964.

4230. De Rochemont, Richard. The pets' cookbook: a layman's comprehensive guide to the feeding of dogs, cats, birds, fish and odd animals around the house. 1st ed. New York, Knopf, 1964.

4231. Falk, J. R. The practical hunter's dog book. New York, Winchester Press, 1971.

4232. Gaddis, Vincent, and Margaret Gaddis. The strange world of animals and pets. 1st ed. New York, Cowles, 1970.

4233. Greer, Milan. The fabulous feline; or, Dogs are passé. New York, Dial, 1961.

4234. Griffen, Jeff. The huntings dogs of America. 1st ed. Garden City, N.Y., Doubleday, 1964.

4235. Hartwell, Dickson. Dogs against darkness: the story of the Seeing Eye. rev. ed. New York, Dodd, Mead, 1968.

4236. Henderson, G. N., and N. St. C. Mead. Cats: an intelligent owner's guide. London, Faber & Faber, 1966.

4237. Levinson, B. M. Pet-oriented child psychotherapy. Springfield, Ill., Thomas, 1969.

4238. ——. Pets and human development. Springfield, Ill., Thomas, 1972.

4239. Mathews, R. K. Wild animals as pets. 1st ed. Garden City, N.Y., Doubleday, 1971.

4240. Mundis, J. J. The guard dogs: maximum protection for you, your home, and your business. New York, McKay, 1970.

4241. Pfaffenberger, C. J. The new knowledge of dog behavior. New York, Howell Book House, 1971.

4242. Podhajsky, Alois. The complete training of horse and rider in the principles of classical horsemanship. Garden City, N.Y.,

Doubleday, 1967. (*Die klassische Reitkunst*, translated by Eva Podhajsky and V. D. Williams.)

4243. Rice, Berkeley. The other end of the leash: the American way with pets. 1st ed. Boston, Little, Brown, 1968.

4244. Scott, T. C. Obedience and security training for dogs. New York, Arco, 1969.

4245. Smith, Heather. A horse in your life: a guide for the new owner. Cranbury, New Jersey, A. S. Barnes, 1966.

4246. Smith, R. C. The complete cat book. New York, Walker, 1963.

4247. Szasz, Kathleen. Petishism: pets and their people in the western world. 1st American ed. New York, Rinehart & Winston, 1969.

4248. Watson, S. D., Jr. Dogs for police service, programming and training. Springfield, Ill., Thomas, 1963.

35

EDUCATION AND TRAINING, VETERINARY MEDICINE AND ALLIED HEALTH

4249. Alexander, A. C. A summary of the types of "paraprofessional training" provided by junior and senior colleges and universities in the area of health, education, and welfare during academic year, 1970-1971. Washington, D.C., Dept. of Health, Education and Welfare, Office of New Careers, 1971.

4250. American Association for Laboratory Animal Science. Manual for laboratory animal technicians. Joliet, Ill., 1967. (AALAS publication no. 67-3.)

4251. Barnum, M. R., et al. A laboratory manual and workbook for audio-tutorial introductory biology: principles. rev. ed. Beverly Hills, Calif., Glencoe, 1969.

4252. Conference on veterinary education. London, 1970. Proceedings. London, Zoological Society of London, 1970.

4253. Council on Education of the American Veterinary Medical Association and the Association of Deans of Colleges of Veterinary Medicine. Survey of veterinary medical education. Chicago, American Veterinary Medical Association, 1958.

4254. Educational Testing Service, Princeton, N.J. Graduate and professional school opportunities for minority students. 3d ed. Princeton, N.J., 1971.

4255. Fein, Rashi, and G. I. Weber. Financing medical education: an analysis of alternative policies and mechanisms. New York, McGraw-Hill, 1971.

4256. Frontiers in comparative medicine. Ed. by W. I. B. Beveridge. v. 1- Minneapolis, Univ. of Minnesota Press, 1972- (Wesley W. Spink lectures on comparative medicine.)

4257. Hubbard, J. P. Measuring medical education: the tests and test procedures of the National Board of Medical Examiners. Philadelphia, Lea & Febiger, 1971.

4258. Innovations in medical education: the Springfield conference. Ed. by H. M. Kimmich. Carbondale, Southern Illinois Univ. Press, 1971.

4259. Reform of medical education: the effect of student unrest. J. R. Krevans and P. G. Condliffe, eds. Washington, D.C., National Academy of Sciences, 1970.

4260. Reynolds, Laura A. Programmed instruction and teaching machines in the field of medical education: an annotated bibliography. McLean, Va., Research Analysis Corp., 1966.

4261. Rudman, Jack. Civil service examination passbook: veterinarian. Plainville, N.Y., National Learning Corp., n.d.

4262. ———. Civil service examination

passbook: **veterinarian trainee**. Plainville, N.Y., National Learning Corp., n.d.

4263. ——. Civil service examination passbook: veterinary medical officer. Plainville, N.Y., National Learning Corp., n.d.

4264. ——. Civil service examination passbook: veterinary science officer. Plainville, N.Y., National Learning Corp., n.d.

4265. Schipper, I. A. Lecture outline of preventive veterinary medicine for animal science students. 4th ed. Minneapolis, Burgess, 1967.

4266. Storey, P. B., J. W. Williamson, and C. H. Castle. Continuing medical education: a new emphasis. Chicago, Division of Scientific Activities, American Medical Association, 1968.

4267. Syllabus for the animal technologist. Ed. by G. R. Collins. Joliet, Ill., American Association for Laboratory Animal Science, 1972.

4268. Symposium on education in veterinary public health and preventive medicine. St. Paul, 1968. Proceedings. Washington, D.C., World Health Organization, 1969.

4269. Symposium on veterinary medical education. 1st- Proceedings. (For specific titles see entry 0783.)

4270. U.S. Division of Allied Health Manpower. Equivalency and proficiency testing: a descriptive compilation of existing testing programs in allied health and other health occupations with an annotated bibliography. Bethesda, Md., 1970?

4271. U.S. National Research Council. Committee on Education. A guide to postdoctoral education in laboratory animal medicine. Washington, D.C., National Academy of Sciences, 1971.

4272. ——. Committee on Veterinary Medical Research and Education. New horizons for veterinary medicine. Washington, D.C., National Academy of Sciences, 1972.

4273. World directory of veterinary schools. Geneva, World Health Organization, 1963.

4274. World trends in medical education: faculty, students, and curriculum. Report of a Macy conference. Ed. by Elizabeth Purcell. Baltimore, published for the Josiah Macy, Jr. Foundation by Johns Hopkins Press, 1971.

PART IV
Laboratory Animals

Chapters 36—40 contain materials pertinent to the study of human disease processes, anomalies, surgical procedures, behavioral problems, and other pathological conditions, through the use of animal subjects as models. Some of the treatises, such as anatomical atlases, textbooks of physiology, and encyclopedic classification guides, are concerned primarily with animals; however, they serve to orient the investigator to animal structure and function in relation to the problem he may be considering. Many of the tests, manuals, and treatises deal with the comparative aspects of a given subject. Some works concerned primarily with specific animal or human medical problems have been included for their coverage of comparative medicine within a particular section or chapter.

In addition to the materials listed in the chapters noted above, the reader is encouraged to examine other parts of this compilation for references that may be germane to his specific investigative interests, e.g., microbiology, neoplasia, or physiology. Because research results are reported first in the periodical literature, in the proceedings of conferences, or in other nonmonographic formats, the investigator is urged to use the indexes, abstracts, and specialized bibliographies as keys to this vast body of literature concerned with experimental animals in biomedical research.

The initial chapter in Part IV includes materials generally applicable to all species of vertebrates which are used as laboratory animals. A number of the citations are relevant to the breeding, nutrition, management, or husbandry of the experimental animal in a controlled laboratory environment. Others pertain to regulations, standards, vivisection, and legal concerns. Some have to do with pathological conditions to which the laboratory animal itself may be subject and which should be taken into consideration before the animal is used experimentally. The investigator who is seeking information on the use of the most appropriate laboratory animal as a model for a specific biomedical research problem will find many references to monographic materials that will be of extensive value to him. Included are texts, treatises, manuals, and atlases of comparative vertebrate anatomy, physiology, and pathology. There are citations to proceedings of meetings and conferences and other publications devoted specifically to the laboratory investigation of a unique human medical problem.

The utilization of primates in the study of many human diseases, in environmental and behavioral considerations, and in developmental and reproductive problems has led to the production of an extensive body of literature which is covered in part in a chapter on primates. Sections within the chapter deal with anatomy, physiology, genetics, behavior, disease, and other aspects of the study of primates and their utilization in biomedical research.

The extensive use of rodents as experimental animals in many disciplines within the biomedical sciences led to the preparation of a chapter dealing with this literature. The importance of the rodent in cancer research, as well as in the study of bacterial, viral, and mycological diseases and of genetic, developmental, and behavioral problems, is well documented. Sections within the chapter include citations to the literature of rodent anatomy and genetics, behavior and ecology, medicine and surgery, and the utilization of rodents as models for research.

Many other vertebrates are suitable for specific investigations. A chapter on this subject cites materials relevant to anatomical and physiological considerations, surgical procedures, and the use of many species of animals as research models.

A final chapter directs attention to the expanding role of invertebrates in the investigation of many biomedical problems related to the well-being of mankind. Invertebrates may be utilized to good advantage in many types of research. Mention may be made of their importance in ecological and behavioral investigations and in the study of neoplastic and related disorders. A few general treatises on anatomy, physiology, and classification have been included, in addition to some publications that emphasize appropriate biomedical relationships.

The need for the discovery of new and suitable vertebrate and invertebrate models for a vast variety of human problems is a pressing one. The testing of biologic agents, the investigation of diseases of metabolic, neoplastic, sensory, or developmental origin, and the many behavioral problems that beset man are challenges to the biomedical investigator who seeks answers through the use of animals as models.

36

EXPERIMENTAL ANIMALS (VERTEBRATES)

General Applications

BREEDING, MANAGEMENT, NUTRITION, AND CARE

4275. Animal Welfare Institute. Basic care of experimental animals. rev. ed. New York, 1958.

4276. Arrington, L. R. Introductory laboratory animal science: the breeding, care and management of experimental animals. Danville, Ill., Interstate, 1972.

4277. Beary, E. G. Laboratory animals; their care and use in research: a bibliography. Natick, Mass., U.S. Army Natick Laboratories, 1968. (U.S. Army Natick Laboratories. Bibliographic series 68-1.)

4278. Coates, Marie E., H. A. Gordon, and B. S. Wostmann. Germ-free animal in research. New York, Academic Press, 1968.

4279. Commonwealth Bureau of Animal Nutrition. Diet in relation to reproduction and the viability of the young. Part I—Rats and other laboratory animals. Aberdeen, Scotland, 1948. (Its Technical communication no. 16.)

4280. Conalty, M. L. Husbandry of laboratory animals. New York, Academic Press, 1967.

4281. Farris, E. J., et al., eds. The care and breeding of laboratory animals. New York, Wiley, 1954.

4282. Genetics in laboratory animal medicine: proceedings of a symposium. Boston, 1968. Washington, D.C., National Academy of Sciences, 1969. (Its publication no. 1724.)

4283. A guide to production, care and use of laboratory animals: an annotated bibliography. Federation proceedings 19(4) pt. 3; 22(2) pt. 3. Washington, D.C., Federation of American Societies for Experimental Biology, 1960-1963. (This has been updated by entry 4401.)

4284. Hafez, E. S. E., ed. Reproduction and breeding techniques for laboratory animals. Philadelphia, Lea & Febiger, 1970.

4285. ———, and I. A. Dyer, eds. Animal growth and nutrition. Philadelphia, Lea & Febiger, 1969.

4286. International symposium on germfree life research. 1st. Nagoya and Inuyama, Japan, 1967. Advances in germfree research and gnotobiology. Ed. by M. Miyakawa and T. D. Luckey. Cleveland, CRC Press, 1968.

4287. Laboratory animal handbooks. v. 1- London, Laboratory Animals Ltd., 1968-
    v. 1 The design and function of laboratory animal houses. Ed. by Ronald Hare and P. N. O'Donoghue.
    v. 2 Dietary standards for laboratory rats and mice. Ed. by M. E. Coates et al.
    v. 3 Transplanted tumours of rats and mice: an index of tumours and host strains. Ed. by D. C. Roberts.
    v. 5 Safety in the animal house. Ed. by John Seamer.

4288. Lane-Petter, William, ed. Animals for research: principles of breeding and management. New York, Academic Press, 1963.

4289. Lane-Petter, William, and A. E. G. Pearson. The laboratory animal: principles and practice. New York, Academic Press, 1971.

4290. Medical Research Council (Great Britain) Laboratory Animals Centre. International index of laboratory animals, giving sources and locations of animals used in laboratories throughout the world. Ed. by M. F. W. Festing. 2d ed. Carshalton, Surrey, England, 1971.

4291. ———. Registers of accredited breeders and recognised suppliers. Carshalton, Surrey, England, 1971.

4292. Notre Dame University. The design, development and study of germfree isolators for use with patients and research. Principal investigator, P. C. Trexler. Notre Dame, Ind., 1962.

4293. Porter, George, and William Lane-Petter, eds. Notes for breeders of common laboratory animals. New York, Academic Press, 1962.

Experimental Animals

4294. Silvan, James. Raising laboratory animals: a handbook for biological and behavioral research. 1st ed. Garden City, N.Y., published for the American Museum of Natural History by Natural History Press, 1966.

4295. Simmons, M. L., and K. F. Burns. The vivarium: management, medicine, and service. Baltimore, Williams & Wilkins, in press 1972.

4296. Symposium on environmental requirements for laboratory animals. Manhattan, Kan., 1971. Proceedings. Manhattan, Kansas State Univ., 1971.

4297. Tavernor, W. D., ed. Nutrition and disease in experimental animals. Baltimore, Williams & Wilkins, 1970.

4298. U.S. Armed Services Technical Information Agency. Gnotobiotics: a report bibliography, 1959-1962. Arlington, Va., 1962.

4299. U.S. National Research Council. Committee on Revision of the Guide for Laboratory Animal Facilities and Care. Guide for laboratory animals facility and care. 3d rev. ed. Washington, D.C., National Institutes of Health, 1968.

4300. Universities Federation for Animal Welfare. Handbook of the care and management of laboratory animals. 3d ed. Baltimore, Williams & Wilkins, 1967.

ANATOMY AND PHYSIOLOGY

4301. Alexander, R. M. Animal mechanics. Seattle, Univ. of Washington Press, 1968.

4302. Andrew, W. Comparative hematology. New York, Grune & Stratton, 1965.

4303. Armstrong, C. N., and A. J. Marshall, eds. Intersexuality in vertebrates, including man. New York, Academic Press, 1964.

4304. Gnotobiotes: standards and guidelines for the breeding, care and management of laboratory animals. Washington, D.C., National Academy of Sciences, 1970.

4305. Daniel, J. C., Jr., ed. Methods in mammalian embryology. San Francisco, Freeman, 1971.

4306. Gordon, M. S., et al. Animal function: principles and adaptation. Toronto, Macmillan, 1968.

4307. Graham-Jones, Oliver, ed. Aspects of comparative ophthalmology. New York, Pergamon Press, 1966.

4308. Hoar, W. S. General and comparative physiology. Englewood Cliffs, N.J., Prentice-Hall, 1966.

4309. Kent, G. C., Jr. Comparative anatomy of the vertebrates. 2d ed. St. Louis, Mosby, 1969.

4310. Kuhlenbeck, Hartwig. The central nervous system of vertebrates. New York, Academic Press, 1967.

4311. Liebow, A. A., and D. E. Smith, eds. The lung. Baltimore, Williams & Wilkins, 1968.

4312. Lockwood, A. P. M. Animal body fluids and their regulation. Cambridge, Mass., Harvard Univ. Press, 1966.

4313. Matousek, Josef. Blood groups of animals. The Hague, Junk, 1965.

4314. Morton, John. Guts: the form and function of the digestive system. New York, St. Martin's Press, 1967.

4315. Patt, D. I., and Gail R. Patt. Comparative vertebrate histology. New York, Harper & Row, 1969.

4316. Polyak, S. L. The vertebrate visual system. Ed. by W. H. Klüver. Chicago, Univ. of Chicago Press, 1957.

4317. Prosser, C. L., and F. A. Brown, Jr. Comparative animal physiology. 2d ed. Philadelphia, Saunders, 1961.

4318. Robinson, Roy. Gene mapping in laboratory mammals. 2 v. New York, Plenum Press, 1971-1972.

4319. Schermer, Siegmund. The blood morphology of laboratory animals. 3d ed. Philadelphia, Davis, 1967.

4320. Schoenborn, H. W., and Harold Gainer. Laboratory exercises in animal physiology. Dubuque, Ia., Brown, 1964.

4321. Sealander, J. A., and C. E. Hoffman.

Laboratory manual of elementary mammalian anatomy, with emphasis on the rat. 2d ed. Minneapolis, Burgess, 1967.

4322. Symposium on comparative pathophysiology of circulatory disturbances. Santa Ynez Valley, Calif., 1971. Comparative pathophysiology of circulatory disturbances: proceedings. Ed. by C. M. Bloor. New York, Plenum Press, 1972. (*Advances in experimental medicine and biology*, v. 22.)

4323. Torrey, T. W. Laboratory studies in developmental anatomy. 2d ed. Minneapolis, Burgess, 1962.

4324. Trautmann, Alfred, and Josef Fiebiger. Fundamentals of the histology of domestic animals. Ithaca, Comstock, 1952.

4325. Weber, R., ed. The biochemistry of animal development. 2 v. New York, Academic Press, 1967.

4326. Weichert, C. K. Anatomy of the chordates. 4th ed. New York, McGraw-Hill, 1970.

4327. Witherspoon, J. D. Functions of life: a laboratory guide for animal physiology. Reading, Mass., Addison-Wesley, 1970.

4328. Yoshikawa, T. Atlas of the brains of domestic animals. University Park, Pennsylvania State Univ., 1967.

Note. The U.S. National Research Council, Institute of Laboratory Animal Resources (ILAR), has issued many useful publications for laboratory animal users dealing with procurement, facilities and care, genetic standards, support personnel, etc. A listing of those available may be found in the *ILAR news*, Washington, D.C.

SURGERY, ANESTHESIA, RADIOLOGY, AND SPECIAL TECHNIQUES

4329. Carlson, W. D. Veterinary radiology. 2d ed. Philadelphia, Lea & Febiger, 1967.

4330. Committee for the Preparation of a Technical Guide for Roentgen Techniques in Laboratory Animals. Roentgen techniques in laboratory animals: radiography of the dog and other experimental animals. Ed. by Benjamin Felson. Philadelphia, Saunders, 1968.

4331. Croft, Phyllis G. An introduction to the anaesthesia of laboratory animals. 2d ed. London, UFAW (Universities Federation for Animal Welfare), 1964.

4332. Demikhov, V. P. Experimental transplantation of vital organs. New York, Consultants Bureau, 1962.

4333. Klemm, W. R. Animal electroencephalography. New York, Academic Press, 1969.

4334. Leonard, E. P. Fundamentals of small animal surgery. Philadelphia, Saunders, 1968.

4335. Lumb, W. V. Small animal anesthesia. Philadelphia, Lea & Febiger, 1963.

4336. Markowitz, J., J. Archibald, and H. G. Downie. Experimental surgery, including surgical physiology. 5th ed. Baltimore, Williams & Wilkins, 1964.

4337. Sawyer, D. C., ed. Experimental animal anesthesiology. San Antonio, Tex., Brooks Air Force Base, School of Aerospace Medicine, 1965.

4338. Smith, R. H. Electrical anesthesia. Springfield, Ill., Thomas, 1963.

4339. Soma, L. R., ed. Textbook of veterinary anesthesia. Baltimore, Williams & Wilkins, 1971.

4340. Westhues, Melchior, and Rudolf Fritsch. Animal anesthesia: local anesthesia. Philadelphia, Lippincott, 1964.

4341. Wilson, J. G., and J. Warkany, eds. Teratology: principles and techniques. Chicago, Univ. of Chicago Press, 1965.

4342. Wright, J. G. Veterinary anesthesia and analgesia. 7th ed. Ed. by L. W. Hall. London, Bailliere, Tindall & Cassens, 1971.

DISEASES AND THERAPEUTICS

4343. Barnes, C. D., and L. G. Eltherington. Drug dosage in laboratory animals: a handbook. Berkeley, Univ. of Calif. Press, 1964.

4344. Cohrs, P., R. Jaffe, and H. Meessen, eds. Pathologie der Laboratoriumstiere. (Pathology of laboratory animals.) 2 v. Berlin, Springer, 1958.

4345. Conference on the pathology of laboratory animals. New York, 1963. The pathology of laboratory animals. Comp. and ed. by W. E. Ribelin and J. R. McCoy. Springfield, Ill., Thomas, 1965.

4346. Estes, R. R., et al. Endoparasites of laboratory animals. San Antonio, Tex., Brooks Air Force Base, School of Aerospace Medicine, 1966.

4347. Illinois Institute of Technology, Research Institute, Chicago, Ill., Technology Center. Susceptibility to infection in irradiated animals: final report, November 1967. B. J. Mieszkuc and A. M. Shefner. Chicago, 1967.

4348. International Committee on Laboratory Animals. Symposium. 2d. The problems of laboratory animal disease. Ed. by R. J. C. Harris. New York, Academic Press, 1962.

4349. Tavernor, W. D., ed. Nutrition and disease in experimental animals. Baltimore, Williams & Wilkins, 1970.

ANIMAL MODELS FOR RESEARCH

4350. Andrew, Warren. The anatomy of aging in man and animals. New York, Grune & Stratton, 1971.

4351. Animal models for biomedical research: proceedings of symposia sponsored by the Institute of Laboratory Animal Resources and the American College of Laboratory Animal Medicine. v. 1- Washington, D.C., National Academy of Sciences, 1967- (I—Its publication no. 1594. II—Its publication no. 1736. III—Its publication no. 1854. IV- Issued as unnumbered publications. Symposia held approximately annually.)

4352. Bourne, G. H., ed. In vivo techniques in histology. Baltimore, Williams & Wilkins, 1967.

4353. Chemical mutagenesis in mammals and man. Ed. by F. Vogel and G. Röhrborn. New York, Springer-Verlag, 1970.

4354. Clarke, W. J., et al. Myeloproliferative disorders of animals and man. Oak Ridge, Tenn., U.S. Atomic Energy Commission, 1970. (AEC symposium series, no. 19.)

4355. Conference on atypical virus infections—possible relevance to animal models and rheumatic disease. Atlanta, 1968. Proceedings. Ed. by C. L. Christian, P. E. Phillips, and R. C. Williams. New York, Arthritis Foundation, 1971.

4356. Conference on the morphology of experimental respiratory carcinogenesis. Gatlinburg, Tenn., 1970. Morphology of experimental respiratory carcinogenesis. Proceedings. Ed. by P. Nettesheim, M. G. Hanna, Jr., and J. W. Deatherage, Jr. Oak Ridge, Tenn., U.S. Atomic Energy Commission, Division of Technical Information, 1970; available from U.S. Dept. of Commerce, Springfield, Va. (AEC symposium series, no. 21.)

4357. Davies, Jack. Survey of research in gestation and the developmental sciences. Baltimore, Williams & Wilkins, 1960.

4358. Davison, A. N., and Alan Peters. Myelination. Springfield, Ill., Thomas, 1970.

4359. Domer, F. R. Animal experiments in pharmacological analysis. Springfield, Ill., Thomas, 1971.

4360. Gauze, G. F. Microbial models of cancer cells. Amsterdam, North-Holland Publishing, 1966. (*Frontiers of biology*, v. 1.)

4361. Gay, W. I., ed. Methods of animal experimentation. 3 v. New York, Academic Press, 1965-1968.

4362. Gross, Ludwik, ed. Oncogenic viruses. 2d ed. New York, Pergamon Press, 1970.

4363. Hamburger, Viktor. A manual of experimental embryology. rev. ed. Chicago, Univ. of Chicago Press, 1960.

4364. A handbook: animal models of human disease. Ed. by T. C. Jones, D. B. Hackel, and George Migaki. Washington, D.C., Universities Associated for Research and Education in Pathology, 1972.

4365. Hinde, R. A. Animal behaviour: a synthesis of ethology and comparative psychology. 2d ed. New York, McGraw-Hill, 1970.

4366. International workshop in teratology.

2d. Kyoto, 1968. Methods for teratological studies in experimental animals and man. Proceedings. Ed. by Hideo Nishimura, J. R. Miller, and Mineo Yasuda. Tokyo, Igaku Shoin, 1969 (distributed in U.S. and Canada by Medical Examination Publishing Co., Flushing, N.Y.).

4367. The laboratory animal in gerontological research: proceedings of a symposium. St. Petersburg, Fla., 1967. Washington, D.C., National Academy of Sciences, 1968. (Its publication no. 1591.)

4368. National conference on research animals in medicine. Washington, D.C., 1972. Official program and abstracts. Washington, D.C., 1972. (DHEW publication no. (NIH) 72-181.)

4369. Panel on experimental animals in cancer research. Tokyo, 1966. Experimental animals in cancer research: proceedings. Ed. by Otto Mühlbock and Tatsuji Nomura. Tokyo, Maruzen, 1968. (Gann monograph, 5.)

4370. Research using transplanted tumours of laboratory animals: cross-referenced bibliography. Ed. by D. C. Roberts. v. 1- London, Imperial Cancer Research Fund, 1964- (Issued annually.)

4371. Ricciuti, E. R. Animals in atomic research. Oak Ridge, Tenn., U.S. Atomic Energy Commission, Division of Technical Information, 1967.

4372. Rich, M. A., ed. Experimental leukemia. New York, Appleton-Century-Crofts, 1968.

4373. Selected abstracts on animal models for biomedical research. Comp. and ed. by C. B. Frank and Marilyn J. Anderson. Washington, D.C., Institute of Laboratory Animal Resources, National Research Council, National Academy of Sciences, 1971.

4374. Selye, Hans. Experimental cardiovascular diseases. New York, Springer-Verlag, 1970.

4375. Stewart, H. L. Transplantable and transmissible tumors of animals. Washington, D.C., U.S. Armed Forces Institute of Pathology, 1959. (U.S. Armed Forces Institute of Pathology. *Atlas of tumor pathology*, Sect. XII, fasc. 40.)

4376. Symposium on the biochemistry of simple neuronal models. Milan, 1969. Biochemistry of simple neuronal models. Ed. by Erminio Costa and Ezio Giacobini. New York, Raven Press, 1970. (*Advances in biochemical psychopharmacology*, v. 2.)

4377. Symposium on the use of subhuman primates in drug evaluation. San Antonio, 1967. Use of nonhuman primates in drug evaluation. Ed. by Harold Vagtborg. Published for the Southwest Foundation for Research and Education, San Antonio. Austin, Univ. of Texas Press, 1968.

4378. Telford, I. R., and Larus Einarson. Experimental muscular dystrophies in animals: a comparative study. Springfield, Ill., Thomas, 1971.

4379. U.S. National Research Council. Committee on Physiological Effects of Environmental Factors on Animals. A guide to environmental research on animals. Washington, D.C., National Academy of Sciences, 1971.

4380. Universities Federation for Animal Welfare. UFAW symposium. Birmingham, England, 1969. The use of animals in toxicological studies. Potters Bar, England, Universities Federation for Animal Welfare, 1969.

4381. Van Tienhoven, Ari. Reproductive physiology of vertebrates. Philadelphia, Saunders, 1968.

4382. Volgyesi, F. A. Hypnosis of man and animals, with special reference to the development of the brain in the species and in the individual. 2d ed. Baltimore, Williams & Wilkins, 1966.

4383. Walter, H. E., and L. P. Sayles. Biology of the vertebrates: a comparative study of man and his animal allies. 2d ed. New York, Macmillan, 1949.

4384. Zarrow, M. X., J. M. Yochim, and J. L. McCarthy. Experimental endocrinology: a sourcebook of basic techniques. New York, Academic Press, 1964.

Note. The agencies or information centers listed below may prove helpful to the investigator who is seeking specific information regarding animal model systems.

U.S. National Academy of Sciences, Institute of Laboratory Animal Resources, Animal Models and Genetic Stock Program, 2101 Constitution Ave., N.W., Washington, D.C. 20418. This agency is described in entry 0989.

The *ILAR news*, Washington, D.C., published quarterly by the U.S. National Research Council, Institute of Laboratory Animal Resources (ILAR), has in each issue an extensive section titled "References on animal models for biomedical research" which cites the pertinent literature, including periodical articles, monographs, and conference proceedings, arranged by body system and/or basic medical research discipline.

U.S. Armed Forces Institute of Pathology, Registry of Comparative Pathology, Washington, D.C. 20305. This national center offers consultative assistance to scientists working in the broad area of comparative pathology.

VIVESECTION

4385. Erbe, N. A. Anti-cruelty laws and scientific use of animals in the United States. London, Research Defense Society, 1966.

4386. Fund for the Replacement of Animals in Medical Experiments. FRAME bibliography of alternatives to laboratory animals. Ed. by M. P. Fowler. London, 1971- (Loose-leaf.)

4387. Godlovitch, Stanley, Roslind Godlovitch, and John Harris, eds. Animals, men and morals: an enquiry into the maltreatment of non-humans. London, Gollancz, 1971.

4388. Niven, C. D. History of the humane movement. Levittown, N.Y., Transatlantic Arts, 1967.

4389. Research Defense Society. The biologist and the experimental animal. London, 1965.

4390. Vyvyan, John. The dark face of science. London, Joseph, 1971.

4391. ———. In pity and in anger: a study of the use of animals in science. London, Joseph, 1969.

LEGAL ASPECTS, REGULATIONS, AND STANDARDS

4392. Leavitt, Emily S. Animals and their legal rights: a survey of American law from 1641 to 1968. New York, Animal Welfare Institute, 1968.

4393. Russell, W. M. S., and R. L. Burch. The principles of humane experimental technique. London, Methuen, 1959.

4394. U.S. National Institutes of Health. Procurement Branch. Standards and Specifications Section. NIH animal care equipment, 1972. Bethesda, Md., 1972. (DHEW publication no. (NIH) 72-17.)

4395. U.S. National Research Council. Institute of Laboratory Animal Resources. Committee on Standards. Standards for the breeding, care and management of laboratory animals. v. 1- Washington, D.C., National Research Council, 1960- (For additional coverage see entry 0793.)

4396. ———. ———. Subcommittee on Genetic Standards. A guide to genetic standards for laboratory animals. Washington, D.C., National Academy of Sciences, 1969.

Note. The U.S. National Research Council, Institute of Laboratory Animal Resources (ILAR), has issued many useful publications for laboratory animal users dealing with regulations and standards. A listing of those available may be found in the *ILAR news*, Washington, D.C.

MISCELLANEOUS

4397. Altman, P. L., ed. Blood and other body fluids. Washington, D.C., Federation of American Societies for Experimental Biology, 1961.

4398. ———, and Dorothy S. Dittmer, eds. Biology data book. 2d ed. Bethesda, Md., Federation of American Societies for Experimental Biology, 1972. (An earlier edition was titled *Handbook of biological data*. Ed. by W. S. Spector, 1956.)

4399. American Association for Laboratory Animal Science. Manual for laboratory animal technicians. Joliet, Ill., American Association for Laboratory Animal Science, 1967. (AALAS publication no. 67-3.)

4400. Carworth Europe. Collected papers. v. 1- Ed. by Annie M. Brown. Alconbury, 1967-
- v. 1 Symposium on the future of the defined laboratory animal. Cambridge, England, 1966. The future of the defined laboratory animal. 1967.
- v. 2 Symposium on the interaction of the laboratory animal with its associated organisms. Cambridge, England, 1967. 1968.
- v. 3 Symposium on uniformity. Cambridge, England, 1968. 1969.
- v. 4 Symposium on the effects of environment. Cambridge, England, 1969. 1970.

4401. Cass, J. S., et al., eds. Laboratory animals: an annotated bibliography of informational resources covering medicine-science (including husbandry)—technology. New York, Hafner, 1971. (This updates entry 4283.)

4402. Fiennes, Richard. Some recent developments in comparative medicine. N.Y., Academic Press, 1966. (Zoological Society of London. Symposia, no. 17.)

4403. Frontiers in comparative medicine. Ed. by W. I. B. Beveridge. v. 1- Minneapolis, Univ. of Minnesota Press, 1972- (Wesley W. Spink lectures on comparative medicine.)

4404. International Committee on Laboratory Animals. Symposium. 1st- Proceedings. 1958- (For specific titles see entry 0757.)

4405. Kay, R. H. Experimental biology: measurement and analysis. New York, Reinhold, 1964.

4406. Short, D. J., and Dorothy P. Woodnott, eds. The I.A.T. manual of laboratory animal practice and techniques. 2d ed. rev. and enl. Springfield, Ill., Thomas, 1971.

4407. Syllabus for the animal technologist. Ed. by G. R. Collins. Joliet, Ill., American Association for Laboratory Animal Science, 1972.

4408. U.S. National Academy of Sciences. Future of laboratory animal resource and research programs. Washington, D.C., 1971. (Proceedings of a conference at the National Academy of Sciences in association with the Institute of Laboratory Animal Resources, Washington, D.C., 1970.)

4409. U.S. National Research Council. Committee on Education. A guide to post-doctoral education in laboratory animal medicine. Washington, D.C., National Academy of Sciences, 1971.

4410. ——. Institute of Laboratory Animal Resources (ILAR). Laboratory animal medical subject headings. Washington, D.C., National Academy of Sciences, 1972.

4411. ——. ——. Laboratory animals. (For specific titles see entry 0792.)

4412. ——. ——. Washington, D.C.
This agency publishes many brochures, treatises, pamphlets and other publications of interest to persons in laboratory animal medicine. A sampling of its many publications follows:
- Principles of laboratory animal care. 1968.
- A guide to genetic standards for laboratory animals. 1969.
- Gnotobiotes: standards and guidelines for the breeding, care and management of laboratory animals. 1970.
- Institute of Laboratory Animal Resources: objectives, committees, activities, publications. 1970.
- Users of laboratory animals in the United States. 9th ed. 1972.

The *ILAR news*, Washington, D.C., is a good source of information about publications, meetings, sources of supply, etc.

4413. ——. ——. Committee on Technical Education. Support personnel for animal research. Washington, D.C., National Academy of Sciences, 1969.

4414. Walter, H. E., and L. P. Sayles. Biology of the vertebrates: a comparative study of man and his animal allies. 2d ed. New York, Macmillan, 1949.

Note. For additional material on experimental animals the investigator is referred to appropriate entries in Chap. 1, "Indexes and

Abstracts"; Chap. 3, "Periodicals"; and Chap. 4, "Review Serials and Serial Monographs.

Some indexes and abstracts of particular relevance are: *Index medicus; Animaux de Laboratorie: Revue bibliographique; Biological abstracts; Excerpta medica; Carcinogenesis abstracts*; and *Virology abstracts*.

Some periodicals are: *Journal of comparative laboratory medicine; Laboratory animals; Laboratory animal science;*
Journal of comparative pathology; American journal of veterinary research;
Society for Experimental Biology and Medicine, *Proceedings; Pathologia veterinaria.*

Some review serials and serial monographs are: *Advances in veterinary science and comparative medicine; Advances in virus research; Monographs in virology;* National Cancer Institute *Monographs*; and *International review of experimental pathology.*

37

PRIMATES

ANATOMY, PHYSIOLOGY, AND GENETICS

4415. Chiarelli, A. B., ed. Comparative genetics in monkeys, apes and man. New York, Academic Press, 1971. (Proceedings of a symposium on comparative genetics in primates and human heredity, Erice, Sicily, 1970.)

4416. Circadian rhythms in nonhuman primates. Ed. by F. H. Rohles. New York, Karger, 1969.

4417. Davis, Ross, and R. D. Huffman. Stereotaxic atlas of the brain of the baboon (Papio). Published for the Southwest Foundation for Research and Education, San Antonio. Austin, Univ. of Texas Press, 1968.

4418. DeLucchi, M. R., et al. A stereotaxic atlas of the chimpanzee brain (Pan satyrus). Berkeley, Univ. of California Press, 1965.

4419. Emmers, Raimond, and Konrad Akert. A stereotaxic atlas of the brain of the squirrel monkey (Saimiri sciureus). Madison, Univ. of Wisconsin Press, 1963.

4420. Hafez, E. S. E., ed. Comparative reproduction of nonhuman primates. Springfield, Ill., Thomas, 1971.

4421. Hartman, C. G., and W. L. Straus, eds. The anatomy of the rhesus monkey (Macaca mulatta). By T. H. Bast et al. New York, Hafner, 1933. (Reprinted with corrections, 1961, 1965.)

4422. Hendrickx, A. G. Embryology of the baboon. Chicago, Univ. of Chicago Press, 1971.

4423. Hill, W. C. Primates: comparative anatomy and taxonomy. v. 1- (For specific titles see entry 1422.)

4424. Huser, H. J. Atlas of comparative primate hematology. New York, Academic Press, 1970.

4425. International symposium on breeding non-human primates for laboratory use. Bern, 1971. Breeding primates: apes, baboons, macaques, New World monkeys; general comments on breeding; research in reproduction. Proceedings. Ed. by W. I. B. Beveridge. New York, Karger, 1972.

4426. Kusama, Toshio, and Masako Mabuchi. Stereotaxic atlas of the brain of Macaca fuscata. Baltimore, University Park Press, 1970.

4427. Levy, B. M., S. Dreizen, and S. Bernick. The marmoset periodontium in health and disease. New York, Karger, 1972. (Monographs in oral science, v. 1.)

4428. Madigan, J. C., Jr., and M. B. Carpenter. Cerebellum of the rhesus monkey: atlas of lobules, laminae, and folia, in sections. Baltimore, University Park Press, 1971.

4429. Malinow, M. R., ed. Biology of the howler monkey (Alouatta carga). New York, Karger, 1968. (*Bibliotheca primatologica*, no. 7.)

4430. Manocha, S. L., and T. R. Shantha. Macaca mulatta: enzyme histochemistry of the nervous system. New York, Academic Press, 1970.

4431. ——, T. R. Shantha, and G. H. Bourne. A stereotaxic atlas of the brain of the Cebus monkey. Oxford, Clarendon Press, 1968.

4432. The primate brain. Ed. by C. R. Noback and William Montagna. New York, Appleton-Century-Crofts, 1970. (*Advances in primatology*, v. 1.)

4433. Schön, M. A. The muscular system of the red howling monkey. Washington, D.C., Smithsonian Institution Press, 1968.

4434. Shantha, T. R., S. L. Manocha, and G. H. Bourne. A stereotaxic atlas of the Java monkey brain (Macaca irus). New York, Karger, 1968.

4435. Snider, R. S., and J. C. Lee. A stereotaxic atlas of the monkey brain. Chicago, Univ. of Chicago Press, 1961.

4436. Stern, J. T., Jr. Functional myology of the hip and thigh of cebid monkeys and its implications for the evolution of erect posture. New York, Karger, 1971. (*Bibliotheca primatologica*, no. 14.)

4437. Szebenyi, E. S. Atlas of Macaca mulatta. Rutherford, N.J., Fairleigh Dickinson Univ. Press, 1969.

4438. Winters, W. D., R. T. Kado, and W. R. Adey. A stereotaxic brain atlas for Macaca nemestrina. Berkeley, Univ. of California Press, 1969.

BEHAVIOR, ECOLOGY, AND EVOLUTION

4439. Altmann, S. A., ed. Social communication among primates. Chicago, Univ. of Chicago Press, 1967.

4440. Altmann, S. A., and Jeanne Altmann. Baboon ecology: African field research. Chicago, Univ. of Chicago Press, 1970.

4441. Bourne, G. H. The ape people. New York, Putnam's, 1971.

4442. Buettner-Janusch, John, ed. Evolutionary and genetic biology of primates. 2 v. New York, Academic Press, 1963-1964.

4443. Burg Wartenstein symposium on functional and evolutionary biology of primates: methods of study and recent advances, 1970. The functional and evolutionary biology of primates. Ed. by Russell Tuttle. Chicago, Aldine-Atherton, 1972.

4444. Carpenter, C. R. Naturalistic behavior of nonhuman primates. University Park, Pennsylvania State Univ. Press, 1964.

4445. Chance, M. R. A., and C. J. Jolly. Social groups of monkeys, apes and men. London, Jonathan Cape, 1970.

4446. Dolhinow, Phyllis, ed. Primate patterns. New York, Holt, Rinehart & Winston, 1972.

4447. Hinde, R. A. Social behavior and its development in subhuman primates. Eugene, Oregon State System of Higher Education, 1972.

4448. Jay, Phyllis C., ed. Primates: studies in adaptation and variability. New York, Holt, Rinehart & Winston, 1968.

4449. Jolly, Alison. The evolution of primate behavior. New York, Macmillan, 1972.

4450. Morris, Ramona, and Desmond Morris. Men and apes. London, Hutchinson, 1966.

4451. Rosenblum, L. A., ed. Primate behavior: developments in field and laboratory research. 2 v. New York, Academic Press, 1970-1971.

4452. Schaller, G. B. The mountain gorilla: ecology and behavior. Chicago, Univ. of Chicago Press, 1968. (c1963.)

4453. Schrier, A. M., et al. Behavior of nonhuman primates: modern research trends. v. 1-4. New York, Academic Press, 1965-1971.

4454. Symposium on cognition. 6th. Carnegie-Mellon University, 1970. Cognitive processes of nonhuman primates. Ed. by L. E. Jarrard. New York, Academic Press, 1971.

4455. Wenner-Gren conference on the systematics of Old World monkeys. Wartenstein, Austria, 1969. Old World monkeys: evolution, systematics, and behavior. Ed. by J. R. Napier and P. H. Napier. New York, Academic Press, 1970.

4456. Williams, Leonard. Man and monkey. Philadelphia, Lippincott, 1968.

DISEASES

4457. Coatney, G. R., et al. The primate malarias. Bethesda, Md., National Institute of Allergy and Infectious Diseases, 1971.

4458. Fiennes, Richard. Zoonoses of primates: the epidemiology and ecology of simian diseases in relation to man. Ithaca, N.Y., Cornell Univ. Press, 1967.

4459. ——, ed. Pathology of simian primates. Part 1- New York, Karger, 1972-

4460. International symposium on infections and immunosuppression in subhuman primates. Rijswijk, Netherlands, Dec. 17-19, 1969. Proceedings. Ed. by H. Balner and W. I. B. Beveridge. Copenhagen, Munksgaard, 1970.

4461. Lapin, B. A., and L. A. Yakovleva. Comparative pathology in monkeys. Springfield, Ill., Thomas, 1963.

4462. Ruch, T. C. Diseases of laboratory primates. Philadelphia, Saunders, 1959.

4463. Sauer, R. M., et al. Care and disease of the research monkey. New York, 1960. (New York Academy of Sciences *Annals*, v. 85, art. 3.)

MODELS FOR RESEARCH

4464. Akademiĩa meditsinskikh Nauk SSSR. Institut eksperimental'noĭ patologii i terapii. Pathogenesis, clinical treatment and therapy of acute radiation sickness based on experiments with monkeys: an All-Union symposium. Ed. by B. A. Lapin. Jerusalem, Israel Program for Scientific Translations, 1967. (Translation of *Patogenez, klinika i lechenie ostroĭ luchevoĭ boleznoi v opytakh na obez'ianakh.*)

4465. Bergwin, C. R., and W. T. Coleman. Animal astronauts: they opened the way to the stars. Englewood Cliffs, N.J., Prentice-Hall, 1963.

4466. Conference on experimental medicine and surgery in primates. 1st- 1967- (For coverage of 1st and 2d congresses see entry 0745.)
　3d. Lyon, 1972. (Proceedings in preparation.)

4467. Histocompatibility workshop on primates. 1st. Rijswijk, 1971. Transplantation genetics of primates. Ed. by Hans Balner and Jon J. van Rood. New York, Grune & Stratton, 1972. (Issued also as *Transplantation proceedings*, v. 4, no. 1, with title "Proceedings of a workshop and symposium on transplantation genetics of primates.")

4468. International symposium on the baboon and its use as an experimental animal. Proceedings. The baboon in medical research. Ed. by Harold Vagtborg.
　1st San Antonio, 1963. Austin, Univ. of Texas Press, 1965.
　2d San Antonio, 1967. Austin, Univ. of Texas Press, 1967.

4469. Symposium on the use of subhuman primates in drug evaluation. San Antonio, 1967. Use of nonhuman primates in drug evaluation. Ed. by Harold Vagtborg. Published for the Southwest Foundation for Research and Education, San Antonio. Austin, Univ. of Texas Press, 1968.

4470. Whitney, R. A., D. J. Johnson, and W. C. Cole. The subhuman primate: a guide for the veterinarian. Edgewood Arsenal, Md., U.S. Dept. of the Army, Edgewood Arsenal Research Laboratories, Medical Research Laboratory, 1967. (Edgewood Arsenal special publication, 100-26.)

MISCELLANEOUS

4471. Bibliotheca primatologica. v. 1- Basel, Karger, 1962- (For specific titles see entry 0737.)

4472. Bourne, G. H., ed. The chimpanzee. v. 1- Baltimore, University Park Press, 1969- (For further coverage see entry 2233.)

4473. The chimpanzee: a topical bibliography. 2d ed. Ed. by F. H. Rohless, Jr. Manhattan, Institute for Environmental Research, Kansas State University, 1972.

4474. Gibbon and siamang. Ed. by D. M. Rumbaugh. v. 1- New York, Karger, 1972-
　v. 1: Evolution, ecology, behavior and captive maintenances. v. 2: Anatomy, dentition, taxonomy, molecular evolution and behavior.

4475. Hofer, Helmut, A. H. Schultz, and

Anatomy and Genetics

D. Starck. Primatologia. Handbuch der Primatenkunde. Handbook of primatology. Manuel de primatologie. 4 v. in 16. New York, Karger, 1956-1971.

4476. International congress of primatology. 1st- Proceedings. 1966- (For coverage of 1st and 2d congresses see entry 0760.) 3d Zurich, 1970. 3 v. New York, Karger, 1972. v. 1—Taxonomy, anatomy, reproduction. Ed. by J. Biegert and W. Leutenegger. v. 2—Neurobiology, immunology, cytology. Ed. by J. Biegert and W. Leutenegger. v. 3—Behavior. Ed. by H. Kummer.

4477. Journal of medical primatology. v. 1- New York, Karger, 1972-

4478. Napier, J. R., and P. H. Napier. A handbook of living primates: morphology, ecology and behaviour of nonhuman primates. New York, Academic Press, 1967.

4479. Primates in medicine. v. 1- New York, Karger, 1968- (For specific titles see entry 0776.)

4480. Reynolds, Vernon. The apes: the gorilla, chimpanzee, orangutan, and gibbon; their history and their world. 1st ed. New York, Dutton, 1967.

4481. Rosenblum, L. A., and R. W. Cooper, eds. The squirrel monkey. New York, Academic Press, 1968.

4482. Southwest Foundation for Research and Education. The baboon: an annotated bibliography 1607-1964. San Antonio, Texas, 1964-
...Supplemental baboon information issued annually, 1963-1965.
...Suppl. 2, 1967. (Issued by Biological Science Communication Project.)
...Suppl. 3, 1969.
...Suppl. 4, 1971.

4483. Symposium on feeding and nutrition of nonhuman primates. Bethesda, Md., 1969. Feeding and nutrition of nonhuman primates. Proceedings. Ed. by R. S. Harris. New York, Academic Press, 1970.

4484. Valerio, D. A., et al. Macaca mulatta: management of a laboratory breeding colony. New York, Academic Press, 1969.

Note. For additional material on primates the investigator is referred to the entries in Chap. 36, "Experimental Animals (Vertebrates): General Applications," many of which have specific chapters or sections dealing with primates as laboratory animals.

38

RODENTS

ANATOMY AND GENETICS

4485. Chiasson, R. B. Laboratory anatomy of the white rat. 2d ed. Dubuque, Ia., Brown, 1969.

4486. Cook, Margaret J. The anatomy of the laboratory mouse. New York, Academic Press, 1965.

4487. Craigie's neuroanatomy of the rat. Rev. and expanded by Wolfgang Zeman and J. R. M. Innes. New York, Academic Press, 1963.

4488. Greene, Eunice C. Anatomy of the rat. New York, Hafner, 1959.

4489. Jackson Laboratory. Handbook on genetically standardized Jax mice. Ed. by E. L. Green. 2d ed. Bar Harbor, Me., 1968.

4490. ———. Standardized nomenclature for inbred strains of mice. 4th ed. Bar Harbor, Me., 1968. (*Cancer research* 28(3), 1968.)

4491. König, J. F. R., and R. A. Klippel. The rat brain: a stereotaxic atlas of the forebrain and lower parts of the brain stem. Baltimore, Williams & Wilkins, 1963.

4492. Kovac, Werner, and Helmut Denk. Der Hirnstamm der Maus: Topographie, Cytoarchitektonik und Cytologie. Wien, Springer-Verlag, 1968.

4493. Luparello, T. J. Stereotaxic atlas of the forebrain of the guinea pig. Baltimore, Williams & Wilkins, 1967.

4494. Mouse news letter. no. 1- Carshalton, Surrey, England, Laboratory Animals Centre, 1949-

4495. Mühlen, Klaus aus der. The hypothalamus of the guinea pig: a topographic survey of its nuclear regions. New York, Karger, 1966.

4496. Pellegrino, L. J., and Anna J. Cushman. A stereotaxic atlas of the rat brain. New York, Appleton-Century-Crofts, 1967.

4497. Robinson, Roy. Genetics of the Norway rat. New York, Pergamon Press, 1965.

4498. Rugh, Roberts. The mouse: its reproduction and development. Minneapolis, Burgess, 1968.

4499. Sealander, J. A., and C. E. Hoffman. Laboratory manual of elementary mammalian anatomy, with emphasis on the rat. Minneapolis, Burgess, 1967.

4500. Sherwood, Nancy M., and P. S. Timiras. A stereotaxic atlas of the developing rat brain. Berkeley, Univ. of California Press, 1970.

4501. Sidman, R. L., J. B. Angevine, Jr., and Elizabeth T. Pierce. Atlas of the mouse brain and spinal cord. Cambridge, Mass., Harvard Univ. Press, 1971.

4502. ———, M. C. Green, and S. H. Appel. Catalog of the neurological mutants of the mouse. Cambridge, Mass., Harvard Univ. Press, 1965.

4503. Smith, Esther M., and M. Lois Calhoun. The microscopic anatomy of the white rat: a photographic atlas. Ames, Iowa State Univ. Press, 1968.

4504. Staats, Joan, comp. Selected references added to the Subject-strain bibliography of inbred strains of mice. Bar Harbor, Me., Jackson Laboratory, 1949- ? (Suppl. to *Mouse news letter.*)

4505. Whitehouse, R. H., and A. J. Grove. The dissection of the rabbit, with an appendix on the rat. 5th ed. London, University Tutorial Press, 1956.

4506. Wirtschafter, Z. T. The genesis of the mouse skeleton: a laboratory atlas. Springfield, Ill., Thomas, 1960

BEHAVIOR AND ECOLOGY

4507. Barnett, S. A. The rat: a study in behavior. Chicago, Aldine, 1963.

4508. Calhoun, J. B. The ecology and sociology of the Norway rat. Washington, D.C., U.S. Public Health Service, 1962. (Its publication no. 1008.)

Note. For additional coverage of rodent behavior the reader is referred to more general treatises of animal behavior, a listing of which will be found in the section on "Behavior" (entries 2222-2323) in Chap. 9.

MEDICINE AND SURGERY

4509. Cotchin, Ernest, and F. J. C. Roe, eds. Pathology of laboratory rats and mice. Philadelphia, Davis, 1967.

4510. Lambert, Rene. Surgery of the digestive system in the rat. Springfield, Ill., Thomas, 1965. (Translated by Brian Julien.)

4511. U.S. National Research Council. Committee on Laboratory Animal Diseases. A guide to infectious diseases of mice and rats: a report. Washington, D.C., National Academy of Sciences, 1971.

MODELS FOR RESEARCH

4512. Farris, E. J., and J. Q. Griffith, eds. The rat in laboratory investigation. 2d ed. New York, Hafner, 1962.

4513. Gibson, D. C., ed. Development of the rodent as a model system of aging. v. 1- Bethesda, Md., National Institute of Child Health and Human Development, 1972- (DHEW publication no. (NIH) 72-121, etc.)

4514. Herrick, C. J. Brains of rats and men: a survey of the origin and biological significance of the cerebral cortex. New York, Hafner, 1963.

4515. Hoffman, R. A., P. F. Robinson, and Hulda Magalhaes. The golden hamster: its biology and use in medical research, with a master bibliography by Hulda Magalhaes and including a stereotaxic atlas of the brain of the golden hamster by K. M. Knigge and Shirley A. Joseph. Ames, Iowa State Univ. Press, 1968.

4516. Hunziker, Nicole. Experimental studies on guinea pig's eczema. New York, Springer-Verlag, 1969.

4517. Korec, Rudolf. Experimental diabetes mellitus in the rat. New York, Appleton-Century-Crofts, 1968.

4518. Magnusson, Bertil, and A. M. Kligman. Allergic contact dermatitis in the guinea pig: identification of contact allergens. Springfield, Ill., Thomas, 1970.

4519. Rafferty, K. A. Methods in experimental embryology of the mouse. Baltimore, Johns Hopkins Press, 1970.

4520. Reid, Mary E. The guinea pig in research: biology—nutrition—physiology. Washington, D.C., Human Factors Research Bureau, 1958.

4521. Stephens, Charlene B. Development of the middle and inner ear in the golden hamster (Mesocricetus auratus): a detailed description to establish a norm for physiopathological study of congenital deafness. Uppsala, Sweden, 1972. (*Acta oto-laryngologica.* Suppl. 296.)

4522. Strong, L. C. Biological aspects of cancer and aging: studies in pure line mice. New York, Pergamon, 1968.

Note. The agencies listed in the Note following entry 4384 serve as sources of information pertinent to all animal model systems including the rodent. Specific information on the biology of the laboratory mouse and rodent model systems is available from Jackson Laboratory, Library Bibliographic Service, Bar Harbor, Maine 04609.

MISCELLANEOUS

4523. Commonwealth Bureau of Animal Nutrition. Diet in relation to reproduction and the viability of the young. Part I— Rats and other laboratory animals. Aberdeen Scotland, 1948. (Its Technical communication no. 16.)

4524. The gerbil: an annotated bibliography. Comp. by Victor Schwentker. West Brookfield, Mass., Tumblebrook Farm, Inc., 1972.

4525. Jackson Laboratory. Biology of the laboratory mouse. 2d ed. Ed. by E. L. Green. New York, McGraw-Hill, 1966.

4526. Rich, S. T. Selected bibliography on gerbillinae. Los Angeles, Division of Laboratory Animal Medicine, Center for the Health Sciences, Univ. of California, 1968.
...Addendum no. 1, 1971.
...Addendum no. 2, 1972.

4527. Rowett, Helen G. Q. The rat as a small mammal. 2d ed. London, Murray, 1965.

4528. Schwentker, Victor. Care and maintenance of the Mongolian gerbil: a basic manual for laboratory technicians. Brant Lake, N.Y., Tumblebrook Farm, Inc., 1969?

4529. Simmons, M. L., and J. O. Brick. The laboratory mouse: selection and management. Englewood Cliffs, N.J., Prentice-Hall, 1970.

Note. For additional material on rodents the investigator is referred to the entries in Chap. 36, "Experimental Animals (Vertebrates): General Applications," many of which have specific chapters or sections dealing with rodents as laboratory animals.

39

OTHER LABORATORY ANIMALS (VERTEBRATES)

ANATOMY AND PHYSIOLOGY

4530. Craigie, E. H. A laboratory guide to the anatomy of the rabbit. 2d ed. Toronto, Univ. of Toronto Press, 1966.

4531. Downs, L. E. Laboratory embryology of the frog. Dubuque, Ia., Brown, 1968.

4532. Hall, L. W. Fluid balance in canine surgery. London, Bailliere, Tindall & Cassell, 1967.

4533. Hoar, W. S., and D. J. Randall, eds. Fish physiology. 6 v. (For specific titles see entry 1929.)

4534. Kemali, M., and V. Braitenberg. Atlas of the frog's brain. New York, Springer-Verlag, 1969.

4535. Love, R. M. The chemical biology of fishes. New York, Academic Press, 1970.

4536. Marshall, A. J., ed. Biology and comparative physiology of birds. 2 v. New York, Academic Press, 1960-1961.

4537. Schebitz, Horst, and H. Wilkens. Atlas of radiographic anatomy of dog and horse. Berlin, Parey, 1968.

4538. Sturkie, P. D. Avian physiology. 2d ed. Ithaca, N.Y., Cornell Univ. Press, 1965.

4539. Underhill, R. A. Laboratory anatomy of the frog. 2d ed. Dubuque, Ia., Brown, 1969.

4540. Whitehouse, R. H., and A. J. Grove. The dissection of the rabbit, with an appendix on the rat. 5th ed. London, University Tutorial Press, 1956.

SURGERY

4541. Harrison, B. M. Dissection of the cat (and comparisons with man): a laboratory manual on Felis domestica. 6th ed. St. Louis, Mosby, 1970.

4542. Leonard, E. P. Fundamentals of small animal surgery. Philadelphia, Saunders, 1968.

4543. Ormrod, A. N. Surgery of the dog and cat: a practical guide. Baltimore, Williams & Wilkins, 1966.

4544. Piermattei, D. L., and R. G. Greeley. An atlas of surgical approaches to the bones of the dog and cat. Philadelphia, Saunders, 1966.

4545. Strande, Anders. Repair of the ruptured cranial cruciate ligament in the dog: an experimental and clinical study (a few experiments in cats included). Baltimore, Williams & Wilkins, 1968.

MODELS FOR RESEARCH

4546. Andersen, A. C., and Loraine S. Good, eds. The beagle as an experimental dog. Ames, Iowa State Univ. Press, 1970.

4547. Anderson, W. D. Thorax of sheep and man: an anatomy atlas. Minneapolis, Dillon Press, 1972.

4548. Animal models for biomedical research: proceedings of symposia sponsored by the Institute of Laboratory Animal Resources and the American College of Laboratory Animal Medicine. v. 1- Washington, D.C., National Academy of Sciences, 1967- (I—Its publication no. 1594. II—Its publication no. 1736. III—Its publication no. 1854. IV- Issued as unnumbered publications. Symposia held approximately annually.)

Good source of information for particular species of animals that are appropriate as models for unique biomedical phenomena. Examples are the use of the armadillo in the study of polyembryony and the pygmy goat in fetal hemodynamics.

4549. Bondy, S. C., and F. L. Margolis. Sensory deprivation and brain development: the avian visual system as a model. Jena, Fischer, 1971. (*Abhandlungen aus dem Gebiet der Hirnforschung und Verhaltenphysiologie*, bd. 4.)

4550. Campbell, J. G. Tumours of the fowl. Philadelphia, Lippincott, 1969.

4551. Ettinger, S. J., and P. F. Suter. Canine cardiology. Philadelphia, Saunders, 1970.

4552. Fitzgerald, T. C. The Coturnix quail: anatomy and histology. Ames, Iowa State Univ. Press, 1969.

4553. Hoerlein, B. F. Canine neurology: diagnosis and treatment. 2d ed. Philadelphia, Saunders, 1971.

4554. Ingle, D., ed. The central nervous system and fish behavior. Chicago, Univ. of Chicago Press, 1968.

4555. International symposium on cetacean research. 1st. Washington, D.C., 1963. Whales, dolphins, and porpoises. Ed. by K. S. Norris. Berkeley, Univ. of California Press, 1966.

4556. Kaplan, H. M. The rabbit in experimental physiology. 2d ed. New York, Scholar's Library, 1962.

4557. Lewis, W. M. Maintaining fishes for experimental and instructional purposes. Carbondale, Southern Illinois Univ. Press, 1963.

4558. Monnier, M., and H. Gangloff. Atlas for stereotaxic brain research on the conscious rabbit. New York, Elsevier, 1961. (*Rabbit brain research*, v. 1.)

4559. Mount, L. E., and D. L. Ingram. The pig as a laboratory animal. New York, Academic Press, 1971.

4560. Pekas, J. C., and L. K. Bustad. A selected list of references on swine in biomedical research. Richland, Wash., Pacific Northwest Laboratory, 1965.

4561. Prince, J. H. The rabbit in eye research. Springfield, Ill., Thomas, 1964.

4562. Schlieper, Carl, ed. Research methods in marine biology. Seattle, Univ. of Washington Press, 1972.

4563. Schwartzman, R. M., and Milton Orkin. Comparative study of skin diseases of dog and man. Springfield, Ill., Thomas, 1962.

4564. Swine in biomedical research: proceedings of a symposium at the Pacific Northwest Laboratory, Richland, Wash., 1965. Ed. by L. K. Bustad, R. O. McClellan and M. P. Burns. Richland, Wash., Biology Dept., Battelle Memorial Institute, Pacific Northwest Laboratory, 1966.

4565. Symposium—Biology of amphibian tumors. New Orleans, 1968. Biology of amphibian tumors. Ed. by Merle Mizell. New York, Springer-Verlag, 1969.

4566. Symposium on marine bio-acoustics. 2d. American Museum of Natural History, 1966. Marine bio-acoustics. Ed. by W. N. Tavolga. New York, Symposium Publications Division, Pergamon Press, 1967.

4567. Symposium on the use of fish as an experimental animal in basic research. Vermillion, S.D., 1968. Fish in research. Ed. by O. W. Neuhaus and J. E. Halver. New York, Academic Press, 1969.

4568. Tumbleson, M. E. A selected list of references on the use of swine in biomedical research. (Loose-leaf.) Columbia, Univ. of Missouri, 1971.

4569. U.S. Chemical Research and Development Laboratories. Pathologic findings in the laboratory goat. By F. W. Light, Jr. 3 parts in 1 v. Edgewood Arsenal, Md., 1965. (CRDL special publication, 2-62, 2-67. Part 3 by C. E. Hopkins and F. W. Light, Jr.)

4570. Walker, W. F., Jr. A study of the cat, with reference to man. 2d ed. Philadelphia, Saunders, 1972.

MISCELLANEOUS

4571. Agricultural Research Council. (Great Britain.) Index on current research on pigs. v. 1- Shinfield, Reading, England, National Institute for Research in Dairying, 1954-

4572. Templeton, G. S. Domestic rabbit production. 4th ed. Danville, Ill., Interstate, 1968.

4573. Thompson, H. V., and A. N. Worden. The rabbit. London, Collins, 1956.

4574. Wells, T. A. G. The rabbit: a practical guide. 2d ed. London, Heinemann Educational Books, 1964.

40

INVERTEBRATES

ANATOMY, PHYSIOLOGY, AND CLASSIFICATION

4575. Barrington, E. J. W. Invertebrate structure and function. Boston, Houghton Mifflin, 1967.

4576. Beklemishev, W. N. Principles of comparative anatomy of invertebrates. Ed. by Z. Kabata. Chicago, Univ. of Chicago Press, 1969. (Translated by J. M. MacLennan.)

4577. Bullock, T. H., and G. A. Horridge. Structure and function in the nervous systems of invertebrates. San Francisco, Freeman, 1965.

4578. Highnam, K. C., and Leonard Hill. The comparative endocrinology of the invertebrates. New York, American Elsevier, 1969.

4579. Kelly, G. F. Endocrine coordination in invertebrates: an introduction to neuroendocrine regulation in animals. 1st ed. New York, Pageant Press, 1967.

4580. Pimentel, R. A. Invertebrate identification manual. New York, Reinhold, 1967.

4581. Ramsay, J. A. Physiological approach to the lower animals. 2d ed. Cambridge, England, The University Press, 1969.

4582. Reverberi, Giuseppe, ed. Experimental embryology of marine and freshwater invertebrates. Amsterdam, North-Holland Publishing, 1971.

4583. Tombes, A. S. An introduction to invertebrate endocrinology. New York, Academic Press, 1970.

4584. Wolken, J. J. Invertebrate photoreceptors: a comparative analysis. New York, Academic Press, 1971.

BIOMEDICAL RELATIONSHIPS

4585. Bliss, Dorothy E., and Dorothy M. Skinner. Tissue respiration in invertebrates. New York, American Museum of Natural History, 1963.

4586. Carthy, J. D. An introduction to the behaviour of invertebrates. New York, Hafner, 1962.

4587. Chemistry of learning: invertebrate research. Proceedings of a symposium sponsored by the American Institute of Biological Sciences. Michigan State University, 1966. Ed. by W. C. Corning and S. C. Ratner. New York, Plenum Press, 1967.

4588. Colloquium on experimental embryology. Clermont-Ferrand, France, 1968. Invertebrate organ cultures. Organized by H. Lutz. New York, Gordon & Breach, 1970. (Documents on biology, v. 2.)

4589. Johnson, Phyllis T., and Faylla A. Chapman. An annotated bibliography of pathology in invertebrates other than insects. Minneapolis, Burgess, 1968.

4590. Steinhaus, E. A., and M. E. Martignoni. An abridged glossary of terms used in invertebrate pathology. 2d ed. Portland, Ore., Pacific Northwest Forest & Range Experiment Station, 1970.

4591. Symposium on neoplasms and related disorders of invertebrate and lower vertebrate animals. Smithsonian Institution, 1968. Neoplasms and related disorders of invertebrate and lower vertebrate animals. Ed. by C. J. Dawe and J. C. Harshbarger. Bethesda, Md., National Cancer Institute, 1969. (National Cancer Institute monograph 31.)

4592. Symposium on neurobiology of invertebrates. Tihany, Hungary, 1967. Neurobiology of invertebrates. Ed. by J. Salankí. New York, Plenum Press, 1968.

4593. Thorpe, W. H. Learning and associated phenomena in invertebrates. London, Bailliere, Tindall & Cassell, 1964.

4594. Vago, C., ed. Invertebrate tissue culture. v. 1- New York, Academic Press, 1971-

MISCELLANEOUS

4595. Barnes, R. D. Invertebrate zoology. 2d ed. Philadelphia, Saunders, 1968.

4596. Dale, R. P., ed. Practical invertebrate zoology: a laboratory manual for the study of the major groups of invertebrates, excluding protochordates. By F. E. G. Cox et al. London, Sidgwick & Jackson, 1969.

4597. Gardiner, Mary S. The biology of invertebrates. New York, McGraw-Hill, 1972.

4598. Journal of invertebrate pathology. v. 7- New York, 1965- (Continues *Journal of insect pathology.*)

4599. Kaestner, Alfred. Invertebrate zoology. New York, Wiley-Interscience, 1967- (Translated and adapted from the 2d German ed. by H. W. Levi and Lorna R. Levi.)

4600. Spotte, S. H. Fish and invertebrate culture: water management in closed systems. New York, Wiley-Interscience, 1970.

INDEXES

AUTHOR INDEX

The Author Index includes the names of organizations as well as individuals who are responsible for the publication of specific numbered entries, followed by a listing of cited conferences, congresses, symposia, and other meetings. Many of the entries in the main body of the work are germane to several subject areas and appear in more than one chapter. In this index an effort has been made to refer to a specific entry only once.

Abbott, Mary, 0145
Abelev, G. I., 2807
Abraham, Ambrus, 2001
Abramoff, P., 3022
Abrams, J. T., 3745, 3746
Abramson, D. I., 1871
Abuladze, K. I., 2968
Académie vétérinaire de France, 0217
Ackerknecht, E. W., 0958
Ackermann, P. G., 1323
Ackroyd, J. F., 3076
Ada, G. L., 3014
Adam, K. M., 2920
Adam, W. S., 1597, 3968
Adams, A. R. D., 3126
Adams, C. E., 0145
Adams, C. W. M., 2399
Adams, F. H., 0788
Adams, J. M., 2847
Adams, J. W., 3882
Adams, O. R., 3198
Adams, R. D., 3490
Adelberg, E. A., 2727
Adey, W. R., 4438
Adinolfi, Matteo, 3023
Adler, F. H., 2003
Adrianov, O. S., 1748
Afzelius, Bjorn, 1598
Aggarwala, A. C., 3360
Agricultural Development Association, 3816
Agricultural Research Council (Great Britain), 3924
Agricultural Research Service. See U.S. Agricultural Research Service
Aidley, D. J., 2002
Aikawa, J. K., 2347
Ainsworth, G. C., 0742, 0877
Aitken, F. C., 0743
Ajello, Libero, 1226, 2876
Akademiiā meditsinskikh Nauk SSSR, 4464
Akademiiā Nauk SSSR, Institute of Higher Nervous Activity, 2026
Akert, Konrad, 1770
Akker, L. M. van den. See Van den Akker, L. M.
Albe-Fessard, D., 1749
Albiston, H. E., 3738
Albricht, J. L., 2265a

Alexander, A. C., 4249
Alexander, Frank, 2542
Alexander, H. L., 3073
Alexander, Nancy J., 2032
Alexander, R. M., 1444, 1705, 4114
Alicata, J. E., 2969
Allen, A. C., 1825
Allen, A. R., 3565
Allen, D. M., 3349
Allen, Edgar, 2196
Allen, Edgar van Nuys, 3257
Allen, J. H., 2471
Allen, R. M., 1150
Allgöwer, M., 3596
Allison, J. B., 2382
Allred, D. M., 2983
Alm, G., 0732
Alston, J. M., 2878
Altman, Joseph, 2222
Altman, P. L., 1075–1079, 2348
Altmann, Jeanne, 4440
Altmann, S. A., 2223, 4440
Alton, G. G., 0800
Altschule, M. D., 2106
Alverdes, Friedrich, 2224
Amberson, Rosanne, 4033
Ambrose, E. J., 1265
American Animal Hospital Association, 0221, 0586, 1750, 3258, 3791
American Association for Laboratory Animal Science, 0146, 4250
American Association of Avian Pathologists, 2841
American Association of Bovine Practitioners, 0587
American Association of Equine Practitioners, 0588, 0921, 3870
American Chemical Society, 0922
American College of Laboratory Animal Medicine, 4351
American College of Veterinary Ophthalmologists, 0589
American Fisheries Society, 4115
American Institute of Biological Sciences, 4587
American Kennel Club, 3984
American Medical Association, 0314, 0865, 0866, 0923

American National Red Cross, 1872
American Pharmaceutical Association, 0472, 2544
American Psychopathological Association, 2241
American Public Health Association, 1176–1179
American Registry of Pathology, 1232
American Society of Animal Production, 3775
American Society of Planning Officials, 3792
American Society of Veterinary Ophthalmology, 3219
American Type Culture Collection, 1180
American Veterinary Medical Association, 0223, 0867, 0925, 3165, 4253
American Veterinary Radiology Society, 0224
Amiel, Jean-Louis, 3626
Aminoff, David, 3031
Amoroso, E. C., 0773, 4083
Ampino, R., 0750
Amromin, G. D., 3556
Andersen, A. C., 3949
Andersen, H. T., 4143
Anderson, E., 1863
Anderson, J. J. B., 3317
Anderson, J. M., 3613
Anderson, J. R., 3052, 3382
Anderson, M. C., 3287
Anderson, Marilyn J., 0197
Anderson, P. J., 1235
Anderson, R. C., 2973
Anderson, W. A. D., 3361, 3362
Anderson, W. D., 1445
Andrew, B. L., 1706, 1908
Andrew, W., 1508, 1599, 1600, 1873
Andrewartha, H. G., 4180
Andrewes, C. H., 2746–2748
Andrews, G. S., 1601
Andrews, R. D., 2749
Andrist, Friedrich, 3871
Angevine, C. D., 2435
Angevine, J. B., Jr., 1812
Animal Health Association. See U.S. Animal Health Association
Animal Welfare Institute, 4275
Ankara üniversite, Veteriner fakültesi, 0561
Anliker, M., 1740
Annan, Gertrude L., 0997
Annis, J. R., 3565
Ansell, G. B., 2409
Anson, B. J., 1434
Anstall, H. B., 3381
Anthony, D. J., 3925
Appel, M. J. G., 0795
Appel, S. H., 2206
Archer, R. K., 1088
Archibald, J., 3571, 4336
Ardrey, Robert, 2225–2227
Arena, Victor, 3695

Arey, L. B., 1509, 1602
Ariens-Kappers, C. U., 1446
Armed Forces Institute of Pathology. See U.S. Armed Forces Institute of Pathology
Armed Forces Medical Library. See U.S. Armed Forces Medical Library
Armitage, P., 2514
Armstrong, C. N., 4303
Armstrong, W. M., 2058
Army Biological Warfare Laboratories. See U.S. Army Biological Warfare Laboratories
Arnow, L. E., 2349
Arnstein, Paul, 4100
Aroeste, Jean L., 1023
Aronson, S. B., 3235
Arquembourg, P. C., 1089, 3462
Arrington, L. R., 4276
Arthur, G. H., 3304
Arthur H. Thomas Company, 1064
Arvy, Lucie, 2400
Asboe-Hansen, Gustav, 2079
Ascenzi, A., 0750
Asdell, S. A., 2156, 2157, 3110
Ash, Lee, 0990
Assali, N. S., 2158, 2187
Assendelft, O. W. van, 1090
Association of American Medical Colleges, 0926
Association of Deans of Colleges of Veterinary Medicine, 4253
Association of Official Agricultural Chemists, 1080
Astwood, E. B., 2080
Atamer, M. A., 3329
Atanasiu, Pascu, 0800
Atomic Energy Commission. See U.S. Atomic Energy Commission
Austin, C. R., 1677, 2159–2161
Austin, George, 1751
Austwick, P. K. C., 0742
Aviado, D. M., 3269
Axelrod, H. R., 4144
Ayers, D. M., 0900

B

Babel, Jean, 1678
Bacharach, A. L., 2098
Back, N., 0732
Baer, J. G., 1447, 2921
Bailey, F. R., 1603
Bailey, J. W., 3842
Bailey, N. T. J., 2515
Bailey, O. T., 0756
Bailey, W. R., 1181
Baillie, A. H., 2421
Bairacli-Levy, Juliette de, 3986
Bajusz, E., 0769, 2004, 2081, 3514
Baker, A. A., 0779

Baker, J. K., 3938
Baker, J. R., 1233
Baker, L. A., 3810
Baker, L. E., 1335
Baker, S. B. De C., 0751
Balassa, Bela, 0186
Balazs, E. A., 1604
Baldwin, Ernest, 2324
Balinsky, B. I., 1510
Ballard, W. W., 1448
Balner, H., 3062, 3066, 4467
Ban, T. A., 2676
Banatvala, J. E., 2750
Bancroft, J. D., 1234
Bander, E. J., 0878
Banga, Ilona, 1707
Banks, W. J., Jr., 1605
Barber, A. A., 1343
Barcroft, Joseph, 1962
Barden, E. S., 0210
Barka, Tibor, 1235
Barner, R. D., 3726
Barnes, C. D., 2630
Barnes, R. D., 4595
Barnett, J. T., 2437
Barnett, K. C., 1752
Barnett, R. N., 1091
Barnett, S. A., 2228, 2229
Barnum, M. R., 4251
Baroldi, G., 0769, 2044
Barone, R., 1449
Barrett-Connor, Elizabeth, 2922
Barrington, E. J. W., 4575
Barron, C. N., 1741
Barrow, Pat, 3593
Barry, H. C., 3199
Bartholomew, Davis, 1325
Barzun, Jacques, 1006
Baslow, M. H., 2545
Basmajian, J. V., 0148, 1391, 1400
Bast, T. H., 1421
Bates, D. V., 3270
Bauchwitz, F. A., 3879
Bauer, G. C. H., 0750
Beadle, G. W., 3088
Beament, J. W. L., 1909, 2039
Beary, E. G., 0149
Beaver, P. C., 2881
Beck, Ernest, 1401
Beck, J. W., 2922
Becker, R. F., 1402
Becker, Walter, 1326
Beckwith, J. R., 1369
Beer, G. R. de. See De Beer, G. R.
Beeson, P. B., 3132
Behbehani, A. M., 0795, 1182, 2822
Behrens, C. F., 3696
Beierwaltes, W. H., 3697
Bekkum, D. W. van, 0786

Beklemishev, W. N., 1450, 4576
Belding, W. N., 2923
Bell, Bettye P., 3163
Bell, D. J., 1910
Bell, Eugene, 1512
Bell, S., 2840
Bellairs, A. A., 4164, 4165
Bellairs, Ruth, 1513
Bellanti, J. A., 2998
Bellville, J. W., 1911
Belschner, H. G., 3811, 3853, 3902, 3926
Bement, A. L., Jr., 2501
Benbrook, E. A., 2924
Bendall, J. R., 1709
Bender, G. T., 1327
Beneke, E. S., 2808
Benirschke, K., 1538, 1624, 3111
Benjamin, Maxine M., 1092
Bennett, J. C., 3200
Bennett, R. B., 4104
Bennington, J. L., 3537
Bensley, B. A., 1403
Benson, Wes, 4228
Bentley, P. J., 2082
Benton, Mildred E., 0095
Benzie, David, 1841
Berenblum, Isaac, 3404
Berg, Gerald, 2801
Berge, Ewald, 3567
Berger, A. J., 1723
Bergsman, D., 0746
Bergwin, C. R., 4465
Beritashvili, I. S., 2230, 2231
Berlyne, G. M., 3305
Berman, A. L., 1753
Bern, H. A., 2097
Bernard, B. de. See De Bernard, B.
Bernard, Claude, 0959, 0960
Berne, R. M., 2045
Berner, R. D., 2835
Bernick, S., 4427
Bernreiter, Michael, 1363
Bernshtein, N. A., 1710
Berntzen, A. K., 1213
Berrens, L., 0770
Berrier, H. H., 3142
Berson, S. A., 2112a
Bertelli, A., 0732, 0751, 2546
Bertke, E. M., 1615
Bertrand, Mireille, 0737
Beshear, Donna R., 1953
Best, C. H., 1912
Besterman, Theodore, 0151
Bethlem, J., 3201
Betts, A. O., 2848
Betuel, H., 3472
Beutler, Ernest, 2351
Bevelander, Gerrit, 1606, 1607
Beveridge, W. I. B., 0776, 3066, 4256, 4425

Bickers, J. N., 3462
Bickford, R. G., 0154
Biegert, J., 4476
Bier, Milan, 1093
Bierer, B. W., 0961
Bierschwal, C. J., 3586
Biester, H. E., 4074
Biggam, Alexander, 3141
Billingham, R. E., 3054, 3069
Binford, C. H., 2812
Bingham, W. G., 0778
Biosciences Information Service, 0096
Birke, G., 0796
Birkett, L., 4130
Bishop, C. W., 1875
Bishop, O. N., 2517
Bisseru, Balideo, 2849, 2879
Bistner, S. I., 0794, 3148
Bjorkman, Nils, 1679
Bjornson, B. F., 2896
Blackith, R. E., 2518
Blacklow, R. S., 3161
Blackwood, W., 3491, 3493
Blahd, W. H., 3698
Blair, J. E., 1183
Blake, J. B., 0802
Blake, T. M., 1364
Blaker, A. A., 1038
Blaker, Gertrude G., 2909
Blandau, R. J., 1514, 1857
Blaxter, K. L., 0747, 2352
Bliss, C. I., 2519
Bliss, Dorothy E., 4585
Bliss, E. L., 2232
Bloch, Bernard, 3586a
Block, S. S., 2908
Blood, D. C., 3713
Bloodworth, J. M. B., 3316
Bloom, B. R., 1608
Bloom, F., 2353
Bloom, W., 1609
Bloor, C. M., 0732
Blount, W. P., 4057
Blum, H. F., 3127
Blumenthal, H. T., 1981
Boalch, D. E., 0097
Boch, Josef, 2925
Bock, H. E., 2997
Bockus, H. L. R., 3283
Boddie, G. F., 3143
Boddington, M. M., 1662
Bogardus, C. R., 3672
Bogart, Ralph, 3119
Bois, C. H. de. See DeBois, C. H.
Bolande, R. P., 3453
Bolander, D. O., 0901
Boley, S. J., 3284
Bolis, Liana, 2468, 2478
Bologna, Camillo V., 3464
Bondy, S. C., 2005

Bone, J. F., 3128
Bonner, J. F., 2472
Bonneville, Mary A., 1697
Booth, E. S., 1404, 1438, 1557
Bootzin, David, 1735
Bordley, James, 3145
Boreva, L. I., 0155
Bossy, Jean, 1754
Bottle, R. T., 0803
Boulle, M., 3244
Bourke, G. J., 2520
Bourne, G. H., 1328, 1544, 1610, 1711, 1755, 1792, 1809, 1990, 2233, 4441
Bowling, Mary C., 1236
Bowsher, David, 1756
Box, P. G., 2866
Boyarsky, Saul, 2076
Boyd, W. C., 2999
Boyd, William, 3363, 3364
Brachet, Jean, 1516, 1612
Bradley, O. C., 1405
Bradley, P. B., 2690
Bradley, R. E., 2977
Bradshaw, L. J., 2473
Brain, E. B., 1237
Brain, W. R., 2210, 3220
Braitenberg, V., 1783
Braley, A. E., 1151
Brambell, F. W., 3055
Brand, D. L., 1288
Brander, G. C., 2553
Brandly, P. J., 2897
Branemark, Per-Ingvar, 1876
Branson, Dorothy, 1184
Brantigan, O. C., 1406
Brauer, Kurt, 1757
Braun, Werner, 2449
Brazier, Mary A. B., 0788, 2007
Breathnach, A. S., 1680
Breazile, J. E., 1913
Breed, R. S., 1223
Breirem, Knut, 3760
Breland, Keller, 2234
Breland, Marian, 2234
Breslav, I. S., 2051
Brewer, G. J., 0732
Brick, J. O., 4529
Brinkhous, K. M., 0756
Brion, A., 2631
Briskey, E. J., 1999
British Cattle Veterinary Association, 3816
British Crop Protection Council, 2580
British Small Animal Veterinary Association, 3794, 3994
British Society for Parasitology, 0739, 1094
British Veterinary Association, 2970, 3927
Britton, C. J. C., 3356
Brock, Samuel, 2211
Brockis, J. Gwynne, 3570
Brodal, A., 1758

Bronner, Felix, 2327
Bronner, Marcell, 2869
Bronner, Max, 2869
Brookes, Murray, 1712
Brookhaven National Laboratory, 1007
Brooks, F. P., 2060
Broom, J. C., 2878
Brown, A. W. A., 2993
Brown, Annie M., 4400
Brown, B., 1392
Brown, C. C., 1329
Brown, Donald, 4153
Brown, F. A., Jr., 1947
Brown, H. D., 2423
Brown, J. H. U., 2487
Brown, Margaret E., 4117
Brown, R. J., 0202
Brown, R. L., 2581
Brown, T. H., 3247
Brown, W. V., 1614, 1615
Brucke, F. Th. von, 2677
Brumlik, Joel, 1385
Bruner, D. W., 2830
Bruyn, G. W., 3239
Bryans, J. T., 0758
Buchanan, R. E., 1207
Buchanan, W. W., 3052
Bucher, Nancy, 3642
Bucherl, Wolfgang, 2604
Buchner, Paul, 2693
Buchthal, Fritz, 1383
Buechner, H. A., 2870
Buechner, H. K., 4214
Buettner-Janusch, John, 3090
Buhlmann, A. A., 2055
Bulbring, Edith, 1713
Bulgren, W. G., 2531
Bullock, T. H., 1759
Bullough, W. S., 1517, 2162
Bundy, C. E., 3939
Bunge, Mary B., 1603
Bunge, R. P., 1603
Bunnell, Fred, 4205
Bunnell, Pille, 4205
Burch, G. E., 3259
Burch, R. L., 4393
Burchenal, J. H., 0767
Burchfield, H. P., 2582
Burdette, W. J., 2521, 3112, 3405
Bureau of Medicine and Surgery. See U.S. Bureau of Medicine and Surgery
Bŭres, Jan, 1330
Burger, Alfred, 2669, 2678
Burke, Shirley R., 1914
Burket, L. W., 3286
Burkhardt, Dietrich, 2235
Burkhardt, Rolf, 1714
Burkitt, D. P., 0767
Burn, J. H., 2009
Burnet, F. M., 1616, 2751, 2752, 2824

Burns, K. F., 4295
Burns, M. P., 3936
Burns, Marca, 3113
Burrell, R. G., 2784, 3078
Burris, R. H., 1142
Burrows, William, 2719
Burstein, A. H., 1720, 2491
Burstone, M. S., 2424
Burton, A. C., 2046
Burton, A. L., 1039
Burton, Maurice, 0817
Burtt, H. E., 2236
Busby, D. W. G., 1185
Busch, H. F. M., 4009
Busch, Harris, 1617, 3407, 3408
Bustad, L. K., 0184, 3935, 3936
Busvine, J. R., 2898, 2985
Byrom, F. B., 3260

C

Caceres, C. A., 2488
Cainiello, E. R., 1808
Caldwell, D. K., 4138
Caldwell, Melba C., 4138
Caldwell, W. L., 3538
Calhoun, J. B., 2237
Calhoun, M. Lois, 1618, 4503
Callander, Robin, 1936
Cameron, I. L., 1518, 1619, 1620
Camougis, George, 1384
Campbell, D. H., 1095
Campbell, D. M., 0972
Campbell, H. J., 2011
Campbell, J. G., 3409, 4071
Campbell, J. W., 2325
Campbell, M. F., 3306
Campbell, P. N., 1621, 2577
Canadian Society for Immunology, 3024
Cantarow, Abraham, 2326
Cappell, D. F., 3382
Caras, R. A., 4057a, 4181, 4182
Card, Leslie E., 4105
Carey, J. E., 3697
Carleton, H. M., 1238
Carlson, B. M., 3642a
Carlson, W. D., 3673
Carpenter, C. R., 0760, 4444
Carpenter, M. B., 1817, 4428
Carpenter, P. L., 2694, 3080
Carpovich, E. A., 0827
Carr, C. J., 2566
Carr, Katherine E., 1172
Carrington, Richard, 4165
Carroll, E. J., 2125
Carroll, W. E., 3943
Carsky, T. R., 1188
Carter, G. R., 1186, 1297, 2695
Carterette, E. C., 0788
Carthy, J. D., 4586

Cartwright, G. E., 1096
Carvalho, A. P. de, 2469
Casarett, G. W., 3689
Case, James, 2012
Casey, Donn, 0927
Casey, E. J., 2454
Casey, H., 3752
Cass, J. S., 0156, 4401
Cassens, R. G., 1999
Castle, C. H., 4266
Catcott, E. J., 3129, 3131, 3795, 3813, 3959
Catty, David, 0770
Causey, Gilbert, 1152
Cawley, L. P., 1097
Cecil, R. L., 3132
Center for Disease Control. See U.S. Center for Disease Control
Central Veterinary Laboratory, 1187
Cesvet, Helen E., 0011, 0066
Chagas, Carlos, 4137
Chamberlain, E. N., 3153
Chambers, K. L., 0738
Champion, R. H., 1826
Chance, M. R. A., 4445
Chapman, Faylla A., 4589
Charnley, John, 3587
Chase, M. W., 2452
Chauveau, A., 1451
Chayen, Joseph, 1239
Cheatle, Esther L., 1196
Chemical Rubber Company, 1081
The Chemical Society (London), 0004
Cheng, T. C., 2928
Cherim, S. M., 098
Cherry, W. B., 188
Cheshier, R. G., 0994
Chiappa, S., 0780, 3699
Chiarelli, A. B., 4415
Chiasson, R. B., 1404, 1407, 1438, 1557, 4071, 4094, 4485
Chiodi, Valentino, 1834
Chowdhuri, A. N. Rai, 2850
Christensen, E. E., 3653
Christensen, G. C., 3969
Christensen, Sherryl E., 3163
Christian, C. L., 2850a
Christie, A. B., 2825
Christie, R. V., 3270
Chrystie, Frances N., 4229
Church, D. C., 2061
Chusid, J. G., 2199
Chvapil, Milos, 1992
Ciba Foundation, 0643, 1622, 2013, 2128, 2474, 2721, 3025
Cinader, Bernhard, 3026
Clairville, A. L., 0828
Clare, N. T., 0742
Clark, G. L., 1153
Clark, George, 1240
Clark, Margaret P., 1240

Clark, R. B., 4222
Clark, R. L., 3423
Clark, W. E., 1623
Clark, W. G., 2679
Clarke, E. G. C., 2587
Clarke, Myra L., 2587
Clarke, W. J., 3348, 4354
Clayden, E. C., 1241
Clegg, A. G., 2085
Clegg, Catherine, 2085
Clegg, H. A., 0122
Clemente, C. D., 0788
Clifford, Margaret A., 3674
Cloudsley-Thompson, J. L., 4166
Clynes, Manfred, 2490
Coates, M. E., 4278, 4287
Coatney, G. R., 4457
Cochran, W. G., 2537
Cockburn, Aidan, 2826
Coffin, D. L., 4072
Cohen, Arnold, 2754
Cohen, Daniel, 2900
Cohen, H. L., 1385
Cohen, J., 1519, 2240
Cohen, S. S., 0741, 2755
Cohrs, P., 1486, 1741, 4344
Colbert, E. H., 1452
Cold Springs Laboratory on Quantitative Biology, 0740
Cole, D. J. A., 0773, 3940
Cole, F. J., 0964
Cole, H. H., 0733, 2164
Cole, R. B., 3273
Cole, V. G., 3904
Cole, W. C., 4470
Cole, W. H., 3414, 3573
Coleman, W. T., 4465
Coles, E. H., 3465
College of American Pathologists, 0868
Collins, C. H., 2696
Collins, G. R., 4267
Collins, R. D., 3170
Collison, R. L., 0100
Colter, J. S., 2797
Columbia University, 0741
Colyer, Frank, 3506
Comar, C. L., 2327
Comfort, Alexander, 1520
Commissie voor Beentumoren, 3482
Committee for the Preparation of a Technical Guide for Roentgen Techniques in Laboratory Animals, 3676
Commonwealth Bureau of Animal Breeding and Genetics, 0164, 3783a
Commonwealth Bureau of Animal Health, 0101, 0742
Commonwealth Bureau of Animal Nutrition, 0743, 4279, 4523
Commonwealth Bureau of Helminthology, 0063, 0744

Communicable Disease Center. See U.S. Center for Disease Control
Comroe, J. H., Jr., 2052
Conalty, M. L., 0757
Conant, N. F., 1191
Condliffe, P. G., 4259
Condon, Patricia A., 0201
Condoyannis, G. E., 0829
Conn, H. J., 1242
Constantine, D. G., 2880
Cook, C. A. G., 3234
Cook, Margaret J., 1408
Cooke, E. I., 0856
Cooke, Ian, 0965
Coombs, R. R. A., 3059
Cooper, J. R., 2680
Cooper, L. Z., 2789
Cooper, R., 1373
Cooper, R. W., 4481
Copenhaver, W. M., 1603
Copley, A. L., 2141, 2458
Cornatzer, W. E., 2371
Cornelius, C. E., 1114
Corning, W. C., 4587
Costa, Erminio, 4376
Costa, M., 2962
Costello, D. P., 4142
Cotchin, E., 0742, 4509
Cottier, H., 3041
Council for International Organizations of Medical Sciences, 0134
Council of Biology Editors, 1040, 1041
Courtice, F. C., 1906
Cowan, H., 3819
Cowan, S. T., 2722
Cowdry, E. V., 1625, 1626
Cowen, H., 2843
Cowie, A. T., 0775, 2107, 2118
Cox, Barbara G., 0732
Cox, F. E. G., 4596
Crabtree, D. G., 2601
Craig, C. F., 2933
Craigie, E. H., 1403, 1409
Crampton, E. W., 3750
Crandall, L. S., 4206
Crasemann, Edgar, 3760
Crawford, L. V., 0781
Crawford, M. A., 0801
Criep, L. H., 3057
Crismon, Cathrine, 2681
Cristofalo, V., 1282
Croft, Phyllis G., 3604
Crofton, H. D., 0744
Crofton, J. W., 3274
Cross, A. G., 3234
Cross, Wilbur, 1008
Crouch, J. E., 1410
Crowe, Alan, 2522
Crowe, Barry, 0855
Crowe, P. K., 4183

Crowle, A. J., 2440
Crowley, L. V., 3366
Cruickshank, Bruce, 1123
Cruickshank, Robert, 2723
Cruse, J. M., 3001
Culling, C. F. A., 1243
Cumings, J. N., 2402
Cumming, Hamish, 1192
Cunningham, C. H., 1193
Cunningham, J. R., 3657
Cupps, P. T., 2164
Currah, Ann, 4034
Curran, R. C., 3454
Currie, D. J., 1042
Curry, T. S., 3653
Cushman, Anna J., 1798
Custer, P. L., 1878
Cuthbertson, D. P., 3751
Cutting, W. C., 2549
Cutts, J. H., 1101
Cyan, Erwin di. See Di Cyan, Erwin
Cyriax, J. H., 3202

D

Dacie, J. V., 1102
Dahlin, D. C., 3483
Dale, R. P., 4596
Dallenbach, F. D., 3539
Dallenbach-Hellweg, G., 3539
Dalton, A. J., 1681
Dameshek, William, 3333, 3334
Damm, H. C., 1082, 1104, 1105, 2455, 2456, 3171
D'Amour, F. E., 1331
Dangerfield, Stanley, 3988
Daniel, J. C., Jr., 1521
Danielli, J. F., 1917
Danowski, T. S., 2086, 3318
D'Arcy, P. F., 2551
Darnell, J. E., 2780
Darwin, Charles, 2242, 2242a
Daudel, Pascaline, 2475, 3412
Daudel, Raymond, 2475, 3412
Dausset, Jean, 1270, 2426, 3617
Davenport, F. M., 2865
Davenport, H. W., 2062
Davidsohn, I., 3172
Davies, Jack, 4357
Davies, M. J., 3516
Davis, B. D., 2697
Davis, D. E., 2243
Davis, H. P., 3989
Davis, J. W., 2827, 2828, 2973, 4073, 4175
Davis, Ross, 1762
Davison, A. N., 0732, 1763, 2403
Davson, Hugh, 1764, 1918, 1955, 2015, 2016
Dawe, C. J., 0785, 4591
Dawes, C. J., 1154
Dawes, G. S., 1982

Dawson, Mary, 2552
Day, E. D., 2441, 2442
Daykin, P. V., 2553
Dayton, D. H., Jr., 3027
Deal, S. J., 2711, 2738
Dean, J. A., 1359
Deatherage, J. W., Jr., 4356
De Bairacli-Levy, Juliette. See Bairacli-Levy, Juliette de
De Beer, G. R., 1522
De Bernard, B., 0750
DeBois, C. H., 3586
Debus, A. G., 0957
De Carvalho, A. P. See Carvalho, A. P. de
Decker, Kurt, 2212
Defendi, Vittorio, 0797
De Graciansky, P. See Graciansky, P. de
De Groot-de-Rook, A. S. See Groot-de-Rook, A. S. de
De Gruchy, G. C., 3173
DeHaan, R. L., 1523
Deichmann, W., 2584
De Jong, R. H., 3605
DeLahunta, Alexander, 1307
DeLand, F. H., 3700
Dellmann, Horst-Dieter, 1627
DeLucchi, M. R., 1765
Del Rio-Hortega, Pio. See Rio-Hortega, Pio del
Demark, N. L. van. See Van Demark, N. L.
Dembeck, Hermann, 2243a
Demikhov, V. P., 3618
Denenberg, V. H., 2244
Denis, Armand, 4035
Denk, Helmut, 1785
Denny, M. R., 2245
Denoix, Pierre, 0767
Dent, Anthony, 3871
Dent, J. B., 3752
Department of Agriculture. See U.S. Department of Agriculture
Department of the Army. See U.S. Department of the Army
Department of the Interior. See U.S. Department of the Interior
DeReuck, A. V. S., 1998
Der Hoeden, Jacob van. See Hoeden, Jacob van der
Dermer, Joseph, 1025
De Robertis, E. D. P. See Robertis, E. D. P. de
De Rochemont, Richard, 3753
Dersal, W. R. van. See Van Dersal, W. R.
Dethier, V. G., 2017, 2246
Detwiler, S. R., 1524
Deuchar, Elizabeth M., 1525
De Vries, A., 2613
De Vries, Louis, 0830, 0831
De Wardener, H. E., 1850
De Weck, A. L., 0770, 0786

Dewhurst, D. J., 1332
Dick, Esther A., 1897
Dicker, S. E., 0775
Dickson, J. F., III, 2494
Di Cyan, Erwin, 1052
DiDio, L. J. A., 1843
Di Fiore, M. S. H. See Fiore, M. S. H. di
Diggins, R. V., 3939
Di Luzio, N. R., 0732
Di Mascio, Alberto, 2682, 2688
Dimona, E. G., 1366
Dimond, S. J., 2247
Dintenfass, Leopold, 1835
DiPalma, J. R., 2554
Disbrey, Brenda D., 1244
Ditmars, R. L., 4167
Dittmer, Dorothy S., 1078, 1079
Division of Allied Health Manpower. See U.S. Division of Allied Health Manpower
Dobberstein, J. C., 1454, 3376
Dobson, Jessie, 0879
Dobzhansky, T., 3100
Dodd, F. H., 3816
Dodds, T. C., 1123, 2699, 2792, 3251, 3455
Dodge, J. D., 1682
Dodge, N. N., 2605
Dodson, A. I., 3619
Dogel', V. A., 4119
Dolhinow, Phyllis, 2248
Dolling, C. H., 3916
Domer, F. R., 2670
Domermuth, C. H., 0181
Domino, E. F., 1748
Domonkos, Anthony, 3242
Donaldson, D. D., 3222
Donaldson, P. E. K., 1333
Donáth, Tibor, 0880
Donovan, B. T., 1862, 2087
Dorfman, R. I., 2088
Dorland, W. A., 0818
Dougherty, R. W., 0765
Dougherty, W. M., 3336
Douglas, Andrew, 3274
Douglas, S. W., 3678, 3679
Dowdy, G. S., Jr., 1844
Downie, H. G., 4336
Downs, L. E., 4531
Doxey, D. L., 3466
Dransfield, J. W., 1479
Drazil, J. V., 0881
Dreizen, S., 4427
Dröscher, V. B., 2249
Drummond, Ann E., 3674
Drury, R. A. B., 1238
Dua-Sharma, Sushil, 1766
Dubbs, Del Rose, 0771
Dubes, G. R., 2758
Dublin, W. B., 3492
Dubos, R. J., 2842
Dubrul, E. L., 1767

Duguid, J. P., 2723
Duijn, C. van, 4120
Duke, K. L., 1482
Duke-Elder, W. S., 1768
Dukes, H. H., 1919
Dumbleton, C. W., 0832
Dummer, G. W. A., 1067
Duncan, D. L., 0743
Duncker, Hans-Rainer, 4075
Dunlop, R. H., 2774
Dunn, A. M., 0158
Dunn, Arnold, 1920
Dunn, L. C., 3100
Dunne, H. W., 3928
Dunphy, J. E., 3648
Durham, R. H., 0819
Dutcher, R. M., 3333
Dutt, R. H., 0733
Duuren, B. L. van. See Van Duuren, B. L.
Dvořák, Jaroslav, 2811
Dyce, K. M., 1411
Dye, W. E., 2730
Dyer, I. A., 2265a, 3842a
Dyke, S. C., 3467
Dystra, R. R., 2901
Dymling, J. F., 0750

E

Eakin, R. M., 1526
Earle, K. M., 1296, 3500
Eastham, R. D., 2356, 3174
Eaton, T. H., Jr., 1455
Ebashi, Setsuro, 1993
Ebbesson, S. O. E., 1795
Eberstein, Arthur, 1386
Ebert, J. D., 1527
Ebert, Myrl, 0805
Ebling, F. J., 1827, 3255
Eccles, J. C., 2019
Eckmann, Leo, 2724
Eddy, B. E., 0795
Eddy, Samuel, 1043
Ede, D. A., 1412
Eden, Murray, 2499
Education Services Press, 1026
Educational Testing Service, 4254
Edwards, J. E., 3517
Edwards, P. R., 1194
Edwards, R. A., 3761
Eggleton, M. Grace, 1955
Eibl-Eibesfeldt, Irenaus, 2250
Eichwald, Claus, 2851
Eik-Nes, K. B., 0772
Einarson, Larus, 3218
Eisenstein, A. B., 1861
Elden, H. R., 1787
Elias, Hans, 1845
Elkan, E., 4131
Elkind, M. M., 1267

Ellenberg, Max, 3319
Ellenberger, Wilhelm, 1457
Elliott, F. A., 2213
Elliott, H. C., 1769
Ellis, George, 0862
Él'piner, I. E., 1393
Eltherington, L. G., 2630
Elveback, Lila R., 2903
Elves, M. W., 1879
Emmens, C. W., 4144
Emmers, Raimond, 1770
Emmett, J. L., 3540
Emmons, C. W., 2812
Endroczi, E., 2413
Engström, Arne, 1683
Enke, C. G., 1345
Enneking, W. F., 3484
Ensminger, M. E., 3777, 3843, 3872, 3917, 3942, 4106
Enzinger, F. M., 0761
Epstein, Ervin, 1828
Epstein, H., 3777a
Equen, Murdock, 3275
Eränkö, Olavi, 0777
Erbe, N. A., 4385
Erichson, R. B., 3341
Ernst, Jenö, 1715
Ershov, V. S., 2974
Escamilla, R. F., 1106
Estes, R. R., 4346
Ethicon, Inc., 3590
Etheridge, J. M., 0854
Etkin, William, 2251
Etter, L. E., 0882
Ettinger, S. J., 3263
European Association for Animal Production, 0747
European Society for Animal Blood Group Research, 0748
European Society for Experimental Surgery, 0751
European Society for the Study of Drug Toxicity, 0749, 0751
European Society of Haematology, 1270
Euzéby, Jacques, 2975
Evans, C. A. L., 1955
Evans, F. G., 1716
Evans, H. E., 1432
Evans, R. W., 3413
Evans, W. C., 2577a
Evans, W. F., 2252
Everett, N. B., 1771
Evers, R. A., 2606
Everson, T. C., 3414
Ewer, R. F., 2253
Ewing, W. H., 2725

F

Fagerström, Rune, 2616

Fairbairn, J. F., II, 3257
Falconer, I. R., 0773, 2330
Falk, J. R., 4231
Faraday Society of London, 2141
Farner, D. S., 4065
Farnes, Patricia, 0785
Farris, E. J., 4281, 4512
Faulkner, L. C., 2165
Faulkner, W. R., 1107
Faust, E. C., 2932, 2933
Fawcett, D. W., 1609, 1684
Fazzini, Eugene, 3367
Federal Advisory Council on Medical Training Aids. See U.S. Federal Advisory Council on Medical Training Aids
Federal Water Quality Administration. See U.S. Federal Water Quality Administration
Federation of American Societies for Experimental Biology, 0934
Federation of European Biochemical Societies, 0752
Federlin, K., 0772
Fein, Rashi, 4255
Felsenfeld, Oscar, 2882, 2902
Felson, Benjamin, 3676
Felter, Jacqueline W., 0997
Fêng, Yuan-chêng, 1718
Fenner, F. J., 0160, 2760, 2852
Fennestad, K. L., 2843
Ferguson, M. M., 2421
Ferner, Helmut, 1435
Fernex, Michel, 1880
Festing, M. F. W., 4290
Fetter, B. F., 2872
Fiebiger, Josef, 1668
Field, H. E., 1413
Fiennes, R. N., 0801, 2883, 3384
Filley, G. F., 3276
Finch, B., 0835
Finco, D. R., 3309
Fine, B. S., 1628
Finean, J. B., 1683
Finerty, J. C. A., 1626
Fink, M., 2690
Finkel, H. E., 3349
Finstein, M. S., 1195
Fiore, M. S. H. di, 1629
Fiore-Donati, L., 0732
Fishbein, Morris, 1044
Fisher, J. W., 2090
Fisher, James, 4184, 4207
Fisher, K. C., 2370
Fisher, R. A., 1010, 2523
Fitzgerald, P. R., 2976
Fitzgerald, T. C., 1414
Fitzpatrick, T. B., 3243
Fleissner, H. K., 3685
Flemans, R. J., 1882
Fleming, George, 0962, 3857

Fleming, Robert, 2534
Flemming, K., 0732
Fletcher, June L., 3051
Fleurent, C. H. A., 0122
Flint, W. P., 2991
Floch, M. H., 2068
Florey, Ernst, 1965
Florey, H. W., 3368
Florkin, Marcel, 2331, 2359
Foley, R. C., 3844
Folk, G. E., Jr., 1921
Folley, S. J., 2091, 2129
Food and Agriculture Organization of the United Nations, 0753, 2829, 2832, 2862, 3742, 3825, 3827, 3833, 3846, 3918
Fooden, Jack, 0737
Ford, D. H., 0777, 2104
Forscher, B. K., 1613
Foulds, Leslie, 3415
Foulk, W. T., 3290
Foundation Center, 1027
Fourman, Paul, 1719
Foust, H. L., 1458, 1630
Fowler, M. P., 4386
Fowler, Maureen J., 0102
Fox, Herbert, 4079
Fox, J. P., 2903
Fox, M. W., 2254–2257, 3957
Fox, Mary A., 1667
Fraenkel, G. S., 2258
Fraenkel-Conrat, H. L., 2761–2763
Frandson, R. D., 1459
Frank, C. B., 0197, 1018, 4373
Frank, E. R., 3575
Frankel, Joseph, 1083
Frankel, Sam, 1108
Frankel, V. H., 1720
Fraschini, F., 2103
Fraser, A. F., 2259
Fraser, Allan, 3906
Fraser, Dean, 2764
Fraser, Margaret N., 3113
Fraser, R. G., 1839a
Frazier, W. C., 2904
Frederickson, D. S., 2390
Freedman, Russell, 1460
Freeman, B. E., 4218
Freeman, B. M., 1910, 2259a, 4093
Freeman, J. A., 1155
Frei, Emil, III, 3346
Frei, Walter, 3369
Frenay, Agnes C., 0902
French, M. H., 3918
Frey, J. R., 0770
Fried, Rainer, 2404
Friedberg, C. K., 3154, 3518
Friede, R. L., 2405
Frieden, Earl, 1966
Friedrich, Heinz, 2260
Friedrich, Hermann, 4145

Fritsch, Rudolf, 361
Fritz, J. C., 4112
Frohlich, E. D., 1944
Frøholm, L. O., 0752
Frost, H. M., 1721, 1722
Fruton, J. S., 2332
Frye, B. E., 2093
Fudenberg, H. H., 3003
Fujita, Tsueno, 1156
Fuller, J. L., 2297
Fund for the Replacement of Animals in Medical Experiments, 4386
Fung, Y. C., 1740
Furchtgott, Ernest, 2683
Fyvie, Audrey, 4177

G

Gaafar, S. M., 0799
Gabe, M., 2406
Gaddis, Margaret, 4232
Gaddis, Vincent, 4232
Gaillard, P. J., 0750
Gainer, Harold, 4320
Gajdusek, D. C., 2765
Galigher, A. E., 1245
Gallagher, C. H., 2361
Galloway, I. A., 2845
Galloway, J. H., 3716
Galton, Mildred M., 1227, 4100
Galuzo, I. G., 2976
Gangloff, H., 4558
Ganong, W. F., 1923, 2112, 2415
Gans, Carl, 1461, 4168
Garber, E. D., 1631
Gardiner, Mary S., 4597
Gardiner-Hill, H., 2095
Gardner, A. F., 3155, 3371
Gardner, D. L., 3203
Gardner, E. D., 1415, 2201
Gardner, M. A., 4036
Gardner, William, 0856
Garn, S. M., 1045
Garner, F. M., 0202, 0790
Garner, R. J., 2587
Garnham, P. C. C., 2934, 2935
Garrison, F. H., 0967
Garrod, L. P., 2642
Garry, Leon, 0110
Gassner, F. X., 0733
Gatz, A. J., 2020
Gauze, G. F., 3417
Gavan, T. L., 1196
Gay, C. W., 2607
Gay, W. I., 4361
Geddes, L. A., 1335, 3156
Gehan, E. A., 2521
Geisler, Rolf, 4121
Geist, Valerius, 2261
Gelatt, K. N., 2558, 3224

Gelder, R. G. van. See Van Gelder, R. G.
Gell, R. G. H., 3059
George, J. C. 4080
Georgi, J. R., 2936
Gerarde, H., 2584
Gerber, H., 0758
Gerencser, V. F., 2711, 2738
Géroudet, Paul, 4195
Gersh, Harry, 4208
Getty, R., 1416, 1458, 1489
Giacobini, Ezio, 4376
Gibbons, W. J., 3717, 3813
Gibbs, B. M., 2698
Gibbs, C. J., 0153
Gibbs, D. D., 3531
Gibbs, Erna L., 1375, 3176
Gibbs, F. A., 1374, 1375, 3176
Gibian, H., 2167
Gibson, D. C., 4513
Giese, A. C., 1632
Gilbert, R. M., 2262
Gilbert, S. G., 3929, 4013
Gillespie, J. H., 0795, 2830
Gillies, R. R., 2699
Gillis, W. T., 2608
Gilman, Alfred, 2560
Giroud, Antoine, 1528
Gitter, K. A., 3227
Glade, P. R., 3077
Glanville, J. N., 1930
Glasser, Otto, 2459
Gleason, M. N., 2588
Gleiser, C. A., 1303
Glick, David, 1246
Glickstein, Mitchell, 1805
Gluckstein, Fritz, 0883
Glynn, L. E., 3060
Godlovitch, Roslind, 4387
Godlovitch, Stanley, 4387
Goin, C. J., 1462, 4169
Goin, Olive B., 1462, 4169
Goldberg, Morris, 0836, 0837
Goldberg, R. E., 1394
Goldberger, Emanuel, 1367
Goldberger, R. F., 2477
Goldblatt, L. A., 2589
Golden, Abner, 3541
Goldman, Leon, 1336
Goldman, M. J., 1368
Goldman, Morris, 1188, 1198
Goldsmith, E. I., 0745
Goldstein, Avram, 2559
Golley, F. B., 4214
Gompel, Claude, 3308
Good, Loraine S., 3949
Good, R. A., 0746, 3004, 3040, 3044
Goodale, R. H., 3178
Goodall, Daphne M., 3873
Goodgold, Joseph, 1386
Goodman, L. S., 2560

Goodman, N. M., 0936
Goodwin, J. F., 3267
Goodwin, P. N., 3655
Gorbman, Aubrey, 2097
Gordon, B. L., 0865, 0866
Gordon, D., 1392
Gordon, H. A., 4278
Gordon, J. E., 0800
Gordon, M. S., 1924, 1967
Gordon, Maxwell, 2684
Gordon, R. F., 2259a, 4093
Gordon, R. M., 2986
Gorham, J. R., 4050
Gorski, R. A., 0788
Goth, Andres, 2561
Gotto, R. V., 4146
Goudie, R. B., 3052
Gould, S. E., 2884, 3519
Gozsy, Bela, 2362
Grabar, P., 0762
Graber, C. D., 2726
Graciansky, P. de, 3244
Grady, H. C., 0756
Graff, H. F., 1006
Graham, E. H., 4203
Graham, Ruth M., 3418
Graham, W. D., 3485
Graham-Jones, Oliver, 1968, 2893, 3994
Gran, F. C., 0752
Grande, Francisco, 0963
Granit, Ragnar, 1994
Grant, J. C. B., 1298, 1418
Grant, Lester, 3401
Grant, Murray, 2905
Grant, R. P., 1369
Graumann, Walther, 2425
Graves, F. T., 3591
Graves, Susan, 1008
Gray, Annie P., 0164
Gray, C. H., 2098, 2590, 3468
Gray, D. F., 3005
Gray, D. J., 1415
Gray, Henry, 1419
Gray, J. R., 3919
Gray, James, 1724
Gray, Peter, 0884, 0914
Gray, T. C., 3606
Graymore, C. N., 2363
Great Britain Department of Scientific and Industrial Research, 0139
Great Britain Ministry of Agriculture, Fisheries and Food, 1200, 4080a
Greeff, R., 1805
Greeley, R. G., 3874, 3976
Green, D. E., 2477
Green, E. L., 4489, 4525
Green, H. N., 3419
Green, M. C., 4502
Greenberg, D. M., 2333
Greenblatt, G. M., 4014

Greene, Eunice C., 4488
Greenfield, J. G., 3493
Greenhalgh, J. F. D., 3761
Greenough, P. R., 3819a
Greenwalt, T. J., 1872, 3006
Greenwood, Maud E., 2460
Greer, Milan, 4233
Gregor, H. P., 3599
Gregory, P. H., 0781
Gregory, R. W., 3920
Gresham, G. A., 3372
Grey, R. M., 2665
Grieg, J. R., 3373
Griffen, Jeff, 4234
Griffenhagen, G. B., 2644
Griffin, D. R., 1464
Griffin, J. P., 2635
Griffith, J. Q., Jr., 4512
Griffith, J. S., 2524
Grinnell, Alan, 1933
Grist, N. R., 2766
Grollman, Arthur, 2562
Grollman, Evelyn F., 2562
Grollman, Sigmund, 1465
Groot-de-Rook, A. S. de, 0140
Gross, C. M., 1419
Gross, Ludwik, 2767
Gross, V. E., 0903
Grossman, C. C., 1395
Grossman, J. D., 1496
Grossman, M. I., 0788
Group of European Nutritionists, 0736
Grove, A. J., 4055
Grubb, R., 0796
Gruchy, G. C. de. See De Gruchy, G. C.
Grumbach, A., 0735
Grundmann, E., 2997
Grzimek, Bernhard, 2263
Gual, C., 0751
Guggisberg, C. A., 4186
Gunn, D. L., 2258
Gunz, Frederick, 3334
Gurr, Edward, 1247
Gurr, G. T., 1248
Gutcho, M., 3755
Guyer, M. F., 1249
Guyton, A. C., 1774, 1925—1927
Gyorgy, Paul, 2340

H

Haan, R. L. de. See DeHaan, R. L.
Habel, Karl, 2768
Habel, R. E., 1466, 3820, 3864
Hackel, D. B., 4364
Hackett, P. L., 3348
Hadek, Robert, 2168
Hadley, Anne, 0905, 1046
Haesler, W. E., Jr., 3007
Hafez, E. S. E., 1857, 2169, 2170, 2181,

2264, 2265, 3756, 4284
Hafs, H. D., 2178
Haguenau, F., 1681, 2757
Haigh, Basil, 3618
Hailman, J. P., 2275
Hainsworth, Marguerite D., 2266
Hale, J. F., 3701
Haley, Leanor D., 1202
Haley, T. J., 0766
Hall, Carrie E., 2903
Hall, E. R., 4223
Hall, L. W., 3958, 4342
Hall, M. C., 1633
Hall, V. E., 0788, 3513
Halliday, W. J., 0885
Halstead, B. W., 4122, 4147
Halstead, J. A., 3476
Halver, J. E., 4567
Ham, A. W., 1634
Hamburger, Jean, 3621
Hamburger, Viktor, 1531
Hamburgh, Max, 2101
Hamerman, David, 2387
Hamilton, Ferelith, 3995
Hamilton, W. J., 2285
Hamkin, R. G., 0743
Hammack, Gloria M., 0103
Hamre, Dorothy, 0771
Hancox, N. M., 1726
Hanlon, J. J., 2906
Hanna, M. G., Jr., 0732, 4356
Hannah, H. W., 3796
Hansel, William, 0733
Hanshaw, J. B., 0795
Hardin, J. W., 2611
Hare, D. B., 1287
Hare, Ronald, 4287
Harell, A., 0750
Hargreaves, Tom, 2063
Harmar, Hilary, 3996
Harmon, R. W., 2612
Harned, Jessie M., 0906
Harper, H. A., 2365
Harris, G. W., 1862
Harris, J. E., 3421
Harris, J. W., 3337
Harris, John, 4387
Harris, L. W., 3750
Harris, Patricia J., 0738
Harris, R. H. C., 0757, 0767, 2769
Harris, R. S., 2340, 4483
Harris, W. C., 4187
Harris, W. H., 3205
Harrison, B. M., 1467, 1468, 1533
Harrison, J. C., 3875
Harrison, J. H., 3306
Harrison, R. G., 1534
Harrison, R. J., 4148
Harrison, T. R., 3134
Harrisson, Barbara, 0776

Harrop, A. E., 3997
Harrow, Benjamin, 2334
Harshbarger, J. C., 4591
Hart, A. H., 3998
Hart, D. M., 2421
Harthoorn, A. M., 4188
Hartman, C. G., 0212, 2195, 4421
Hartwell, Dickson, 4235
Harvard University, 2174
Harvey, A. M., 3135, 3145
Harvey, D. G., 3757
Harvey, E. B., 4081
Harvey, William, 0968
Harwood, R. F., 2988
Hasegawa, Toshio, 3542
Hashimoto, Ken, 3245
Hassall, A., 0007, 0008
Hastings, G. W., 3586a
Hathaway, D. E., 0161
Haugen, Jostein, 4051
Haupt, R. E., 4030
Haurowitz, Felix, 2443
Hawk, P. B., 2366
Hawkins, C. F., 1047
Hawkins, Linda L., 2644
Hawley, G. G., 0886
Haxhe, J. J., 0751
Hay, E. D., 4082
Hayat, M. A., 1157
Hayes, M. H., 3876–3878
Hayhoe, F. G. J., 1882
Haymaker, W., 1863
Haynes, Williams, 0857
Hayt, W. H., 1337
Hayward, C. L., 1469
Haywood, B. J., 0166
Hazard, J. B., 0756
Hazen, Elizabeth L., 1204
Heaney, R. P., 3205
Heath, J. S., 3797
Heath, R. G., 0789
Hedgecock, L. W., 2646
Hediger, Heini, 2267–2269
Hedinger, C. E., 0767
Heftman, Erich, 1109
Heidrich, H. J., 3322
Heidrich, J. F., 3718
Heldemann, W. H., 0786
Helminthological Society of Washington, 0395
Helwig, E. B., 0756
Henderson, G. N., 3719, 4236
Henderson, Isabella F., 0820
Henderson, J. A., 3713
Henderson, J. R., 2771
Henderson, W. D., 0820
Hendrickx, A. G., 4422
Hendry, W. S., 2872
Henley, Catherine, 4142
Henning, M. W., 2831

Henning, Norbert, 3532
Henry, J. B., 3172
Henry, J. P., 2047
Henry, R. J., 2367
Henry, S. M., 2700
Herbert, W. J., 3009, 3720
Herman, Robert, 2939
Herrath, Ernst von, 1635
Herrick, C. J., 2270, 4514
Hershey, D., 2489
Hess, W. R., 0795
Hickman, John, 3622
Hicks, E. A., 4030
Highnam, K. C., 4578
Hildebrand, Milton, 1302
Hildemann, W. H., 3094
Hill, A. B., 2525, 2526
Hill, A. V., 0969
Hill, D. W., 2495, 3607
Hill, Leonard, 4578
Hill, R. B., 3379
Hill, W. C., 1422
Hinde, R. A., 2270a, 4447
Hine, G. J., 1338
Hinshaw, L. B., 0732
Hirano, Asao, 3499
Hirohata, Kazushi, 1687
Hirsch, E. F., 2048
Hirsch, J. G., 2871
Hirsch, W. J., 3557
Hirschhorn, H. H., 0838
Hisaw, F. L., 0738
Hoar, W. S., 1970, 4123
Hoare, C. A., 2937
Hobday, F. T. G., 3577
Hoch, F. L., 2385
Hodos, William, 4086
Hoeden, Jacob van der, 2885
Hoerlein, B. F., 3959
Hoerr, N. L., 0821
Hofer, Doris J., 0907
Hofer, H. O., 0760, 4475
Hofer, Helmut, 4475
Hoffman, C. E., 4321
Hoffman, G. L., 4124
Hoffman, R. A., 4515
Hoffman, W. S., 3180
Hofstad, M. S., 4074
Höhn, E. O., 2100
Hoijer, Dorothy T., 2408
Holborrow, E. J., 3060
Holdeman, Lillian V., 2739
Holečková, Emma, 1282
Hollaender, Alexander, 2583
Hollander, J. L., 3207
Hollinshead, W. H., 1423
Holman, H. H., 1012
Holmes, R. L., 1636
Holmes, W. L., 0732, 1883
Holt, S. H., 3181

Homburger, F., 0778
Hope, A. B., 2461
Hopkins, C. E., 3913
Hopkins, H. U., 3159
Hornaday, W. T., 4189, 4190
Horning, E. C., 0772
Hornykiewicz, O., 2677
Horridge, G. A., 1776, 4577
Horsbrugh-Porter, Andrew, 3877
Horsfall, F. L., 2854
Horsfall, W. R., 2987
Hort, W., 0769
Horton, E. W., 0772, 2111
Horton-Smith, C., 4083
Hoskins, H. P., 3953, 3954
Hoskins, J. M., 2770
Hotchin, John, 0771
Houck, J. C., 0751
House, E. L., 1777
House, W., 2753
Howard, E. B., 3348
Howell, A. B., 1728
Howell, Elsworth, 3988
Howell, P. G., 0795
Hoyle, L., 0795
Hraba, I., 0770
Hsia, D. Yi-Yung, 2369
Hsiung, G. D., 2771
Hsu, T. C., 1538, 4116
Hubbard, J. P., 4257
Hubinont, P. E., 2102
Hudson, J. R., 0753
Hudson, R. E. B., 3520
Huettner, A. F., 1470
Huffman, G. C., 2072
Huffman, R. D., 4417
Hugh, A. E., 1930
Hughes, D. E., 0781
Hughes, D. O., 2462
Hughes, E. C., 0888
Hughes, G. M., 1938
Hughes, H. V., 1479
Hughes, Helena E., 3455
Hughes, J. T., 3240
Huhn, Dieter, 1688
Huisman, T. H. J., 1112, 1885
Hull, R. N., 0795
Humason, Gretchen L., 1268
Humphrey, J. H., 3011
Hungerford, T. G., 3722, 4084
Hunt, H. F., 2241
Huntsberger, D. V., 2527
Hunziker, Nicole, 4516
Hurley, Rosalinde, 2875
Hurry, S. W., 1637
Huser, H. J., 4424
Hutch, J. A., 1852
Hutchison, H. E., 3339
Hutt, F. B., 3116, 3117
Hütter, Ralf, 0735

Hutyra, Ference, 3373
Huxley, J. S., 3422
Hyman, H. T., 3157

I

Igarashi, Shiro, 1471
Ignatieff, E., 3946
Illingworth, Charles, 3643
Illinois Institute of Technology, Research Institute, 4347
Ingle, D., 4118
Ingram, D. L., 3934
Ingram, V. M., 0741
Innes, J. R. M., 3494, 4487
Inoue, Hajime, 1156
Institut français de la fievre aphteuse, 2855
Institut national de recherches vétérinaires, Bibliotheque, 0104
Institut Pasteur, 0067, 0406
Institute of Animal Technicians, 0242
Institute of Biology, 0661, 1444, 1484, 1746, 1846, 2027
Institute of Laboratory Animal Resources, 4408, 4548
Institute of Microbiology and Hygiene of Montreal University, 2362
International Academy of Pathology, 0756
International Anatomical Nomenclature Committee, 0869, 0870
International Association of Veterinary Food Hygienists, 0604a
International Atomic Energy Agency, 3703
International Cell Research Organization, 0996
International Committee on Laboratory Animals, 0757
International Committee on Veterinary Anatomical Nomenclature, 1426
International Dairy Federation, 3822
International Society for Cell Biology, 0669, 3048
International Society of Blood Transfusion, 1270
International Society of Haematology, 0599
International Society of Hemorheology, 1888
International Society of Psychoneuroendocrinology, 2411
International Study Group for Steroid Hormones, 0707
International Union Against Cancer, 0767, 0871
International Veterinary Reference Service, 0036
Ireland, Norma O., 0940
Isenbart, Hans-Heinrich, 3879
Israel, M. S., 3397
Israels, M. C. G., 3561
Issitt, P. D., 1111

Ivanov-Smolenskiĭ, A. G., 2026
Ivens, Virginia, 3828

J

Jablonski, Stanley, 0840, 0858
Jackson, E. R., 3816
Jackson, G. J., 3067
Jackson, George, 2939
Jackson, H. L., 3705
Jackson Laboratory, 4489, 4490, 4525
Jacobs, H. L., 3955
Jacobs, J. E., 2487
Jacobson, H. G., 3687
Jacobson, Marcus, 1540
Jacobson, S. A., 3486
Jaffe, H. L., 3208
Jaffe, R., 4344
Jain, N. C., 3832
James, M. T., 2988
Jamieson, G., 1872
Jamroz, Casimir, 4087
Jane, J. A., 3495
Jansen, Jan, 1476
Japanese Cancer Association, 0754
Jaroslow, B. N., 3668
Jarrard, L. E., 4454
Jasmin, G., 0769, 3321
Jawetz, Ernest, 2727
Jay, J. M., 2907
Jay, Phyllis C., 4448
Jeanrenaud, B., 2078
Jeffrey, H. C., 2940
Jellison, W. L., 2873
Jenkin, Penelope M., 2105
Jenkins, E. D., 1667
Jenkins, Frances B., 0807
Jenkins, T. W., 3961
Jennings, A. R., 3375
Jennings, H. S., 4025
Jennings, J. B., 3758
Jensen, H. E., 3228
Jensen, Rue, 3824
Joest, Ernst, 3724
Johansson, Ivar, 3118
Johns, H. E., 3657
Johnsgard, P. A., 2272
Johnson, A. D., 2176
Johnson, Charles, 2243a
Johnson, D. E., 2582
Johnson, D. J., 4470
Johnson, Patricia H., 3880
Johnson, Phyllis T., 4589
Johnson, R. N., 2273
Johnstone, P. N., 1541
Joint FAO/WHO Expert Committee on Brucellosis, 0753
Joklik, W. K., 2717
Jollie, M. T., 4216
Jolly, Alison, 4449

Jolly, C. J., 4445
Jones, A. F., 3995, 3999
Jones, B. V., 3794
Jones, Clyde, 0737
Jones, E. G., 4000
Jones, I. C., 1864
Jones, J. D., 2054
Jones, L. M., 0800, 2564
Jones, T. C., 1303, 3394, 4364
Jones, W. E., 3859
Jong, R. H. de. See De Jong, R. H.
Jonxis, J. H. P., 1112
Jordan, D. S., 4149
Jordan, F. T. W., 3809
Joseph, Shirley A., 4515
Joshua, J. O., 4017
Joslin, E. P., 3323
Joyce, C. R. B., 2685
Jubb, K. V. F., 3377
Jude, A. C., 4040
Judy, W. L., 3798
Juel-Jensen, B. E., 2857
Juergens, J. L., 3257
Juergenson, E. M., 3938
Juhasz, S. E., 2776
Julien, Brian, 4510
Jung, R. C., 2932, 2933
Jungerman, P. F., 2816

K

Kabat, E. A., 2444, 2445
Kabata, Z., 4119, 4576
Kado, R. T., 4438
Kaestner, Alfred, 4599
Kahn, A., 3438
Kahn, R. H., 1272
Kaiser, Frances E., 0141
Kaiser, H. E., 4081
Kalbus, Barbara H., 3930, 4018
Kaldor, George, 2385
Kalinová, Eva, 2811
Kalter, Harold, 1781
Kamath, S. H., 1113
Kamiya, Toshiro, 1471
Kammlade, W. G., 3920
Kampmeier, O. F., 1890
Kaneko, J. J., 2372
Kaniuga, Z., 0752
Kaplan, A. S., 0795
Kaplan, H. M., 4052
Karczmar, A. G., 2565
Karlson, Peter, 2336
Karpovich, E. A., 0841
Karpovich, P. V., 1995
Karstad, L. H., 4174
Karten, H. J., 4086
Kato, Laszlo, 2362
Katsuta, Hajim, 3424
Kattus, A. A., 0788

Kauffmann, Fritz, 2728
Kavanau, J. L., 1641
Kawamura, Akiyoshi, 1210
Kay, D. E., 1158
Kay, R. H., 4405
Kays, D. J., 3881
Kazal, L. A., 2151
Kedrevskiĭ, B. V., 1642
Keele, C. A., 1961
Keitzer, W. A., 2072
Kelen, Eva M. A., 2622
Keller, R., 0770
Kellermeyer, R. W., 3337
Kelley, E. C., 0970
Kellogg, A. R., 4204
Kelly, Elizabeth, 1398
Kelly, G. F., 4579
Kelly, W. R., 3146
Kelson, K. R., 4223
Kemali, M., 4155
Kemmerly, J. E., 1337
Kemp, L. A. W., 3658
Kendall, Florence P., 1996
Kendall, H. O., 1996
Kendall, S. B., 2956
Kenedi, R. M., 2504
Kennedy, A. H., 4058
Kennedy, P. C., 3377
Kenneth, J. H., 0820
Kenshalo, D. R., 1778
Kent, G. C., Jr., 1473, 1474
Kenworthy, Walter, 1481
Kerker, Ann E., 0808
Kerley, Peter, 3693
Kerr, A. H., 0859
Kestens, P. J., 0751
Keyes, H. M.R., 0939
Keyes, J. W., Jr., 3697
Kheĭsin, E. M., 2941
Kiddy, C. A., 2178
Kienitz, M., 0735
Kihlman, B. A., 1643
Kiloh, L. G., 1376
Kimmich, H. M., 4258
King, D. W., 2479
King, E. J., 2394
King, E. R., 3696
King, J. O. L., 3759
King, J. R., 4065
King, J. W., 3171
King, Judith E., 4148
King, L. S., 1048
King, R. C., 0889
Kingsbury, John, 2615
Kirk, H., 3147
Kirk, R. W., 0794, 2634, 3725, 3731
Kirkham, Hadley, 3543
Kiss, F., 1427
Kissmeyer-Nielsen, F., 0786
Kit, Saul, 0771

Kitay, J. I., 2106
Kleerekoper, Herman, 4126
Kleiber, Max, 2373
Klemm, W. R., 1377
Kleynenberg, S. Y., 4150
Klienenberger-Nobel, Emmy, 2729
Kligman, A. M., 3252, 4518
Klingberg, M. A., 0732
Klingener, David, 4052a
Klinghammer, Erich, 2250
Klintworth, G. K., 2872
Klip, Willem, 2464
Klippel, R. A., 4491
Klopfer, P. H., 2275
Klostermann, G. F., 3230
Klumbies, Gerhard, 2321
Klüver, W. H., 1801
Knigge, K. M., 4515
Knight, C. R., 1049
Knight, Julie, 1283, 1998, 2128, 2793
Knight, Maxwell, 2203
Knowles, F. G. W., 2108
Knudsen, J. W., 1050
Kobayashi, Mitsunao, 1396
Kochva, E., 2613
Koenig, Elizabeth, 3443
Koepke, J. A., 3470
Kohn, R. R., 1984
Koldovsky, O., 1973
Koldovsky, P., 0780
Kon, S. K., 2107
König, J. F. R., 4491
Kooi, K. A., 1378
Kopech, Gertrude, 0179, 0180
Kopin, I. J., 3324
Koppányi, Theodore, 2565
Koprowsky, Hilary, 2782
Korec, Rudolf, 4517
Korenchevsky, V., 1544
Kosakai, Nozomu, 1115
Koss, L. G., 1644
Kotliarevskiĭ, L. I., 2026
Kovac, Werner, 4492
Kowalski, E., 0752
Kozloff, E. N., 1245
Kradjian, R. M., 3537
Král, Frank, 3978
Krantz, J. S., 2566
Kratochvil, C. H., 0776
Kremer, M., 2402
Krevans, J. R., 4259
Kreyberg, Leiv, 0761
Krider, J. L., 3943
Krieg, A. F., 1116
Kritchevsky, David, 0797
Krogh, A., 1836, 4127
Krull, W. H., 2943
Krushinski, L. V., 2276
Kruzas, A. T., 0931
Kubica, G. P., 1117

Kubitschek, H. E., 2703
Kuhlenbeck, Hartwig, 1786
Kummer, H., 4476
Kurstak, Edouard, 2783
Kurtz, S. M., 1689
Kusama, Toshio, 4426
Kusama, Yoshio, 0842
Kwapinski, J. B., 2446, 3081

L

Labadan, R. M., 2994
Laborit, Henri, 1645
Lacroix, J. V., 3953
LaDu, B. N., 2567
Laguens, R. P., 1690
Lahunta, Alexander de. See DeLahunta, Alexander
Laing, J. A., 0753, 2179
Lajtha, Abel, 0777, 2412
Laki, Koloman, 1729, 1891
Laland, S. G., 0752
Lambert, Rene, 4510
Lamela, Alberto, 1119
Lamming, G. E., 0773
Lampe, K. F., 2616
Land, F. H. de. See DeLand, F. H.
Landy, Maurice, 3030, 3033, 3045, 3074
Lane-Petter, William, 4288, 4289, 4293
Lang, E. K., 3682, 3683
Langebartel, D. A., 4156
Langley, L. L., 1646, 1931
Langman, Jan, 1546
Lannigan, Robert, 3521
Lansford, E. M., Jr., 0919
Lanyon, W. E., 4107
Lapage, Geoffrey, 2944
Lapin, B. A., 4461, 4464
Largiadèr, Felix, 3624
Larimer, James, 1932
Larner, Joseph, 2374
Larsell, Olof, 1476
Lasley, J. F., 3783
Last, R. J., 1820
Latner, A. L., 2375
Laufman, Harold, 3341
Lauslahti, K., 0735
La Via, M., 3022, 3379
Lavine, I., 1074
Lavoipierre, M. M. J., 2986
Lawrence, C. A., 2908
Lawrence, H. S., 3033
Lawrie, R. A., 0773
Leach, R. M., 2940
Leach, W. J., 1477
Leader, R. W., 0891
Leahy, J. R., 3799
Lear, Erwin, 2686
Leavell, B. S., 3342
Leaverton, P. E., 2527

Leavitt, Emily S., 4392
Leclercq, Marcel, 2989
Leden, Hans von. See Von Leden, Hans
Ledley, R. S., 2529
Lee, D. G., 3869
Lee, H. H., 2994
Lee, Henry, 3594
Lee, J. C., 4435
Lee, Julius, 2108
Lee, L. W., 1341
Lee, R. V., 0907
Lee-Delisle, Dora, 0843
Lees, R., 1211
Leeson, C. R., 1648
Leeson, T. S., 1648
Leghissa, Silvano, 1478
Lehmann-Grube, F., 0795
Lehninger, A. L., 2376
Leider, Morris, 0892
Lejeune, Fritz, 0844
Lenihan, J. M., 1342
Lenkeit, Walter, 3760
Lenman, J. A. R., 1388, 3185
Lennert, Karl, 1693
Lennette, E. H., 0763, 2692, 2777
Lentz, T. L., 1547, 1691
Leonard, A. H., 4059
Leonard, E. P., 3962, 3963
Leone, C. A., 3931
Leopold, S. S., 3159
Leoschke, W. L., 4060
Lépine, Pierre, 0845
Leutenegger, W., 4476
Levedahl, B. H., 1933
Levenson, S. M., 3651
Lever, W. F., 3507, 3508
Levi, H. W., 4599
Levi, Lorna R., 4599
Levi, W. M., 4108
Levine, B. S., 0840
Levine, Laurence, 1649
Levine, N. D., 2834, 2889, 2945−2947
Levinson, B. M., 4237, 4238
Levinson, S. A., 1120
Leviton, A. E., 4170
Levy, B. M., 4427
Levy, H. B., 2779
Lewis, A. E., 2530
Lewis, D., 0765
Lewis, E. F., 3925
Lewis, Marianna O., 1028
Lewis, S. M., 3332
Lewis, W. M., 4128
Leyhausen, P., 4041
Library of Congress. See U.S. Library of Congress
Licht, S. H., 1389
Lichtenstein, Louis, 3211, 3487
Liebig, Justus, 0971
Liebow, A. A., 0756

Liener, I. E., 2625
Light, F. W., Jr., 3913
Lillie, R. D., 1252, 1253
Lillie, R. J., 2592
Lim, R. K. S., 3964
Lin, K'o-shêng, 3964
Lindahl-Kiessling, K., 0732
Linde, R. W. te. See Te Linde, R. W.
Lindsley, D. B., 0788
Linford, J. H., 2377
Ling, N. R., 2144
Lingeman, C. H., 0790
Link, R. P., 2606
Linné, J. J., 1121
Lipkin, M. J., 0965
Lipner, Harry, 2092
Lissák, Kálmán, 2413
Littig, K. S., 2896
Little, L. M., 1013
Litwack, Gerald, 2083
Liu, Chan-nao, 3964
Livestock Sanitary Association. See U.S. Livestock Sanitary Association
Livingood, C. S., 3253
Ljungqvist, Ulf, 3965
Lloyd, L. E., 1014
Lobitz, W. C., 1830
Lockley, R. M., 2277
Lockwood, A. P. M., 4312
Lodge, G. A., 0743
Lodge, G. E., 0773
Lodge, R. W., 2617
Lodwick, G. S., 3684
Loewy, A. G., 1650
Lofts, Brian, 2027
Long, Cyril, 2337
Longacre, J. J., 3643a
Longree, Karla, 2909
Loosli, J. K., 3763, 3846
Lopashov, G. V., 1548
Lorenz, D., 0990
Lorenz, K. Z., 2278−2281
Loutit, J. F., 3659
Love, R. M., 4129
Loveland, R. P., 1159
Low, D. G., 3974
Lowbeer, Leo, 3536
Lowe, W. C., 3533
Lucas, A. M., 4087
Lucas, I. A. M., 0743
Lucchi, M. R. de. See DeLucchi, M. R.
Luckey, T. D., 4286
Luisada, A. A., 1833, 3277
Lumb, J. W., 3829
Lumb, W. V., 4335
Lumsdaine, A. A., 0788
Luna, L. G., 1232
Lungwitz, Anton, 3882
Luparello, T. J., 4493
Lupulescu, A., 1692

Author Index

Luria, S. E., 2780
Lush, I. E., 3107
Lutz, H., 4588
Luyendijk, W., 0777
Luzio, N. R. di. See Di Luzio, N. R.
Lyght, C. E., 1015
Lyle, T. K., 3234
Lynch, H. T., 3095
Lynch, M. J., 1122
Lyons, H. A., 3529
Lyons, Kathleen, 2921

M

Mabuchi, Masako, 4426
McAllister, H. A., 2704
McAllister, R. A., 3471
McBride, G., 0779
MacBryde, C. M., 3161
MacCallum, F. J., 3819a
MacCallum, F. O., 3240
McCarthy, J. L., 2134
McCauley, J. E., 0738
McCauley, W. J., 1974
McClellan, R. O., 3936
McClung, C. E., 1160
McCluskey, R. T., 3401
McComas, A. J., 1376
McCombs, R. P., 3136
McCoy, J. J., 4042
McCoy, J. R., 4345
McCready, V. R., 3704
McCunn, James, 3577, 3898
McDermott, Walsh, 3132
McDiarmid, A., 0801
McDonald, G. A., 3344
McDonald, I. W., 3846
MacDonald, J. R., 1185, 2753
McDonald, L. E., 2109
McDonald, Peter, 3761
MacDonald, Virginia, 0196
McFadden, H. W., Jr., 1196
McFadyean, J., 1479
McFarland, D. J., 2282
Macfarlane, R. G., 0801
MacFate, R. P., 1120, 1124
McGilvray, James, 2520
McGrath, J. T., 3966, 3967
McGree, W. R., 3883, 3884
Machado, M. A., 2858, 2887
McIlwain, Henry, 2414
Mack, Roy, 0872
MacKay, R. S., 2497
Mackenzie, D. H., 3562
Mackenzie, David, 3907
Mackenzie, P. Z., 4179
McKerns, K. W., 1856, 2094, 2110
Mackey, D. R., 3824
Mackintosh, N. J., 2311
Macklem, P. T., 3524

McLaughlin, C. A., 4053
McLean, F. C., 1997
MacLennan, J. M., 4576
McLeod, W. M., 3829
MacMahon, Brian, 2910
McMinn, R. H., 3644
MacNalty, A. S., 0816
McNaught, Ann B., 1936
McNeer, Gordon, 3534
McQuown, A. L., 2818
Macy, R. W., 2948
Maddin, Stuart, 3247
Madigan, J. C., Jr., 1791
Maegraith, B. G., 3126
Magalhaes, Hulda, 4515
Magalini, Sergio, 0893
Magnusson, Bertil, 4518
Magrane, W. G., 3232
Mahaffey, L. W., 3885
Maher, J. F., 2071
Mahlum, D. D., 3667
Maibach, H. I., 3645
Maier, Barbara M., 2284
Maier, N. R. F., 2283
Maier, R. A., 2284
Maizels, M., 3188, 3345
Makelä, O., 0786, 3034
Malek, I., 2702
Malinin, Theodore, 1271
Malinow, M. R., 4429
Malmstadt, H. V., 1345
Malt, R. A., 3642
Maltry, Emile, 3546
Manclark, C. R., 1214
Mandel, H. G., 2567
Manko, H. H., 1051
Manner, H. W., 1551
Manning, R. T., 3162
Manninger, Rudolph, 3373
Manocha, S. L., 4430, 4431, 4434
Manson, L. A., 0797
Manson-Bahr, P. H., 3137
Mantegazza, P., 2111
Manton, S. M., 1492, 4043
Manuel, Y., 3472
Manuila, A., 0846
Manwill, M. C., 3886
Maplesden, D. C., 3762
Maramorosch, Karl, 2781-2783
Marble, Alexander, 3323
Marcial-Rojas, R. A., 2978
Marcus, A. J., 2146
Marek, Joseph, 3373
Margolis, F. L., 4549
Marinacci, A. A., 1390
Mark, D. D., 1125
Markell, E. K., 2949
Markowitz, J., 4336
Marks, Charles, 3580
Marler, E. E. J., 0860

Marler, P. R., 2285
Marrable, A. W., 3932
Marriott, H. J. L., 1371
Marsh, B. B., 1999
Marsh, Hadleigh, 3909
Marshall, A. J., 4088, 4303
Marshall, I. H. A., 2180
Marshall, P. T., 1938
Martignoni, M. E., 4590
Martin, A. E., 1346
Martin, A. W., 1978
Martin, E. W., 2568
Martin, Hubert, 2580
Martini, G. A., 2859
Martini, L., 2103, 2112, 2116
Mascio, Alberto di. See DiMascio, Alberto
Mascoli, Carmine C., 2784, 3078
Mason, Charlene, 0998
Mason, E. E., 2531
Mason, I. L., 0894, 3777a
Mason, M. M., 0174
Masoro, E. J., 2385
Masui, Kiyoshi, 4089
Mathé, G., 0780, 0786, 3626
Mather, J. F., 3541
Mathews, C. A., 3834
Mathews, Christopher, 2705
Mathews, R. K., 4239
Mathews, W. W., 1552a
Matousek, Josef, 0748
Matsumoto, Teruo, 3595
Mattenheimer, Hermann, 1126
Matthews, C. A., 3912
Matthews, L. H., 2203
Matthews, P. B. C., 0775
Mattingly, P. F., 2990
Mattingly, R. F., 3638
Matzen, P. F., 3685
Matzke, H. A., 1793
Maule, J. P., 3783a
Maumenee, A. E., 3233
Maupin, Bernard, 1893
Mauro, Alexander, 0751
Mawdesley-Thomas, L. E., 0801
Maxwell, M. H., 2379
May, J. W., 1289
May, N. D. S., 3908
Mayer, Edmund, 1431
Mayer, Karl, 3953
Mayer, W. V., 1939
Mayersbach, H. von, 1665
Maynard, L. A., 3763
Mazia, Daniel, 1651
Mazur, Abraham, 2334
Mead, N. St. C., 4236
Mech, L. D., 4191
Mecherles, T. F., 3349
Medical Library Association, 0802, 0944, 0997
Medical Research Council, 2817, 4290, 4291

Medway, William, 3473
Meek, E. S., 0780
Meek, G. A., 1161
Meessen, H., 4344
Mehregan, A. H., 3254
Meites, Joseph, 2116
Mellon, Deforest, Jr., 2028
Mellon, M. G., 0809
Melmon, K. L., 2633
Melnick, J. L., 2727, 2775
Meloan, C. E., 1347, 1348
Menczel, J., 0750
Mendelsohn, E. I., 1941
Mengel, C. E., 3346
Mercer, E. H., 1162
Merchant, D. J., 0790, 1272
Merchant, I. A., 1215, 2706, 2835
Merck and Co., 0177, 2569, 3149, 4109
Merillat, L. A., 0972
Mering, T. A., 3946
Merino-Rodriguez, Manuel, 2931
Merkow, L., 0778
Merler, Ezio, 2439
Merritt, H. H., 2204
Merritt, Lynne L., 1359
Mertens, Robert, 4171
Mertz, Walter, 2371
Metcalf, C. L., 2991
Metz, C. B., 1553
Meyboom, F., 0847
Meyer, J. S., 0777
Meyer, M. C., 2950
Meyler, L., 2555, 2574
Meynell, Elinor, 2707
Meynell, G. G., 2707
Miale, J. B., 3190
Miatello, V. R., 1855
Michaelis, L. S., 3685
Micklem, H. S., 3627
Midwest Regional Medical Library, 0929
Miescher, P. A., 0746, 0762, 3040, 3479
Mieszkuc, B. J., 4347
Migaki, George, 2897, 4364
Mihich, Enrico, 3429
Miles, A. A., 2714
Miles, A. E. W., 1829
Milhaud, G., 0750
Milhorn, H. T., 1942
Millard, Max, 3474
Miller, A. T., 2380
Miller, F. N., Jr., 3386
Miller, J. N., 1217
Miller, J. R., 4366
Miller, M. E., 3969, 3970
Miller, O. L., Jr., 0790
Miller, S. E., 3475
Miller, W. C., 0815, 3800
Mills, C. F., 2398
Mills, Harlan, 1554
Milovanov, V. K., 3784

Minckler, Jeff, 2465, 3381, 3496
Minckler, T. M., 3381
Minton, Madge R., 4157
Minton, S. A., Jr., 2618, 4157
Mirand, E. A., 0732
Mirsky, A. E., 1612
Mitchell, H. H., 3764
Mitchell, Ralph, 2911
Mittwoch, Ursula, 3096
Miya, T. S., 2570
Miyakawa, M., 4286
Mizell, Merle, 0780, 4161
Modell, Walter, 2556, 2557
Moffitt, R. L., 3964
Moghissi, K. S., 2181
Möller, Göran, 0786
Moloney, J. B., 0790
Moltz, Howard, 2287
Monaco, A. P., 2546, 3633
Monlux, A. W., 3388
Monlux, W. S., 3388
Monnier, M., 2029, 4558
Mönnig, H. O., 2959, 3729
Monroy, Alberto, 1553
Montagna, William, 0731, 1480, 1481, 1830, 4432
Montagu, M. F. A., 1500
Monteith, J. L., 0781
Montgomery, G. L., 3510
Montgomery, Hamilton, 3509
Moon, H. D., 0756
Moon, H. W., 2774
Moore, F. D., 3278
Moore, H. B., 2731
Moore, K. E., 2687
Moor-Jankowski, J., 0745
Moreland, Sara L., 0929
Moreno, Esteban, 2978
Morgagni, G. B., 1433
Morgan, J. P., 3686
Morgan, J. T., 0773
Morgan, R. H., 3655
Mori, Yoshitaka, 1693
Morimoto, Kazuo, 1687
Morrelli, H. F., 2633
Morris, Desmond, 2288-2290, 4450
Morris, H. P., 0754
Morris, Henry, 1434
Morris, M. L., 3765
Morris, Ramona, 4450
Morrison, Elsie B., 3766
Morrison, F. B., 3766
Morrison, S. H., 3766
Morriss, J. E., 1460
Morton, John, 1846
Morton, L. T., 0967
Moser, R. H., 1052
Moss, Emma S., 2818
Mossman, H. W., 2183
Mostofi, F. K., 0756

Motta, M., 2103
Moulton, J. E., 3730
Mount, L. E., 0775, 2114, 3933, 3934, 4559
Movat, H. Z., 3024, 3070
Moyer, D. L., 2186
Moyer, J. H., 2671
Mozgovoï, A. A., 2979
Mrosovsky, Nicholas, 2115
Mudd, Stuart, 2836, 3071
Mueller-Eberhard, H. J., 3479
Muffley, H. C., 1735
Mühlbock, Otto, 4369
Mühlen, Klaus aus der, 4495
Muir, Robert, 3382
Muller, A. F., 2073
Muller, E. E., 0751
Muller, G. H., 3249, 3250
Müller, J., 0772
Müller, M. E., 3596
Muller, R., 0739
Mulligan, H. W., 2980
Mundis, J. J., 4240
Munro, H. N., 2382
Murphy, H. T., 0808
Murphy, Robert, 4192
Murphy, W. H., Jr., 1272
Murray, E. G. D., 2712
Murray, Margaret R., 0179, 0180
Murray, R. O., 3687
Musumeci, R., 3699
Muth, O. H., 2648, 3769
Muybridge, Eadweard, 1731, 1732
Myers, J. A., 2844, 3830
Myren, J., 2067
Myrvik, Q. N., 3020

N

Nachman, Ralph, 3346
Naether, C. A., 4061
Nagaishi, Chuzo, 1840
Nagano, Yasuiti, 2865
Naib, Z. M., 3431
Nakahara, Waro, 3434
Nakai, Junnosuke, 1794
Nalbandov, A. V., 2184
Napier, J. R., 4455, 4478
Napier, P. H., 4455, 4478
Narin, R. C., 1127
Natelson, Samuel, 1128
National Academy of Sciences. See U.S. National Academy of Sciences
National Agricultural Library. See U.S. National Agricultural Library
National Audiovisual Center. See U.S. National Audiovisual Center
National Cancer Institute. See U.S. National Cancer Institute
National Council on Radiation Protection and

Measurements. See U.S. National Council on Radiation Protection and Measurements
National Federation of Science Abstracting and Indexing Services, 0106
National Institute of Allergy and Infectious Diseases. See U.S. National Institute of Allergy and Infectious Diseases
National Institute of General Medical Sciences. See U.S. National Institute of General Medical Sciences
National Library of Medicine. See U.S. National Library of Medicine
National Mastitis Council, Research Committee, 3831
National Museum. See U.S. National Museum
National Referral Center for Science and Technology. See U.S. National Referral Center for Science and Technology
National Research Council. See U.S. National Research Council
National Science Foundation. See U.S. National Science Foundation
Naumenko, A. I., 2030
Nauta, W. J. H., 1795, 1863
Naval Medical School. See U.S. Naval Medical School
Neal, K. G., 3930, 4018
Nealon, T. F., 3597
Necheles, T. F., 3349
Necker, Claire, 4044
Needham, Dorothy M., 1733
Needham, Joseph, 1986
Neil, Eric, 1961
Neil, M. W., 2338
Neimark, H., 0186
Nelsen, O. E., 1555
Nesheim, M. C., 4105, 4111
Neter, Erwin, 2733
Netter, F. H., 1053
Nettesheim, P., 4356
Neuhaus, O. W., 4140
Neumann, A. L., 3847
Neumann, Karlheinz, 2425
Neurath, A. R., 0771
Neveu-Lemaire, Maurice, 2951
Neville, Kris, 3594
New, D. A., 1273
Newell, R. C., 2952
Newill, V. A., 2983
Newman, H. H., 1483
Newsom, I. E., 3909
Newth, D. R., 1484
New York Academy of Sciences, 0497, 4463
New York Heart Association, 2794, 3522
New York Society of Electron Microscopists, 1163
Nezlin, R. S., 2447
Ng, A. B. P., 3548

Nickel, Richard, 1485
Nicol, J. A. C., 4151
Nieberle, Karl, 3383
Nieburgs, H. E., 1274
Niemer, W. T., 4027
Niewiarowski, S., 0752
Nikolskiĭ, G. V., 4130
Nishikawa, Y., 3860
Nishimura, Hideo, 4366
Niven, C. D., 4388
Noback, C. R., 4432
Nobel, A., 0848
Nobelstiftelsen, Stockholm, 0976
Noble, E. R., 1219, 2953
Noble, G. A., 1219, 2953
Nodine, J. H., 2671
Noe, Lee, 1027
Nomura, Tatsuji, 4369
Norman, J. C., 1308
Norrby, E., 0795
Norris, J. R., 2734
Norris, K. S., 4125
North, Pamela H., 2619
Nossal, G. J., 3036
Nottingham Easter School in Agricultural Science, 0773
Nowinski, W. W., 1656
Nowotny, Alois, 2448
Nunn, A. S., Jr., 2058
Nunn, F. J., 3606
Nunn, J. F., 3608
Nunnally, James, 3653
Nuzzolo, Lucio, 1129
Nybakken, O. E., 0908

O

Oakley, W. G., 3325
O'Brien, Donough, 1130
Ochs, Sidney, 2031
O'Connor, Meave, 2796
Odell, W. D., 2186
Odlaug, T. O., 1557
O'Donoghue, P. N., 4287
Oehme, F. W., 3732
Office international des épizooties, 0576
Ogilvie, Colin, 3153
Ogilvie, P. W., 3931
Ohno, S., 0772
Oldham, P. D., 2532
Olitzki, A., 0735
Oliver, C. P., 1043
Oliver, R., 3658, 3660
Olmsted, E. H., 0959
Olmsted, J. M. D., 0959
Olsen, Gjerding, 1943
Olsson, Sten-Erik, 3688
O'Malley, C. D., 0788
O'Mary, C. C., 3842a
O'Neal, L. W., 3628
Oppenheim, I. A., 1131

Oppenheimer, Jane M., 1592
O'Rahilly, R., 1415
Orbison, J. L., 0756
Orejas-Miranda, Braulio, 4172
Orkin, Milton, 3512
Orlov, N. P., 2981
Ormrod, A. N., 3582
Orr, R. T., 4217
Orten, J. M., 2339
Osbaldiston, G. W., 1132
Osborne, C. A., 3974
Osborne, W. D., 3887
Oser, B. L., 2366
Osol, A., 0821, 2571
Osselton, J. W., 1373, 1376
Oswaldo-Cruz, E., 4053a
Otčenášek, Miloš, 2811
Ottaway, C. W., 3898
Ouchterlony, Örjan, 2450
Owen, L. N., 3488

P

Pace, W. R., 3279
Pack, G. T., 3534
Packer, R. A., 2706
Padmore, G. R. A., 1134
Palladin, A. V., 2417
Palma, J. R. di. See DiPalma, J. R.
Palmer, A. C., 2205
Palmer, E. D., 3292
Palminteri, A., 3637
Palocsay, G. P., 1255
Pampiglione, Giuseppe, 1379
Pan, H. P., 0789
Pan American Sanitary Bureau, 0755
Panchen, A. L., 4222
Panoya, E. M., 0155
Pansky, Ben, 1777
Pantelouris, E. M., 1975
Pantin, C. F. A., 1164
Paoletti, R., 0732
Papez, R., 0977
Paradiso, J. L., 4227
Parakkal, P. F., 2032
Paranchych, William, 2797
Paré, J. A. P., 2053
Parish, H. J., 2657
Parker, H. W., 4171
Parker, R. C., 1276
Parker, W. H., 3787a
Parkes, A. S., 2180
Parnas, Józef, 0735
Parsons, D. F., 1165
Paschkis, K. E., 2117
Paton, Alexander, 3293
Patt, D. I., 1652
Patt, Gail R., 1652
Patten, B. M., 1558–1561
Patten, J. W., 3888
Patten, S. F., 3547

Patton, A. R., 2384
Patton, H. D., 1950
Paul, J. R., 1277, 2912
Paul, S., 3878
Paulson, Moses, 3294
Pavlovskiĭ, E. N., 2889
Pavlovskiy, Y. N., 2890
Pearce, J. H., 0781
Pearsall, Nancy N., 3020
Pearse, A. G. E., 2428
Pearson, A. E. G., 4289
Pearson, Ronald, 4091
Pearson, W. N., 2340
Pease, D. C., 0788, 1256
Pease, P. E., 2736
Pecile, A., 0751
Pederson, C. S., 2913
Peele, T. L., 2219
Peer, L. A., 3629
Peery, T. M., 3386
Pegg, D. E., 3563
Pekas, J. C., 3935
Pellegrino, L. J., 4496
Pellérdy, L. P., 2954
Penfield, Wilder, 3498
Penner, L. R., 2950
Pepys, J., 0770
Percival, G. H., 3510
Pereira, H. G., 2748
Pernkopf, Eduard, 1436
Perrone, N., 2505
Perry, E. J., 3785
Perry, J. S., 0768
Perry, T. W., 3767
Petchesky, Rosalind P., 0185
Peters, Alan, 1695, 1763
Peters, Hannah, 1550
Peters, J. A., 0896, 4172
Pethica, B. A., 2478
Petrak, Margaret L., 4092
Petráň, Mojmír, 1330
Petrov, I. R., 3665
Petrushevskiĭ, G. K., 4119
Pettit, G. D., 3975
Pfaffenberger, C. J., 2291
Phillips, G. B., 0204
Phillips, P. E., 4355
Phillips, R. W., 3735
Phillipson, A. T., 0765, 3282
Physiological Society of London, 0775
Pi, J. S., 0737
Piazza, Marcello, 2861
Pick, Joseph, 1800
Pickett, M. J., 2731
Pickford, J. H., 3792
Pierce, Elizabeth T., 4501
Piermattei, D. L., 3631, 4024
Pillsbury, D. M., 3252, 3253
Pimentel, R. A., 4580
Pincus, Gregory, 2119

Pindborg, J. J., 0761
Pinkava, J., 1135
Pinkus, Hermann, 3254
Pinniger, R. S., 3794
Pirchner, Franz, 3097
Pitt-Rivers, Rosalind, 1867
Pitts, R. F., 2074
Pitzschke, Horst, 2851
Pizzarello, D. J., 3666
Platt, W. R., 3350
Plescia, O. J., 2449
Plotz, E. J., 2167
Plous, F. K., Jr., 2889, 2941
Plowright, W., 0795
Plum Island Animal Disease Laboratory, 0108, 0186
Plummer, Gordon, 2776
Podhajsky, Alois, 4242
Podhajsky, Eva, 4242
Polezhaev, L. V., 1562
Pollard, C. B., 2612
Pollard, Morris, 0774
Pollister, A. W., 1350
Polson, C. J., 2593
Polyak, S. L., 4316
Polyanski, Y. I., 4119
Pontén, J., 0795
Poon, T. P., 3499
Popesko, Peter, 1489
Porter, George, 4293
Porter, K. R., 1697
Porter, Ruth, 2084
Potts, W. H., 2980
Poyer, J., 1340
Prankerd, T. A. J., 3345
Pratt, H. D., 2896
Preece, Ann, 1257
Prévot, A. R., 2708
Price, L. W., 1351
Prier, J. E., 3473, 3581
Priest, J. H., 3098
Priester, W. A., 0874
Prince, J. H., 1802, 1803, 4054
Pritchard, J. J., 3644
Prosser, C. L., 1947
Prywes, Moshe, 3015
Public Health Service. See U.S. Public Health Service
Pugh, D. M., 2553
Pugh, T. F., 2910
Purcell, Elizabeth, 4274
Purpura, D. P., 1868
Putt, F. A., 3457
Pyke, D. A., 3325

Q

Queensland University, 0779
Quick, A. J., 1895, 3352
Quimby, C. W., 3609
Quimby, Edith, 3655, 3708

R

Rack, J. H., 1244
Radeleff, R. D., 2594
Raekallio, Jyrki, 3647
Rafferty, K. A., 1563
Rall, J. E., 3324
Ramachandran, G. N., 2344
Ramón y Cajal, Santiago, 1564, 1805
Ramsay, Alexandra, 2299
Ramsay, J. A., 4581
Ramsey, D. M., 0788
Rand, Elias, 1397
Rand, R. W., 3589
Randall, D. J., 1929
Rang, Mercer, 1734
Ransom, J. H., 3889
Rapaport, S. I., 3353
Rapp, Fred, 2856
Raspé, Gerhard, 1578
Ratcliffe, F. N., 2852
Ratner, S. C., 2245, 4587
Ravel, Richard, 3192
Raven, C. P., 1565, 1566, 1987
Raven, R. W., 3433a
Ray, J. G., Jr., 3084
Rayner, R. W., 2819
Read, C. P., 2955
Reade, Winwood, 4209
Reagan, J. W., 3548
Rebuck, J. W., 0756
Rech, R. H., 2687
Reed, F. C., 2814
Reemtsma, Keith, 3636
Regamey, R. H., 0782
Reichenbach-Klinke, H. H., 4131
Reid, Eric, 3436
Reid, Mary E., 4520
Reid, W. M., 0810
Reighard, J. E., 4025
Reiner, J. M., 2480
Reith, E. J., 1655
Remenchik, A. P., 3138
Rendel, Jan, 3118
Rendel, John, 3999
Renetzky, Alvin, 1023
Renk, W., 3718
Rensch, B., 0741
Reproduction Research Information Service, 0145, 0188
Reuck, A. V. S. de. See DeReuck, A. V. S.
Reuterskiöld, A. G., 2075
Revel, J. P., 4082
Reverberi, Giuseppe, 4582
Revillard, J. P., 3038, 3472
Rey, Barbara, 3879
Reyment, R. A., 2518
Reynolds, H. H., 0776

Reynolds, Laura A., 4260
Reynolds, M., 2129
Reynolds, Vernon, 4480
Rezek, P. R., 1309
Rheingold, Harriet L., 2292
Rhinesmith, H. S., 2481
Rhodes, A. J., 2787
Rhodin, J. A., 1698
Ribbons, D. W., 2734
Ribelin, W. E., 4345
Ricciuti, E. R., 4371
Rice, Berkeley, 4243
Rice, V. A., 3787
Rich, M. A., 0790, 3564
Rich, S. T., 0190
Richards, J. D. M., 3345
Richardson, G. S., 2123
Richardson, I. W., 2466
Richardson, U. F., 2956
Richer, Paul, 1054
Richter, Derek, 2418
Rickett, H. W., 2621
Ridgway, S. H., 4132
Riegel, J. A., 1976a
Riemann, Hans, 2914
Rifkin, Harold, 3319
Riggs, D. S., 2533
Rigler, L. G., 3691
Ríha, I., 3028
Rinfret, A. P., 3589
Ring, A. M., 1136
Rio-Hortega, Pio del, 3500
Riser, W. H., 3671
Ristic, Miodrag, 3355
Rita, Geo, 2772
Ritchie, A. E., 3209
Rittenhouse, J. G., 0181
Rivers, T. M., 2854
Rob, Charles, 3583
Robb, Jane S., 2049
Robbins, S. L., 3387
Robertis, E. D. P. de, 1656
Roberts, Catherine, 3125
Roberts, D. C., 0189, 4287
Roberts, Ffrangcon, 0909
Roberts, J. A. F., 3099
Roberts, J. C., 3262
Roberts, S. J., 3584
Robinson, M. C., 1438
Robinson, P. F., 4515
Robinson, Roy, 3108, 3121, 3122
Robinson, T. J., 3910
Rocha-Miranda, C. E., 4053a
Rochemont, Richard de. See De Rochemont
Rodahl, Kåre, 1806
Rodman, N. F., Jr., 0756
Roe, F. J. C., 4509
Rogers, A. L., 2808
Rohles, F. H., Jr., 0156a, 1915
Röhrborn, G., 1587

Röhrer, Heinz, 2788
Rokitskiĭ, P. F., 0849
Roland, C. G., 1048, 1055
Rolands, I. W., 0801
Romanoff, A. J., 1569, 1570
Romanoff, A. L., 1568—1570
Romer, A. S., 1439, 4159
Rood, J. J. van, 3062, 4467
Rook, A. J., 2040, 3255
Rooney, J. R., 1310, 2220, 3862
Roos, Charles, 0802
Root, W. S., 2573
Rooyen, C. E. van, 2787
Roscoe B. Jackson Memorial Laboratory.
  See Jackson Laboratory
Rose, A. H., 0781, 1948
Rose, G. G., 1278
Rose, S. M., 1571
Rosen, L., 0795
Rosen, Robert, 1016
Rosenberg, Eugene, 2482
Rosenberg, L. E., 4081
Rosenberg, Nancy, 2789
Rosenblum, L. A., 4451, 4481
Rosenfeld, Gastäo, 0191
Ross, C. F., 1312
Ross, G., 0788
Ross, K. F., 1166
Ross, M. H., 1655
Rossier, P. H., 2055
Rossoff, Irving, 2659
Roth, L. M., 2992
Rothblat, G. H., 0797, 1279
Rothfield, L. I., 2483
Rothman, Stephen, 2041
Rothstein, Howard, 1949
Rouiller, Charles, 1847, 1854
Round, M. C., 0744
Rouse, J. E., 3848
Rovee, D. T., 3645
Rowett, Helen G. Q., 4527
Rowlands, I. W., 0768, 0801
Rowlands, W. T., 3911
Roy, D. N., 2993
Roy, J. H. B., 3849
Royal Army Veterinary Corps, 0263
Royal Microscopical Society, 0464
Royal Society of London, 0018, 0019
Royal Society of Medicine, 0862
Royal Society of Tropical Medicine and Hygiene, 0515
Royer, Pierre, 1719
Rubin, B. A., 0771, 0778
Rubin, E. H., 3280
Rubin, Philip, 3489, 3689
Rubinstein, L. J., 3501
Ruch, T. C., 1950, 4462
Rudin, S. C., 1352
Rudman, Jack, 4261—4264
Ruffer, F. G., 0854

Rugh, Roberts, 1572—1575
Ruiz Torres, F., 0850
Rumbaugh, D. M., 4474
Runnells, R. A., 3388
Rupp, Charles, 2202
Ruščák, Michal, 2419
Ruščáková, Dagmar, 2419
Rushmer, R. F., 1838, 2498
Russell, Dorothy S., 3501
Russell, F. E., 0192, 2614
Russell, G. V., 3500
Russell, P. F., 2933
Russell, P. S., 3633
Russell, W. M. S., 4393
Rutstein, D. D., 2499
Rüttimann, A., 1889
Rycroft, P. V., 3623
Rytömaa, Tapio, 1582

S

Sack, W. O., 1311
Sadleir, R. M., 2293
Saez, F. A., 1656
Said, A. H., 3649
Sainsbury, D. B. W., 3801
Sainsbury, Peter, 3801
St. Pierre, R. L., 3051
Salankí, J., 4592
Salisbury, G. W., 2190
Salle, A. J., 2709
Salter, R. B., 3213
Salvaggio, J. E., 3462
Salzman, N. P., 2768
Samter, Max, 3073
Samuelsson, G., 0796
Sanborn, W. R., 0193
Sandborn, E. B., 1167
Sanders, Murray, 0763
Sandritter, Walter, 3458, 3459
San Pietro, A., 1660
Saphir, Otto, 1313, 3389
Sarin, L. K., 1394
Sarma, K. P., 3549
Sarner, Harvey, 3802
Saslow, C., 2294
Satchell, G. H., 4133
Sauer, R. M., 4463
Saunders, G. C., 3890
Saunders, J. T., 1492
Saunders, J. W., Jr., 1576, 1577
Saunders, L. Z., 3494
Saunders, Leonard, 2534
Saunders, P. R., 2614
Savitt, Sam, 3891
Sawin, C. T., 2124
Sawyer, C. H., 0788
Sawyer, D. C., 4337
Sax, N. I., 2595
Sayles, L. P., 4383

Scadding, J. G., 3390
Scanga, F., 1259
Scarff, R. W., 0761
Schadé, J. P., 1807
Schaeffer, Morris, 0763
Schaible, P. J., 4110
Schaller, G. B., 4452
Schalm, O. W., 2125, 3326, 3354, 3832
Scharffenberg, R. S., 0192
Scharrer, Berta, 0741
Scharrer, Ernst, 0741
Schebitz, H., 3690
Scheer, B. T., 1977, 2359
Schepartz, Bernard, 2326
Schermer, Siegmund, 4319
Scheuer, P. J., 3296
Schiff, Leon, 3297
Schindler, Rudolf, 1281, 3298
Schinz, H. R., 3691
Schipper, I. A., 4265
Schlieper, Carl, 4134
Schmid, Elisabeth, 1493
Schmid, F., 1661
Schmidt, A. J., 3650
Schmidt, G. H., 2191
Schmidt, J. A., 0194
Schmidt, J. E., 0822, 0823, 0910, 0911, 0980
Schmidt, Nathalie J., 2777
Schmutz, E. M., 2624
Schneirla, T. C., 2283
Schoenborn, H. W., 4320
Schön, M. A., 4433
Schorn, Julius, 3459
Schothorst, M. van, 2919
Schottelius, B. A., 1958
Schreiber, M. H., 3692
Schrier, A. M., 4453
Schroeder, W. A., 1885
Schubert, Maxwell, 2387
Schull, W. J., 3091
Schultz, A. H., 4475
Schultz, D. P., 2296
Schultz, D. R., 0770
Schwabe, C. W., 2891, 2892
Schwan, H. P., 2500
Schwarte, L. H., 4074
Schwartz, R. S., 0778
Schwartz, S. S., 3284
Schwartzman, R. M., 2816, 3511, 3512
Schwarze, E., 1495
Schwarzenberg, Leon, 3626
Schwentker, Victor, 0163a, 4528
Scomazzoni, Guiseppe, 3515
Scott, E. G., 1181
Scott, G. R., 3833
Scott, J. P., 2297
Scott, J. W., 3550
Scott, M. L., 4111
Scott, T. C., 4244
Scotti, T. M., 3362

Scowen, E. F., 3701
Scrimshaw, N. S., 0800
Sealander, J. A., 1441
Seamer, John, 4287
Searcy, R. L., 2388
Searle, A. G., 3109
Searles, Aysel, 1031
Sebeok, T. A., 2298, 2299
Sebrell, W. H., Jr., 2340
Seddon, H. R., 3738
Seegers, W. H., 2389
Seiden, R., 2660
Seitz, J. F., 2431
Sela, Michael, 3015
Selkurt, E. E., 1952
Selye, Hans, 1019, 1899, 2126, 3268, 3391
Semb, L. S., 2067
Seneviratna, P., 4097
Serth, G. W., 3893
Severi, Lucio, 3425, 3526
Shader, R. I., 2682, 2688
Shanks, S. C., 3693
Shantha, T. R., 4430, 4431, 4434
Shapton, D. A., 2698
Sharf, J. M., 1176
Sharma, K. N., 3955
Sharma, R. M., 3360
Sharp, J. T., 2737
Shaw, J. C., 1373
Shearman, R. P., 2127
Sheehy, T. W., 3299
Shefner, A. M., 4347
Shehadi, W. H., 3677
Shelesnyak, M. C., 2192
Shelley, W. B., 3252, 3256
Sherlock, Sheila, 3300
Sherman, W. B., 3039
Sherwood, Nancy, 4500
Shifrine, M., 0186
Shilling, C. W., 0095, 0916
Skhol'nik-ĪAarros, Ekaterina G., 1811
Short, D. J., 4406
Shtark, M. B., 2033
Shubin, Herbert, 3140
Shugar, D., 0752
Sidman, R. L., 3490, 4501, 4502
Siegel, P. D., 2385
Siegelman, S. S., 3280
Siegert, R., 2859
Siegler, P. E., 2671
Siegmund, O. H., 3150
Siekevitz, Philip, 1650
Sigel, M. M., 0785, 3044
Sigg, E. B., 2677
Sikov, M. R., 3667
Silvan, James, 4294
Silverberg, S. G., 3308
Silvers, Willys, 3054
Silverstein, A. M., 3478
Silverstein, Alvin, 1736
Silverstein, Virginia B., 1736

Silvestri, L. G., 3428
Simionescu, Nicoale, 3553
Simkiss, K., 2193
Simmonds, Sofia, 2332
Simmons, Arthur, 1137
Simmons, M. L., 4295, 4529
Simon, Noel, 4195
Simon, William, 2535
Simonton, Wesley, 0998
Simpson, D. C., 2502
Simpson, G. G., 0741, 4225
Simpson, L. L., 2596
Simpson, R. M., 4179
Sindermann, C. J., 4135
Singer, Ira, 3067
Singer, Marcus, 3980
Sinkovics, J. G., 3421
Sinning, W. E., 1995
Sinnott, E. W., 3100
Sircom, Geoffrey, 2268, 4210
Sisson, J. A., 3392, 3393
Sisson, S., 1496
Sjöstrand, F. S., 1168
Skerman, V. B. D., 0199, 2710
Skinner, Dorothy M., 4585
Skinner, F. A., 2698
Skinner, J. E., 2207
Slack, J. M., 2711, 2738
Slack, R., 2575
Slager, Ursula E. T., 3502
Slater, E. C., 0752
Slifkin, M., 0778
Sliosberg, A., 0834
Slonim, N. B., 1138
Sloss, Margaret W., 2957
Sluckin, Wladyslaw, 2300
Smialowski, Arthur, 1042
Smillie, I. S., 3214
Smith, C. A. B., 2536
Smith, D. A., 1354
Smith, D. D., 2301
Smith, D. E., 0756
Smith, D. T., 2717
Smith, Esther M., 4503
Smith, F. E., 4221
Smith, H. A., 3394
Smith, Harry, 0781
Smith, Heather, 4245
Smith, Ivor, 1139
Smith, K. M., 2790
Smith, Louis D. S., 2739
Smith, Martin, 3065
Smith, N. R., 2712
Smith, P. F., 2740
Smith, R. C., 0810, 4046
Smith, R. H., 4338
Smith, R. N., 3694
Smith, R. T., 0746, 3040, 3074
Smith, Rodney, 3583
Smithcors, J. F., 0981, 0982, 3813

Smithers, D. W., 3554
Smorto, M. P., 1391
Smyth, J. D., 0744, 2958
Smythe, R. H., 0983, 2034, 2302, 2303, 3585, 3894—3896, 3981
Snapp, R. R., 3847
Snapper, I., 3438
Snedecor, G. W., 2537
Snell, W. H., 0897
Snider, R. S., 0766, 4027, 4435
Snieszko, S., 4115, 4136
Snively, W. D., Jr., 3395
Soave, O. A., 3803
Sober, H. A., 2341
Società italiana delle scienze veterinarie, 0602
Society for Developmental Biology, 0710
Society for Endocrinology, 0711
Society for Experimental Biology and Medicine, 0265
Society for General Microbiology, 0781
Society for the Study of Fertility, 0768
Society for Veterinary Ethology, 2304
Society of American Bacteriologists, 1223
Society of Chemical Industry, 0004, 2597
Sodeman, W. A., 3396
Sodeman, W. A., Jr., 3396
Söderström, Nils, 3598
Sojka, W. J., 0742
Sokal, R. R., 2538
Sokoloff, Leon, 3215
Soltys, M. A., 2837
Soma, L. R., 4339
Sommer, Robert, 2305
Sommerville, J. C., 3491
Somogyi, J. C., 0736
Sophian, John, 3311
Sorkin, E., 3029
Soulsby, E. J. L., 0799, 2959, 2961
South Africa Veterinary Medical Association, 0266
Southwest Foundation for Research and Education, 0200
Southwick, C. H., 2306
Spector, W. G., 2576
Spector, W. S., 1078, 1087, 2598
Spedding, C. R. W., 3921
Spencer, E. S., 1140
Spencer, Frank, 2484
Spencer, Herbert, 3527
Spencer, Marjory C., 0203
Spiegel, E. A., 0764
Spiers, F. W., 3709
Spittell, J. A., Jr., 3257
Spotte, S. H., 4600
Sprent, J. F., 2960
Spriggs, A. I., 1662
Staats, Joan, 4504
Stableforth, A. W., 2845
Stacy, R. W., 2539

Stainsby, G., 2141
Stamey, T. A., 3312
Stamp, J. T., 3906
Stanbury, J. B., 2390
Stanley, W. M., 2752
Stansbury, R. L., 0794
Starck, D., 4475
Stark, Lawrence, 2487, 2503, 2510
Starling, E. H., 1955
Startup, F. G., 3982
Starzl, T. E., 0786, 3634
Stauffer, J. F., 1142
Stave, Uwe, 1988
Stebbins, W. C., 2307
Steel, K. J., 2722
Steele, J. H., 2888
Steen, E. B., 0863
Steen, J. B., 2056
Steffanides, G. F., 0873
Stein, G. N., 3535
Stein, J., 1661
Steinberg, Hannah, 2689
Steinhaus, E. A., 4590
Stellar, Eliot, 2246
Stent, G. S., 2791
Stephen, L. E., 0742
Stephens, Charlene B., 4521
Stern, J. T., Jr., 4436
Sterzl, J., 3028
Stevens, C. F., 2036
Stevens, G. W. W., 1169
Stevens, P. G., 3922
Stevenson, H. W., 2308
Stevenson, L. G., 0984
Stewart, G. T., 2893a
Stewart, H. L., 3439
Stibitz, G. R., 2540
Stich, Walther, 1884
Stiers, Georges, 0104
Stiles, C. W., 0007, 0008
Stiles, K. A., 1663
Stoker, M. G. P., 0781
Stoker, Michael, 0797
Stokes, A. W., 2309
Stokes, E. Joan, 2713
Stokoe, W. M., 1497
Stonehouse, Bernard, 4196
Storey, P. B., 4266
Storm, D. F., 3796
Storm, R. M., 0738, 2310
Storr, J. F., 1461
Storz, Johannes, 2742
Stowe, E. C., 1132
Stowell, R. E., 0756
Straatsma, B. R., 0788
Strande, Anders, 3983
Stratton, John, 3719
Straus, Reuben, 3262
Straus, W. L., 4421
Street, Phillip, 4197, 4212

Stroeva, O. G., 1548
Strömberg, Berndt, 3866
Strong, C. L., 3897
Strong, L. C., 4522
Stuart, A. E., 1900
Stuart-Harris, C. H., 2863
Stubbs, George, 3898
Studer, A., 3041
Stünzl, H., 3369
Sturkie, P. D., 4098
Sugano, Haruo, 0754
Sulman, F. G., 0772
Sulzberger, M. B., 3253
Summerhays, R. S., 3899
Sumner, K. C., 3804
Sunderland, Sydney, 2221
Sunderman, F. W., 3315, 3460
Sunderman, F. W., Jr., 3315, 3460
Sunshine, Irving, 2599
Supperer, R., 2925
Surgenor, D. M., 3330
Sussman, Maurice, 1581
Suter, P. F., 3956
Sutherland, N. S., 2262, 2311
Swain, R. H. A., 2723, 2792
Swan, H. J. C., 3513
Swansburg, R. C., 3164
Swanson, C. P., 1664
Swenson, M. J., 1963
Swett, W. W., 3912
Swiss Society of Microbiology, 0735
Sykes, George, 1224
Symington, Thomas, 3327
Symmers, W. St. C., 3400
Szabó, D., 1170
Szasz, Kathleen, 4247
Szebenyi, E. S., 4437
Szentagothai, J., 1427

T

Taber, C. W., 0825
Taboada, Oscar, 2995
Takahashi, Masayoshi, 3441
Taliaferro, Lucy G., 3668
Taliaferro, W. H., 3668
Täljedal, I. B., 0796
Talso, P. J., 3138
Tamm, Igor, 2854
Tansley, Katharine, 2038
Tavernor, W. D., 4349
Tavolga, W. N., 2312, 4139
Taylor, Angela R., 0739, 1094
Taylor, C. E., 0800
Taylor, J. A., 1499
Taylor, J. H., 3101
Taylor, K. E., 2897
Taylor, K. W., 3325
Taylor, Mary E., 4012
Taylor, N. B., 1912

Taylor, R. G. W., 1666
Taylor, R. M., 2798
Taylor, W. H., 2393
Tedeschi, C. G., 3503
Teir, Harald, 1582
Telford, I. R., 4378
Te Linde, R. W., 3638
Templeton, G. S., 4062
Tepperman, Jay, 2130
Texter, E. C., Jr., 2069
Thal, A. P., 3139
Theodor, Oskar, 2962
Thévenin, René, 2313
Thienes, C. H., 2600
Thimann, K. V., 2119
Thomas, A. K., 1189
Thomas, Carolyn, 3171
Thomas, J. A., 1285
Thomas, Picton, 2662
Thomas, R. C., 0854
Thomas Company, Arthur H. See Arthur H. Thomas Company
Thompson, H. V., 4573
Thompson, P. D., 2799
Thompson, R. B., 0786
Thompson, R. H. S., 2394
Thompson, S. W., 1261, 1667
Thomson, J. F., 3669
Thomson, R. G., 3835
Thomson, W., 0743
Thomson, W. A., 0826
Thorek, Philip, 3601
Thorndike, E. L., 2314
Thorne, Charles, 1056
Thornton, Horace, 2915
Thorpe, D. R., 1315
Thorpe, Sylvia A., 1805
Thorpe, W. H., 2315, 4593
Thorpe, W. V., 2342
Thorsby, E., 0786
Thorup, O. A., Jr., 3342
Thrasher, J. D., 1518
Threadgold, L. T., 1701
Threlfall, W., 0744
Tienhoven, Ari van. See Van Tienhoven, Ari
Tietz, N. W., 2343
Tietz, W. J., 1945
Timiras, P. S., 4500
Tinbergen, Niko, 2316—2318
Tindal, J. S., 0775
Titus, H. W., 4112
Tocantins, L. M., 1901
Todd, K. S., Jr., 2941
Toivanen, Auli, 0735
Toivanen, P., 0735
Tokunaga, Junichi, 1156
Tolansky, S., 1171
Tolman, E. C., 2319
Tombes, A. S., 4583

Tompsett, D. H., 1316
Toner, P. G., 1172, 1848
Top, F. H., 2838
Topley, W. W. C., 2714
Torrey, T. W., 1585, 4323
Traber, Y. A., 1173
Trainer, D. O., 4174
Transplantation Society, 3617
Trautman, J. C., 1999
Trautmann, Alfred, 1668
Travis, B. V., 2994
Trease, G. E., 2577a
Tregear, R. T., 2042
Treherne, J. E., 2039, 2984
Trentin, J. J., 3058
Trexler, P. C., 4292
Tricker, B. J. K., 1742
Tricker, R. A. R., 1742
Triggle, D. J., 2420
Trinca, J. S., 3739
Trippensee, R. E., 4198
Tritsch, G. L., 1669
Tromp, S. W., 2506
Troshin, A. S., 1640
Trotter, D. M., 3829
Trotter, W. R., 2120
Truant, J. P., 2692
Truex, R. C., 1817
Truhaut, René, 0767
Tsuboi, Eitaka, 3528
Tuchmann, Marcel, 3158
Tuchmann-Duplessis, H., 1549
Tucker, R. K., 2601
Tudhope, G. R., 2131
Tumbleson, M. E., 4568
Tunevall, Gösta, 0107
Turek, S. L., 3639
Turk, J. L., 3018
Turner, C. D., 2132
Turner, Helen, 3923
Turner, J. P., 1456
Tuttle, Russell, 4443
Tuttle, W. W., 1958
Tyler, Albert, 1651
Tyrrell, D. A. J., 0795
Tyson, Edward, 0985

U

Udall, D. H., 3740
Udenfriend, Sidney, 1356
Ugolev, A. M., 2070
Uhler, Katherine, 0203
Uhr, J. W., 3045
Ulmer, M. J., 4030
Umbreit, W. W., 1142
Umeda, Noritsugu, 3196
Underhill, R. A., 4162
Underwood, E. J., 2395
Union of International Associations, 0864

U.S. Agricultural Research Service, 0269, 0270, 1225
U.S. Animal Health Association, 0604
U.S. Armed Forces Institute of Pathology, 0202, 1262, 1317, 4375
U.S. Armed Forces Medical Library. See U.S. National Library of Medicine
U.S. Armed Services Technical Information Agency, 4298
U.S. Army Biological Warfare Laboratories, 0204
U.S. Atomic Energy Commission, 3348, 3667
U.S. Bureau of Animal Industry, 0007, 0008
U.S. Bureau of Medicine and Surgery, 2627
U.S. Center for Disease Control (formerly U.S. Communicable Disease Center), 0286, 1226, 2846, 2894, 4100
U.S. Chemical Research and Development Laboratories, 3913
U.S. Communicable Disease Center. See U.S. Center for Disease Control
U.S. Defense Documentation Center, 0205
U.S. Department of Agriculture, 0950, 2663, 2917, 4050, 4087
U.S. Department of Agriculture, Animal and Plant Health Inspection Service, 0206
U.S. Department of the Army, 1228
U.S. Department of the Interior, 0207, 0789, 4200, 4201
U.S. Division of Allied Health Manpower, 4270
U.S. Federal Advisory Council on Medical Training Aids, 1057
U.S. Federal Water Quality Administration, 2801
U.S. Livestock Sanitary Association, 0604
U.S. National Academy of Sciences, 4202, 4282, 4408
U.S. National Agricultural Library, 0115
U.S. National Audiovisual Center, 1058
U.S. National Cancer Institute, 0532, 0790, 1032, 2802, 3543, 4591
U.S. National Cancer Institute, Epizootiology Section, 0874
U.S. National Cancer Institute, Viral Oncology Program, 0999
U.S. National Council on Radiation Protection and Measurements, 3670
U.S. National Institute of Allergy and Infectious Diseases, 1287, 1288, 3046, 3084
U.S. National Institute of General Medical Sciences, 1143
U.S. National Institutes of Health, 0951, 1033–1035, 4394
U.S. National Library of Medicine, 0116–0118, 0135, 0203, 0209, 0811, 1059
U.S. National Museum, 4172
U.S. National Referral Center for Science and Technology, 1001–1003

U.S. National Research Council, Committee on Animal Health, 0986
U.S. National Research Council, Committee on Animal Nutrition, 0791
U.S. National Research Council, Committee on Education, 4409
U.S. National Research Council, Committee on Laboratory Animal Diseases, 4511
U.S. National Research Council, Committee on Pathology, 3445
U.S. National Research Council, Committee on Physiological Effects of Environmental Factors on Animals, 4379
U.S. National Research Council, Committee on Revision of the Guide for Laboratory Animal Facilities and Care, 4299
U.S. National Research Council, Committee on Trauma, 3651
U.S. National Research Council, Committee on Veterinary Medical Research and Education, 4272
U.S. National Research Council, Food Protection Committee, 1229
U.S. National Research Council, Institute of Laboratory Animal Resources, 0792, 0793, 4410, 4412
U.S. National Research Council, Institute of Laboratory Animal Resources, Committee on Technical Education, 4413
U.S. National Research Council, Institute of Laboratory Animal Resources, Subcommittee on Genetic Standards, 4396
U.S. National Research Council, Subcommittee on Avian Diseases, 4101
U.S. National Research Council, Subcommittee on Prenatal and Postnatal Mortality, 3851
U.S. National Science Foundation, 1036
U.S. Naval Medical School, 2995
U.S. Public Health Service, 0208, 0212, 0874, 0952, 1034, 1035, 1226, 1227, 1230, 2802
Universities Associated for Research and Education in Pathology, 3446
Universities Federation for Animal Welfare, 2673, 3788, 4300
University Microfilms, 0109
University of Maine at Orono, Department of Animal and Veterinary Sciences, 0210
Unseld, D. W., 0851
Ursprung, Heinrich, 1523
Uskavitch, Robert, 0108, 0186
Uslenghi, C., 3699
Utmann, J. E., 0780
Utz, J. P., 2812

V

Vago, C., 4594
Vagtborg, Harold, 4468, 4469

Vaillancourt, Pauline M., 0812
Valdes-Dapena, A. M., 3535
Valdés-Dapena, Maria A., 1670
Valentine, G. H., 3124
Valerio, D. A., 4484
Van Assendelft, O. W. See Assendelft, O. W. van
Van Bekkum, D. W. See Bekkum, D. W. van
Vanbreuseghem, R., 2877
Vandel, Albert, 4218
Van Demark, N. L., 0733, 3850
Van den Akker, L. M., 4103
Van der Hoeden, Jacob. See Hoden, Jacob van der
Van Dersal, W. R., 4203
Van Duijn, C. See Duijn, C. van
Van Duuren, B. L., 0778
Van Gelder, R. G., 1939, 4219
Vann, Edwin, 2541
Van Rooyen, C. E. See Rooyen, C. E. van
Van Schothorst, M. See Schothorst, M. van
Van Tienhoven, Ari, 4381
Van Winkle, Walton, Jr., 3648
Vaughan, J. T., 3836
Vaughan, Janet M., 2000
Vaughan-Jones, R., 0751
Veillon, Emmanuel, 0852
Veldman, F. J., 3729
Verhaart, W. J. C., 1501
Vernberg, F. J., 2320
Vernberg, Winona B., 2320
Verts, B. J., 4220
Vesey-FitzGerald, B. S., 4047
Vestal, Annie L., 1230
Veterinary Research Laboratory (Onderstepoort, South Africa), 3742
Vevers, Gwynne, 4145, 4209
Via, M. La. See La Via, M.
Viamonte, Manuel, 1887
Vida, J. A., 2578
Vileck, J., 0795
Villee, C. A., 1586, 2174, 4221
Villemin, M., 0853
Vincent, W. S., 0790
Vine, Louis L., 4002, 4003
Vinken, P. J., 3239
Visual Science Information Center, 0119
Voge, Marietta, 2949
Vogel, F., 4353
Völgyesi, F. A., 2321
Vollman, R., 2195
Von Brucke, F. Th. See Brucke, F. Th. von
Von Herrath, Ernst. See Herrath, Ernst von
Von Leden, Hans, 3588, 3589
Von Mayersbach, H. See Mayersbach, H. von
Vredenbregt, J., 2496
Vries, A. de. See De Vries, A.
Vries, Louis de. See De Vries, Louis
Vyvyan, John, 4390, 4391

W

Waddington, C. H., 0741
Wadsworth, Gladys E., 1996
Wagner, B. M., 0756
Wagner, H. N., 3700
Wahi, P. N., 0761
Wain, Harry, 0912
Wainio, W. W., 1671
Waksman, B. H., 3480
Waksman, S. A., 2821
Waldenström, Jan, 3448
Walford, A. J., 0813
Walford, R. L., 3047
Walker, D. F., 3836
Walker, E. P., 4226, 4227
Walker, W. F., Jr., 1502, 1503, 4221
Wallach, Jeffrey, 0878
Wallington, E. A., 1238
Walshe, F. M. R., 3241
Walter, H. E., 4383
Walter, J. B., 3397
Walter, Pat L., 1380
Walton, G. S., 2040
Walton, J. N., 2209, 3220
Wamberg, Kjeld, 0918
Ward, C. O., 2674
Wardener, H. E. de. See De Wardener, H. E.
Ware, Martin, 0122
Warkany, J., 4341
Warner, Nancy E., 3555
Warren, K. S., 0214
Warren, Katherine B., 3048
Warren, Shields, 3328
Wartenweiler, J., 2496
Wartman, W. B., 1280
Wasley, G. D., 1289
Waterman, A. J., 1504
Waterson, A. P., 2743, 2804
Watson, J. D., 2486
Watson, J. M., 0800
Watson, S. D., Jr., 4248
Watterson, R. L., 1589
Waxman, B. D., 2539
Way, E. L., 2567
Way, R. F., 3837, 3869
Waymouth, Charity, 0785, 1290
Weaver, A. D., 3819a
Weaver, C. S., 3603
Weber, G. I., 4255
Weber, R., 2396
Weber, T. B., 1340
Weck, A. L. de. See De Weck, A. L.
Weed, R. I., 1896
Weichert, C. K., 4326
Weigle, W. O., 0786, 3072
Weil, M. H., 3140
Weinman, David, 2839
Weir, D. M., 3019, 3086

Weiser, R. S., 3020
Weiss, D. W., 3449
Weiss, E., 3369
Weiss, K. E., 0795
Weiss, Leon, 3049
Weiss, P. A., 1591
Weitbrecht, Josias, 1745
Weller, J. M., 3475
Wells, B. B., 3476
Wells, P. N. T., 3197
Wells, T. A. G., 4063
Welsh, J. J., 1060
Welty, J. C., 4113
Wenner, H. A., 0795
Wenner-Gren Center Foundation, 0796
Wensing, C. J. G., 3817
Wepler, Wilhelm, 3536
West, Billy, 0215
West, E. S., 2345
West, G. P., 0815, 2863a, 3800
West, J. B., 2057
Western Society of Naturalists, 1978
Westhues, Melchior, 3567, 3611
Westphal, O., 2997
Westphal, U., 0772
Whalen, R. E., 0788
Whitaker, J. R., 1149
Whitby, L. E. H., 3356
White, Abraham, 2346
White, D. O., 1291, 2760
White, E. G., 3809
White, F. Dianne, 2801
White, M. J. D., 3102
White, P. R., 1292
White, R. G., 3011
White, Wilma L., 1145
Whitehead, G. K., 2321a
Whitehouse, R. H., 4055
Whitfield, I. C., 1358, 2507
Whitlock, J. H., 2963
Whitlock, Ralph, 3838
Whitmore, G. F., 3654
Whitmore, R. L., 2470
Whitney, R. A., 4470
Whitteridge, Gweneth, 0987
Whittow, G. C., 1979
Whitty, C. W. M., 3240
Widmann, Frances K., 3178
Wied, G. L., 2433
Wiener, A. S., 2152
Wiener, Norbert, 2508
Wiesinger, K., 2055
Wilcocks, Charles, 3137
Wildhirt, Egmont, 3536
Wildlife Disease Association, 0279
Wildy, Peter, 0771
Wilkens, H., 3865
Wilkie, D. R., 1746
Wilkie, David, 3103
Wilkinson, D. S., 3255

Wilkinson, G. T., 4032
Wilkinson, J., 2877
Wilkinson, J. S., 3473
Wilkinson, P. C., 3009
Willard, H. H., 1359
Willenegger, H., 3596
Williams, C. A., 2452
Williams, C. H., 0933
Williams, F. L., 2667
Williams, G., 3312a
Williams, L. F., Jr., 3284
Williams, Leonard, 4456
Williams, M. Ruth, 1147
Williams, N. E., 1083
Williams, R. C., 4355
Williams, R. H., 2133
Williams, R. J., 0919
Williams, Roger, 3065
Williams, V. D., 4242
Williams, W. J., 3338
Williams, W. L., 3313
Williamson, H. D., 3678, 3679
Williamson, J. W., 4266
Williamson, M. E., 1672
Williamson, Stanton, 0794
Willier, B. H., 1592
Willis, A. T., 2716, 2744, 3398, 3652
Willis, E. R., 2992
Willis, R. A., 1593, 2716, 3451
Willmer, E. N., 1293
Wills, M. R., 3314
Wilner, B. I., 2805
Wilson, David, 3021
Wilson, G. B., 1673
Wilson, G. S., 2714
Wilson, J. A., 1959
Wilson, J. G., 1595, 4341
Wilson, Kit, 4048
Wilson, Marjorie K., 2279
Wilt, F. H., 1294
Winburne, J. N., 0898
Winchell, Constance M., 0814
Windhager, E. E., 1360
Windle, W. F., 1674
Wineburgh, Margaret, 1381
Winer, B. J., 1022
Winge, Øjvind, 4004
Winkle, Walton van, Jr. See Van Winkle, Walton, Jr.
Winner, H. I., 2875
Winqvist, G., 0796
Winsor, Travis, 3259
Winstead, Martha, 1361
Winter, H., 0779, 3839, 3914
Winters, W. D., 4438
Winters, W. L., Jr., 1837
Wintrobe, M. M., 1148
Wirth, David, 3151
Wirtschafter, Z. T., 4506
Wischnitzer, Saul, 1174, 1321

Wiseman, R. F., 3900
Wismar, Beth L., 1675
Wistar Institute of Anatomy & Biology, 0797
Witcofski, R. L., 3666
Witherspoon, J. D., 4327
Witte, Siegfried, 3532
Witten, D. M., 3540
Wogan, G. N., 2815
Wojtczak, L., 0752
Wolf, George, 3711
Wolf, J. K., 1819
Wolff, Etienne, 1584
Wolff, Eugene, 1820
Wolff, H. S., 2509
Wolken, J. J., 4584
Wolman, Moshe, 3399
Wolstenholme, G. E. W., 1283, 1622, 2084, 2793, 2796, 3042
Wood, D. W., 1960
Wood, E. J. F., 4141
Woodbine, Malcolm, 0773
Woodburne, R. T., 1322
Woodford, F. P., 1041
Woodliff, H. J., 1295
Woodnott, Dorothy P., 4406
Woodruff, A. W., 2840, 3302
Woods Hole, Mass., Marine Biological Laboratory, 4142
Worden, A. N., 4573
Work, Kresten, 2964
World Association for Animal Production, 2398
World Association for Buiatrics, 0798
World Association for the Advancement of Veterinary Parasitology, 0799
World Association of Veterinary Food Hygienists, 0604a, 2919
World Health Organization, 0718, 0800, 1175, 2668, 2806
World Veterinary Association, 1061
Wostmann, B. S., 4278
Wrathall, A. E., 0742
Wright, A. H., 4173
Wright, Anna A., 4173
Wright, C. A., 2966
Wright, F. J., 3141
Wright, G. P., 3400
Wright, J. G., 3744
Wright, S. E., 3915
Wright, Samson, 1961
Wurtman, R. J., 1870
Wyatt, H. V., 0803
Wycis, H. T., 0764
Wyngaarden, J. B., 2390
Wynn, R. M., 1676
Wynne-Edwards, V. C., 2322

Y

Yahr, M. D., 1868

Yakovleva, L. A., 4461
Yamada, Hiroshi, 2511
Yamaguti, Satyu, 2967
Yamamoto, Tadashi, 0754
Yanof, H. M., 2512
Yapp, W. B., 1443, 1980
Yashon, David, 3495
Yasuda, Mineo, 4366
Yates, Frank, 2523
Yochim, J. M., 2134
Yoffey, J. M., 1905, 1906
York, C. J., 2848
Yoshida, Tomizo, 0754
Yoshikawa, T., 4328
Young, Clara G., 0899
Young, J. Z., 1822
Young, R. J., 4111
Young, Sydney, 3923
Young, W. C., 2196
Yu, P. N., 3267

Z

Zacarian, S. A., 3602
Zachar, Jozef, 1330, 1747

Zacks, S. I., 3504
Zajonc, R. B., 2323
Zaman, V., 2920
Zamboni, Luciano, 1703
Zarafonetis, C. J. D., 3340
Zarrow, M. X., 2134
Zelickson, A. S., 1832
Zeman, Wolfgang, 4487
Zemjanis, R., 3790
Zil'ber, L. A., 3452
Zimmer, Arthur, 1125
Zimmerman, H. M., 3499
Zimmerman, O. T., 1074
Ziswiler, Vinzenz, 4205
Zmijewski, C. M., 3051
Zollinger, R. M., 1062
Zoological Society of London, 0801
Zubin, Joseph, 2241
Zucker, M. H., 1362
Zuckerman, A. J., 3303
Zuckerman, Solly, 0876
Zugibe, F. T., 1264
Zülch, K. J., 3505
Zweifach, B. W., 3401
Zweig, Gunter, 1149, 2603

Conferences, congresses, symposia, and other meetings the proceedings of which are cited individually as noted

Brook Lodge Conference on Inflammation, 3641
Colloquium on Experimental Embryology, 4588
Colloquium on Veterinary Education in Radiology, Radiobiology, and Radioisotope Utilization, 0783
Conference and Workshop on Histocompatibility Testing, 0591
Conference on Atypical Virus Infections, 2850a
Conference on Biologic Aspects and Clinical Uses of Immunoglobulins, 2439
Conference on Biology of Hard Tissue, 0592
Conference on Comparative Mammalian Cytogenetics, 1624
Conference on Experimental Medicine and Surgery in Primates, 0745
Conference on Female Reproduction in Farm Animals, 0733
Conference on Fetal Homeostasis, 0593
Conference on Genetics, 3091
Conference on Murine Leukemia, 0790
Conference on Parturient Paresis in Dairy Animals, 3317
Conference on the Morphology of Experimental Respiratory Carcinogenesis, 4356

Conference on the Pathology of Laboratory Animals, 4345
Conference on the Secretory Immunologic System, 3027
Conference on Veterinary Education, 4252
European Symposium on Calcified Tissues, 0750
International Cancer Congress, 0767, 3423
International Conference on Tissue Culture in Cancer Research, 3424
International Conference on Biological Membranes, 2478
International Conference on Equine Infectious Diseases, 0758
International Conference on Foot and Mouth Disease, 2856
International Conference on Leukemia-Lymphoma, 3560
International Conference on Lung Tumours in Animals, 3526
International Conference on Tetanus, 2724
International Conference on the Morphological Precursors of Cancer, 3425
International Congress for Microbiology, 0595

International Congress of Chemotherapy, 0759
International Congress of Endocrinology, 0751
International Congress of Entomology, 2039
International Congress of Histochemistry and Cytochemistry, 2426
International Congress of Immunology, 3012
International Congress of Lymphology, 1887
International Congress of Neuropathology, 0597
International Congress of Primatology, 0760, 4476
International Congress of Ultrasonography in Ophthalmology, 3227
International Congress on Animal Reproduction and Artificial Insemination, 0598
International Congress on Hormonal Steroids, 2103
International Congress on Muscle Diseases, 0751
International Convocation on Immunology, 0662
International Corneo-Plastic Conference, 3623
International Neurochemical Conference, 2409
International Neurochemical Symposium, 2410, 2418
International Nuclear Medicine Symposium, 3704
International Research Conference, 2202
International Seminar on Biomechanics, 2496
International Seminar on Reproductive Physiology and Sexual Endocrinology, 2102
International Symposium on Adjuvants of Immunity, 0782
International Symposium on Animal and Plant Toxins, 2613
International Symposium on Animal Toxins, 2614
International Symposium on Antilymphocyte Serum, 0782
International Symposium on BCG Vaccine, 0782
International Symposium on Biological Assay Methods, 0782
International Symposium on Biotechnical Developments in Bacterial Vaccine Production, 0782
International Symposium on Blood and Tissue Antigens, 3031
International Symposium on Breeding Nonhuman Primates for Laboratory Use, 4425
International Symposium on Brucellosis, 0782
International Symposium on Canine Heartworm Disease, 3960
International Symposium on Cetacean Research, 4125

International Symposium on Combined Vaccines, 0782
International Symposium on Continuous Cultivation of Microorganisms, 1208
International Symposium on Cytoecology, 1640
International Symposium on Diseases in Zoo Animals, 4178
International Symposium on Enterobacterial Vaccines, 0782
International Symposium on Foot-and-Mouth Disease, 0782
International Symposium on Freeze-Drying, 1339
International Symposium on Germfree Life Research, 4286
International Symposium on Growth Hormone, 0751
International Symposium on Health Aspects of the International Movement of Animals, 3723
International Symposium on Immunological Methods of Biological Standardization, 0782
International Symposium on Immunopathology, 0762
International Symposium on Infections and Immunosuppression in Subhuman Primates, 4460
International Symposium on Inflammation, Biochemistry and Drug Interaction, 0751
International Symposium on Interferon and Interferon Inducers, 0782
International Symposium on Interferons, 2772
International Symposium on Laboratory Animals, 0782
International Symposium on Lymphology, 1889
International Symposium on Medical and Applied Virology, 0763
International Symposium on Mycotoxins in Foodstuffs, 2815
International Symposium on Natural Mammalian Hibernation, 2370
International Symposium on Neurovirulence of Viral Vaccines, 0782
International Symposium on Pertussis, 0782
International Symposium on Pseudotuberculosis, 0782
International Symposium on Rabies, 0782
International Symposium on Rubella Vaccines, 0782
International Symposium on Selenium in Biomedicine, 2648
International Symposium on Steroencephalotomy, 0764
International Symposium on the Baboon and Its Use as an Experimental Animal, 4468
International Symposium on the Newer Trace Elements in Nutrition, 2371

International Symposium on the Nucleolus: Its Structure and Function, 0790
International Symposium on the Physiology of Digestion in the Ruminant, 0765
International Symposium on the Response of the Nervous System to Ionizing Radiation, 0766
International Symposium on the Skin Senses, 1778
International Symposium on Vectorcardiography, 1370
International Workshop in Teratology, 1539
Lepetit Colloquia on Biology and Medicine, 2778
Leucocyte Culture Conference, 0601
Lister Centenary Scientific Meeting, 3643
National Biomedical Sciences Instrumentation Symposium, 1349
National Conference on Histoplasmosis, 2874
National Conference on Research Animals in Medicine, 4368
National Rabies Symposium, 2860
Panel on Current Problems of Bone-marrow Cell Transplantation with Special Emphasis on Conservation and Culture, 1275
Panel on Experimental Animals in Cancer Research, 4369
Panel on Radiation and the Control of Immune Response, 3037
Panel on Radiation Sterilization of Biological Tissues for Transplantation, 3662
Panel on the Effects of Various Types of Ionizing Radiations from Different Sources of Haematopoietic Tissue, 3663
Panel on the Radiation Sensitivity of Toxins and Animal Poisons, 2620
Panel on the Use of Isotopes in Studies of Mineral Metabolism and Disease in Domestic Animals, 3706
Panel on the Use of Nuclear Techniques in Studies of Mineral Metabolism and Disease in Domestic Animals, 3707
Rock Island Arsenal Biomechanics Symposium, 1735
Schering Symposium on Intrinsic and Extrinsic Factors in Early Mammalian Development, 1578
School on Neural Networks, 1808
Seminar-Workshop in Biomaterials, 2501
Seminars in Arthritis and Rheumatism, 3064
Seminars in Hematology, 1896
Sigrid Jusélius Symposium, 1582, 3034
Sloan Symposium on Uveitis, 3235

Study Group on Hormones and the Immune Response, 3042
Symposia Series in Immunobiological Standardization, 0782
Symposium—Biology of Amphibian Tumors, 4161
Symposium on Aging in Cell and Tissue Culture, 1282
Symposium on Anticholinergic Drugs and Brain Functions in Animals and Man, 2690
Symposium on Applied Virology, 0763
Symposium on Avian Leukosis, 4099
Symposium on Biomechanics and Related Bio-engineering Topics, 2504
Symposium on Biophysics and Physiology of Biological Transport, 2468
Symposium on Body Composition in Animals and Man, 1498
Symposium on Candida Infections, 2875
Symposium on Cellular Biology of Myxovirus Infections, 2793
Symposium on Clinical Education: Professional, Graduate and Continuing, 0783
Symposium on Coccidioidomycosis, 2876
Symposium on Cognition, 4454
Symposium on Comparative Bioelectrogenesis, 2469
Symposium on Comparative Genetics in Primates and Human Heredity, 3104
Symposium on Comparative Pathophysiology of Circulatory Disturbances, 4322
Symposium on Cross-Disciplinary Sciences in Biomedical Research, 1020
Symposium on Cross-Species Transplantation, 3636
Symposium on Diseases and Husbandry of Aquatic Mammals, 4138
Symposium on Diseases of Fishes and Shellfishes, 4115
Symposium on Education in Veterinary Public Health and Preventive Medicine, 4268
Symposium on Emergencies in Veterinary Practice, 0794
Symposium on Environmental Requirements for Laboratory Animals, 4296
Symposium on Equine Bone and Joint Diseases, 3217
Symposium on Feeding and Nutrition of Nonhuman Primates, 4483
Symposium on Feline Medicine, 0794
Symposium on Fundamental Cancer Research, 0714
Symposium on Gastrointestinal Medicine and Surgery, 0794
Symposium on Graduate Education in Veterinary Medicine, 0783
Symposium on Growth Control in Cell Cultures, 1283
Symposium on Interferon, 2794

Symposium on Lactogenesis, 2129
Symposium on Marine Bio-acoustics, 4139
Symposium on Medical Applications of Plastics, 1355
Symposium on Myotatic, Kinesthetic and Vestibular Mechanisms, 1998
Symposium on Neoplasms and Related Disorders of Invertebrate and Lower Vertebrate Animals, 4591
Symposium on Neurobiology of Invertebrates, 4592
Symposium on Oncogenicity of Virus Vaccines, 2795
Symposium on Orthopedic Surgery in Small Animals, 0794
Symposium on Physical Diagnosis in Small Animals, 0794
Symposium on Plastics in Surgical Implants, 3600
Symposium on Practice Management and Hospital Design, 0794
Symposium on Preprofessional Veterinary Education—Student Selection and Curriculum, 0783
Symposium on Rhythmic Research, 1665
Symposium on Strategy of the Viral Genome, 2796
Symposium on the Biochemistry of Simple Neuronal Models, 4376
Symposium on the Effects of Environment, 4400
Symposium on the Evaluation of Anthelminthics, 0799
Symposium on the Foundations and Objectives of Biomechanics, 1740
Symposium on the Future of the Defined Laboratory Animal, 4400
Symposium on the Interaction of Drugs and Subcellular Components in Animal Cells, 2577
Symposium on the Interaction of the Laboratory Animal with Its Associated Organisms, 4400
Symposium on the Molecular Biology of Viruses, 2797
Symposium on the Physiology and Biochemistry of Muscle as a Food, 1999
Symposium on the Use of Fish as an Experimental Animal in Basic Research, 4567
Symposium on the Use of Radioisotopes in Animal Nutrition and Physiology, 1957
Symposium on the Use of Subhuman Primates in Drug Evaluation, 2672
Symposium on Ultrasound in Biology and Medicine, 1398
Symposium on Uniformity, 4400
Symposium on Veterinary Medical Education, 0783
Symposium: Sulfur in Nutrition, 2392
Technical Conference on Artificial Insemination and Reproduction, 0603
Texas University Annual Clinical Conference on Cancer, 0784
Wenner-Gren Conference on the Systematics of Old World Monkeys, 4455
Wesley W. Spink Lectures on Comparative Medicine, 4256
Working Conference on Rabies, 2865
Workshop in Animal Technician Training, 0792
Workshop in Teratology, 1595
Workshop on the Status of Research and Training in Biomaterials, 2510
World Conference on Animal Production, 3789
World Congress on Animal Feeding, 3773
World Congress on Fertility and Sterility, 0605
World's Poultry Congress, 0607, 2866
World Veterinary Congress, 0606, 0988

SUBJECT INDEX KEY

The Subject Index Key contains the inclusive numbers of the entries within a given subject division of the text, using headings drawn from the Table of Contents. This key may be used to advantage in conjunction with the Subject Index in order to effect more complete access to materials included under form designation (e.g., Serial Monographs or Handbooks) and subject discipline (e.g., Pathology or Animal Management).

0001-0086	Indexes and Abstracts	1382-1391	Electromyography and Electroneurography
0087-0094	Current Awareness Services		
0095-0122	Bibliographies of Periodicals	1392-1399	Ultrasonics
0123-0137	Bibliographies of Conferences and Congresses	1400-1506	Anatomy, General and Comparative
0138-0144	Translation Sources	1507-1596	Anatomy, Developmental
0145-0216	Specialized Subject Bibliographies	1597-1676	Histology and Cytology
		1677-1704	Ultrastructure
0217-0607	Periodicals	1705-1906	Anatomy by Systems
0608-0730	Review Serials	1907-1980	Physiology, General and Comparative
0731-0801	Serial Monographs		
0802-0814	Guides to the Literature	1981-1989	Physiology, Developmental
0815-0853, 0877-0899, 0913-0919	Dictionaries and Encyclopedias	1990-2154	Physiology by Systems
		2155-2197	Physiology, Reproductive
		2198-2221	Neurology, General and Clinical
0854-0864	Abbreviations and Synonyms	2222-2323	Behavior
0865-0876	Nomenclature	2324-2346	Biochemistry, General
0900-0912	Medical Terminology and Word Structure	2347-2398	Physiological Chemistry
		2399-2420	Neurochemistry
0920-0957	Biographical Tools and Directories	2421-2434	Histochemistry
		2435-2452	Immunochemistry
0958-0988	Histories and Sourcebooks	2453-2470	Biophysics
0989-1004	Libraries and/or Information Centers and Their Services	2471-2486	Molecular Biology
		2487-2513	Bioengineering
1005-1022	Research Methodology	2514-2541	Biomathematics
1023-1036	Research Support	2542-2579	Pharmacology
1037-1062	Medical Writing, Speaking, Illustration, and Audiovisual Materials	2580-2603	Toxicology
		2604-2627	Poisonous Plants and Animals
		2628-2668	Therapeutics
1063-1074	Industrial Information	2669-2674	Drug Testing
1075-1087	Biomedical Handbooks	2675-2690	Psychopharmacology
1088-1149	Clinical Laboratory Methods	2691-2717	Microbiology, General
1150-1174	Microscopy and Microphotography	2718-2745	Medical Bacteriology
		2746-2807	Virology
1175-1231	Microbiological and Parasitological Cultures, Techniques, and Classification	2808-2821	Mycology
		2822-2895	Infectious Diseases
		2896-2919	Epidemiology and Public Health
1232-1264	Histological and Histochemical Technique, Stains and Staining	2920-2982	Parasitology and Parasitic Diseases
		2983-2995	Medical Entomology
1265-1295	Cell and Tissue Culture	2996-3087	Immunology
1296-1322	Dissection, Autopsy, Necropsy, and Specimen Preparation	3088-3125	Genetics
		3126-3141	Internal Medicine, General
1323-1362	Instrumentation, General	3142-3197	Internal Medicine, Diagnosis
1363-1372	Electrocardiology	3198-3359	Internal Medicine, Diseases by Systems
1373-1381	Electroencephalography		

Subject Index Key

3360-3401	Pathology, General
3402-3452	Neoplasia, General
3453-3459	Histopathology
3460-3476	Pathology, Clinical
3477-3480	Immunopathology
3481-3564	Pathology and Neoplasia by Systems
3565-3602	Surgery and Obstetrics, General
3603-3612	Anesthesia
3613-3640	Surgery, Special
3641-3652	Wound Healing
3653-3694	Radiology
3695-3711	Radioisotopes and Radiotherapy
3712-3744	Veterinary Medicine and Surgery
3745-3774	Animal Nutrition
3775-3790	Animal Production and Management
3791-3809	Veterinary Medical Practice
3810-3851	Cattle
3852-3900	Horses
3901-3923	Sheep and Goats
3924-3944	Swine
3945-4004	Dogs
4005-4048	Cats
4049-4063	Furbearing Animals
4064-4113	Birds and Poultry
4114-4151	Fish and Aquatic Mammals
4152-4173	Reptiles and Amphibians
4174-4212	Wildlife and Zoo Animals
4213-4227	Zoology
4228-4248	Companion Animals
4249-4274	Veterinary Medicine, Education and Training
4275-4349, 4392-4414	Laboratory Animals, General
4350-4384	Animal Models for Research
4385-4391	Vivesection
4415-4484	Laboratory Animals, Primates
4485-4529	Laboratory Animals, Rodents
4530-4574	Laboratory Animals, Vertebrates
4575-4600	Laboratory Animals, Invertebrates

SUBJECT INDEX

The main subject headings, subheadings, and cross-references used in this index conform as far as possible with the National Library of Medicine, *Medical subject headings* 1973 (MESH). Because of the emphasis of the text, it has not been possible to use MESH headings exclusively.

abortion
   livestock, 2165
actinomyces, 2821, 2869
adaptation. See also behavior, spatial
   animals, domestic, 2264
   primates, 2271
adaptation, physiological, 2004
adenoviruses
   serial monographs, 0778
adipose tissue
   handbooks, 1928
   metabolism, 2078
adrenal glands, 1861, 1864, 2086. See also endocrine glands
   diseases, 3327
   physiology, 2094
   serial monographs, 0756
   surgery, 3628
aflatoxins. See mycotoxins
African horse sickness
   bibliographies, 0186
African swine fever
   bibliographies, 0186
   serial monographs, 0795
aggression, 2280, 2306
   serial monographs, 0788
aggression, comparative, 2273
aging. See geriatrics
agriculture
   bibliographies, 0097
   dictionaries, 0898
   handbooks, 0950
   indexes and abstracts, 0046, 0047, 0089
   serial monographs, 0773
   statistics, 2523
air microbiology, 2880
   serial monographs, 0781
air pollution. See also environmental health
   animals, domestic, 2592
alimentary tract. See gastrointestinal system
allergy, 3024, 3057. See also hypersensitivity
   guinea pigs, 3068
   indexes and abstracts, 0078
   periodicals, 0287, 0407
   review serials, 0692, 0703

allergy (cont.)
   serial monographs, 0770
amines, 2111, 2112a
amino acids, 3757
amphibia, 4170, 4171
   anatomy, 4162, 4531
   biology, developmental, 1525
   biology, molecular, 4153
   brain, atlases, 1783
   chromosomes, 4067
   classification, 4172
   embryology, 1589
   neoplasms, 0780
analgesia
   veterinary, 3612
anatomy. See also dissection; neuroanatomy
   amphibia, 4162, 4531
   animals, domestic, 1454, 1457-1459, 1485, 1489, 1490, 1496, 1499
      musculoskeletal system, 1479
   apes, 4474
   art, 1043, 1049, 1054, 1401, 1456
   atlases, 1405, 1410, 1413, 1416-1418, 1427, 1436, 1456, 1458, 1489
   biliary tract, 1844
   birds, 1412
      respiratory system, 4075
   bladder, 1852
   blood cells, 1876
   bone, 1716
   cats, 1404, 1410, 1413, 1417, 1437, 1502
      atlases, 4030
   cattle, 1411, 1429, 1744, 3834
   chickens, 1438
   chimpanzees, 2233
   Coturnix quail, 1414
   dictionaries, 0879, 0880
   dogs, 1405, 1432
      radiography, 1440
   erythrocytes, 1872
   eye, 1490, 1764, 1768, 1820
   fish and aquatic animals, 4114
   gastrointestinal system, 1475, 1843, 1846, 3287
   goats, 3834
   gonads, 1856
   handbooks, 0973
   horses, 1401, 1442, 3894, 3898

Subject Index

anatomy: horses (cont.)
    radiography, 1440
  human, 1400, 1415, 1418, 1419, 1427,
      1434, 1436, 1445
  indexes and abstracts, 0032
  kidney, 1851
  liver, 1845
  lung, 1840
    radiography, 1839a
  mammals, 1465
  manuals, 1299, 1316, 1424
  medicine, clinical, 1402, 1406
  mice, 1408
  microcirculation, 1836
  minks, 4052a
  monkeys, 1421
  nomenclature, 0869, 1426
  periodicals, 0288, 0293, 0303, 0316,
      0420
  pigeons, 4071
  primates, 1422, 4476
  rabbits, 1403, 1409, 1428
  rats, 1407, 1420, 4499
  retina, 1805
  sheep, 1430
    respiratory system, 1445
  snakes, 4156
  teeth, 1829
  testis, 1853
  urethra, 1852
  vertebrates, 1439, 1441, 1443, 1456,
      1464, 1473, 1504, 1505
    heart, 1834
  veterinary, 1416, 1466, 1495
anatomy, comparative, 0962, 1444, 1448,
    1468, 1480, 1497
  animals, domestic, 1451
  atlases, 1461, 1493, 1506
  cerebellum, 1476
  eye, 1491
  history, 0964, 1500
  invertebrates, 1450
  mammals, 1449, 1477
  manuals, 1481
  nervous system, 1501
  vertebrates, 1447, 1455, 1462, 1474, 1478
    nervous system, 1446
anatomy, developmental, 1509. See also
    growth
  eye, 1548
  geriatrics, 1508, 1515
  veterinary, 1484
anatomy, pathological, 1433, 3724
anatomy, surgical, 3580
  cats, 1488
  dogs, 1488
  human, 1423
androgens. See hormones
anesthesia
  animals, laboratory, 3604

anesthesia (cont.)
  electronics, medical, 3607
  periodicals, 0317, 0341
  review serials, 0644, 0720
  veterinary, 3610, 3611, 3612, 4335, 4337
anesthesia, electrical, 4338
anesthesia, general, 3606
anesthesia, local
  pharmacology, 3605
  physiology, 3605
anesthesiology, 3609
  bibliographies, 0042
  indexes and abstracts, 0032
  instrumentation, 1911
  respiratory system, 3608
angiocardiography, 3517. See also heart
animal behavior. See behavior
animal diseases. See also medicine, vet-
    erinary; names of specific diseases
  nomenclature, 0867, 0874
animal hospitals. See hospitals, veterinary
animal husbandry. See husbandry
animal models, 4364
  arthritis and rheumatism, 2850a
  baboons, 4468
  bibliographies, 0197
  birds, vision, 2005
  brain, 1822
  dogs, 3949
  fish and aquatic mammals, 4128, 4140
  gestation, 4357
  goats, 3913
  guinea pigs, 0770
  indexes and abstracts, 1018
  information retrieval, 0989
  invertebrates, 4587
  mice
    geriatrics, 4513, 4522
    neoplasia, 4522
  neurochemistry, 4376
  pharmacology, experimental, 2670, 2671
  primates, 4377, 4425
  rats, 4512
    geriatrics, 4513
  research, biomedical, 1018, 1021, 4548
  swine, 3934-3937
animal production. See also husbandry
  congresses, 3787
  periodicals, 0318
  research, 3775
  serial monographs, 0747
animals, domestic. See also livestock;
    ruminants; vertebrates
  adaptation, 2264
  air pollution, 2592
  anatomy, 1454, 1457-1459, 1485, 1489,
      1490, 1496, 1499
    musculoskeletal system, 1479
  anatomy, comparative, 1451
  bacterial diseases, 3738

animals, domestic (cont.)
  behavior, 2265
  biochemistry, clinical, 2372
  brain, atlases, 1821
  coccidiosis, 2941, 2981
  ecology, 2293
  eye, 1490
  feeds and feeding, 3762
  fertility, 2179
  genetics, 3117
  genitourinary diseases, 3313
  histology, 1490, 1668
  infectious diseases, 2830, 2831, 2835
  insemination, 2188
  mammary glands, diseases, 3718
  minerals, 2383
  nematoda, 2945
  neoplasms, 3410, 3430
  nutrition, 3762
    comparative, 2381
  ophthalmology, 3228
  origin, 3777a
  parasitic diseases, 2975, 3738
  parasitology, 2924
  pathology, 1486
  physiology, 1459, 1919, 1963
  protozoology, 2946
  radioisotopes, 2383, 3706
  reproduction, 2164
  reproduction, comparative
    physiology, 2184
  teratology, nervous system, 1542
  trace elements, metabolism, 3706
  viral diseases, 3738
animals, exotic. See exotic animals
animals, heterothermic
  physiology, 1978
animals, laboratory, 1011, 4271, 4289, 4390,
    4391, 4939. See also opossums; primates; rabbits; reptiles; rodents
  anesthesia, 3604
  bibliographies, 0149, 0156, 0165, 0189,
    0209
  blood cells, morphology, 1898
  breeds and breeding, 4276, 4281, 4284,
    4288, 4293
  care and management, 4275-4277, 4281,
    4283, 4287, 4288, 4294, 4295, 4299,
    4300, 4388, 4393-4395, 4406, 4412
  catalogs, 4290, 4291
  classification, 4410
  conferences, 4400
  diseases, 0757, 4297
  environment, 4400
  environmental health, 4296
  genetics, 3108, 3114, 4396, 4412
  geriatrics, 1545
  husbandry, 4280
  immunology, 0782
  indexes and abstracts, 0025

animals, laboratory (cont.)
  legislation, 4385
  medicine, nuclear, 4371
  neoplasia, 4369
  neoplasms, 0754, 3437
  nutrient requirements, 0791
  nutrition, 4279, 4297
  parasitic diseases, 4346
  pathology, 4344, 4345, 4509
  periodicals, 0251, 0568a
  radiation effects, 4347
  radiography, 3676
  reproduction, 4284
  research, 4277, 4386, 4408
    biological, 4389
    biomedical, 4368
  serial monographs, 0735, 0792
  space medicine, 4465
  standards, 4395, 4396, 4412
  teratology, 1539
    nervous system, 1542
  therapeutics, 2630
animals, poisonous, 2604, 2605, 2609, 2610,
    2627. See also snakes; venoms
animals, tropical. See tropical animals
anthelminthics
  serial monographs, 0799
anthropology, 2289
  indexes and abstracts, 0032
antibiotics, 2642, 2646
antibodies, 2452, 3021, 3026, 3028, 3034.
    See also immunoglobulins
  biochemistry, 2447
  biosynthesis, 2443
  immunochemistry, 2443
anticoagulants. See also blood coagulation
  rabbits, 2645
antigens, 2452, 3014, 3031
  neoplasms, 0767
  periodicals, 0527
  serial monographs, 0780, 0786
  toxoids, 2657
antiserums, 2657
apes, 4441. See also primates
  anatomy, 4474
  behavior, 4445, 4474
  cerebellum, 1787
  classification, 4474
  ecology, 4474, 4480
  evolution, 4474
  genetics, comparative, 3104
  history, 4480
apparatus. See equipment
aquatic mammals. See fish and aquatic
    mammals
art. See also photography
  anatomy, 1043, 1049, 1054, 1401
  biology, 1050
  cardiovascular system, 1053
  endocrine glands, 1053

art (cont.)
    gastrointestinal system, 1053
    genitourinary system, 1053
    nervous system, 1053
arteries. See blood vessels
arteriosclerosis, comparative, 3262
arthritis and rheumatism, 3207
    animal models, 2850a
    bibliographies, 0209
    indexes and abstracts, 0002, 0032, 0065
    periodicals, 0327
    viral diseases, 2850a
arthropods, 2987. See also insects
ASCA Topics, 0991
astronauts, animal. See space medicine
audiovisual aids
    directories, 1037
    film catalogs, 1057, 1058, 1061
    indexes and abstracts, 1059
Aujeszky's disease (pseudorabies), 0795
autoimmune diseases, 2736
automation
    bibliographies, 0209
autopsy. See necropsy

B

baboons. See also primates
    animal models, 4468
    bibliographies, 4482
    brain, atlases, 1762
    conferences, 4468
    ecology, 0737
    embryology, 1537
bacteria. See also enterobacteriaceae;
        Escherichia coli; salmonella
    classification, 1180, 1207, 1221-1223,
        2704
    manuals, 2722
    symbiosis, 2693
bacteria, pathogenic, 2719
    animals, 2837
    human, 2837
bacterial diseases
    animals, domestic, 3738
    birds, 4101
    fish and aquatic mammals, 4136
    human, 2842
    manuals, 1181, 1199, 1214
bacteriology, 2709, 2714, 2716. See also
        cultures; microbiology; stains and
        staining
    indexes and abstracts, 0032, 0085
    manuals, 1117, 1132, 1184, 1186, 1214,
        1215, 1223, 1230, 2699
    periodicals, 0406, 0421, 0423
    review serials, 0640
    veterinary, 1186, 1215, 2695, 2706
bacteriology, anaerobic
    pathogenic, 2739, 2744

bacteriology, clinical, 2713
bacteriology, experimental, 2707
bacteriophage, biochemistry, 2705
bats. See also biospeleology
    bibliographies, 0209
BCG vaccination
    serial monographs, 0782
bedsonial diseases, 2742, 2822
behavior, 2035, 2225, 2244, 2317. See also
        aggression; communication: animals;
        instinct; intelligence; psychology;
        psychopathology; stress; telemetry
    animals, 2035, 2222, 2224, 2234, 2243,
        2246, 2247, 2263, 2266, 2272, 2278,
        2285, 2302, 2304, 2309, 2312, 2316,
        2322
        abnormal, 2257, 2276
        drugs, 2689
        feedback, 2282
        neurology, 2270
    animals, domestic, 2265
    apes, 4445, 4474
    bibliographies, 0150, 0167, 0209
    biology, 2250
    cats, 4038, 4041
    chimpanzees, 0776
    dogs, 2254-2256, 2279, 2291
    evolution, 2275
    fish and aquatic mammals, 1929, 2239,
        4117
    gorillas, 4194
    horses, 3896
    invertebrates, 4586
    mammals, 2253, 2301
        maternal, 2292
    monkeys, 4445, 4455
    periodicals, 0225, 0232, 0329, 0367, 0401,
        0434, 0504, 0508
    primates, 0150, 0691, 0760, 2238, 2248,
        2295, 4447, 4451, 4476, 4478
        evolution, 2274
    rats, 2229, 2237
    reproduction, 2182, 2293
        animals, 2259
    review serials, 0631, 0634, 0691, 0699
    senses, 2296
    serial monographs, 0738, 0764, 0768,
        0776, 0779, 0788
    sheep, 2261
    vertebrates, 2230, 2287
    wildlife, 2254
    wolves, 2254, 2286
behavior, comparative, 2242, 2242a, 2251,
        2260, 2267, 2281, 2284, 2288, 2319,
        4445, 4450, 4456
    neonatal, 2308
behavior, social, 2224, 2226, 2227, 2247,
        2251
    dogs, 2297
    genetics, 2232, 2297

behavior, social (cont.)
    instinct, 2232
    monkeys, 2240
behavior, spatial, 2227, 2305
    poultry, 2259a
    zoos, 2269
bile
    metabolism, 2063
bilharziasis. See schistosomiasis
biliary tract, 1844
biliary tract diseases, 3300
biochemistry, 2326, 2331, 2332, 2334, 2336, 2339, 2342, 2345, 2346, 2349, 2356, 2365, 2366, 2368, 2385. See also cytology, biochemical; embryology, biochemical; psychopharmacology, biochemical; research, biochemical
    antibodies, 2447
    bacteriophage, 2705
    bibliographies, 0207
    birds, 0789
        embryology, 1570
    blood
        cats, 2353
        dogs, 2353
    blood cells, 2149
    bone, 1719, 1990
    bone marrow, 2149
    connective tissue, 1604, 2387, 2435
    diseases, 2394
    encyclopedias, 0919
    erythrocytes, manuals, 2351
    eye, 1764, 2363
    fish and aquatic mammals, 1935
    handbooks, 1085, 1086, 1103, 1104
    hormones, 2083, 2119
    indexes and abstracts, 0032, 0088
    invertebrates, 2325
    kidney, 1854
    kidney failure, 2397
    lactation, 2091
    liver, 1847
    mammals, 2330
    manuals, 1009, 1130, 1142
    medicine, clinical, 2356, 2368
    muscles, 0794, 1729
    neoplasia, 3407, 3436
    nervous system, 2417
        diseases, 2402
    periodicals, 0235, 0331-0333, 0353, 0366, 0424, 0440
    poultry, 1910
    review serials, 0614, 0635, 0641, 0656a, 0677
    serial monographs, 0752
    sheep, 3915
    skin, 2041
    teeth, 1829
    testis, 1853.
    vertebrates, 2325, 2338

biochemistry (cont.)
    viruses, 2779
biochemistry, clinical, 1113
    animals, domestic, 2372
biochemistry, comparative, 2324, 2325
biochemistry, developmental, 1516, 1553, 1567, 1570, 1590
biochemistry, diagnostic, 3195
biocompatible materials, 2501, 2510, 2511. See also plastics
bioelectricity
    fish and aquatic mammals, 2469
biological clocks, 2027
    cytology, 1665
    primates, 1915
    serial monographs, 0737
biological products, 1068, 1073, 3805, 4101. See also vaccines
    tuberculin, 2657
    veterinary, 2663
biological standardization
    serial monographs, 0782
biological transport, 2058
biology. See also immunobiology; microbiology; neurobiology; radiobiology; research, biological; zoology
    art, 1050
    behavior, 2250
    birds, 1937, 4065, 4107, 4113
    bone, 1726
    cells, 2482
    computers, 2529, 2534
    dictionaries, 0820, 0832, 0845, 0846, 0884, 0891, 0900
    directories, 0928, 0934, 1003
    education, 4251
    encyclopedias, 0913, 0914
    fish and aquatic mammals, 4132
    furbearing animals, 4220
    guides to the literature, 0803, 0808, 0810
    guinea pigs, 4520
    hamsters, 4515
    handbooks, 1076, 1078, 1087
    indexes and abstracts, 0029, 0032, 0047, 0068, 0089
    information retrieval, 0992
    invertebrates, 4595, 4597, 4599
    lactation, 2191
    mammals, 4219
    manuals, 1083
    mathematics, 2515, 2522, 2534, 2536, 2538, 2540
    mice, 4525
    monkeys, 4429
    periodicals, 0265, 0320, 0328, 0373, 0387, 0441, 0486
    primates, 3090, 4443
    reptiles, 4168
    skin, 1826

biology (cont.)
  statistics, 2517, 2519, 2523, 2527, 2530, 2534, 2541
  ultrastructure, 1683
    atlases, 1682
  vertebrates, 1483, 4217
biology, comparative
  vertebrates, 4383
biology, developmental, 1294, 1512, 1527, 1577, 1580, 1581, 1582, 1594. See also anatomy, developmental; growth
  amphibia, 1525
  biochemistry, 1516, 1525, 1567, 1590
  hamsters, 4521
  hormones, 2101
  mice, 1574
  mammals, 1518, 1578
  review serials, 0651, 0710
  vertebrates, 1513
biology, marine, 2609, 2610, 2952, 4132, 4141, 4143, 4145, 4151, 4582
  embryology, 4142
  pharmacology, 2545
  research, 4134
  review serials, 0618
biology, molecular, 2471, 2472, 2474, 2477, 2481, 2484
  amphibia, 4153
  genetics, 2486
  manuals, 2473
  muscles, 1993
  neoplasia, 2475, 2485
  reptiles, 4153
  virology, 2763
  viruses, 2764, 2791, 2797
biomathematics. See mathematics; statistics
biomechanics, 1705, 1710, 1718, 1720, 1722, 1724, 1728, 1731, 1732, 1735, 1740, 1742, 2496, 2502, 2504
  dogs, 1737
  horses, 3862, 3876, 3894, 3895
  muscles, 1709
biomedical engineering, 2487, 2489, 2490, 2494, 2498-2500, 2504, 2509. See also electronics, medical
  bibliographies, 0185
  equipment, 1067
  indexes and abstracts, 0032, 0070
  neurology, 2208
  periodicals, 0336, 0337, 0425, 0426
  review serials, 0610, 0611, 0636
  tissue replacement, 2501
biomedical research. See research, biomedical
biophysics, 1942, 1950, 2454, 2459, 2462, 2464, 2466, 2468. See also physics
  blood, 2141
  circulation, 2046
  genitourinary system, 2072, 2076
  handbooks, 1103, 1104

biophysics (cont.)
  manuals, 2460
  membranes, 2461
  muscles, 1715
  periodicals, 0503
  review serials, 0609, 0636
biopsy, 1314
  liver, 3194
  neoplasms, 3528
  respiratory system, 3528
biospeleology, 4218
birds. See also Coturnix quail; pigeons; poultry
  anatomy, 1412
    respiratory system, 4075
  bacterial diseases, 4101
  bibliographies, 0207
  biochemistry, 0789
  biology, 1937, 4065, 4107, 4113
  brain, 1797
  breeds and breeding, 4104
  care and management, 4072, 4104
  chromosomes, 4067
  diseases, 0210, 0231, 4079, 4092
  embryology, 1568, 1569, 4080a
    biochemistry, 1570
  eye, ultrastructure, 1536
  hematologic diseases, 4099
  hematology, atlases, 1892
  indexes and abstracts, 0036
  infectious diseases, 2827
  muscles, 1723
  mycotic diseases, bibliographies, 4102
  parasitic diseases, 2827
  parasitology, 0744, 2962
  periodicals, 0231
  physiology, 1956
    comparative, 1937
  psychology, 2236
  sense organs, 2005
  serial monographs, 0789
  sex determination, 1858
  vision, animal models, 2005
birth defects. See also metabolism, inborn errors of; teratology
  indexes and abstracts, 0003
bladder. See also genitourinary system; urology
  anatomy, 1852
  neoplasms, 3538, 3546, 3549
    diagnosis, radiologic, 3544
  physiology, 1852
blood. See also body fluids; hematology; hematopoietic system; hemoglobin
  atlases, 1878
  biochemistry
    cats, 2353
    dogs, 2353
  biophysics, 2141
  bone, 1712

blood (cont.)
  diagnosis, laboratory, 1144, 1894, 3167
  handbooks, 1075
  hormones, 2098
  infectious diseases, 2839
  periodicals, 0338, 0339
  rheology, 2141, 2463
  tissue culture, 1295
  ultrastructure, 1688
blood cells, 1883. See also bone marrow;
    erythrocytes; leukocytes; lympho-
    cytes; platelets
  anatomy, 1876
  animals, laboratory
    morphology, 1898
  atlases, 1882
  biochemistry, 2149
blood circulation. See circulation
blood coagulation, 1901, 2148, 3351, 3352
  anticoagulants, 2645
  fibrinogen, 1891
blood diseases. See hematologic diseases
blood fluids, 4397
blood groups, 1146
  periodicals, 0226
  serial monographs, 0748
blood pressure, 3152, 3156
blood transfusion, 1270, 3626
  serial monographs, 0734
blood vessels, 1871
  innervation, 2001
  serial monographs, 0756
body composition
  human, 1498
  veterinary, 1498
body fluids, 1914, 2135. See also blood
    fluids; cerebrospinal fluid
  animals, 1934
  metabolism, 1934
  physiology, 2074
body temperature. See temperature
bone, 1721, 2086. See also calcification;
    orthopedics; skeleton
  anatomy, 1716
  archaeology, 1493
  bibliographies, 0173, 0203
  biochemistry, 1719, 1990
  biology, 1726
  blood, 1712
  growth, 1734
  histopathology, atlases, 1714
  metabolism, diseases, 1532
  neoplasms, 3482, 3483, 3485, 3487
    radiology, 3684
    veterinary, 3488
  neoplasms, comparative, 3486
  nutrition, 1738
  pathology, classification, 3489
  periodicals, 0427-0429
  phosphorous, 2086

bone (cont.)
  physiology, 1716, 1990, 1997, 2000
  radiology, atlases, 3482
  regeneration
    cytology, 3650
    vitamin K, 3649
  serial monographs, 0736
bone diseases, 3199, 3208, 3211, 3214
  horses, 3217, 3862
  metabolism, 1532
  ultrastructure, 1687
bone marrow, 1904. See also myeloprolifer-
    ative disorders
  atlases, 1878
  biochemistry, 2149
  histopathology, atlases, 1714
  pathology, atlases, 3561
  serial monographs, 0734, 0779
  sheep, 0779
  tissue culture, 1295
  transplantation, 1275, 3563, 3626
  ultrastructure, 1688
borna disease
  bibliographies, 0186
borrelia infections, 2882
bovine. See cattle
brain. See also brain stem; cerebellum;
    electroencephalography
  amphibia, atlases, 1783
  animal models, 1822
  animals, domestic
    atlases, 1821
  animals, hibernating, 2033
  baboons, atlases, 1762
  bibliographies, 0187
  birds, 1797
  cats, atlases, 1753, 1816, 4009
  chemistry, 2405
  chimpanzees, atlases, 1765
  dogs, 2256
    atlases, 1748, 1766, 1789, 1813
  drugs
    animals, 2690
    human, 2690
  guinea pigs, atlases, 1790
  hamsters, atlases, 4515
  human, 1775
  injuries, 2211
  mammals, atlases, 1757
  mice, atlases, 1812
  monkeys, atlases, 1770, 1792, 1809,
    1815, 4426, 4438
  necropsy, 1296
  neoplasms, 0778, 3505
    ultrastructure, 1696
  opossums, 4053a
  periodicals, 0340, 0378
  pigeons, atlases, 1782
  primates, 1796
  rabbits, atlases, 4558

brain (cont.)
  rats, 1775
    atlases, 1749, 1784, 1798, 1810
    research, 1775, 1804
    serial monographs, 0732, 0764, 0777, 0778, 0788
  vertebrates, 1471
brain stem
  atlases, 1771
  mice, 1785
  neurochemistry, 1819
breast. See mammary glands
breeds and breeding, 2175, 3097, 3787.
    See also genetics; insemination; reproduction
  animals, laboratory, 4276, 4281, 4284, 4288, 4293
  birds, 4104
  cats, 0793, 4043
  chickens, 0793
  Coturnix quail, 0793
  dogs, 0793, 2172, 3110, 3984, 3999
  guinea pigs, 0793
  hamsters, 0793
  horses, 3871, 3876, 3881, 3884, 3888, 3892
  indexes and abstracts, 0024
  mice, 0793
  monkeys, 4484
  primates, 0793, 4425
  rabbits, 0793
  rats, 0793
  rodents, 0793
  sheep, 3916
  swine, 0793
brucellosis
  manuals, 1175
  serial monographs, 0735, 0753, 0782, 0800
Burkitt's lymphoma
  serial monographs, 0767

C

calcification, 3391. See also bone
  periodicals, 0351, 0592
  serial monographs, 0750
calcium, 1859, 2086
  deficiency, 3317
  physiology, 2327
cancer. See neoplasia
canine. See dogs
capillaries. See blood vessels
carbohydrate metabolism. See metabolism, carbohydrate
carcinogens. See also neoplasia
  indexes and abstracts, 0050
  serial monographs, 0731, 0749, 0767, 0778

cardiology, 3267. See also electrocardiology; heart
  dogs, 3263
  medicine, veterinary, 3258
cardiology, clinical
  therapeutics, 1833
cardiology, comparative, 2049
cardiovascular diseases, 0032, 0788, 3257, 3260, 3268, 3514, 3518. See also thrombosis; vascular diseases
  congenital, 3513, 3517
  diagnosis, 3522
    clinical, 3154
  nomenclature, 3522
  nutrition, 3266
cardiovascular surgery. See surgery, cardiovascular
cardiovascular system, 1838. See also blood vessels; heart
  art, 1053
  diagnosis, radiologic, 3693
  encyclopedias, 1833
  embryology, 1541
  indexes and abstracts, 0032
  pathology, 3513, 3519-3521
  periodicals, 0302, 0361, 0363, 0492
  physiology, 2045
  review serials, 0621, 0721
cats, 4035, 4044, 4236, 4246
  anatomy, 1404, 1410, 1413, 1417, 1437, 1502
    atlases, 4030
  anatomy, surgical, 1488
  behavior, 4038, 4041
  blood biochemistry, 2353
  brain, atlases, 1753, 1816, 4009
  breeds and breeding, 0793, 4043
  care and management, 4033, 4039, 4042, 4046, 4047
  diagnosis, clinical, 3147
  dictionaries and encyclopedias, 4034, 4048
  diseases, 4017, 4032
  dissection, 1301, 1304
  emergencies
    diagnosis, radiologic, 3688
  eye, 1750
    diseases, 3219
  genetics, 3121, 4040, 4043
  indexes and abstracts, 0036
  medicine, veterinary, 3130, 4029
  muscles, 1725
  neuroanatomy, 1780
  nutrient requirements, 3994
  nutrition, 4036
  periodicals, 0240
  skin diseases, 3511
  surgery, 3130, 3579, 3582
    atlases, 1488
    diseases, 3577

cats (cont.)
  surgery, orthopedic, 3625, 3631, 3635
  urology, 3309
cattle, 3842-3844, 3848. See also livestock; ruminants
  anatomy, 1411, 1429, 1744, 3834
  bibliographies, 0186
  diseases, 0186, 0753, 0798, 3811, 3824
  fertility, 2156
  indexes and abstracts, 0036, 0055
  insemination, 2190
  leptospirosis, 2843
  losses, 3851
  mammary gland diseases, 1869
  management, 3810, 3847
  mastitis, 1218, 2125, 3816, 3822
  medicine, 3129
  muscular diseases, 3819a
  musculoskeletal system, 1741
  nutrient requirements, 0791
  nutrition, 3849
  ovary, 0779
  parasitic diseases, 2970
  periodicals, 0372, 0399, 0435, 0587
  reproduction, 2190
  serial monographs, 0753, 0779
  sterility, 2156
  surgery, 3129
    genitourinary, 3640
cell culture. See tissue culture
cells, 1649, 1650, 1664, 1673. See also cytology; mast cells; nucleolus; nucleus
  biology, 2482
  development, 1620
  drugs, 2552
  metabolism, 1619
  morphology, 1612
    neuroglia, 1794
  nervous system, pathology, 1653
  pharmacology, 1917, 2552
  physiology, 1632, 1646, 1917, 1940, 2002
  research, 1661
    directories, 0996
cellular immunity. See immunity, cellular
cerebellum. See also brain
  anatomy, comparative, 1476
  apes, 1787
  circulation, physiology, 2030
  dogs, 1379
  histology, comparative, 1476
  monkeys, 1791
  serial monographs, 0777
cerebrospinal fluid
  physiology, 2015
cerebrovascular diseases. See also vascular diseases
  indexes and abstracts, 0051
cervix uteri. See uterus

cestoda, 2958, 2968
  serial monographs, 0744
chemicals
  catalogs, 2569
  toxicology, 2584
chemistry. See also biochemistry; histocytochemistry; immunochemistry; mutagens, chemical; neurochemistry; parasitology, chemical; protozoology, chemical; zoology, chemical
  brain, 2405
  dictionaries, 0886, 0890
  encyclopedias, 0895
  guides to the literature, 0809
  handbooks, 1081
  hormones, 2578
  indexes and abstracts, 0052, 0091
  information retrieval, 0993
  instrumentation, 1350
  invertebrates, 2359
  manuals, 1080
  synonyms, 0856, 0857, 0860
  veterinary, 2575
chemistry, clinical, 2343. See also biochemistry
  history, 0971
  instrumentation, 1327
  manuals, 1098, 1110, 1128, 1141, 1145
  review serials, 0613, 0712
  statistics, 1134
  veterinary, 1114
chemistry, pathological, 2349
chemotherapy, 2642
  indexes and abstracts, 0049
  neoplasia, 2654
  review serials, 0626
  serial monographs, 0759
chick embryo, 1533, 1558, 1589
chickens. See also birds; fowl; poultry
  anatomy, 1438
  breeds and breeding, 0793
  feeds and feeding, 4112
  histology, gastrointestinal system, 1618
  nutrition, 4111
chimpanzees. See also primates
  anatomy, 2233
  behavior, 0776, 2233
  bibliographies, 0156a
  brain, atlases, 1765
  diseases, 2233
  ecology, 4480
  history, 4480
  immunology, 0776
  nervous system, 0776
  physiology, 2233
  reproduction, 2233
  restraint, 2233
  serial monographs, 0737, 0776
  space medicine, 4465
chlamydial diseases, 2742, 2822

cholera, 2902
chromatography, 1149
   bibliographies, 0166
   manuals, 1109, 1139
   serial monographs, 0772
chromosomes, 3124. See also cytogenetics; genetics
   birds, 3111
   classification, 3111
   fish and aquatic mammals, 3111
   mammals, 1538
   reptiles, 3111
   serial monographs, 0785
   sex, 3096
circadian rhythms. See biological clocks
circulation, 0968, 0987, 1835. See also microcirculation; pulmonary circulation; rheology
   biophysics, 2046
   cerebellum, physiology, 2030
   congresses, 1888
   fish and aquatic mammals, 1929, 4133
   handbooks, 1077, 1928
   heart
      normal, 2044
      pathology, 2044
   kidney, 1851
   lung, 3269
   pathology, 4323
   physiology, 2046, 2047, 4322
   serial monographs, 0732
CLASS, 0992
clinical chemistry. See chemistry, clinical
clinical diagnosis. See diagnosis, clinical
clinical laboratories. See laboratories, clinical
clinical medicine. See medicine, clinical
clinical pathology. See pathology, clinical
clostridia, 2745
coccidiosis, 2876, 2954. See also poultry: diseases
   animals, domestic, 2941, 2981
   bibliographies, 0177
   ruminants, 2947
cold. See cryogenics; freezing and freeze-drying
collagen, 1707, 2344. See also connective tissue
communicable diseases, 1188, 2838. See also infectious diseases
communication, 2508. See also behavior; speech
   animals, 0147, 2235, 2252, 2298, 2299
   chemistry, 2240a
   primates, 2223
   serial monographs, 0788
communication, comparative, 4456
computers. See also diagnosis, laboratory: computer assisted
   biology, 2529, 2534

computers (cont.)
   medicine, 1340, 2529, 2531
   nutrition, 3752
   pharmacy, 2534
   research, biomedical, 2539
conferences and congresses
   bibliographies, 0123-0137
connective tissue. See also collagen; mast cells
   biochemistry, 1604, 2387, 2435
   cytology, 1604
   hormones, 2079
   immunochemistry, 2435
   immunology, 3064
   nutrition, 1738
   periodicals, 0368
   physiology, 1992
   review serials, 0663
   serial monographs, 0736, 0756
connective tissue diseases, 2435, 3200, 3203, 3562
contagious bovine pneumonia
   bibliographies, 0186
   serial monographs, 0753
cornea
   transplantation, 3629
Coturnix quail. See also birds
   anatomy, 1414
   breeds and breeding, 0793
   histology, 1414
cryogenics. See also freezing and freeze-drying
   periodicals, 0369
   surgery, 3588, 3589
      dermatology, 3602
cultures, 2703. See also bacteriology; tissue culture
   catalogs, 1180
   manuals, 1208
cybernetics, 2480, 2503, 2508. See also feedback
cytochemistry. See histocytochemistry
cytogenetics, 1614, 1631, 1654, 3103. See also chromosomes; genetics
   periodicals, 0371
cytogenetics, comparative
   mammals, 1624
cytology, 1598, 1615, 1621, 1640, 1656, 1666, 2480, 3453, 3455. See also cells; histology; ultrastructure, cytological
   connective tissue, 1604
   diagnosis, laboratory
      uterus, 3547
   instrumentation, 1350
   manuals, 1167, 1233, 1251
   neoplasia, 3418
   neoplasms, 3450
      atlases, 3441
      nervous system, 3495

cytology: neoplasms (cont.)
    uterus, 3548
  periodicals, 0379, 0430, 0526
  regeneration, bone, 3650
  respiratory system, 1671
  review serials, 0612, 0649, 0664, 0669, 0674, 0694
  uterus, 1676
cytology, biochemical, 1612, 1642, 1643, 1660
cytology, diagnostic, 1601, 1644, 1662, 3431
  gastrointestinal system, atlases, 3532
  stomach, 3531
cytology, histopathological, 1644
cytology, pathological, 1625
cytology, physiological, 1612, 1622, 1645, 1650, 1651
cytomegaloviruses
  serial monographs, 0795

D

dairy products. See also milk
  microbiology, 1178
deformities. See teratology
dentistry. See also oral manifestations; teeth
  veterinary, 3506
    bibliographies, 0209
dermatitis. See skin diseases
dermatology, 3242, 3243, 3247, 3252, 3255, 3256. See also skin
  atlases, 3244
  dictionaries, 0892
  indexes and abstracts, 0032, 0064
  manuals, 3253
  medicine, veterinary, 3250
  periodicals, 0455
  review serials, 0722
  surgery, cryogenic, 3602
dermatology, comparative, 3512
dermis. See skin
diabetes, 3319, 3323, 3325, 3328. See also endocrine diseases; pancreas
  indexes and abstracts, 0056
diabetes, experimental
  rats, 4517
diagnosis. See also biochemistry, diagnostic; cytology, diagnostic; electrodiagnosis; immunochemistry, diagnostic
  cardiovascular diseases, 3522
  manuals, 2651
  medicine, veterinary, 3142
  neoplasms, 3433a
  uterus, atlases, 3550
  veterinary manuals, 2653
  viral diseases, 2771, 2777

diagnosis, clinical, 3145, 3153, 3159, 3161, 3162, 3164, 3192
  atlases, 3158
  cardiovascular diseases, 3154
  cats, 3147
  dogs, 3147
  handbooks, 3157
  serial monographs, 0794
diagnosis, laboratory, 1120, 2356, 2368, 2388, 3171, 3178, 3181, 3184, 3464
  blood, 1144, 1894, 3167
  chemistry, clinical, 1110
  cytology, uterus, 3547
  computer assisted, 1116, 1143
  directories, 0942
  endocrine glands, 3166
  endocrinology, 1106
  hematology, 1894
  manuals, 1096, 1099, 1108, 1136, 3170
diagnosis, radiologic, 3165, 3692
  cardiovascular system, 3693
  cats, emergencies, 3688
  dogs, emergencies, 3688
  eye, 3693
  gastrointestinal system, 3535, 3693
  genitourinary system, 3540, 3693
  neoplasms
    bladder, 3544
    kidney, 3545
  nervous system, 3691
  obstetrics and gynecology, 3693
  respiratory system, 3693
  skeleton, 3691, 3693
dietetics. See also feeds and feeding; nutrient requirements; nutrition; vitamins
  veterinary, 3745, 3759
digestion. See also gastric juice; gastrointestinal system
  livestock, 3758
  pathology, 2070
  physiology, 2070
  ruminants, 0765
dirofilariasis
  dogs, 2977
disease vectors, 2781, 2798, 2881, 2985, 2990-2992. See also insects
disinfection. See infection control
dissection, 1322, 1458, 1506. See also necropsy
  cats, 1301, 1304
  dogs, 1307
  horses, 1311
  rabbits, 1319
  ruminants, 1300
  swine, 1554, 1583, 3930
  vertebrates, 1305, 1318, 1458
distemper
  serial monographs, 0795
dogs
  anatomy, 1405, 1432

dogs: anatomy (cont.)
    radiography, 1440
  anatomy, surgical, 1488
  animal models, 3949
  behavior, 2254-2256, 2279, 2291
  behavior, social
    genetics, 2297
  bibliographies, 0171, 0174
  biomechanics, 1737
  blood biochemistry, 2353
  brain, 2256
    atlases, 1748, 1766, 1789, 1813
  breeds and breeding, 0793, 2172, 3110, 3984, 3999
  cardiology, 3263
  cerebellum, 1379
  diagnosis, clinical, 3147
  dictionaries and encyclopedias, 3988, 3989, 3995
  dirofilariasis, 2977
  diseases, 0795
  dissection, 1307
  electroencephalography, 1379
  emergencies
    diagnosis, radiologic, 3688
  eye, 1750, 1752
    diseases, 3219, 3237
  genetics, 3113, 3125
  hemorrhage, 3343
  histology, atlases, 1597
  indexes and abstracts, 0036
  intervertebral disc displacement, 3212
  medicine, veterinary, 3131, 3998, 4002, 4003
  neuroanatomy, 1780
  neurology, 2215, 2218
  nutrient requirements, 0791, 3994
  nutrition, 3986, 3993
  ophthalmology, 3232
  pediatrics, 3957
  radiography, 3676
  reproduction, 2173
  skin diseases, 3249, 3511, 3512
  spinal diseases, 3967
  surgery, 3131, 3571, 3579, 3582
    atlases, 1488, 3565
    diseases, 3577
    water-electrolyte balance, 2364
  surgery, orthopedic, 3625, 3631, 3635
  urology, 3309
dogs, guard, 4240, 4244, 4248
dogs, hunting, 4231, 4234
dogs, Seeing Eye, 4235
domestic animals. See animals, domestic
drug-induced diseases, 2551, 2555
drug interactions
  serial monographs, 0749, 0751
drugs. See also chemotherapy; formularies; pharmacology; psychopharmacology; therapeutics; toxicology

drugs (cont.)
  adverse effects, 2568, 2574, 2688
  behavior, 2683
    animals, 2689
  brain
    animals, 2690
    human, 2690
  catalogs, 2543, 2556, 2557, 2569, 2629, 2638, 2655, 2656, 2658
  cells, 2552
  classification, 2575
  handbooks, 2644
  metabolism, 2567, 2577
  nervous system, 2573, 2678
  nomenclature, 2668
  psychotherapeutic, 2677, 2684, 2688
  standards, 2544, 2572
  toxicity, 2584, 2669, 2671, 2672
    bibliographies, 2681
  veterinary
    catalogs, 2659, 2660, 2664, 2665
    standards, 2547
drug therapy. See chemotherapy; therapeutics

E

ear. See also otorhinolaryngology; sense organs
  hamsters, 4521
  manuals, 1326
eccrine sweat glands, 0731
echoviruses
  serial monographs, 0795
ecology. See also adaptation; environment; migration; populations
  animals, domestic, 2293
  apes, 4474, 4480
  baboons, 0737
  chimpanzees, 4480
  fish and aquatic mammals, 4130, 4600
  gorillas, 4194, 4480
  invertebrates, 4600
  primates, 4478
  rats, 2237
  wildlife, 4182, 4183, 4184, 4189, 4193, 4195, 4196, 4205, 4214
  wolves, 2286
economics, veterinary, 3844, 3845
  poultry, diseases, 4093
education, 4249
  biology, 4251
  directories, 0945
  support, 1026, 1027, 1030
education, graduate, 4254
  support, 1024, 1029, 1031, 1035
education, medical, 4257-4260, 4266, 4274
  directories, 0926, 0954
  periodicals, 0458
  serial monographs, 0788

education, medical (cont.)
  support, 4255
education, veterinary medical, 4252, 4253,
    4268, 4271
  directories, 0955
  serial monographs, 0783, 0792
electrical anesthesia, 4338
electrical safety
  bibliographies, 0209
electrocardiography, 1363, 1364, 1371
  interpretation, 1365-1369, 1372
  periodicals, 0437, 0600
  symposia, 1370
electrodiagnosis
  manuals, 1385
  muscles, 1382, 1383, 1387-1390
  nervous system, 1391
  neuromuscular diseases, 1386
electroencephalography
  bibliographies, 0154, 0213, 0216
  dogs, 1379
  handbooks, 2023
  indexes and abstracts, 0058
  interpretation, 1374-1376
  manuals, 1373, 1378
  periodicals, 0374
  veterinary, 1377
electromyography. See electrodiagnosis
electron microscopy, 1152, 1155-1157,
    1161, 1162, 1172, 1174. See also
    ultrastructure
  atlases, 1699
  bibliographies, 0182
  manuals, 1154, 1158, 1165, 1168, 1256
electroneurography. See electrodiagnosis
electronics, medical, 1323, 1325, 1337,
    1345, 1350, 2495, 2507, 2512. See
    also instrumentation
  anesthesia, 3607
  equipment, 1067, 1071, 1333
  handbooks, 1329, 1330, 1351, 1362
  surgery, 3607
electrophoresis, 1093
  immunochemistry, 1133, 2438
  immunology, 1089
  manuals, 1139, 1149
electrophysiology, 1747
  manuals, 1330, 1350, 1358, 1384
embryology, 1510, 1514, 1519, 1560, 1580,
    1630, 2158, 2160. See also anatomy,
    developmental
  amphibia, 1589
  atlases, 1552a, 1670
  baboons, 1537
  biology, marine, 4142
  birds, 1568, 1569, 4080a
    biochemistry, 1570
  cardiovascular system, 1541
  chick, 1533, 1558, 1589
  handbooks, 1079

embryology (cont.)
  human, 1561
    abnormal, 1546
    normal, 1546
  indexes and abstracts, 0032
  mammals, 1521, 1550
    manuals, 1526, 1573
  mice, 1563
    atlases, 4506
  morphogenesis, 1534
  neoplasia, 1535
  nervous system, 1524, 1540, 1564
  nomenclature, 0870
  nutrition, 1528
  pathology, 1593
  periodicals, 0438, 0593
  radiobiology, 3667
  serial monographs, 0732, 0749, 0768
  swine, 1533, 1552, 1554, 1557, 1559,
    1583, 1589, 3929-3931
  vertebrates, 1470, 1526, 1564, 1573, 1575
embryology, biochemical, 1986
embryology, comparative, 1448
  vertebrates, 1551, 1555
embryology, experimental, 1591, 1592
  invertebrates, 4582
  manuals, 1531, 1572
  mice, 1563
emergencies. See also shock; trauma
  cats, radiologic diagnosis, 3688
  dogs, radiologic diagnosis, 3688
  serial monographs, 0794
  veterinary, 3148, 3804
encephalitis, Venezuelan equine, 0205, 0209
endocrine diseases, 3316, 3318, 3555. See
    also diabetes; mastitis; parturient
    paresis
endocrine glands. See also adrenal glands;
    hormones; pancreas; parathyroid
    glands; pineal body; pituitary gland;
    thymus gland; thyroid gland
  art, 1053
  diagnosis, laboratory, 3166
  enzymology, histochemical, 2400
  fish and aquatic mammals, 1929
  metabolism, 2130
  physiology, 2082, 2124, 2130
    comparative, 2100
  transplantation, 3629
endocrinology, 2095, 2112a, 2132, 2133,
    3166, 3321. See also neuroendocrin-
    ology
  handbooks, 1928
  indexes and abstracts, 0032, 0045, 0059
  invertebrates, 4583
  manuals, 1106
  periodicals, 0241, 0289, 0375, 0431, 0439
  review serials, 0652, 0711, 0724
  serial monographs, 0751, 0772
  vertebrates, 1966

endocrinology (cont.)
  veterinary, 2109
endocrinology, clinical, 2080, 2086, 2117
endocrinology, comparative, 2097, 2121
  invertebrates, 4578
endocrinology, experimental, 4384
endocrinology, neurochemical, 2081, 2413, 2415
endometrium. See uterus
energy transfer, 2373, 2376, 2377, 2380, 2384. See also metabolism
  mammals, 2385
enterobacteriaceae, 2728
  manuals, 1194
entomology, 2986-2989, 2994, 2995. See also insects
  bibliographies, 0103
  manuals, 2983
  periodicals, 0349, 0436, 0459
  veterinary, 2993
environment. See also ecology
  animals, 2320
    laboratory, 4400
  handbooks, 1076
  microbiology, 1195
  research, 4379
environmental health
  animals
    domestic, 2592
    laboratory, 4296
  indexes and abstracts, 0032, 0080, 0089
environmental physiology. See physiology, environmental
enzymes, 2375
  histocytochemistry, 2424, 2429
  review serials, 0615
  serial monographs, 0741
enzymology
  animals, 2361
enzymology, histochemical
  endocrine glands, 2400
ephemeral fever
  bibliographies, 0186
epidemiology, 2893a, 2903, 2910, 2912. See also public health
  indexes and abstracts, 0054
  infectious diseases, 2825
  manuals, 2900
  periodicals, 0305
  zoonoses, 2894
epidermis. See skin
epilepsy. See also neuromuscular diseases
  indexes and abstracts, 0032
epithelium. See skin
equine. See horses
equipment. See also instrumentation
  biomedical engineering, 1067, 1071
  catalogs, 1064, 1065, 1072, 1074
  electronics, medical, 1333, 1334
  handbooks, 1354

erythrocytes, 1875, 1896
  anatomy, 1872
  biochemistry, manuals, 2351
  biosynthesis, 1881
  metabolism, 1881, 2351
  pathology, 1881
  physiology, 1872
  serial monographs, 0732
Escherichia coli
  serial monographs, 0742
ethology. See behavior
evolution, 4436
  apes, 4474
  behavior, 2275
  horses, 3876
  monkeys, 4455
  primates, 4442, 4443
    behavior, 2274
  serial monographs, 0741
  sheep, 2261
  vertebrates, 1452
exotic animal diseases, 2829, 3742. See also foot-and-mouth disease
  bibliographies, 0186
  indexes and abstracts, 0028
exotic animals. See also tropical animals
  indexes and abstracts, 0036
  pets, 4228
eye. See also ophthalmology; retina; sense organs; vision
  anatomy, 1490, 1764, 1768, 1820
    comparative, 1491
    developmental, 1548
  animals, domestic, 1490
  biochemistry, 1764, 2363
  birds, ultrastructure, 1536
  cats, 1750
  diagnosis, radiologic, 3693
  dogs, 1750, 1752
  histology, 1628
  injuries, 1768
  physiology, 1764, 1768, 2003, 2016
  rabbits, research, 4054
  surgery, 3623
    cornea transplantation, 3629
eye diseases, 1768, 3222, 3234. See also uveitis
  cats, 3219
  dogs, 3219, 3237

F

farm animals. See animals, domestic
farriery. See horseshoeing
feed additives, 3772
feedback. See also cybernetics
  behavior, animal, 2282
feeds and feeding, 3755, 3757, 3758, 3766, 3767, 3773, 3842a. See also nutrient requirements; nutrition

feeds and feeding (cont.)
  analysis, 2603
  animals, domestic, 3762
  bacterial control, 1225
  chickens, 4112
  furbearing animals, 0743
  horses, 3881
  livestock, 0743
  periodicals, 0388
  pets, 3753
  primates, 4483
  toxins, 2625
feline. See cats
fertility. See also gonadotropins; reproduction; sterility
  animals, domestic, 2179
  cattle, 2156
  periodicals, 0389, 0410, 0476, 0605
fertilization, 2159. See also reproduction
  mammals, 2181
  ultrastructure, 1677
    mammals, 1530, 1596
fertilization, comparative
  biochemistry, 1553
  immunology, 1553
  morphology, 1553
fetal membranes, 1586. See also placenta
fetus. See embryology
fibrinogen, 1891. See also blood coagulation
fish and aquatic mammals, 4148
  anatomy, 4114
  animal models, 4128, 4140
  bacterial diseases, 4136
  behavior, 1929, 2239, 4117
  bibliographies, 0172
  biochemistry, 1935
  bioelectricity, 2469
  biology, 4132
  care and management, 4144
  chromosomes, 4067
  circulation, 1929, 4133
  classification, 4149
  diseases, 0801, 4115, 4120, 4121, 4131, 4132, 4135, 4138
  ecology, 4130, 4600
  endocrine glands, 1929
  husbandry, 4138
  indexes and abstracts, 0043
  metabolism, 4117
  morphology, 4150
  nervous system, 2239
  parasitology, 0739, 2938, 4119
  physiology, 1929, 4117
  research, 4125, 4140
  respiration, 1929
  sense organs, 4126, 4139
  symbiosis, 4146
  water-electrolyte balance, 4127
fluid balance. See water-electrolyte balance

flukes, 2966, 2967
fluorescence, 1356. See also immunofluorescence
food. See also nutrition
  mycotoxins, 2591
  toxins, 2625
food additives
  analysis, 2603
food inspection, 2897, 2901, 2909, 2914, 2915. See also public health
  congresses, 2919
  handbooks, 1211
  manuals, 1229, 2917
  microbiology, 1176, 2904, 2907, 2913
  standards, 2899
foot-and-mouth disease, 2855, 2858, 2864, 2887
  bibliographies, 0186
  congresses, 2856
  periodicals, 0594
  serial monographs, 0755, 0782
foreign bodies, 3275
formularies, 2544, 2547, 2572, 2631, 2667
fowl. See also birds; chickens; poultry; turkeys
  care and management, 4198
  neoplasms, 3409
  physiology, 4083
foxes. See also wildlife
  nutrient requirements, 0791
freezing and freeze-drying. See also cryogenics
  bibliographies, 0215
  symposia, 1340
frogs. See amphibia
fungi. See also mycology; mycotic diseases
  animals, 2811, 2868
fungi, pathogenic
  animals, 2837
  classification, 2817
  human, 2837
  manuals, 1204
furbearing animals, 4057a. See also minks; rabbits; wildlife
  biology, 4220
  care and management, 4198
  feeds and feeding, 0743

G

gamma globulin
  serial monographs, 0796
gastric juice. See also digestion
  physiology, 2067
gastroenterology, 3283, 3292, 3294
  indexes and abstracts, 0032, 0061
  periodicals, 0392, 0393
gastrointestinal diseases, 3284, 3299, 3302. See also liver diseases

gastrointestinal system. See also biliary tract; digestion; liver; stomach
  anatomy, 1475, 1843, 1846, 3287
  art, 1053
  chickens, histology, 1618
  cytology, diagnostic
    atlases, 3532
  diagnosis, radiologic, 3535, 3693
  histopathology, 3535
  instrumentation, 1353, 1357
  mammals, peptides, 2354
  medicine, 0794
    veterinary, 3637
  metabolism, 2058
  neoplasms, 0784, 3533
  pathology, 3535
  photography, 1357
  physiology, 1843, 1846, 1928, 2060-2062, 2068, 2069, 3287
    comparative, 1973
  ruminants, 1841
  surgery, 0794, 1843, 3287
    rats, 4510
  ultrastructure, atlases, 1702
genetics, 2175, 2796, 3088, 3097, 3099, 3100, 3116. See also breeds and breeding; chromosomes; cytogenetics; immunogenetics; RNA
  animals
    domestic, 3117
    laboratory, 3108, 3114, 4396, 4412
  behavior, social, 2232
  bibliographies, 0164
  biology, molecular, 2486
  cats, 3121, 4040, 4043
  congresses, 3091
  dictionaries, 0889, 3095
  directories, 0938
  diseases, 2390
  dogs, 3113, 3125
    behavior, social, 2297
  horses, 3119
  indexes and abstracts, 0032, 0062
  information retrieval, 0989
  livestock, 3120
  mammals, 3109, 3112
  mice, 4494
    bibliographies, 4504
    catalogs, 4502
    standards, 4489
  neurochemistry, 2409, 2410
  periodicals, 0396, 0448, 0488
  primates, 0776, 3090
  rats, 3122
  review serials, 0616, 0664
  serial monographs, 0734, 0740, 0741, 0768, 0776, 0797
  sheep, 3923
  transplantation, primates, 3062
  vertebrates, 3107

genetics, comparative
  apes, 3104
  human, 3104
  monkeys, 3104
genetics, molecular, 3101
genetics, physiological
  animals, 1975
genitourinary diseases
  animals, domestic, 3313
genitourinary system. See also bladder; kidney; ovary; placenta; testis; uterus
  art, 1053
  biophysics, 2072, 2076
  diagnosis, radiologic, 3540, 3693
  physiology, 1852, 2075
  review serials, 0623
  surgery, 3619
    cattle, 3640
    horses, 3640
gerbils. See also animals, laboratory
  bibliographies, 0163a, 0190
  care and management, 4528
geriatrics, 1520, 1544
  anatomy, developmental, 1508, 1515
  animals, laboratory, 1545
  human, 1508
  immunology, 1588, 3044
  indexes and abstracts, 0032
  mammals, 1543
  mice, animal models, 4513, 4522
  neoplasia, 4522
  nervous system, 1981
  periodicals, 0381, 0446
  rats, animal models, 4513
  research, 1545, 4513
  review serials, 0617
  serial monographs, 0731
  veterinary, 1508
germ-free life, 4278, 4292, 4298, 4304
  research, 4286
  serial monographs, 0732, 0793
gerontology. See geriatrics
gestation. See also obstetrics and gynecology; reproduction
  animal models, 4357
  pathology, 2187
  physiology, 2158, 2187
  research, 4357
gnotobiology. See germ-free life
goats, 3918. See also ruminants
  anatomy, 3834
  animal models, 3913
  husbandry, 3907
gonadotropins. See also hormones
  serial monographs, 0733
gonads, 2086. See also ovary; testis
  anatomy, 1856
  serial monographs, 0788
gorillas. See also primates
  behavior, 4194

gorillas (cont.)
 ecology, 4194, 4480
 history, 4480
 serial monographs, 0737
growth. See also anatomy, developmental
 animals, 2265a
 bone, 1734
 handbooks, 1079
 horses, 3870
guard dogs, 4240, 4244, 4248
guinea pigs. See also animals, laboratory
 allergy, 3068
 animal models, 0770
 biology, 4520
 brain, atlases, 1790
 breeds and breeding, 0793
 hypothalamus, 1866
 nutrition, 4520
 physiology, 4520
 research, 4520
 skin diseases, 3068, 3246
gynecology. See obstetrics and gynecology

H

hair
 serial monographs, 0731
hamsters. See also animals, laboratory
 bibliographies, 0168
 biology, 4515
 biology, developmental, 4521
 brain, atlases, 4515
 breeds and breeding, 0793
 ear, 4521
 neoplasms, kidney, 3543
 pathology, 0778
health occupations, 4265, 4270
health physics, 3653, 3657, 3658, 3660, 3672. See also radiation protection
hearing. See ear; otorhinolaryngology; sense organs
heart. See also cardiology
 innervation, 2001, 3516
 normal, circulation, 2044
 pathology
  angiocardiography and, 3517
  circulation, 2044
  conducting tissue, 3516
 vertebrates
  anatomy, 1834
  innervation, 2048
heart diseases. See cardiovascular diseases
heart function tests, 3163
heartworm disease. See dirofilariasis
helminthology, 2951, 2959, 2978. See also parasitology
 atlases, 1209
 bibliographies, 0158
 indexes and abstracts, 0063
 periodicals, 0395, 0447

helminthology (cont.)
 serial monographs, 0744
 veterinary, 2930
hematologic diseases, 3302, 3329, 3356. See also leukemia
 birds, 4099
 horses, 3855
 manuals, 3339
 surgery, 3341
hematology, 2131, 3173, 3188, 3336, 3338, 3342, 3346, 3353. See also blood
 atlases, 1882, 3350
 birds, atlases, 1892
 diagnosis, laboratory, 1894
 handbooks, 1102
 immunology, 2142, 2154
 indexes and abstracts, 0032, 0060
 manuals, 1088, 1096, 1101, 1123, 1137
 periodicals, 0290, 0343, 0382, 0394, 0599
 primates, atlases, 1886
 review serials, 0656, 0660, 0708, 0709
 serial monographs, 0734, 0801
 veterinary, 1897
hematology, clinical
 handbooks, 1148
hematology, comparative, 1873
hematopoietic system. See also reticuloendothelial system
 diseases, 1895
 neoplasms, 0790, 3438, 3448
 radiation effects, 3663
hemoglobin
 bibliographies, 0178
 biosynthesis, 2137, 2143, 3349
 function, 1885
 manuals, 1090, 1112, 3339
 metabolism, 2137
 mice, 0178
 physiology, 2143
 serial monographs, 0734, 0741
hemorrhage
 dogs, 3343
hepatitis. See also infectious diseases; liver diseases
 viral, 2861
herbicides. See pesticides
herbivores. See mammals
herd management programs, 3810, 3847
heredity. See genetics
hermaphroditism, 4303
 serial monographs, 0768
herpes virus, 2857
 serial monographs, 0795
herpetology, 4169. See also snakes
 dictionaries, 0896
heterologous transplantation, 3636
heterothermic animals, 1978
hibernation, 2115
 animals, 2033
 mammals, 2370

histamine. See also hypersensitivity
  serial monographs, 0775
histocompatability antigens
  serial monographs, 0780, 0786
histocompatability testing, 1270, 1287,
    1288. See also transplantation
histocytochemistry, 1239, 1250, 2422, 2423,
    2425, 2428, 2433. See also enzymol-
    ogy, histochemical
  bibliographies, 1235
  congresses, 2426
  enzymes, 2424, 2429
  manuals, 1234, 1235, 1246, 1253, 1261,
    1264, 1293
  monkeys, 4430
  neoplasms, 2424
  neural transmission, 0777
  periodicals, 0397, 0398, 0449
  review serials, 0694
  steroids, 2421
  wound healing, 2429
histology, 1603, 1606, 1607, 1609, 1626,
    1634, 1637, 1648, 1674. See also
    cytology; stains and staining
  animals, domestic, 1490, 1668
  atlases, 1597, 1605, 1629, 1635, 1655,
    1675
  chickens, gastrointestinal system, 1618
  classification, 1667
  Coturnix quail, 1414
  dogs, atlases, 1597
  eye, 1628
  handbooks, 1663
  human, 1599, 1602, 1623, 1629, 1635
  indexes and abstracts, 0032
  lung, 1840
  manuals, 0761, 1232, 1237, 1238, 1241,
    1244, 1245, 1255-1257, 1260, 1262,
    1263, 1328
  musculoskeletal system, 1633
  neoplasms, 0761, 3413
  nervous system, 1633
  periodicals, 0535
  rats, atlases, 4503
  review serials, 0664
  serial monographs, 0761
  veterinary, 1605, 1627, 1630
histology, comparative, 1600
  cerebellum, 1476
  vertebrates, 1487
histology, neonatal
  atlases, 1670
histology, physiological
  manuals, 1610, 1636
histopathology, 1281, 2479
  atlases, 3454, 3458, 3459
  bone, atlases, 1714
  bone marrow, atlases, 1714
  classification, 1667
  cytology and, 1644

histopathology (cont.)
  gastrointestinal system, 3535
  liver, 3536
  manuals, 1236, 1243, 1251, 1253, 1254,
    1258, 1261, 1280
  neoplasms, 1672
  nervous system, 3500
  skin, 3251, 3254, 3508
    atlases, 3510
  thyroid gland, 3553
  uterus, 3539
histoplasmosis, 2874
hormones, 2124, 2196. See also endocrin-
    ology; gonadotropins; steroids
  animals, 2108
    comparative, 2105
  bibliographies, 0198
  biochemistry, 2083, 2119
  biology, developmental, 2101
  blood, 2098
  chemistry, 2578
  congresses, 2103
  connective tissue, 2079
  immunology, 2128
  invertebrates, 4578
  kidney, 2090
  mammals, 2085
  periodicals, 0400, 0401
  pharmacology, 2578
  physiology, 2119
  reproduction, 2160
  research, 2088
  review serials, 0705, 0717
  serial monographs, 0751, 0788
  therapeutics, 2119
  vertebrates, 2093
hormones, comparative, 2100
horsemanship, 3872, 3875, 4242
horses, 3872, 3873, 3878-3880, 3883, 3887,
    3889-3891, 4242, 4245. See also
    stable management
  anatomy, 1401, 1442, 3894, 3898
    radiography, 1440
  behavior, 3896
  bibliographies, 0186, 0205, 0209
  biomechanics, 3862, 3876, 3894, 3895
  bone diseases, 3217, 3862
  breeds and breeding, 3871, 3876, 3881,
    3884, 3888, 3892
  dictionaries and encyclopedias, 3899
  diseases, 0186, 0205, 0209, 0758, 3853,
    3893
  dissection, 1311
  evolution, 3876
  feeds and feeding, 3881
  genetics, 3119
  growth, 3870
  hematologic diseases, 3855
  indexes and abstracts, 0033, 0036
  joint diseases, 3217, 3897

Subject Index

horses (cont.)
  medicine, 3128
  muscular diseases, 3198, 3862, 3897
  necropsy, 1310
  neurology, clinical, 2220
  nutrient requirements, 0791
  parasitic diseases, 3856
  periodicals, 0238, 0588
  reproduction, 2185
  review serials, 0671
  roaring, 3857
  serial monographs, 0758
  surgery, 3128
    genitourinary system, 3640
    tendons and ligaments, 1739
horseshoeing, 0979, 3874, 3882, 3886, 3900
hospitals, veterinary, 3791, 3792, 3795, 3798
  periodicals, 0221
  serial monographs, 0794
hunting dogs, 4231, 4234
husbandry, 3777, 3784, 3787, 3801, 3843. See also animal production
  animals, laboratory, 4280
  fish and aquatic mammals, 4138
  goats, 3907
  poultry, 4105, 4106, 4109
  serial monographs, 0757
  sheep, 3902, 3906, 3917, 3920-3922
  swine, 3925, 3927, 3938-3940, 3942, 3943
hygiene, 2898. See also food inspection; public health
  indexes and abstracts, 0032, 0041
  periodicals, 0450, 0515
hypersensitivity, 3039, 3070. See also allergy; histamine
  serial monographs, 0749, 0770
hypnosis, 2321
hypocalcemia, 3317. See also calcium
hypothalamus, 1863, 2086, 2115. See also nervous system
  guinea pigs, 1866
  neurochemistry, 2116

I

iatrogenic diseases, 2551, 2555
illustration. See art; photography; photomicrography
immune serums, 2657
immunity, 2714, 3022, 3025, 3030, 3040, 3041, 3052, 3055, 3060, 3072
  manuals, 3077
  neoplasia, 2654, 3058
  parasitic diseases, 2939
  radiobiology, 3037, 3043
  serial monographs, 0746, 0782
immunity, cellular, 1608, 1616, 1647, 3014, 3024, 3034, 3038, 3048, 3049, 3070

immunobiology
  review serials, 0782
  transplantation, 3054
immunochemistry, 2441, 2442, 2445, 2448
  antibodies, 2443
  connective tissue, 2435
  electrophoresis, 1133, 2438
  handbooks, 2452
  periodicals, 0402
  research, 2446
immunochemistry, diagnostic, 2440
immunochemistry, experimental, 2444
immunofluorescence. See also fluorescence
  bibliographies, 0193
  manuals, 1188, 1198, 1210
immunogenetics, 3003, 3006, 3010
immunoglobulins, 3063. See also antibodies
  congresses, 2439
immunologic diseases, 3073, 3085. See also autoimmune diseases
  serial monographs, 0786
immunology, 1608, 2437, 2445, 2997-2999, 3001, 3004, 3005, 3011, 3015, 3018-3020, 3023, 3057, 3059, 3074, 3080. See also allergy; antibodies; antigens; immunity, cellular; serology
  animals, laboratory, 0782
  bibliographies, 0032, 0107
  chimpanzees, 0776
  congresses, 3012, 3024, 3027
  connective tissue, 3064
  dictionaries, 0885, 0887
  electrophoresis, 1089
  geriatrics, 1588, 3044
  hematology, 2142, 2154
  hormones, 2128
  infectious diseases, 2836, 4347
  liver, 3065
  lymphocytes, 1905
  manuals, 1095, 2452, 3076
  neoplasia, 2807, 3419, 3421, 3434
  neoplasms, 3449
  nucleic acids, 2449
  periodicals, 0362, 0364, 0376, 0403, 0405, 0407, 0451, 0452, 0517, 0591, 0596
  primates, 0776, 4476
  research, 2446
  review serials, 0647, 0653, 0662, 0679, 0695
  serial monographs, 0731, 0732, 0734, 0735, 0740, 0770, 0776
  skin, 3069
  veterinary, 3720
immunology, developmental
  serial monographs, 0746
immunology, experimental, 3078, 3085
  manuals, 3086
  serial monographs, 0762, 0772, 0774
  uveitis, 3233

immunosuppression, 3029, 3045, 3046
  infectious diseases, 3066
  primates, 3066
implantation, 1579, 2189. See also insemination
inborn errors of metabolism, 2369, 2390
infection control, 2908, 2909. See also wound infection
  bibliographies, 0146, 0204, 0209
  manuals, 1224
infectious diseases, 2824, 2826, 2838, 2840. See also borrelia infections; brucellosis; chlamydial diseases; cholera; coccidiosis; communicable diseases; contagious bovine pneumonia; hepatitis; histoplasmosis; influenza; leptospirosis; mycoplasma; Newcastle disease; pertussis; pseudorabies; rabies; rickettsial diseases; rinderpest; schistosomiasis; spirochete infections; tetanus; toxoplasmosis; trichinosis; trichomonas foetus; trypanosomiasis; tuberculosis; Venezuelan equine encephalitis; vesicular exanthema; vibrio fetus
  animals, 2845
    domestic, 2830, 2831, 2835
  birds, 2827
  blood, 2839
  epidemiology, 2825
  horses, 0758
  immunology, 2836, 4347
  immunosuppression, 3066
  indexes and abstracts, 0085
  mice, 4511
  periodicals, 0405, 0453, 0518
  prevention and control, 2774
  primates, 0760, 3066, 3384
  rats, 4511
  serial monographs, 0758, 0760
  wildlife, 2828
infertility. See fertility; sterility
inflammation, 1613, 2362, 3401
  pharmacology, 2576
  serial monographs, 0751
influenza, 2863
  serial monographs, 0795
information services, 0989, 0991, 0992, 1000, 1001
  libraries, 0990, 0993, 0995
  neoplasia, 0999
insecticides. See pesticides
insects, 2898, 2991. See also disease vectors; entomology
  control, 2898
  neurophysiology, 2039
  physiology, 1909, 1933
    sense organs, 2017
insemination, 3783a. See also breeds and breeding; implantation

insemination (cont.)
  animals, domestic, 2188
  cattle, 2190
  congresses, 3782a
  livestock, 3784
  manuals, 3782
  periodicals, 0598, 0603
  semen, 3782, 3783a
instinct, 2318, 2321a. See also behavior
  animals, 2228, 2258, 2277, 2315
  behavior, social, 2232
  human, 2228
instrumentation, 1323, 1325, 1332, 1335, 1340-1342, 1346, 1361. See also computers; electrical safety; electrocardiography; electrodiagnosis; electroencephalography; electronics, medical; equipment
  chemistry, 1350
  clinical, 1327
  cytology, 1350
  gastrointestinal system, 1353, 1357
  handbooks, 1329
  indexes and abstracts, 0032
  laboratories, clinical, 1323
  manuals, 1359
  medicine, nuclear, 1338
  periodicals, 0590
  physiology, 1352
  radioisotopes, 3708
  serial monographs, 0764, 0772
  symposia, 1349
  technology, medical, 1347, 1348
intelligence. See also learning
  animals, 2228, 2314
  human, 2228
  primates, 4454
interferon, 2749, 2772, 2794
  serial monographs, 0782, 0795
internal medicine. See medicine, internal
interneurons. See neurons
intersexuality. See hermaphroditism
intervertebral disc displacement in dogs, 3212
invertebrates
  anatomy, comparative, 1450
  animal models, 4587
  behavior, 4586
  biochemistry, 2325
  biology, 4595, 4597, 4599
  chemistry, 2359
  classification, 4580
  ecology, 4600
  embryology, experimental, 4582
  endocrinology, 4583
    comparative, 4578
  hormones, 4578
  learning, 4587, 4593
  manuals, 4596
  neoplasms, 4591

invertebrates (cont.)
  nervous system, 1786, 1788
  neuroanatomy, 1547, 1759
  neurobiology, 4592
  neuroendocrinology, 4579
  neurophysiology, 1759
  parasitology, 0739
  pathology
    bibliographies, 4589
    dictionaries, 4590
    periodicals, 4598
  physiology, 4575, 4579, 4581
  respiration, 4585
  serial monographs, 0739
  temperature, 1979
  tissue culture, 4588, 4594
  vision, 4584
invertebrates, poisonous, 2610
isoenzymes. See enzymes

J

joint diseases, 3208, 3211, 3215
  horses, 3217, 3897
  ultrastructure, 1687
joints. See also orthopedics
  neoplasms, radiology, 3684
  periodicals, 0427-0429

K

kennels. See hospitals, veterinary
kidney, 1850, 2071. See also genitourinary system; urine; urology
  anatomy, 1851
  biochemistry, 1854
  circulation, 1851
  hormones, 2090
  micromanipulation, 1360
  morphology, 1854
  neoplasms, 3537
    diagnosis, radiologic, 3545
    hamsters, 3543
  physiology, 1854, 2074
    comparative, 1976a
  regeneration, 3642
  serial monographs, 0756
  surgery, 1851
  transplantation, 3061
  ultrastructure, 1681
    atlases, 1690
kidney, artificial
  indexes and abstracts, 0044
kidney diseases, 3305, 3312, 3451
  atlases, 3312a
  pregnancy, 3311
kidney failure
  biochemistry, 2397

L

laboratories, clinical
  computer assisted, 1143
  indexes and abstracts, 0053
  instrumentation, 1323
  manuals, 1099, 1105, 1107, 1108, 1125, 1126
  statistics, 1091
laboratory animals. See animals, laboratory
laboratory diagnosis. See diagnosis, laboratory
lactation, 2129. See also milk
  biochemistry, 2091
  biology, 2191
  physiology, 2091, 2118
  serial monographs, 0772, 0773
large animals. See cattle; goats; horses; sheep; swine
lasers, 1336
law. See legislation
learning. See also intelligence
  animals, 2243a, 2262, 2311, 2315
  bibliographies, 0157, 0176
  invertebrates, 4587, 4593
  mammals, 2301
  serial monographs, 0788
  vertebrates, 2231
learning, comparative, 2300
legislation
  animals, laboratory, 4385
  vivisection, 4385
legislation, veterinary, 3796, 3803, 4392
leptospirosis
  animals, 2878
  cattle, 2843
  human, 2878
  manuals, 1227
leukemia, 3333, 3334, 3340, 3556. See also hematologic diseases; myeloproliferative disorders; neoplasia
  indexes and abstracts, 0069
  serial monographs, 0734, 0780, 0784, 0790, 0796
leukemia, experimental, 3564
leukocytes. See also blood cells
  periodicals, 0601
  serial monographs, 0754
libraries, medical, 0994, 0997
ligaments. See tendons and ligaments
light, 2027
  diseases, 3127
lipids in mammals, 2385
liver
  anatomy, 1845
  biochemistry, 1847
  biopsy, 3194
  histopathology, 3536
  immunology, 3065

liver (cont.)
  metabolism, 2063
  morphology, 1847
  physiology, 1847
  regeneration, 3642
  review serials, 0696
  serial monographs, 0779
  transplantation, 0786, 3634
liver diseases, 3290, 3293, 3297, 3300, 3303. See also gastrointestinal diseases; hepatitis
livestock. See also animals, domestic
  abortion, 2165
  care and management, 3788
  dictionaries and encyclopedias, 0894
  digestion, 3758
  diseases, 3722, 3729
  feeds and feeding, 0743
  genetics, 3120
  insemination, 3784
  nutrition, 3751
  parasitic diseases, 2974
  plants, poisonous, 2606, 2617, 2624, 2626
  reproduction, 2170
locomotion. See biomechanics
lumpy skin disease
  serial monographs, 0795
lung. See also pulmonary circulation; respiration
  anatomy, 1840
    radiography, 1839a
  circulation, 3269
  diseases, 0770
  histology, 1840
  mycotic diseases, 2870
  neoplasms, 0761
    atlases, 3528
    veterinary, 3526
  pathology, 3269, 3281, 4311
  physiology, 2057, 3269, 3279
    radiography, 1839a
  serial monographs, 0756, 0761
  vascular diseases, 3529
lymphatic system, 1871, 1906
  morphology, comparative, 1890
  neoplasms, 0767, 0780, 0784, 0786
  periodicals, 0487
  serial monographs, 0732
  ultrastructure, atlases, 1693
lymphocytic choriomeningitis virus
  serial monographs, 0795
lymphocytes, 1879, 2144
  biosynthesis, 1905
  congresses, 1887, 1889
  immunology, 1905
  serial monographs, 0756
lymphomas, 3340
  Burkitt's lymphoma, 0767
  radiotherapy, 3699

M

macaca. See monkeys
macromolecular systems, 2481
magnesium
  metabolism, comparative, 2347
malaria, 2934, 2990
  primates, 2971
malnutrition. See nutrition
mammals, 4226
  anatomy, 1465
    comparative, 1449, 1477
  behavior, 2253, 2301
    maternal, 2292
  biochemistry, 2330
  biology, 4219
    developmental, 1518, 1578
  brain, atlases, 1757
  chromosomes, 1538
  classification, 4222-4225, 4227
  cytogenetics, comparative, 1624
  dictionaries, 0817
  embryology, 1521, 1550
  energy transfer, 2385
  fertilization, 2181
    ultrastructure, 1530, 1596
  gastrointestinal system, peptides, 2354
  genetics, 3109, 3112
  geriatrics, 1543
  hibernation, 2370
  hormones, 2085
  learning, 2301
  lipids, 2385
  musculoskeletal system, 1449
  mutagens, chemical, 1587
  neoplasia, 0742
  neuroanatomy, 1471
  neuroendocrinology, 2087
  nucleic acids, 2385
  oviduct, 1857
  parasitology, 0744
  periodicals, 0457
  physiology, 1465, 1916, 1938, 1939, 4214
    neonatal, 1988
  proteins, 2385
    metabolism, 2382
  radiation protection, 3669
  reproduction, 0801, 2157, 2160, 2167, 2195
  skin, 1827
  teratology, 1549
  tissue culture, 1669
  water-electrolyte balance, 2385
mammary glands, 2107
  diseases, 3322
    animals, domestic, 3718
    cattle, 1869
  neoplasms, 0761
marine biology. See biology, marine
marine microbiology, 4141

mast cells, 1880, 1899
   serial monographs, 0770
mastitis in cattle, 1218, 2125, 3816, 3822
materials. See biocompatible materials
mathematics, 2464. See also statistics
   biology, 2515, 2522, 2534, 2536, 2538, 2540
   medicine, 2515, 2540
      nuclear, 2528
   morphology, 2518
   neurophysiology, 2021
   pharmacy, 2534
   physiology, 2533, 2535
   radiology, 2528
meat inspection. See food inspection
medical education. See education, medical
medical electronics. See electronics, medical
medical illustration. See art; photography; photomicrography
medical research. See research, biomedical
medical technology. See technology, medical
medicine
   computers, 1340, 2529, 2531
   mathematics, 2515, 2540
   statistics, 2520, 2523, 2525-2527, 2532
medicine, clinical
   biochemistry, 2356, 2368
   indexes and abstracts, 0090
   periodicals, 0456
   statistics, 2521
medicine, comparative, 4256
   periodicals, 0354
   serial monographs, 0801
medicine, experimental, 0959, 0960, 0963
   periodicals, 0383, 0442, 0462, 0483, 0516
   review serials, 0632
   serial monographs, 0732, 0751
medicine, internal, 3132, 3134-3136, 3138
   indexes and abstracts, 0032
medicine, nuclear, 2383, 3672, 3696, 3698
   animals, laboratory, 4371
   atlases, 3700
   dictionaries, 0882
   indexes and abstracts, 0032, 0073
   instrumentation, 1338
   manuals, 3697
   mathematics, 2528
   periodicals, 0468
   review serials, 0726
medicine, preventive
   veterinary, 3743, 4268
medicine, tropical, 2840, 3126, 3137, 3141, 3302
   periodicals, 0515
   review serials, 0668
medicine, veterinary, 3713, 3716, 3740, 3784a, 3800, 3807, 4261, 4263-4265

medicine, veterinary (cont.)
   cardiology, 3258
   cats, 3130, 4029
   dermatology, 3250
   diagnosis, 3142
      clinical, 3143, 3144, 3146, 3151
   dogs, 3131, 3998, 4002, 4003
   gastrointestinal system, 3637
   human health, 3537
   nutrition, 3752
   orthopedics, 3206
   radiation protection, 3670
   wildlife, 4179
MEDLARS, 1000
membranes, 1641, 2478, 2983. See also fetal membranes
   biophysics, 2461
   periodicals, 0463
   review serials, 0642
   serial monographs, 0797
   ultrastructure, 1681
metabolism, 2333, 2348, 2374, 2376, 2377, 2379, 2380. See also energy transfer
   adipose tissue, 2078
   animals, 2373
   bibliographies, 0161
   bile, 2063
   body fluids, 1934
   bone, diseases, 1532
   drugs, 2567, 2577
   endocrine glands, 2130
   erythrocytes, 1881, 2351
   fish and aquatic mammals, 4117
   gastrointestinal system, 2058
   hemoglobin, 2137
   liver, 2063
   mammals, proteins, 2382
   manuals, 1142
   minerals, 2327, 3754
   nitrogen, 2325
   periodicals, 0431, 0491
   radioisotopes, 3709
   reptiles, 4166
   review serials, 0619
   ruminants, 0765, 2352
   serial monographs, 0765
   steroids, 2110, 2628
   trace elements
      animals, 2398, 3706
metabolism, carbohydrate, 1978
metabolism, comparative
   magnesium, 2347
metabolism, inborn errors of, 2369, 2390
meteorology, 2506
mice. See also animals, laboratory
   anatomy, 1408
   animal models
      geriatrics, 4513, 4522
      neoplasia, 4522
   bibliographies, 0178

mice (cont.)
  biology, 4525
    developmental, 1574
  brain, atlases, 1812
  brain stem, 1785
  breeds and breeding, 0793
  care and management, 4529
  control, 2896
  embryology
    atlases, 4506
    experimental, 1563
  genetics, 4495
    bibliographies, 4504
    catalogs, 4502
    standards, 4489
  hematology, 0178
  infectious diseases, 4511
  mutagens, 2206
  neoplasms, 4287
  nomenclature, 4490
  nutrient requirements, 4287
  pathology, 4509
  reproduction, 1574
  spinal cord, atlases, 1812
microbes, serial monographs, 0781
microbiology, 2694, 2697, 2717, 2719, 2723.
    See also bacteriology
  bibliographies, 0107, 0199
  environment, 1195
  food inspection, 2904, 2907, 2913
  indexes and abstracts, 0032, 0071
  manuals, 1181, 1183, 1186, 1190, 1196,
    1197, 1199, 1220
  periodicals, 0291, 0292, 0295, 0322,
    0323, 0330, 0355, 0419, 0443, 0460,
    0595
  review serials, 0608, 0620, 0653
  serial monographs, 0735, 0781
  water, 1177, 1179, 2911
microbiology, clinical
  manuals, 2692
microbiology, experimental, 2711, 2738
microbiology, marine, 4141
microbiology, medical, 2726, 2727, 2733,
    2743
microcirculation. See also circulation
  anatomy, 1836
  physiology, 1836
  symposia, 1837
micromanipulation
  kidney, 1360
  manuals, 1245, 1249
microorganisms. See bacteria
microphotography. See photomicrography
microscopy. See also electron microscopy
  atlases, 1151
  dictionaries, 1153
  interference, 1166, 1171
  manuals, 1160, 1164, 1167
  periodicals, 0464

microscopy (cont.)
  phase contrast, 1166
  slit-lamp, 1151
migration, 2227, 2242a, 2258, 2277, 2310,
    2313, 2322. See also populations
milk, 2107. See also dairy products; lactation
milk fever, 3317
minerals. See also calcium; magnesium;
    phosphorous; selenium; sulfur; trace
    elements
  animals, domestic, 2383
  metabolism, 2327, 3754
minks. See also furbearing animals
  anatomy, 4052a
  care and management, 4059
  diseases, 4050, 4058
  nutrient requirements, 0791
  nutrition, 4060
  parasitology, 4050
mitosis, 1649, 1673
models for research. See animal models
molecular biology. See biology, molecular
molecular genetics, 3101
mollusca, 2966
monitoring systems, 1344, 2488
monkeys, 4481. See also primates
  anatomy, 1421
  atlases, 4437
  behavior, 4445, 4455
    social, 2240
  biology, 4429
  brain, atlases, 1770, 1792, 1809, 1815,
    4426, 4438
  breeds and breeding, 4484
  care and management, 4463
  cerebellum, 1791
  classification, 4455
  diseases, 4463
  evolution, 4455
  genetics, comparative, 3104
  histocytochemistry, 4430
  mouth, 4427
  muscles, 4436
  musculoskeletal system, 4433
  nervous system, 4430
  pathology, comparative, 3378
  radiation injuries, 4464
  serial monographs, 0737
monsters. See teratology
morphogenesis, 1512, 1517, 1523, 1550,
    1553, 1565, 1566, 1576, 1584
  embryology, 1534
  vertebrates, 1585
morphology, 1431
  animals, laboratory
    blood cells, 1898
  cells, 1612
  fish and aquatic mammals, 4150
  handbooks, 1079

morphology (cont.)
  kidney, 1854
  liver, 1847
  manuals, 1492
  mathematics, 2518
  neoplasia, experimental, 3525
  neuroglia, 1794
  periodicals, 0293, 0465
  primates, 4478
  reptiles, 4168
  review serials, 0622
  serial monographs, 0732, 0769, 0790
  vertebrates, 1472
morphology, comparative
  lymphatic system, 1890
  ovary, 1482
motor neurons. See neurons
mouth
  monkeys, 4427
  review serials, 0624
movement. See biomechanics
multiple sclerosis
  indexes and abstracts, 0014
muscle contraction, 1747
  serial monographs, 0775
muscles, 1713, 1746, 1994
  biochemistry, 0974, 1729
  biology, molecular, 1993
  biomechanics, 1709
  biophysics, 1715
  birds, 1723
  cats, 1725
  electrodiagnosis, 0148, 1383, 1387-1390
  function, 1383, 1384, 1387
  innervation, 1706
  monkeys, 4436
  neurophysiology, 1706
  nutrition, 1999
  pharmacology, 1711
  physiology, 0148, 1711, 1995, 1996
  regeneration, 3642a
muscular diseases, 3201. See also neuro-
    muscular diseases
  cattle, 3819a
  horses, 3198, 3862, 3897
  serial monographs, 0751
muscular dystrophy, 3218
  indexes and abstracts, 0072
musculoskeletal system, 1736, 1998. See
    also bone; connective tissue
  animals, domestic
    anatomy, 1479
  cattle, 1741
  diseases, 3213
  histology, 1633
  injuries, 3213
  mammals, 1449
  monkeys, 4433
  pathology, 3484
  radiology, 3687

musculoskeletal system (cont.)
  reptiles, 4159
  transplantation, 3629
  turkeys, 1727
mutagens, 3091
  mice, 2206
mutagens, chemical, 2583
  human, 1587
  mammals, 1587
myasthenia gravis. See nervous system
    diseases
mycology, 2808, 2812
  animals, 2811, 2868
  atlases, 2818, 2819
  classification, 2817
  dictionaries, 0877, 0897
  indexes and abstracts, 0032, 0038, 0040
  manuals, 1191, 1202, 1226
  serial monographs, 0742
  veterinary, 1186, 2695, 2816
mycoplasma, 2721, 2729, 2737, 2740
  bibliographies, 0181, 0186
mycotic diseases, 2873, 2875
  animals, 2868, 2877
  birds, bibliographies, 4102
  human, 2842, 2877
  lung, 2870
  nervous system, 2872
mycotoxins, 2589
  food, 2591
myelin sheath, 1763
myeloproliferative disorders, 3348, 3561.
    See also leukemia
myxovirus, 2793, 2852

N

necropsy. See also dissection; specimen
    handling
  dictionaries, 0905
  horses, 1310
  manuals, 1296-1298, 1303, 1309, 1312,
    1313
  ruminants, 1320
nematoda
  animals, domestic, 2945
  human, 2945
  serial monographs, 0744, 0770
neonate. See also pediatrics
  histology, atlases, 1670
  physiology, 1982, 1988, 2158
  radiobiology, 3667
neoplasia, 1265, 3422. See also carcino-
    gens; chemotherapy; leukemia
  animals, laboratory, 4369
  bibliographies, 0196, 0208, 0209, 0812
  biochemistry, 3407, 3436
  biology, molecular, 2475, 2485
  chemotherapy, 2654
  congresses, 3423

neoplasia (cont.)
  cytology, 3418
  education, 3423, 3433a
  embryology, 1535
  etiology, 3417, 3423, 3433a
  geriatrics, 4522
  immunity, 2654, 3058
  immunology, 2807, 3419, 3421, 3434
  indexes and abstracts, 0032, 0050
  information services, 0999
  mice, animal models, 4522
  periodicals, 0342, 0359, 0360, 0391, 0409, 0499, 0532
  research, 0780, 1269, 3404, 3408, 3433a
  review serials, 0706, 0714
  serial monographs, 0754, 0767, 0784, 0790
  skin, 3602
  virology, 0754, 2807, 3434
  viruses, 2747, 2767, 2778, 2795, 2802, 2807, 3405
neoplasia, comparative
  bibliographies, 3442
neoplasia, developmental, 1535
neoplasia, experimental, 1638
  morphology, 3525
neoplasm regression, 3414
neoplasms
  amphibia, 0780
  animals, domestic, 3410, 3430
  animals, laboratory, 0754, 3437
  antigens, 0767
  atlases, 3445, 3446
  bibliographies, 0189
  biopsy, 3528
  bladder, 3538, 3546, 3549
    diagnosis, radiologic, 3544
  bone, 3482, 3483, 3485, 3487
    radiology, 3684
    veterinary, 3488
  brain, 0778, 3505
    ultrastructure, 1696
  cytology, 3450
    atlases, 3441
  diagnosis, 3433a
  etiology, 3415
  fowl, 3409
  gastrointestinal system, 0784, 3533
  hematopoietic system, 0790, 3438, 3448
  histocytochemistry, 2424
  histology, 0761, 3413
  histopathology, 1672
  immunology, 3449
  indexes and abstracts, 0049
  invertebrates, 4591
  joints, radiology, 3684
  kidney, 3537
    diagnosis, radiologic, 3545
    hamsters, 3543
  lung, 0761

neoplasms: lung (cont.)
    atlases, 3528
    veterinary, 3526
  lymphatic system, 0767, 0780, 0784, 0786
  mammary glands, 0761
  mice, 4287
  nervous system, 3500, 3501
    cytology, 3495
  nomenclature, 0871
  ovary, 0784
  pathology, 3451
  radioisotopes, 3704
  rats, 4287
  research, 0778
  respiratory system, 3525
  review serials, 0693
  serial monographs, 0778
  skin, 3245
  stomach, 3534
  therapeutics, 3433a
  thymus gland
    radiotherapy, 3552
  thyroid gland, 0767, 3553, 3554
  transplantation, 3437, 3439, 3629
  uterus, 0784
    cytology, 3548
  viruses, ultrastructure, 1681
neoplasms, comparative
  bone, 3486
nerve cells, 1794
nerve endings, 2019
nervous system, 1807, 1994
  anatomy, comparative, 1501
  animals, domestic and laboratory
    teratology, 1542
  art, 1053
  biochemistry, 2417
  chimpanzees, 0776
  diagnosis, radiologic, 3691
  drugs, 2573, 2678
  electrodiagnosis, 1391
  embryology, 1524, 1540, 1564
  fish and aquatic mammals, 1929, 2239
  function, 1384
  geriatrics, 1981
  histology, 1633
  histopathology, 3500
  indexes and abstracts, 0032, 0045, 0058, 0075
  injuries, 2221
  invertebrates, 1786, 1788
  monkeys, 4430
  mycotic diseases, 2872
  neoplasms, 3500, 3501
    cytology, 3495
  pathology, 3496
    cells, 1653
  pharmacology, 2680
  radiation effects, 0766
  serial monographs, 0756, 0776

nervous system (cont.)
    tissues, 1806
    toxicology, 2596
    transplantation, 3629
    ultrastructure, 1678, 1695
    vertebrates, 1786
nervous system diseases, 1763, 2026, 3220, 3239, 3241, 3552
    biochemistry, 2402
    viral diseases, 3240
neural transmission
    congresses, 1808
    histocytochemistry, 0777
neuroanatomy, 1755, 1756, 1769, 1774, 1777, 1793, 2199, 2202, 2219
    atlases, 1471, 1754, 1771
    cats, 1780
    dogs, 1780
    human, 1460, 1817
    invertebrates, 1547, 1759
    mammals, 1471
    manuals, 1240
    rats, 1823
    research, 1795
    vertebrates, 1460
neuroanatomy, clinical, 1758, 1773
neuroanatomy, comparative, 1800
neurobiology, 1786
    invertebrates, 4592
    periodicals, 0494
    primates, 4476
    review serials, 0667
neurobiology, developmental, 1779
neurochemistry, 1755, 2399, 2403, 2414, 2420. See also endocrinology, neurochemical
    animal models, 4376
    bibliographies, 0169
    brain stem, 1819
    conferences, 4587
    congresses, 2409, 2410
    genetics, 2409, 2410
    handbooks, 2412
    hypothalamus, 2116
    manuals, 2404
neurochemistry, comparative, 2418
neurochemistry, experimental, 2419
neuroendocrinology, 2029, 2112
    congresses, 2104
    invertebrates, 4579
    mammals, 2087
    periodicals, 0495
    review serials, 0658
    serial monographs, 0741
neuroglia
    morphology, 1794
neurohistochemistry. See neurochemistry
neurology, 2199, 2201, 2204, 2209, 2465
    behavior, animals, 2270
    bibliographies, 0209

neurology (cont.)
    biomedical engineering, 2208
    dogs, 2215, 2218
    indexes and abstracts, 0032
    manuals, 2207
    periodicals, 0247, 0384, 0466, 0496
    review serials, 0686, 0699
    serial monographs, 0777
    veterinary, 2205
neurology, clinical, 2210, 2212, 2213, 2219
    horses, 2220
neurology, comparative, 0977, 1786, 1800
neuromuscular diseases. See also muscular diseases; nervous system diseases; spinal diseases
    electrodiagnosis, 1386
    epilepsy, 0032
    multiple sclerosis, 0014
    Parkinson's disease, 0075, 0183, 0209, 0211
neurons, 1776, 1811, 1994
    serial monographs, 0788
neuro-ophthalmology, 1768
neuropathology, 3490, 3492, 3493, 3502
    atlases, 3491, 3504
    manuals, 1240, 1284
    periodicals, 0294, 0466, 0597
    review serials, 0700
neuropathology, comparative, 3494
neurophysiology, 0965, 1755, 1756, 1774, 2007, 2011, 2026, 2029, 2031, 2036
    bibliographies, 0159
    handbooks, 1928
    insects, 2039
    invertebrates, 1759
    mathematics, 2021
    muscles, 1706
    pharmacology, 2009
neurophysiology, clinical, 1773, 2023
neurosecretion, 2406
neurosurgery
    periodicals, 0467
    spinal cord, 1751
neurosurgery, comparative, 1800
Newcastle disease, 2853, 2866
nitrogen
    metabolism, 2325
Nobel prize
    physiology, 0976, 0984
nuclear medicine. See medicine, nuclear
nucleic acids
    immunology, 2449
    mammals, 2385
    review serials, 0701
    RNA, 0788
    viruses, 2758
nucleolus, 1617
    serial monographs, 0790
nucleus
    ultrastructure, 1681

nursing, veterinary, 3794, 3797
nutrient requirements. See also amino
    acids; minerals; proteins; trace ele-
    ments; vitamins
  animals, laboratory, 0791
  cats, 3994
  cattle, 0791
  dogs, 0791, 3994
  foxes, 0791
  horses, 0791
  mice, 4287
  minks, 0791
  poultry, 0791
  rabbits, 0791
  rats, 4287
  serial monographs, 0791
  sheep, 0791
  swine, 1791
nutrition. See also feeds and feeding;
    food
  animals, domestic, 3762
  animals, laboratory, 4279, 4297
  bibliographies, 0209
  bone, 1738
  cardiovascular diseases, 3266
  cats, 4036
  cattle, 3849
  chickens, 4111
  computers, 3752
  connective tissue, 1738
  dogs, 3986, 3993
  embryology, 1528
  guinea pigs, 4520
  indexes and abstracts, 0074
  livestock, 3751
  medicine, veterinary, 3752
  minks, 4060
  muscles, 1999
  periodicals, 0344, 0469, 0478
  poultry, 0773, 4110
  primates, 4483
  radioisotopes, 1957
  rats, 4279
  review serials, 0687, 0719
  ruminants, 0765, 2061, 3846
  serial monographs, 0736, 0743, 0765,
    0773, 0800, 0801
  sheep, 0743
  sulfur, 2392
  swine, 0773, 3941, 3944
  trace elements, 2371
    animal, 2395
    human, 2395
  veterinary, 2265a, 2361, 3745, 3746,
    3750, 3752, 3759-3761, 3763,
    3765
  wildlife, 0801
nutrition, comparative
  animals, domestic, 2381
  human, 2381

O

obstetrics and gynecology, 3638. See
    also gestation; pregnancy; reproduc-
    tion
  diagnosis, radiologic, 3693
  dictionaries, 0888
  indexes and abstracts, 0032
  pathology, 3308
  periodicals, 0306, 0523
  review serials, 0645
  ultrasonics, 1396
  veterinary, 3304, 3310, 3586
oncology. See neoplasia
ophthalmology, 1768. See also eye
  animals, domestic, 3228
  dogs, 3232
  indexes and abstracts, 0032
  periodicals, 0380, 0412, 0589
  review serials, 0727
  ultrasonics, 3227
  veterinary, therapeutics, 2558
ophthalmology, comparative, 1463, 3224
opossums. See also animals, laboratory
  brain, 4053a
optics. See vision
oral manifestations, 3286. See also dentis-
    try; mouth; teeth
  diseases, 3155
organ culture. See tissue culture
organ transplantation. See transplantation
orthopedics, 3202, 3639. See also bone;
    calcification; joints; surgery, ortho-
    pedic
  medicine, veterinary, 3206
  radiology, veterinary, 3686
  review serials, 0646
osmoregulation. See water-electrolyte bal-
    ance
otorhinolaryngology. See also ear; sense
    organs
  bibliographies, voice, 0147
  indexes and abstracts, 0032
  manuals, 1326
ovary. See also endocrine glands; gonads;
    ovulation; reproduction
  cattle, 0779
  morphology, comparative, 1482
  neoplasms, 0784
  physiology, 2123
  serial monographs, 0756, 0779
  transplantation, 0145
oviducts
  mammals, 1857
ovine. See sheep
ovulation. See also reproduction
  control, 2127, 2174
ovum. See also embryology; fertilization;
    morphogenesis; reproduction
  implantation, 1579, 2189

P

pancreas. See also diabetes
  handbooks, 1928
  serial monographs, 0796
parasitic diseases, 2979. See also coccidiosis; malaria; trichinosis; trypanosomiasis
  animals, domestic, 2975, 3738
  animals, laboratory, 4346
  birds, 2827
  cattle, 2970
  horses, 3856
  human, 2969
  immunity, 2939
  livestock, 2974
  primates, 3384
  serial monographs, 0739
  veterinary, 2969
  wildlife, 2973, 4177
parasitology, 2834, 2922, 2923, 2928, 2933, 2935, 2949, 2953, 2960, 2961, 2989. See also cestoda; helminthology; nematoda
  animals, domestic, 2924
  birds, 0744, 2962
  dictionaries, 0833
  fish and aquatic mammals, 0739, 2938, 4119
  indexes and abstracts, 0032, 0085
  mammals, 0744
  manuals, 1094, 1200, 1213, 1216, 1219, 1231
  minks, 4050
  periodicals, 0321, 0385, 0470, 0500
  review serials, 0625
  serial monographs, 0738, 0739, 0799
  sheep, 0744
  veterinary, 1200, 1216, 1231, 2921, 2925, 2936, 2943, 2944, 2955, 2957, 2963
    manuals, 2927
  wildlife, 2962, 2973
parasitology, chemical, 2359
parasitology, experimental
  manuals, 2948
parathyroid glands, 2086
Parkinson's disease
  bibliographies, 0183, 0209, 0211
  indexes and abstracts, 0075
parturient paresis, 3317
parturition
  serial monographs, 0768
pathology, 2716, 3361-3364, 3368, 3371, 3379, 3381, 3382, 3386, 3387, 3392, 3397, 3431, 3453. See also biopsy; histopathology; neuropathology; psychopathology, comparative
  anatomy, 1433, 3724
  animals, domestic, 1486
  animals, laboratory, 4344, 4345, 4509

pathology (cont.)
  atlases, 3365, 3372
  bone, classification, 3489
  bone marrow, atlases, 3561
  cardiovascular system, 3513, 3519-3521
  chemistry, 2349
  circulation, 4323
  cytology, 1625
  dictionaries, 0905
  digestion, 2070
  embryology, 1593
  erythrocytes, 1881
  gastrointestinal system, 3535
  gestation, 2187
  hamsters, 0778
  heart
    circulation, 2044
    conducting tissue, 3516
  history, 0971
  human, 1925
  indexes and abstracts, 0020, 0032
  invertebrates
    bibliographies, 4589
    dictionaries, 4590
    periodicals, 4598
  lung, 3269, 3281, 4311
  manuals, 1309, 1317, 3366, 3367
  mice, 4509
  musculoskeletal system, 3484
  neoplasms, 3451
  nervous system, 3496
    cells, 1653
  nomenclature, 0868
  obstetrics and gynecology, 3308
  periodicals, 0275, 0295, 0296, 0307, 0325, 0342a, 0377, 0454, 0471
  physiology, 1944, 1953, 1954
  pigments, 3399
  primates, 3384
  protozoology, 2978
  radiation injuries, 4464
  rats, 4509
  review serials, 0665, 0681, 0688, 0689, 0729
  serial monographs, 0756, 0778
  stomach, 1353
  surgery, 3367
  technology, radiologic, 3674
  veterinary, 3369, 3373, 3375-3377, 3383, 3388, 3394
    manuals, 3360
pathology, clinical, 3467, 3468, 3471, 3475, 3476
  manuals, 1092, 1122
  periodicals, 0246, 0304, 0433
  review serials, 0729
  veterinary, 1100, 3189, 3466
pathology, comparative
  dictionaries, 0891
  monkeys, 3378

pathology, comparative (cont.)
  periodicals, 0248
  review serials, 0650
  serial monographs, 0769
  skin, 2040
pathology, systemic, 3389, 3393, 3400
pediatrics. See also neonate
  dogs, 3957
  indexes and abstracts, 0032
peptides, 2111
  mammals, gastrointestinal system, 2354
perfusion, 1308. See also therapeutics
periodicals
  bibliographies, 0095-0122
pertussis
  serial monographs, 0782
pesticides. See also toxicology
  analysis, 2580-2582, 2603
  indexes and abstracts, 0076, 0077
  pharmacology, 2581
  serial monographs, 0789
  therapeutics, 2581
  veterinary, 2597
  wildlife, 2601
pets, 4232, 4238, 4243, 4247
  care and management, 4229
  exotic animals, 4228
  feeds and feeding, 3753
  psychology, 4237
  wildlife, 4239
pharmacognosy, 2577a
pharmacology, 2548, 2554, 2559-2563,
    2566, 2571, 2573. See also chemo-
    therapy; drugs; psychopharmacology;
    therapeutics
  anesthesia, local, 3605
  biology, marine, 2545
  cells, 1917, 2552
  encyclopedias, 0895
  handbooks, 2549
  hormones, 2578
  indexes and abstracts, 0032, 0091, 0094
  inflammation, 2576
  manuals, 2570
  muscles, 1711
  nervous system, 2680
  neurophysiology, 2009
  periodicals, 0297, 0356, 0473, 0528
  pesticides, 2581
  research, 2669
  review serials, 0626, 0638, 0682, 0690
  synonyms, 0860
  transplantation, 2546
  veterinary, 2542, 2553, 2564
    ophthalmology, 2558
pharmacology, experimental, 2565, 2579
  animal models, 2670, 2671
pharmacopoeias. See formularies
pharmacy
  computers, 2534

pharmacy (cont.)
  mathematics, 2534
  periodicals, 0472
  statistics, 2534
phonation, 0147. See also communication;
    speech
phosphorous
  bone, 2086
photography, 1731. See also art; photo-
    micrography
  gastrointestinal system, 1357
  handbooks, 1038, 1039, 1042
  wildlife, 4187
photomicrography, 1150, 1159, 1169, 1170,
    1173
photosensitization
  serial monographs, 0742
physical diagnosis. See diagnosis, clinical
physics. See also biophysics; psycho-
    physics
  handbooks, 1081
  radioisotopes, 3708
physiological adaptation, 2004
physiological chemistry. See biochemistry
physiology, 1918, 1923, 1927, 1931, 1936,
    1949, 1950, 1952, 1958. See also
    electrophysiology; neurophysiology
  adrenal glands, 2094
  anesthesia, local, 3605
  animals, domestic, 1459, 1919, 1963
    reproduction, comparative, 2184
  animals, heterothermic, 1978
  bibliographies, 0969
  birds, 1956
  bladder, 1852
  body fluids, 2074
  bone, 1716, 1990, 1997, 2000
  calcium, 2327
  cardiovascular system, 2045
  cells, 1632, 1646, 1917, 1940, 2002
  cerebrospinal fluid, 2015
  chimpanzees, 2233
  circulation, 2046, 2047, 4322
    cerebellum, 2030
  connective tissue, 1992
  cytology, 1612, 1622, 1645, 1650, 1651
  digestion, 2070
  endocrine glands, 2082, 2124, 2130
    comparative, 2100
  erythrocytes, 1872
  eye, 1764, 1768, 2003, 2016
  fish and aquatic mammals, 1929, 4117
  fowl, 4083
  gastric juice, 2067
  gastrointestinal system, 1843, 1846, 1928,
    2060-2062, 2069, 3287
  genetics, 1975
  genitourinary system, 2075
  gestation, 2158, 2187
  guinea pigs, 4520

physiology (cont.)
  handbooks, 1928
  hemoglobin, 2143
  histology, 1610, 1636
  history, 0971
  hormones, 2119
  human, 1925, 1926, 1955
  indexes and abstracts, 0015
  insects, 1909, 1933
  instrumentation, 1352
  invertebrates, 4575, 4579, 4581
  kidney, 1854, 2074
  lactation, 2091, 2118
  liver, 1847
  lung, 2057, 3269, 3279
    radiography, 1839a
  mammals, 1465, 1916, 1938, 1939, 4214
  manuals, 1009, 1299, 1331, 1343, 1424
  mathematics, 2533, 2535
  microcirculation, 1836
  muscles, 0148, 1711, 1995, 1996
  Nobel prize, 0976, 0984
  ovary, 2123
  periodicals, 0235, 0298, 0299, 0308, 0356, 0366, 0422, 0440, 0444, 0474, 0502, 0504, 0510, 0513
  pineal body, 2106
  platelets, 2146
  poultry, 1910, 4083
  radioisotopes, 1957
  reproduction, 1859, 2180, 2186, 2190
  respiration, 2052, 2055
  review serials, 0614, 0639, 0656a
  ruminants, 0765
  sense organs, 2012, 2024, 2028
    insects, 2017
  serial monographs, 0765, 0775
  skin, 2041, 2042
  surgery, experimental, 4336
  swine, 2114
  synapses, 2019
  testis, 1853
  urethra, 1852
  vertebrates, 1443, 1464, 1504, 1938, 1943, 1974
    reproduction, 2194
  veterinary, 1913, 1924, 1932, 1945, 1959, 1960, 1967, 1975, 1980
    manuals, 4320, 4327
  vision, 1768
physiology, clinical, 1911, 1912, 1961
physiology, comparative, 1962, 1965, 1970, 1976, 1977
  birds, 1937
  gastrointestinal system, 1973
  kidney, 1976a
  respiration, 2054
  respiratory system, 2056
  skin, 2040
  veterinary, 1947

physiology, developmental, 1987
physiology, environmental
  handbooks, 1928
  mammals, 1921
physiology, experimental, 1908
  rabbits, 1972
  veterinary, 1920
physiology, neonatal, 1982, 2158
  mammals, 1988
physiology, pathological, 1944, 1953, 1954
pig. See swine
pigeons, 4108. See also birds
  anatomy, 4071
  brain, atlases, 1782
  diseases, 4084
  spinal cord, 1818
pigments
  pathology, 1399
pineal body, 1870, 2086. See also endocrine glands
  physiology, 2106
pituitary gland, 1862, 2086
placenta, 1586. See also obstetrics and gynecology
  ultrastructure, 1679
plants
  manuals, 2621
plants, poisonous, 2605, 2607, 2608, 2611, 2615, 2616. See also toxins
  livestock, 2606, 2617, 2624, 2626
  manuals, 2619
plastics, 1306. See also biocompatible materials
  surgery, 1355, 3586a, 3595, 3600
    orthopedic, 3587
plastic surgery. See surgery, plastic
platelets, 3343. See also blood cells; blood coagulation; hematology
  bibliographies, 0175
  physiology, 2146
  serial monographs, 0752, 0756
poisonous animals. See animals, poisonous
poisonous invertebrates, 2610
poisonous plants. See plants, poisonous
polyoma virus
  serial monographs, 0795
populations. See also ecology; migration
  animals, 4180
  wildlife, 4182, 4184, 4202
porcine. See swine
poultry. See also birds; chickens; fowl
  behavior, spatial, 2259a
  biochemistry, 1910
  congresses, 2866
  diseases, 0742, 2853, 2866, 4074, 4084, 4097, 4100
    economics, veterinary, 4093
  husbandry, 4105, 4106, 4109
  nutrient requirements, 0791
  nutrition, 0773, 4110

poultry (cont.)
  periodicals, 0280, 0348, 0507, 0607
  physiology, 1910, 4083
  salmonella, 2841
PPLO. See mycoplasma
practice management, 3802, 3844. See also economics, veterinary; herd management programs; legislation, veterinary
  periodicals, 0249, 0273, 0586
  review serials, 0628
  serial monographs, 0794
pregnancy. See also obstetrics and gynecology
  kidney diseases, 3311
primates, 2672, 4470. See also apes; baboons; chimpanzees; gorillas; monkeys
  adaptation, 2271
  anatomy, 1422, 4476
  animal models, 4377, 4425
  behavior, 0150, 0691, 0760, 2238, 2248, 2295, 4447, 4451, 4476, 4478
    evolution, 2274
  bibliographies, 0150, 0170
  biological clocks, 1915
  biology, 3090, 4443
  brain, 1796
  breeds and breeding, 0793, 4425
  classification, 1422, 4476
  communication, 2223
  diseases, 0760, 4462
  ecology, 4478
  evolution, 4442, 4443
  feeds and feeding, 4483
  genetics, 0776, 3090
    transplantation, 3062
  handbooks, 4475
  hematology, atlases, 1886
  immunology, 0776, 4476
  immunosuppression, 3066
  infectious diseases, 3066, 3384
  intelligence, 4454
  malaria, 2971
  morphology, 4478
  neurobiology, 4476
  nomenclature, 0876
  nutrition, 4483
  parasitic diseases, 3384
  pathology, 3384
  periodicals, 0390, 0461, 0508
  psychology, 2271
  reproduction, 2169, 4476
  research, 4451, 4453
  review serials, 0691
  serial monographs, 0737, 0745, 0760, 0776, 0792
  virology, 0795
  zoonoses, 2883
prostaglandins, 2111. See also drugs
  bibliographies, 0209

prostaglandins (cont.)
  serial monographs, 0772
proteins. See also nutrition
  biosynthesis, 2430
  mammals, 2385
    metabolism, 2382
  manuals, 1127
  serial monographs, 0773
  urine, 2378
protozoology, 2959
  animals, domestic, 2946
  atlases, 1209
  handbooks, 0975
  human, 2946
  periodicals, 0475
  veterinary, 2920, 2956
protozoology, chemical, 2359
protozoology, pathological, 2978
pseudorabies
  serial monographs, 0795
pseudotuberculosis. See sarcoidosis
psittacosis. See chlamydial diseases
psychology. See also behavior
  animals, 2283
    zoos, 2267, 2268
  birds, 2236
  pets, 4237
  primates, 2271
psychology, comparative, 2245, 2270a, 2303
psychology, social
  animals, 2323
psychopathology, comparative, 2241. See also aggression; behavior
psychopharmacology, 2676, 2679, 2683, 2685, 2687. See also drugs: psychotherapeutic; pharmacology; tranquilizing agents
  handbooks, 2682
psychopharmacology, biochemical, 2675
psychophysics, 2035, 2294
public health, 2905, 2906, 4268. See also environmental health; epidemiology; food inspection; hygiene
  indexes and abstracts, 0032
  periodicals, 0309, 0357, 0509
pulmonary circulation, 2051, 2057. See also lung

Q

quail, Coturnix. See Coturnix quail

R

rabbits, 4061, 4573
  anatomy, 1403, 1409, 1428
  anticoagulants, 2645
  brain, atlases, 4558
  breeds and breeding, 0793
  care and management, 4062, 4063

rabbits (cont.)
  diseases, 4057
  dissection, 1319
  eye research, 4054
  nutrient requirements, 0791
  physiology, experimental, 1972
rabies, 1189, 2849, 2860, 2880. See also pseudorabies; zoonoses
  animals, 2851, 2863a
  bibliographies, 0209
  congresses, 2865
  human, 2851, 2863a
  serial monographs, 0782, 0800
radiation
  transplantation, 3627, 3662
radiation effects, 3665
  animals, laboratory, 4347
  hematopoietic system, 3663
  nervous system, 0766
  toxins, 2620
  venoms, 2620
radiation injuries, 3689
  monkeys, 4464
  pathology, 4464
  therapeutics, 4464
radiation injuries, experimental, 4347
radiation protection. See also health physics
  mammals, 3669
  medicine, veterinary, 3670
radiation sickness. See radiation injuries
radiobiology, 1267, 3666, 3695
  embryology, 3667
  encyclopedias, 0916
  immunity, 3037, 3043
  neonate, 3667
radiography
  animals, laboratory, 3676
  dogs, 3676
    anatomy, 1440
  horses, anatomy, 1440
  lung
    anatomy, 1839a
    physiology, 1839a
  veterinary, 3694
radioisotopes, 3711
  animal research, 3703
  animals, domestic, 2383, 3706
  instrumentation, 3708
  manuals, 3703
  metabolism, 3709
  neoplasms, 3704
  nutrition, 1957
  physics, 3708
  physiology, 1957
radiology, 3655, 3657, 3658, 3660, 3672. See also diagnosis, radiologic; health physics; medicine, nuclear; technology, radiologic
  atlases, 1730

radiology (cont.)
  bibliographies, 0187
  bone, atlases, 3482
  dictionaries, 0882, 1046
  indexes and abstracts, 0032
  mathematics, 2528
  musculoskeletal system, 3687
  neoplasms
    bone, 3684
    joints, 3684
  orthopedics, veterinary, 3686
  periodicals, 0224, 0310, 0408, 0411, 0413, 0511, 0512
  review serials, 0654, 0704
  veterinary, 3673, 3678, 3679
radiology, clinical, 2212
radiotherapy
  lymphomas, 3699
  neoplasms, thymus gland, 3552
  serial monographs, 0780
rats, 4527
  anatomy, 1407, 1420, 4499
  animal models, 4512
    geriatrics, 4513
  behavior, 2229, 2237
  brain, 1775
    atlases, 1749, 1784, 1798, 1810
  breeds and breeding, 0793
  control, 2896
  diabetes, experimental, 4517
  ecology, 2237
  genetics, 3122
  histology, atlases, 4503
  infectious diseases, 4511
  neoplasms, 4287
  neuroanatomy, 1823
  nutrient requirements, 4287
  nutrition, 4279
  pathology, 4509
  surgery, gastrointestinal system, 4510
red blood cells. See erythrocytes
regeneration, 1571
  bone
    cytology, 3650
    vitamin K, 3649
  kidney, 3642
  liver, 3642
  muscles, 3642a
  serial monographs, 0751
  skeleton, 1532
  vertebrates, 0194
  veterinary, 1562
renal diseases. See kidney diseases
reoviruses
  serial monographs, 0795
reproduction, 2102. See also abortion; animal production; breeds and breeding; fertility; gestation; insemination; obstetrics and gynecology; parturient paresis; pregnancy; semen; sex; sterility

reproduction (cont.)
  animals
    diagnosis, 2197
    therapeutics, 2197
  animals, domestic, 2164
  animals, laboratory, 4284
  behavior, 2182, 2293
    animals, 2259
  bibliographies, 0145, 0188, 0212
  cattle, 2190
  chimpanzees, 2233
  directories, 0927
  dogs, 2173
  handbooks, 1079
  hormones, 2160
  horses, 2185
  indexes and abstracts, 0027
  livestock, 2170
  mammals, 2157, 2160, 2167, 2195
  mice, 1574
  periodicals, 0335, 0476, 0598, 0603
  physiology, 1859, 2180, 2186, 2190
  primates, 2169, 4476
  review serials, 0627
  serial monographs, 0733, 0738, 0743, 0768, 0773, 0801
  sheep, 0743, 3910
  swine, 2114, 3941
  vertebrates, 2162, 2163
    physiology, 2194
  veterinary, 2109
reproduction, comparative
  animals, physiology of domestic, 2184
reptiles, 4164, 4165, 4167, 4170, 4171. See also amphibia; herpetology; snakes; venoms
  biology, 4168
    molecular, 4153
  chromosomes, 4067
  classification, 4172
  metabolism, 4166
  morphology, 4168
  musculoskeletal system, 4159
  water-electrolyte balance, 4166
research
  animals, laboratory, 4277, 4386, 4408
  biology, marine, 4134
  brain, 1775, 1804
  directories, 0946
  environment, 4379
  eye, rabbits, 4054
  fish and aquatic mammals, 4125, 4140
  geriatrics, 1545, 4513
  germ-free life, 4286
  gestation, 4357
  guinea pigs, 4520
  handbooks, 0948
  hormones, 2088
  immunochemistry, 2446
  immunology, 2446

research (cont.)
  methods, 1006, 1007, 1010, 1014, 1022
  neoplasia, 0780, 1269, 3404, 3408, 3433a
  neoplasms, 0778
  pharmacology, 2669
  primates, 4451, 4453
  support, 1023, 1025, 1027, 1028, 1032-1034, 1036
  swine, 0093
  toxicology, animals, 2673
  wildlife, 4201
research, biochemical
  animal models, 1018, 1021, 4548
research, biological, 4405
  animals, laboratory, 4389
  methods, 1012, 1013, 1016, 1019
research, biomedical
  animal models, 1018
  animals, laboratory, 4368
  computers, 2539
  directories, 0922, 0937
  indexes and abstracts, 0087
  methods, 1005, 1008, 1011, 1015, 1020, 1021
  review serials, 0675
  serial monographs, 0776, 0777, 0788, 0792
  statistics, 2514
  swine, 3935, 3936, 3937
respiration
  fish and aquatic mammals, 1929
  handbooks, 1077, 1928
  invertebrates, 4585
  physiology, 2052, 2055
    comparative, 1971
respiratory function tests, 3163, 3270
respiratory insufficiency, 3276, 3278
respiratory system. See also lung
  anesthesiology, 3608
  biopsy, 3528
  birds, anatomy, 4075
  cytology, 1671
  diagnosis, radiologic, 3693
  indexes and abstracts, 0032
  neoplasms, 3525
  periodicals, 0315, 0513
  physiology, comparative, 2056
  viral diseases, 2863
respiratory tract diseases, 3273, 3274, 3280. See also lung: diseases; lung: mycotic diseases; tuberculosis
  human, 3277
  veterinary, 3277
restraint
  animals, 3593
  chimpanzees, 2233
  wildlife, 4188
reticuloendothelial system, 1900, 3562. See also hematopoietic system

reticuloendothelial system (cont.)
   periodicals, 0514
   serial monographs, 0732, 0752
retina. See also eye
   anatomy, 1805
   diseases, 1768
   serial monographs, 0788
rheology. See also circulation
   blood, 2141, 2463, 2470
rheumatism. See arthritis and rheumatism
rhinoviruses
   serial monographs, 0771, 0795
rickettsial diseases, 2822
   animals, 2848
   handbooks, 1182, 1212
   human, 2854
rinderpest virus, 2862
   serial monographs, 0795
RNA. See also genetics; nucleic acids
   serial monographs, 0788
roaring
   horses, 3857
rodents. See also gerbils; guinea pigs; hamsters; mice; rats
   breeds and breeding, 0793
rubella virus
   serial monographs, 0782, 0795
ruminants. See also cattle; goats; sheep
   coccidiosis, 2947
   digestion, 0765
   dissection, 1300
   gastrointestinal system, 1841
   metabolism, 0765, 2352
   necropsy, 1320
   nutrition, 0765, 2061, 3846
   serial monographs, 0765

S

salmonella
   feeds and feeding, 1225
   poultry, 2841
sarcoidosis, 3390
   bibliographies, 0209
   serial monographs, 0782
scar tissue. See wound healing
schistosomiasis
   bibliographies, 0214
   serial monographs, 0800
scientific writing. See writing
scrapie virus
   bibliographies, 0153
sebaceous glands
   serial monographs, 0731
Seeing Eye dogs, 4235
selenium, 2648
semen, 3782, 3783a. See also insemination; reproduction; spermatogenesis
sense organs, 1998. See also ear; eye
   atlases, 1754

sense organs (cont.)
   birds, 2005
   fish and aquatic mammals, 4126, 4139
   insects, physiology, 2017
   physiology, 2012, 2024, 2028
   review serials, 0648
   serial monographs, 0740
   skin, 1778
   ultrastructure, 1678
   vertebrates, 1760, 1761
senses
   animals, 2203, 2249
   behavior, 2296
serology, 1111, 3080. See also technology, medical
   indexes and abstracts, 0032
   manuals, 1118, 1129
   periodicals, 0322
serums, immune, 2657
sex, 2196. See also reproduction
   chromosomes, 3096
sex determination
   birds, 1858
sex ratio, 2178
sheep. See also ruminants
   anatomy, 1430
      respiratory system, 1445
   behavior, 2261
   biochemistry, 3915
   bone marrow, 0779
   breeds and breeding, 3916, 3922
   diseases, 3902, 3904, 3906, 3909, 3911
   evolution, 2261
   genetics, 3923
   husbandry, 3902, 3906, 3917, 3920-3922
   indexes and abstracts, 0036
   nutrient requirements, 0791
   nutrition, 0743
   parasitology, 0744
   reproduction, 0743, 3910
   thorax, atlases, 1839
shellfish, 2966
   microbiology, 1177
shock, 3139, 3140. See also emergencies; trauma
   serial monographs, 0732
skeleton. See also bone
   diagnosis, radiologic, 3691, 3693
   regeneration, 1532
skin, 1825, 1830. See also dermatology
   biochemistry, 2041
   biology, 1826
   histopathology, 3251, 3254, 3508
      atlases, 3510
   human, ultrastructure, 1680
   immunology, 3069
   mammals, 1827
   neoplasia, 3602
   neoplasms, 3245
   pathology, comparative, 2040

skin (cont.)
  physiology, 2041, 2042
    comparative, 2040
  sense organs, 1778
  serial monographs, 0731, 0756, 0787
  surgery, 1828
  ultrastructure, 1704
  vertebrates, 2032
  wound healing, 3645
skin diseases, 0795, 2616, 3248.
  cats, 3511
  dogs, 3249, 3511, 3512
  guinea pigs, 3068, 3246
  human, 3512
skull
  vertebrates, 1522
slow virus diseases
  serial monographs, 0771
small animals. See cats; dogs; furbearing animals
smell. See sense organs
snakes. See also herpetology; reptiles; venoms
  anatomy, 4156
  bibliographies, 0192
  classification, 4172, 4173
  manuals, 2627
social behavior. See behavior, social
social psychology, animal, 2323
space, medicine
  animals, laboratory, 4465
  bibliographies, 0155
  chimpanzees, 4465
  serial monographs, 0788
spatial behavior. See behavior, spatial
specimen handling, 1271, 1297, 1299, 1302, 1306. See also dissection; tissue preservation
  manuals, 1314, 1316, 1317
spectrophotometry
  instrumentation, 1348
  manuals, 1090
speech, 0147, 1047, 1051, 1060, 1062, 1767. See also communication
spermatogenesis. See also semen
  serial monographs, 0768
spinal cord
  injuries, 2211
  mice, atlases, 1812
  neurosurgery, 1751
  pigeons, 1818
spinal diseases
  dogs, 3967
spirochete infections
  manuals, 1217
stable management, 3875, 3877, 3881, 3885, 3890. See also horses
stains and staining
  manuals, 1232, 1237, 1240-1242, 1247, 1248, 1252, 1258, 1262

stains and staining (cont.)
  periodicals, 0521
statistics, 2537. See also mathematics
  agriculture, 2523
  biology, 2517, 2519, 2523, 2527, 2530, 2534, 2541
  laboratories, clinical, 2516
  medicine, 2520, 2523, 2525-2527, 2532
    clinical, 2521
  pharmacy, 2534
  research, biomedical, 2514
stereoencephalotomy
  serial monographs, 0764
sterility. See also fertility; reproduction
  cattle, 2156
  periodicals, 0389, 0605
  serial monographs, 0733
sterilization. See infection control
steroids, 2102. See also hormones
  congresses, 2103
  histocytochemistry, 2421
  metabolism, 2110, 2628
  review serials, 0629, 0707
  serial monographs, 0772, 0788
  therapeutics, 2662
stomach. See also digestion; gastric juice; gastrointestinal system
  cytology, diagnostic, 3531
  neoplasms, 3534
  pathology, 1353
stress, 1645. See also behavior
sulfur
  nutrition, 2392
surgery, 3570, 3573, 3583, 3585, 3597. See also anatomy, surgical; anesthesia; neurosurgery; transplantation; wound infection
  adrenal glands, 3628
  atlases, 3601
  cats, 3130, 3579, 3582
    atlases, 1488
    diseases, 3577
  cattle, 3129
  cryogenic, 3588, 3589
  dermatology, 3602
  dogs, 3131, 3157, 3579, 3582
    atlases, 1488, 3565
    diseases, 3577
    water-electrolyte balance, 2364
  electronics, medical, 3607
  eye, 3623
  gastrointestinal system, 0794, 1843, 3287
    rats, 4510
  genitourinary system, 3619
    cattle, 3640
    horses, 3640
  hematologic diseases, 3341
  horses, 3128
  indexes and abstracts, 0032

surgery (cont.)
  kidney, 1851
  manuals, 3367, 3590
  pathology, 3367
  periodicals, 0311, 0326, 0345, 0370,
    0477, 0505, 0522, 0523
  plastics, 1355, 3586a, 3595, 3600
  review serials, 0672, 0683, 0685, 0702,
    0713, 0725, 0728
  skin, 1828
  veterinary, 3567, 3575, 3581, 3637
  wound healing, 3643a, 3648
surgery, cardiovascular
  indexes and abstracts, 0032
  review serials, 0685, 0721
surgery, experimental
  periodicals, 0383
  physiology, 4336
  primates, 0745
  serial monographs, 0745
surgery, orthopedic, 3596
  cats, 3625, 3631, 3635
  dogs, 3625, 3631, 3635
  indexes and abstracts, 0032
  plastics, 3587
  review serials, 0728
  serial monographs, 0794
surgery, plastic
  indexes and abstracts, 0032
  periodicals, 0505
  review serials, 0728
surgery, thoracic
  indexes and abstracts, 0032
sweat glands
  serial monographs, 0731
swine. See also African swine fever;
    trichinosis; vesicular exanthema
  animal models, 3934-3937
  breeds and breeding, 0793
  diseases, 0186, 0742, 3925-3928
  dissection, 1554, 1583, 3930
  embryology, 1533, 1552, 1554, 1557,
    1559, 1583, 1589, 3929-3931
  husbandry, 3925, 3927, 3938-3940, 3942,
    3943
  indexes and abstracts, 0036, 0093
  nutrient requirements, 0791
  nutrition, 0773, 3941, 3944
  physiology, 2114
  reproduction, 2114, 3941
  research, 0093
    biomedical, 3935-3937
symbiosis, 2776
  bacteria, 2693
  fish and aquatic mammals, 4146
synapses
  physiology, 2019
systemic pathology, 3389, 3393, 3400

T

taste. See sense organs
technology, medical, 1100, 1113, 1124, 1131,
    1147, 3171, 3464
  instrumentation, 1347, 1348
  manuals, 1092, 1098, 1105, 1107, 1115,
    1119, 1121, 1122, 1135, 1145
  periodicals, 0456, 0490
  review serials, 0697
technology, radiologic
  pathology, 3674
teeth, 3506. See also dentistry
  anatomy, 1829
  biochemistry, 1829
telemetry, 1344, 2488
temperature, 1978
  invertebrates, 1979
  physiological effect, 1640, 1941, 1948
  vertebrates, 1979
tendons and ligaments
  horses, 1739
  human, 1745
teratology, 1595. See also birth defects;
    metabolism, inborn errors of
  animals, domestic
    nervous system, 1542
  animals, laboratory, 1539
    nervous system, 1542
  human, 1546
  indexes and abstracts, 0032
  mammals, 1549
  periodicals, 0524
  review serials, 0630
teratology, comparative
  methods, 1556
testis. See also endocrine glands; gonads
  anatomy, 1853
  biochemistry, 1853
  physiology, 1853
tetanus, 2724
thalamus, 1868. See also nervous system
therapeutics, 2556, 2557, 2560, 2562, 2563,
    2657, 2658, 3805. See also antibi-
    otics; chemotherapy; drug interactions;
    drugs; perfusion; radiotherapy; tran-
    quilizing agents
  animals, laboratory, 2630
  hormones, 2119
  manuals, 2651
  neoplasms, 3433a
  periodicals, 0473
  pesticides, 2581
  radiation injuries, 4464
  reproduction, animal, 2197
  review serials, 0682, 0723
  serial monographs, 0759
  steroids, 2662

therapeutics (cont.)
  veterinary, 2564, 2632, 2634, 2665, 3373, 3739
    manuals, 2653
    ophthalmology, 2558
  water-electrolyte balance, 2393
thoracic surgery, 0032
thorax. See also lung; respiratory system
  sheep, atlases, 1839
thrombosis, 1895, 1901, 3351
thymus gland, 2084, 2096. See also endocrine glands
  neoplasms, radiotherapy, 3552
thyroid gland, 1867, 2086, 2112a, 2131. See also endocrine glands
  diseases, 2131
  histopathology, 3553
  neoplasms, 0767, 3553, 3554
  serial monographs, 0756
  ultrastructure, 1692
tissue culture, 1192, 1265, 1267, 1269, 1272, 1273, 1276-1279, 1281, 1285, 1290, 1291
  atlases, 1657
  bibliographies, 0179, 0180
  blood, 1295
  bone marrow, 1295
  indexes and abstracts, 0011, 0021, 0066, 0081
  invertebrates, 4588, 4594
  mammals, 1669
  manuals, 1268, 1289, 1292-1294
  periodicals, 0404
  serial monographs, 0785, 0790, 0795, 0797
  symposia, 1282, 1283
  vertebrates, 1657
tissue preservation, 1271, 1308. See also specimen handling
tissue repair. See wound healing
tissues
  classification, 1667
  nervous system, 1806
tissue therapy, 1661
tissue typing. See histocompatibility testing
toxicology, 2560, 2585, 2588. See also animals, poisonous; drugs: toxicity; pesticides; plants, poisonous; venoms
  bibliographies, 0201
  chemicals, 2584
  directories, 1002
  handbooks, 1070, 1084, 2598
  indexes and abstracts, 0032, 0082
  manuals, 2599
  nervous system, 2596
  periodicals, 0297, 0365, 0528, 0529
  research, animals, 2673

toxicology (cont.)
  review serials, 0655, 0684
  serial monographs, 0749
  veterinary, 2587, 2594
toxicology, clinical, 2593, 2600
toxins, 2613. See also mycotoxins
  animals, 2614
  feeds and feeding, 2625
  food, 2625
  radiation effects, 2620
toxoids, 2657. See also antigens
toxoplasmosis, 2964, 2976
trace elements
  metabolism
    animals, 2398, 3706
  nutrition, 2371
    animals, 2395
    human, 2395
tranquilizing agents, 2686
transfusion. See blood transfusion
translations
  bibliographies, 0138-0144
transplantation, 3613, 3617, 3618, 3624, 3633. See also histocompatibility testing; tissue preservation
  bibliographies, 0145
  bone marrow, 1275, 3563, 3626
  cornea, 3629
  endocrine glands, 3629
  genetics
    primates, 3062
  immunobiology, 3054
  indexes and abstracts, 0032
  kidney, 3061
  liver, 0786, 3634
  manuals, 1287
  musculoskeletal system, 3629
  neoplasms, 3437, 3439, 3629
  nervous system, 3629
  ovary, 0145
  periodicals, 0530, 0531
  pharmacology, 2546
  radiation, 3627, 3662
  serial monographs, 0734, 0786
transplantation, heterologous, 3636
transplantation antigens. See histocompatibility antigens
transport of animals, 3723
trauma. See also emergencies; shock
  periodicals, 0479
  review serials, 0728
trematoda, 2966, 2967
trichinosis, 2884
trichomonas foetus
  serial monographs, 0753
trophoblast. See embryology
tropical animals. See also exotic animals; wildlife

tropical animals (cont.)
  diseases, 0022, 0083, 0313, 0321, 0350
  indexes and abstracts, 0022, 0083
  periodicals, 0268, 0300, 0313, 0321, 0350
tropical medicine. See medicine, tropical
trypanosomiasis, 2937, 2980
  serial monographs, 0742
tuberculin, 2657. See also biological products
tuberculosis, 1230, 2844, 2888. See also respiratory tract diseases
  indexes and abstracts, 0032
  serial monographs, 0782
tumors. See neoplasms
turkeys. See also fowl; poultry
  musculoskeletal system, 1727

U

udder. See mastitis
ultrasonics
  diagnosis, 1394, 1395, 1397, 1398, 1399
  dictionaries, 0882
  manuals, 1392
  obstetrics and gynecology, 1396
  ophthalmology, 3227
  therapeutics, 1393, 1394, 1398
ultrastructure, 1689. See also electron microscopy
  atlases, 1530, 1682, 1698
  blood, 1688
  bone diseases, 1687
  bone marrow, 1688
  eye, birds, 1536
  fertilization, 1677
  gastrointestinal system, atlases, 1702
  joint diseases, 1687
  kidney, 1681
    atlases, 1690
  lymphatic system, atlases, 1693
  manuals, 1683
  membranes, 1681
  neoplasms
    brain, 1696
    viruses, 1681
  nervous system, 1678, 1695
  nucleus, 1681
  periodicals, 0480
  placenta, 1679
  sense organs, 1678
  serial monographs, 0769
  skin, 1704
    human, 1680
  thyroid gland, 1692
ultrastructure, cytological, 1697, 1701
  atlases, 1684, 1685, 1691
urethra. See also genitourinary system
  anatomy, 1852
  physiology, 1852

urine, 3312. See also kidney
  manuals, 1140
  proteins, 2378
  serial monographs, 0775
urogenital system. See genitourinary system
urology, 3306. See also bladder; genitourinary system; kidney
  cats, 3309
  dogs, 3309
  indexes and abstracts, 0032
  periodicals, 0414, 0481
  review serials, 0730
uterus
  cytology, 1676
    diagnosis, laboratory, 3547
    diagnosis, atlases, 3550
    histopathology, 3539
  neoplasms, 0784
    cytology, 3548
uveitis, 1768, 3235. See also eye diseases
  immunopathology, 3233

V

vaccines, 2657. See also biological products
  serial monographs, 0782, 0790
  viruses, 2789
vascular diseases, 3522. See also cardiovascular diseases; cerebrovascular diseases
  lung, 3529
vectorcardiography. See electrocardiography
veins. See blood vessels
venereology
  indexes and abstracts, 0032
Venezuelan equine encephalitis
  bibliographies, 0205, 0209
venoms, 0192, 2604, 2613, 2618, 4157
  bibliographies, 0191, 0192, 2612
  radiation effects, 2620
vertebrates
  anatomy, 1439, 1441, 1443, 1456, 1464, 1473, 1504, 1505
    heart, 1834
  anatomy, comparative, 1447, 1455, 1462, 1474, 1478
  nervous system, 1446
  behavior, 2230, 2287
  biochemistry, 2325, 2338
  biology, 1483, 4217
    comparative, 4383
    developmental, 1513
  brain, 1471
  dissection, 1305, 1318, 1458
  embryology, 1470, 1526, 1564, 1573, 1575
    comparative, 1551, 1555
  endocrinology, 1966
  evolution, 1452
  genetics, 3107

vertebrates (cont.)
   heart, innervation, 2048
   histology, comparative, 1487
   hormones, 2093
   learning, 2231
   manuals, 1469
   morphogenesis, 1585
   morphology, 1472
   nervous system, 1786
   neuroanatomy, 1460
   physiology, 1443, 1464, 1504, 1938, 1943, 1974
   regeneration, 1494
   reproduction, 2162, 2163
      physiology, 2194
   sense organs, 1760, 1761
   skin, 2032
   skull, 1522
   temperature, 1979
   tissue culture, 1657
   viruses, 2748
   vision, 1801, 1814, 2038
   water-electrolyte balance, 2082
vesicular exanthema
   bibliographies, 0186
vesicular stomatitis virus
   bibliographies, 0186
veterinary assistants, 3794, 3795, 3797, 3808, 4250, 4262, 4267, 4413
   periodicals, 0242
   serial monographs, 0792
veterinary economics. See economics, veterinary
veterinary law. See legislation, veterinary
veterinary medical education. See education, veterinary medical
vibrio fetus
   serial monographs, 0753
viral diseases, 2765, 2788, 2822, 2847, 2859
   animals, 2848
      domestic, 3738
   arthritis and rheumatism, 2850a
   diagnosis, 2777
   handbooks, 1182, 1212
   hepatitis, 2861
   human, 2854
   liver, 2867
   nervous system diseases, 3240
   respiratory system, 2863
viral interference, 2749. See also interferon
virology, 1192, 2754, 2760, 2780, 2782, 2787
   bibliographies, 0160, 0162, 0208
   biology, molecular, 2763
   congresses, 2775
   indexes and abstracts, 0032, 0084
   manuals, 1185, 1193, 1201, 1203, 1205, 1206, 1228, 2768, 2777
   neoplasia, 0754, 2807, 3434
   periodicals, 0301, 0324, 0445, 0482, 0533

virology (cont.)
   primates, 0795
   review serials, 0633, 0676, 0680, 0698
   serial monographs, 0754, 0763, 0771, 0774, 0795
   veterinary, 2706, 2751, 2804
virology, clinical, 2750, 2792
virology, comparative, 2783
virology, diagnostic, 2771
virology, experimental, 2769, 2784
virus cultivation. See also tissue culture
   serial monographs, 0790
viruses, 2746, 2762, 2781, 2790, 2798, 2799. See also names of specific viruses
   animals, 2752
   bibliographies, 0153, 0186, 0205, 0209
   biochemistry, 2779
   biology, molecular, 2764, 2791, 2797
   classification, 2805
   directories, 1004
   genetics, 2796
   neoplasia, 2747, 2767, 2778, 2795, 2802, 2807, 3405
   neoplasms, ultrastructure, 1681
   nucleic acids, 2758
   plants, 2752
   serial monographs, 0740, 0752, 0771, 0778, 0781, 0782, 0795
   vaccines, 2789
   vertebrates, 2748
   water, 2801
vision, 1764. See also eye; sense organs
   birds, animal models, 2005
   invertebrates, 4584
   physiology, 1768
   vertebrates, 1801, 1814, 2038
vitamins, 2340, 3649. See also nutrient requirements
   review serials, 0717
   serial monographs, 0744
vivarium. See animals, laboratory
vivisection, 4387, 4388, 4390, 4391, 4393
   legislation, 4385
voice. See also communication; otorhinolaryngology; speech
   bibliographies, 0147

W

water
   microbiology, 1177, 1179, 2911
   viruses, 2801
water-electrolyte balance, 2379
   dogs, surgery, 2364
   fish and aquatic mammals, 4127
   mammals, 2385
   reptiles, 4166
   therapeutics, 2393
   vertebrates, 2082
weather, 2506

wildlife, 0789, 4181, 4186, 4204. See also foxes; furbearing animals; wolves; zoos
  behavior, 2254
  care and management, 4198
  diseases, 0209, 0278, 0279, 0801, 4079, 4177, 4178
  ecology, 4182-4184, 4189, 4193, 4195, 4196, 4205, 4214
  infectious diseases, 2828
  medicine, veterinary, 4179
  nutrition, 0801
  parasitic diseases, 2973, 4177
  periodicals, 0278, 0279
  pesticides, 2601
  pets, 4239
  photography, 4187
  populations, 4182, 4184, 4202
  protection, 4183, 4189, 4190, 4192, 4197, 4203, 4205
  research, 4201
  restraint, 4188
  zoos, 4206
wolves
  behavior, 2254, 2286
  ecology, 2286
wound healing, 3643, 3644, 3646, 3651
  histocytochemistry, 2429
  serial monographs, 0731
  skin, 3645
  surgery, 3634a, 3648
wound infection, 2745. See also infection control

writing, 1040-1042, 1044, 1045, 1047, 1048, 1052, 1055, 1056, 1062

X

x-ray, diagnostic. See diagnosis, radiologic; radiography
x-ray therapy. See radiotherapy

Z

zoo animals. See wildlife; zoos
zoology, 4221. See also biology; protozoology; wildlife
  indexes and abstracts, 0008, 0086
  periodicals, 0358, 0416
  review serials, 0666
zoology, chemical, 2359
zoonoses, 2879, 2881, 2885, 2889-2893. See also brucellosis; chlamydial diseases; communicable diseases; leptospirosis; rabies; trichinosis
  epidemiology, 2894
  periodicals, 0285, 0286, 0576, 0594
  primates, 2883
  serial monographs, 0755
zoos, 2269, 4207, 4208, 4211, 4212
  behavior, spatial, 2269
  periodicals, 0250
  psychology, animals, 2267, 2268
  review serials, 0670
  wildlife, 4206